Lecture Notes in Computer Science 5870

Commenced Publication in 1973
Founding and Former Series Editors:
Gerhard Goos, Juris Hartmanis, and Jan

Robert Meersman Tharam Dillon
Pilar Herrero (Eds.)

On the Move to Meaningful Internet Systems: OTM 2009

Confederated International Conferences
CoopIS, DOA, IS, and ODBASE 2009
Vilamoura, Portugal, November 1-6, 2009
Proceedings, Part I

Springer

Volume Editors

Robert Meersman
Vrije Universiteit Brussel (VUB), STARLab
Bldg G/10, Pleinlaan 2, 1050 Brussel, Belgium
E-mail: meersman@vub.ac.be

Tharam Dillon
Curtin University of Technology, DEBII - CBS
De Laeter Way, Bentley, WA 6102, Australia
E-mail: t.dillon@curtin.edu.au

Pilar Herrero
Universidad Politécnica de Madrid, Facultad de Informática
Campus de Montegancedo S/N, 28660 Boadilla del Monte, Madrid, Spain
E-mail: pherrero@fi.upm.es

Library of Congress Control Number: 2009937359

CR Subject Classification (1998): H.2, H.3, H.4, C.2, H.5, D.2.12, I.2, K.4

LNCS Sublibrary: SL 3 – Information Systems and Application, incl. Internet/Web and HCI

ISSN 0302-9743
ISBN-10 3-642-05147-2 Springer Berlin Heidelberg New York
ISBN-13 978-3-642-05147-0 Springer Berlin Heidelberg New York

springer.com

Printed in Germany

Typesetting: Camera-ready by author, data conversion by Scientific Publishing Services, Chennai, India
Printed on acid-free paper SPIN: 12779802 06/3180 5 4 3 2 1 0

Volume Editors

Robert Meersman
Tharam Dillon
Pilar Herrero

CoopIS 2009

Ted Goranson
Hai Zhuge
Moira C. Norrie

DOA 2009

Mark Little
Jean-Jacques Dubray
Fabio Panizeri

IS 2009

Stefanos Gritzalis
Xueqi Cheng

ODBASE 2009

Avigdor Gal
Annika Hinze
Sharma Chakravarthy

General Co-chairs' Message for OTM 2009

The OnTheMove 2009 event in Vilamoura, Portugal on 1-5 November, further consolidated the growth of the conference series that was started in Irvine, California in 2002, and held in Catania, Sicily in 2003, in Cyprus in 2004 and 2005, in Montpellier in 2006, a first time in Vilamoura in 2007, and in Monterrey Mexico in 2008. The event continues to attract a diversified and representative selection of today's worldwide research on the scientific concepts underlying new computing paradigms, which, of necessity, must be distributed, heterogeneous and autonomous yet meaningfully collaborative. Indeed, as such large, complex and networked intelligent information systems become the focus and norm for computing, there continues to be an acute and even increasing need to address and discuss face to face in an integrated forum the implied software, system and enterprise issues as well as methodological, semantical, theoretical and applicational issues. As we all know, email, the Internet, and even video conferences are not sufficient for effective and efficient scientific exchange.

The OnTheMove (OTM) Federated Conferences series has been created to cover the scientific exchange needs of the community/ies that work in the broad yet closely connected fundamental technological spectrum of Web-based distributed computing. The OTM program every year covers data and Web semantics, distributed objects, Web services, databases, information systems, enterprise workflow and collaboration, ubiquity, interoperability, mobility, grid and high-performance computing.

OnTheMove aspires to be a primary scientific meeting place where all aspects of the development of such Internet- and Intranet-based systems in organizations and for e-business are discussed in a scientifically motivated way. This eighth edition of the OTM Federated Conferences event again provided an opportunity for researchers and practitioners to understand and publish these developments within their individual as well as within their broader contexts. Originally the federative structure of OTM was formed by the co-location of three related, complementary and successful main conference series: DOA (Distributed Objects and Applications, since 1999), covering the relevant infrastructure-enabling technologies, ODBASE (Ontologies, DataBases and Applications of SEmantics, since 2002) covering Web semantics, XML databases and ontologies, and CoopIS (Cooperative Information Systems, since 1993) covering the application of these technologies in an enterprise context through, e.g., workflow systems and knowledge management, and in 2007 IS was added (Information Security). In 2006 GADA (Grid computing, high-performAnce and Distributed Applications) was added to this as a main symposium but unfortunately this year attracted too few submissions to guarantee both representativity and quality; a new GADA edition is, however, planned for 2010. Both IS and GADA started as successful workshops at OTM, the first covering the issues of security in complex

Internet-based information systems, the second covering the large-scale integration of heterogeneous computing systems and data resources with the aim of providing a global computing space.

Each of these four conferences encourages researchers to treat their respective topics within a framework that incorporates jointly (a) theory, (b) conceptual design and development, and (c) applications, in particular case studies and industrial solutions.

Following and expanding the model created in 2003, we again solicited and selected quality workshop proposals to complement the more "archival" nature of the main conferences with research results in a number of selected and more "avant-garde" areas related to the general topic of Web-based distributed computing. For instance, the so-called Semantic Web has given rise to several novel research areas combining linguistics, information systems technology, and artificial intelligence, such as the modeling of (legal) regulatory systems and the ubiquitous nature of their usage. We were glad to see that ten of our earlier successful workshops (ADI, CAMS, EI2N, SWWS, ORM, OnToContent, MONET, SEMELS, COMBEK, IWSSA) re-appeared in 2008 with a second, third or even fifth edition, sometimes by alliance with other newly emerging workshops, and that no fewer than three brand-new independent workshops could be selected from proposals and hosted: ISDE, ODIS and Beyond SAWSDL. Workshop audiences productively mingled with each other and with those of the main conferences, and there was considerable overlap in authors.

We were also happy to see that in 2009 the number of quality submissions for the OnTheMove Academy (OTMA, formerly called Doctoral Consortium Workshop), our "vision for the future" in research in the areas covered by OTM, took off again. We must thank the new OTMA Dean, Erich Neuhold, and his team of collaborators led by Peter Spyns and Anja Schanzenberger, for their gallant efforts in implementing our interactive formula to bring PhD students together: research proposals are submitted for evaluation; selected submissions and their approaches are (eventually) presented by the students in front of a wider audience at the conference, and intended to be independently and extensively analyzed and discussed in public by a panel of senior professors.

As said, all four main conferences and the associated workshops shared the distributed aspects of modern computing systems, and the resulting application-pull created by the Internet and the so-called SemanticWeb. For DOA 2009, the primary emphasis stayed on the distributed object infrastructure; for ODBASE 2009, it became the knowledge bases and methods required for enabling the use of formal semantics; for CoopIS 2009, the focus as usual was on the interaction of such technologies and methods with management issues, such as occur in networked organizations, and for IS 2008 the emphasis was on information security in the networked society. These subject areas overlap in a scientifically natural fashion and many submissions in fact also treated an envisaged mutual impact among them. As for the earlier editions, the organizers wanted to stimulate this cross-pollination by a "shared" program of famous keynote speakers: this year we were proud to announce Wolfgang Prinz in Computing Science at the

University of Bonn, Santosh Shrivastava in Computing Science at the University of Newcastle upon Tyne, Kai Hwang in Electrical Engineering and Computer Science and Director of Internet and Cloud Computing Lab at the University of Southern California (USC) and last but not least Alejandro Buchmann of the Department of Computer Science at Technische Universität Darmstadt where he heads the Databases and Distributed Systems Group.

The registration fee structure again wanted to strongly encourage multiple event attendance by providing *all* main conference authors with free access or discounts to *all* other conferences and workshops (workshop authors paid a small extra fee to attend the main conferences).

We received a total of 234 submissions for the four main conferences and 131 submissions in total for the workshops. The numbers are about 25% lower than for 2008, not unexpected because of the prevailing economic climate. But, not only may we indeed again claim success in attracting an increasingly representative volume of scientific papers, many from the USA, Central and South America, but these numbers of course allow the Program Committees to compose a high-quality cross-section of current research in the areas covered by OTM. In fact, in spite of the number of submissions, the Program Chairs of each of the three main conferences decided to accept only approximately the same number of papers for presentation and publication as in 2007 and 2008 (i.e., average 1 paper out of 3-4 submitted, not counting posters). For the workshops, the acceptance rate varies but the aim was to stay as strict as before, consistently about 1 accepted paper for 2-3 submitted. We have separated the proceedings into three books with their own titles, two for the main conferences and one for the workshops, and we are grateful to Springer for their suggestions and collaboration in producing these books and USB sticks. The reviewing process by the respective Program Committees was again performed very professionally, and each paper in the main conferences was reviewed by at least three referees, with arbitrated email discussions in the case of strongly diverging evaluations. It may be worthwhile emphasizing that it is an explicit OnTheMove policy that all conference Program Committees and Chairs make their selections completely autonomously from the OTM organization itself. The OnTheMove Federated Event organizers again made all proceedings available on a CDROM to all participants of the conferences and workshops, independently of their registration to a specific conference or workshop. Paper proceedings were on request this year, and incurred an extra charge.

The General Chairs are once more especially grateful to the many people directly or indirectly involved in the setup of these federated conferences. Few people realize what a large number of people have to be involved, and what a huge amount of work, and in 2009 certainly also financial risk, the organization of an event like OTM entails. Apart from the persons in their roles mentioned above, we therefore wish to thank in particular our 17 main conference PC Co-chairs: DOA 2009: Mark Little, Jean-Jacques Dubray, Fabio Panizeri, ODBASE 2009: Avigdor Gal, Annika Hinze, Sharma Chakravarthy, CoopIS 2009: Ted Goranson, Hai Zhuge, Moira C. Norrie, IS 2009: Gritzalis Stefanos, Xueqi Cheng; and the Workshop PC Co-chairs: Stefan Jablonski, Olivier Curé, Christoph Bussler, Annika Hinze, George Buchanan, Hervé Panetto, Ricardo Goncalves, Peter Bernus, Ted Goranson, Alok Mishra, Deepti Mishra, Ozlem Albayrak, Lawrence Chung, Nary Subramanian, Manuel Noguera, José Luis Garrido, Patrizia Grifoni, Fernando Ferri, Irina Kondratova, Arianna D'Ulizia, Paolo Ceravolo, Mustafa Jarrar, Andreas Schmidt, Matt-Mouley Bouamrane, Christophe Gravier, Frederic Cuppens, Jacques Fayolle, Simon Harper, Saturnino Luz, Masood Masoodian, Terry Halpin, Herman Balsters, Tharam S. Dillon, Ernesto Damiani, Elizabeth Chang, Chen Wu, Amandeep Sidhu, Jaipal Singh, Jacek Kopecky, Carlos Pedrinaci, Karthik Gomadam, Maria Maleshkova , Reto Krummenacher, Elena Simperl, Françoise Baude, Philippe Merle, Ramonville Saint-Agne, Pieter De Leenheer, Martin Hepp, Amit Sheth, Peter Spyns, Erich J. Neuhold and Anja Schanzenberger.

All, together with their many PC members, performed a superb and professional job in selecting the best papers from the harvest of submissions. We are all grateful to Ana Cecilia Martinez-Barbosa and to our extremely competent and experienced Conference Secretariat and technical support staff in Antwerp, Daniel Meersman, Ana-Cecilia (again), and Jan Demey, and last but not least to our two editorial teams, one in Perth (DEBII-Curtin University) and one in Madrid (Quoriam Ingenieros).

The General Chairs gratefully acknowledge the academic freedom, logistic support and facilities they enjoy from their respective institutions, Vrije Universiteit Brussel (VUB), Curtin University, Perth Australia, and Universitad Politécnica de Madrid (UPM), without which such an enterprise would not be feasible. We do hope that the results of this federated scientific enterprise contribute to your research and your place in the scientific network... We look forward to seeing you again at next year's event!

August 2009

Robert Meersman
Tharam Dillon
Pilar Herrero

Organization

OTM (On The Move) is a federated event involving a series of major international conferences and workshops. These proceedings contain the papers presented at the OTM 2009 Federated conferences, consisting of four conferences, namely CoopIS 2009 (Cooperative Information Systems), DOA 2009 (Distributed Objects and Applications), IS 2009 (Information Security) and ODBASE 2009 (Ontologies, Databases and Applications of Semantics).

Executive Committee

General Co-chairs	Robert Meersman (VU Brussels, Belgium) Tharam Dillon (Curtin University of Technology, Australia) Pilar Herrero (Universidad Politécnica de Madrid, Spain)
CoopIS 2009 PC Co-chairs	Ted Goranson (Earl Research, USA) Hai Zhuge (Chinese Academy of Sciences, China) Moira C. Norrie (ETH Zurich, Switzerland)
DOA 2009 PC Co-chairs	Mark Little (Red Hat, UK) Jean-Jacques Dubray (Premera, Mountlake Terrace, WA, USA) Fabio Panizeri (University of Bologna, Italy)
IS 2009 PC Co-chairs	Stefanos Gritzalis (University of the Aegean, Greece) Xueqi Cheng (Chinese Academy of Science, China)
ODBASE 2009 PC Co-chairs	Avigdor Gal (Technion, Israel Institute of Technology) Annika Hinze (University of Waikato, New Zealand) Sharma Chakravarthy (The University of Texas at Arlington, USA)
Local Organizing Chair	Ricardo Goncalves (New University of Lisbon, Portugal)
Publication Chair	Houwayda Elfawal Mansour (DEBII, Australia)
Publicity-Sponsorship Chair	Ana-Cecilia Martinez Barbosa (DOA Institute, Belgium)
Logistics Team	Daniel Meersman (Head of Operations) Ana-Cecilia Martinez Barbosa Jan Demey

CoopIS 2009 Program Committee

Ghaleb Abdulla
Anurag Agarwal
Marco Aiello
Antonia Albani
Elias Awad
Joonsoo Bae
Zohra Bellahsene
Salima Benbernou
Djamal Benslimane
M. Brian Blake
Klemens Böhm
Christoph Bussler
James Caverlee
Yiling Chen
Meng Chu Zhou
Francisco Curbera
Vincenzo D'Andrea
Ke Deng
Xiaoyong Du
Schahram Dustdar
Johann Eder
Rik Eshuis
Opher Etzion
Renato Fileto
Paul Grefen
Michael Grossniklaus
Amarnath Gupta
Mohand-Said Hacid
Geert-Jan Houben
Zhixing Huang
Patrick Hung
Paul Johannesson
Epaminondas Kapetanios
Dimka Karastoyanova
Rania Khalaf
Hiroyuki Kitagawa
Akhil Kumar
Allen Lee
Frank Leymann
Ling Li
Ling Liu
Sanjay K. Madria
Tiziana Margaria
Leo Mark
Maristella Matera
Massimo Mecella
Ingo Melzer
Mohamed Mokbel
Jörg Müller
Nirmal Mukhi
Miyuki Nakano
Werner Nutt
Andreas Oberweis
Gérald Oster
Hervé Panetto
Cesare Pautasso
Frank Puhlmann
Lakshmish Ramaswamy
Manfred Reichert
Stefanie Rinderle-Ma
Rainer Ruggaber
Duncan Ruiz
Radhika Santhanam
Kai-Uwe Sattler
Ralf Schenkel
Jialie Shen
Aameek Singh
Xiaoping Sun
Wei Tang
Edison Tse
Susan Urban
Ricardo Valerdi
Willem-Jan Van den Heuvel
Maria Esther Vidal
John Warfield
Mathias Weske
Li Xiong
Li Xu
Jian Yang
Leon Zhao
Aoying Zhou

DOA 2009 Program Committee

Giorgia Lodi
Subbu Allamaraju
Mark Baker
Judith Bishop
Gordon Blair
Harold Carr
Geoffrey Coulson
Frank Eliassen
Patrick Eugster
Pascal Felber
Benoit Garbinato
Medhi Jazayeri
Eric Jul
Nick Kavantzas
Joe Loyall
Frank Manola
Gero Mühl
Nikola Milanovic
Graham Morgan
Rui Oliveira
Jose Orlando Pereira
Francois Pacull
Fernando Pedone
Arno Puder
Michel Riveill
Luis Rodrigues

IS 2009 Program Committee

Alessandro Acquisti
Gail-Joon Ahn
Vijay Atluri
Joonsang Baek
Manuel Bernardo Barbosa
Ezedin Barka
Elisa Bertino
Yu Chen
Bruno Crispo
Gwenael Doerr
Josep Domingo Ferrer
Nuno Ferreira Neves
Simone Fischer-Huebner
Clemente Galdi
Aiqun Hu
Jiankun Hu
Hai Jin
Christos Kalloniatis
Maria Karyda
Stefan Katzenbeisser
Hiroaki Kikuchi
Spyros Kokolakis
Wei-Shinn Ku
Kwok-Yan Lam
Costas Lambrinoudakis
Xiaodong Lin
Ling Liu
Evangelos Markatos
Sjouke Mauw
Chris Mitchell
Yi Mu
Barry Clifford Neuman
Yi Pan
Jong Hyuk Park
Guenther Pernul
Milan Petkovic
Frank Piessens
Bhanu Prasad
Bart Preneel
Rodrigo Roman
Pierangela Samarati
Biplab K. Sarker
Haiying (Helen) Shen
Weisong Shi
Mikko T. Siponen
Diomidis Spinellis
Pureui Su
Luis Javier Garcia Villalba
Cheng-Zhong Xu
Yixian Yang
Alec Yasinsac
Moti Yung
Wei Zou
Andre Zuquete

ODBASE 2009 Program Committee

Karl Aberer
Harith Alani
María Auxilio Medina
Renato Barrera
Sonia Bergamaschi
Leopoldo Bertossi
Alex Borgida
Mohand Boughanem
Paolo Bouquet
Christoph Bussler
Silvana Castano
Paolo Ceravolo
Oscar Corcho
Ernesto Damiani
Aldo Gangemi
Benjamin Habegger
Mounira Harzallah
Bin He
Andreas Hotho
Jingshan Huang
Farookh Hussain
Prateek Jain
Maciej Janik
Vana Kalogeraki
Dimitris Karagiannis
Uladzimir Kharkevich
Manolis Koubarakis
Maurizio Lenzerini
Juanzi Li
Alexander Löser
Li Ma
Vincenzo Maltese
Maurizio Marchese
Gregoris Metzas
Riichiro Mizoguchi
Peter Mork
Ullas Nambiar
Anne Ngu
Sandeep Pandey
Adrian Paschke
Peter R. Pietzuch
Axel Polleres
Wenny Rahayu
Rajugan Rajagopalapillai
Sudha Ram
Satya Sahoo
Pavel Shvaiko
Sergej Sizov
Il-Yeol Song
Veda C. Storey
Umberto Straccia
Eleni Stroulia
Heiner Stuckenschmidt
Vijayan Sugumaran
York Sure
Robert Tolksdorf
Susan Urban
Yannis Velegrakis
Guido Vetere
Kevin Wilkinson
Baoshi Yan
Laura Zavala
Jose Luis Zechinelli
Roberto Zicari

Supporting Institutions

OTM 2009 was proudly supported by Vrije Universiteit Brussel in Belgium, Curtin University of Technology in Australia and Universidad Politechnica de Madrid in Spain.

Sponsoring Institutions

OTM 2009 was proudly sponsored by *algardata S.A.* in Portugal, Collibra as a spin-off of STARLab at the Vrije Universiteit Brussel, *Lecture Notes in Computer Science* by Springer and the *Universidade do Algarve* in Portugal.

Table of Contents – Part I

CoopIS 2009 Keynote

DOA 2009 Keynote

Cooperative Information Systems (CoopIS) International Conference 2009

Workflow

Process Models

Ontology Challenges

Network Complexity 1

Network Complexity 2

Modeling Cooperation

Information Complexity

Infrastructure

Information

Distributed Objects and Applications (DOA) International Conference 2009

Aspect-Oriented Approaches for Distributed Middleware

Distributed Algorithms and Communication Protocols

Distributed Databases and Transactional Systems

Distributed Infrastructures for Cluster and Grid Computing

Object-Based, Component-Based, Resource-Oriented, Event-Oriented and Service-Oriented Middleware

Peer to Peer and Decentralized Infrastructures

Performance Analysis of Distributed Computing Systems

Reliability, Fault Tolerance, Quality of Service, and Real Time Support

Self* Properties in Distributed Middleware

Software Engineering for Distributed Middleware Systems

Security and Privacy in a Connected World

Ubiquitous and Pervasive Computing

Table of Contents – Part II

IS 2009 Keynote

ODBASE 2009 Keynote

International Symposium on Information Security (IS) International Conference 2009

Information Systems Security

Privacy and Authentication

Security Policies and Verification

Ontologies, DataBases, and Applications of Semantics (ODBASE) International Conference 2009

Managing Ontologies

Using Ontologies

Event Processing

Dealing with Heterogeneity

Building Knowledge Bases

XML and XML Schema

Developing Collaborative Working Environments and What Can We Learn from Web 2.0

Wolfgang Prinz

Fraunhofer FIT in Bonn - Germany

Short Bio

Professor Wolfgang Prinz, PhD studied informatics at the University of Bonn and received his PhD in computer science from the University of Nottingham. He is deputy head of Fraunhofer FIT in Bonn, division manager of the Collaboration systems research department in FIT, and Professor for cooperation systems at RWTH Aachen.

He is carrying out research in the area of Cooperative Systems, Social Web and Pervasive Games. He participated in and managed several national research and international research projects and he is currently coordinator of a large European research project on collaborative work environments.

Talk

"Developing Collaborative Working Environments and what can we learn from Web 2.0"

As developers of collaborative work environments, we may consider the rising wave of new ideas and applications within the Web 2.0 domain as a hazard that sweeps us away or we can learn to surf this wave. In my presentation, I'll first distill Web 2.0 applications and technologies to identify the basic concepts. Then I'll discuss how these concepts can be used for the design and development of cooperative work applications.

R. Meersman, T. Dillon, P. Herrero (Eds.): OTM 2009, Part I, LNCS 5870, p. 1, 2009.

Third Party Services for Enabling Business-to-Business Interactions

Santosh Shrivastava

University of Newcastle upon Tyne - UK

Short Bio

Santosh Shrivastava was appointed a Professor of Computing Science, University of Newcastle upon Tyne in 1986. He received his Ph.D. in Computer Science from Cambridge University in 1975.

His research interests are in the areas of computer networking, middleware and fault tolerant distributed computing. Current focus of his work is on middleware for supporting inter-organization services where issues of trust, security, fault tolerance and ensuring compliance to service contracts are of great importance. From 1985-98 he led a team that developed the Arjuna distributed object transaction system. Arjuna transaction service software is now an integral part of JBoss application sever middleware from Red Hat Inc. So far he has supervised 25 PhD students.

Talk

"Third party services for enabling business-to-business interactions".

Abstract. Business-to-business (B2B) interactions concerned with the fulfilment of a given business function (e.g., order processing) requires business partners to exchange electronic business documents and to act on them. This activity can be viewed as the business partners taking part in the execution of a shared business process, where each partner is responsible for performing their part in the process. Naturally, business process executions at each partner must be coordinated at run-time to ensure that the partners are performing mutually consistent actions (e.g., the seller is not hipping a product when the corresponding order has been cancelled by the buyer). A number of factors combine to make the task of business process coordination surprisingly hard:

(i) Loose coupling: B2B interactions take place in a loosely coupled manner, typically using message oriented middleware (MoM) where business partners are not required to be online "at the same time". Shared business process coordination in such a setting is inherently difficult, as interacting partners rarely have an up-to-date information on the state of other partners, so there is a danger of partners getting out of synchrony with each other (state misalignment).

R. Meersman, T. Dillon, P. Herrero (Eds.): OTM 2009, Part I, LNCS 5870, pp. 2–3, 2009.

(ii) Timing and validity constraints: Business document exchange protocols (e.g., RosettaNet PIPs) have stringent timing and validity constraints: a business message is accepted for processing only if it is timely and satisfies specific syntactic and semantic validity constraints. Such constraints can be yet another cause of state misalignment between the partners. For example, if a message is delivered but not taken up for processing due to some message validity condition not being met at the receiver, the sender‘s and the receiver‘s views could divert (the sender assumes that the message is being processed whereas the receiver rejected it).

(iii) Faulty environment: Business interactions encounter software, hardware and network related problems (e.g., clock skews, unpredictable transmission delays, message loss, incorrect messages, node crashes etc.).

In summary, one can appreciate that there is plenty of scope for business partners to misunderstand each other leading to disputes. Partner misbehaviour adds additional complications. Within this context, we will explore the possibility of developing third party services for coordination, fair exchange, exception resolution, contract monitoring that can be utilized by business partners to simplify the task of performing B2B interactions.

CoopIS 2009 – PC Co-chairs' Message

Welcome to the 17th International Conference on Cooperative Information Systems (CoopIS 2009).

This year CoopIS was held in Vilamoura, Algarve, Portugal, during November 3–5, 2009.

Cooperative information systems (CISs) are the basic component of the information management infrastructure that supports much of what we do in the world. It happens to inherit challenging problems from many disciplines, the solutions to which are early in the food chain of innovation in computer science. The CIS domain includes enterprise integration and interoperability, distributed systems technologies, middleware, business process management (BPM), and Web and service computing architectures and technologies.

The CoopIS conferences provide a long-lived forum for exchanging ideas and results on scientific research from a variety of areas, such as business process and state management, collaborative systems, electronic commerce, human–computer interaction, Internet data management and strategies, distributed systems, and software architectures. As a prime international forum for cooperative information systems and applications, we encourage the participation of both researchers and practitioners, aimed at facilitating exchange and cross-fertilization of ideas and support of technology deployment and transfer.

We are very pleased to share the proceedings comprising the exciting technical program with you. This year's conference included seven full-paper research sessions, two short-paper research sessions and one panel. The program covered a broad range of topics in the design and development of cooperative information systems: business process technologies, business process tracing, distributed process management, Web services, e-service management, schema matching, workflow, and business applications. We were very pleased to have Wolfgang Prinz as our keynote speaker.

This high-quality program would not have been possible without the authors who chose CoopIS as a venue to submit their publications to, and the Program Committee members who dedicated their efforts and time to the review and the online PC meeting discussions. We received about 80 submissions from 35 countries. Every paper received at least three independent reviews. Through a careful two-phase review process, consisting of PC members' independent reviews and online PC meeting discussions, 20 full papers and 9 short papers were selected and included in this year's technical program. We are grateful for the dedication and excellent job of the CoopIS 2009 PC members, who are experts in the field.

We would like to take this opportunity to express our sincere thanks to Daniel Meersman and Jan Demey for their hard work, enthusiasm, and almost constant availability. Daniel and Jan were critical in facilitating the paper management process and making sure that the review process stayed on track. We would also like to thank the General Chairs of OTM, Robert Meersman, Tharam Dillon and Pilar Herrero, for their support. They provided leadership and infrastructure for

R. Meersman, T. Dillon, P. Herrero (Eds.): OTM 2009, Part I, LNCS 5870, pp. 4–5, 2009.

the consolidated conference/workshop enterprise. CoopIS benefits greatly from the larger mature structure of OTM. Tharam additionally handled the duties of marshalling the paper review process, and is due special thanks.

The Publicity and Sponsorship Chair, Ana-Cecilia Martinez, has our appreciation for her efforts. The OTM support team (Daniel Meersman, Jan Demey) worked behind the scenes performing the many tasks that move things along. Our deepest thanks go to the members of the PC, who worked tight deadlines, providing the technical guidance that can be clearly seen in the quality of the papers.

And finally, we wish to thank the authors. This challenging and important field is contributing to the general well-being and health of many enterprises. Your work is appreciated.

August 2009

Ted Goranson
Hai Zhuge
Moira C. Norrie

Resolution of Compliance Violation in Business Process Models: A Planning-Based Approach

Ahmed Awad, Sergey Smirnov, and Mathias Weske

Business Process Technology Group
Hasso Plattner Institute at the University of Potsdam
D-14482 Potsdam, Germany
{Ahmed.Awad,Sergey.Smirnov,Mathias.Weske}@hpi.uni-potsdam.de

Abstract. Keeping business processes compliant with regulations is of major importance for companies. Considering the huge number of models each company possesses, automation of compliance maintenance becomes essential. Therefore, many approaches focused on automation of various aspects of compliance problem, e.g., compliance verification. Such techniques allow localizing the problem within the process model. However, they are not able to resolve the violations. In this paper we address the problem of (semi) automatic violation resolution, addressing violations of execution ordering compliance rules. We build upon previous work in categorizing the violations into types and employ automated planning to ensure compliance. The problem of choosing the concrete resolution strategy is addressed by the concept of context.

Keywords: Business process modeling, compliance checking, process model parsing, process model restructuring.

1 Introduction

In today's business being compliant with regulations is vital. Process models provide enterprises an explicit view on their business. Thus, it is rational to employ process models for compliance checking. Companies hire experts to audit their business processes and to evidence process compliance to external/internal controls. Keeping processes compliant with constantly changing regulations is expensive [12].

Compliance requirements (rules) originate from different sources and address various aspects of business processes. For instance, a certain order of execution between activities is required. Other rules force the presence of activities under certain conditions, e.g., reporting banking transactions to a central bank, if large deposits are made. Violations of compliance requirements originating from regulations, e.g., the Sarbanes-Oxley Act of 2002 (see [1]), lead to penalties, scandals, and loss of reputation. Several approaches have been proposed to handle the divergent aspects of compliance on the level of process models. Most of them are focused on model compliance checking, i.e., on model verification problem [2,7,11,15].

Although the problem of compliance violation resolution was discussed in literature, it is usually perceived as a human expert task. However, it would be possible to (semi-) automate the task of resolving compliance violations. This would be valued as an aid to the human expert to speed up the process of ensuring compliance. In [3], we made the

R. Meersman, T. Dillon, P. Herrero (Eds.): OTM 2009, Part I, LNCS 5870, pp. 6–23, 2009.

first step towards resolving violations of compliance rules regarding execution ordering of activities by identifying the different violation patterns.

In this paper we show how automated planning techniques can be used for resolution of compliance violations. We present resolution algorithms for violation patterns identified in [3] and explain the role of *resolution context*. The developed approach assumes that compliance violations can be resolved sequentially, one after another. This implies that there are no contradictions between compliance rules: for any two rules r_1 and r_2, resolution of r_1 violation does not lead to violation of r_2 and vice versa.

The rest of the paper is organized as follows. Section 2 provides the necessary formalism. Section 3 describes a motivating example. Section 4 discusses a set of resolution algorithms for the different violation patterns. The related work is presented in Section 5. Section 6 concludes the paper and discusses future work.

2 Preliminaries

In this section we introduce the basic concepts, supporting the violation resolution approach. As resolutions are realized on a structural level, we introduce a supporting formalism—the concept of process structure trees. Further, we show how the problem domain can be described by a *resolution context* and the task of violation resolution can be interpreted in terms of automated planning.

2.1 Process Structure Tree

Correction of compliance violations assumes analysis and modification of business process models on the structural level. Hence, we need a technique efficiently supporting these tasks. We rely on the concept of a process structure tree (PST), which is the process analogue of abstract syntax trees for programs. The concept of a PST is based on the unique decomposition of a process model into fragments. Fragments, which are the decomposition result, are organized into a hierarchy according to the nesting relation. This hierarchy is called a process structure tree. PSTs can be constructed using various algorithms. One approach is a decomposition into *canonical single entry single exit (SESE) fragments*, formally described in [25]. Informally, SESE fragments can be defined as fragments with exactly one incoming and one outgoing edge. The node sets of two canonical SESE fragments are either disjoint, or one contains the other. Following [25], we consider the maximal sequence of nodes to be a canonical SESE fragment. As we assume process models to be structured workflows, the SESE decomposition suits the task well.

Definition 1. A *process structure tree* $P = (N, r, E, type)$ is a tree, where:

- N is a finite set of nodes, where nodes correspond to canonical SESE fragments
- $r \in N$ is the root of the tree
- $E \subseteq (N \times (N \setminus \{r\}))$ is the set of edges. Let tree nodes $n_1, n_2 \in N$ correspond to SESE fragments f_1 and f_2, respectively. An edge leads from n_1 to n_2 if SESE fragment f_1 is the direct parent of f_2
- $type : N \rightarrow \{\texttt{act}, \texttt{seq}, \texttt{and}, \texttt{xor}, \texttt{or}, \texttt{loop}\}$ is a function assigning a type to each node in N: `act` corresponds to activities, `seq`—sequences, `and`, `xor`, `or`—blocks of corresponding type, `loop`

- $N_{<type>} \subseteq N$ denotes the set of nodes with specific type, e.g., N_{seq} are the nodes of type seq.

A process model may contain several occurrences of one activity (e.g., activity A). Then, the model's PST has the set of nodes which correspond to occurrences of A. To address such a set of nodes we denote it with $N_a \subseteq N$. Since a tree has no cycles, a path between two nodes is a unique sequence of nodes.

Definition 2. A *path* between two nodes $n_0, n_k \in N$, is a sequence of nodes $path(n_0, n_k) = (n_0, n_1, \ldots, n_k)$ where $(n_i, n_{i+1}) \in E, 0 \leq i < k$. If there is no path between n_0 and n_k we denote it with $path(n_0, n_k) = \perp$. We write $n \in path(n_0, n_k)$ to express the fact that n lies on the path from n_0 to n_k.

Definition 3 formalizes the notion of the least common ancestor of two nodes.

Definition 3. The *least common ancestor* of two nodes $n, m \in N$ in the $P = (N, r, E, type)$ is a node $lca(n, m) = \{p : p \in N \wedge p \in path(r, n) \wedge p \in path(r, m) \wedge \nexists p' (p' \neq p \wedge p' \in path(r, n) \wedge p' \in path(r, m) \wedge p \in path(r, p'))\}$.

Depending on the type of the least common ancestor of two nodes, we can determine the behavioral relation between them, either sequence, choice, parallel, etc. If a node is of type seq the execution order for its direct children is defined.

Definition 4. The *order* of execution of a node $n \in N$ with respect to node $p \in N_{seq}$ is a function $order : N_{seq} \times N \rightarrow \mathbb{N}$, defined if $(p, n) \in E$ and where the first argument is the parent node and the second—its child.

To reflect the effect of data on formulating conditions for branches, we define a function $condition$ as follows.

Definition 5. *Let Pr be a set of predicates representing data conditions. A function $condition : N_{seq} \cup N_{loop} \rightarrow 2^{Pr}$ associates with each sequence or loop fragment a condition. An empty condition is evaluated as $true$.*

2.2 Catalog of Violations

We are concerned with resolving violations to execution ordering rules. Generally, execution ordering rules can be divided into *leads to* and *precedes* rules [2]. Informally, a rule *A leads to B* requires that after every execution of A, B must *eventually be executed*. On the other hand, a rule *A precedes B* requires that before executing B, A must have been executed before. Variants of these rules can be derived [4]. In this paper, we develop algorithms for resolving the *leads to* rules. These algorithms can be symmetrically applied to the *precedes* case and adapted to the variants. In general, four types of violation can be identified:

Splitting Choice. Activity A executes and B can be skipped due to the choice of an alternative branch.
Different Branches. If A and B are on different threads of the process.
Inverse Order. If A and B appear in order different then specified by the rule.
Lack of Activity. Depending on the rule type, for instance *A leads to B*, if a process model lacks activity B.

Since a process model might contain more than one occurrence of the activities A and B under investigation, we assume a priori knowledge about the pairing of such occurrences. In this paper we study in detail the different violations to the *leads to* rules. All results achieved can be symmetrically applied to the case of *precedes* rules.

2.3 The Resolution Context

The resolution context represents a global process independent description of the business activities. This context describes various relations between activities. We call these relations $aspects$ of the context. With the notion of the context, we try to simulate the knowledge needed while process models are first composed or later on modified.

Definition 6. The *resolution context C* is a 7-tuple $(N_{act}, A, T, asptype, con_{t\in T}, pre_{t\in T}, post_{t\in T})$, where:

- N_{act} is the set activities
- A is the set of objects, which define model aspects
- T is the set of aspect types
- $asptype : A \rightarrow T$ is the function relating each object to a particular aspect
- $con : N_{act} \times N_{act}$ is the relation between two activities, indicating contradiction
- $pre_{t\in T} \subseteq N_{act} \times 2^{\{a:\forall a\in A, type(a)=t\}}$ is the relation defining the prerequisites of activity execution in terms of aspects
- $post_{t\in T} \subseteq N_{act} \times 2^{\{a:\forall a\in A, type(a)=t\}}$ is the relation defining the result of activity execution in terms of aspects. Postconditions of an activity can be divided into positive and negative, i.e.,$post_{t\in T} = post^{+}_{t\in T} \cup post^{-}_{t\in T}$

One *aspect* of a context is a tuple $(N_{act}, A_t, t, asptype, con_t, pre_t, post_t)$, where $t \in T \wedge A_t = \{a : a \in A, asptype(a) = t\}$.

Definition 6 captures aspects with three elements: sets A and T and function $asptype$. Set A is the set of objects, describing the business environment from a certain perspective, e.g., dependencies of activities on data objects or their semantic annotations. Set T consists of the object types; an example is $T = \{activity, data, semantic\ Annotation\}$. Function $asptype$ specifies a type (element of set T) for an element of A. Typification of objects allows distinguishing aspects, e.g., distinguishing data flow from semantic description of a process.

The three basic relations are pre, $post$, and con. They respectively describe preconditions, postconditions, and contradiction relations. The pre relation describes what objects with different (types) aspects that are required for a certain activity in order to execute it. Similarly, the $post$ relation specifies what are the effects of executing a certain activity. For each activity there might be different sets of pre/post conditions to resemble the notion of alternation. Taking data as an aspect to describe such relations, pre_{data} would describe the precondition of each activity in terms of data elements. The con relation describes contradictions between activities. If two activities are known to be contradicting, at most one is allowed to appear in any process instance.

We assume process models to be consistent with the context: at least one precondition for any activity in the model must be satisfied in the process model. Also, activities are assumed to produce the effect, post condition(s), as described in the context. Moreover, for any two contradicting activities, there must not be a chance to execute both of them in a single instance.

Definition 7. *A process model and its PST $P = (N, r, E, type)$ are consistent with a resolution context $C = (N_{act}, A, T, asptype, con_{t\in T}, pre_{t\in T}, post_{t\in T})$ if:*

- *Preconditions are satisfied:* $\forall n \in N_{act} \wedge \forall t \in T \wedge pre_t \neq \emptyset \wedge \forall a \in A \wedge asptype(a) = t \wedge (n, a) \in pre_t : n \in N \rightarrow \exists m \in N_{act} : m \in N \wedge (m, a) \in post_t \wedge ((type(lca(m, n)) = seq \wedge order(lca(m, n), m) < order(lca(m, n), n)) \vee (type(lca(m, n)) = loop \wedge$ *m is on the mandatory part of the loop*$)$
- *No two contradicting execute in the same instance:* $\forall n, m \in N_{act} \wedge (n, m) \in con : N_n = \emptyset \vee N_m = \emptyset \vee \forall n' \in N_n \forall m' \in N_m type(lca(n', m')) = xor$

Similarly, the requirement imposed by a compliance rule has to be consistent with the context. For instance, a compliance rule must not impose order between two activities that are known to be contradicting according to the context. In this paper, we consider only rules that are consistent with the context; while might be violated by some process models. The resolution context plays central role assuring model consistency once changes have been applied to make them compliant.

2.4 Automated Planning

A violation resolution often implies that a business process logic is changed. The task is always to come up with a compliant model, fulfilling the business goal. This implies that a process should be reorganized to assure that requirements are satisfied. Given that the set of activities required to construct a compliant process is available in the context, this task can be approached with techniques of automated planning [21].

The problem of automated planning can be described as follows. Given a system in an initial state it is required to come up with a sequence of actions, bringing the system to the goal state. The sought sequence of actions is called a plan. A system can be represented as a state-transition system which is a 3-tuple $\Sigma = (S, A, \gamma)$, where S is a finite set of states, A is a finite state of actions, and $\gamma : S \times A \rightarrow 2^S$ - a state transition function. A planning task for system $\Sigma = (S, A, \gamma)$, an initial state s_0, and a subset of goal states S_g is to find a sequence of actions $\langle a_1, a_2, \ldots, a_k \rangle$ corresponding to a sequence of transitions $(s_0, s_1, \ldots, s_k)$ such that $s_1 \in \gamma(s_0, a_1), s_2 \in \gamma(s_1, a_2), \ldots, s_k \in \gamma(s_{k-1}, a_k)$ and $s_k \in S_g$.

To formalize the resolution problem in terms of automated planning we need to explain what are $\Sigma = (S, A, \gamma)$, s_0, and S_g. The system Σ is a business environment, where a business process is executed and which evolves as the next activity completes. Hence, actions in the planning task are associated with instances of activities described by the business context, while system states—with the states of the environment where a process executes. Function γ defines transition rules in the planning domain. In the process domain the context defines the preconditions and effects of activities, which aligns with the transition function. A transition from the current state to the next state via application of an activity results in removing of all the effects defined by $post^-$ relation and adding the effects defined by $post^+$ relation. The current state reflects the effects of all the activities which have taken place. Initial state s_0 corresponds to the state of the environment before a certain activity of the process took place. Set S_g consists of the states in which the business goal of the process is fulfilled and a compliance rule is not violated. The states can be described in terms of first order logic (notice that a compliance rule can be described in first order logic as well). Finally, we argue that the resulting plan corresponds to one instance of the business process. To retrieve a process model, all possible plans satisfying the initially stated goal should be considered. One possible approach enabling the construction of a model from several plans is process mining [23].

We propose to avoid construction of a business process from scratch. The preferable strategy is to identify a process fragment whose update is enough for achieving compliance. In [3] we have developed an approach enabling violation handling on the structural level. According to this approach, violations are classified into 4 categories and for each category an appropriate resolution technique is applied. In Section 4, we will demonstrate how automated planning techniques are employed in each case.

3 Motivating Example

We introduce an example, to be used throughout the paper to illustrate the ideas. The example includes the resolution context and business process fragments. The resolution context is defined by the tuple $(N_{act}, A, T, type, con_{t \in T}, pre_{t \in T}, post_{t \in T})$. The set of activities N_{act} is formed by *Go to checkout, Notify customer, Pay by credit card, Pay by*

Table 1. Pre and post relations of the example resolution context

Activity	Precondition	Postcondition	
		Negative	Positive
Go to checkout	order [init]	order [init]	order [conf] payment method [card]
Go to checkout	order [init]	order [init]	order [conf] payment method [transfer]
Notify customer	order [conf] payment [received]	notification [init]	notification [sent]
Pay by credit card	payment method [card] card data [filled]	payment [init]	payment [received]
Pay by bank transfer	payment method [transfer] bank data [filled]	payment [init]	payment [received]
Provide bank data	payment method [transfer] bank data [unfilled]	bank data [unfilled]	bank data [filled]
Provide credit card data	payment method [card] card data [unfilled]	card data [unfilled]	card data [filled]
Prepare goods	order [conf]	goods [init]	goods [prepared]
Send goods	address [filled] payment [received] goods [prepared]	goods [prepared]	goods [sent]
Provide shipping address	order [conf]		address [filled]
Cancel order	order [init]	order [init]	order [canceled]
Archive order	order [conf] goods [sent] payment [received]	order [conf] order [conf] order [conf]	order [archived] order [archived] order [archived]
Archive order	order [canceled]	order [canceled]	order [archived]

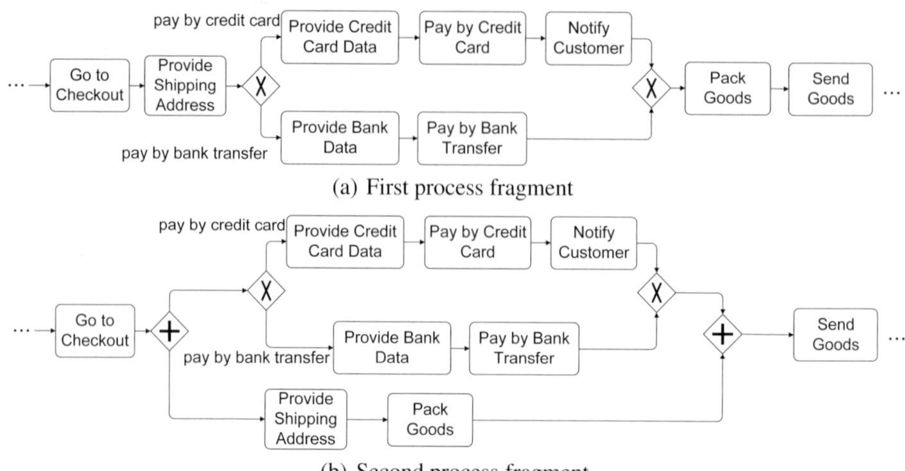

(a) First process fragment

(b) Second process fragment

Fig. 1. Two process fragments consistent with the context

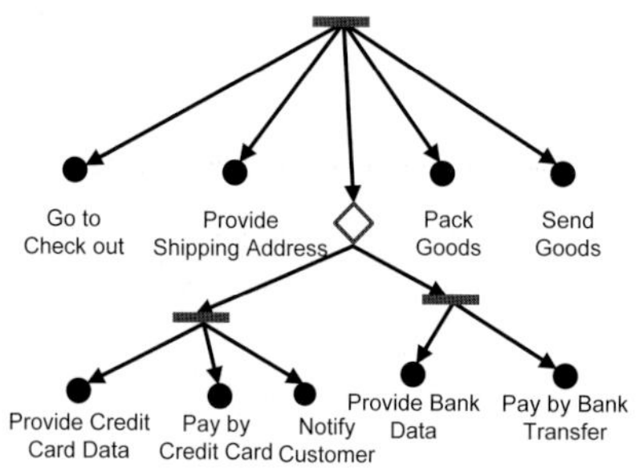

Fig. 2. The PST of the process model in Fig. 1(a)

bank transfer, *Provide bank data*, *Provide credit card data*, *Prepare goods*, *Send goods*, *Provide shipping address*, *Cancel order*, *Archive order*. In the example we consider one aspect—data flow. Hence, set T contains one element *data object*. The set of objects A is the set of data objects *address*, *bank data*, *card data*, *goods*, *order*, *notification*, *payment*, *payment method*. Subsequently, function $type$ relates each of the data objects to type *data object*. Table 1 captures $pre_{t \in T}$ and $post_{t \in T}$ relations. An activity may have more than one pre- or postcondition. For instance, activity *Archive order* expects either a confirmed order, received payment, and sent goods, or it expects the canceled order. In this way it is possible to express disjunctive preconditions. Activities *Pay by credit card* and *Pay by bank transfer* are the only contradicting activities.

Based on the above resolution context, the two process fragments shown in Fig. 1 represent two different ways of composing the activities, where the composition is consistent with the context. The tree representation of the process in Fig. 1(a) is shown in Fig. 3. In that figure, horizontal bars represent *sequence* blocks, diamond represents the choice block, and black circles represent the activities.

4 Resolving Violation

In this section we explain how compliance rule violations can be resolved. We base the resolution strategy on the violation patterns discussed in Section 2.2. The resolution technique to a large extent exploits automated planning.

Obviously, there are more than one possible way to resolve a certain violation [3]. However, not every resolution could be acceptable. For instance, one way of resolving a violation for a rule *A leads to B* would be by simply removing occurrences of A from the process model. Obviously, this is not acceptable as the resulting process might not be consistent with the context, i.e., it has broken dependencies. Thus, for each violation pattern, we discuss resolution strategies that guarantees both consistency and compliance, if possible.

Recalling the discussion about planning in Section 2.4, we use the function $findPlan(init, goal, condition, context)$ in our algorithms to encapsulate the call for an AI planner. The parameter $init$ describes the initial state for the planner. The $goal$ parameter determines which activity(ies) that have to be executed as goals for the planner. Moreover, the $condition$ parameter might be used to express extra constraints on the goal state of the planner. Finally, the $context$ parameter is the encoding of the resolution context. Resolution context is the domain knowledge used by the planner to find a plan.

Before we discuss violation resolution, we explain how the initial and goal states are calculated for the $findPlan$ function. Generally, for a rule *Source leads to Destination*; the $initial$ parameter reflects the execution of a set of activities from starting of the process up to and including the $source$ activity. $source$ is an occurrence of the Source activity in the process, i.e., $source \in N_{Source}$. We use the notion $source^-$ to reflect the execution history before $source$.

As discussed in Section 2.4, there might be more than one $initial$ state to check depending on the number of alternative branches. For instance, calling $findPlan($ $Pack\ Goods^-, Archive\ Order, true, context)$, regarding the process in Fig. 1(a), would constitute two initial states; reflecting the alternating branches. To succeed with resolving violation, a plan has to be found in each case.

4.1 Splitting Choice Violation

A violation of a compliance rule *A* `leads to` *B* is categorized as *splitting choice* if:

- a process model contains occurrences of activities A and B;
- there is a pair of nodes a and b, which are occurrences of A and B, respectively;
- b belongs to the path from a to a process end;
- there is a path from a to a process end, which does not contain occurrences of B.

Informally speaking, the model allows execution of A giving an option to skip activity B. The cause of splitting choice violation is the split node allocated on the path between occurrences of A and B.

Let us turn to the example process fragment in Fig. 1. The process fragment violates the compliance rule *Go to checkout* `leads to` *Notify customer*. *Go to checkout* activity is succeeded by the choice block. While one branch of the choice block contains activity *Notify Customer*, the other does not.

```
input : m—process structure tree
input : a, b ∈ N—the occurrences of activities A and B, respectively
input : c—resolution context
output: m—updated process structure tree
1  s = lca (a,b) ;
2  if type(s) = loop then
3      plan = findPlan (a^-, b, exitCondition (s), c);
4      if plan = ∅ then
5          return null;
6      insert plan into m exactly after the loop exit;
7  else if type(s) = seq then
8      x is the choice block containing b;
9      forall branch is a branch of x with no b do
10         if branch has activities contradicting b then
11             remove branch from m ;
12         else
13             plan = findPlan (a^-, b, condition (branch), c) ;
14             if plan = ∅ then
15                 remove branch from m;
16             else
17                 add plan to m merging it into branch;
18 return m;
```

Algorithm 1. Resolution of a *splitting choice* violation

A violation resolution implies that a process model is modified in such a way that activity B is always executed after a. We aim at introduction of local modifications to the model. Hence, we first identify the smallest fragment of a model, whose modification can be sufficient for the violation resolution. Afterwards, the fragment is modified in the way that execution of B is assured. Algorithm 1 provides a deeper insight to this approach. Initially, a block enclosing occurrences of activities A and B is sought (see line 1 in Algorithm 1). If the enclosing block is a loop, the algorithm has to assure that B is executed after the loop. Automated planning attempts to construct a suitable plan containing activity B. If the plan is constructed, it is inserted exactly after the loop block. Otherwise, the resolution cannot be performed. If the block is not a loop, we seek inside it for a choice block containing b on its branch (line 7). For each block branch missing B, we check if it has activities contradicting B, with respect to the resolution context C. If a contradiction exists, the branch is removed (line 11). In case of no contradictions, we use automated planning based on the information of the resolution context to find a path from a to b under the branch condition (see line 13). If no plan could be found, the branch is removed from the model. Otherwise, the found plan is merged to the branch to enforce compliance.

In the context of the example in Fig. 1(a) an occurrence of activity *Notify customer* is added to the branch where it was missing. The occurrence is added to the branch after activity *Pay by Bank Transfer*.

We argue that the proposed algorithm is correct, i.e., it resolves the violation of type splitting choice and delivers a consistent model free of contradictions.

Theorem 1. *If a process model contains a violation of type* splitting choice, *Algorithm 1 resolves this violation and delivers a model which is consistent with the context.*

Proof. To prove the theorem we have to show that:

- in the resulting process model the violation is resolved;
- the model does not have inconsistencies or contradictions.

Algorithm 1 localizes changes in the fragment s which is the least common parent of a and b. We focus the analysis on fragment s as well.

If fragment s is of type $loop$, a plan containing an occurrence of B is created and inserted exactly after the loop. This assures the violation resolution. The resulting plan is consistent with the process model, as this is one of the requirements to the planner.

If fragment s is a sequence, we identify a choice block within s, let it be x. Block x contains an occurrence of B on at least one of the branches. New occurrences of B are added to the branches where B is missing. If a branch has an activity contradicting B, the branch is removed. If a there is no contradicting activities, a plan containing B is designed by the planner and added to the branch. At this stage every branch in the choice contains B and, there is no way to skip B execution in the block. Thus, we have shown the first statement of the theorem.

Every branch which initially contained B is consistent and free of contradictions. Let us look at those branches where we add occurrences of B and their prerequisites. If such a branch has a contradiction or introduces an inconsistency, it is deleted. Removal of the branches does not lead to inconsistencies. Activities preceding a branch to be removed do not depend on activities of this branch. Activities succeeding the removed branch expect the effects of execution of at least one branch in the choice. While at least one branch of the original process is preserved, no inconsistencies can be caused by the branch removal. Thus, the resulting model is free of inconsistencies or contradictions. Thereby, we proved the second statement and completed the proof. □

4.2 Different Branches

A violation is of type *different branches* if a process model contains a and b (occurrences of activities A and B, respectively) and there is no path leading from a to b. A violation of this type takes place if the two activities are allocated on different branches of a block. One can notice that the violation occurs independent of a block type, i.e., in an OR, XOR, or AND block. However, the resolution strategy varies depending on the block type. Before we turn to a discussion of resolution strategies, let us illustrate different branches violation by an example. *Different branches* violation is shown in Fig. 1(b), where *Pay by credit card* and *Pack goods* activities are executed in parallel. Once the company policy requires to pack the goods after receiving the payment, the business process becomes non-compliant.

In case of an AND block, the resolution strategy aims at sequentializing A and B. To achieve the sequential execution of A and B, we move an occurrence of B from a block branch to the position exactly after the block. However, such a manipulation with an occurrence of B might introduce inconsistencies into the process model: there might be activities on the branch expecting B in the initial place. Hence, we move not only an occurrence of B, but the set of activities succeeding B on the branch and depending on B. An alternative strategy is to allocate a and move it, together with preceding activities on which a depends, exactly before the block. The preference to one of these strategies can be given basing on the number of activities to be moved.

```
input : m—process structure tree
input : a, b ∈ N—the occurrences of activities A and B, respectively
input : c—resolution context
output: m—updated process structure tree
s = lca (a,b) ;
switch type(s) do
    case AND
        PRE_a is the set of all nodes that execute before a within the same thread;
        POST_b is the set of all nodes that execute after b within the same thread;
        if |PRE_a| <|POST_b| then
            move PRE_a before the parallel block;
        else
            move POST_b after the parallel block;
    case XOR
        forall branch is a branch of x with a, but no b do
            resolve Lack of activity violation;
    case OR
        restructure the model;
        call to other resolution strategies;
return m;
```

Algorithm 2. Resolution of a *different branches* violation

The resolution of a different branches violation in a XOR block is reducible to the resolution of lack of activity violation. Indeed, the branch containing a misses an occurrence of B. Thus, the technique for lack of activity violation resolution is applicable for this branch (see Section 4.4).

Resolution of different branches violation in an OR block profits from the other resolution strategies. First, we replace an OR block with a combination of AND and XOR blocks, exhibiting the same behavior. The restructuring of the process relies on the approach introduced in [5]. Fig. 3 illustrates the solution for a trivial case of an OR block with two alternative branches. Once the OR block is replaced, other resolution strategies can be invoked.

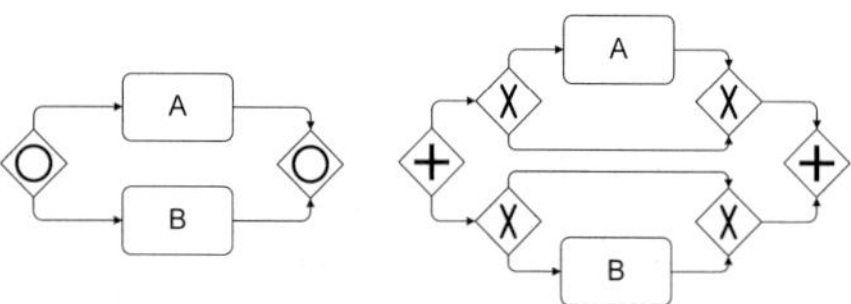

(a) Initial fragment (b) Transformed fragment

Fig. 3. Example of overwriting mechanism

Returning to the running example in Fig. 1(b), the resolution algorithm moves activity *Pack Goods* from the lower branch of the parallel block to the position between the AND join and *Send Goods* activity.

Theorem 2. *If a process model contains a violation of type* different branches, *Algorithm 2 resolves this violation and delivers a model which is consistent with the context.*

Proof. Similar to the case described in Theorem 1, it is required to show that in the resulting process model a violation is resolved and the resulting model does not have inconsistencies or contradictions.

Let us consider the AND block, XOR block, and OR block one by one. In case of an AND block, an occurrence of B is moved to the position directly after the block. After this modification B is always executed after A, as B directly succeeds the parallel block containing A. The compliance requirement holds. The initial model is free of contradictions and inconsistencies. Moving the occurrence of B together with dependent activities does not introduce contradictions, not inconsistencies. This is true, since the performed sequentialization only restricts the initial model.

An OR block is reduced to a combination of XOR and AND blocks. A transition to XOR and AND blocks and the resolution strategies for these two blocks define the properties of OR block resolution. The transformation from an OR block to the combination of XOR and AND blocks does not introduce inconsistencies and contradictions, since the new construct exhibits the same behavior. We have already argued about the properties of an AND resolution. For the XOR case we employ the resolution of *lack of activity* violation. In Section 4.4 we will argue that lack of activity violation meets the stated requirements. Thus, for the OR and XOR blocks the desired properties hold. □

4.3 Inverse Order

A process model violates a compliance rule A `leads to` B if it contains a and b (occurrences of activities A and B, respectively) connected with a path, but this path leads from b to a. The inverse order violation can be illustrated by the process fragment in Fig. 1(a), where the company sends a notification with an order summary to a customer. Afterwards, the company contacts its logistics partner to pack and send goods. New business conditions might require the company to include the delivery information in the notification, i.e., first the goods should be packed. This requirement is captured in the rule *Pack goods* `leads to` *Notify customer* rule. This is an *inverse order* violation.

To resolve an *inverse order* violation, we propose to analyze the fragment from activity B to A. The main idea is to reorder A and B and achieve model compliance. Reordering of the activities A and B is feasible, if there are no dependencies of activity A on B. If a dependency exists, reordering introduces inconsistencies into a model. To answer if reordering is feasible, we attempt to construct a model fragment from the process start to the point when A is executed. To come up with this fragment, we construct a plan. In contrast to the initial model, activity B should not appear in this plan. The plan construction is approached with automatic planning techniques. The initial state of the planning task reflects the process state directly before activity B is executed. The goal state describes the process after activity A is executed. If the planner comes up with the plan, the reordering is feasible.

Once reordering turns out to be feasible, the designed plan must be complemented to assure execution of B. Again the planning task is carried out by the planner. The initial state of this new plan corresponds to the goal state of the previous step. The new goal state describes the process state directly after an execution of B in the initial process. The designed plan has to be inserted into the model. If after insertion of the plan, the model is free of contradictions and inconsistencies the resolution is completed. In the opposite case we say that the automatic resolution is not feasible.

In the example of *Pack goods* `leads to` *Notify customer* rule violation the resolution strategy moves the occurrence of *Notify Customer* activity to the position after *Pack Goods* activity.

Theorem 3. *If a process model contains a violation of type* inverse order, *the proposed resolution strategy resolves this violation and delivers a model which is consistent with the context.*

Proof. We have to show that in the resulting process model a violation is resolved and the resulting model does not have inconsistencies or contradictions.

If the planner succeeds with plan construction, the resulting plan contains B. As the plan is inserted into the process model exactly after A, A `leads to` B holds. The modifications to the model are limited to adding new fragments constructed by the planner. As the planner uses the context, the designed plans are free of contradictions and inconsistencies. Adding the plan into the model, we check if it has any contradictions with the rest of the model. If it is the case, the plan is not accepted. Hence, the resulting model (if produced) is free of contradictions and inconsistencies. □

4.4 Lack of Activity

A process model contains a violation of type A `leads to` B if it contains at least one occurrence of A and no occurrence of B. Consider a compliance rule *Go to checkout* `leads to` *Archive order*. Checking the process model fragment in Fig. 1(a) against this rule, we see that this rule is violated, since *Archive order* is missing in that fragment.

To resolve a violation of this type we introduce an occurrence of B into the process model exactly after an occurrence of A. If the process model does not contain activities contradicting to B, we construct a plan using *findPlan(A, B, true, context)*. The plan is merged into the process model directly after an occurrence of A. In case there is an activity contradicting to B, let it be C, the resolution requires extra actions. The actions depend on the relations between occurrences of A and C:

- occurrences of A and C are allocated on different branches of a choice block;
- occurrences of A and C are allocated on different branches of a parallel block;
- an occurrence of C is allocated before A;
- an occurrence of C is allocated after A.

In the first case, B can be added to the process model exactly after A. As the branch with an occurrence of A does not contain activities contradicting B, B can be introduced to this branch without any conflicts. In the latter three cases a process model contains occurrences of activities contradicting B. We propose to introduce an occurrence of B into the model in such a way that B and C appear on alternative branches. We first seek for a SESE fragment containing an occurrence of C and activities tightly coupled with C. Such a process fragment contains activities which are transitively dependent on C or on which only C transitively depends. The fragment is preceded by an activity, let it be pre, and succeeded by an activity—$post$. We aim at complementing the process model with a branch, alternative to the identified fragment with C and containing B. We can obtain such a sequence as a result of a planning task, requiring it to fit between pre and $post$ and containing B. Finally, we add the choice block with the two branches into the model: the branch with C and the branch with plan containing B.

Identification of the activities dependent on C is based on the analysis of the resolution context and can be found as the closure of all the activities transitively depending on C. Similarly, the activities on which only C transitively depends can be found. Once

```
input : m—process structure tree
input : a ∈ N—is the occurrences of A for which the violation has to be resolved
input : b ∈ N—a new occurrence of B to be added to the tree
input : c—resolution context
output: m—updated process structure tree
CON_b is the set of activities contradicting b based on c;
if CON_b ≠ ∅ then
    forall x ∈ CON_b do
        if type(lca(x,a)) = choice then
            if findPlan (a⁻, b, true, c) ≠ ∅ then
                insert findPlan (a⁻, b, true, c) after a in m;
            else
                return null;
        else
            Find pre, post;
            if pre = null ∨ post = null then
                return null;
            FRAG_x contains activity x and its tightly coupled activities;
            FRAG_b contains activity b and the results of findPlan (pre⁻, b, true,
            c) and findPlan (b⁻, post, true, c);
            insert FRAG_x and FRAG_b in a choice block between pre and post;
else
    if findPlan (a⁻, b, true, c) ≠ ∅ then
        insert findPlan (a⁻, b, true, c) after a in m;
    else
        return null;
return m;
```

Algorithm 3. Resolution of a *lack of activity* violation

the model fragment is identified, it is possible to learn its first preceding activity—*pre* and the first succeeding—*post*.

As the result of the described model transformation, the violation type is no longer *lack of activity*. Instead, it changes either to *inverse order*, *splitting choice*, or *different branches* violation. Notice that a *lack of activity* violation can be reduced to *different branches* violation and vice versa. However, there is no mutual dependency between them, as we reduce these violations to not intersecting subcases of violations. Algorithm 3 summarizes the approach.

Returning to the example with a compliance rule *Go to checkout* `leads to` *Archive Order* and the process fragment in Fig. 1(a), according to the resolution strategy and the resolution context, activity *Archive Order* is added after goods are sent.

The resolution strategy for *lack of activity* violation reduces the violation to the previous three cases: *inverse order*, *splitting choice*, and *different branches*. Hence, its properties, i.e., model compliance, freedom of contradictions and inconsistencies originate from the properties of resolution methods for the named violation types. However, we have already shown that the resolutions for the three violations satisfy the stated requirement. Fig. 4 illustrates how a lack of activity violation is reduced to other violation types and is resolved.

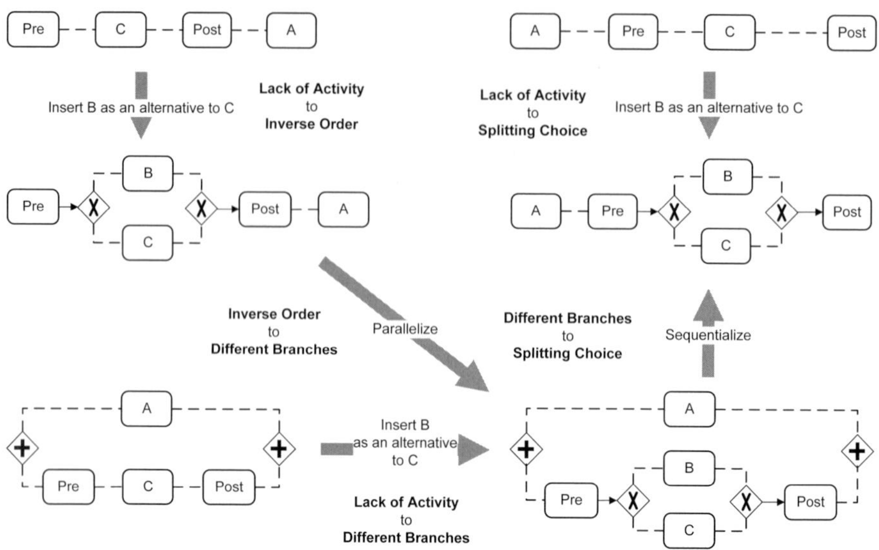

Fig. 4. Resolution of *lack of activity* violation in case of contradictions

4.5 The Overall Resolution Process

According to the discussion above, it might be the case that to resolve some violations is to first transform it from a violation type to another. For instance, in some cases of lack of activity violation, we transform it first to a splitting choice violation. Later on, we apply the algorithm of splitting choice to completely resolve the violation. Thus, the compliance violation resolution is of iterative nature. In each iteration, the violation type is recognized and the appropriate violation resolution algorithm is applied. This process is repeated until either no further violations are recognized or at one step the resolution algorithm either fails to resolve the violation or to transform it to another type. Algorithm 4 summarizes our approach to resolve compliance violations.

Utilizing the priori of pairs of activity occurrences, the identification of violation type is achieved in a polynomial time. The presented approach is prototypically implemented in our compliance management tool chain.

5 Related Work

An essential problem of compliance is identification of violations. Today a large body of research on compliance checking of business process models is published. Our primary interest is the work on execution order compliance checking. The research in this area can be divided into two directions: compliance by design and compliance checking of existing models. Compliance by design is to enforce process model compliance already at the stage of design. [8,9,13,19,20] show how this approach can be realized. A limitation of such approach is the need to recheck a process model once a rule is changed or newly introduced. Thus, it lacks the ability to resolve violations. The other branch of research employs model checking techniques to verify that existing process models satisfy the compliance rules. [2,6,15,27] consider this problem. Following this

```
input  : m—process structure tree
input  : A Leads to B
input  : c—the resolution context
output: m—violation free process structure tree
violation=getViolationType (m,A,B);
while violation ≠ none do
    switch violation do
        case Splitting Choice
            m = resolveSplittingChoice(m, a, b, c);
        case Different Branches
            m =resolveDifferentBranches (m,a,b,c);
        case Inverse Order
            m =resolveInverseOrder (m,a,b,c);
        case Lack of Activity
            m =resolveLackActivity (m,a,b,c);
    if m =null then
        return null;
    violation=getViolationType (m,A,B);
return m;
```

Algorithm 4. Compliance violation resolution

approach the authors of [10,11] use business contracts as the source of compliance rules. To formalize compliance rules Formal Contract Language (FCL) is used. [22] separates the process modeling from control objectives; it also uses FCL to express control requirements over annotated process models. These approaches are capable of *verifying* processes against compliance requirements. Again, the capability of resolving violations by means of changing the structure of the process through adding removing sets of activities is not addressed.

Resolution of violations can be driven by the severity of violations. In this sense an interesting method was introduced in [14] to measure the *degree* of compliance. Once a compliance rule is specified, the approach tells the degree of compliance for a given process model on the scale from 0 to 1.

Recently, several requirements frameworks for business process compliance management have been proposed. In [16] the authors formulate requirements for compliance tools. The requirements address the issues of lifetime compliance. Among other requirements, the authors discuss compliance enforceability. In this context they perceive the violation resolution as a human task, i.e., only human experts are responsible for taking remedy actions when a violation is discovered.

It should be noticed that [7] discussed possible strategies for resolving compliance violations. However, the discussion was very high level.

In this paper, we demonstrated that automated resolution of violations is achievable. This is to be considered as a step forward in supported automated compliance checking.

Automated planning techniques [21] have been used for service composition, e.g., in [18,17] for realizing service oriented architecture SOA. Unlike our approach, compositions are made from scratch. I.e., the user needs behind the service composition define the goal state to be reached where they start from a *clean* initial state. In our approach, we used the notion of cumulative effect to use planning to compose parts of the process that is related to the compliance rule. Thus, we developed more sophisticated techniques to construct the initial state for the planning problem.

6 Conclusions and Future Work

In this paper we have addressed the problem of compliance rule violation resolution. We focused on a special class of violations—execution order compliance rule violations. In order to cope with these violations we introduced a violation catalog, describing 4 violation types and their resolution methods. Once a violation is categorized according to the catalog, a proper resolution strategy is employed. The resolution strategies are based on automated planning techniques and an extensive description of a business domain—resolution context.

A violation resolution implies that a process model structure is changed. To control model structural modifications we employ the concept of SESE fragments. As a consequence, the approach is limited to process models, which are block-structured. To neglect this limitation advanced decomposition techniques, as described in [24], can be employed. This is the first direction of future work.

The proposed resolution strategy assumes that we know the nodes resulting in a violation. If each activity has a unique occurrence in the process model, identification of nodes causing a violation is straightforward. When activities have multiple occurrences, a mechanism for identification of occurrence pairs leading to a violation is required. A naive approach is to consider all combinations of node pairs. However, this problem is not trivial and is considered by us as another direction of future work.

The presented algorithms were dedicated to resolve violation to *leads to* rules. Adaptation of these algorithms to resolve violation to other rules is a future work.

Although the changes introduced by the resolution algorithms result in consistent and compliant processes models, an updated model might look unnatural for a human reader. For instance, sequentialization of two branches in case of different branches violation, leaves only one branch in the parallel block. Thus, a resulting process model needs refactoring to develop a naturally looking process model. Process refactoring techniques proposed in [26] can facilitate the problem solution.

References

1. Sarbanes-Oxley Act of 2002. Public Law 107-204 (116 Statute 745), United States Senate and House of Representatives in Congress (2002)
2. Awad, A., Decker, G., Weske, M.: Efficient Compliance Checking Using BPMN-Q and Temporal Logic. In: Dumas, M., Reichert, M., Shan, M.-C. (eds.) BPM 2008. LNCS, vol. 5240, pp. 326–341. Springer, Heidelberg (2008)
3. Awad, A., Smirnov, S., Weske, M.: Towards Resolving Compliance Violations in Business Process Models. In: GRCIS, vol. 459. CEUR-WS.org (2009)
4. Awad, A., Weske, M.: Visualization of compliance violation in business process models. In: 5th Workshop on Business Process Intelligence BPI 2009. Springer, Heidelberg (to appear, 2009)
5. Decker, G., Kopp, O., Leymann, F., Pfitzner, K., Weske, M.: Modeling Service Choreographies Using BPMN and BPEL4Chor. In: Bellahsène, Z., Léonard, M. (eds.) CAiSE 2008. LNCS, vol. 5074, pp. 79–93. Springer, Heidelberg (2008)
6. Förster, A., Engels, G., Schattkowsky, T., Van Der Straeten, R.: Verification of Business Process Quality Constraints Based on VisualProcess Patterns. In: TASE, pp. 197–208. IEEE Computer Society, Los Alamitos (2007)
7. Ghose, A., Koliadis, G.: Auditing Business Process Compliance. In: Krämer, B.J., Lin, K.-J., Narasimhan, P. (eds.) ICSOC 2007. LNCS, vol. 4749, pp. 169–180. Springer, Heidelberg (2007)

8. Goedertier, S., Vanthienen, J.: Compliant and Flexible Business Processes with Business Rules. In: BPMDS. CEUR Workshop Proceedings, vol. 236. CEUR-WS.org (2006)
9. Goedertier, S., Vanthienen, J.: Designing Compliant Business Processes from Obligations and Permissions. In: Eder, J., Dustdar, S. (eds.)BPD 2006. LNCS, vol. 4103, pp. 5–14. Springer, Heidelberg (2006)
10. Governatori, G., Milosevic, Z.: Dealing with Contract Violations: Formalism and Domain Specific Language. In: EDOC, pp. 46–57. IEEE Computer Society Press, Los Alamitos (2005)
11. Governatori, G., Milosevic, Z., Sadiq, S.: Compliance Checking between Business Processes and Business Contracts. In: EDOC, pp. 221–232. IEEE Computer Society Press, Los Alamitos (2006)
12. Hartman, T.: The Cost of Being Public in the Era of Sarbanes-Oxley. Foley&Lardner, Chicago, Ill (2006)
13. Lu, R., Sadiq, S., Governatori, G.: Compliance Aware Business Process Design. In: ter Hofstede, A.H.M., Benatallah, B., Paik, H.-Y. (eds.) BPM Workshops 2007. LNCS, vol. 4928, pp. 120–131. Springer, Heidelberg (2008)
14. Lu, R., Sadiq, S., Governatori, G.: Measurement of Compliance Distance in Business Processes. Inf. Sys. Manag. 25(4), 344–355 (2008)
15. Lui, Y., Müller, S., Xu, K.: A Static Compliance-checking Framework for Business Process Models. IBM Systems Journal 46(2), 335–362 (2007)
16. Ly, L.T., Göser, K., Rinderle-Ma, S., Dadam, P.: Compliance of Semantic Constraints – A Requirements Analysis for Process Management Systems. In: GRCIS, vol. 339. CEUR-WS.org (2008)
17. Marconi, A., Pistore, M., Traverso, P.: Automated Composition of Web Services: the ASTRO Approach. IEEE Data Eng. Bull. 31(3), 23–26 (2008)
18. Meyer, H., Weske, M.: Automated Service Composition Using Heuristic Search. In: Dustdar, S., Fiadeiro, J.L., Sheth, A.P. (eds.) BPM 2006. LNCS, vol. 4102, pp. 81–96. Springer, Heidelberg (2006)
19. Milosevic, Z., Sadiq, S., Orlowska, M.: Translating Business Contract into Compliant Business Processes. In: EDOC, pp. 211–220. IEEE Computer Society Press, Los Alamitos (2006)
20. Namiri, K., Stojanovic, N.: Pattern-Based Design and Validation of Business Process Compliance. In: Meersman, R., Tari, Z. (eds.) OTM 2007, Part I. LNCS, vol. 4803, pp. 59–76. Springer, Heidelberg (2007)
21. Nau, D., Ghallab, M., Traverso, P.: Automated Planning: Theory & Practice. Morgan Kaufmann Publishers Inc., San Francisco (2004)
22. Sadiq, S., Governatori, G., Namiri, K.: Modeling Control Objectives for Business Process Compliance. In: Alonso, G., Dadam, P., Rosemann, M. (eds.) BPM 2007. LNCS, vol. 4714, pp. 149–164. Springer, Heidelberg (2007)
23. van der Aalst, W., Weijters, T., Maruster, L.: Workflow Mining: Discovering Process Models from Event Logs. IEEE Trans. on Knowl. and Data Eng. 16(9), 1128–1142 (2004)
24. Vanhatalo, J., Völzer, H., Koehler, J.: The Refined Process Structure Tree. In: Dumas, M., Reichert, M., Shan, M.-C. (eds.) BPM 2008. LNCS, vol. 5240, Springer, Heidelberg (2008)
25. Vanhatalo, J., Völzer, H., Leymann, F.: Faster and More Focused Control-Flow Analysis for Business Process Models Through SESE Decomposition. In: Krämer, B.J., Lin, K.-J., Narasimhan, P. (eds.) ICSOC 2007. LNCS, vol. 4749, pp. 43–55. Springer, Heidelberg (2007)
26. Weber, B., Reichert, M.: Refactoring process models in large process repositories. In: Bellahsène, Z., Léonard, M. (eds.) CAiSE 2008. LNCS, vol. 5074, pp. 124–139. Springer, Heidelberg (2008)
27. Yu, J., Manh, T.P., Han, J., Jin, Y., Han, Y., Wang, J.: Pattern Based Property Specification and Verification for Service Composition. In: Aberer, K., Peng, Z., Rundensteiner, E.A., Zhang, Y., Li, X. (eds.) WISE 2006. LNCS, vol. 4255, pp. 156–168. Springer, Heidelberg (2006)

A Two-Stage Probabilistic Approach to Manage Personal Worklist in Workflow Management Systems*

Rui Han[1,2,3], Yingbo Liu[1,2,3], Lijie Wen[1,2,3], and Jianmin Wang[1,2,3]

[1] School of Software, Tsinghua University, Beijing, P.R. China, 100084
[2] Key Laboratory for Information System Security, Ministry of Education, P.R. China, 100084
[3] Tsinghua National Laboratory for Informativasron Science and Technology, P.R. China, 100084
{hanr07,lyb01,wenlj00}@mails.tsinghua.edu.cn, jimwang@tsinghua.edu.cn

Abstract. The application of workflow scheduling in managing individual actor's personal worklist is one area that can bring great improvement to business process. However, current deterministic work cannot adapt to the dynamics and uncertainties in the management of personal worklist. For such an issue, this paper proposes a two-stage probabilistic approach which aims at assisting actors to flexibly manage their personal worklists. To be specific, the approach analyzes every activity instance's continuous probability of satisfying deadline at the first stage. Based on this stochastic analysis result, at the second stage, an innovative scheduling strategy is proposed to minimize the overall deadline violation cost for an actor's personal worklist. Simultaneously, the strategy recommends the actor a feasible worklist of activity instances which meet the required bottom line of successful execution. The effectiveness of our approach is evaluated in a real-world workflow management system and with large scale simulation experiments.

Keywords: workflow management system, workflow scheduling, personal worklist management, probability.

1 Introduction

Workflow management systems (WFMSs) improve business processes by automating tasks and getting the right information to the right place for a specific job function in time [1]. One crucial area that WFMSs can bring great improvement to business processes is the application of scheduling in workflow [2, 3]. Existing techniques on workflow scheduling [2, 4, 5, 6, 7, 8] mainly concern about allocating tasks (activity instances) among multiple resources (mainly refers to actors, i.e., activity executors or

* Supported by the NSFC (90718010), National Basic Research Program (973 Plan) under grant No. 2009CB320700, National High-Tech Development Program (863 Plan) under grant No. 2008AA042301 and 2007AA040607, and Program for New Century Excellent Talents in University.

R. Meersman, T. Dillon, P. Herrero (Eds.): OTM 2009, Part I, LNCS 5870, pp. 24–41, 2009.

workflow participants). However, people are actually the driving force of workflow [9]. It is necessary to apply workflow scheduling to assist actors to manage their personal worklists, which contain activity instances (i.e., work items, this paper uniformly calls activity instances for convenience) that would be executed by the actors. Therefore, personal worklist management should be an important part of workflow scheduling and a necessary complement of current workflow scheduling techniques.

At present, there is little research on personal worklist management in WFMSs. In [10, 11], authors present preliminary work in this area. For an activity instance in the personal worklist, they analyze its several states of satisfying deadline together with discrete probabilities. Based on these deterministic analysis results, they sort activity instances in the personal worklist under rigorous assumptions (specified in Section 6).

Nevertheless, the dynamic nature of workflows causes highly uncertainties in the personal worklist management. From the aspect of process instance, two uncertainties, which are variations in activity (execution) durations and the unpredictability of execution paths, cause activity instances' continuous probabilities of satisfying deadlines. For example, in Figure 1, activity instance a_6's probability of satisfying deadline is 96.31%, and that probability is only 5~10% for activity instance a_5. From the aspect of personal worklist, new activity instances are continuously added to the personal worklist, and activity instances from multiple systems and process instances compete for execution during the same time interval. Hence some activity instances may have extremely low probabilities of satisfying the given deadlines, because these activity instances' executions are delayed by their preceding activity instances. For example, in Figure 1's personal worklist, probabilities of satisfying deadlines for activity instances a_1, a_2, a_4, a_7 to a_{10}, a_{12}, a_{14}, and a_{15} are all below 5%.

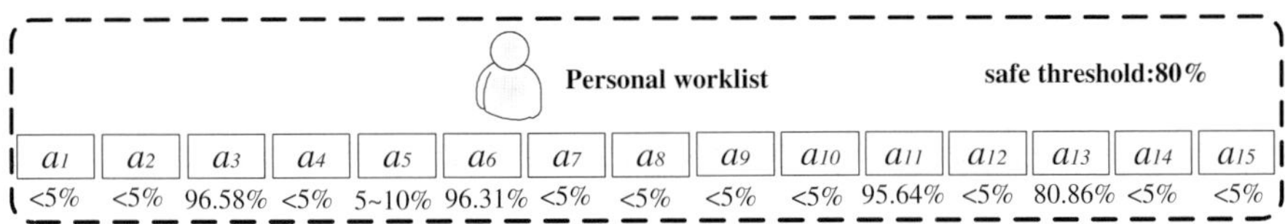

Fig. 1. An example of personal worklist

As stated above, serious deadline violations may occur in an actor's personal worklist, which cause considerable exception-handling (e.g., deadline-based escalation [12]) costs in the system. Moreover, the possibilities of violating deadline are stochastic due to the complex issues involved in the process instance and personal worklist. Therefore, traditional deterministic management of personal worklist is too rigorous.

In this paper, we proposed a two-stage probabilistic approach to provide stochastic and flexible management of personal worklist. Specifically, at the first process analysis stage, the approach analyzes the continuous probability of satisfying deadline for every activity instance. This probability is denoted as successful execution ratio and it provides a precise estimation to guide personal worklist scheduling. At the second worklist scheduling stage, an innovative scheduling strategy [13] is proposed to manage an actor's personal worklist. When new activity instances are added to the personal worklist, this strategy maintains a feasible worklist (named worklist *S*) and an infeasible worklist (named worklist *U*). The worklist *S* involves activity instances whose joint successful execution ratio meets the required bottom line of successful execution, i.e., a probability value denoted as safe threshold (e.g., 80% for Figure 1's personal worklist). The worklist *U* is comprised of other deadline-violating activity instances. The objective of our scheduling strategy is to minimize the total deadline violation cost for activity instances in worklist *U*, thus achieving improved performance of WFMSs.

The rest of the paper is organized as follows. Section 2 gives the overview of our two-stage probabilistic approach. Section 3 explains the first process analysis stage and Section 4 introduces the second worklist scheduling stage. Section 5 presents the implementation of our approach in a real-world WFMS and utilizes large scale simulation experiments to evaluate the effectiveness of our scheduling strategy. Section 6 discusses related work. Finally, Section 7 presents the conclusions and future work.

2 A Two-Stage Probabilistic Worklist Management Approach

In this section, we present the overview of our two-stage probabilistic worklist management approach. As shown in Table 1, the approach includes two stages and each stage is comprised of two steps.

Table 1. A two-stage probabilistic worklist management approach

Overview	**Input**: Process instance, workflow event log. **Method**: Two-stage probabilistic worklist management approach. **Output**: Feasible worklist *S*, infeasible worklist *U*.
Stage 1: Process analysis	**Step 1.1**: Modeling process instance with PTCWF-nets.
	Step 1.2: Analyzing every activity instance's successful execution ratio.
Stage 2: Worklist scheduling	**Step 2.1**: Adding a new activity instance to an actor's personal worklist.
	Step 2.2: Scheduling worklist *S* to meet the safe threshold, while minimizing the total deadline violation cost in infeasible worklist *U*.

At the process analysis stage, when a new process instance is initiated for execution, step 1.1 models this process instance with Probabilistic Time Constraint WorkFlow Nets (PTCWF-nets). PTCWF-nets extend classical Petri nets and Workflow nets (WF-nets) [14] with time information. Based on PTCWF-nets, step 1.2 analyzes activity instances' successful execution ratios and then allocates these activity instances. Generally, an activity instance is directed to an actor for execution [10, 15].

At the worklist scheduling stage, when new activity instances are added to an actor's personal worklist, the approach schedules the worklist. To be specific, for a newly added activity instance, step 2.1 inserts it into the feasible worklist *S*. Then, step 2.1 recalculates the joint successful execution ratio for activity instances in worklist *S*. Next, step 2.2 schedules worklist *S* to make the ratio meet the safe threshold by eliminating some activity instances. The eliminated activity instances are added to infeasible worklist *U* for exception handling. Step 2.1 and 2.2 would iterate several times if multiple activity instances are added. The approach performs immediately when the actor is idle based on two assumptions: 1) the execution of activity instance is non-preemptive [13]; 2) the scheduling takes a short time with respect to activity instances' execution time. Note that our approach recommends the actor a list of activity instances which ensure successful execution, while other deadline-violating activity instances' exception-handling costs are minimized. However, the real execution of activity instances is always up to the actors themselves.

3 Analysis of Process Instance

In this section, we introduce step 1.1 and 1.2 of our approach in subsection 3.1 and 3.2, respectively.

3.1 Modeling Process Instance with PTCWF-Nets

Inheriting and developing from Stochastic Petri nets (SPN) [16], PTCWF-nets model the process instance by adding time constraint C_P, time parameter C_T, and time parameter C_F to classical WF-nets' [14] set of places P, set of transitions T, and set of arcs F.

In PTCWF-nets, transitions represent activity instances. Therefore, time constraint C_P describes temporal relations between activity instances. Two time parameters separately represent two uncertainties in business processes (i.e., variations of activity execution durations and paths): C_T describes continuous distributed durations, and C_F represents probabilities of selecting execution paths. These time constraint and parameters can be obtained from workflow event log by statistical analysis [1, 17].

Definition 1. (Time constraint C_P). *Time constraint C_P is a finite set of pairs ($E_{\min}$, $E_{\max}$). For any place p ($p \in P$), c_p ($c_p \in C_P$) denotes token's available time for p's output transitions. To be specific, from the moment when p's all input transitions complete and a token arrives at p, $c_p.E_{\min}$ and $c_p.E_{\max}$ ($c_p.E_{\max} \geq c_p.E_{\min} \geq 0$) are separately token's earliest and latest available time for p's output transitions.*

For a place p, $c_p.E_{\min}$ represents the minimum time interval between the predecessor and successor activity instances. Similarly, $c_p.E_{max}$ represents the maximum interval. For example, in a process instance of altering contract, when activity instance *Countersign* is completed, its successor activity instance *Review* has to wait 2 hours (the minimum time interval) before being executable and should be finished within 58 hours (the maximum time interval).

Definition 2. (Time parameter C_T). *Time parameter C_T is a finite set of random variables which obey normal distribution $N(\mu, \sigma^2)$. For any transition t ($t \in T$), c_t ($c_t \in C_T$) denotes t's firing duration which is not instantaneous where $c_t \sim N(\mu_t, \sigma_t^2)$ and $\mu_t, \sigma_t^2 \in R^+$.*

Time parameter C_T represents the probability distribution of activity duration. Without losing generality, we assume that all activity durations follow the normal distribution model $N(\mu, \sigma^2)$, where μ is the expected value, σ^2 is the variance, and σ is the standard deviation [17]. Those follow non-normal distribution models can be treated by normal transformation [18].

Definition 3. (Time parameter C_F). *Time parameter C_F is a finite set of probability values. For any arc f ($f \in F$), c_f ($c_f \in C_F$) denotes the probability of selecting f where $c_f \in R^+$. In addition, $c_f = 1$ iff f belongs to a sequential or parallel path which is certainly be selected. $0 \le c_f \le 1$ iff f belongs to a selective or iterative path which is one of the n ($n \ge 1$) possible execution paths: $\{f_1, f_2, \ldots, f_n\}$, where $\sum_{i=1}^{n} c_{f_i} = 1$.*

Definition 4. (PTCWF-nets). *A PTCWF-net is a six tuple ($P, T, F; \theta_P, \theta_T, \theta_F$), where <P, T, F > is the classical WF-net. θ_P is **place** function. It is defined from P into C_P such that: $\forall p \in P$, $[\theta_P(p) \in C_P]$. θ_T is **transition** function. It is defined from T into C_T such that: $\forall t \in T$, $[\theta_T(t) \in C_T]$. θ_F is **arc** function. It is defined from F into C_F such that: $\forall f \in F$, $[\theta_F(f) \in C_F]$.*

PTCWF-nets have the property of safe and free-choice [14]. Also, PTCWF-nets are isomorphic to continuous time Markov chains [16]. The number of states of the Markov chain corresponds to the number of reachable markings of the PTCWF-nets. In addition, PTCWF-nets include four types of workflow control structures: sequential, parallel, selective, and iterative (Table 2). These four structures can be combined to model majority of typical business processes in various domains, such as manufacture factory, hospital, bank, and so on [14].

Table 2. Four types of workflow control structures in PTCWF-nets

Workflow control structure	Figure
Sequential	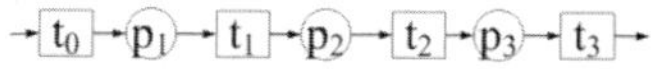 $t_0 \to p_1 \to t_1 \to p_2 \to t_2 \to p_3 \to t_3$
Parallel	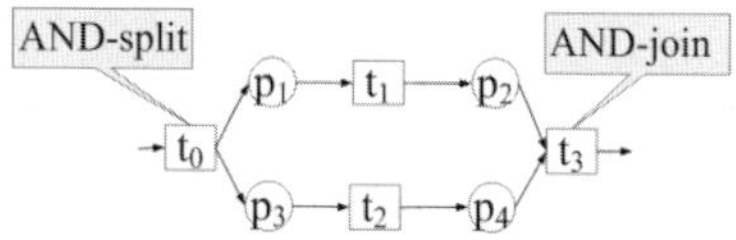 AND-split, AND-join; t_0, p_1, t_1, p_2, p_3, t_2, p_4, t_3
Selective	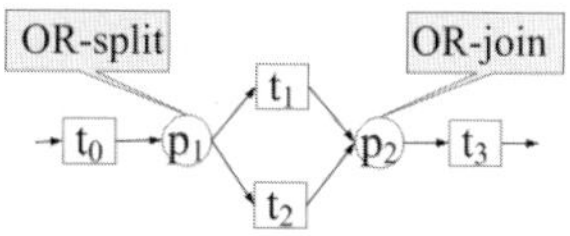 OR-split, OR-join; t_0, p_1, t_1, t_2, p_2, t_3
Iterative	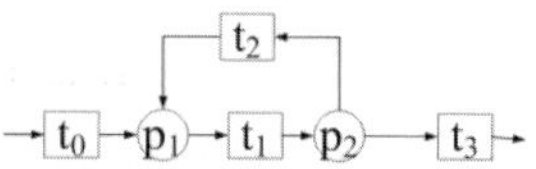 t_2; t_0, p_1, t_1, p_2, t_3

3.2 Analyzing Every Activity Instance's Successful Execution Ratio

After modeling the process instance with a PTCWF-net, the approach analyzes transition's successful firing ratio to represent activity instance's successful execution ratio. Figure 2 shows the pseudocode of the analyzing algorithm.

Input: The PTCWF-net with time constraints and time parameters
Output: Every transition's successful firing ratio

Method
1. a^i := c; // a^i is the token's arrival time of source place i, and c is a time constant
2. t := T. getTransitionByBFT();// obtain the first transition
3. WHILE $t \neq \emptyset$ DO
   BEGIN
4. $d^t_{min} := \mu_t - 3\sigma_t$; // minimum firing duration
5. $d^t_{max} := \mu_t + 3\sigma_t$; // maximum firing duration
6. e^t_{min} := Max{$a^p + \theta_P(p).E_{min} \mid p \in \bullet t$ }; // earliest enabled time
7. e^t_{max} := Min{$a^p + \theta_P(p).E_{max} \mid p \in \bullet t$}; // latest enabled time
8. $s^t := e^t_{min}$; // start time
9. r^t := P($d^t_{min} < \theta_T(t) < e^t_{max} - s^t$); // successful firing ratio
10. FOR EVERY p ($p \in t\bullet$) DO
    BEGIN
11. p.calculateTokenArrivalTime();
    // calculate p's token's arrival time
    END
12. t := T.getTransitionByBFT(); // obtain a new transition
   END

Function getTransitionByBFT()
// return a new transition through BFT of the PTCWF-net; return $\emptyset$ if there is no new transition in the PTCWF-net
Function calculateTokenArrivalTime()
// calculate place's token's arrival time

Fig. 2. Analyzing algorithm of successful firing ratio

Assumed that in the PTCWF-net, set T has n transitions, set P has m places, and set F has k arcs (n, m, $k \geq 1$). The algorithm in Figure 2 utilizes the basic breadth-first traversal (BFT) strategy to select transition from the PTCWF-net for analyzing its successful firing ratio (line 2 and line 12). The time complexity of this BFT strategy is $O(n+m+k)$ for directed graph PTCWF-net [19]. In addition, all operations in each analysis of a transition's successful firing ratio (line 4 to line 11) are linear time complexity. Therefore, the time complexity of the whole algorithm is $O(n+m+k)$.

As shown in Figure 2, to analyze a transition t's successful firing ratio (line 9), the algorithm first needs to calculate t's possible interval of duration ($d^t_{\min}, d^t_{\max}$) (line 4 and line 5), its enabled time interval ($e^t_{\min}, e^t_{max}$) (line 6 and line 7), and its start time s^t (line 8).

We employ the "3σ" rule to define transition t's possible interval of duration ($d^t_{\min}, d^t_{\max}$) where t's duration $\theta_T(t) \sim N(\mu_t, \sigma_t^2)$. This rule describes that for any sample comes from normal distribution model, it has a probability of 99.73% to fall into the range of [μ-3σ, μ+3σ] which is a systematic interval of 3 standard deviations around the expected value [17]. Therefore, t's minimum duration is defined as $d^t_{\min} = \mu_t - 3\sigma_t$, and the maximum duration is $d^t_{\max} = \mu_t + 3\sigma_t$.

Transition t is enabled if and only if all its input places have available tokens. For any place p ($p \in \bullet t$), its time constraint is $\theta_P(p)$ and its token's arrival time is a^p. According to definition 1, in place p, token's earliest available time for t is $a^p + \theta_P(p).E_{\min}$ and the latest time is $a^p + \theta_P(p).E_{max}$. Therefore, t's earliest enabled time $e^t_{\min}$ is the maximum one of token's earliest available time in t's input places: $e^t_{\min} = \mathrm{Max}\{a^p + \theta_P(p).E_{\min} | p \in \bullet t\}$. Similarly, t's latest enabled time e^t_{max} is the minimum one of token's latest available time in t's input places: $e^t_{max} = \mathrm{Min}\{a^p + \theta_P(p).E_{\max} | p \in \bullet t\}$. e^t_{max} is also t's deadline.

When transition t is enabled at time $e^t_{\min}$, it starts at time s^t after a time period Δt where $s^t = e^t_{\min} + \Delta t$. There are three situations for s^t : 1) t starts as soon as it is enabled if $s^t = e^t_{\min}$; 2) when t is enabled, it starts after a period of time if $s^t = e^t_{\min} + \Delta t$ and $e^t_{\min} < s^t \le e^t_{max}$; 3) t cannot start if $s^t > e^t_{max}$. In this paper, we discuss situation 1 and leave out situation 3. The corresponding discussion for situation 2 is similar as situation 1, because situation 2's definition style of start time ($s^t = e^t_{\min} + \Delta t$) is similar to that of situation 1 ($s^t = e^t_{\min}$).

As stated above, transition t's completion time is $(s^t + \theta_T(t)) \in i_1 = (s^t + d^t_{\min}, s^t + d^t_{\max})$. If t's enabled time interval is $i_2 = (e^t_{\min}, e^t_{max})$, the firing of t is considered as satisfying deadlines only if t completes before its maximum enabled time e^t_{max}, namely, $(s^t + \theta_T(t)) \in i_3 = (s^t + d^t_{\min}, e^t_{max})$, as shown in Figure 3. Therefore, t's successful firing ratio is the probability integral of random variable $(s^t + \theta_T(t))$ over the interval of i_3.

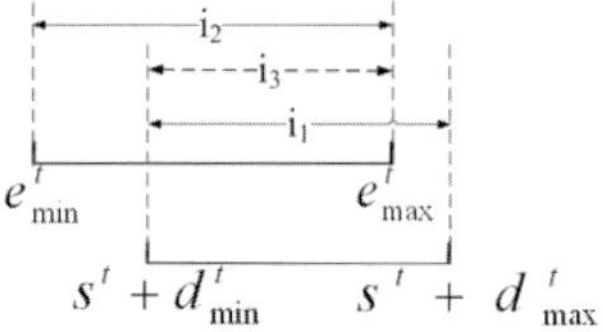

Fig. 3. Analyzing transition t's successful firing ratio

Definition 5. (Transition's successful firing ratio). *For a transition t in PTCWF-nets where $\theta_T(t) \sim N(\mu_t, \sigma_t^2)$, its successful firing ratio:*

$$
\begin{aligned}
&r^t = P(s^t + d^t_{\min} < s^t + \theta_T(t) < e^t_{max}), \\
&i.e.,\ r^t = P(d^t_{\min} < \theta_T(t) < e^t_{max} - s^t)
\end{aligned}
\tag{1}
$$

This ratio can be easily obtained if $\theta_T(t) \sim N(\mu_t, \sigma_t^2)$, because for a given random variable $x \sim N(\mu, \sigma^2)$, x's probability integral over interval (a, b) is [17]: $P(a<x<b)=$

$\Phi((b-\mu)/\sigma)$- $\Phi((a-\mu)/\sigma)$. Function $\Phi(y)$ is the standard normal distribution function where $y\sim N(0, 1)$.

Moreover, in Figure 2, the calculations of time variables $e^t_{\min}$, e^t_{max}, and s^t (line 6 to line 8) are related to token's arrival time of place p ($p\in \bullet t$). Given that, in the PTCWF-net, token's arrival time of the source place is a time constant (line 1). The algorithm calculates token's arrival time of other places by means of token's arrival time of their preceding places (line 11). In four workflow control structures, token's arrival time is calculated differently. In Table 3, we take place p_2 as our target to explain the calculation of token's arrival time.

Table 3. Token's arrival time of place p_2 in four workflow control structures

Workflow control structure	Token's arrival time of place p_2:
Sequential or Parallel	$a^{p_1} + \theta_P(p_1).E_{min} + \mu_{t_1}$
Selective	$\theta_F(f_1)\times(a^{p_1} + \theta_P(p_1).E_{min} + \mu_{t_1}) +$ $\theta_F(f_2)\times(a^{p_1} + \theta_P(p_1).E_{min} + \mu_{t_2})$
Iterative	$a^{p_1} + \theta_P(p_1).E_{min} + \mu_{t_1}$ for transition t_2, $a^{p_1} + (\theta_P(p_1).E_{min} + \mu_{t_1})\times(\theta_F(f_1)/\theta_F(f_2)+1)$ $+(\theta_P(p_2).E_{min} + \mu_{t_2})\times(\theta_F(f_1)/\theta_F(f_2))$ for transition t_3

– In **sequential** or **parallel** control structure, a^{p_2} is transition t_1's expected completion time ($s^{t_1} + \mu_{t_1}$) where $s^{t_1} = e^{t_1}_{\min} = a^{p_1} + \theta_P(p_1).E_{\min}$.

– In **selective** control structure, a^{p_2} is the probability synthesis of transition t_1 and t_2's expected completion time. t_1's probability of firing is $\theta_F(f_1)$ and its expected completion time is ($a^{p_1} + \theta_P(p_1).E_{\min} + \mu_{t_1}$). t_2's probability of firing is $\theta_F(f_2)$ and its expected completion time is ($a^{p_1} + \theta_P(p_1).E_{\min} + \mu_{t_2}$).

– In **iterative** control structure, a^{p_2} is different for p_2's output transition t_2 which is inside the loop body and t_3 which is outside the loop body. Statistically, each time t_3 fires, t_1 fires ($\theta_F(f_1)/\theta_F(f_2)$+1) times and t_2 fires ($\theta_F(f_1)/\theta_F(f_2)$) times. Besides, in each t_1's firing, t_1's holding time includes its waiting time before firing $\theta_P(p_1).E_{\min}$ and its expected firing duration μ_{t_1}. Similarly, t_2's holding time is ($\theta_P(p_2).E_{\min} + \mu_{t_2}$) per firing. For t_2, a^{p_2} is t_1's first expected completion time: $a^{p_1} + \theta_P(p_1).E_{\min} + \mu_{t_1}$. For t_3, a^{p_2} is the time when the iterative firings of t_1 and t_2 complete.

4 Scheduling of Personal Worklist

In this section, we introduce step 2.1 and 2.2 of our approach in subsection 4.1 and 4.2, respectively.

4.1 Adding a New Activity Instance to an Actor's Personal Worklist

To perform scheduling on an actor's personal worklist, the newly added activity instance *a* is associated with six time-related variables: start time s^a, expected duration μ_a, deadline $e^a_{\max}$, successful execution ratio r^a, fixed deadline violation cost c^a_f, and probabilistic deadline violation cost c^a_p. The first four time variables can be obtained from the process analysis stage of the approach. c^a_f is the specified exception-handling cost of activity instance *a* when *a* violates its deadline. c^a_f is set as a constant here with consideration of various conditions in systems (e.g., context, resource utilization) and process instances (e.g., importance, urgency). c^a_f can also be a cost function such as exponential function. Detail discussion about the cost function is beyond the scope of this paper. c^a_p is the estimated deadline violation cost of activity instance *a*, which is the probability synthesis of (1- r^a) and $c^a_p := c^a_f \times (1\text{-}\ r^a)$.

Assumed that, before the insertion of activity instance *a*, worklist *S* contains (*n*-1) activity instances: $\{a_1, a_2, \ldots, a_{n-1}\}$ (*n*>1). After insertion, the execution order of activity instances in worklist *S* are rearranged according to activity instances' start time. Given that after rearrangement, activity instance *a* is the ith one in worklist *S*, thus there are (*i*-1) activity instances $\{a_1, a_2, \ldots, a_{i-1}\}$ $(1<i<n)$ whose execution orders are before *a*, and (*n-i*) ones whose execution orders are later.

For any activity instance a_j $(i \le j \le n)$ which is sorted equal or after the newly inserted activity instance in worklist S, a_j's start time and successful execution ratio should be recalculated. Given that, a_j's previous activity instance is a_{j-1}. a_{j-1} starts at time s^{a-1}, its expected duration is μ_{a-1}, its deadline is $e^{a-1}_{\max}$, so it completes before or equal to its deadline: Min$\{s^{a-1} + \mu_{a-1}, e^{a-1}_{\max}\}$. a_j's start time is decided by the later one of its the original start time and a_{j-1}'s completion time: s^a=Max$\{s^a$, Min$\{s^{a-1} + \mu_{a-1}, e^{a-1}_{\max}\}$ $(i \le j \le n)$. Besides, a_j's successful execution ratio r^{a_j} should be updated along with a_j's new start time by means of formula 1 (definition 5, Section 3.2).

After updating activity instances' successful execution ratios in worklist *S*, the approach recomputes the joint successful execution ratio of activity instances in worklist *S*.

Definition 6. (Joint successful execution ratio of activity instances in worklist *S*). *There are n activity instances: $\{a_1, a_2, \ldots, a_n\}$ (n> 0) in worklist S after the insertion of a new activity instance. The joint successful execution ratio of these activity instances is the product of their successful execution ratios:*

$$r^s = \prod_{i=1}^{n} r^{a_j} \tag{2}$$

4.2 Scheduling of Worklist S

After recalculation of the joint successful execution ratio in worklist *S*, the approach compares this ratio with the safe threshold. If the ratio is greater than or equal to the safe threshold, the worklist *S* is maintained for execution without any scheduling. Otherwise, a scheduling strategy is proposed to eliminate some activity instances from worklist *S* until the ratio meets the safe threshold. This safe threshold also represents the required stability of actor's execution.

The safe threshold is set through a win-win negotiation between the system manager and actor, which considers both requirements of systems (e.g., performance, quality of service, stability, and so on) and conditions of actor (e.g., proficiency level, payment rate, workload, project experience, and so on).

The proposed scheduling strategy first selects the activity instance with the smallest successful execution ratio to eliminate, so this strategy is named Smallest Ratio First (SRF) strategy. Figure 4 shows the pseudocode of our strategy, whose time complexity is $O(n^2)$ given that worklist S contains *n* activity instances.

There are other four fundamental scheduling strategies [13]: First In First Out (FIFO), Latest Deadline First (LDF), Smallest Cost First (SCF), and Longest Excepted Duration First (LEDF). Strategy FIFO first removes the activity instance which is lastly added to the personal worklist, and Strategies LDF, SCF, and LEDF first remove the activity instance with the latest deadline, smallest fixed deadline violation cost, and longest expected duration, respectively. The effectiveness of our strategy with respect to other four strategies is evaluated in the next section.

In the scheduling of worklist *S*, the removed activity instances are added to infeasible worklist *U* for exception handling, such as adjusting deadline or reemploying other actors. How exception handling should be triggered is outside the scope of this paper and it is often related to some other overall aspects of the WFMSs such as quality of service [12].

```
Input: feasible worklist S, infeasible worklist U, safe
threshold r^Safe
Output: Scheduled Worklist S and U
Method
1. num := n;    // initial worklist S has n activity instances
2. While worklist S's joint successful execution ratio r^S < r^Safe
   BEGIN
3.   FOR i := 1 TO num DO
     BEGIN
4.     Find the activity instance a_i with smallest successful
       firing ratio r^{a_i};
     END
5.   S := S \ {a_i};    // remove activity a_i from worklist S
6.   num= num-1;
     // feasible worklist S has (num-1) activity instances
7.   U := U ∪ {a_i}; // add activity a_i to end of worklist U
8.   IF r^S ≥ r^Safe THEN RETURN;    //exit the scheduling strategy
  END
```

Fig. 4. Smallest Ratio First (SRF) scheduling strategy

After scheduling of worklist *S*, the approach recalculates the total deadline violation cost of activity instances in worklist *U*. The approach takes the probabilistic deadline violation cost as the measure of violation cost, because for an activity instance a_i, $c_f^{a_i}$ comprehensively considers both a_i's possibility of violating deadline $(1 - r^{a_i})$ and a_i's fixed deadline violation cost $c_f^{a_i}$.

Definition 7. (Total deadline violation cost of activity instances in worklist *U*). *There are m activity instances: {a_1, a_2, ..., a_m} (m>1) in worklist U after the scheduling strategy. The total deadline violation cost of activity instances in worklist U is the sum of their probabilistic violations costs:*

$$c^U = \sum_{i=1}^{m} (c_f^{a_i} \times (1 - r^{a_i})). \tag{3}$$

5 System Implementation and Simulation Experiment

In this section, we implement our approach as a prototype system in a WFMS called Tsinghua InfoTech Product Lifecycle Management (TiPLM) workflow [20], and demonstrate two stages of our approach by means of concrete examples. The referenced time constraints and time parameters in these examples are extracted from TiPLM event log through applying statistical analysis [17], or set by consulting related workflow documents. Moreover, we evaluate the effectiveness of our scheduling strategy through large scale simulation experiments whose results are independent of specific platforms.

The first process analysis stage of the approach is realized by revising the TiPLM workflow model editor. In the model editor, a workflow process is described by a directed graph that has four different kinds of nodes: activity nodes, route nodes, start node, and end node. Nodes are connected by directed links to define different kinds of control flows. Figure 5 shows a screenshot of the TiPLM workflow model editor.

From TiPLM workflow, we extract a process instance of alteration of contract (Figure 6) to demonstrate step 1.1 and 1.2 of our approach. Step 1.1 converts the example process instance into a PTCWF-net, as depicted in Figure 7. The congruent relationship of elements between the example process instance and PTCWF-net is shown in Table 4. In addition, Table 5 presents time constraints and parameters for

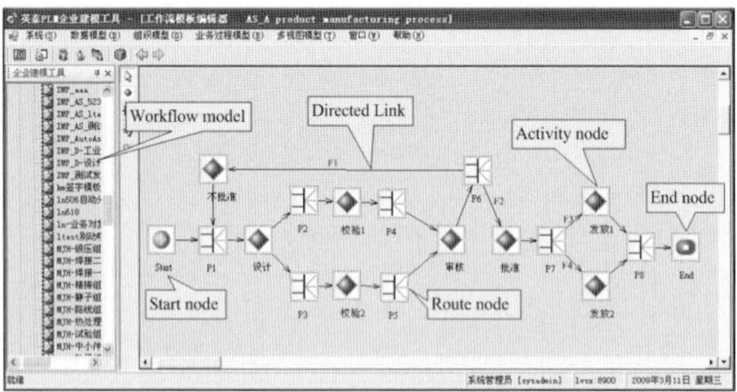

Fig. 5. TiPLM workflow model editor

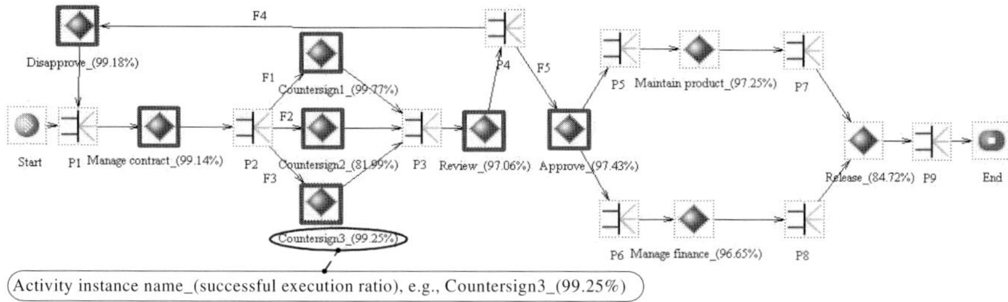

Fig. 6. Example process instance of alteration of contract

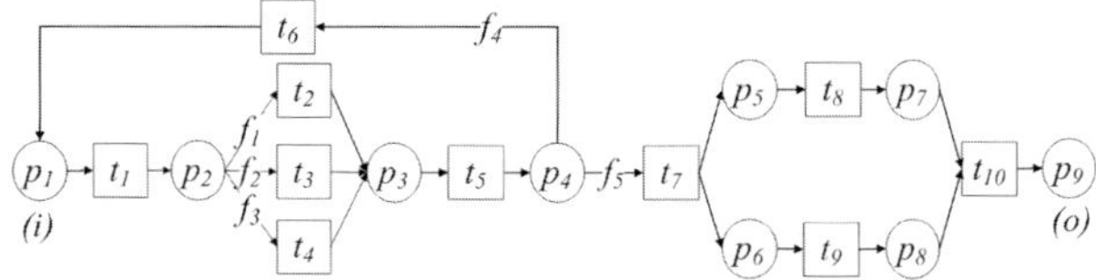

Fig. 7. Modelling the example process instance with a PTCWF-net

Table 4. Congruent relationship of elements between the example process and PTCWF-net

Route node	Place	Activity Node	Transition	Directed link	Arc
P_1	p_1	Manage contract	t_1	F_1	f_1
P_2	p_2	Countersign1	t_2	F_2	f_2
P_3	p_3	Countersign2	t_3	F_3	f_3
P_4	p_4	Countersign3	t_4	F_4	f_4
P_5	p_5	Review	t_5	F_5	f_5
P_6	p_6	Approve	t_6		
P_7	p_7	Disapprove	t_7		
P_8	p_8	Maintain product	t_8		
P_9	p_9	Manage finance	t_9		
		Release	t_{10}		

Table 5. Time constraints and parameters in the example PTCWF-net

p	$\theta_F(f).E_{\min}$	$\theta_F(f).E_{\max}$	T	μ_t	σ_t,	f	$\theta_F(f)$
p_1	1.02	6.98	t_1	5.52	0.18	f_1	0.33
p_2	5.11	68.27	t_2	57.11	1.97	f_2	0.24
p_3	1.47	57.86	t_3	61.26	2.06	f_3	0.43
p_4	2.74	12.85	t_4	58.12	2.02	f_4	0.36
p_5	3.17	17.34	t_5	52.73	1.92	f_5	0.64
p_6	3.24	15.72	t_6	9.52	0.24		
p_7	1.12	36.05	t_7	9.52	0.32		
p_8	1.72	37.27	t_8	13.24	0.48		
p_9	-	-	t_9	11.72	0.41		
			t_{10}	33.49	1.17		

Table 6. Example 15 activity instances and their related time variables

Activity instance name	Start time	Expected duration	Deadline	Successful execution ratio (%)	Fixed deadline violation cost	Probabilistic deadline violation cost
a_1	30.79	5.34	36.46	96.69	70.63	2.34
a_2	28.35	6.25	35.07	98.54	30.96	0.45
a_3	61.12	6.63	68.16	96.58	85.05	2.91
a_4	5.74	9.44	15.93	98.81	60.11	0.72
a_5	15.37	7.00	22.95	99.11	26.82	0.24
a_6	7.91	7.78	16.33	99.17	17.61	0.15
a_7	80.07	11.59	92.76	99.58	95.96	0.41
a_8	33.29	8.42	42.33	98.23	72.12	1.27
a_9	69.24	14.71	84.82	96.31	13.39	0.49
a_{10}	3.59	5.81	9.90	99.43	45.98	0.26
a_{11}	2.41	7.85	10.71	95.64	17.73	0.77
a_{12}	86.42	9.48	96.53	97.38	54.82	1.24
a_{13}	22.53	14.89	38.28	95.66	66.29	2.88
a_{14}	82.48	11.03	94.29	98.22	62.16	1.11
a_{15}	12.10	8.02	20.60	95.83	16.16	0.67

the example PTCWF-net, and the time unit is hour (referring to transitions and places). For arcs, there are only five time parameters of arc f_1 to f_5 are given because other arcs' time parameters are 1. Step 1.2 analyzes every activity instance's successful execution ratio in the example process instance, as shown in Figure 6.

Next, we use Figure 1's personal worklist (Section 1) to illustrate the second worklist scheduling stage of our approach. Table 6 shows 15 activity instances in this worklist and their six related time variables. Time unit for 'Start time', 'Expected duration', and 'Deadline' is a same time unit, for 'Successful execution ratio' is percentage, and for the last two variables is a same cost unit. For the convenience of focusing on the scheduling results of five strategies, in Table 6, these 15 activity instances' start time, expected durations, and successful execution ratios are assumed to randomly range from 0 to 100, 5 to 15, and 95% to 100%, respectively (these assumptions are still applicable in the following simulation experiments).

Given that, in the example the safe threshold is 80%. The initial personal worklist is empty and 15 activity instances are added to the worklist from a_1 to a_{15}.

Figure 8 shows the scheduling results of five scheduling strategies after adding the first two activity instances a_1 and a_2. When a_1 is inserted to worklist S, worklist S only contains one activity instance whose successful execution ratio is 96.69% and this ratio is above the safe threshold. When a_2 is inserted to worklist S, the joint successful execution ratio of a_1 and a_2 is below 5%, so worklist S needs scheduling. In the scheduling, strategy SRF and LDF eliminate a_1 and the other three strategies remove a_2. Hence in strategy SRF and LDF, total deadline violation cost of activity instances in worklist U is a_1's probabilistic violations cost: 2.34. Similarly, in the other three strategies, this cost is a_2's probabilistic violations cost: 0.45.

	Worklist S	Worklist U	Total violation cost of worklist U
SRF	a_2	a_1	2.34
FIFO	a_1	a_2	0.45
LDF	a_2	a_1	2.34
SCF	a_1	a_2	0.45
LEDF	a_1	a_2	0.45

Fig. 8. Example scheduling results after adding the first two activity instances

Figure 9 shows the scheduling results of five strategies after adding all 15 activity instances. Among five scheduling strategies, worklist S in SRF strategy has the largest number of activity instances (sorted according to their start time), hence activity instances in worklist U have the smallest total deadline violation cost.

Strategy	Worklist	Activity instances	Total deadline violation cost
SRF	Worklist S	a_{11}, a_{15}, a_{13}, a_3, a_9	Total deadline violation cost:8.19
	Worklist U	a_1, a_6, a_5, a_8, a_7, a_4, a_{10}, a_2, a_{12}, a_{14}	
FIFO	Worklist S	a_{11}, a_{15}, a_3	Total deadline violation cost:11.19
	Worklist U	a_2, a_1, a_6, a_5, a_9, a_8, a_7, a_{10}, a_4, a_{14}, a_{13}, a_{12}	
LDF	Worklist S	a_{15}, a_{13}, a_{14}	Total deadline violation cost:11.26
	Worklist U	a_1, a_3, a_2, a_5, a_6, a_7, a_9, a_8, a_4, a_{11}, a_{10}, a_{12}	
SCF	Worklist S	a_{15}, a_{13}, a_{14}	Total deadline violation cost:11.26
	Worklist U	a_2, a_1, a_6, a_5, a_9, a_4, a_8, a_3, a_7, a_{11}, a_{10}, a_{12}	
LEDF	Worklist S	a_{15}	Total deadline violation cost:15.25
	Worklist U	a_2, a_1, a_4, a_6, a_5, a_9, a_7, a_8, a_{11}, a_3, a_{10}, a_{12}, a_{13}, a_{14}	

Fig. 9. Example scheduling results after adding all 15 activity instances

In the following, we compare our SRF strategy with the other four strategies through large scale simulation experiments. Each experiment includes 15 steps, and every step adds an activity instance to the personal worklist. Hence from step 1 to step 15, the personal worklist contains 1 to 15 number of activity instances, which reflect the workload of actor from light to heavy. In each step, five strategies are applied to schedule worklist S to meet the given safe threshold. Then the total deadline violation cost of activity instances in worklist U is recalculated. In addition, we set four safe thresholds: 60%, 70%, 80%, and 90%. These safe thresholds represent the stability requirements of actor's execution from low to high. To investigate the statistic performance, for each safe threshold, the experiment is executed for 1000 times. Therefore, in our simulation, a large scale of 4000 independent experiments have been executed. For fairness, the total deadline violation cost in each step of the experiment is specified as the average value of all 1000 ones.

Figure 10 (a) to Figure 10 (d) show the experiment results for safe threshold 60% to 90%, respectively. Experiment results show that for all four safe thresholds, the total deadline violation cost of our strategy is smaller than that of the other four strategies. Besides, when the worklist becomes larger by inserting more activity instances, the gap of total violation cost also becomes larger between our strategy and the other four strategies. Furthermore, our strategy performs even better when safe threshold becomes smaller (i.e., from 90% to 60%), which means our strategy can reduce more deadline violation costs than other strategies when actors are required lower stability of execution. To conclude, our scheduling strategy is almost always advantageous to minimize deadline violation cost than other commonly used scheduling strategies, especially for worklists with heavy workload and low stability.

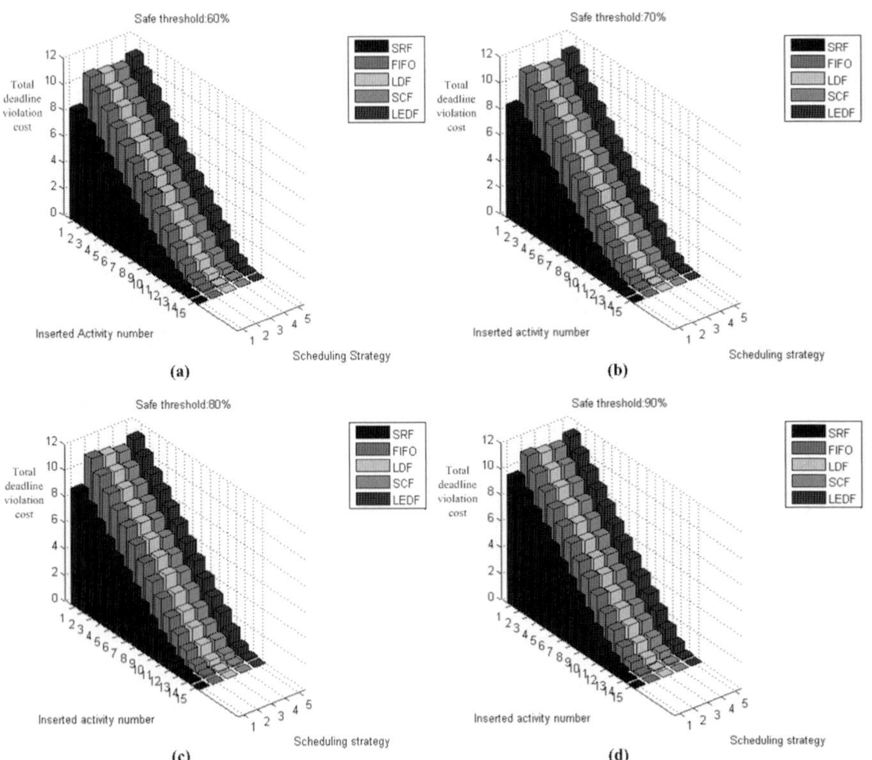

Fig. 10. Simulation experiments of five scheduling strategies

6 Related Work

In this section, we review some related work on workflow scheduling and personal worklist management. To the best of our knowledge, the work that is most similar to ours is presented in [10, 11, 21]. In [10], Eder et al. discuss main problems in the management of personal worklist. Their approach first employs duration histograms [21] to model workflow process and generate time plan for activity instance. Based on time plan, it performs a simple Earliest Deadline First (EDF) scheduling strategy to sort activity instances in the personal worklist. Furthermore, in [11], they envision two new scheduling strategies to sort activity instances without providing much detail.

Our work differs from Eder et al.'s work [10, 11, 21] mainly in two aspects. First, their work introduces the notion of time histograms to describe a number of activity durations together with their probabilities. Hence, they present deterministic analysis result as several satisfaction states and their probabilities. In contrast, our work considers continuous distributed activity durations and analyzes activity instance's stochastic probabilities of satisfying deadline. Secondly, Eder et al.'s scheduling strategy makes two rigorous assumptions: 1) worklist is safe only if its probability of satisfying deadlines is 1.0, i.e., the safe threshold is 100%; 2) execution intervals of activity instances cannot overlap in the worklist. In comparison, our approach relaxes the first assumption by setting the safe threshold through a negotiating process between the system manager and actor. We also consider execution intervals' overlap in our scheduling strategy. Consequently, our approach can provide flexible management of personal worklist which adapts to dynamic workflow environment.

Distinguishing from our work which views workflow scheduling from individual workflow actors' viewpoint, recent research mainly concerns macro-level scheduling from workflow system's point of view, i.e., they concentrate on assigning tasks among multiple actors [22]. In [2, 4], Baggio et al. discuss problems of applying existing scheduling techniques in WFMSs. They use a "Guess and Solve" scheduling approach and other basic scheduling strategies to minimize the late process instances in WFMSs. In [5, 6, 7], Combi et al. focus on temporalities in the conceptual organizational model and task assignment policies. They propose a temporal organizational model, which extends traditional organizational models, to describe different constraints of resources such as availability constraints and deadline constraints. Based on these constraints, they design a scheduling algorithm to evaluate the priority of tasks according to their expected deadlines. In addition, in [8], Senkul et al. mention workflow scheduling in a single process instance. They propose an architecture to model resource information and resource allocation constraints. Then, they employ a scheduler model to incorporate a constraint solver, so as to find proper resource assignments and to schedule workflows with resource allocation constraints.

7 Conclusions

In this paper, we have proposed a two-stage probabilistic approach, which provides stochastic scheduling to flexibly manage individual actor's personal worklist. To be specific, at the first stage, the approach analyzes every activity instance's continuous probability of satisfying deadline. This probability acts as an accurate guidance for personal worklist scheduling. At the second stage, an innovative scheduling strategy is proposed. Every time new activity instances are added to an actor's personal worklist, this strategy maintains a feasible worklist S according to the negotiated safe threshold, and an infeasible worklist U which is comprised of deadline-violating activity instances. The strategy targets at minimizing the overall deadline violation cost for activity instances in worklist U, thus achieving improved system performance.

Moreover, we use concrete examples and large scale simulation experiments to verify the practicability of our work in facilitating personal worklist management under dynamic workflow environment.

In our future work, based on the probabilistic approach proposed in this paper, we plan to investigate the impact of managing individual actor's worklist on other actors' worklists, or even the overall system performance. Besides deadline violation cost, we will discuss other performance measures, such as throughput, turnaround time, and so on.

References

1. Eder, J., Panagos, E.: Managing Time in Workflow Systems. In: Workflow Handbook 2001, Future Strategies INC, in association with Workflow Management Coalition (2000)
2. Baggio, G., Wainer, J., Clarence, E.: Applying scheduling techniques to minimize the number of late jobs in workflow systems. In: Proceedings of the 2004 ACM symposium on Applied computing, pp. 1396–1409. ACM Press, Nicosia (2004)
3. van der Aalst, W.M.P., van Hee, K.M.: Workflow Management: Models, Methods, and Systems. The MIT Press, Cambridge (2002)
4. Tramontina, G.B., Wainer, J.: Modeling the behavior of dispatching rules in workflow systems: A statistical approach. Groupware: Design, Implementation, and Use, 208–215 (2005)
5. Combi, C., Pozzi, G.: Temporal conceptual modelling of workflows. In: Song, I.-Y., Liddle, S.W., Ling, T.-W., Scheuermann, P. (eds.) ER 2003. LNCS, vol. 2813, pp. 59–76. Springer, Heidelberg (2003)
6. Combi, C., Pozzi, G.: Towards temporal information in workflow systems. In: Olivé, À., Yoshikawa, M., Yu, E.S.K. (eds.) ER 2003. LNCS, vol. 2784, pp. 13–25. Springer, Heidelberg (2003)
7. Combi, C., Pozzi, G.: Task scheduling for a temporal workflow management system. In: Proceedings of the Thirteenth International Symposium on Temporal Representation and Reasoning (TIME 2006), pp. 61–68. IEEE Press, Budapest (2006)
8. Senkul, P., Toroslu, I.H.: An architecture for workflow scheduling under resource allocation constraints. Information Systems 30, 399–422 (2005)
9. Moore, C.: Common Mistakes in Workflow Implementations. Giga Information Group, Cambridge (2002)
10. Eder, J., Pichler, H., Gruber, W., Ninaus, M.: Personal schedules for workflow systems. In: van der Aalst, W.M.P., ter Hofstede, A.H.M., Weske, M. (eds.) BPM 2003. LNCS, vol. 2678, pp. 216–231. Springer, Heidelberg (2003)
11. Eder, J., Eichner, H., Pichler, H.: A probabilistic approach to reduce the number of deadline violations and the tardiness of workflows. In: Meersman, R., Tari, Z., Herrero, P. (eds.) OTM 2006 Workshops. LNCS, vol. 4277, pp. 5–7. Springer, Heidelberg (2006)
12. van der Aalst, W.M.P., Rosemann, M., Dumas, M.: Deadline-based escalation in process-aware information systems. Decision Support Systems 43, 492–511 (2007)
13. Brucker, P.: Scheduling algorithms. Springer, Heidelberg (2007)
14. van der Aalst, W.M.P.: The application of Petri nets to workflow management. Journal of Circuits Systems and Computers 8, 21–66 (1998)
15. Russell, N., Ter Hofstede, A., Edmond, D., van der Aalst, W.M.P.: Workflow resource patterns. Technical Report, Eindhoven Univ. of Technology (2004)
16. van der Aalst, W.M.P., van Hee, K.M., Reijers, H.: Analysis of discrete-time stochastic Petri nets. Statistica Neerlandica 54, 237–255 (2000)

17. Stroud, K.A.: Engineering Mathematics, 6th edn. Palgrave Macmillan, New York (2007)
18. Law, A.M., Kelton, W.D.: Simulation Modelling and Analysis, 4th edn. McGraw-Hill, New York (2007)
19. Zhou, R., Hansen, E.A.: Breadth-first heuristic search. Artificial Intelligence 170, 385–408 (2006)
20. Introduction to Infotech Product Lifecycle Management Solution (in Chinese), `http://www.thit.com.cn/chanpinshijie/TiPLM.htm`
21. Eder, J., Pichler, H.: Duration histograms for workflow systems. In: Proc. of the Conf. on Engineering Information Systems in the Internet Context 2002, pp. 239–253. Kluwer Academic Publishers, Dordrecht (2002)
22. Yingbo, L., Jianmin, W., Jiaguang, S.: Using Decision Tree Learning to Predict Workflow Activity Time Consumption. In: Proceedings of the Ninth International Conference on Enterprise Information Systems (ICEIS 2007), pp. 69–75 (2007)

Flaws in the Flow: The Weakness of Unstructured Business Process Modeling Languages Dealing with Data

Carlo Combi and Mauro Gambini

Dipartimento di Informatica – Università di Verona
Strada Le Grazie, 15, Verona, Italy
{carlo.combi,mauro.gambini}@univr.it

Abstract. Process-Aware Information Systems (PAISs) need more flexibility for supporting complex and varying human activities. PAISs usually support business process design by means of graphical graph-oriented business process modeling languages (BPMLs) in conjunction with textual executable specifications. In this paper we discuss the flexibility of such BPMLs which are the main interface for users that need to change the behavior of PAISs. In particular, we show how common BPMLs features, that seem good when considered alone, have a negative impact on flexibility when they are combined together for providing a complete executable specification. A model has to be understood before being changed and a change is made only when the benefits outweigh the effort. Two main factors have a great impact on comprehensibility and ease of change: concurrency and modularity. We show why BPMLs usually offer a limited concurrency model and lack of modularity; finally we discuss how to overcome these problems.

Keywords: process-aware information systems, unstructured business process modeling languages, concurrency, modularity, refactoring.

1 Introduction

Business process modeling involves the understanding of the observed or the desired reality and the design of a consistent specification which can be used for driving a Process-Aware Information System (PAIS) [1] to streamline the business. The description of a business process has to be shared among all the stakeholders that participate in the design activity to reach an agreement on how the business works and how the information system will be implemented.

PAISs support business process design by means of graphical Business Process Modeling Languages (BPMLs) in conjunction with textual executable specifications that contain additional data necessary at run-time. These languages are usually graph-oriented and they allow multiple connections among different parts of the same model, for instance connecting the body of two distinct loops.

R. Meersman, T. Dillon, P. Herrero (Eds.): OTM 2009, Part I, LNCS 5870, pp. 42–59, 2009.

The business process management community recognizes several limits of PAISs, in particular the lack of flexibility [2,3]: the demand for more flexible systems is not surprising, especially for supporting complex and varying human activities. Flexibility is a pervasive concept in the design of a PAIS and it can be related to several aspects, for instance it may refer to the deployment of new processes or the ability to change process instances at run-time or else the allocation of human resources. In this paper we discuss the flexibility of BPMLs which can be considered the main interface of PAISs: they are used by process designers for capturing business logic, by developers for implementing a working system and by common users, especially if they need to perform some changes during the enactment of a process.

A software system has to be comprehended before being changed, at least for avoiding unpredictable results. Only by understanding the consequences of a change we can estimate the effort needed to make it and a change is made only when the benefits outweigh the effort. For improving BPMLs flexibility we need to enhance comprehensibility and reduce the effort needed to change business process models. Two main intertwined factors have a great impact on comprehensibility and ease of change, namely *concurrency* and *modularity*. In this paper we will bring to light frequent design flaws of graph-oriented BPMLs that severely limiting their ability to cope with concurrency and modularity, producing negative impacts on PAISs flexibility. For this purpose we introduce a typical BPML supporting (1) unstructured control-flow design with token-based semantics, (2) hierarchical decomposition of tasks, (3) parameter passing among local task variables and (4) asynchronous message passing. We show that such flaws are not related to a specific language aspect but they arise when different features are put together for obtaining a complete executable specification. In particular, unstructured control-flow features mixed with shared task variables offer a limited concurrency model; we explain why asynchronous message passing is not a viable alternative to shared variables communication because in conflict with hierarchical decomposition; finally we argue that the mixing of these language abstractions reduce modularity, severely limiting the ability to express and change large business process models.

The remainder of this paper is organized as follows: in Section 2 we introduce some related work. In Section 3 we clarify some basic notions such as concurrency, comprehensibility and modularity in the context of executable BPMLs. In Section 4 we propose a core modeling language for capturing the main features of unstructured BPMLs. In Section 5 we show the limits of business process modeling based only on control-flow constructs and we discuss the issues that arise when data are not carefully modeled during design. In Section 6 we show why asynchronous message passing cannot be used instead of shared data communication among tasks. In Section 7 we point out the difficulties to overcome for transforming a model preserving its semantics. In Section 8 we discuss plausible solutions for the exposed issues and we explain why there are no quick fixes. Finally, in Section 9 we summarise the results of this paper and we discuss open problems.

2 Related Work

Several interesting techniques have been proposed to gain flexibility in PAISs, for instance exception handling can be used for modeling unpredictable events [4]; the adaptability of business process models can be improved by supporting change patterns [5] or by allowing the composition of blocks starting from simple fragments [6]; data-driven and declarative approaches can be used to adapt the structure of a model at run-time [7,8].

At the best of our knowledge, the solutions to gain flexibility in PAISs fall into two categories: (1) solutions based on innovative techniques that go beyond the current practice like declarative ones and (2) solutions that improve existing systems and modeling languages by adding new features; these solutions are often related to a particular aspect of PAISs such as control-flow design. On the contrary this paper focuses on the interaction among common core features of existing BPMLs that are put together for providing a complete executable model. We will show that such mixing of features often goes against well-established software engineering principles like modularity and program comprehensibility [9,10] and hence against flexibility.

A business process model may be sound or not depending on the chosen soundness criteria. Many soundness criteria have been proposed, each one attempts to capture good models without being too restrictive [11]. Such criteria are usually defined through Petri Net Theory ignoring data-flow aspects, even though such aspects are relevant for obtaining executable specifications. The aim of this paper is not to define yet another soundness criteria but to explain some fundamental flaws in the design of BPMLs. For such purpose we use simple business process models that can be sound respect to particular criteria but are poor from a program comprehension perspective. These models are expressed using an abstract BPML for capturing the core features of common modeling languages; a broad comparison among features of existing BPMLs has been already done through the workflow patterns initiative [12,13].

3 Background

We define the data-flow of a system as the set of connections among components needed to exchange data. In a process we can distinguish different aspects and study them nearly independently; such aspects are called *perspectives* and mainly concern control-flow logic, data management and resource management [1]. Such distinction cannot be freely transferred to an executable BPML, because control-flow and data-flow concerns cannot be easily separated: we cannot ignore data dependencies if we would obtain correct specifications [14]. Some BPMLs neglect data-flow aspects, but their specification is only postponed when processes are implemented in a PAIS. For these reasons we consider jointly control-flow and data-flow aspects when we will discuss about BPMLs.

As mentioned in Section 1, for improving flexibility we have to enhance comprehensibility. Comprehensibility is a property of a system to be easily grasped in all of its parts. Comprehensibility may refer to different subjects such as business processes, business process models or BPMLs; it is necessary to clarify these distinct but interrelated meanings: for a human, understanding a process means building a clear *mental* model about how it works. To understand a business process and share knowledge about it, we use a *concrete* model expressed in a certain BPML: a tool for tackling the inherent complexity of the observed reality. Therefore, clear semantics and comprehensibility are important requirements for BPMLs, but to be useful they must be able to manage complexity. As stated previously, two main intertwined factors have a great impact on comprehensibility of BPMLs, namely concurrency and modularity.

Reasoning about concurrent entities is difficult: for instance, nontrivial multi-threaded programs become soon incomprehensible to humans [15]. Zapf et al. argue that it is not trivial to increase business process productivity by exploiting parallel tasks, because the needed coordination efforts reduce expected efficiency gains [16]. Nevertheless, parallelism cannot be avoided by removing it from models, since the modeled reality is inherently concurrent: not surprisingly one goal of a PAIS is to reduce coordination efforts into a truly concurrent environment, where agents need to interact with each others. Graph-oriented BPMLs adopt a token-based semantics which is typically formalized through Petri Nets Theory. Petri Nets are good for describing true concurrency: unfortunately the concept of token does not differ from the concept of thread when it is associated with the wrong data model. Not surprisingly, current systems impose properties like safeness [12] or they adopt structured control-flow constructs when possible.

Modularity is one of the few means to face complexity. We define modularity of a BPML as the ability to decompose its models in small interrelated components which in turns can be recombined in different configurations [9]. The lack of modularity makes large models hard to understand and change, since a slight modification in one part may affect the model as a whole with the risk to introduce new errors [9]. Modularity enhances the ability to change a model; anyway it is worth noting that adding features like hierarchical decomposition of tasks to a BPML does not necessary increase modularity: a large model decomposed into some compound tasks may be more comprehensible but more hard to change. Quantifying the modularity of a language is not a simple matter because we need to compare the size of each change with the efforts necessary to apply it. Nevertheless, the effort to make a change in a model must be proportional to the semantics gap between the current and the new version: as a consequence, small semantics changes should be easy to perform and semantics-preserving ones should be effortless, as for the creation of a sub-process in order to reduce the size of the original model and to enhance comprehensibility.

4 A Typical Unstructured Graph-Oriented BPML

In this section we introduce a typical graph-oriented BPML which supports the definition of unstructured models with multiple connections among components; in the following we will refer to it as $\mathcal{A}$ language.

Adopting the approach in [17], $\mathcal{A}$ is a *model* of existing systems like YAWL [18], BPMN with WS-BPEL executable semantics [19], and jBPM from the JBoss Enterprise Middleware [20] which resembles UML Activity Diagrams [21]. The aim is twofold: first, we gain generality by capturing relevant aspects of graph-oriented languages without restricting to a specific implementation and second, we base the discussion on a language with a well-defined behavior, since several languages propose their own particular constructs and expose a slightly different semantics often given only informally.

As the name suggests, the syntax of graph-oriented languages can be described with the Graph Theory, while its semantics can be given through the Petri Nets Theory. The formalization of a complete executable modeling language is not a simple matter: for our purposes we need only core constructs which were already analyzed in literature through the YAWL initiative, workflow control-flow and data-flow patterns [22,13,12].

The graphical symbols of the language are depicted in Fig. 1, whereas the main elements are summarized in the UML Class Diagram depicted in Fig. 2, where the in-degree and out-degree of each element are expressed within square brackets for not cluttering the diagram.

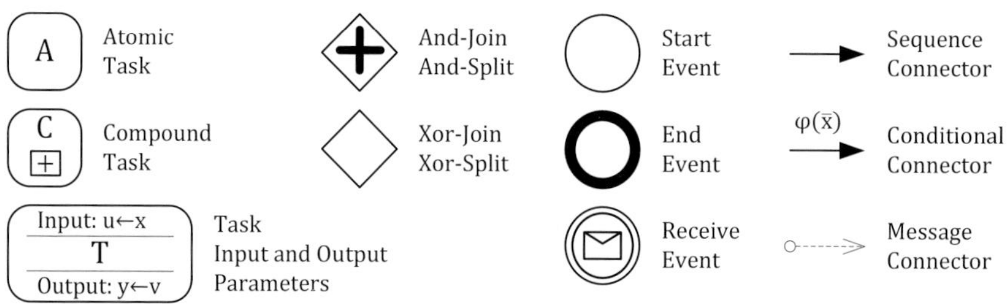

Fig. 1. BPMN symbols for $\mathcal{A}$ language constructs

Symbols in Fig. 1 are taken from the BPMN standard [19] except for the representation of input and output parameters of a generic task T, where x and y are external task variables, u and v are internal ones and the left arrow denote an assignment. When the internal implementation of a task is not relevant, we will denote only external variables affected by such task and whenever input and output are evident from control-flow path, we will drop the initial labels. A set of variables is denoted with a vector like $\overline{x}$, while a boolean expression over a set of variables $\overline{x}$ takes the form $\varphi(\overline{x})$. Two examples of a model expressed in $\mathcal{A}$ are depicted in Fig. 3 and Fig. 4.

For simplifying the language, out-degree of split gateways and in-degree of join gateways are limited to two: such limits do not compromise the freedom in the use of connectors, neither the generality of results. Control elements with higher degree are obtainable by composing basic constructs: for instance, inclusive Or-Split Or-Join pair can be simulated by And-Split followed by a Xor-Split for each parallel branch; this also holds for tasks and events with multiple entering

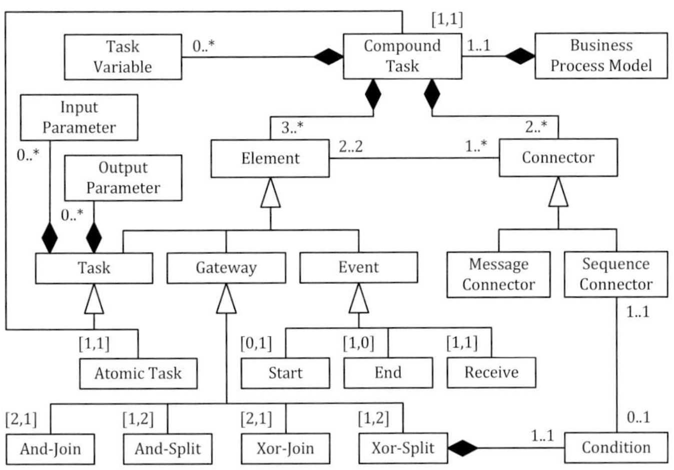

Fig. 2. The elements of the $\mathcal{A}$ modeling language

and leaving connectors. Arbitrary loops with many entry and exit points can be obtained with the available constructs.

The $\mathcal{A}$ language exposes a token-based semantics typical of unstructured BPMLs: a token is produced by a start event and consumed by an end event. Any model expressed in $\mathcal{A}$ could have more start and end events; we assume that all start events begin together as if they were connected backwards with one or more subsequent And-Splits. During the execution of an instance of the model, namely a *case*, every start event produces a token that follows the control-flow net defined by tasks, sequence connectors and gateways control elements. In particular, an And-Split construct consumes one token from the incoming branch and produces a new token for every outgoing branch, while the And-Join waits one token for each incoming branch before producing one outgoing token. With only two outgoing branches, a Xor-Split does not differ from an if-then-else construct, while the Xor-Join is interpreted as a simple merge [12] which replicates to the outgoing branch all the incoming tokens without synchronization. Every final event consumes tokens that arrive to it by removing them from the net and when all tokens have been removed the case terminates according to the implicit termination control-flow pattern [12]. When a token reaches a task, a new instance is enabled and loaded in the work-list ready to be performed; when the corresponding activity is finished, the task is closed and the token is released through the outgoing connector. This also holds for compound tasks, with the difference that a token is placed in every contained start event and it terminates when all the contained tokens have been removed.

Any compound task can declare one or more *task variables* of a certain type and the whole set of variables of a task instance is called *store*. A task instance is considered *stateless* because its store is not retained after the task conclusion. The scope of a variable coincides with the compound task where it is declared and its visibility does not extend to task instances invoked in the same scope: in this manner $\mathcal{A}$ implements a form of encapsulation where each instance has its own local variables. The language does not support explicit termination [12], since it causes many troubles in concurrent context: a single end event may interrupt all parallel tasks possible leaving the stores into an inconsistent state.

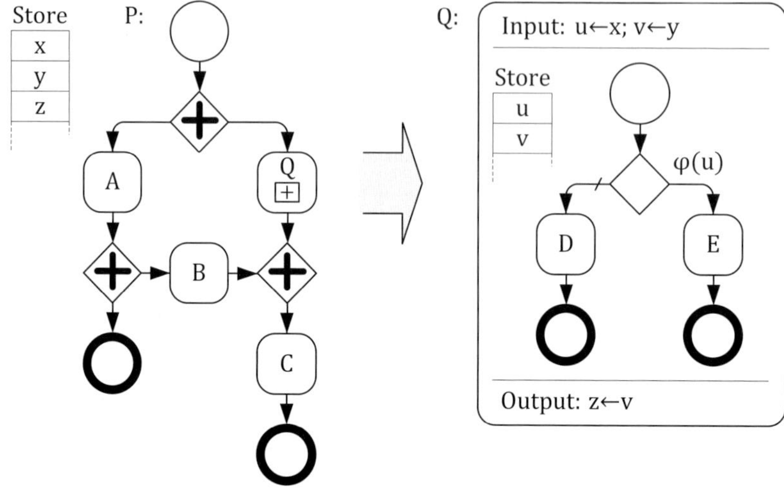

Fig. 3. Local Shared Variables (LSV) communication between tasks

A task can declare *input parameters* and *output parameters*; when a task is enabled, it reads the variables from the current scope and copy their content into its internal variables; when the task has completed, external variables are updated by means of output parameters as shown in Fig. 3. Indeed, from a data-flow perspective, any task T can be seen as a function between its input and output parameters, therefore we denote with $T : \overline{x} \mapsto \overline{y}$ such relation, where $\overline{x}$ is the set of variables read by T and $\overline{y}$ is the set of variables updated by T; for instance the task Q in Fig. 3 is denoted with $Q : \{x, y\} \mapsto \{z\}$. Moreover, we abstract from the computational aspects of the language, since any needed function $f : \overline{x} \mapsto \overline{y}$ can be implemented by an atomic task F with the aid of external programs or human intervention.

Tasks, even in a different scope, can communicate among them through input and output parameters and shared variables without breaking encapsulation. In the following, we will refer to this type of exchange as *Local Shared Variables* (LSV) communication. Beyond LSV, $\mathcal{A}$ supports *Asynchronous Message Passing* (AMP) communication for representing interactions among tasks. For sake of simplicity, we consider only point-to-point message exchanges from atomic send tasks to intermediate receive events: a message exchange between generic tasks means that the source task contains at least one atomic send task and the destination task contains at least one receive intermediate event. A receive intermediate event is implemented as message queue stored in the same scope in which it is placed. As will become clear in the next sections, LSV remain the first choice for inter-task communication because AMP cannot be used in all situations.

5 Tokens and Shared Variables

In this section we expose the problems of LSV into a truly concurrent environment and the difficulty to take advantage of state invariants in presence of token-based semantics.

5.1 Control-Flow Design Hides Race Conditions

If we consider only control-flow concerns during the design of a business process, we may obtain a perfectly sound but unusable model, especially if such model takes advantage of parallelism. Sooner or later data-flow concerns must be taken into account and they could be in contrast with control-flow logic.

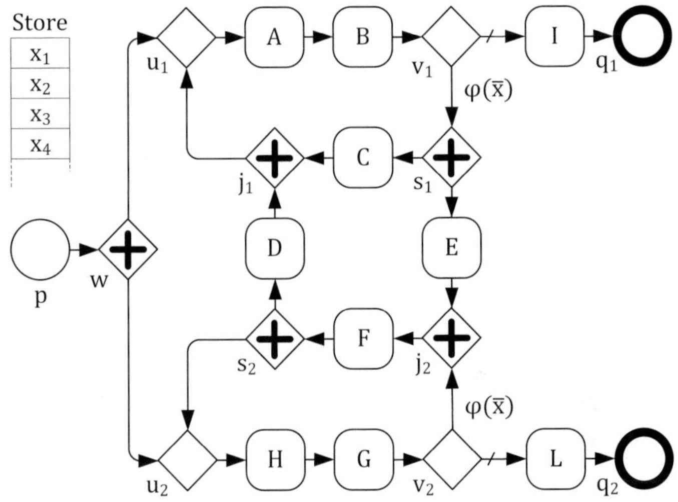

Fig. 4. An unstructured model with LSV communication strategy

For instance, let us consider the use of conditions in Xor-Split constructs using LSV communication strategy among tasks and in particular let us examine the use of loop conditions $\varphi(\overline{x})$ in the model depicted in Fig. 4. The model was used in [23] for proving the existence of well-behaved unstructured models that do not have a structured form. The model contains two main loops $(u_1, A, B, v_1, s_1, C, j_1)$ and $(u_2, H, G, v_2, j_2, F, s_2)$ that run in parallel and synchronize each other at every iteration. Abstracting from data-flow details, the model is free from deadlocks, but with LSV strategy it almost certainly reaches an invalid state due to race conditions: there is no guarantee that the two conditions $\varphi(\overline{x})$ in the two loops are evaluated at the same time, so one loop may exit before the other.

We assume that the state underlying $\overline{x}$ changes during the execution, otherwise we would fall into two trivial cases: $\varphi(\overline{x})$ is always *true* and the process never terminates or $\varphi(\overline{x})$ is always *false* and the loops are never executed entirely. Supposing that initially $\varphi(\overline{x})$ evaluates to *true*, if the state is mutable then there exists at least one task t that modifies one or more variables of the subset $\overline{x}$ by means of output parameters. For sake of simplicity we assume that exactly one task modifies the state, thereby for each $t \in \{A, \ldots, H\}$ excluding F we can exhibit an execution that lead to deadlock. Tasks I and L are not considered since they are placed outside the loops. To become convinced of this fact let us note that one of the two loops does not require to wait the end of t for reaching its Xor-Split gateway. For instance, let $t = A$, the trace $p \to w \to (u_1, u_2) \to (A, H) \to (A, G) \to (A, v_2) \to (A, j_2) \to (B, j_2) \to (v_1, j_2) \to (I, j_2) \to (q_1, j_2)$ leads to deadlock, in similar way a deadlock may occur for $t \in \{B, H, G\}$. For $t \in \{C, E\}$ the model may reach a deadlock in j_1 and j_2 respectively, since the branch terminating in G may still running when C and E finish. For $t = D$ the system can evolve from s_2 in $s_2 \to (u_2, D) \to (H, D) \to (G, D) \to (v_2, D) \to (j_2, D) \to (j_2, j_1) \to (j_2, u_1) \to (j_2, A) \to (j_2, B) \to (j_2, v_1) \to (j_2, I) \to (j_2, q_1)$. On the contrary, if the change of the state is made only by the task F the process does not run into a deadlock because F is correctly synchronized between the two loops.

5.2 Tokens Invalidate State Invariants

In languages like $\mathcal{A}$ we cannot rely on state invariants about a portion of the model to express business logic, as after a conditional gateway or inside the body of a loop: potentially, any assertion about a set of variables can be invalidated by a task instance enabled by another token that flows in the same branch.

Let us consider the model in Fig. 5. At least one task between A and B updates the state underlying $\overline{x}$, otherwise $\varphi(\overline{x})$ is always constant and the choice u is useless; if there exists an update, we can state that $\overline{x}_A \cup \overline{x}_B \neq \emptyset$, where $\overline{x}_A$ and $\overline{x}_B$ denote the subsets of variables of $\overline{x}$ modified by tasks A and B, respectively.

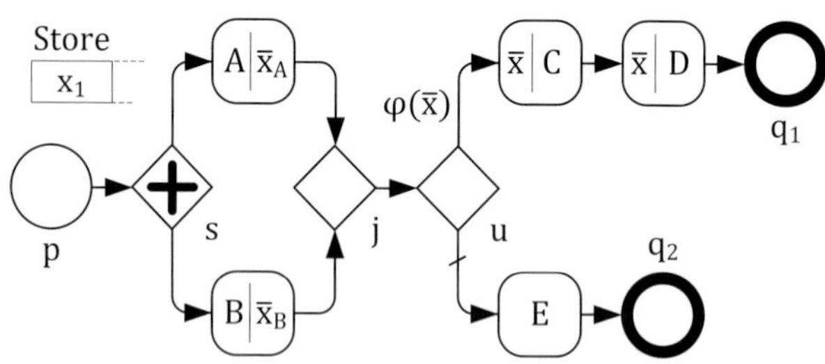

Fig. 5. Tasks C and D cannot rely on state invariant over $\overline{x}$

We discard trivial race conditions supposing $\overline{x}_A \cap \overline{x}_B = \emptyset$, so A and B tasks update different variables; we also suppose that $\varphi(\overline{x})$ is evaluated differently when the two tokens produced by the And-Split s reach the Xor-Split u, so we exclude concurrent activations of the same outgoing branch.

When A completes, a token reaches the Xor-Split u; if $\varphi(\overline{x})$ is true, C is enabled and reads the variables $\overline{x}$. At this point, if the state underlying $\overline{x}$ is updated by B before the conclusion of C, the task D will read different values than C such that $\varphi(\overline{x})$ is false. As a consequence, D cannot rely on the state invariant asserted by $\varphi(\overline{x})$ even if it is below a branch guarded by such assertion. Similar considerations hold if B finishes before A and in special cases when $\overline{x}_A = \emptyset$ or $\overline{x}_B = \emptyset$.

For instance, A and B may be activities counting the number of items into two warehouses $\mathcal{W}_A$ and $\mathcal{W}_B$, where the result is stored in $\overline{x}_A = \{x_1\}$ and $\overline{x}_B = \{x_2\}$ variables. Any time a counting activity finishes, the condition $x_1 \geq x_2$ is evaluated for deciding in which warehouse there are more items. If $x_1 \geq x_2$ the system schedules the task C which sends a notice by mail and then it enables task D which reads x_1 and x_2 to move $(x_1 - x_2)/2$ items from $\mathcal{W}_A$ to $\mathcal{W}_B$. When D is enabled, x_2 may have been already updated by B and so the computation $(x_1 - x_2)/2$ performed by D could be wrong.

6 Message Receivers Are Ambiguous

In this section we will explain why LSV cannot be easily replaced by AMP strategy. We have shown that LSV is not a good communication strategy, so one could ask why the AMP facility offered by $\mathcal{A}$ cannot be exploited in concurrent situations. Unfortunately AMP strategy is not a viable alternative for inter-task communication due to two main shortcomings: first, a task instance is stateless and it does not exist before its activation; second, the model does not guarantee that such instance is unique, because multiple instances may be enabled at the same time with a different dynamically assigned identifier. Thus, from a model it is not clear who is the receiver of a message, unless it is a specific intermediate event into an already enabled task. For example, in BPMN messages are used for modeling the interactions among instances of different process models, while within the same process, AMP among tasks like in Fig. 6 is forbidden in favour of strategies like parameter passing: in this manner BPMN guarantees that there exists only one process instance that receives a message.

To clarify the problem let us consider the model depicted in Fig. 6, where P_1 is the main process and P_2 is a compound task used by the first one. Due to the loop (v_1, s_1, A, B, u_1), P_1 can produce an unbounded number of tokens that flow through the right branch of s_1, so it can invoke P_2 multiple times. We suppose that the tasks of P_2 are highly interrelated and they need to exchange data using LSV strategy; as a consequence we have made P_2 an isolated compound task with its local store, so each instance of P_2 in the same case runs without interfere each other. An instance of P_2 can terminate only if e_2 receives a message from the external process P_1, that in turn guarantees the existence of one message for

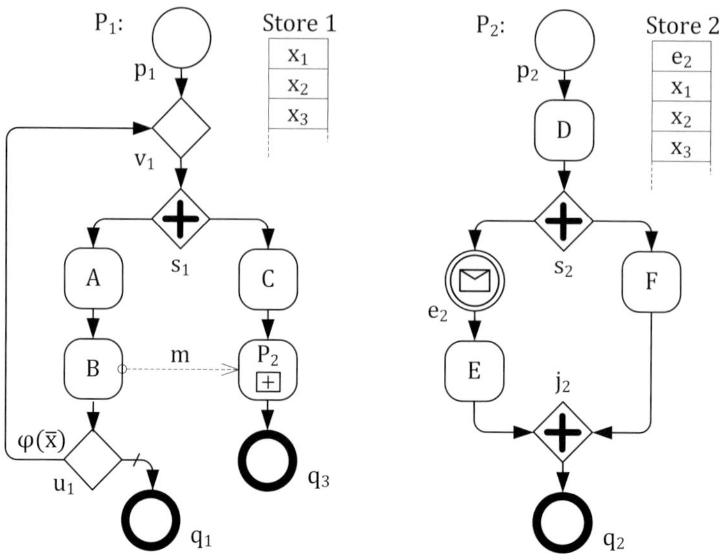

Fig. 6. An ill-defined message passing interaction

each instance, because they are started by the same loop (v_1, s_1, A, B, u_1) which afterwards produces a message m with the B send task.

The simplicity of the model hides a really complex execution semantics. First, there is no guarantee that an instance of P_2 exists when the first message m is sent, so intermediate event queues like e_2 are not sufficient to store the messages. As a consequence an external buffering strategy is needed and it cannot be as simple as a message queue: due to AMP semantics, the loop (v_1, s_1, A, B, u_1) may run multiple times and exit before a single instance of P_2 is created. The scheduling of these instances cannot be predicted because it depends on the order with which C instances are performed. As a result, to maintain the correct association between messages and receivers, the use of a message queue for each communication channel is not sufficient.

7 Unstructured Models and Language Modularity

In this section we discuss the modularity of unstructured BPMLs, in particular we point out which obstacles have to be overcome in order to make a common refactoring transformation useful for reducing the complexity of a model and gain comprehensibility.

7.1 Single Parts Are Complex as the Whole

We have argued that in executable models control-flow and data-flow concerns cannot be easily separated and we have pointed out the problems of LSV and AMP communication techniques in languages like $\mathcal{A}$. A different separation can

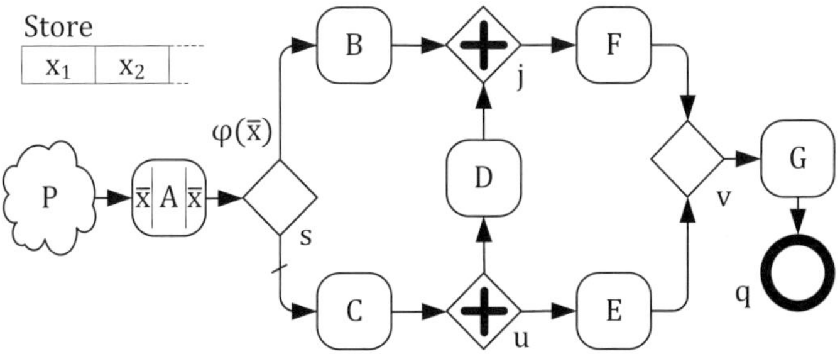

Fig. 7. An unstructured model with a first order improper nesting

be performed by isolating a portion of a model where a particular property holds; unfortunately, it is really difficult to state local properties in unstructured models due to arbitrary control-flow connections and token-based semantics.

Let us consider the model depicted in Fig. 7 and suppose for a moment that the sub-graph P is a single start event. The resulting model exhibits a first order improper nesting [24], because (s, j) and (u, v) are both corresponding pair of constructs that differ in their type and these pairs are not properly nested. This construction is considered erroneous since it leads to a deadlock for each of the two possible executions, because only one token can reach j in both cases.

In general, the correctness of the same pattern placed into an existing model cannot be determined without understanding the other parts. For instance, we reconsider the model in Fig. 7 but now preceded by an arbitrary graph P, and we suppose for simplicity that every time A is executed, the variables in $\overline{x}$ are updated to obtain a different evaluation of $\varphi(\overline{x})$, so the branches of the Xor-Split s are chosen alternatively: the A task can obtain this effect by reading the previous value of $\overline{x}$. If the number of tokens that exit from the sub-graph P is even, then this part of the model is free from deadlocks. The evenness of tokens exiting from P is a property of the entire graph: to infer something about a small part of the model one must consider the graph as a whole; even worse, the number of tokens into a graph is a property hard to derive statically from the model because it may vary in each case.

7.2 Chosen Constructs Kill Modularity

We have pointed out that one of the main goals of a model is to face the inherent complexity of the observed reality and we have stated that modularity is an essential property for managing complexity. In this section we argue that languages like $\mathcal{A}$ lack of modularity. In particular we reason about the difficulties encountered when we want to extract some compound tasks from a model and recombine them together to obtain a new model semantically equivalent to the original one: such transformation does not alter the semantics of the model but only the way in which it is represented for enhancing comprehensibility, therefore it must be effortless.

Let us reconsider the unstructured model in Fig. 4; because it is not so clear, we want to transform it in the model of Fig. 8, where P and Q are compound tasks that encapsulated the two main loops $(u_1, A, B, v_1, s_1, C, j_1)$ and $(u_2, H, G, v_2, j_2, F, s_2)$; moreover, we choose to place E in P and D in Q, obtaining two sub-graphs composed with $\{A, B, C, E\}$ and $\{H, G, F, D\}$. The details of this decomposition are reported in Fig. 9.

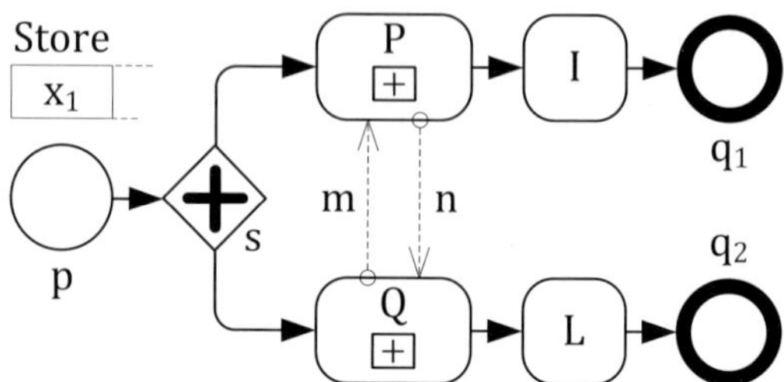

Fig. 8. A hierarchical decomposition between two interrelated tasks

A distinctive feature of unstructured BPMLs is allowing arbitrary connections among tasks; for this reason any control-flow sub-graph may have multiple entry and exit points. This feature, worsened by permissive token-based semantics, makes the extraction of a sub-graph a nontrivial operation. The control-flow dependencies among tasks cannot be discarded along with language constructs, so they must be kept by adding different constructs.

In the model of Fig. 4 only two main tokens can flow, one for each loop, while the tokens that flow through D and E are used for synchronization purpose, then it is fairly simple to remove crossing control-flow constructs. Direct control-flow dependences are given by $E \rightarrow F$ and $D \rightarrow A$, which can be replaced with two dummy messages m and n depicted in Fig. 8. We have just discussed the problems of AMP communication strategy: fortunately, in this example the main model assures the existence of one and only one receiver for each dummy message.

After the creation of a new compound task from a pre-existing control-flow sub-graph, data-flow relations must be adjusted since each compound task has its own local variables. This operation is nothing but trivial: on one hand, it is necessary to understand the order in which shared variables are updated and on the other hand, parameter passing must be replaced with message exchanges. For instance, let us consider the models in Fig. 4 and Fig. 8, and suppose that B updates a variable x_1 used by F: this is conceivable because B runs always before F for the control-flow dependencies $B \rightarrow E \rightarrow F$; for the same reason, let us suppose that D updates a variable x_2 used by A. In short, we can state $B : \emptyset \mapsto \{x_1\}$, $F : \{x_1\} \mapsto \emptyset$, $D : \emptyset \mapsto \{x_2\}$ and $A : \{x_2\} \mapsto \emptyset$. If we describe such relations through LSV parameter passing we have $P : \{x_2\} \mapsto \{x_1\}$ and $Q : \{x_1\} \mapsto \{x_2\}$, therefore these two compound tasks cannot run in parallel neither in sequence due to data-flow dependencies. As a consequence, AMP strategy must be adopted instead of LSV parameter passing which in turn does not have a clear semantics because of stateless tasks and multiple instances.

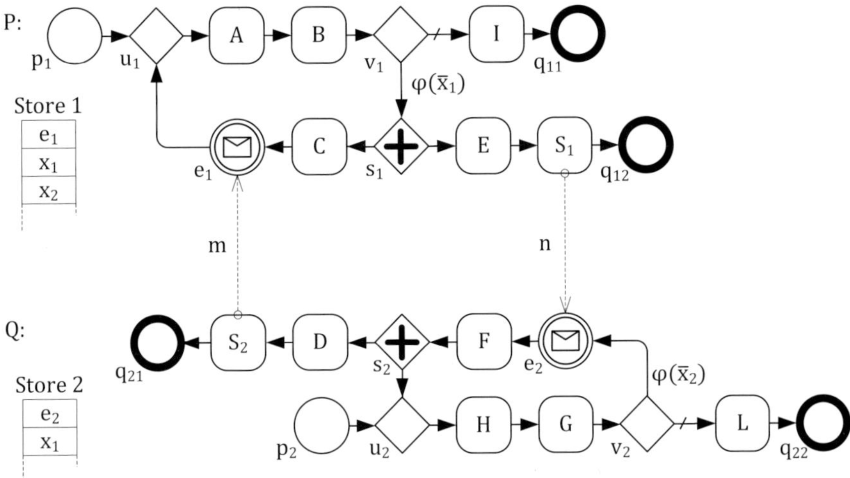

Fig. 9. Details of the hierarchical decomposition

This kind of transformations is common during the design and maintenance of a business process model whose size grows naturally for adjusting new requirements and special cases. Altogether the language abstractions chosen by languages like $\mathcal{A}$ for modeling business processes make these changes hard to perform.

8 Improving BPMLs

The primary aim of this paper is to clarify the interaction between control-flow and data-flow features that often undermine unstructured BPMLs flexibility: understanding these shortcomings is a required step for designing better BPMLs. In this section we summarize the obtained results and we discuss open problems and some potential solutions.

In the current practice data-flow aspects are usually ignored during business process modeling; as a consequence, in order to provide sufficient expressiveness, BPMLs need to support unstructured control-flow design with arbitrary connections among tasks [23]. However, data-flow concerns have to be considered for obtaining executable models able to drive a PAIS.

For existing BPMLs, the use of some verification methods seems the only viable solution for ensuring the absence of errors in the models or at least for warning about strange conditions. These methods may operate simultaneously on control-flow and data-flow relations to exclude common errors [14], however they may not be feasible on a full featured BPML, because detecting race conditions is an NP-Hard problem for any language able to express mutual exclusion [25]. In any case, whatever is the verification method used, ill-defined models have to be adjusted with the risk to introduce new errors or to lose the alignment between high level business logic and low level specification.

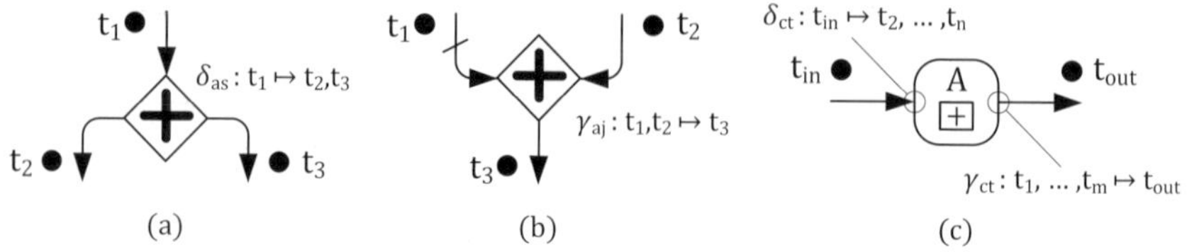

Fig. 10. Dispatch and combine extension points for TLV

These reasons justify the research of better data models for designing more flexible unstructured BPMLs that take care of concurrency and modularity issues. On one hand, LSV strategy offers a limited concurrency model and cannot scale for expressing complex business processes without AMP. On the other hand, AMP cannot be used in all situations, due to the stateless nature of tasks.

One possible solution is using a local store for each token, instead of a single store for each task. In this context a task receives tokens of a certain type, reads and updates all associated variables and releases them when completed. This behaviour is similar to parameter passing used in LSV strategy, but it does not have side effects due to shared variables; we refer to this strategy as Token Local Variables (TLV) communication.

In TLV every token has a unique identifier and its type provides information about the declared variables. Tokens are dynamically created by start event places or And-Split constructs, and they can be initialized using extension points as described in the following. Extension points are essentially functions for transferring data among tokens and they can be used for dispatching the data of one token to multiple ones, or for combining the data of many tokens to a single one. TLV requires advanced And-Split and And-Join constructs for supporting the extension points mentioned above:

- An And-Split has to support a dispatch extension point δ_{as} that receives the data of the incoming token t_1 and transforms them for producing the data associated to each outgoing token t_2 and t_3, as depicted in Fig. 10.a. By default, this extension point simply replicates the incoming data on each outgoing branch. Data may be only logically replicated through a lazy strategy, as they need to be physically copied only during an update.
- An And-Join has to support a combine extension point γ_{aj} that integrates the data of the two incoming tokens t_1, t_2 and resolves possible conflicts for producing the new outgoing token t_3, as in Fig. 10.b. If no combine function is provided, only the token that arrives from the default incoming branch is replicated in output after join. The default incoming branch can be represented as in Fig. 10.b.
- A compound task may have multiples start or end events; we have explained in Section 4 that all start events may be considered as connected backwards by an And-Split. Therefore, the creation of inner tokens does not differ from the And-Join semantics explained above, except for the use of a generalized function δ_{ct} that produces several tokens $t_1, \ldots, t_n$ from the incoming one t_{in}. Tokens $t_1, \ldots, t_m$ that reach end events are collected and when the compound

task is completed an extension point γ_{ct} is used for producing the outgoing token t_{out}, as in Fig. 10.c.

The TLV communication strategy can be a good data model for executable specifications with unstructured control-flow, at least for avoiding common concurrency issues. However, for modeling purposes, TLV presents some of the limits of LSV strategy: TLV associates data with tokens that are dynamically generated during the execution and are not statically visible from the model. Moreover, TLV strategy does not substantially improve the modularity of the language which largely depends on control-flow constructs.

9 Conclusions

In this paper we have introduced a typical modeling language, named $\mathcal{A}$, for capturing relevant features of existing graph-oriented BPMLs and we have considered jointly control-flow and data-flow concerns to have a broad perspective on modeling and execution issues.

We have argued that the essential features of languages like $\mathcal{A}$ do not fit well together, undermining their ability to express complex business process logic. Therefore, we have discussed some fundamentals language flaws related to (1) control-flow design into LSV context, (2) state invariants in presence of permissive token-based semantics, (3) the impedance between stateless tasks and AMP communication strategy, (4) the difficulty to verify a portion of a model isolated from the remaining ones and (5) the lack of modularity. Finally, we have considered the possibility to use a TLV strategy, where each token has its own local store. This solution offer a better data model for unstructured BPMLs, since it prevents some common race conditions. Unfortunately, it does not resolve all the problems highlighted for LSV and AMP strategies, because some of these are directly related to the choice of an unstructured control-flow.

We conjecture that the need for more flexible PAISs, pursued by increasingly large and varying business processes, brings to a fundamental paradigm shift in modeling, especially for expressing concepts related to concurrency. The flaws described in this paper are fundamental in nature and they cannot be overcame by simply adding new language constructs to existing unstructured BPMLs, while adding restrictions to tokens or adopting a more structured control-flow simply distort the peculiarities of this kind of languages. We hope that the presented analysis can help the design of the next generation of BPMLs.

Acknowledgments. We are grateful to Marcello La Rosa for his comments and suggestions that helped us to improve the quality of the original manuscript.

References

1. Dumas, M., van der Aalst, W.M., ter Hofstede, A.H.: Process-Aware Information Systems: Bridging People and Software Through Process Technology. Wiley-Interscience, Hoboken (2005)
2. Bandara, W., Indulska, M., Chong, S., Sadiq, S.: Major issues in business process management: an expert perspective. BPTrends (October 2007)

3. Mutschler, B., Reichert, M., Bumiller, J.: Unleashing the effectiveness of process-oriented information systems: Problem analysis, critical success factors, and implications. IEEE Transactions on Systems, Man, and Cybernetics, Part C 38(3), 280–291 (2008)
4. Adams, M., ter Hofstede, A.H.M., van der Aalst, W.M.P., Edmond, D.: Dynamic, extensible and context-aware exception handling for workflows. In: OTM Conferences (1), pp. 95–112 (2007)
5. Weber, B., Reichert, M., Rinderle-Ma, S.: Change patterns and change support features - enhancing flexibility in process-aware information systems. Data Knowl. Eng. 66(3), 438–466 (2008)
6. Sadiq, S.W., Orlowska, M.E., Sadiq, W.: Specification and validation of process constraints for flexible workflows. Inf. Syst. 30(5), 349–378 (2005)
7. Müller, D., Reichert, M., Herbst, J.: A new paradigm for the enactment and dynamic adaptation of data-driven process structures. In: Bellahsène, Z., Léonard, M. (eds.) CAiSE 2008. LNCS, vol. 5074, pp. 48–63. Springer, Heidelberg (2008)
8. Pesic, M., van der Aalst, W.M.P.: A declarative approach for flexible business processes management. In: Business Process Management Workshops, pp. 169–180 (2006)
9. Reijers, H.A., Mendling, J.: Modularity in process models: Review and effects. In: Dumas, M., Reichert, M., Shan, M.-C. (eds.) BPM 2008. LNCS, vol. 5240, pp. 20–35. Springer, Heidelberg (2008)
10. von Mayrhauser, A., Marie Vans, A.: Program comprehension during software maintenance and evolution. Computer 28(8), 44–55 (1995)
11. Weske, M.: Business Process Management: Concepts, Languages, Architectures. Springer-Verlag New York, Inc., Secaucus (2007)
12. Russell, N., Ter Hofstede, A.H.M., Mulyar, N.: Workflow control-flow patterns: A revised view. BPM center report BPM-06-22, bpmcenter.org. Technical report (2006)
13. Russell, N., ter Hofstede, A.H.M., Edmond, D., van der Aalst, W.M.P.: Workflow data patterns: Identification, representation and tool support. Conceptual Modeling - ER, 353–368 (2005)
14. Sadiq, S., Orlowska, M., Sadiq, W., Foulger, C.: Data flow and validation in workflow modelling. In: ADC 2004: Proceedings of the 15th Australasian database conference, Darlinghurst, Australia, pp. 207–214. Australian Computer Society, Inc., Australia (2004)
15. Lee, E.A.: The problem with threads. Computer 39(5), 33–42 (2006)
16. Zapf, M., Lindheimer, U., Heinzl, A.: The myth of accelerating business processes through parallel job designs. Inf. Syst. E-Business Management 5(2), 117–137 (2007)
17. Ouyang, C., Dumas, M., Breutel, S., ter Hofstede, A.H.M.: Translating standard process models to BPEL. In: Dubois, E., Pohl, K. (eds.) CAiSE 2006. LNCS, vol. 4001, pp. 417–432. Springer, Heidelberg (2006)
18. van der Aalst, W.M.P., Aldred, L., Dumas, M., ter Hofstede, A.H.M.: Design and implementation of the YAWL system. In: Persson, A., Stirna, J. (eds.) CAiSE 2004. LNCS, vol. 3084, pp. 142–159. Springer, Heidelberg (2004)
19. Object Management Group (OMG). Business Process Modeling Notation (BPMN) version 1.1. (January 2008)
20. JBoss Enterprise Middleware Red Hat. JBoss jBPM (2008)
21. OMG. Unified modeling language: Superstructure, version 2.1.1. Technical Report formal/2007-02-03, Object Management Group (2007)

22. Dijkman, R.M., Dumas, M., Ouyang, C.: Semantics and analysis of business process models in BPMN. Inf. Softw. Technol. 50(12), 1281–1294 (2008)
23. Kiepuszewski, B., ter Hofstede, A.H.M., Bussler, C.J.: On structured workflow modelling. In: Wangler, B., Bergman, L.D. (eds.) CAiSE 2000. LNCS, vol. 1789, pp. 431–445. Springer, Heidelberg (2000)
24. Liu, R., Kumar, A.: An analysis and taxonomy of unstructured workflows. In: van der Aalst, W.M.P., Benatallah, B., Casati, F., Curbera, F. (eds.) BPM 2005. LNCS, vol. 3649, pp. 268–284. Springer, Heidelberg (2005)
25. Netzer, R.H.B., Miller, B.P.: On the complexity of event ordering for shared-memory parallel program executions. In: Proceedings of the 1990 International Conference on Parallel Processing, pp. 93–97 (1990)

Maintaining Compliance in Customizable Process Models

Daniel Schleicher, Tobias Anstett, Frank Leymann, and Ralph Mietzner

Institute of Architecture of Application Systems,
University of Stuttgart
Universitätsstraße 38, 70569 Stuttgart, Germany
{schleicher,anstett,leymann,mietzner}@iaas.uni-stuttgart.de

Abstract. Compliance of business processes has gained importance during the last years. The growing number of internal and external regulations that companies need to obey has led to this state. This paper presents a practical concept of ensuring compliance during design time of customizable business processes.

We introduce the concept of a business process template that implicitly contains compliance constraints as well as points of variability. We further present an algorithm that ensures that these constraints cannot be violated. We also show how these algorithms can be used to check whether a customization of this process template is valid regarding these compliance constraints. So the designer of a business process, in contrast to the template designer, does not have to worry about compliance of the eventual process.

In a final step we show how these general concepts can be applied to WS-BPEL.

1 Introduction

Today compliance is a more important quality of service (QoS) requirement for organizations than it was in the past. As a consequence from financial and other scandals, organizations are forced to ensure the compliance to more and more regulations. These regulations are not only imposed by laws such as the Sarbanes–Oxley Act (SOX) [15] which regulates financial transactions, but also corporate self-commitments and social expectations such as realizing a *green IT*.

Regulations apply to the way a company does business by means of how their business processes are executed. Thus being compliant requires rethinking of the way business processes are managed in respect to regulations. This may include the installation of new roles in the organizational structure such as compliance-officers responsible for analyzing compliance requirements to capture the ones relevant for the organization. It also may require changes to the IT infrastructure executing the business process to provide enough evidence to (external) auditors responsible for assessing the compliance. Furthermore there is a need for specialized process designers which are able to map the identified regulations to abstract business process models which allow further customizations by

R. Meersman, T. Dillon, P. Herrero (Eds.): OTM 2009, Part I, LNCS 5870, pp. 60–75, 2009.

domain experts. During these customizations the real business logic is added. These abstract business process models are further referred as *compliance templates* which basically realize reference process implementations of one or more compliance constraints.

The complexity of real world business processes and the poor tool support today makes it hard for process designers to build processes in a way that is common for traditional programming languages like Java. Thus humans should not need to bother with compliance constraints besides the business logic of a process. What is needed is support of the process designer during design-time to automatically check compliance concerns so that the designer could be absolutely sure, the eventual process will be compliant to certain compliance constraints. Further requirements are performance and scalability of the algorithms, which check the compliance of a process. It is not reasonable for the process designer to wait 5 minutes after inserting a new activity into the process for the validation to finish.

A company may employ their own compliance-officers and specialized process designers to develop compliant processes but also has the option to outsource these tasks. 3rd party standardization organizations as well as software as a service providers may offer compliance templates which may be further customized to execute them in their own infrastructure or in the cloud. One big problem of customizations of a predefined compliance template is that the implicitly contained compliance constraints might be compromised which would cause a violation of the identified regulations at execution time. Thus there is a need to detect those malicious customizations and restrict the modeler of being able to use them.

In this paper we present an algorithm that ensures that compliance constraints of (customizable) compliance templates can not be compromised. We introduce a motivating example in Section 2 which illustrates the problems we address and which is used throughout the paper. We then present an approach how to ensure compliance using compliance templates comprising an abstract business process, variability descriptors and compliance descriptors in Section 3. In Section 4 we present formal definitions of compliance assurance rules which are part of compliance descriptors to implement a compliance validation algorithm. The introduced concepts are then applied to WS-BPEL [9], the standard for modeling and executing business processes in service oriented environments, in Section 5. Finally we compare the presented approach with related work in this field and finish with a conclusion and an outlook to future work.

2 Motivating Example

In this Section we show the motivation for this paper by means of a SaaS (Software as a Service) [14] scenario. We assume a fictional SaaS provider, *John Doe IT Service* offers a template for a loan approval process. This template will be completed by the customers of John Doe. The template implicitly comprises some compliance rules, like "two approve activities which must be executed in sequence" or "a credit check activity must be executed at the end of the process". After the finalization the fully functional business process is deployed on John

Does infrastructure. During the process of completing the template, activities, which are being inserted, could violate some of the contained compliance rules. For example an inserted activity could end the process before a credit check activity was executed.

In this paper we provide an approach to support the process designer in order to complete the template to a fully compliant business process.

3 Controlling Compliance with Process Templates

Compliance of a business process can be verified in numerous ways. There are approaches, which use monitoring to ensure the compliant execution of a business process [12]. Another approach [3] uses deontic logic to model obligations and permissions, which can then be used in the design phase of a business process to verify the compliance of the process. Our approach differs from the approaches above in two ways. First of all we provide an abstract business process, which already contains all activities important for compliance. This approach is very powerful because one can model a big set of compliance constraints within an abstract business process. This abstract process then will be completed by the process designer. Secondly to ensure that no activity, inserted by the process designer, violates the compliance of the abstract process, on the one side we constrain the set of activities, which can be inserted, and on the other side we provide a set of rules, which ensure, that an inserted activity cannot violate the compliance. In this section we show how to ensure compliance of customizable business processes by means of compliance templates (see Figure 1). A compliance template comprises three parts, (i) an *abstract business process*, (ii) *variability descriptors* and (iii) *compliance descriptors*. The abstract business process implicitly contains compliance rules. These are the rules the resulting business process should comply with after being completed by a process designer. The business activities contained within the abstract business process implement these rules. Additionally the abstract business process defines placeholders that need to be completed before the abstract process can be executed. Placeholders are described and restricted in more detail by the variability descriptors that are included in the compliance template. These variabilities are independent from the abstract business process and can differ for different application domains. Compliance descriptors describe additional compliance rules that customizations of the abstract process must comply with and that cannot be specified in the abstract process directly. In the following each of the parts of a compliance template are described in detail.

3.1 Abstract Business Processes

An abstract business process is a business process that contains so-called *opaque activities* and *compliance activities*. Opaque activities are used as placeholders that need to be replaced with concrete constructs during the customization to fill the abstract business process with concrete business logic and make the process executable. Compliance activities are activities in the business process that are not allowed to be removed or altered otherwise. They implement the compliance requirements which the compliance template should fulfill.

Figure 1 shows an abstract business process as a template for a loan approval process, a compliance descriptor, and a variability descriptor of a compliance template. The dotted arrows show exemplarily what alternative, compliance link, and compliance assurance rules is applied to which activity. The activities with label *1st decision* and *2nd decision* are compliance activities. Opaque activities are labeled *Opaque*.

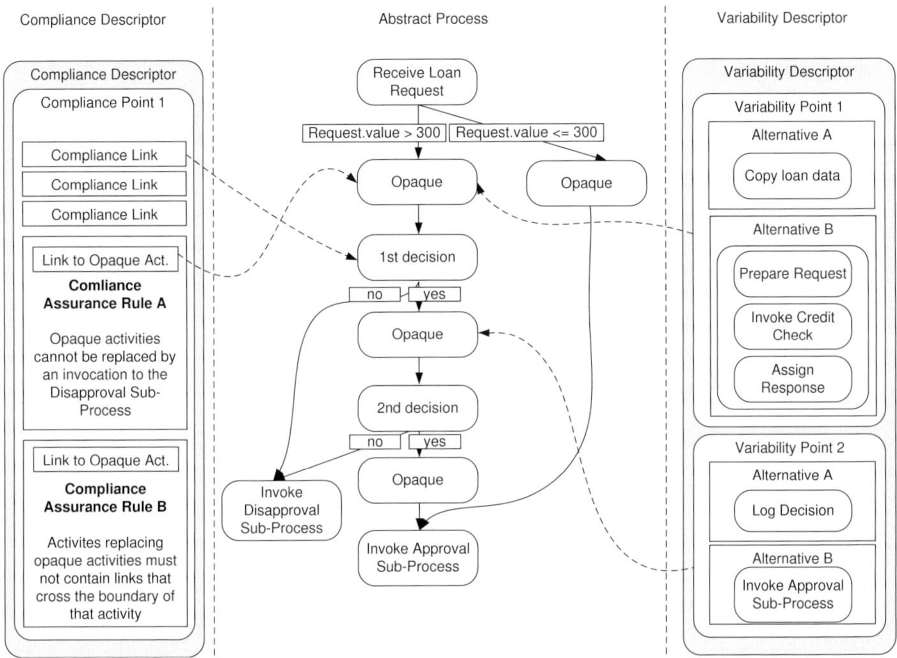

Fig. 1. Loan approval process with compliance descriptor and variability descriptor

3.2 Variability Descriptors

Variability descriptors [7,8] are a means to describe variability for applications. Unlike other approaches [13,5,11] they are not focused on process models alone but whole applications. Variability descriptors can be used to describe variability across all layers of the application ranging from the GUI over the workflow layer down to individual services and database schemes.

Besides documenting variability, variability descriptors can be transformed into workflows [8] that can be used to guide users through the customization of applications. They present the user with choices and abstract from the concrete implementation making it easier for non-developers to customize the applications.

In short a variability descriptor consists of a set of *variability points*. A variability point consists of a set of *locators* and a set of *alternatives*. Locators point into artifacts of the application such as a BPEL file or an HTML file which should be affected by that variability point. Alternatives describe alternative values for that variability point. Alternatives can be one of the following types.

- *explicit*, meaning that the concrete value is specified,
- *empty*, meaning that the code the locator points to is deleted,
- *free*, meaning that a user is prompted to enter any text that should go where the locator points to, or
- *expression*, meaning that the value where the expression points to or that is calculated during the evaluation of the expression, is inserted at the location the locator points to. An expression could be for example "insert the value inserted at another point in the document times 2".

Variability points are connected by *dependencies*. Dependencies describe the order in which variability points must be bound. Additionally to describe more advanced dependencies, so-called *enabling conditions* can be used to enable certain activities based on a condition. Using dependencies and enabling conditions it is possible to describe complex dependencies such as "after alternative A of variability point V1 has been bound only alternative C and not alternative D of variability point V2 can be bound".

3.3 Compliance Descriptors

Similar to variability descriptors explained in Section 3.2 *compliance descriptors* are defined. Compliance descriptors are specified independently from the abstract business process they belong to. This facilitates the reuse of compliance descriptors. For example an abstract business process, which implements a certain separation of duty constraint, can be reused several times in different compliance templates because the compliance descriptors are not tied to this abstract business process.

Compliance descriptors (see Figure 1) consist of so-called *compliance points*. Every compliance point expresses one compliance requirement for the business process. A compliance point consists of one or more *compliance links* and *compliance assurance rules*. Compliance links are used to mark the activities of an abstract process, which belong to a compliance point. These activities are called *compliance activities*. Activities marked this way should not be altered in any way in order to preserve the compliance of the abstract business process. When editing a process with a graphical tool, compliance activities, with a compliance link pointing on them, can for example be coloured red to show that these activities can not be altered in any way. Listing 1.1 shows the XML representation of a compliance link. The *document*-function of XSLT is used to point to the file where the corresponding activity is located. In this case it is an XML document that describes the abstract business process (e.g., a BPEL file). Additional XPath functionality is used to navigate inside the document to the designated activity.

A compliance assurance rule consists of the rule itself and a set of links to opaque activities of the abstract business process contained in the compliance

Listing 1.1. Compliance Link

```
<ComplianceLink>
   document("loanApproval.bpel")//bpel:opaqueActivity[@name="payment"]
</ComplianceLink>
```

template. In this way the rule is applied to a number of opaque activities. Compliance assurance rules describe which alternatives are not allowed to be put into an opaque activity. This ensures that the corresponding abstract process template stays compliant after inserting an alternative from a variability descriptor.

A compliance descriptor has one compliance point. The duty of this compliance point is to make sure, that a loan request is approved by two different persons. So there are two compliance links to two activities in the abstract business process, which implement the approve functionality of the process. These activities are now marked as unchangeable. Further there is a set of compliance assurance rules. These rules are written to make sure, that no activity inserted into an opaque activity can bypass one of the approve activities. For example one of these rules would permit that links, from activities replacing opaque activities, bypass activities marked by compliance links. Further examples for compliance assurance rules follow in Section 4.

4 Checking Variability against Compliance Rules

There are two rule sets. Compliance rules implicitly defined in the abstract business process of a compliance template and *compliance assurance rules* which are part of a compliance descriptor. Compliance assurance rules ensure that a compliance template could not be completed in a way that destroys one or many compliance rules which are implicitly implemented in the abstract business process of a compliance template. The following Section introduces formal definitions which are then used to compose a formal representation of each compliance assurance rule. This formal representation is then used in Section 4.1 to implement a compliance validation algorithm.

4.1 Formal Definitions

In this section we define sets and functions for use in the compliance validation algorithm presented in Section 4.2. We also show a list of compliance assurance rules, which are essential to preserve compliance of a compliance template. For every compliance assurance rule a formal description is provided. This formal description facilitates the implementation of the rule in an algorithm.

In [4] the abstract syntax of WS-BPEL 2.0 is described. In the following we use some definitions from this document shown in short in the list below. Although the formal representation of a business process used in this paper is based on BPEL it is generic enough to be adapted for other modelling languages.

- $\mathcal{A}$ is the set of all activities in a process.
- $\mathcal{A}_{type}$ is the set of all activities of a certain type.
- $\mathcal{V}$ is the set of all variables of a process.
- $\mathcal{L}$ is the set of control links in a process.
- $\mathcal{C}$ is the transition condition of a link.
- LR denotes a link between two activities, a link can contain a transition condition:

$$\mathsf{LR} \subset \mathcal{A} \times \mathcal{L} \times \mathcal{C} \times \mathcal{A} \tag{1}$$

  For further information see [4].
- The function $\mathsf{descendants} : \mathcal{A} \rightarrow 2^{\mathcal{A}}$ returns the set of all descendants of an activity. Descendants of an activity $a \in \mathcal{A}$ are the activities that are nested within this activity.
- The function $\mathsf{partnerLink}_{\mathcal{CO}}$ assigns a partner link to a message construct.

Definition 1 (Set of Replacement Activities)
The set of replacement activities $\mathcal{A}_{replacement}$ consists of business activities, which replace an opaque activity in the process of gaining an executable completion [9] out of the compliance template. An opaque activity can only be replaced by one *replacement activity. This is necessary because an opaque activity can have one or more incoming and one or more outgoing links. These links need to be mapped to the activity replacing the opaque activity. However $\mathcal{A}_{replacement}$ could contain several other activities. For example, in figure 1 $\mathcal{A}_{replacement}$ in alternative A consists only of the activity* Copy loan data.

Definition 2 (Replace Function)
$\mathcal{A}_{opaque}$ is the set of all opaque activities in a BPEL process. $\mathcal{E}$ is the set of all activities in a BPEL process except the opaque activities. The function $\mathsf{replace} : \mathcal{A}_{opaque} \rightarrow \mathcal{E}$ *with $\mathcal{E} = \mathcal{A} \setminus \{\mathcal{A}_{opaque}\}$ replaces an opaque activity with a non opaque activity in order to become an executable completion of the abstract business process of a compliance template. The variability descriptor, shown in figure 1, shows all activities, which are available for replacement of opaque activities of the abstract business process. If an opaque activity can be replaced by a particular alternative of a variability descriptor, is verified by the algorithm presented in section 4.2.*

Definition 3 (Variable Read Relation)
The relation $VR \subseteq \mathcal{A} \times \mathcal{V}$ shows that there is a read relation between an activity $a \in \mathcal{A}$ and a variable $v \in \mathcal{V}$.

In other words it shows that a variable v is read by an activity A.

Definition 4 (Variable Write Relation)
The relation $VW \subseteq \mathcal{A} \times \mathcal{V}$ shows that there is a write relation between an activity $a \in \mathcal{A}$ and a variable $v \in \mathcal{V}$.

In other words it shows that a variable v is written by an activity A.

Definition 5 (Set of Compliance Activities)
The set of compliance activities $\mathcal{A}_{compliance}$ comprises all activities with a compliance link pointing to. In figure 1 the activity labeled 1st decision *is such a compliance activity because a compliance link from compliance point 1 is pointing to it.*

Definition 6 (Compliance Template)
A compliance template $\mathcal{T}$ comprises three parts, (i) an abstract business process, *(ii)* variability descriptors *and (iii)* compliance descriptors *as described in section 3.3 and shown in figure 1.*

Compliance assurance rules can be divided into two kinds. Rules which have to be applied to all compliance templates and rules which have to be applied to specific compliance templates to ensure certain requirements. Compliance Assurance rule 1 is an example for the first kind of rules.

The following list describes the compliance assurance rules which are applied to all compliance templates and provides a formal description of each rule. We do not argue that the provided rules are complete as compliance assurance rules can differ strongly for any application domain. However, these rules are standard rules applied to every compliance template. A violation of these rules would allow to circumvent or deactivate compliance activities at run-time.

1. **Opaque activities must not be replaced with exit activities.** This rule ensures that the execution of a business process could not be stopped by replacement activities and thus compliance rules could not be violated. This rule applies for every compliance template because a compliance rule, implicitly defined in an abstract business process of a compliance template, could easily be violated by inserting an exit-activity like the *exit*-activity in WS-BPEL.

$$\forall o \in \mathcal{A}_{opaque} : \mathsf{replace}(o) = \mathcal{A} \setminus \{\mathcal{A}_{exit}\} \tag{2}$$

For instance, if any opaque activity in figure 1 would be replaced by an exit activity the loan approval process could be terminated abnormally. This is not intended because the compliance rules, implicitly contained within the abstract process, could then be violated.
2. **An alternative replacing an opaque activity must not contain links that cross the boundary of that opaque activity.** This rule ensures that compliance activities could not be bypassed. Links are only allowed between sub activities of replacement activities. This rule must also be applied to every compliance template because any compliance activity could be bypassed with a link from an inserted activity.

$$\begin{aligned} &\forall r \in \mathcal{A}_{replacement}, \forall a \in \mathsf{descendants}(r), \forall a' \in \mathcal{A} \setminus \mathsf{descendants}(r) : \\ &\neg\exists l \in \mathcal{L} : (a, l, c, a') \in \mathsf{LR} \vee (a', l, c, a) \in \mathsf{LR} \end{aligned} \tag{3}$$

In figure 2 the irrelevant parts of the loan approval process are drawn transparent. As you can see an opaque activity has been replaced by the replacement activity *Copy loan data*. We assume that the alternative A containing this activity also contains a link to another activity. This link is labeled *Cross Boundary Link* in figure 2. It is not allowed to insert this link into the abstract process because it bypasses the activity labeled *2nd decision* and another opaque activity. Thus the compliant execution of the process template could no longer be guaranteed.

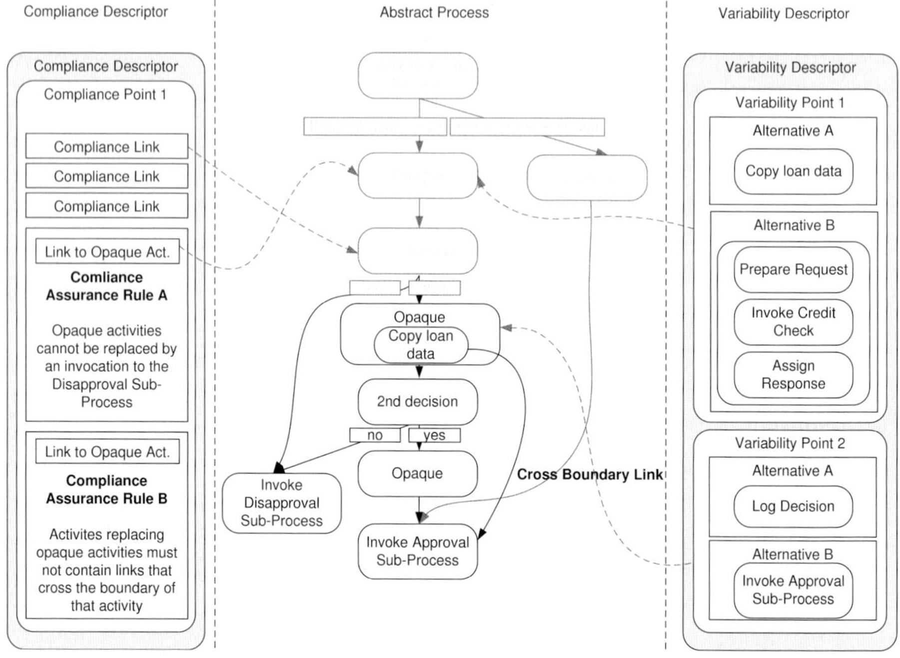

Fig. 2. Example for the use of links crossing the boundary of an opaque activity

3. **Variables that are read by one of the compliance activities must not be modified by replacement activities.** The possibility of data corruption through replacement activities is prevented with this rule. $\mathcal{V}_t$ denotes all variables which are part of the abstract business process of a compliance template. This rule must not necessarily be included in a compliance template. Structural compliance rules, like *activity A must be executed before activity B*, can not be violated by any write access of business activities to process variables.

   The fact that write access to such variables is denied, is a small restriction for the process designer. In most processes it won't be necessary to write to a variable belonging to compliance activities, because then the process designer would want to change the execution behavior of the resulting process. It will be more likely that an extra set of variables is introduced, which is used by the replacement activities. However, it is possible to read from a Variable in $\mathcal{V}_t$.

$$\forall a \in \mathcal{A}_{compliance}, \forall r \in \mathcal{A}_{replacement}, \forall v \in \mathcal{V}_t : \exists VR \Rightarrow \neg\exists VW \qquad (4)$$

As an example, in mind we add a variable with name *loan amount* to the abstract process shown in figure 1. This variable is written by the activity *Receive Loan Request*. If this variable is later in the process modified by a replacement activity (e.g. from the value 200 to value 500) some compliance rules, implicitly defined in the compliance template, could be violated. The process execution would take another branch than intended.

4. **Services which are being invoked from compliance activities must not be invoked from replacement activities.** This rule prevents violation of compliance rules through invocation of services which are used by compliance activities. This rule also must not be included in every compliance template because not every business process invokes other services.

$$\forall a \in \mathcal{A}_{compliance,invoke}, r \in \mathcal{A}_{replacement,invoke} : \\ \mathsf{partnerLink}_{\mathcal{CO}}(a) \neq \mathsf{partnerLink}_{\mathcal{CO}}(r) \quad (5)$$

In figure 1, the activity *Invoke Approval Sub-Process* invokes a sub-process to approve the loan request. This sub-process is only invoked once during the execution of the loan approval process. If a replacement activiy would invoke the *Approval Sub-Process* before the activity *Invoke Approval Sub-Process* is executed, it could bypass compliance rules implicitly defined in the compliance template.

4.2 Validation Algorithm

A valid replacement for an opaque activity must not violate compliance rules defined in a compliance template. In this section we present an algorithm to verify if a compliance assurance rule is being violated by replacing an opaque activity with a certain alternative. The algorithm is written in Java-style pseudocode. It is divided into two parts to reduce complexity and support readability.

An enterprise might have a lot of variability descriptors with even more alternatives for all kinds of purposes. The goal of the algorithm is now to select the valid alternatives for each opaque activity which do not violate compliance rules implicitly defined in the process template. So the process designer does not need to think about the validity of alternatives to be inserted into an opaque activity. Thus with this algorithm the compliance of the resulting executable completion is preserved by the design tool and must not be considered by the process designer at design time. The designer can then fully concentrate on the business objectives rather than struggling with compliance concerns.

Algorithm 1 shows how opaque activities, alternatives, and compliance assurance rules are linked in data structures. The prerequisite for the algorithm is a list of opaque activities. Furthermore it shows how these data structures are traversed in order to validate an alternative. Essentially it iterates over all opaque activities, gets the alternatives for each opaque activity and validates each alternative against the rules which apply for the current opaque activity. One opaque activity object contains a list of alternatives and a list of compliance assurance rules.

Algorithm 1. Calculate valid alternatives

```
//An opaque activity contains a list of alternatives and compliance assurance rules
Require: List opaqueActivities;
for opaqueActivity in opaqueActivities do
    List alternatives = opaqueActivity.getAlternatives();
    List complAssurRules = opaqueActivity.getAssurRules();
    for alternative in alternatives do
        for complianceAssuranceRule in complianceAssuranceRules do
            if not (complianceAssuranceRule.validate(alternative)) then
                alternatives.remove(alternative);
            end if
        end for
    end for
end for
```

Algorithm 2 is the continuation of algorithm 1. It shows a sample implementation for the validation of compliance assurance rule 1 ”‘Opaque activities must not be replaced with exit and opaque activities”’ from section 4.1. In order to check if a business activity is an exit activity, we have to iterate over all business activities of an alternative inserted into the abstract business process.

Algorithm 2. Example implementation of method *validate* (algorithm 1) to check Compliance Assurance Rule 1.

```
Require: alternative;
Array activities = alternative.getActivities();
for activity in activities do
    if activity.getType == 'Exit' then
        return false;
    end if
end for
return true;
```

5 Application to WS-BPEL

In the specification of WS-BPEL so called *abstract BPEL processes* are defined. Abstract BPEL processes are used to define business process templates and thus can be used analogue to our definition of an abstract process. An abstract BPEL process can include so-called *opaque activities*, which during the customization will be filled with replacement activities from the provided alternatives.

To meet the constraint, that one opaque activity can only be replaced by exactly one replacement activity (see section 4.1), so called structured activities such as the scope or flow activity can be used. In BPEL a structured activity can contain several other activities. Besides this it has a number of other functions. The same is true for a sequence, while, repeatUntil, and forEach activity.

In the following we show how every compliance assurance rule can be applied to a business process written in BPEL.

- **Rule 1 (Opaque activities must not be replaced with exit and opaque activities):** Exit and opaque activities are included in the BPEL specification.
- **Rule 2 (An alternative replacing an opaque activities must not contain links that cross the boundary of that opaque activity):** This rule must be considered when using BPEL because in BPEL links for example can cross the boundary of certain activities like a Scope activity.
- **Rule 3 (Variables which are read by one of the compliance activities must not be modified by replacement activities):** Since with the BPEL language one can define variables, this rule can be applied.
- **Rule 4 (Services which are being invoked from compliance activities must not be invoked from replacement activities):** This rule must be considered, because BPEL is intended to be used above the Web service layer, so Web service invocations are an integral capability of BPEL.

The algorithms presented in section 4.2 can easily be applied when using BPEL, because they are written in an process language independent form.

6 Implementation Aspects

As a proof of concept we have implemented a prototype, which implements the algorithms presented in Section 4.2. It further shows how the algorithms and formal definitions from Section 4.1 could be used to verify alternatives being inserted into a business process. The prototype therefore implements the situation where a human wants to insert an alternative into a business process. The design of the prototype reflects two important aspects, extensibility and applicability in a graphical workbench.

Figure 3 shows design of the prototype in an UML class-diagram and the dependencies of the classes. For the reason of applicability in a graphical workbench, the init-method in the *VariabilityChecker*-class does the job the workbench later should do. It initializes the installed compliance assurance rules present in the system and loads a Variability Descriptor from the file system. Everything is saved in a new object of the type *OpaqueActivity*. This object represents an opaque activity of an abstract process of a compliance template. Then one alternative included in the Variability Descriptor is loaded into an object of the type *Alternative*.

After initialization, the validation of the loaded alternatives is triggered by invoking the *validateAlternative* method. In the future this method is the interface to the graphical process modeling tool. The implementation of this method is kept as similar as possible with the algorithms 1 and 2. Within this method the *validate*-method of the class *ExitComplAssurRule*, which implements compliance assurance rule 1, is invoked. At this point new compliance assurance rules can easily be added, because a single compliance assurance rule is represented

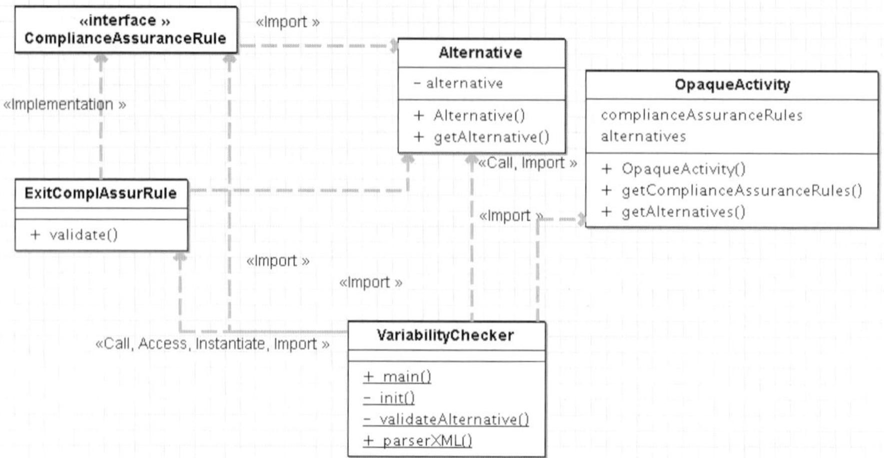

Fig. 3. Class diagram with dependencies of the prototype classes

by one Java class. Classes, which implement additional compliance assurance rules, have to implement the *ComplianceAssuranceRule* interface. After that the class can be added to the implementation of the validateAlternative method. Every class, which implements the ComplianceAssuranceRule interface, needs to implement the method *validate*. Thus there is no sole method where every compliance assurance rule is verified but a number of classes where this is done. This eases maintainability of core component, the implementation of the validation algorithms. Every class implements the validation algorithm for a single compliance assurance rule. With this approach it is easy to add, delete and modify implementations of compliance assurance rules.

7 Related Work

Our approach is different from related work as it combines the modelling of variability with the need for compliance in business processes. Other approaches such as configurable EPCs [11] focus on the modelling and description of variability in process models. The same approach is followed by [5] where points-of-variability in BPEL processes are introduced. However there is no means to verify whether the alternatives which are modelled do not contradict compliance rules which need to be followed by the process. Another approach, VXBPEL [13] extends BPEL with variability making it possible to model alternatives which need to be bound before execution.

Other approaches which deal with run-time adaptation of workflows (such as [10]) also need to take compliance violations into account as these might occur during the adaptation of the workflow. Therefore in [10] compliance rules are enforced during the adaptation which prevent the user to adapt the workflow in a way that these rules are violated. Our approach is different and does not require the extension of existing workflow middle-ware as it allows a modeller to check before the workflow is deployed, whether the variability the modeller has added to the process model is contradictory to the compliance rules. However the algorithms presented in this paper can also be applied in a run-time adaptation scenario because compliance descriptors and variability descriptors can be used to annotate a running BPEL process. In addition to that our approach allows the process designer to annotate different compliance requirements to the same process model thus making the process more flexible and adaptable to different requirements.

Business processes need to be consistent with internal regulations of enterprises and financial markets, laws and customer needs. This leads to a strong need for flexibility of business processes because of changing legislation, customer needs and market conditions [1]. But it is also crucial not to violate compliance rules which apply to a business process when adapting it at design time or at run-time.

There is a considerable amount of work done in annotation of business processes with compliance requirements. In [12] an approach is described which uses the formal contract language (FCL [2]) to describe compliance constraints on business processes. These compliance rules then are used as annotations for the business processes. Furthermore the notions of *compliance distance* and *control tags* are introduced. Compliance distance denotes the effort to make a process model compliant to certain rules. Control tags are used to model control objectives in a process model. In our work we use so called compliance descriptors to annotate business processes with compliance constraints. One advantage of compliance descriptors is, that you can reuse them for different business processes. This is made possible by the fact, that every compliance descriptor defines one atomic compliance constraint which then can be composed with other compliance descriptors to more complex compliance constraints.

Another approach for integration of semantic constraints in process management systems is shown in [6]. The main focus lies here on the process management system and its capabilities to verify semantic constraints of business processes when they are modified at run-time. Here only dependency and mutual exclusion constraints are considered. In our work the compliance constraints are implicitly defined by an abstract process template. In this template a infinite number of constraints can be defined.

8 Conclusions and Outlook

In this paper we presented an approach to prevent process designers from bothering with compliance constraints during design time. This approach is based upon the fact that in a compliance template most kinds of compliance constraints can be implicitly included. It is then the duty of the algorithms of the process design workbench to check if alternatives inserted into the template violate these constraints.

In the future we will work on the implementation of a design time tool with the concepts of this paper built in. Furthermore some aspects of these concepts can also be applied to a tool to adapt an already running business process. Here we want to extend a BPEL engine to support validation of run-time adaptations against compliance rules.

Our work will also include research on dependencies between alternatives and dynamic compliance rules. It could for example be possible that when a certain alternative is inserted into a process, another has to be inserted, too. By dynamic compliance rules we understand that a rule, which has not be satisfied during execution at the beginning of a process, must be satisfied at the end of a process. One example for this is that a credit check activity must be executed at least once during process execution. For example due to a Web service timeout a credit check activity could not be executed. For this reason an extra credit check activity has to be inserted at the end of the process in order to verify if the credit check has already been executed. If not, the credit check must be executed in order to fulfill the compliance requirement. This extra check could be inserted automatically if a credit check alternative is inserted into the process by a process designer.

Acknowledgments

The work published in this article was partially funded by the MASTER project[1] under the EU 7th Research Framework Programme Information and Communication Technologies Objective (FP7-216917).

References

1. Goedertier, S., Vanthienen, J.: Designing Compliant Business Processes from Obligations and Permissions. In: Eder, J., Dustdar, S. (eds.) BPD 2006. LNCS, vol. 4103, pp. 5–14. Springer, Heidelberg (2006)
2. Governatori, G., Milosevic, Z.: A Formal Analysis of a Business Contract Language. Int. J. Cooperative Inf. Syst. 15(4), 659–685 (2006)
3. Kharbili, M.E., Stein, S., Markovic, I., Pulvermüller, E.: Towards a Framework for Semantic Business Process Compliance Management. In: GRCIS 2008 (June 2008)
4. Kopp, O., Mietzner, R., Leymann, F.: Abstract Syntax of WS-BPEL 2.0. Technical report, University of Stuttgart, IAAS, Germany (2008)
5. Lazovik, A., Ludwig, H.: Managing process customizability and customization: Model, language and process. In: Benatallah, B., Casati, F., Georgakopoulos, D., Bartolini, C., Sadiq, W., Godart, C. (eds.) WISE 2007. LNCS, vol. 4831, pp. 373–384. Springer, Heidelberg (2007)
6. Ly, L.T., Rinderle, S., Dadam, P.: Integration and Verification of Semantic Constraints in Adaptive Process Management Systems. Data Knowl. Eng. 64(1), 3–23 (2008)
7. Mietzner, R.: Using Variability Descriptors to Describe Customizable SaaS Application Templates. Technical Report Computer Science 2008/01, University of Stuttgart, Faculty of Computer Science, Electrical Engineering, and Information Technology, Germany, University of Stuttgart, Institute of Architecture of Application Systems (January 2008)

[1] http://www.master-fp7.eu/

8. Mietzner, R., Leymann, F.: Generation of BPEL Customization Processes for SaaS Applications from Variability Descriptors. In: IEEE SCC, pp. 359–366. IEEE Computer Society Press, Los Alamitos (2008)
9. OASIS. Web Services Business Process Execution Language Version 2.0 - Oasis Standard (2007)
10. Reichert, M., Rinderle-Ma, S., Dadam, P.: Flexibility in Process-Aware Information Systems. T. Petri Nets and Other Models of Concurrency 2, 115–135 (2009)
11. Rosemann, M., van der Aalst, W.M.P.: A Configurable Reference Modelling Language. Inf. Syst. 32(1), 1–23 (2007)
12. Sadiq, W., Governatori, G., Namiri, K.: Modeling control objectives for business process compliance. In: Alonso, G., Dadam, P., Rosemann, M. (eds.) BPM 2007. LNCS, vol. 4714, pp. 149–164. Springer, Heidelberg (2007)
13. Sun, C.-A., Aiello, M.: Towards Variable Service Compositions Using VxBPEL. In: Mei, H. (ed.) ICSR 2008. LNCS, vol. 5030, pp. 257–261. Springer, Heidelberg (2008)
14. Turner, M., Budgen, D., Brereton, P.: Turning Software Into a Service. Computer 36(10), 38–44 (2003)
15. United States Code. Sarbanes-Oxley Act of 2002, PL 107-204, 116 Stat 745. Codified in Sections 11, 15, 18, 28, and 29 USC (July 2002)

Measuring the Compliance of Processes with Reference Models

Kerstin Gerke, Jorge Cardoso, and Alexander Claus

SAP AG, SAP Research, CEC Dresden,
Chemnitzer Str. 48, 01187 Dresden, Germany
mail@kerstin-gerke.de, jcardoso@dei.uc.pt, mail@alexander-claus.de

Abstract. Reference models provide a set of generally accepted best practices to create efficient processes to be deployed inside organizations. However, a central challenge is to determine how these best practices are implemented in practice. One limitation of existing approaches for measuring compliance is the assumption that the compliance can be determined using the notion of process equivalence. Nonetheless, the use of equivalence algorithms is not adequate since two models can have different structures but one process can still be compliant with the other. This paper presents a new approach and algorithm which allow to measure the compliance of process models with reference models. We evaluate our approach by measuring the compliance of a model currently used by a German passenger airline with the IT Infrastructure Library (ITIL) reference model and by comparing our results with existing approaches.

1 Introduction

Reference models have gained increasing attention, because they make a substantial contribution to design and execute processes efficiently. Obviously, reference models are useful, but to which extent are these best practices adopted and implemented in a specific business context? Process mining algorithms [1,17] have shown a considerable potential for assessing the compliance of instances with reference models. The instances are typically recorded by process-aware IS and serve as a starting point for reconstructing an as-is process model. The derived model can be compared with other models (e.g. reference models) using existing algorithms to determine the equivalence of processes. Nevertheless, the results of a former compliance analysis using process mining and equivalence algorithms are not sufficient [11]. Our previous studies have evaluated the compliance of an as-is process model of a passenger airline with a reference model, which had incorporated the fundamentals of ITIL [15]. We found that the techniques available yield low values of compliance which could not be confirmed by the passenger airline. This difference was mainly due to: (1) different levels of details, (2) partial view of process mining, and (3) overemphasis of the order of activities. First, the level of detail characterizing a process differs widely when comparing a reference model with an as-is or to-be process model. Second, the derived as-is model only partially represents the processes of the airline. The execution of the processes

R. Meersman, T. Dillon, P. Herrero (Eds.): OTM 2009, Part I, LNCS 5870, pp. 76–93, 2009.

does not only result in log files but it also results in written record files, manual activities as well as human knowledge. Information outside the reach of process mining algorithms may compromise the results of compliance. Finally, reference models typically do not state whether dependencies between activities are compulsory. During our former studies [11] on compliance using existing equivalence algorithms, we have changed the order of activities in a reference model. While the compliance should remain the same since the reference model did not enforce a specific order for the execution of the activities, the compliance yielded different results.

This paper motivates the reader for the importance of measuring the compliance of process models with reference models. We also discuss the differences between process equivalence and process compliance and argue for the need of specific algorithms to measure the compliance between processes. We show that two models can have different structures but one process can still be compliant with the other. Furthermore, we develop a new approach and algorithm to overcome the drawbacks identified. We measure the compliance of an as-is process model of a German passenger airline with a reference model. To validate our methodology, we compare our compliance results with two existing approaches and explain why current algorithms are not suitable to evaluate the compliance.

The remainder of our paper is organized as follows. Section 2 introduces the fundamentals of reference models. Section 3 explains our methodology to measure compliance. The following section investigates the requirements for determining compliance. Sect. 5 presents and evaluates our rational and concept to develop a new algorithm. Sect. 6 describes the main related work. Finally, Sect. 7 formulates our conclusions based on our findings.

2 The Importance of Reference Models

Reference models offer a set of generally accepted processes which are sound and efficient. Their adoption is generally motivated by the following reasons. First, they significantly speed up the design of process models by providing reusable and high quality content. Second, they optimize the design as they have been developed over a long period and usually capture the business insight of experts [25]. Third, they ease the compliance with industry regulations and requirements and, thus, mitigate risk. Fourth, they are an essential mean to create a link between the business needs and IT implementations [25].

Reference models can be differentiated along their scope, their granularity, and the views, which are depicted in the model [25]. We distinguish (1) reference models focusing on capturing domain-specific best practices like ITIL, COBIT, and SCOR, and (2) configurable reference models, such as SAP Solution Manager [18], which aim at capturing the functionalities of a software system. Although the focus of this paper is on the first class of models, we explain both classes shortly with respect to their characteristics and their contribution to compliance.

The Information Technology Infrastructure Library (ITIL) is a set of guidance published as a series of books by the Office of Government Commerce. These books describe an integrated best practice approach to managing and controlling IT services [15]. The Control Objectives for Information and related Technology (COBIT) has been developed by the IT Governance Institute to describe good practices, to provide a process framework and to present activities in a manageable and logical structure. The Supply Chain Operations Reference Model (SCOR) provides a unique framework, which links business process and technology features into a unified structure to support communication among supply chain partners and to improve the effectiveness of supply chains [19].

A process is compliant in terms of the introduced reference models if the process is implemented as described by the reference model and the process and its results comply with laws, regulations and contractual arrangements [21]. Other popular reference models include the APQC Process Classification Framework SM (PCF) [2] and the Capability Maturity Model Integration (CMMI) [6].

The SAP Solution Manager of SAP NetWeaver [18] provides configurable reference models for business scenarios. Their usage ensures quality of the IT solution and enables traceability of all changes and, thus, compliance to the organizational needs. Most of the ERP vendors have similar approaches to support the configuration and implementation procedure of an IS landscape.

3 Methodology to Analyze Compliance

Based on our experiences with business processes of the air travel industry, we devised a generic approach and methodology to analyze the compliance between processes. The methodology identifies 5 entities, illustrated in Fig. 1, which need to be considered when measuring the compliance with reference models: the meta reference model M_0, the adopted reference model M_1, the to-be process model M_2, the instances of a process model M_2, and the as-is process model M_3. Depending on the scope, a meta reference model M_0 may provide either generally accepted processes or a set of abstract guidelines. In both cases, and particularly in the latter case, the reference model M_1 needs to be adapted to the needs of an organization yielding a set of processes M_2. The execution of the processes generates a set of instances. The analysis of these instances provides an as-is process model M_3 which reflects how a process M_2 was executed. The level of compliance can be measured by analyzing process models M_0, M_1, M_2, and M_3. Since M_0 is generally specified in natural language, we will concentrate our study on analyzing models M_1, M_2, and M_3.

Model M_1 and M_2 are mainly constructed manually, whereas M_3 is usually inferred from log files. These log files serve as a starting point for process mining algorithms, which aim at the automatic extraction of process knowledge. Various algorithms [1,17] have been developed and implemented in ProM [16] to discover different types of process models, for instance Petri nets [22] or Event-driven Process Chains (EPCs) [26]. ProM is a process mining workbench offering algorithms to discover and verify process models [26].

The level of compliance is expressed by a quality indicator, which can be incorporated into a maturity model, e.g. the COBIT maturity model “Manage

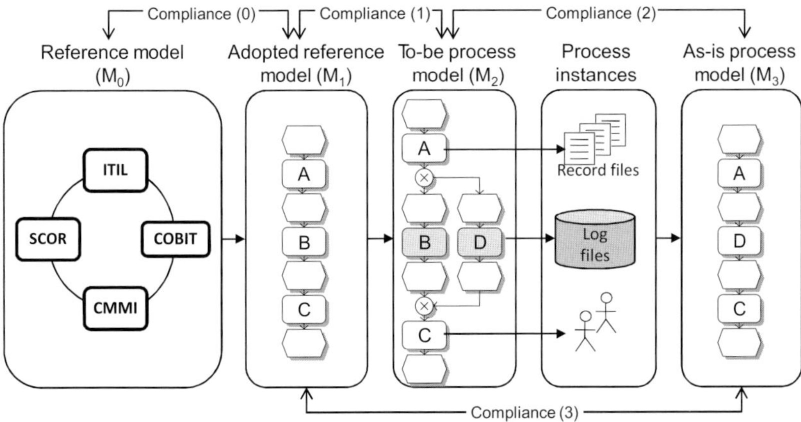

Fig. 1. Entities of a compliance analysis with reference models

Service Desk and Incidents" [21]. Such a model classifies the degree to which a process is aligned with a reference process. The level of compliance measured by the comparison of model M_2 or M_3 with M_1 serves as an initial estimate of the as-is compliance maturity. Opposing the as-is maturity and the to-be maturity supports the identification of potential improvements and contributes to determine alternative actions.

4 Requirements for a Compliance Analysis

We define process compliance as the degree to which a process model behaves in accordance to a reference model. The behavior is expressed by the instances, which can be generated by the model.

Figure 2 shows two EPCs capturing similar functionalities. Both are taken from the complaint handling process of a German passenger airline. The process is supported by the application "Interaction Center" (IAC) of the SAP Customer Relationship Management (CRM) system. The IAC facilitates the processing of interactions between business partners. Each interaction is registered as an activity. Besides a complaint description, further information, such as associated documents (e.g. e-mails), may be related to activities. Based on the characteristics of a complaint, an activity of the categories "Cust. Relations" or "Cust. Payment" is established. For example, complaints associated with payments are processed by the "Cust. Payment" department.

The EPC in the center of the figure shows model M_1, which depicts three activities: *Create incident*, *Categorize incident*, and *Prioritize incident*. The EPC on the right-hand side of the figure shows model M_2. Processing starts with an incoming complaint. Customers can complain by sending an e-mail or by filling an online form. In the latter case, the customer has to classify the complaint. In the former case, an employee has to read the e-mail to understand the complaint and determine the category manually. To measure the compliance, we need to discuss characteristics of business and reference models.

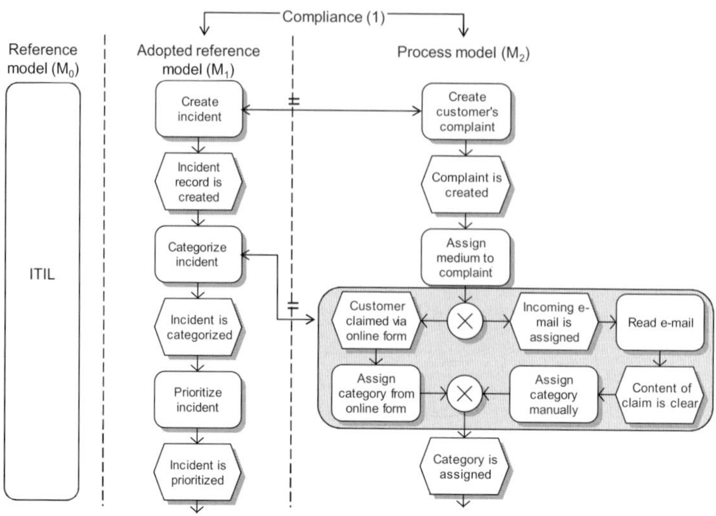

Fig. 2. The complaint handling process of a German passenger airline

Compliance Maturity and Degree. Our case study has identified two major concerns when it comes to evaluating compliance with reference models. First, the passenger airline wanted to learn if its processes followed the behavior recommended by the reference model. Second, the airline wanted to inquire if all the behavior recommended by the reference model was being implemented. In the context of compliance, we refer to the former as compliance degree and we denote the latter as compliance maturity. Let us consider the processing of incoming customer complaints. Model M_1 may recommend accepting complaints either via e-mail, letter or phone. If the airline accepts complaints via the first two mentioned communication channels only a part of the recommendations is implemented. We say that the airline is partially mature with respect to compliance maturity. But the ones currently being implemented (e-mail and letter) correspond to what the reference model M_1 recommends. In such a case, we say that the airline is fully compliant with respect to compliance degree.

Granularity of Models. Having two models M_1 and M_2 it may happen that the granularity characterizing the level of detail of activities varies. For example, in Fig. 2, activity *Prioritize incident* exists in model M_1, but no such activity exists in model M_2. Furthermore, it is possible that compliance applies to a set of activities, rather than individual activities. For example, activity *Categorize incident* of model M_1 corresponds to a set of activities in model M_2 highlighted in Fig. 2. In order to account for the granularity we have to identify the correspondence of activities. Correspondence is a mapping between activities of model M_2 to activities of model M_1 where the functionality of the activities is the same. Existing approaches, for example schema or semantic matching [26,9], assume that the correspondence can be established automatically based on the labels. The examples of our use case show that it is not realistic to only assume that equivalent activities may be identified by considering similarities of labels. For example, the activities *Create incident* in model M_1 and *Create customer's complaint* in model M_2 have the same functionality, but they have different labels. Since the automatic mapping is not applicable, we favor the manual mapping.

Customization of the Reference Model. It is often important to treat parts of model M_1 in a special way when measuring compliance. For example, since reference models do not typically state if the activities have to be executed exactly in a specified order, the order may not always be important. We refer to these special parts as partitions. A partition is a user-selected set of activities with a type, which can be "Order" or "Exclusion". Figure 3 shows that activities *Categorize incident* and *Prioritize incident* in partition P_1 may be executed in an arbitrary order. A partition of type "Exclusion" allows the definition of activities, which need to be excluded from the compliance analysis. Consider partition P_2. In our use case, the preprocessing of an incident is not supported by the IS right now. However, a manual activity corresponding to the functionality expressed by activity *Preprocess incident* is executed. To prevent the missing activity to erroneously affect the compliance, the activity is excluded.

Iteration. A special circumstance is the case in which an activity is part of an arbitrary cycle in process M_2 while it is not in model M_1. This means that this activity can be executed repetitively, while in model M_1 it must be performed correctly in only one iteration. For example, in our use case, the activities *Search for a solution* and *Inform Customer* are performed repeatedly until the customer accepts the processing of the claim. The existence of the cycle increases the quality of the process and contributes to a higher degree of the customer satisfaction. Thus, even if ITIL does not explicitly recommend a cycle, the airline feels that this cycle in model M_2 does not affect the compliance with model M_1 - a contrast with a cycle, which purely means to redo work. The latter cycle negatively affect the efficiency of a process. What makes it even more complicated is the fact that various reference models neither contain cycles nor state a precise number of recommended iterations. Without knowing the semantics of cycles it is not possible to state in general its effect on compliance.

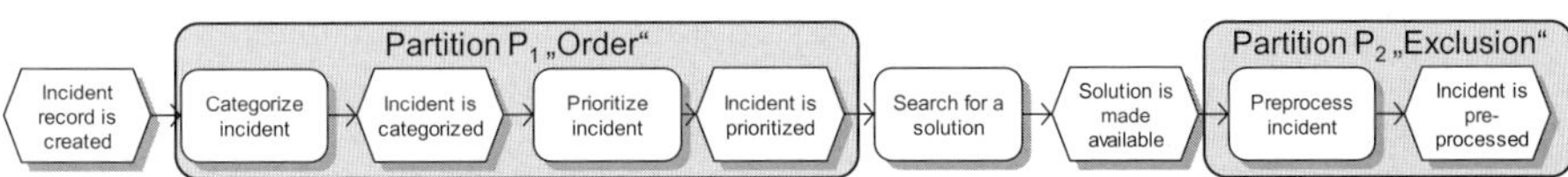

Fig. 3. Customization of reference model M_1

5 Sequence-Based Compliance

Based on requirements from Sect. 4 we have developed an algorithm to measure the compliance of model M_2 or M_3 with model M_1. Its main characteristic is that two models can have different structures but the algorithm can still judge one process to be compliant with the other. Figure 4, for example, clearly illustrates that the process models are different, but we will show that they are compliant.

5.1 Theoretical Foundations

Previous sections have used the EPC language to model processes since it is easy to understand and it is widely used in the industry (e.g. the common language of our use case). We use a more formal approach based on WF-nets [22] for the design of the compliance algorithm. It is a formalism well suited to analyze processes since there is a vast amount of research done in this area. We define the degree of compliance based on the firing sequences of WF-nets.

Definition 1 (WorkFlow net)
A WorkFlow net (WF-net) *is a tuple* $M = (P, T, F, i, o)$ *such that:*

- P *is a finite set of* places,
- T *is a finite set of* transitions,
- $P \cap T = \emptyset$,
- $F \subseteq (P \times T) \cup (T \times P)$ *is a set of arcs,*
- $i \in P$ *is the unique* source *place such that* $\bullet i = \emptyset$,
- $o \in P$ *is the unique* sink *place such that* $o\bullet = \emptyset$,
- *Every node* $x \in P \cup T$ *is on a path from* i *to* o,

where for each node $x \in P \cup T$ *the set* $\bullet x = \{y|(y, x) \in F\}$ *is the* preset *of* x *and* $x\bullet = \{y|(x, y) \in F\}$ *is the* postset *of* x.

Transitions represent the activities of an instance. The input place (i) and the output place (o) of the WF-net express the entry point when instances are created and the exit point when instances are deleted. The last requirement ensures that there are no transitions and places which do not contribute to processing.

Definition 2 (Firing sequence)
Let $M = (P, T, F, i, o)$ *be a WF-net and let* $t \in T$ *be a transition of* M.

- *A* marking $K : P \to \mathbb{N}$ *is a mapping defining the number of tokens per place.*
- t *is* enabled *in a marking* K *if* $(\forall p \in \bullet t)\, K(p) \geq 1$.
- t fires from *marking* K to *marking* K', *denoted by* $K[t\rangle K'$, *if* t *is enabled in* K *and* $(\forall p \in \bullet t)\, K'(p) = K(p) - 1$ *and* $(\forall p \in t\bullet)\, K'(p) = K(p) + 1$.
- $\sigma = \langle t_1, t_2, \ldots, t_n \rangle \in T^*$ *is a* firing sequence *leading from a marking* K_1 *to a marking* K_{n+1}, *denoted by* $K_1[\sigma\rangle K_{n+1}$, *if there are markings* $K_2, \ldots, K_n$, *such that* $K_1[t_1\rangle K_2[t_2\rangle \ldots K_n[t_n\rangle K_{n+1}$.

To capture relevant behavior we restrict ourselves to firing sequences representing process instances, which are terminated properly.

Definition 3 (Complete sound firing sequences). *Let* $M = (P, T, F, i, o)$ *be a WF-net and* $\sigma \in T^*$.

- K_i *is the* initial marking *with* $K_i(i) = 1$ *and* $(\forall p \neq i)\, K_i(p) = 0$.
- K_o *is the* final marking *with* $K_o(o) = 1$ *and* $(\forall p \neq o)\, K_o(p) = 0$.

- *σ is a* complete sound firing sequence, *if $K_i[\sigma\rangle K_o$.*
- *Let us use $S(M)$ to denote the set of all complete sound firing sequences.*

This definition ignores unsound behavior, for instance process instances running into a deadlock or a livelock. When no ambiguity occurs, we simply refer to σ as a firing sequence.

Since WF-nets can be considered as directed graphs, where $P \cup T$ is the set of nodes and F is the set of arcs, we use the standard graph-theoretical notion of a cycle.

Definition 4 (Cycle). *A* cycle *in a WF-net $M = (P, T, F, i, o)$ is a sequence of nodes $(x_1, \ldots, x_n) \in (P \cup T)^*$, such that $(\forall 1 \leq i < n)\,(x_i, x_{i+1}) \in F$ and $x_1 = x_n$.*

The existence of cycles causes the set $S(M)$ to be in general infinite. Therefore, we restrict the number of unroll factors for cycles by a variable parameter[1]. We end up with a finite subset of $S(M)$ denoted by $S'(M)$. The set $S'(M)$ grows exponentially in the number of transitions $|T|$. However, Sect. 5.5 will show that our approach can be used in practice. Our strategy to deal with cycles and their contribution to compliance among competing requirements (see Sect. 4) is to equate cycles having no correspondence in model M_1 with the action of redoing work. The superfluous work may have a negative effect on the compliance values.

5.2 Measuring Compliance

To account for the special characteristics of compliance with reference models, which we have identified in Sect. 4, we use several parameters to our algorithm.

Definition 5 (Granularity mapping). *Let be $M_1 = (P_1, T_1, F_1, i_1, o_1)$ and $M_2 = (P_2, T_2, F_2, i_2, o_2)$ two WF-nets where we refer to M_1 as the* reference model *and to M_2 as the* process model. *We use a mapping $\mathcal{G} : T_2 \rightarrow T_1$ to map activity labels in the process model to activity labels in the reference model. Since $\mathcal{G}$ can be non-injective, this mapping can handle granularity differences between the two models. Let us use the term* granularity mapping *for $\mathcal{G}$.*

Definition 6 (User-selected partition). *Let M_1 be a reference model as stated in Def. 5. A* user-selected partition *of M_1 is a set of transitions $p \subseteq T_1$ which can be of type* exclusion *or* order. *User-selected partitions of type exclusion are represented with $\bar{p}$ and those of type order with $\check{p}$. M_1 can have associated with it at most one user-selected partition of type exclusion and an arbitrary finite number of user-selected partitions of type order. Let us use $\mathcal{P}$ to denote the set of all user-selected partitions associated with M_1.*

Now that we have defined the parameters we deduce the compliance measures.

[1] We omit the parameter here and in subsequent equations since it has no significant effect to the equations and we want to keep them readable.

Definition 7 (Extended firing sequence set, Mapped firing sequence set). *Let M_1 and M_2 be the reference model and the process model as stated in Def. 5. Let $\mathcal{P}$ be the set of all user-selected partitions related to M_1 and let $\mathcal{G}$ be the granularity mapping between M_1 and M_2. Let $\sigma_1 \in T_1^*$ and $\sigma_2 \in T_2^*$.*

- $\sigma_1^{\mathrm{ext}}(\mathcal{P})$ *is the* set of extended firing sequences *of σ_1, which is derived from σ_1 by applying two actions to σ_1: (1) remove the transitions in $\bar{p}$ from σ_1 and (2) generate the permutations of $\sigma_1 \setminus \bar{p}$ for all user-selected partitions $\check{p}$.*
- *Let us use $|\sigma_1|_{\mathrm{ext}} = |\sigma_1'|$ $(\sigma_1' \in \sigma_1^{\mathrm{ext}}(\mathcal{P}))$ to denote the length of an arbitrary extended firing sequence σ_1' of σ_1.*
- $\sigma_2^{\mathrm{map}}(\mathcal{G})$ *is the* set of mapped firing sequences *of σ_2, which is derived from σ_2 by applying $\mathcal{G}$ to all transitions of σ_2, whereas for each subsequence of transitions of σ_2, which are mapped to the same transition $t_1 \in T_1$ only one occurrence of t_1 is placed in the resulting sequences, but possibly at different positions resulting in several mapped sequences.*
- *Let us use $|\sigma_2|_{\mathrm{map}} = |\sigma_2'|$ $(\sigma_2' \in \sigma_2^{\mathrm{map}}(\mathcal{G}))$ to denote the length of an arbitrary mapped firing sequence σ_2' of σ_2.*

Note, that $|\sigma_1|_{\mathrm{ext}}$ is well defined. The length of all extended sequences $\sigma_1' \in \sigma_1^{\mathrm{ext}}(\mathcal{P})$ is equal since they differ only in the order of transitions. The same holds for $|\sigma_2|_{\mathrm{map}}$. Removing transitions by $\bar{p}$ guarantees $|\sigma_1|_{\mathrm{ext}} \leq |\sigma_1|$ and the mapping of possible multiple transitions to one transition ensures $|\sigma_2|_{\mathrm{map}} \leq |\sigma_2|$.

Definition 8 (Compliance measures). *Let M_1, M_2, $\mathcal{G}$ and $\mathcal{P}$ as stated in the definitions above. Let $\sigma_1 \in T_1^*$ and $\sigma_2 \in T_2^*$.*

- *The* firing sequence compliance (fsc) *of σ_2 w.r.t. σ_1 is:*

$$\mathrm{fsc}(\sigma_2, \sigma_1, \mathcal{P}, \mathcal{G}) = \max\{\mathrm{lcs}(s, s') | s \in \sigma_1^{\mathrm{ext}}(\mathcal{P}), s' \in \sigma_2^{\mathrm{map}}(\mathcal{G})\} \ . \tag{1}$$

- *The* firing sequence compliance degree (fscd) *of σ_2 w.r.t. σ_1 is:*

$$\mathrm{fscd}(\sigma_2, \sigma_1, \mathcal{P}, \mathcal{G}) = \frac{\mathrm{fsc}(\sigma_2, \sigma_1, \mathcal{P}, \mathcal{G})}{|\sigma_2|_{\mathrm{map}}} \ . \tag{2}$$

- *The* firing sequence compliance maturity (fscm) *of σ_2 w.r.t. σ_1 is:*

$$\mathrm{fscm}(\sigma_2, \sigma_1, \mathcal{P}, \mathcal{G}) = \frac{\mathrm{fsc}(\sigma_2, \sigma_1, \mathcal{P}, \mathcal{G})}{|\sigma_1|_{\mathrm{ext}}} \ . \tag{3}$$

- *The* compliance degree (cd) *of M_2 w.r.t. M_1 is given by:*

$$\mathrm{cd}(M_2, M_1, \mathcal{P}, \mathcal{G}) = \frac{\sum_{\sigma_2 \in S'(M_2)} \max_{\sigma_1 \in S'(M_1)}\{\mathrm{fscd}(\sigma_2, \sigma_1, \mathcal{P}, \mathcal{G})\}}{|S'(M_2)|} \ . \tag{4}$$

- *The* compliance maturity (cm) *of M_2 w.r.t. M_1 is given by:*

$$\mathrm{cm}(M_2, M_1, \mathcal{P}, \mathcal{G}) = \frac{\sum_{\sigma_1 \in S'(M_1)} \max_{\sigma_2 \in S'(M_2)}\{\mathrm{fscm}(\sigma_2, \sigma_1, \mathcal{P}, \mathcal{G})\}}{|S'(M_1)|} \ . \tag{5}$$

Function lcs in (1) calculates the length of the longest common subsequence of two firing sequences, thereby finding the maximum number of identical activities while preserving the activity order. The greater the value returned, the more similar the firing sequences are. See [4] for details on lcs. Since the firing sequences σ_1 and σ_2 can have various structures manifesting in their extended and mapped firing sequence sets, (1) will select the variation of σ_1 and σ_2 which will yield a greater similarity of σ_1 and σ_2. The compliance degree (2) of σ_2 indicates the extent to which the transitions of σ_2 are executed according to the specifications of a reference model expressed with σ_1. The compliance maturity (3) of a firing sequence σ_2 points at the extent to which the specification of a reference model expressed with σ_1 is followed by σ_2. In (4), (5), the degree and maturity of compliance express the ratio of instances, which can be produced by one model that can also be produced by the other model. From the viewpoint of compliance degree the process model is related to the reference model; from maturity vice versa. These compliance measures return a value in interval $[0, 1]$. For example, if the compliance degree is 1, the compliance is the highest since all firing sequences of model M_2 can also be produced by model M_1.

5.3 Industrial Application

This section applies the sequence-based compliance analysis to the case study introduced in Sect. 4 and compares the results with two existing approaches available in ProM: "Structural Precision/Recall" and "Footprint Similarity". We have chosen these two approaches since they are sometimes used to determine the compliance between models. We discuss the results of our study in Sect. 5.4.

Measuring Sequence-Based Compliance. Fig. 4 shows the starting point for the compliance analysis in ProM: two WF-nets. The left-hand side model portrays the reference model M_1, which was adopted from ITIL. Initially created as an EPC in the ARIS toolset, it has been converted into a WF-net and imported into ProM. The right-hand side model illustrates the as-is model M_3, which represents the complaint handling process of the passenger airline. It was extracted with the ProM plugin "Heuristic Miner" [16] from a log file containing 4,650 cases and 44,006 events being observed over a period of one year.

To adapt the reference model to the needs of the airline, model M_1 was customized as follows. The activity *Identify responsible employee* was excluded because the activity was not recorded by the IS. The airline assumes that the activities *Inform customer* and *Preprocess incident* may be executed in an arbitrary order. As a result, the airline has agreed on a user-selected partition of type exclusion ($\bar{p}$ = {*Identify responsible employee*}) as well as on a partition of type order ($\check{p}$ = {*Inform customer, Preprocess incident*}). Besides the user-selected partitions, the left-hand side of Fig. 5 shows the granularity mapping. Please note, that the figure denotes the as-is model M_2. During the mapping, we found typical characteristics in the airline process discussed in Sect. 4: missing and additional activities and activities with different levels of detail. For example, the activity *Prioritize incident* is missing in model M_3 and the activities

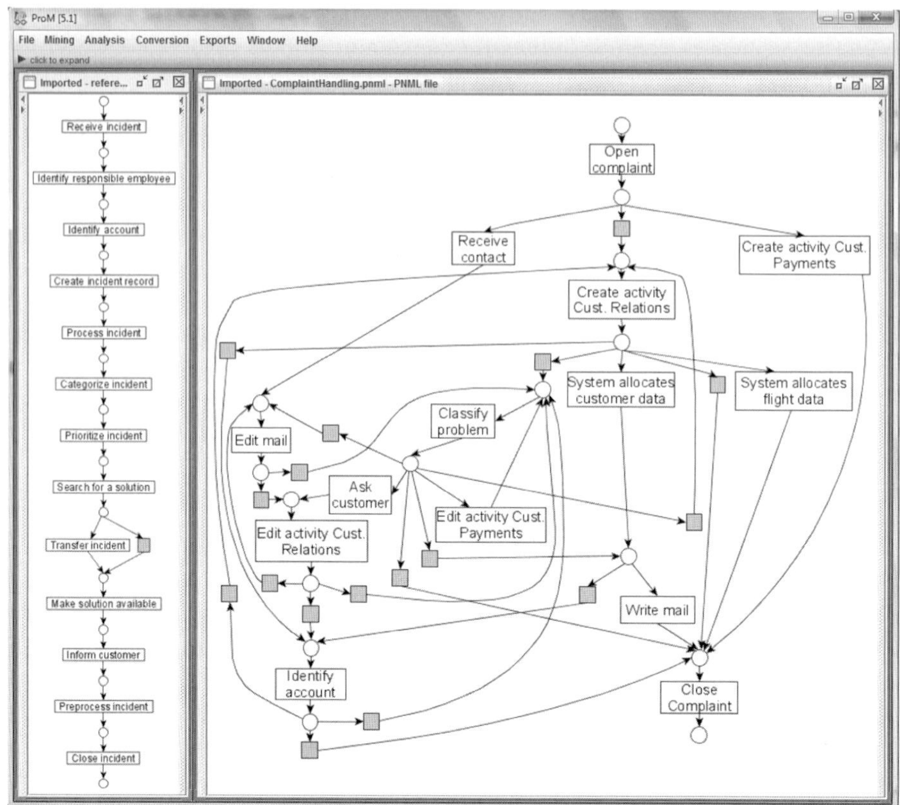

Fig. 4. Process models in ProM used for compliance analysis

Create activity Cust. Relations and *Create activity Cust. Payments* of model M_3 correspond to the activity *Create incident* in model M_1. Figure 4 shows that the airline uses iterations: model M_3 has cycles. Since the cycles are seen as quality improvement, the limit for cycle unrolling is set to 1. This limit ensures that all activities are considered but that the iteration of activities is not punished.

The right-hand side of Fig. 5 illustrates the results of our compliance analysis. Visible are the compliance degree and compliance maturity, which were computed according to Equations (4 and 5) per passed cycle as well as the extended firing sequences $\sigma_1^{\text{ext}}(\mathcal{P})$ of model M_1 and the firing sequences $\sigma_3^{\text{map}}(\mathcal{G})$ of model M_3. Unrolling a cycle once, yields the compliance degree $cd(M_3, M_1, \mathcal{P}, \mathcal{G})$ of 0.82 and the compliance maturity $cm(M_3, M_1, \mathcal{P}, \mathcal{G})$ of 0.52. To explain these values, we study the first line of the sequences σ_1 and σ_3, respectively. We consider the following extended firing sequence $\sigma'_{1-1} = \langle$*Receive incident, Identify account, Create incident record, Process incident, Categorize incident, Prioritize incident, Search for a solution, Make solution available, Inform customer, Preprocess incident, Close incident*$\rangle$ and $\sigma''_{1-1} = \langle$*Receive incident, Identify account, Create incident record, Process incident, Categorize incident, Prioritize*

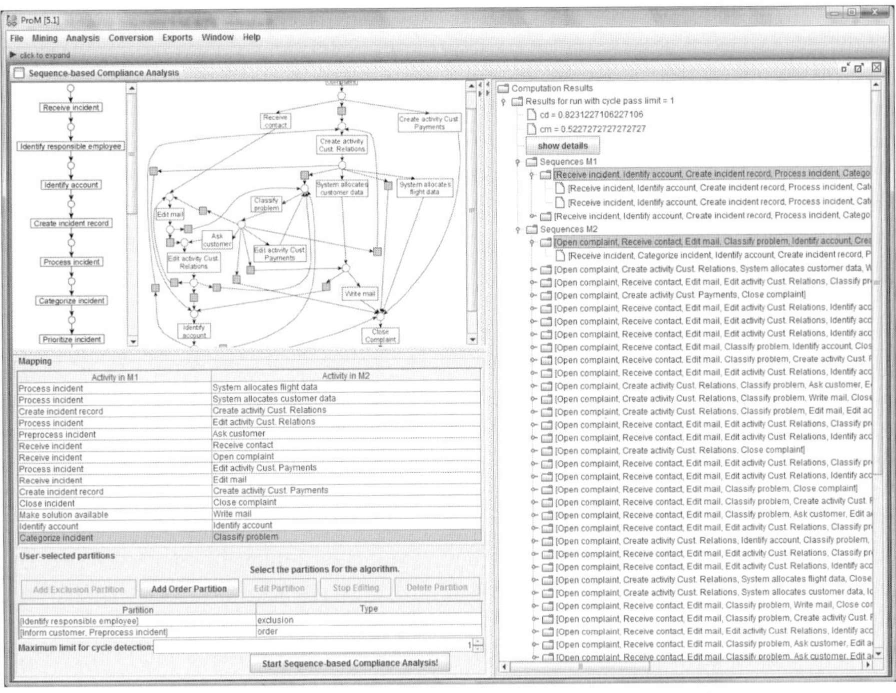

Fig. 5. Sequence-based compliance analysis plugin

incident, Search for a solution, Make solution available, Preprocess incident, Inform customer, Close incident⟩, $\sigma'_{1-1}, \sigma''_{1-1} \in \sigma_1^{\text{ext}}(\mathcal{P})$. Let us also consider the firing sequence σ_{3-1} = ⟨*Open complaint, Receive contact, Edit mail, Classify problem, Identify account, Create activity Cust. Relations, System allocates flight data, Close complaint*⟩, which results in the firing sequence σ'_{3-1} = ⟨*Receive incident, Categorize incident, Identify account, Create incident record, Process incident, Close incident*⟩ $\in \sigma_3^{\text{map}}(\mathcal{G})$. Since the maximum common longest subsequence of σ'_{1-1} and σ''_{1-1} with σ'_{3-1} corresponds to ⟨*Receive incident, Identify account, Create incident record, Process incident, Close incident*⟩, the firing sequence compliance $fsc(\sigma_{3-1}, \sigma_{1-1}, \mathcal{P}, \mathcal{G})$ is 5. The firing sequence compliance degree $fscd(\sigma_{3-1}, \sigma_{1-1}, \mathcal{P}, \mathcal{G})$ is $\frac{5}{6}$. This means that the instance σ_{3-1} of the as-is process model follows the order of the reference model with an overlap of 83%. The firing sequence compliance maturity $fscm(\sigma_{3-1}, \sigma_{1-1}, \mathcal{P}, \mathcal{G})$ is $\frac{5}{11}$. This means that only 45% of instance σ_{1-1} prescribed by the reference model are being followed by instance σ_{3-1} of the as-is process model. The result of the compliance degree of 82% indicates that the processes executed by the airline correspond to the recommendations of the reference model. We can say that, although the models M_3 and M_1 look different, the model M_3 is highly compliant with reference model M_1. The compliance maturity of 52% indicates that there are recommendations in reference model M_1 which are not implemented by the airline. Nonetheless, because of the maturity value of 52% we can conclude that model M_3 is also partially mature with reference model M_1.

Measuring Precision and Recall. In [24], the authors introduce the structural precision and recall. $Precision^S(M_1, M_2)$ is the fraction of connections in M_2 that also appear in M_1. If this value is 1, the precision is the highest because all connections in the second model exist in the first model. $Recall^S(M_1, M_2)$ is the fraction of connections in M_1, which also appear in M_2. If the value is 1, the recall is the highest because all connections in the first model exist in the second model. To analyze the compliance, model M_1 and M_3 of our use case need to be represented by a heuristic net. Therefore, we have converted model M_1, originally represented by an EPC, into a Heuristic net using ProM. Since the ProM plugin expects same labels, we have renamed the labels of model M_3 according to model M_1 and carried out the mapping depicted in Fig. 5. The structural precision obtained was 3% and the recall was 8%.

Measuring Causal Footprint. The causal footprint [26] is the second approach we have compared with our algorithm. The footprint identifies two relationships between activities: look-back and look-ahead links. This paper does not elaborate on the corresponding equation due to its complexity. We refer interested readers to [26]. Since the analysis of the causal footprint is based on comparing two EPCs, we have converted model M_3 into an EPC using a conversion plugin in ProM. The mapping was manually performed in accordance to the mapping shown in Fig. 5. To analyze the causal footprint, the ProM plugin "Footprint Similarity" was used and yielded a result of 27%.

5.4 Evaluation

This section discusses the compliance values, which we yielded in Sect. 5.3 based on the requirements from Sect. 4.

Precision and recall rely on the notion of equivalence and expect process models, which need to be compared, to be equal in their structure. This is the reason why the values obtained are relatively low: 3% and 8%, respectively. Similar to our approach these two measures allow to analyze the compliance from the perspectives compliance degree (i.e. precision) and compliance mature (i.e. recall). By contrast the approach neither offers a mapping functionality nor accounts for the necessary customization of the reference model: ordering or exclusion of activities. Expressing the behavior of a model in terms of connections results in the loss of information whether two connected transitions are part of a cycle and neglects the control flow of process models. However, these are relevant information when measuring the compliance with reference models.

The causal footprint also relies on the notion of equivalence. However, the approach assumes that process models with different structures may be similar. Therefore, the result of 27% is closer to the values obtained when using the algorithm we have developed (i.e. 82% and 52%). Since the formula is symmetric, measuring the compliance of model M_3 with model M_1 or of model M_1 with model M_3 yields the same value. It is clear that this situation is perfectly aligned with the notion of equivalence but fails to meet the requirements of determining compliance from the perspectives degree and maturity. Like our approach the notion of mapping is included. However, a non-injective mapping is not supported. Since the algorithm accounts for the ordering of activities, it partially fulfills the requirements for customization of reference models. Nonetheless, it does not account for the exclusion of activities. The authors [26] do not state the behavior of their formula with respect to cycles.

Using algorithms with the notion of equivalence, we are tempted to infer that the processes are not compliant. In contrast to the sequence-based compliance, the recall and precision and the causal footprint yield a value, which is little expressive and hard to explain. It is not possible to trace the missing or dissent instances. The solution proposed in this paper obtains two different values for compliance (i.e. degree and maturity) and also calculates intermediate results from instance compliance. This enables process designers to trace back which instances are affecting positively or negatively the compliance of the processes under analysis. The industrial application shows that the notion of equivalence cannot be used with satisfactory results to evaluate the compliance of processes with a reference model.

5.5 Feasibility Study

The sequence-based compliance algorithm is based on the generation of sets of firing sequences to describe the behavior of a process model. Unfortunately, in general, the size of these sets can grow exponentially with the size of the WF-net in terms of activities. This section shows the applicability of our algorithm in spite of its exponential complexity. Like Dijkman [8], we used a sample of EPCs of the SAP reference model to test whether our algorithm can be applied in practice by showing that the computation times are acceptable. The SAP reference model has been described in [20,10] and is referred to in many research papers (e.g. [8]). Since it is among the most comprehensive reference models covering over 600 business processes,we assume that these models can be regarded as a representative example. The study is performed by applying the sequence-based compliance algorithm to a subset of 126 pairs of EPCs from the SAP reference model, which we have converted to WF-nets. The pairs are put together based on their similarity computed by the ProM plugin "EPC Similarity Calculator". Our pairs are characterized with a similarity greater than 50%. Figure 6 shows the percentage of model pairs for which the compliance can be computed within a given number of milliseconds on a regular desktop computer. Ninety percent of the process models analyzed with our compliance algorithm took less than 62 milliseconds. In the experiment, the runtime of the algorithm takes on average 50.5 milliseconds with a standard deviation of 9.3 milliseconds. Figure 7 shows the runtime per activities in the processes of a model pair. The average number of activities in these processes is 16. We only found a weak correlation between runtime and the number of activities of a process. Therefore, we conclude that for the number of activities, which we found in the SAP reference models, the

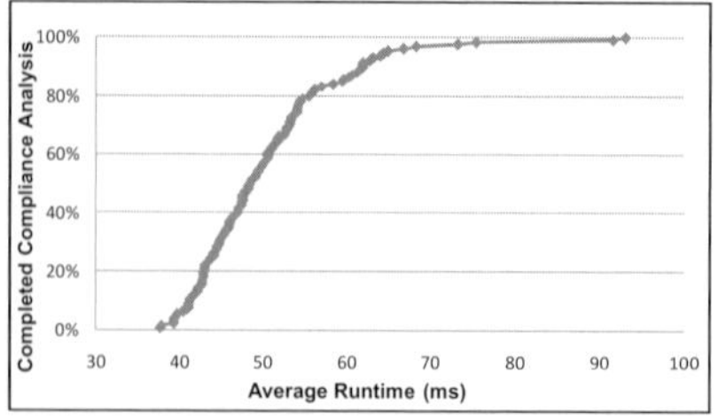

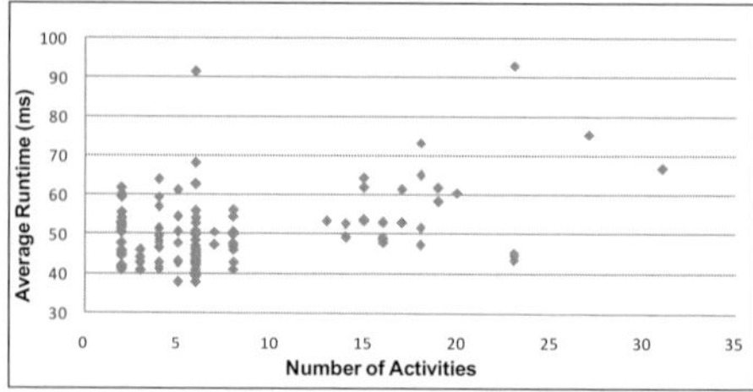

Fig. 6. Average runtime

Fig. 7. Runtime as function of activities

sequence-based compliance analysis is applicable. These results show that, in theory we are confronted with exponential runtime when the complexity is measured in terms of the input size only, i.e. activities. However, in practice there are natural boundaries, e.g. the number of activities per process model is between a lower bound and an upper bound. Hence, the algorithm can be used in practice despite its exponential complexity.

An alternative to address complexity with regard to the input size of the algorithm is to capture the behavior of a model using the state space of a WF-net. A state space corresponds to the set of reachable markings of a WF-net [3]. The resulting graph is denoted as the reachability graph. Buchholz et al. [5] present a method focusing on optimizing the generation of the reachability graph of large Petri nets. The central idea is to decompose a net, to generate reachability graphs for the parts and to combine them. Furthermore, there exist various techniques for state space reduction [8], which may be exploited to improve the efficiency of the underlying algorithm of the sequence-based compliance algorithm. Corresponding approaches are referred to reduction rules. These rules aim at reducing the size of the state space by reducing the number of places and transitions preserving information relevant for analysis purpose. For example, it is possible to account for the significance of transitions. Transitions, which are rarely executed, can be left out using abstraction or encapsulation. Again, we found arguments for the applicability of state spaces in the context of the input size. For example, Verbeek et al. [28] argue that state spaces generating a reachability graph are often feasible for systems up to 100 transitions.

6 Related Work

Our work can be related to various research areas, namely process discovery and verification, process integration, and behavior inheritance.

Measuring compliance assumes the presence of a given model. Therefore, process mining, which aims at the discovery of such a model, is related to the work presented in this paper. Various algorithms have been developed to discover process models based on a log file [17,1].

In the literature, we have identified two ways to verify the compliance between processes and supporting IS: log-based verification and inter-model verification.

Since it is possible to verify if a model and a log file fit together, measuring the compliance can be seen as a very specific form of log-based verification. Thus, our paper is related to the work of Cook et al. [7] who have introduced the concept of process validation. They propose a technique comparing the event stream coming from the process model with the event stream from the execution log based on two different string distance metrics. The notion of compliance has also been discussed in the context of genetic mining [1]. Compliance checking is applied by using fitness, behavioral precision and recall. All these compliance measures propose some kind of replay of the instances in a Petri net. However, the applicability of the log-based verification presumes the existence of log files which are not always available. In the context of the inter-model verification van der Aalst introduces the delta analysis, which compares the real behavior of an IS with the expected behavior (e.g. a reference model) [23]. Different notions of equivalence of process models being subject to verification, such as trace equivalence [27], bisimulation [27], and behavioral equivalence [24], have been developed. The classical equivalence notions are defined as a verification property which yields yes or no, but do not provide a degree of equivalence [26]. Notions searching for behavioral similarity, for instance causal footprint [26] and structural appropriateness [17,24] are applicable in the context of process mining. However, they do not account for the characteristics of compliance with reference models. We introduced them in Sect. 5.3. For a detailed overview we refer to [24].

From a conceptual viewpoint, process integration and process inheritance are similar to our work. Comparing two process models in order to measure compliance in terms of corresponding behavior implies that there are distinctions. Common integration approaches for process models show how these distinctions can be integrated, for example to harmonize processes after an organizational merger [14]. In [8], Dijkman has categorized differences related to control flow, resource assignment, and activity correspondence and has presented a technique to diagnose these differences between process models. Juan [13] applied a string comparison approach of the firing sequences embedded in each process model to identify differences between process models. These works are complementary to our approach and can be considered together during the compliance analysis to locate the exact position of a difference between the models and analyze the type of a difference in the process models. However, since process integration approaches are designed for similar business situations, they typically focus on very similar processes on the same level of abstraction. Basten and van der Aalst [3] have introduced the relations of behavioral inheritance, which can also be used to identify commonalities and differences in process models. The approach is motivated by improving reusability and adaptivity of process models and concentrates on applying the idea of inheritance known from object-oriented modeling. The relations are based on labeled transition systems and branching bisimulation and correspond to the algebraic principles of encapsulation and abstraction [3]. Process inheritance assumes that process models originate from common sources and, therefore, are different yet very similar. Thus, notions of inheritance do not account for different level of granularities.

7 Conclusion and Future Work

Reference models provide valuable recommendations for the implementation of business processes. However, methods and solutions to determine how these guidelines are implemented in practice are non-existing. Known algorithms to evaluate the equivalence of processes have proven to be insufficient to measure compliance since many factors and characteristics related to compliance are ignored. In this paper, we have investigated the characteristics of compliance and we have devised a generic approach to analyze the compliance of process models with reference models. Our main contribution is an algorithm, called sequence-based compliance, which is based on the observation that process models can have different structures but one process can still be compliant with the other.

In order to validate our approach and our algorithm we have measured the compliance of a complaint handling process of a German passenger airline. The passenger airline has obtained transparency of its current customer support processes by carrying out process mining on their log files. Nonetheless, the next step, which needed to be executed, was to determine to which extent the process were aligned with a reference model (i.e. ITIL). This second step has been addressed in this paper.

We have further evaluated our methodology by comparing the results with two existing approaches. The validation was not trivial since we applied process mining and equivalence algorithms on real data. The results have shown that the sequence-based compliance yields more insightful values when compared to the results of existing algorithms based on analyzing the equivalence of processes.

In the future, we are planning to apply our approach and algorithm to other business and industry domains. We also aim to learn which additional types of customization of reference models are important and study how traceability can be incorporated into compliance analysis to enable organizations to quickly identify problematic parts of their running processes.

References

1. Alves de Medeiros, A.K., Weijters, A.J.M.M., van der Aalst, W.M.P.: Genetic Process Mining: A Basic Approach and its Challenges. In: Bussler, C.J., Haller, A. (eds.) BPM 2005. LNCS, vol. 3812, pp. 203–215. Springer, Heidelberg (2006)
2. APQC, American Productivity & Quality Center, http://www.apqc.org/pcf
3. Basten, T., van der Aalst, W.M.P.: Inheritance of Behavior. Journal of Logic and Algebraic Programming 47(2), 47–145 (2001)
4. Bergroth, L., Hakonen, H., Raita, T.: A Survey of Longest Common Subsequence Algorithms. In: 7th IEEE Intl. Symposium on String Processing Information Retrieval, pp. 39–48. IEEE Press, Los Alamitos (2000)
5. Buchholz, P., Kemper, P.: Hierarchical Reachability Graph Generation for Petri Nets. Form. Methods Syst. Des. 21(3), 281–315 (2002)
6. CMMI, Software Engineering Institute, http://www.sei.cmu.edu/cmmi
7. Cook, J.E., He, C., Ma, C.: Measuring Behavioral Correspondence to a Timed Concurrent Model. In: 17th IEEE Intl. Conf. on Software Maintenance, p. 332. IEEE Press, Los Alamitos (2001)

8. Dijkman, R.: Diagnosing Differences between Business Process Models. In: Dumas, M., Reichert, M., Shan, M.-C. (eds.) BPM 2008. LNCS, vol. 5240, pp. 261–277. Springer, Heidelberg (2008)
9. Ehrig, M., Koschmider, A., Oberweis, A.: Measuring Similarity between Semantic Business Process Models. In: 4th Asia-Pacific Conf. on Conceptual Modeling, pp. 71–80 (2007)
10. Teufel, T., Keller, G.: SAP R/3 Process Oriented Implementation: Iterative Process Prototyping. Addison-Wesley, Reading (1998)
11. Gerke, K., Tamm, G.: Continuous Quality Improvement of IT Processes based on Reference Models and Process Mining. In: 15th Americas Conf. on Information Systems (2009)
12. IDS Scheer AG, `http://www.ids-scheer.com`
13. Juan, Y.C.: A String Comparison Approach to Process Logic Differences between Process Models. In: 9th Joint Conference on Information Sciences (2006)
14. Mendling, J., Simon, C.: Business Process Design by View Integration. In: Eder, J., Dustdar, S. (eds.) BPM Workshops 2006. LNCS, vol. 4103, pp. 55–64. Springer, Heidelberg (2006)
15. Official Introduction to the ITIL Service Lifecycle. Stationery Office Books, London (2007)
16. ProM, `http://www.processmining.org`
17. Rozinat, A., Veloso, M., van der Aalst, W.M.P.: Evaluating the Quality of Discovered Process Models. In: Bridewell, W., et al. (eds.) 2nd Intl. Workshop on the Induction of Process Models, Antwerp, Belgium, pp. 45–52 (2008)
18. SAP AG, `http://www.sap.com`
19. Supply-Chain Council, Supply Chain Operations Reference Model, SCOR (2006)
20. Ladd, A., Curran, T., Keller, G.: SAP R/3 Business Blueprint: Understanding the Business Process Reference Model. Prentice Hall PTR Enterprise Resource Planning Series, Upper Saddle River (1997)
21. The IT Governance Institute, COBIT 4.1 (2007)
22. van der Aalst, W.M.P.: Verification of Workflow Nets. In: Azéma, P., Balbo, G. (eds.) ICATPN 1997. LNCS, vol. 1248, pp. 407–426. Springer, Heidelberg (1997)
23. van der Aalst, W.M.P.: Business Alignment: Using Process Mining as a Tool for Delta Analysis and Conformance Testing. Requir. Eng. 10(3), 198–211 (2005)
24. van der Aalst, W.M.P., Alves de Medeiros, A.K., Weijters, A.J.M.M.: Process Equivalence: Comparing Two Process Models Based on Observed Behavior. In: Dustdar, S., Fiadeiro, J.L., Sheth, A.P. (eds.) BPM 2006. LNCS, vol. 4102, pp. 129–144. Springer, Heidelberg (2006)
25. van der Aalst, W.M.P., Dreiling, A., Gottschalk, F., Rosemann, M., Jansen-Vullers, M.: Configurable Process Models as a Basis for Reference Modeling. In: Bussler, C.J., Haller, A. (eds.) BPM 2005. LNCS, vol. 3812, pp. 512–518. Springer, Heidelberg (2006)
26. van Dongen, B.F., Dikman, R., Mendling, J.: Measuring Similarity between Business Process Models. In: Bellahsène, Z., Léonard, M. (eds.) CAiSE 2008. LNCS, vol. 5074, pp. 450–464. Springer, Heidelberg (2008)
27. van Glabbeek, R.J., Peter Weijland, W.: Branching Time and Abstraction in Bisimulation Semantics. Communications of the ACM 43(3), 555–600 (1996)
28. Verbeek, H.M.W.: Verification and Enactment of Workflow Management Systems. PhD thesis, University of Technology, Eindhoven, The Netherlands (2004)

Formalized Conflicts Detection Based on the Analysis of Multiple Emails: An Approach Combining Statistics and Ontologies

Chahnez Zakaria[1], Olivier Curé[1], Gabriella Salzano[1], and Kamel Smaïli[2]

[1] Université Paris-Est, IGM Terre Digitale, Marne-la-Vallée, France
{chahnez.zakaria,olivier.cure,gabriella.salzano}@univ-mlv.fr
[2] Loria, Campus Scientifique, BP 239 54506 Vandoeuvre Lès-Nancy, France
smaili@loria.fr

Abstract. In Computer Supported Cooperative Work (CSCW), it is crucial for project leaders to detect conflicting situations as early as possible. Generally, this task is performed manually by studying a set of documents exchanged between team members. In this paper, we propose a full-fledged automatic solution that identifies documents, subjects and actors involved in relational conflicts. Our approach detects conflicts in emails, probably the most popular type of documents in CSCW, but the methods used can handle other text-based documents. These methods rely on the combination of statistical and ontological operations. The proposed solution is decomposed in several steps: (i) we enrich a simple negative emotion ontology with terms occuring in the corpus of emails, (ii) we categorize each conflicting email according to the concepts of this ontology and (iii) we identify emails, subjects and team members involved in conflicting emails using possibilistic description logic and a set of proposed measures. Each of these steps are evaluated and validated on concrete examples. Moreover, this approach's framework is generic and can be easily adapted to domains other than conflicts, e.g. security issues, and extended with operations making use of our proposed set of measures.

1 Introduction

Multinational enterprises have developed well since the emergence of globalization, i.e. the process by which local, regional or national phenomena become integrated on a global scale. In the early 90s, the total number of multinational entreprises exceeded 37.000 and they had more than 170.000 affiliates abroad [9]. This has led to the creation of virtual or geographically distributed teams that overcome the problems of distance by using Computer Supported Cooperative Work (CSCW) tools. However it is still difficult for a team leader to remotely manage the emotions of its members and the conflicts that may arise between them. Such situations can complicate communication and cooperation between them, and it affects their work efficiency.

R. Meersman, T. Dillon, P. Herrero (Eds.): OTM 2009, Part I, LNCS 5870, pp. 94–111, 2009.

During the experiments of Hawthorne [3], Elton Mayo studied the importance of emotions in the professional environment. This is opposed to the classical School of management, especially Taylor's model which created the symbol of the work dehumanization. Mayo proved that good horizontal and/or vertical relationships (i.e. between colleagues and/or between employ and his employer), in a professional environment, have a major influence on overall satisfaction provided by the work and personal productivity.

The constitution of virtual teams has accentuated the difficulty of understanding an employee's behaviour. Nevertheless, the team leader can overcome this situation with the data generated by the CSCW tools, especially through the analysis of emails which allow to generate important textual corpora due to its large exploitation in professional environments [23]. But the number of emails exchanged between team members on a daily basis can be so important that it may not be possible for a team leader to read them all. In fact we studied an e-collaborative work of educational content mediatization team[1]. Its members communicate using email and they put their leader in copy for all emails exchanged, allowing him to monitor their collaboration. This has generated a minimum of 40 emails daily.

Our solution, named Handling Conflict Email (HaCoEma), consists in the automatic detection of conflicts in emails exchanged between team members. Hence it enables team leaders to intervene and manage conflicts before they lead to irreversible situations. HaCoEma solves the task of conflict detection by classifying emails, according to a domain ontology of relational conflicts.

Using topic identification (TID) techniques may allow to identify conflict emails. The main objective of topic identification is to assign one or several topic labels to a set of textual data. Labels are chosen from a set of topics fixed a priori. Several approaches have been proposed at the end of 90's [26]. All these techniques use a metric which compares the document under processing with a list of topics. Our purpose here is to detect conflict in emails by considering a conflict as a specific topic which one has to identify.

In [2], we addressed the issue of email routing. It has been also considered as an identification problem and we showed the difficulty to process emails. Indeed, they have specific features which make them different from newspaper documents (which are in general the material raw for TID). In opposition to newspapers, it is not easy to find special email corpora. Obviously, everyone has a considerable list of emails in his mailbox, but for our purpose we need company's emails which are unfortunately not available for obvious confidentiality reasons. E-mails are often noisy which makes their interpretation uncertain. Hence, it is difficult to process them automatically in order to retrieve the most relevant information. Which information should be kept? Firstly, should all the headings which constitute the structure of an email be removed? Some of them could be very relevant for detecting the topic as subject, date, sender. Then, which likelihood can be attributed to them. In addition, emails are often ungrammatical, punctuation is usually missed, abbreviations are widely used, foreign words are utilized, images,

[1] http://ufc.dz/

web pages may be present, etc. All these problems make detecting conflicts in emails an interesting challenge to raise.

HaCoEma uses the TFIDF (Term Frequency-Inverse Document Frequency) principle and the SVM (Support Vector Machine) model to classify emails according to the concepts of an ontology of relational conflicts. Our study also addresses the issue of building an ontology of negative emotions, which is made up of two phases. First we conceptualize the domain by hand, then we enrich the ontology by using a trigger-based model which finds terms corresponding to different conflicts in corpora.

In many contexts, analysis of emails supports decision making, for instance to produce cooperative software, such as Bugzilla[2], an open bug tracking system, where the emails report bugs affecting Internet browsers, e.g. Mozilla, Firefox, Thunderbird. Therefore, we analyze the emails in the conceptual framework of Documents for Action (DofA) [27]. With this approach, we focus on the collective dimension of the writing process, to analyze emails in an asynchronous communication process between several agents sharing common interests.

The remainder of the paper is organized as follows. Section 2 presents concepts and technologies used in HaCoEma. Section 3 describes our conceptualization approach of the conflicts domain in two stages. Section 4 describes the models which we have developed and used for classifying emails based on the concepts of our ontology. It also validates the classification methods exploited with experimental results. Section 5 provides a solution to the analysis of multiple emails based on a set of proposed measures and possibilistic description logics. Section 6 discusses related work. Finally, Section 7 concludes with a discussion of future directions.

2 Background

2.1 Possibilistic Description Logic

In information technology, an ontology provides a shareable, reusable piece of knowledge about a specific domain and can be specified more or less formally in order to create an agreed-upon vocabulary. We have selected Description Logics (DL) [1] as a mean to represent ontologies in the context of this work. The choice of DL is motivated by the important number of available DL-based ontologies hence enabling cooperation between them and the increasing amount of associated tools, e.g. reasoners, editors, APIs.

DL corresponds to a family of knowledge representation formalisms allowing to present and reason over domain knowledge in a formally and well-understood way. Central DL notions are concepts (unary predicates), relationships, also called roles (binary predicates), and individuals. A standard DL knowledge base is usually defined as $\mathcal{K} = \langle \mathcal{T}, \mathcal{A} \rangle$ where $\mathcal{T}$ (or TBox) and $\mathcal{A}$ (or ABox) consist respectively of a set of concept descriptions (resp. concept and role assertions).

[2] http://www.bugzilla.org/

The following concept description:

$$NegativeEmotion \sqsubseteq Thing \sqcap \forall hasForm.String \sqcap \forall hasSynonym.String \sqcap \forall hasCause.String$$

states that a concept named *NegativeEmotion* is described as the set of objects which have forms, causes and synonyms that are strings of characters. Concept assertions, for instance, are *Person(paul)* and *Email(email1)*; an example of a role assertion is *sentBy(email1,paul)*. Based on $\mathcal{K}$, some standard DL reasoning tasks are concept satisfiability, knowledge base consistency, concept subsumption and instance checking which are detailed in [1].

Possibilistic logic, or possibility theory, [8] provides an efficient solution for handling uncertain or prioritized formulas and coping with inconsistency. In this logic, each formula is associated to a real value in [0,1]. The notion of possibility distribution π, defined as $\pi : \Omega \rightarrow [0,1]$, where Ω represents the set of all classical interpretations, is fundamental in defining the logic's semantics. From this possibility distribution, two important measures can be computed: (i) the possibility degree of a formula ϕ, defined as $\Pi(\phi)= \max\{\pi(\omega) : \omega \models \phi\}$, where $\omega(\phi)$ is the degree of compatibility of interpretation ω with available beliefs. (ii) the certainty degree of a formula ϕ, defined as $N(\phi)$=1 - $\Pi(\neg\phi)$.

In possibilistic DL, a possibilistic formula is a pair (ϕ, α) where ϕ is a standard DL axiom, i.e. TBox or ABox axiom, and α expresses a degree of certainty. A set of possibilistic formulas, also called a possibilistic knowledge base ($\mathcal{PK}$), consists of a possibilistic TBox ($\mathcal{PT}$) and ABox ($\mathcal{PA}$). The classical knowledge base ($\mathcal{K}$) associated with ($\mathcal{PK}$) corresponds to $\{\phi_i | (\phi_i, \alpha_i) \in \mathcal{PK}\}$. A $\mathcal{PK}$ is consistent iff its $\mathcal{K}$ is consistent.

Given a $\mathcal{PK}$ and $\alpha \in [0,1]$, the α-cut of $\mathcal{PK}$, denoted $\mathcal{PK}_{\geq\alpha}$, is defined as $\mathcal{PK}_{\geq\alpha}= \{ \phi \in \mathcal{K} | (\phi, \beta) \in \mathcal{PK}$ and $\beta \geq \alpha\}$. The inconsistency degree of $\mathcal{PK}$, denoted $Inc(\mathcal{PK})$, is defined as $Inc(\mathcal{PK}) = \max\{\alpha_i : \mathcal{PK}_{\geq\alpha}$ is inconsistent$\}$.

Example: Consider a possibilistic DL knowledge base $\mathcal{PK} = \langle \mathcal{PT}, \mathcal{PA} \rangle$ where $\mathcal{PT}$= {(Email $\sqsubseteq$ ConflictEmail $\sqcup$ NonConflictEmail, 1), (ConflictEmail $\sqsubseteq \neg$ NonConflictEmail,1)} and $\mathcal{PA}$ = {(ConflictEmail(email1), 0.7), NonConflictEmail(email1),0.3)}. The TBox $\mathcal{PT}$ states that it is certain that an email is either an email with or without conflicts and that conflict emails are disjoint from non conflict emails. The ABox $\mathcal{PA}$ states that the email identified by *email1* is more likely to be a conflict email (certainty of 0.7). Let $\alpha = 0.3$, we then have $\mathcal{PK}_{\geq 0.3} = \langle \mathcal{PT}_{\geq 0.3}, \mathcal{PA}_{\geq 0.3} \rangle$ where: $\mathcal{PT}_{\geq 0.3}$ ={Email $\sqsubseteq$ ConflictEmail $\sqcup$ NonConflictEmail, ConflictEmail $\sqsubseteq \neg$ NonConflictEmail} and $\mathcal{PA}_{\geq 0.3}$ ={ConflictEmail(email1), NonConflictEmail(email1)}. It is clear that $\mathcal{PK}_{\geq 0.3}$ is inconsistent. Now let $\alpha = 0.7$. Then $\mathcal{PK}_{\geq 0.7} = \langle \mathcal{PT}_{\geq 0.7}, \mathcal{PA}_{\geq 0.7} \rangle$ where $\mathcal{PT}_{\geq 0.7}$ ={Email $\sqsubseteq$ ConflictEmail $\sqcup$ NonConflictEmail, ConflictEmail $\sqsubseteq \neg$ NonConflictEmail} and $\mathcal{PA}_{\geq 0.7}$ ={ConflictEmail(email1)}. So $\mathcal{PK}_{\geq 0.7}$ is consistent. Moreover, $Inc(\mathcal{PK})$=0.3.

2.2 Statistical Models

Several approaches are proposed for building ontologies from corpora. They can be grouped into two categories: (i) structural approaches based on the use of formal grammar and (ii) non-structural approaches, such as statistical approaches which must use important enough corpora, in order to have reliable measures and find out interesting relationships between terms [13].

The acquisition of terms based on statistical approach exists since several decades: Enguehard and Pantera (1995) [10], Dias (2002) [7], etc. It consists on the idea that words of the same area tend to often occur together. Similarity measures are used to identify recurrent associations of terms. The correlated terms recurrences are extracted by using different kind of measures [21] like Mutual Information. It is a measure of distance stemming from the information theory, which allows to measure the degree of association between two events. The mutual information $MI(x, y)$ represents the importance of the relationship between two events x and y. The non-weighted MI is given below:

$$MI(x, y) = \log \frac{P(x, y)}{P(x)P(y)}$$

where $P(x)$ (resp. y) is the marginal probability of x (resp. y) and $P(x, y)$ is the joint probability of x and y.

In general, a classification model consists of two tasks: modeling the document using a model of representation, as the vector model [20], and his assignment to the topic that concerns through a classifier like SVM [4], or a distance measure like Salton's cosine [19].

The essential idea of SVM is to use kernel functions (such as the polynomial, the Gaussian, etc.) to transform not linearly separable data into linearly separable ones using a representation of higher dimension spaces. So the goal is to find a function F, to learn from observation of input-output and to predict other events. The function attempts to minimize the errors of learning while maximizing the margin separating the categories of data [4].

The representation model describes emails or documents with the terms of the vocabulary or the terms involved in the classification's topics. There are several weight functions that represent the importance of each term in the email, e.g. the occurrence frequency of the term in the email or the TFIDF measure. TFIDF is used to evaluate how important a term is to an email in a corpus. The importance increases proportionally to the number of times a term appears in the email but is offset by the frequency of the term in the corpus. It is calculated as follows [18]:

$$TFIDF(w_i, M) = TF(w_i, M) \times IDF(w_i) \quad where \quad IDF(w_i) = \log \frac{T}{t_i}$$

where $TF(w_i, M)$ is the frequency of the term w_i in the email M. T is the size of the corpus and t_i is the number of emails in which the term w_i occurs.

2.3 Evaluation Measure of Retrieval Systems

The combination of recall, precision and F-measure [13] is a popular evaluation for information retrieval systems. Recall is defined as the fraction of relevant emails that are retrieved by the system, precision is defined as the fraction of retrieved emails that are in fact relevant and F-measure characterizes the combined performance of recall and precision. These measures are calculated as follows:

$$Recall = \frac{Number\ of\ relevant\ emails\ retrieved}{Number\ of\ emails\ to\ retrieve}$$

$$Precision = \frac{Number\ of\ relevant\ emails\ retrieved}{Number\ of\ emails\ retrieved}$$

$$F\text{–}measure = 2 \times \frac{Precision \times Recall}{Precision + Recall}$$

There are two other measures that estimate the performance of a system from its errors, namely the False Acceptance (FA), where an email is wrongly considered as conflictual, and the False Rejection (FR), where an email is wrongly rejected. These measures are calculated as follows [13]:

$$FA = \frac{Number\ of\ False\ Acceptances}{Number\ of\ emails\ retrieved} \qquad FR = \frac{Number\ of\ False\ Rejections}{Number\ of\ emails\ to\ retrieve}$$

3 Ontology Construction

3.1 Manual Creation of the Skeleton of the Ontology

To the best of our knowledge, no ontology describes the domain of conflicts. Since such an ontology is required in HaCoEma, we had to design one. We achieved this task by focusing on the litterature on emotions and considering that conflicts are generally associated to negative emotions. To conceptualize the conflict domain, we based our work on the taxonomy of Michelle Larivey [14], but we changed the separation criteria of emotions. We can see in Figure 1 that a first criterion separates emotions according to the degree of conflict, the first category represents emotions that can produce substantial conflicts as *disgust* and *hatred*, the second one leads to anticipate some indirect conflicts as *indifference*. A second criterion separates personal from social emotions, in fact it distinguishes social emotions from other emotions. This is due to the fact that it is very difficult to determine a personal emotion. For instance, the *sadness* emotion may be social when this feeling is due to the behaviour of another colleague or friend, and may be personal when the person did not succeed to reach an objective; however, *jealousy* can easily be classified as social emotions.

Conceptualizing the field of conflict and the classification of emails were made in the French language. Figure 1 presents a translation into the English language of an excerpt of our ontology. In the next section we present the statistical model that we used to enrich our ontology from corpora.

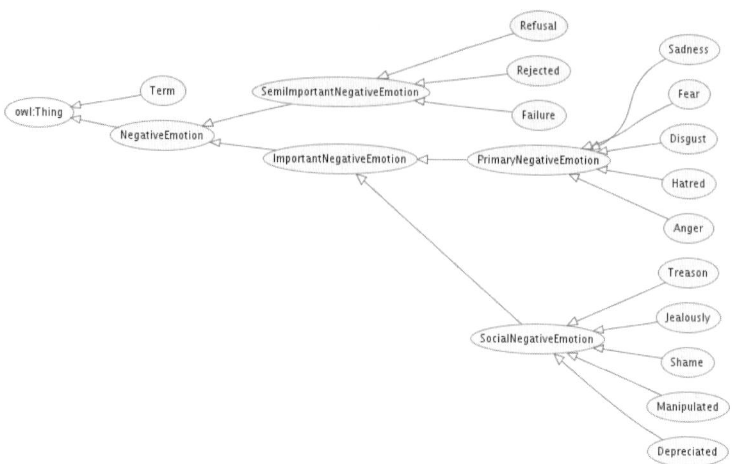

Fig. 1. Conflict ontology

3.2 Automatic Enrichment of the Ontology

Development of statistical language models is historically related to the construction of the first significant linguistic corpora [6]. For these models, a corpus represents a raw material, it is used to learn a maximum of linguistic events (n-grams, part of speech, etc.) [13]. In other words statistical processing of corpora allows to get knowledge by studying recurrent phenomenon. A corpus should be large in order to model statistically a maximum of reliable constructions. The more a corpus is important, the better the events are modeled [13]. For machine translation or speech recognition, it is not surprising to train the language model on a corpus with 300 million words. Classical n-grams models are often enriched by language models based on triggers which are used in several domains, e.g. in translation, and are exploited to build multilingual dictionaries [15].

We use a trigger approach [13] to enrich the ontology. Our aim is to find terms that are semantically related to the terms of the ontology, then to integrate them into the ontology, for a better description of its concepts. The triggers focus on terms that often appear together. That means we can predict the term w_j when w_i occurs (it can be written as: $w_i \rightarrow w_j$). For instance the term *"insult"* will probably predict the term *"humiliation"*. The triggers are determined by calculating for each ontology term its Mutual Information with each term in the dictionary. Then, only terms with a high Mutual Information are kept and used as triggered terms. We use this principle of trigger to enrich the ontology at the level of emotion concepts, because emotion is represented by just a few words which are synonyms, e.g. *"sadness"* emotion. Triggers also allow to collect several non synonym terms. Therefore we create properties to link them to the ontology and hence improve the classification of emails. Each triggered term will be manually associated to concepts of the ontology via *synonym* (regrouping some synonyms of the term representing the concept), *form* (regrouping terms that indicate the expression of emotion) or *cause* (regrouping the reasons which may justify the expression of emotion) properties. For instance, the trigger model enables to enrich *Sadness* concept with the terms *grief, sorrow, etc.* as synonyms, the terms *suffer, endure, etc.* as forms and the terms *annoy, offend, etc.* as causes.

4 Classification of Emails

HaCoEma solves the task of detecting conflicts in emails by classifying them. It consists in identifying the concepts to which an email belongs to and therefore to recognize the emotion expressed in this email. The domain of classification is made up of two distinct approaches: supervised and unsupervised learning. The distinction between these two approaches comes from the knowledge or not of categories. Indeed, supervised classification learns to assign instances to predefined categories, while unsupervised classification is a task, which learns classification from the data, because categories are unknown. For the purposes of this paper we will focus on supervised learning. We classify emails according to concepts of the ontology, i.e. the categories of the classification are emotions of ontology.

4.1 TFIDF Classifier

Each email (E_i) to classify is encoded by a vector according to the terms of a concept (C_i). Then a similarity is calculated to quantify the semantic proximity between the email (its representation by the concept vector) and an emotion. This process is repeated for each emotion. Once all similarities are calculated, the classification process associates to each email the emotion with the highest similarity value. We introduce the following notations: $C_i = \{c_{i1} ..., c_{ij}, ..., c_{in}\}$, where c_{ij} is the weight of the term w_j in the *ith* concept, and n is the number of terms in the concept which varies from one concept to another. $E_i = \{e_{i1}, ..., e_{ij}, ..., e_{in}\}$, where e_{ij} is the weight of the term w_j in the *ith* concept. Weights are estimated using the TFIDF. The classification is done by calculating for each pair (C_i, E_i) the cosine of the angle between vectors C_i and E_i defined as follows [19]:

$$Cos(C_i, E_i) = \frac{\sum_{j=1}^{n} c_{ij} e_{ij}}{\sqrt{\sum_{j=1}^{n} {c_{ij}}^2 \sum_{j=1}^{n} {e_{ij}}^2}}$$

4.2 SVM Classifier

The SVM are a class of algorithms inspired by the theory of statistical learning of Vapnik [4], it is a recent alternative for the classification and has been used in many applications such as face recognition, and bioinformatic. The SVM were originally designed as a binary classifier, however, they were generalized (SVMs) [22] [5] for a multi-class learning. For classifier with SVM, we used the Thorsten Joachims tool[3]. We represented our corpus in the input format of this tool, as follows:

[3] http://svmlight.joachims.org/svm_multiclass.html

```
<line>.=.<target> <feature>:<value>...<feature>:<value>
<target>.=.<integer>
<feature>.=.<integer>
<value>.=.<float>
```

The target value indicates the category of the email one of the classification categories, for example, the tuple: (3 1:0.43 3:0.12 9284:0.2), specifies an email of category 3 for which feature (term) number 1 has the value 0.43, feature (term) number 3 has the value 0.12 and feature (term) number 9284 has the value 0.2.

4.3 Evaluation

As explained before, it is difficult to get a corpus of emotions; in addition only those, which are subject to create conflicts between people, interest us. In order to evaluate our approach, we create an emotion corpus by extracting it from forum discussions. In fact, people in forums can exchange very hard words. To achieve that, we use our ontology as an index, which permits to collect all exchanges containing words predisposed to provoke a quarrelling between people. We then get 2.138 messages split into eight different emotions. In average each category contains 267 messages with a standard deviance of 16. Table 1 present the results of classification by using TFIDF and SVM. The results are presented in terms of recall, precision, F-measure, False acceptance and false reject. The first conclusion is that TFIDF outperforms SVM on our corpus. TFIDF achieves a F-measure of 0.93 whereas SVM gets 0.86. This could be very surprising in comparison to other works. This is due to the size of the corpus and the nature of the messages which are different from what we find classically in other works, e.g. processing of natural language. In fact, texts in our corpus are polluted making the frequency of words very low. It seems that TFIDF is less sensible to low frequency than SVM. Note that TFIDF allows getting a F-measure of 1 for *Hatred* concept and SVM achieves better results than TFIDF for *Anger* category.

Table 1. Performance of the TFIDF and SVM classifiers

Concept	**Recall**		**Precision**		**F-Measure**		**FA**		**FR**	
	TFIDF	SVM	TFIDF	SVM	TFIDF	SVM	TFIDF	SVM	TFIDF	SVM
Anger	0.92	1.0	0.96	0.93	0.94	0.96	0.04	0.07	0.08	0.0
Hatred	1.0	0.84	1.0	0.75	1.0	0.79	0.0	0.25	0.0	0.16
Treason	0.92	0.96	1.0	0.96	0.96	0.96	0.0	0.04	0.07	0.04
Jealousy	0.85	0.88	1.0	0.92	0.92	0.9	0.0	0.08	0.14	0.12
Fear	0.97	0.92	0.82	0.72	0.89	0.81	0.17	0.28	0.03	0.08
Depreciated	0.92	0.88	1.0	0.92	0.96	0.9	0.0	0.08	0.08	0.12
Sadness	0.92	0.52	0.92	0.87	0.92	0.65	0.08	0.13	0.08	0.48
Shame	0.96	0.88	0.89	0.88	0.92	0.88	0.11	0.12	0.04	0.12
Average on all the concepts	0.93	0.86	0.94	0.87	0.93	0.86	0.01	0.02	0.01	0.03

5 Analysis in the Context of Multiple Emails

In this section, we present a model for conflict management based on the analysis and classification of multiple emails, a set of measures and a possibilistic DL approach dealing with uncertainty.

5.1 Model Description

The *Email* class is the core of the UML model of Figure 2. An email is characterized by a date of writing, a subject and a body, which is composed by one or more fragments, having a suitable granularity. An email can be an answer to a previous email, or a forwarded email. In these cases, only the newly provided fragments of the body will be analyzed. Similarly, we do not analyze the attached file(s).

Many relations link the *Email* and *Agent* classes. In our application area, agents represent employees preparing the educational supports and the team leader. An email has one and only one sender and one or more direct receivers. An email can have one or more copy-receivers, who may be visible or not.

Depending on the presence of conflict terms, we classify the emails in two classes: *ConflictEmail* and *NonConflictEmail*. A conflict email is an email such that its body contains fragments related to some negative emotions, specified in the conflict ontology (Section 3). As the conflict emails are used to warn a conflict, *ConflictEmail* and *Conflict* classes are related. In practice, we also associate to a conflict email the emails having the same subject or subjects derived from it, such as Re(⟨subject⟩), Fwd(⟨subject⟩), Fwd(Re⟨subject⟩), and so on.

5.2 Required Measures to Warn a Conflict

The model in Figure 2 enables to warn about a potential conflict. In fact, we can compute some indicators, like as in the Table 2.

The indicators 1.3 and 1.4 give a global idea of the "width" and the "depth" of a conflict email, while the indicators 4.3 to 4.6 inform about the "width", "length", "density" and "urgency" of a conflict. These indicators enable the team manager to give a value to the *status* attribute of a conflict (current or completed), and to the *exit* value (satisfactory or unsatisfactory). We evaluate the "email agent activity" with indicators 5.6 to 5.8. A high conflict activity for an agent can imply his high implication in preventing conflicts (see for example the team manager, or an agent often involved because of his expertise). Moreover, an agent role in a conflict can be deduced from the relation (sentBy, sentTo, copyTo, blindCopyTo in the UML model of Figure 2) associating emails and agents. The indicators of Table 2 can be used to filter emails (respectively subjects, agents and conflicts), verifying some conditions based on the analysis and classification of the emails. Table 3 lists some examples of such conditions, where the values of α, β, γ and δ are chosen by the team leader.

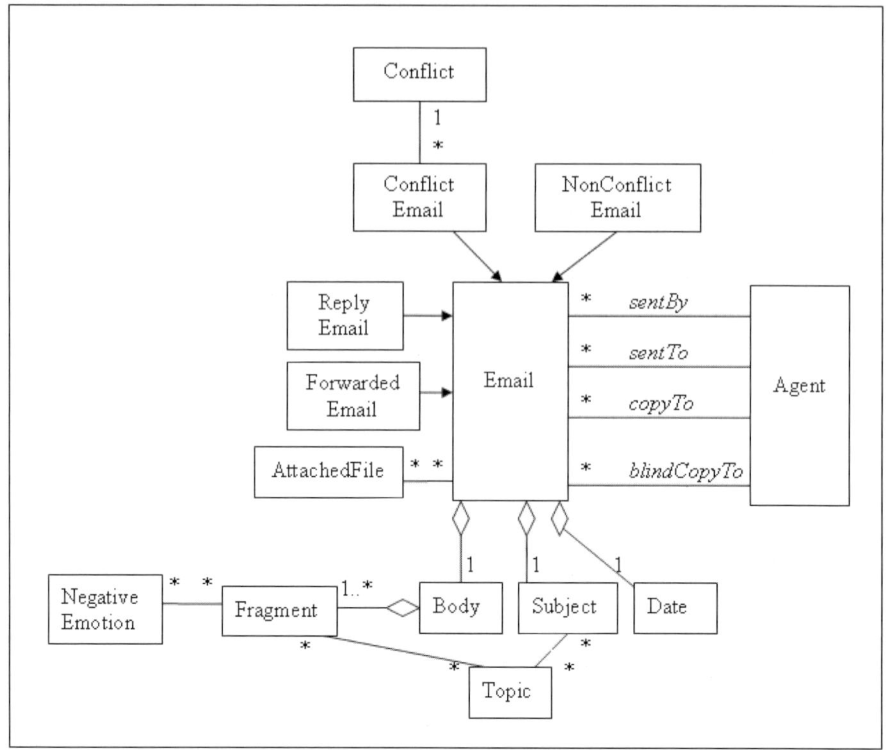

Fig. 2. UML model for conflict management

5.3 Ontology-Based Architecture

Since we are exploiting an ontology to represent conflicts, it seems relevant to develop an ontology-based approach to detect and warn conflicts between agents. This approach has several assets: (i) interoperability since every representation of knowledge exploits a DL formalism, (ii) representation of knowledge related to the fragment of social conflicts can benefit from the exploitation of ontologies developed in social networks, (iii) availability of efficient and user-friendly tools to design, maintain and process ontologies using standards of the Semantic Web.

Given Figure 2, we have everything related to conflicts and their classification represented in our ontology. The agent aspect of this model is already represented in the FOAF[4] RDF vocabulary standard. Moreover, this ontology can be used to represent relationships between (groups of) agents as well as information concerning projects they are working on. Hence, HaCoEma creates an ontology which contains some concepts relevant in our domain, e.g. *Email* and *DofA*, and relating them to the imported FOAF and Conflict ontology, e.g. $Email \sqsubseteq DofA$

[4] http://www.foaf-project.org/

Table 2. Indicator based on the analysis and the classification of the emails (TN stands for Total Number)

1. Email	1.1 TN of conflict terms 1.2 TN of fragments 1.3 TN of agents that receive email (directly or not) 1.4 TN Depth of conflict within an email = (TN of conflict terms) / (TN of fragments)
2. Date	2.1 TN of conflict emails at this date
3. Subject	3.1 TN of conflict emails with this subject (or a subject derived by applying Re and / or FWD) 3.2 TN of emails with this subject or a similar subject 3.3 TN of conflict emails with this subject / TN of emails with this subject
4. Conflict	4.1 TN of conflict emails related to it 4.2 TN of all emails related to it 4.3 TN of agents involved in this conflict =(TN of distinct agents who are senders of conflict email related to this conflict) + (TN of distinct agents who are (direct or not) receivers of a conflict email related to this conflict) 4.4 TN of days for this conflict = (Date of the last conflict email in this conflict) - (Date of the first conflict email in this conflict) 4.5 Density number (for a conflict) = (TN of conflict emails related to it) / (TN of all emails related to it)
5. Agent	5.1 TN of sent emails 5.2 TN of sent conflict emails 5.3 TN of received emails 5.4 TN of received conflict emails with a "send to" 5.5. TN of received conflict emails with a "blind copy to" 5.6 TN of emails in which sehe(he is involved = (TN of sent emails) + (TN of received emails) 5.7 TN of conflict emails in which she/he is involved = (TN of conflict emails that she/he sent) + (TN of conflict emails that she/he received) 5.8 Conflict activity =(TN of conflict emails in which she/he is involved) / (TN of emails in which she/he is involved)

and $DofA \sqsubseteq foaf : Document$. Finally datatype properties have been created to store values associated to the measures presented in Table 2.

Associated to this ontology, i.e. TBox, we can now create an ABox which stores all information according to a given state of emails exchanged between agents. It is also important to stress that we enable a team leader to enrich the ABox manually, i.e. he can assert that an agent or subject is conflicting and attach to this belief a certainty degree. Then, a set of emails, together with the agents sending and receiving them, can be envisioned as a graph.

Table 3. Examples of reports based on the analysis of the conflicts

Report	Conditions
Email	α_1 * total number of agents that receive it (directly of not) (1.3) + β_1 * Depth of conflict within an email (1.4) $\geq \delta_1$
Subject	α_2 * total number of conflict emails with the same subject (3.1) $\geq \delta_2$
Agent	Conflict activity (5.8) $\geq \delta_3$
Conflict	α_4 * total number of agents involved in this conflict (4.3) + β_4 * density number (4.5) + γ_4 * urgency number (4.6) $\geq \delta_4$

5.4 Use Cases Exploiting the Measures

We now present three use cases exploiting the measures proposed in Table 2. They enable to identify (i) agents which are involved in conflicts, (ii) subjects associated to conflicts and (iii) emails containing conflicts of certain types. These approaches exploit a subset of our proposed measures, features of possibilistic DL and its graph-based representation. We present the approaches exploited in the three use cases through an example whose's scenario is the following.

Scenario. A project leader supervises several persons which are exchanging emails on a daily basis. An extract of the information concerning these emails is represented in the graph of Figure 3: 3 persons (*paul* , *mary* and *peter*) and 3 emails are displayed. Each email has a subject property relating to the subject of the email and 3 datatype properties (one for each social, primary and semi importance type of emotions) storing a value in [0,1] and representing the certainty of appearance of such a conflict.

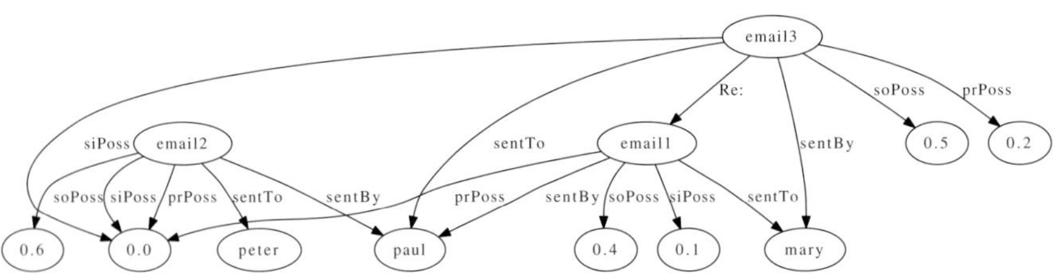

Fig. 3. Extract of an email graph

All use cases can be applied to predefined time periods. That is the emails' date have to satisfy some conditions, e.g. searching for conflicting emails between march 1st, 2009 and may 1st, 2009. This approach enables team leaders to search for agents, subjects and emails with conflicts over a given period of time.

Use Case 1: Agent Identification. This use case requires to compute the Conflict Activity (CA), i.e. 5.8 in Table 2. This is easily performed using a query language adapted to graph navigation, e.g. SPARQL's graph pattern matching. Then for each agent, we assert a possibility formula stating that an agent is an instance of the concept *ConflictingAgent* with a certainty degree corresponding to his CA value. For instance, suppose that in our scenario, the agent denoted *paul* is involved in 5 emails and 3 of them have conflicts. Then the CA of *paul* is 0.6 and we can assert that: ConflictAgent(*paul*,0.6). Our ABox ($\mathcal{PA}$) also contains {ConflictAgent(*peter*, 0.2),(ConflictAgent(*mary*, 0.3), Agent(*paul*,1), Agent(*peter*,1), Agent(*mary*,1)} and an extract of the TBox ($\mathcal{PT}$) consists of {(ConflictAgent $\sqsubseteq$ Agent, 1)}. Then the inconsistency degree of $\mathcal{PK}$ is $Inc(\mathcal{PK}) = 0$ and all 3 agents are conflicting to a certain degree. Now suppose that the team leader has defined that *mary* is certainly not an instance of *ConflictAgent* $(\neg ConflictAgent(mary), 0.8)$. Then $Inc(\mathcal{PK}) = 0.3$ and only *paul* can be considered to be conflicting.

Additionally to this possibility theory approach, we also exploit some user defined preferences corresponding to the values α, β, γ and δ of Table 5. In the context of agent identification, δ_3 is a relevant parameter that enables to restrict the set of agents identified as conflicting. For instance in our running example, if $\delta_3 \geq 0.7$ then no agents are displayed to the team leader.

Use Case 2: Subject Identification. In this use case, for each subject with at least 2 emails, the system: (i) computes the measure 3.3 in Table 2, denoted ψ, by navigating into the graph. (ii) creates a possibilistic instance with type *ConflictSubject* for this subject. (iii) sets ψ as the certainty degree of this concept. Identifying conflicting subjects is then similar to the approach presented for agent identification.

Use Case 3: Email Identification. In this use case, the team leader selects the type of conflict he is interested in, e.g. social, primary or semi-important. Then the possibility value associated to this type, respectively *soPoss,prPoss* and *siPoss* in the email graph, are used to defined *ConflictEmail* possibilitic assertions for each email. The operations used to identify emails is then similar to the ones presented in the agent and subject approaches.

Summary. We have presented 3 use cases based on measures related to the emergence of conflicts. We believe that they correspond to the fundamental metrics expected from team leaders willing to minimize and prevent conflict situation in their teams. The architecture adopted for these 3 use cases is quite general and can easily be reused and extended to define finer grained solutions based on the measures of Table 2.

6 Related Work

In this section we provide a succinct overview over related works. In the first part, we analyze HaCoEma with respect to a system visualizing discussion content and participant behavior, while in the second part we interpret emails in the conceptual framework of Document for Action. In a third part we present different types of work on the building of emotion ontologies for classification.

6.1 Communication Garden

Systems that improve Computed Mediated Communication (CMC) should work on the archives of emails on different aspects (discussion content, participant behavior, social network between participants) and at different layers (information representation, classification or visualization). Generally, only some aspects and layers are treated by the existing systems. With respect to this framework, our research focalizes on the first two layers (conflict email representation and classification), and it enables the analysis of participant behaviors in a social network. The treatment of email content is deferred to our future works (Section 7).

To evaluate a computer mediated organization, the "Communication Garden System" [28] visualizes discussion content and participant behavior. It makes use of the very suggestive "flower metaphor", to represent graphically the liveliness of the threads of communication and the persons' activities. A thread is represented as a flower, where the number of petals equals the number of messages posted for this thread and the number of leaves represents the number of persons participating to the discussion. The starting time and the topic area are displayed by the localization of the thread. Similarly, the "person flower" represents some statistics (number of messages, number of discussions, duration). The visualization system allows suggesting easily "the current hot topic" in the community or the "most active" participants, by using statistics similar to the indicators shown in Table 2.

6.2 Documents for Action

In accordance with [27], emails can be considered as documents used to mediate the coordination of a "community of actions", i.e. a widely distributed group committed to working towards a common goal. Manuel Zacklad distinguishes two kinds of goals, services goals and integration goals : the first tend to be reached as the result of epistemic transactions, while the second rely on relational transactions. Operational and strategic activities are associated to the services goals, while relational and integrative activities are associated to the integration goals. In our application area, we associate the service goals and the operational activities (making educational supports) to the contributor agents. Through the email activity (relational activity), these agents and team leader participate of the integration goals. The team leader plays a special role, through his integrative activities, because he regulates the organization by analyzing conflict emails and solving conflict situations. In this context, emails participate to collective activities, transmit and disseminate information quickly, help in decision making and demonstrate situations. Then an email has all the properties of a Document for Action (DofA) : (a) sustainability, due to the participants commitment ; (b) fragmentation, due to the body structure, Replay and Forwarded emails ; (c) non trivial relations between the email, producers and receivers ; (d) extended state of incompleteness (chain of emails). In concordance with the conceptual framework of DofA, the email model (Figure 2) determine some dependencies between emails and agents, as listed on Table 4. An automatic visualization of these dependencies can be a useful support for the team leader.

Table 4. Examples of dependencies between emails and agents

Type of dependencies	If	Then
Email/Email	e_1 replies (or forwards) to e_2	e_1 depends on e_2
Email/Agent	e_1 is sent by a_1	e_1 depends on a_1
Agent/Email	e_1 is sent to (or copied to) a_2	a_2 depends on e_1
Agent/Agent	e_1 is sent by a_1 and e_1 is sent to (or copied to) a_2	a_2 depends on a_2

6.3 Emotions for Classification

Until recently researchers have ignored the emotional message behind the communication. However, the understanding and expression of emotion is not only important for humans, but is also critical for human-computer interaction. It is studied in different areas, such as psychology, neurology and sociology. Although several description models of emotions exist, categorical [16] [25] and dimensional [11] [12] models are the most commonly encountered. Most of the research on emotion analysis has been done on the applications of machine learning to emotion classification. In [24], affective lexicon ontology is constructed to classify emotional texts with SVM, it includes 10.200 entries and it is used to analyse the text from three different levels : words, sentences and discourses. In [25], Chinese emotion ontology is used in classifying the emotion of the actors in sentences. It is built from HowNet, and it contains just under 5.500 verb concepts covering 113 different emotion categories. [11] proposed an emotional ontology, where each emotional concept is defined in terms of a range of values along three emotional dimensions, corresponding to evaluation, activation and power. Classification and ontology are used to provide particular rules from emotional text. These rules provide configuration parameters for a system for emotional voice synthesis. The classification is used in the other types of emotion expression, [17] present an approach to affective sensing, in spoken language and facial expressions, using a generic model of affective communication and a set of ontologies to assist in the analysis of concepts and the recognition process.

7 Conclusion and Future Works

In this paper, we address an original topic: how to detect conflicts between people in an e-collaborative work. The idea is to stop verbal rise in emails by the manager of an e-learning platform. We proposed to treat this problem as a classification issue. Exchanged emails are analysed to detect the birth of a conflict. We manually developed from scratch a negative emotion ontology, which has been enhanced by using the notion of triggers collected from corpora. Then with this new ontology we collected corpus related to the appropriate emotions. The corpus has been extracted from discussion forums. We got 2.138 polluted messages, which have been split into eight different emotion categories. Classification has been handled by two methods TFIDF and SVM. For this particular corpus, we obtained good results and TFIDF outperforms SVM. Several tracks are under investigation in order to increase more easily the size of the corpus. We are also working on other classification methods. The objective is to have a list of methods in order to take advantage of each of them. In fact, our experience in topic identification showed that each classification method could succeed in the identification of some concepts and fail for others, that is why in general several classifications methods are combined to get better results.

Moreover, the possibilistic DL approach, together with the set of measures we have proposed, enable to derive new solutions based on the use cases proposed in Section 5.4.. We are currently working on an efficient interface to configure declaratively such solutions. We want to go further in the use of the model for conflict management (Figure 2) and perform content analysis of the emails, based on linguistic relations between *Fragment, Subject* and *Topic* classes. This enables us to identify the largest set of emails related to a potential conflict.

References

1. Baader, F., Calvanese, D., McGuinness, D.L., Nardi, D., Patel-Schneider, P.F. (eds.): The Description Logic Handbook: Theory, Implementation, and Applications. Cambridge University Press, Cambridge (2003)
2. Bigi, B., Brun, A., Paul Haton, J., Smaili, K., Zitouni, I.: A comparative study of topic identification on newspaper and e-mail. In: Proceedings of the String Processing and Information Retrieval Conference, SPIRE 2001 (2001)
3. Broches, R.S.: Unraveling the Hawthorne Effect: An Experimental Artifact 'Too Good to Die'. PhD thesis, University of Wesleyan (April 2008)
4. Cortes, C., Vapnik, V.: Support vector networks. Machine Learning, 273–297 (1995)
5. Crammer, K., Singer, Y., Cristianini, N., Shawe-taylor, J., Williamson, B.: On the algorithmic implementation of multiclass kernel-based vector machines. Journal of Machine Learning Research 2, 2001 (2001)
6. Denoual, E.: Méthodes en caractères pour le traitement automatique des langues. PhD thesis, University of Joseph Fourier (2006)
7. Dias, G.: Extraction automatique d'associations lexicales à partir de corpora. PhD thesis, University of Orleans (December 2002)
8. Dubois, D., Lang, J., Prade, H.: Possibilistic logic. In: Gabbay, D., Hogger, C., Robinson, J., Nute, D. (eds.) Handbook of Logic in Artificial Intelligence and Logic Programming, vol. 3, pp. 439–513. Oxford University Press, Oxford (1994)
9. Eden, L.: Multinationales en Amrique du Nord. illustrated (1994)
10. Enguehard, C., Pantera, L.: Automatic natural acquisition of a terminology. Journal of Quantitative Linguistics 2(1), 27–32 (1995)
11. Francisco, V., Gervás, P., Peinado, F.: Ontological reasoning to configure emotional voice synthesis. In: Marchiori, M., Pan, J.Z., Marie, C.d.S. (eds.) RR 2007. LNCS, vol. 4524, pp. 88–102. Springer, Heidelberg (2007)
12. Garcia-Rojas, A., Vexo, F., Thalmann, D., Raouzaiou, A., Karpouzis, K., Kollias, S.: Emotional Body Expression Parameters In Virtual Human Ontology. In: Proceedings of 1st Int. Workshop on Shapes and Semantics, pp. 63–70 (2006)
13. Haton, J., Cerisara, C., Fohr, D., Laprie, Y., Smaili, K.: Reconnaissance automatique de la parole. Du signal à son interprétation. Dunod (2006)
14. Larivery, M.: Les genres d'émotions. La lettre du psy 2(7) (July 1998)

15. Lavecchia, C., Smaili, K., Langlois, D., Haton, J.-P.: Using inter-lingual triggers for machine translation. In: Eighth conference INTERSPEECH (2007)
16. Mathieu, Y.Y.: Annotation of emotions and feelings in texts. In: Tao, J., Tan, T., Picard, R.W. (eds.) ACII 2005. LNCS, vol. 3784, pp. 350–357. Springer, Heidelberg (2005)
17. McIntyre, G., Göcke, R.: Towards Affective Sensing. In: Jacko, J.A. (ed.) HCI 2007. LNCS, vol. 4552, pp. 411–420. Springer, Heidelberg (2007)
18. Robertson, S.E., Jones, S.K.: Relevance weighting of search terms. Journal of the American Society for Information Science 27(3), 129–146 (1976)
19. Salton, G., Mcgill, M.J.: Introduction to Modern Information Retrieval. McGraw-Hill, Inc., New York (1986)
20. Salton, G., Wong, A., Yang, C.S.: A vector space model for automatic indexing. Commun. ACM 18(11), 613–620 (1975)
21. Voltz, R., Oberle, D., Staab, S., Motik, B.: Kaon server - a semantic web management system. In: Alternate Track Proceedings of the Twelfth International World Wide Web Conference, Budapest, Hungary, May 2003, pp. 139–148. ACM, New York (2003)
22. Weston, J., Watkins, C.: Multi-class support vector machines (1998)
23. Whittaker, S., Sidner, C.: Email overload: exploring personal information management of email. In: CHI 1996: Proceedings of the SIGCHI conference on Human factors in computing systems, pp. 276–283. ACM Press, New York (1996)
24. Xu, L., Lin, H.: Ontology-driven affective chinese text analysis and evaluation method. In: Paiva, A.C.R., Prada, R., Picard, R.W. (eds.) ACII 2007. LNCS, vol. 4738, pp. 723–724. Springer, Heidelberg (2007)
25. Yan, J., Bracewell, D.B., Ren, F., Kuroiwa, S.: The creation of a chinese emotion ontology based on hownet. Engineering Letters 16(1), 166–171 (2008)
26. Yang, Y., Liu, X.: A re-examination of text categorization methods (1999)
27. Zacklad, M.: Communities of action: a cognitive and social approach to the design of cscw systems. In: GROUP 2003: Proceedings of the 2003 international ACM SIGGROUP conference on Supporting group work, pp. 190–197. ACM Press, New York (2003)
28. Zhu, B., Chen, H.: Communication-garden system: Visualizing a computer-mediated communication process. Decis. Support Syst. 45(4), 778–794 (2008)

Semantic Annotations and Querying of Web Data Sources

Thomas Hornung[1] and Wolfgang May[2]

[1] Institut für Informatik, Universität Freiburg
hornungt@informatik.uni-freiburg.de
[2] Institut für Informatik, Universität Göttingen
may@informatik.uni-goettingen.de

Abstract. A large part of the Web, actually holding a significant portion of the useful information throughout the Web, consists of views on hidden databases, provided by numerous heterogeneous interfaces that are partly human-oriented via Web forms ("Deep Web"), and partly based on Web Services (only machine accessible). In this paper we present an approach for annotating these sources in a way that makes them citizens of the Semantic Web. We illustrate how queries can be stated in terms of the ontology, and how the annotations are used to selected and access appropriate sources and to answer the queries.

1 Introduction

The Web is today a major source for retrieving *data*. The term *Web Data Sources* covers Web sources that allow access to an underlying database-style repository, and that exhibit at least implicitly a table-like schema. Fortunately, this is the case for virtually all *structured, Web-accessible* data sources, ranging from *human-readable Deep Web* [10] sources – databases that are accessible via filling out Web forms – to data-centric *Web Services* that are machine-accessible via low-level technologies such as REST [9] or SOAP (Web Services). These sources could be very useful for declarative query answering and data workflows, but the task of combining information from different such sources is cumbersome, and is currently an actively researched area [4].

In this paper we propose an approach to describe Web Data Sources in a unified way by assigning semantic annotations to them. The annotations cover the technical handling of (wrapped) sources, the description of their input and output characteristics in terms of *tags*, and the relationship between these tags in terms of the underlying domain ontology. We describe how these annotations are exploited to evaluate SPARQL [28] queries that are stated in terms of the domain ontology.

Structure of the Paper. Section 2 gives a short overview of RDF as the data model underlying the Semantic Web and of SPARQL as the most common RDF query language. In Section 3 we introduce the application scenario, which deals with querying for flight connections. Section 4 develops the concepts used for

R. Meersman, T. Dillon, P. Herrero (Eds.): OTM 2009, Part I, LNCS 5870, pp. 112–129, 2009.

semantic annotation of Web Data Sources. These concepts are formalized in an RDF ontology in Section 5. Section 6 shows how the annotations are used for evaluating queries w.r.t. the annotated sources. Section 7 discusses related work and we conclude with Section 8.

2 Preliminaries: RDF and SPARQL

The *Resource Description Framework (RDF)* [25] is the data model underlying the Semantic Web. In RDF, information is represented by a labeled directed graph where nodes represent *resources*, identified by URIs, and literals, and the labeled edges represent the relationships between them. The edge labels are also URIs, which makes up the salient feature of the Semantic Web: properties and classes can also have properties, which allows for expressing metadata information. In this paper, we represent RDF data in its Turtle [29] notation. Shortly,

res_1 a cl; prop_1 res_2; prop_2 res_3,res_4 . res_4 prop_3 [prop_4 "foo"].

sketches the syntax varieties of Turtle: the resource denoted by res_1 is of class cl ("a" is a shorthand for rdf:type), has properties prop_1, resulting in res_2, and it is related via prop_2 to resources res_3 and res_4, where res_4 in turn is related by property prop_3 to an unnamed resource that has a property prop_4 with value "foo". Instead of writing URIs completely, the usage of prefixes analogous to namespaces in XML is common. We will omit the (repeated) declarations of the common prefixes for rdf:, rdfs: (RDFS [26]), owl: (OWL [20]), and xsd: (XML Schema [32]).

SPARQL [28] is a query language for RDF data. Its design is based on SQL-like clauses using triple patterns with logical variables (join variables) of the form

select *variables* from *sources/files*
where *dot-separated-extended-triple-patterns-and-filters*

where *dot-separated-extended-triple-patterns-and-filters* are patterns similar to the above Turtle notation. Here, positions can optionally be replaced by variables (of the forms "?X" or "$X"). Additionally, filter conditions (e.g. of the form ?X $\leq$?Y) can be stated. Variables occurring twice or more act as *join variables.* A subset of the variables can be declared in the select clause to be the *answer variables*, whose matches with resources and literals are the answers. The result of a query is thus a set of tuples of variable bindings.

3 Application Scenario

We illustrate the framework with a well-known situation: querying for flight connections. Consider four sample data sources that provide information about flights:

(A) Flight portals like http://www.travenjoy.com/ can be queried with a departure airport, a destination airport (by their IATA codes, e.g., FRA and CDG), and the intended dates (roundtrip) and return (possible composite) connections between these airports by flight code/flight codes, departure time, arrival time, and price. For connected flights, they usually succeed only if the connection is by a single airline (or alliance).

(B) Websites of a single airline like http://www.lufthansa.de can be queried by entering a departure airport, a destination airport, and the intended traveling dates (roundtrip), returning a list of connections with departure time, arrival time, and price.

(C) Sources like http://www.theairdb.com/ can be queried with a departure airport and return a list of destination airports and the respective airlines that offer flights to these destinations. In combination with sources of type (B), this can be useful to search for possible connections when the destination airport is not yet fixed (e.g., to choose amongst several airports in a certain area), or when no composite connection by a single airline or alliance exists.

(D) A third type of sources (e.g., the schedule of Frankfurt airport at http://www.airportcity-frankfurt.de/) provides three queries: ($D1$) given a destination airport, a date, and optionally desired departure time (abbreviated dDepT below), get airlines and flight numbers that provide direct connections, ($D2$) symmetrically, select an origin airport and ask for connections to Frankfurt. ($D3$) given a timepoint and a date (e.g., 20.5.2009, 9:00) yields all flights (code and destination) that depart during some hours after that time to anywhere. Note that it does not handle availability and prices at all; for this, another data source of type (B) must be used.

The formalization of the notions of the *domain ontology* is later used to correlate the Web Data Sources with the notions of the application domain. The ontology of the required fragment of the traveling ontology consists of the following classes:

- travel:Airport with property travel:code.
- travel:Airline with properties travel:name and travel:website.
- travel:FlightConnection which is the disjoint union of (i) direct flights (travel:Flight) and (ii) connected flights. Connections have the properties travel:from and travel:to, relating them to instances of travel:Airport, travel:deptTime and travel:arrTime indicating their departure time and arrival time, respectively.
- Flights have additionally the properties travel:operatedBy, relating each flight to an airline, and travel:flightCode. Flight here means the abstract notion of e.g. flight number "LH123" operating on weekdays from Frankfurt to Paris at 12:30h.
- Connections that represent connected flights have additionally the (multi-valued) property travel:consistsOfFlight, relating them to flights.
- travel:BookableConn: Bookable connections are instances of connections (travel:instanceOfConn), i.e., flights or connected flights, on a given day. They are assigned an actual price for which a ticket can be booked *now*, i.e., the extension of this predicate is dynamic, and is obtained at runtime from a Web Data Source.

Some sample data triples are given in Figure 1.

```
@prefix travel: <http://www.semwebtech.org/domains/2006/travel#> .
@prefix airport: <http://www.semwebtech.org/domains/2006/travel/airports/> .
@prefix airline: <http://www.semwebtech.org/domains/2006/travel/airline/> .
@prefix flight: <http://www.semwebtech.org/domains/2006/travel/flights/> .
airport:fra a travel:Airport; travel:code "FRA" .
airport:cdg a travel:Airport; travel:code "CDG" .
airline:lh a travel:Airline; travel:name "Lufthansa";
    travel:website <http://www.lufthansa.com> .
flight:lh123 a travel:Flight; travel:operatedBy airline:lh; travel:flightCode "LH123";
    travel:from airport:fra; travel:to airport:cdg ;
    travel:deptTime "12:30"; travel:arrTime "13:50".
flight:lh123-20-05-2009 a travel:BookableConn;
    travel:instanceOfConn flight:lh123; travel:date "20-05-2009"; travel:price 300.00.
```

Fig. 1. Sample Data of the Application Scenario

4 Annotation of Web Data Sources

The annotation of Web Data sources concerns three levels. The *technical level* describes how to address the source (or its wrapper, respectively). The *signature level* corresponds to the *signature* of query services, specifying their input and output parameters. In contrast, the *semantical level* relates the query services with the underlying domain terminology. We propose WDSDL (Web Data Source Description Language) to describe Web Data Sources. WDSDL consists of a lower, signature level that covers the plain signature, and of a second, semantical level that relates it with the notions of the actual domain ontology.

4.1 Characteristics of Web Data Sources

Conceptually, every Web Data Source can be seen as an n-ary predicate $q(\overline{x}) = q(x_1, \ldots, x_n)$ (its *characteristic predicate*, which contains all input/output mappings) over variables $\{x_1, \ldots, x_n\}$.

The different interaction patterns (i.e., forms or Web Service invocations) with a Web Data Source can be regarded as predefined views over its characteristic predicate, which can also not be queried in general, but only via a restricted access pattern, i.e., certain input arguments must be given to return the corresponding output values. The modeling associates each view v with a unique identifying URI (which is not the URL of the corresponding Web form, but "simply" some RDF URI). For each view, some attributes $\overline{qin} = \{xin_1, \ldots, xin_k\} \subseteq \{x_1, \ldots, x_n\}$ act as inputs, others $\overline{qout} = \{xout_1, \ldots, xout_m\} \subseteq \overline{x} \setminus \overline{qin}$ act as outputs. In the remainder of the paper we call this the *signature* of the view, and denote it by $\overline{qout} \leftarrow v(\overline{qin})$.

4.2 Technical Level

As mentioned in the introduction, Web Data Sources can be distinguished into two types: *Deep Web Sources* where the primary interface is given by a Web form, and *Web Services* that are machine-accessible via protocols like REST or SOAP, but do not provide a declarative interface in any query language. For both types, generic wrapper interfacing languages have been designed, called *DWQL (Deep Web Query Language)* that allows to pose queries against Deep Web Sources, and *WSQL (Web Service Query Language)* that supports the generic querying of (REST- and SOAP-based) Web Services. To the outside, DWQL and WSQL provide a uniform set-oriented interface of the generic form $\overline{qout} \leftarrow v(\overline{\iota n})$: given a set r of tuples of variable bindings over the input variables $\overline{q\iota n}$ for that view, the wrapper returns the set of tuples $\pi[\overline{qout} \cup \overline{q\iota n}](q \bowtie r)$ (which is the same as $\pi[\overline{qout} \cup \overline{q\iota n}](q \ltimes r)$ since $\mathsf{Var}(r) = \overline{q\iota n} \subseteq \overline{x} = \mathsf{Var}(q)$) where q is the characteristic predicate of the source as introduced above[1].

The actual wrappers that map the Web Data Sources to DWQL or WSQL have to be programmed manually; as has been done for the above Sources (A)-(D). The annotations start above this level and annotate the wrapped source.

4.3 Signature Level

The first step of assigning a *signature* consists of *naming* the variables of the characteristic predicate by so-called *tags*, similar to the way they are used in social bookmarking sites.

Example 1. *Consider the sample data sources from above that provide information about flight schedules. The sources (A) and (B) provide nearly the same schema (and similar views over it), while source (B) is implicitly restricted to flights of Lufthansa. Thus, both sources can uniformly be tagged as follows (the chosen tag names are not necessarily the same as the ontology notions – they will be correlated explicitly later):*

(A) connection(from, to, date, airline, fcode, deptTime, arrTime, price).
view: (airline, fcode, deptTime, arrTime, price) ← travenjoyConns(from, to, date).
(B) connection(from, to, date, fcode, deptTime, arrTime, price).
view: (fcode, deptTime, arrTime, price) ← LHConnections(from, to, date).

Source (C) provides only a restricted characteristic predicate and view (which is actually a projection of the above, where all concrete information about the flights is missing):

(C) flight(airline, from, to).
view: (airline, to) ← flies(from).

For source (D), the reference airport, Frankfurt (FRA), is fixed. The formal characterization is less simple than above, where the characteristic predicate could be seen as the contents of an actual database: due to the possibility to enter a desired departure/arrival time in the views which is compared *as an additional condition to the data, the characteristic predicate is an infinite view over the possibly materialized database. We start with the view definitions:*

[1] π denotes relational projection, $\bowtie$ and $\ltimes$ denote natural join and left semijoin, respectively.

(D1) (airline, fcode, depT, arrT) $\leftarrow$ fromFRAtoX(to, date, dDepT).
(D2) (airline, fcode, depT, arrT) $\leftarrow$ fromXtoFRA(from, date, dArrT).
(D3) (to, airline, fcode, depT, arrT) $\leftarrow$ fromFRA(date, dDepT).

The underlying characteristic predicate in this case is the universal relation, i.e., the join of all views, after extending (D1) and (D3) with from: "FRA", *(D2) with* to: "FRA", *and has thus the format*

(D) flight$_D$(from, to, date, airline, fcode, dDepT, depT, dArrT, arrT).

As motivated above, it is defined as an intensional predicate over the database that represents all valid answers:

flight$_D(f, t, d, a, c, dDT, dT, dAT, aT)$ *holds whenever there is an actual flight* (f, t, d, a, c, dT, aT) *such that* $dT > dDT$ *and* $aT > dAT$.

The above description is human-readable since intuitive names are used, but it is not machine-usable since the tag names are not (yet) formally associated to the domain notions. Even more, it does not specify the actual correlation between the values of the result tuple. Formally, it can be considered as a *schema*, similar to an SQL schema, that assigns names to columns of the relation defined by the characteristic predicate, e.g., for source (D): flight$_D$(from, to, date, airline, fcode, dDepT, depT, dArrT, arrT). The semantics what existence of a tuple $(a, b, c, d, e, f, g, h, i) \in$ flight$_D$ means in terms of the domain ontology is not yet specified.

4.4 Semantical Level

Assuming a relational modeling of the domain, assertions about the characteristic predicate can be expressed in terms of the ontology. As usual for ontologies formulated in RDF, the class names are unary predicates, and the property names are binary predicates. For instance, flight$_D$ can be axiomatized as

$$(*)\quad \mathsf{flight}_D(a,b,c,d,e,f,g,h,i) \Leftrightarrow \exists v,w,x,y,z:$$
$$\mathsf{Flight}(v) \wedge \mathsf{flightCode}(v,e) \wedge \mathsf{Airline}(w) \wedge \mathsf{operatedBy}(v,w) \wedge \mathsf{name}(w,d) \wedge$$
$$\mathsf{Airport}(x) \wedge \mathsf{from}(v,x) \wedge \mathsf{code}(x,a) \wedge$$
$$\mathsf{Airport}(y) \wedge \mathsf{to}(v,y) \wedge \mathsf{code}(y,b) \wedge$$
$$\mathsf{deptTime}(v,g) \wedge f < g \wedge \mathsf{arrTime}(v,i) \wedge h < i \wedge$$
$$\mathsf{BookableConn}(z) \wedge \mathsf{instanceOfConn}(z,v) \wedge \mathsf{date}(z,c).$$

which is a *local-as-view* [14] mapping. The mappings of the individual views of a source are obtained as projections.

For applications in the *Semantic Web*, underlying features of the Semantic Web and its data model, RDF, can be exploited:

- Source annotations in RDF format wrt. the agreed target ontology of the application domain can be provided in a homogeneous, web-wide format,
- Description Logic reasoners are available, which implement –depending on the chosen DL– expressive, but decidable fragments of first-order logic.

The following section introduces the WDSDL ontology for expressing Web Data Source Descriptions themselves in RDF.

5 WDSDL: The Ontology of Web Data Source Descriptions

5.1 Signature Level in RDF

The central notion of Web Data Sources is, as introduced above, the characteristic predicate, where the Web Data Source provides one or more views upon. The WDSDL ontology first provides notions for describing the characteristic predicate by enumerating the tags used by the source. In RDF terminology, the tags are just resources that have a name, and that can be annotated by datatypes, and optionally by dimensions (i.e., "time", "price", etc.), formats (e.g, "HH:mm" for times), and units (meters, miles, \$, € etc.) (cf. [12]). The WDSDL vocabulary continues with describing the views provided by the source, and for each view which tags are used in it as input or output. On this level, there is not yet a relationship between the tags and the domain notions.

The WDSDL ontology on the signature level is given in Figure 2, a sample WDSDL Source Description for source (D) is given in Figure 3.

```
@prefix : <http://www.semwebtech.org/languages/2008/wdsdl#>.
:WebDataSource a owl:Class.
:baseURL rdfs:domain :WebDataSource; rdfs:range xsd:anyURI.
:DeepWebSource rdfs:subClassOf :WebDataSource.
:WebServiceSource rdfs:subClassOf :WebDataSource.
:View a owl:Class.
:providesView rdfs:domain :WebDataSource; rdfs:range :View.
:Tag a owl:Class.
:hasTag rdfs:domain :WebDataSource; rdfs:range :Tag.
  # property: tags (called variables in the logic-oriented setting) have a name
:hasInputVariable rdfs:domain :View; rdfs:range :Tag.
:hasOutputVariable rdfs:domain :View; rdfs:range :Tag.
```

Fig. 2. The WDSDL Ontology Part I: Technical and Signature Level

Example 2 (Web Data Source Description). *Consider the WDSDL Source Description for source (D) given in Figure 3. The* RDF blank nodes *of the form "_:xxx" act only internally as tag identifiers. To the outside, only the tag names are known as illustrated by the SPARQL query*

```
prefix : <http://www.semwebtech.org/languages/2008/wdsdl#>
select ?N
from   RDF data given in Figure 3
where { <bla://views/travel/fra/fromFRAtoX> :hasOutputVariable [ :name ?N ] }
```

that can be used to query the names of the tags that are output variables of the view. It yields the variable bindings N/"to", N/"airline", N/"flightCode", N/"deptTime", N/"arrTime".

```
@prefix : <http://www.semwebtech.org/languages/2008/wdsdl#> .
@prefix travel: <http://www.semwebtech.org/domains/2006/travel#> .

<bla://views/travel/fra> a :WebDataSource;
  :baseURL <http://www.airportcity-frankfurt.de/>;
  :providesView <bla://views/travel/fra/fromFRAtoX>,
                <bla://views/travel/fra/toFRAfromX>,
                <bla://views/travel/fra/fromFRA>;
  :hasTag _from,_:to, _:date, _:airline, _:fcode, _:depT, _:arrT, _:dDepT, _:dArrT.

_:from a :Tag; :name "from"; :datatype xsd:string.
_:to a :Tag; :name "to"; :datatype xsd:string.
_:date a :Tag; :name "date"; :datatype xsd:date; :format "dd.MM.yyyy".
_:airline a :Tag; :name "airline"; :datatype xsd:string.
_:fcode a :Tag; :name "flightCode"; :datatype xsd:string.
_:depT a :Tag; :name "deptTime"; :datatype xsd:time; :format "HH:mm".
_:arrT a :Tag; :name "arrTime"; :datatype xsd:time; :format "HH:mm".
_:dDepT a :Tag; :name "desiredDeptTime"; :datatype xsd:time; :format "HH:mm".
_:dArrT a :Tag; :name "desiredArrTime"; :datatype xsd:time; :format "HH:mm".
<bla://views/travel/fra/fromFRAtoX> a :View;
  :hasInputVariable _:to, _:date, _:dDepT;
  :hasOutputVariable _:airline, _:fcode, _:depT, _:arrT.
<bla://views/travel/fra/toFRAfromX> a :View;
  :hasInputVariable _:from, _:date, _:dArrT;
  :hasOutputVariable _:airline, _:fcode, _:depT, _:arrT.
<bla://views/travel/fra/fromFRA> a :View;
  :hasInputVariable _:date, _:dDepT;
  :hasOutputVariable _:to, _:airline, _:fcode, _:depT, _:arrT.
```

Fig. 3. WDSDL Source Description for Source (D): Frankfurt Airport

5.2 Semantical Level: Correlating to the Domain Ontology

The description on the semantical level refers to the description of the sources on the signature level, i.e., the tags (corresponding to the variables of the *characteristic predicate*), and relates its components to the actual domain notions.

The correlation has to cover what has been expressed by the predicate logic formula (*) above. The basic idea is to express the relationships between objects of the application domain in terms of a relevant prototype fragment of the RDF graph, and to associate the tags with the corresponding nodes.

Example 3. *For Source (D), the relationship between the values of from, to, date, airline, fcode, dDepT, depT, dArrT, and arrT is induced by instances of the class travel:Flight as follows: fcode, depT, arrT are the respective attributes of the flight itself, while from and to are actually the airport codes of the origin and destination airport, airline is the name of the airline, there must be an instance of the flight on the given date, and depT/arrT must be later than dDepT/dArrT. Fig. 4 depicts this situation for sample data. Note that Source (D) does not cover all aspects of the ontology, e.g., the price of the connection is not available in it.*

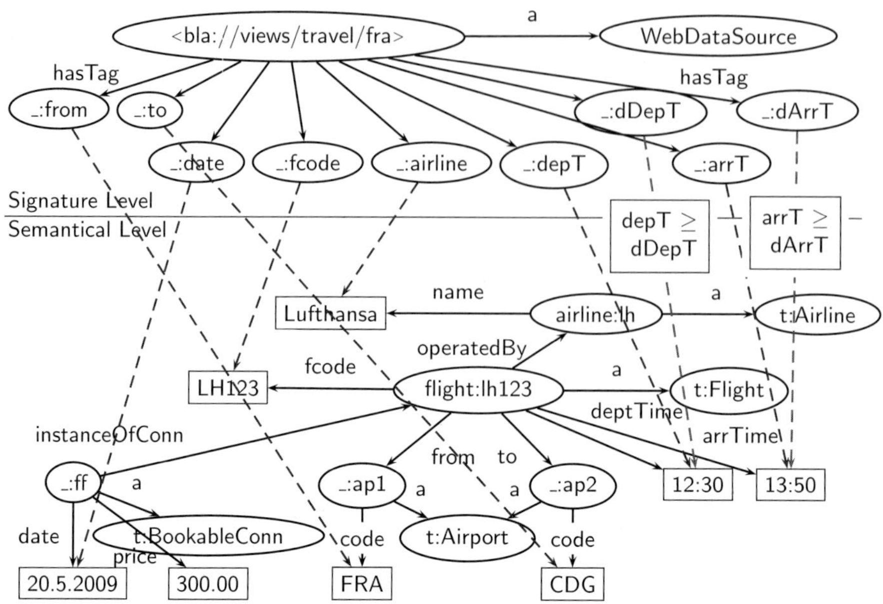

Fig. 4. Semantic Mapping from Source (D) to the Ontology

Temptation: Canonical Instance as Prototype Triples. At a first glance, this correlation could be represented by RDF triples where tag names and existential variables are replaced by literal constants and resource constants, respectively (similar to the idea of the definition of a canonical instance of a conjunctive query in relational database theory [7]). The canonical instance as triples for Source (D) is given in Figure 5 where the replacements are highlighted by boxes. Since the canonical instance is only concerned with the structural aspects, the additional conditions on desired departure/arrival time are not part of it.

```
prefix t : <http://www.semwebtech.org/domains/2006/travel#>.
_:flightR a t:Flight; t:operatedBy [ a t:Airline; t:name "airlineTag" ];
  t:from [a t:Airport; t:code "fromTag" ];
  t:to [a t:Airport; t:code "toTag" ] ;
  t:flightCode "fcodeTag" ; t:deptTime "depTTag" ; t:arrTime "arrTTag" .
_:bconnR a t:BookableConn; t:instanceOfConn _:flightR ; t:date "dateTag" .
```

Fig. 5. Canonical instance of Source (D) as RDF triples

Problems. The first problem is, how this prototype description of the pattern is related in RDF to the WDSDL source description. It can be put in a file at a url u, and a triple (<bla://views/travel/fra> wdsdl:hasPrototypePattern u) can be added. This file is then intended for separated use as pattern, and does not make the description itself part of the Semantic Web.

Nevertheless, since the file at URI u is accessible in the Web, it implicitly *contributes* its triples to the "world-wide RDF database" in which also the triples from the ontology in Section 3 can be found.

If some RDF application (e.g., a Semantic Web search engine) encounters these triples, it cannot know that they are a prototype that is intended for separate use only, and will add them to its knowledge base. This leads to wrong facts (e.g., the resource with URI local0815:flightR is considered to be an instance of the class travel:Flight and is operated by a blank node local0815#id4711 that represents an airline that has a travel:name property with value "airlineTag"). Obviously, having the canonical instance in the Web together with the actual instances is a problem in the RDF world. Instead, an RDF-based language for *describing* patterns has to be defined and used.

Source Annotation Statements. The annotation on the semantic level consists of a set of statements that *refer* to the tags and optionally also refer to additional placeholders acting as existential variables (the RDF *blank nodes* in Figure 5). The statements *describe* the graph pattern that is asserted to hold for all answers returned by any query against the data source. For this, WDSDL makes use of the mechanism of *reification*: it talks *about* statements by wdsdl:AnnotationStatements. These statements relate

- tags (which have already been introduced as resources of the class wdsdl:Tag),
- local variables,
- constants if required,
- the notions of the application domain.

Additionally, constraints (using the operators from the XPath Functions and Operators namespace) are expressed by wdsdl:AnnotationConstraints[2]. Figure 6 shows the second part of the WDSDL ontology that covers the annotations on the semantical level.

Example 4. *The WDSDL source annotation of Source (D) is given in Figure 7. It has to be added to the source description on the signature level that has been given in Figure 3 since it refers to the same local tag resource URIs (the tag URIs are indicated by frames). Note that it additionally contains local identifiers (of the form _:xxxV) for the blank nodes. Note also that each of the views has an additional constraint that specifies _:from and _:to, respectively, to be "FRA".*

If the AnnotationStatements were materialized, they would exactly result in the canonical instance given in Section 5.2. Note that the same reification strategy

[2] The comparison operators used in these constraints are the same as for XPath/XQuery. As the W3C documents do explicitly list these operators like op:time-less-than, but do not assign a namespace URI to the op namespace prefix. For that, we temporarily assign it within the wdsdl namespace.

```
@prefix : <http://www.semwebtech.org/languages/2008/wdsdl#> .
:Annotation a owl:Class.
:hasAnnotation rdfs:domain [ owl:unionOf (:WebDataSource :View)] ;
  rdfs:range :Annotation.
:AnnotationStatement a owl:Class.    :AnnotationConstraint a owl:Class.
:hasAnnotationStatement rdfs:domain :Annotation; rdfs:range :AnnotationStatement.
:hasAnnotationConstraint rdfs:domain :Annotation; rdfs:range :AnnotationConstraint.
```

Fig. 6. The WDSDL Ontology Part II: Semantical Level

```
@prefix wdsdl: <http://www.semwebtech.org/languages/2008/wdsdl#> .
@prefix travel: <http://www.semwebtech.org/domains/2006/travel#> .
<bla://views/travel/fra/> wdsdl:hasAnnotation
[ wdsdl:localVar _:flightV, _:airlineV, _:airp1V, _:airp2V, _:bconnV ;
  wdsdl:hasAnnotationStatement
  [ rdf:subject _:flightV; rdf:predicate rdf:type; rdf:object travel:Flight],
  [ rdf:subject _:flightV; rdf:predicate travel:operatedBy; rdf:object _:airlineV ],
  [ rdf:subject _:airlineV; rdf:predicate rdf:type; rdf:object travel:Airline],
  [ rdf:subject _:airlineV; rdf:predicate travel:name; rdf:object _:airline ],
  [ rdf:subject _:flightV; rdf:predicate travel:from; rdf:object _:airp1V],
  [ rdf:subject _:flightV; rdf:predicate travel:to; rdf:object _:airp2V],
  [ rdf:subject _:flightV; rdf:predicate travel:flightCode; rdf:object _:code ],
  [ rdf:subject _:airp1V; rdf:predicate travel:code; rdf:object _:from ],
  [ rdf:subject _:airp2V; rdf:predicate travel:code; rdf:object _:to ],
  [ rdf:subject _:flightV; rdf:predicate travel:deptTime; rdf:object _:depT ],
  [ rdf:subject _:flightV; rdf:predicate travel:arrTime; rdf:object _:arrT ],
  [ rdf:subject _:bconnV; rdf:predicate rdf:type; rdf:object travel:BookableConn],
  [ rdf:subject _:bconnV; rdf:predicate travel:instanceOfConn; rdf:object _:flightV ],
  [ rdf:subject _:bconnV; rdf:predicate travel:date; rdf:object _:date ];
  wdsdl:hasAnnotationConstraint
  [ rdf:subject _:depT ; wdsdl:comparator wdsdl:time-less-than ; rdf:object _:dDepT ],
  [ rdf:subject _:arrT ; wdsdl:comparator wdsdl:time-less-than ; rdf:object _:dArrT ] ].
<bla://views/travel/fra/toFRAfromX/> wdsdl:hasAnnotation
  [ wdsdl:hasAnnotationConstraint [ wdsdl:onVariable _:to ; wdsdl:hasValue "FRA"] ].
<bla://views/travel/fra/fromFRAtoX/> wdsdl:hasAnnotation
  [ wdsdl:hasAnnotationConstraint [ wdsdl:onVariable _:from ; wdsdl:hasValue "FRA"] ].
<bla://views/travel/fra/fromFRA/> wdsdl:hasAnnotation
  [ wdsdl:hasAnnotationConstraint [ wdsdl:onVariable _:from ; wdsdl:hasValue "FRA"] ].
```

Fig. 7. WDSDL Source Annotation for Source D: Frankfurt Airport

is followed e.g. in OWL-2 [21] when asserting that a statement does *not* hold – describe the statement and annotate it.

On the other hand, the triples described in Figure 7 also represent the formula (*) given in Section 4.4. For instance, all statements about _:flightV (that stands for the variable v in (*)) are equivalent to

$\mathsf{Flight}(v) \wedge \mathsf{operatedBy}(v, w) \wedge \mathsf{from}(v, x) \wedge \mathsf{to}(v, y) \wedge \mathsf{flightCode}(v, e) \wedge \mathsf{deptTime}(v, g) \wedge \mathsf{arrTime}(v, i)$.

6 Source Selection and Query Answering

As already mentioned above, a SPARQL query basically specifies a graph pattern that yields answer bindings by matching the pattern against RDF data. SPARQL queries to be evaluated are submitted to a Query Broker together with the input variable bindings. The task of the query broker is to identify appropriate sources to query them in a suitable order, and to combine the answers. For this, the Query Broker is aware of the WDSDL descriptions of the sources that it can use to answer the query. The identification of Web Data Sources and their views that can contribute to the answer to a SPARQL query can then be reduced to finding suitable structural overlappings between the query and the views provided by Web Data Sources.

We will first illustrate the problem a bit more. Then we analyze the relationship with the areas of query answering using views [30,11,14] and querying data under access limitations [16,15,19,33,6].

6.1 Motivational Examples

Example 5 (Coverage between Sources and a Query). *Consider the query "where can we go from MUC on 21.5.2009 for less than 50€?". The SPARQL query is (without redundant class constraints)*

```
prefix  t: <http://www.semwebtech.org/domains/2006/travel#>
select  ?DEST ?P
where {?C a t:Flight; t:from [t:code "MUC"]; t:to [ t:code ?DEST] .
        ?BC t:instanceOfConn ?C; t:date "21-05-2009"; t:price ?P .
        filter (?P < 50) }
```

The prototype pattern of the query is shown in the upper part of Figure 8. The black portion is the prototype, while the gray portion is the part of the ontology that is not relevant to the query.

Consider first source (A) whose prototype graph is shown in the lower part of Figure 8: obviously, (A)'s prototype covers the query prototype (and additionally departure, arrival, and code of the flight connection). Analogously, also source (B) covers the query. Sources (C) and (D) do not cover the query since they do not contain the price (for (D) compare the upper part of Figure 8 with the mapping shown in Figure 4).

The above example illustrates the basic idea how to map queries onto the *annotated* sources, by using their canonical RDF instances. Nevertheless, there is not necessarily any source that completely covers the query. Additionally, in presence of limited access patterns, complete coverage does not guarantee that the source can be used for actually answering the query, but the input-output-characteristics must also be checked for compatibility.

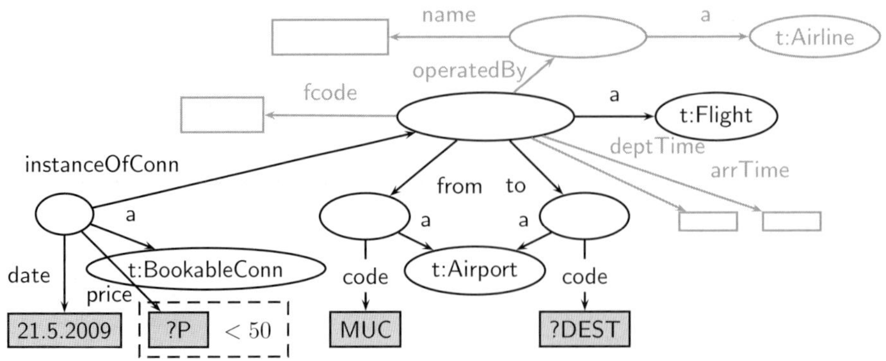

Above: Query Prototype as a Fragment of the Ontology given in Figure 4

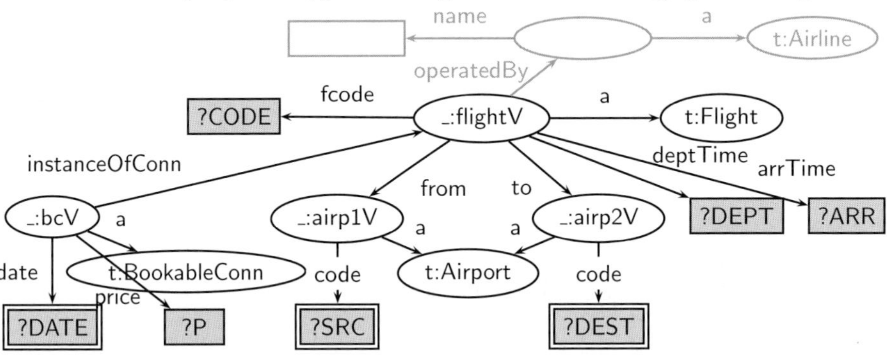

Fig. 8. Query Prototype as a Fragment of the Ontology (upper part) and coverage of the view provided by source (A) of the ontology (lower part). Input variables are double-framed.

Example 6 (Access Limitations). *Reconsider Figure 8 that illustrates that Source (A) covers the notions required in the query. Nevertheless, Source (A) provides the only view*

(airline, fcode, deptTime, arrTime, price) $\leftarrow$ travenjoyConns(from, to, date)

*that requires the destination (*to*) to be an input parameter. Similarly, the only view on Source (B) also declares* to *to be an input parameter. Both cannot be used immediately, since no destination is known for answering the query.*

Furthermore, the views provided by Source (D) require from *or* to *to be "FRA" (so only view* <bla://views/travel/fra/toFRAfromX/> *would only have a small overlap with the flights relevant for the query, namely those from "MUC" to "FRA").*

Source (C) is applicable, which is able to return values for to*, and also for* airline*. The WDSDL Source Annotation is given in Figure 9; its relationship with the query prototype is depicted in Figure 10. Thus, it is able to return the* *destination airports reachable from Munich, and additionally returns knowledge (the name of the airline) which is not required by the query, but can also be used for the next step.*

After that, the input requirements of Sources (A) and (B) are satisfied (i.e., date, start and now also the destinations are known), but (B) is applicable only for connections by Lufthansa. Both will be queried, and the resulting connections that are below 50€ will be returned. Note that this implicitly makes use of currency conversion described in [12] between $ and € if necessary.

```
@prefix travel: <http://www.semwebtech.org/domains/2006/travel#> .
<bla://views/travel/airdb/> :hasAnnotation
[   :localVar _:flightV, _:airlineV, _:airp1V, _:airp2V ;
  :hasAnnotationStatement
  [ rdf:subject _:flightV; rdf:predicate rdf:type; rdf:object travel:Flight],
  [ rdf:subject _:flightV; rdf:predicate travel:operatedBy; rdf:object _:airlineV ],
  [ rdf:subject _:airlineV; rdf:predicate rdf:type; rdf:object travel:Airline],
  [ rdf:subject _:airlineV; rdf:predicate travel:name; rdf:object [_:airline] ],
  [ rdf:subject _:flightV; rdf:predicate travel:from; rdf:object _:airp1V],
  [ rdf:subject _:flightV; rdf:predicate travel:to; rdf:object _:airp2V],
  [ rdf:subject _:airp1V; rdf:predicate travel:code; rdf:object [_:from] ],
  [ rdf:subject _:airp2V; rdf:predicate travel:code; rdf:object [_:to] ] ].
```

Fig. 9. WDSDL Source Annotation for Source (C): airdb

6.2 Query Evaluation

The actual query evaluation can then make use of solutions of two well-investigated issues: *query answering using views* for obtaining possible rewritings, and *querying data under access limitations* for checking which rewritings are compatible with the access limitations, and obtaining executable query plans.

Query Answering Using Views is e.g. investigated in [30,11,14]. Here, *GAV (global-as-view)* and *LAV (local-as-view)* approaches can be distinguished. Intuitively, LAV is the more obvious way: given a global schema, the local schemas are expressed as views over it. In contrast, GAV expresses the global schema in terms of the local ones.

In our setting, the global schema is given as an RDF schema, i.e., unary and binary predicates, and the sources are given as predicates, i.e., a relational schema. As already mentioned in Section 4.4, our approach is based on a *LAV* mapping. The LAV mappings like (*) in Section 4.4 can be derived automatically from the WDSDL Source Annotations. The usual algorithms, like MiniCon [24] construct a union of conjunctive queries over the available views that yields all obtainable answers.

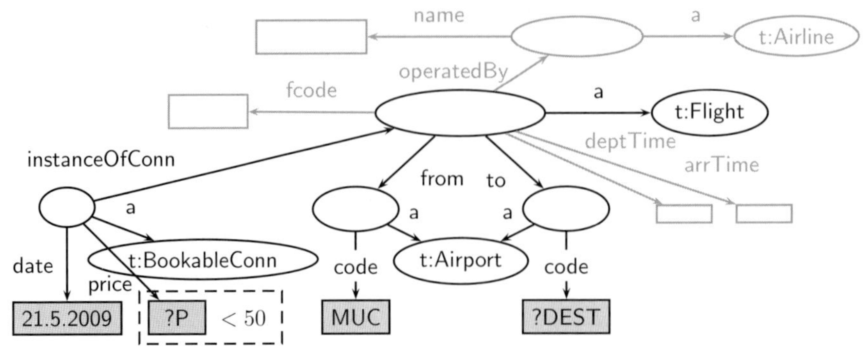

Above: Query Prototype as a Fragment of the Ontology given in Figure 4

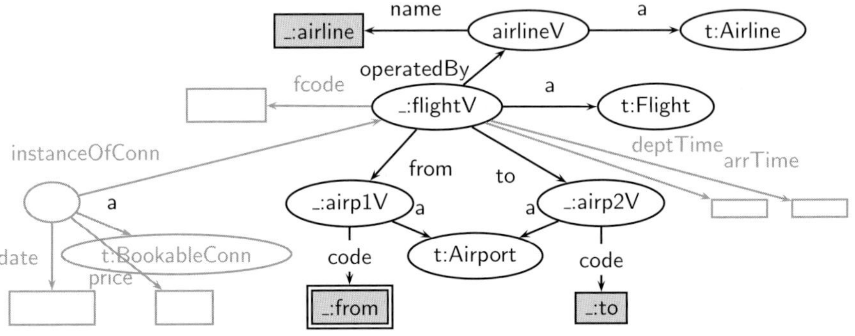

Fig. 10. Query Prototype as a Fragment of the Ontology (upper part) and coverage of the view provided by source (C) of the ontology (lower part)

Example 7. *In the running example, the original SPARQL query is seen as a predicate* q(To,Price) *where the following conjunctive queries contribute results:*

1.) q(To, Price) :-
connection$_A$("MUC", To, "21.05.2009", _Airl, _FCode, _DT, _AT, Price).
2.) q(To, Price) :- flight$_C$(_Airl, "MUC", To),
connection$_A$("MUC", To, "21.05.2009", _Airl, _FCode, _DT, _AT, Price).
3.) q(To, Price) :- flight$_C$("Lufthansa", "MUC", To),
connection$_B$("MUC", To, "21.05.2009", _FCode, _DT, _AT, Price).

Note that the second one is –on this stage– redundant as it is a refinement of the first one. Also the third one is a subset of the second one (the Lufthansa flights).

In the given setting with limited access capabilities, one must not eliminate the redundant rewritings yet, since some of them may turn out not to be executable.

Querying Data under Access Limitations is concerned with the case where the reformulation and mapping to the sources is already done, and a concrete plan is to be found how to do the evaluation. Concretely, two algorithms have been presented in [33,6].

They have to be applied to each of the found rewritings, resulting in an executable plan for those rewritings where such a plan exists. Only after that, redundant rewritings/plans can be removed.

Example 8. *In the running example, from the above rewritings, (1.) is not executable since Source (A) expects* To *as an input. (2.) and (3.) are executable since the subquery containing* flight_C *is supported by (C)'s only view, yielding the possible solutions for* To*. With these, Sources (A) (for all airlines) and (B) (for Lufthansa) can be queried. After that, (3.) can be dropped since it is redundant.*

The query broker executes the obtained query plans and returns the resulting bindings as answers of the original SPARQL query.

7 Related Work

The core related aspects of *Query Answering Using Views* and *Querying Data under Access Limitations* have been discussed in Section 6.2.

Our work is also related to the field of *Semantic Web Services.* There, several different formalisms for enriching Web Services with semantic annotations have been proposed. In [8] a layered language architecture (similar to the standard Semantic Web layer cake [13]) is introduced to enable automatic Web Service discovery, composition and execution. The framework considers different aspects of Web Services, such as the heterogeneity in data representations and the pre- and postconditions of Web Service invocations. DAML-S [5] enhances the WSDL descriptions of input and output messages to cover abstract types for each message part and also comprises a process model ontology, which can be used to support automated Web Service invocation, composition, and interoperation. [17] describes how the sucessor of DAML-S, OWL-S [18], fits into the picture of the "Semantic Annotations for WSDL and XML Schema" (SAWSDL), a standard set by the W3C. SAWSDL defines a fixed set of WSDL extension attributes, which allow to use elements of an external Semantic Web framework, e.g. for mapping message parts to abstract types. Finally, [1] proposes the "Semantic Web Process Description Language" (SWPDL), which is based on an OWL ontology and allows to describe (composite) Web Services, as well as (collections of) Web pages.

Our approach is also based on several layers, which have a clear cut purpose and each level builds neatly on top of the lower level. All semantic annotations are given in standard RDF and OWL constructs and are thus Semantic Web documents. Additionally, we support a more generic and abstract notion of Web Data Sources, which allows us to represent and use heterogeneous data sources, ranging from Deep Web Sources to Web Services. This is possible because we do not rely on a specific low level service specification, such as WSDL.

One of the main incentives for the semantic annotations of Web Services is the ability to do *automatic selection or matchmaking* of these services. Current approaches to this problem usually employ some kind of *degree of match* metric to find relevant services [31,23,3,22]. For the composition of the thus found services sometimes additional user interaction is required [27,2]. In our approach we are only concerned with querying, and with *exact* matches for the query description.

8 Conclusion

The WDSDL ontology serves for annotating Web Data Sources to relate them to the terminology of their domain ontology. These annotations enable a WDSDL-based query broker to handle SPARQL queries that are stated in terms of the domain ontology by selecting appropriate sources to stepwise process the complete query or portions of it, and to combine their answers to answer the original query. The approach is currently under implementation.

References

1. Agarwal, S.: Specification of Invocable Semantic Web Resources. In: ICWS, pp. 124–131. IEEE Computer Society Press, Los Alamitos (2004)
2. Agarwal, S., Studer, R.: Automatic Matchmaking of Web Services. In: ICWS, pp. 45–54. IEEE Computer Society Press, Los Alamitos (2006)
3. Benatallah, B., Hacid, M.-S., Rey, C., Toumani, F.: Request Rewriting-based Web Service Discovery. In: Fensel, D., Sycara, K., Mylopoulos, J. (eds.) ISWC 2003. LNCS, vol. 2870, pp. 242–257. Springer, Heidelberg (2003)
4. Braga, D., Ceri, S., Daniel, F., Martinenghi, D.: Optimization of Multi-domain Queries on the Web. PVLDB 1(1), 562–573 (2008)
5. Ankolekar, A., Burstein, M., Hobbs, J.R., Lassila, O., Martin, D., McDermott, D., McIlraith, S.A., Narayanan, S., Paolucci, M., Payne, T.R., Sycara, K.: DAML-S: Web service description for the semantic web. In: Horrocks, I., Hendler, J. (eds.) ISWC 2002. LNCS, vol. 2342, pp. 348–363. Springer, Heidelberg (2002)
6. Calì, A., Martinenghi, D.: Conjunctive query containment under access limitations. In: Li, Q., Spaccapietra, S., Yu, E., Olivé, A. (eds.) ER 2008. LNCS, vol. 5231, pp. 326–340. Springer, Heidelberg (2008)
7. Chandra, A.K., Merlin, P.M.: Optimal Implementation of Conjunctive Queries in Relational Data Bases. In: STOC, pp. 77–90. ACM Press, New York (1977)
8. de Bruijn, J., Lausen, H., Polleres, A., Fensel, D.: The Web Service Modeling Language WSML: An Overview. In: Sure, Y., Domingue, J. (eds.) ESWC 2006. LNCS, vol. 4011, pp. 590–604. Springer, Heidelberg (2006)
9. Fielding, R.T.: Architectural Styles and the Design of Network-based Software Architectures. PhD thesis, University of California, Irvine, California (2000)
10. Florescu, D., Levy, A.Y., Mendelzon, A.O.: Database Techniques for the World-Wide Web: A Survey. SIGMOD Record 27(3), 59–74 (1998)
11. Halevy, A.Y.: Answering queries using views: A survey. VLDB J. 10(4), 270–294 (2001)
12. Hornung, T., May, W.: Deep web queries in a semantic web environment. In: Workshop on Advances in Accessing Deep Web (ADW). LNBIP, vol. 37, pp. 39–50. Springer, Heidelberg (2009)
13. Horrocks, I., Parsia, B., Patel-Schneider, P.F., Hendler, J.A.: Semantic Web Architecture: Stack or Two Towers. In: Fages, F., Soliman, S. (eds.) PPSWR 2005. LNCS, vol. 3703, pp. 37–41. Springer, Heidelberg (2005)

14. Lenzerini, M.: Data integration: A theoretical perspective. In: PODS, pp. 233–246. ACM Press, New York (2002)
15. Li, C.: Computing complete answers to queries in the presence of limited access patterns. VLDB J. 12(3), 211–227 (2003)
16. Li, C., Chang, E.Y.: Answering queries with useful bindings. ACM Trans. Database Syst. 26(3), 313–343 (2001)
17. Martin, D., Paolucci, M., Wagner, M.: Bringing Semantic Annotations to Web Services: OWL-S from the SAWSDL Perspective. In: Aberer, K., Choi, K.-S., Noy, N., Allemang, D., Lee, K.-I., Nixon, L.J.B., Golbeck, J., Mika, P., Maynard, D., Mizoguchi, R., Schreiber, G., Cudré-Mauroux, P. (eds.) ASWC/ISWC 2007. LNCS, vol. 4825, pp. 340–352. Springer, Heidelberg (2007)
18. Martin, D.L., Burstein, M.H., McDermott, D.V., McIlraith, S.A., Paolucci, M., Sycara, K.P., McGuinness, D.L., Sirin, E., Srinivasan, N.: Bringing Semantics to Web Services with OWL-S. In: World Wide Web, pp. 243–277 (2007)
19. Nash, A., Ludäscher, B.: Processing Unions of Conjunctive Queries with Negation under Limited Access Patterns. In: Bertino, E., Christodoulakis, S., Plexousakis, D., Christophides, V., Koubarakis, M., Böhm, K., Ferrari, E. (eds.) EDBT 2004. LNCS, vol. 2992, pp. 422–440. Springer, Heidelberg (2004)
20. OWL Web Ontology Language (2004), http://www.w3.org/TR/owl-features/
21. OWL 2 Web Ontology Language (2009), http://www.w3.org/2007/OWL/wiki/OWL_Working_Group (work in progess)
22. Pantazoglou, M., Tsalgatidou, A., Athanasopoulos, G.: Quantified Matchmaking of Heterogeneous Services. In: Aberer, K., Peng, Z., Rundensteiner, E.A., Zhang, Y., Li, X. (eds.) WISE 2006. LNCS, vol. 4255, pp. 144–155. Springer, Heidelberg (2006)
23. Paolucci, M., Kawamura, T., Payne, T.R., Sycara, K.P.: Semantic Matching of Web Services Capabilities. In: Horrocks, I., Hendler, J. (eds.) ISWC 2002. LNCS, vol. 2342, pp. 333–347. Springer, Heidelberg (2002)
24. Pottinger, R., Halevy, A.Y.: Minicon: A scalable algorithm for answering queries using views. VLDB J. 10(2-3), 182–198 (2001)
25. Resource Description Framework, RDF (2000), http://www.w3.org/RDF
26. Resource Description Framework (RDF) Schema specification (2000), http://www.w3.org/TR/rdf-schema/
27. Sirin, E., Parsia, B., Hendler, J.A.: Filtering and Selecting Semantic Web Services with Interactive Composition Techniques. IEEE Intelligent Systems 19(4), 42–49 (2004)
28. SPARQL Query Language for RDF (2006), http://www.w3.org/TR/rdf-sparql-query/
29. Turtle - Terse RDF Triple Language, http://www.dajobe.org/2004/01/turtle/
30. Ullman, J.D.: Information integration using logical views. In: Afrati, F.N., Kolaitis, P.G. (eds.) ICDT 1997. LNCS, vol. 1186, pp. 19–40. Springer, Heidelberg (1996)
31. Vaculín, R., Chen, H., Neruda, R., Sycara, K.P.: Modeling and Discovery of Data Providing Services. In: ICWS, pp. 54–61. IEEE Computer Society Press, Los Alamitos (2008)
32. XML Schema (1999), http://www.w3.org/XML/Schema
33. Yang, G., Kifer, M., Chaudhri, V.K.: Efficiently ordering subgoals with access constraints. In: PODS, pp. 183–192. ACM, New York (2006)

An Extended Petri-Net Based Approach for Supply Chain Process Enactment in Resource-Centric Web Service Environment

Xiaodong Wang[2], Xiaoyu Zhang[2], Hongming Cai[1], and Boyi Xu[3]

[1] School of Software, Shanghai JiaoTong University, Shanghai, China
[2] BIT Institute, University of Mannheim, Mannheim, Germany
[3] Antai College of Economic&Management, Shanghai JiaoTong University, Shanghai, China
{wangxd.sjtu,xiaoyu.zhang.cn}@googlemail.com,
{hmcai,byxu}@sjtu.edu.cn

Abstract. Enacting a supply-chain process involves variant partners and different IT systems. REST receives increasing attention for distributed systems with loosely coupled resources. Nevertheless, resource model incompatibilities and conflicts prevent effective process modeling and deployment in resource-centric Web service environment. In this paper, a Petri-net based framework for supply-chain process integration is proposed. A resource meta-model is constructed to represent the basic information of resources. Then based on resource meta-model, XML schemas and documents are derived, which represent resources and their states in Petri-net. Thereafter, XML-net, a high level Petri-net, is employed for modeling control and data flow of process. From process model in XML-net, RESTful services and choreography descriptions are deduced. Therefore, unified resource representation and RESTful services description are proposed for cross-system integration in a more effective way. A case study is given to illustrate the approach and the desirable features of the approach are discussed.

Keywords: supply chain, RESTful web service, Petri-Net, XML-Net, WADL.

1 Introduction

Supply Chain Process is a special business process which interrelates production, logistics and information. Distinguishing from other business processes, supply chain processes span usually departments and organizations. They are generally performed by different independent firms in order to produce and deliver a specified range of goods and services. A supply chain process is normally very complex, while it often involves lots of people, data, materials, products, rules and the relationships among them. The supporting IT systems are accordingly always distributed and heterogeneous. Furthermore, unstructured applications and data also make it difficult to integrate supply chain processes with business entities and services provided by various sources.

Service Oriented Architecture (SOA) [1] is now accepted as a popular construction approach for inter-organizational business process integration and composition of

R. Meersman, T. Dillon, P. Herrero (Eds.): OTM 2009, Part I, LNCS 5870, pp. 130–146, 2009.

applications across systems. In the paradigm of SOA, business services are mapped to IT services, which are mainly implemented based on SOAP (Simple Object Access Protocol) protocol. However, SOAP services are generally complicated. Hence discovering and matching services for integration of business processes are arduous work. Moreover, the representation for the semantics of business entities in SOA framework is still missing.

Currently, REST (Representational State Transfer) [2] receives increasing attention. Compared with SOA, RESTful architecture requires much less work for development. The Resource-Oriented Architecture (ROA) is based on the concept of the resource, which is an abstraction of information [3] and is referenced with a global identifier (URI) containing the name and the address of the resource. From the resource-oriented viewpoint, the entities engaged in business process could be abstracted as resources. Based on this standpoint, business process will be regarded as interactions among resources and resources' states transformation. Accordingly, in a resource-oriented environment, the access to resources is achieved through RESTful Web Services using the Hypertext Transfer Protocol (HTTP).

Integrating supply chain process across organizational and heterogonous systems requires the unified resource modeling and representation. Correlatively, widely accepted modeling tools for describing interaction processes of resources are indispensable. Finally, universal description for RESTful services and their compositions should be generated according to resource models, so as to encapsulate the various implementations from different systems.

In this paper, an extended Petri-net based approach is proposed for modeling and integrating supply chain process in RESTful service environment. To obtain unified definitions of resources, a resource meta-model is built to represent the basic information elements of resources, so as to conceal the incompatibilities among resources from heterogeneous applications. According to this resource meta-model, XML schemas and documents are derived, which describe resources and their states in Petri nets. As self-described messages, the information encapsulated in XML could be simply exchanged through HTTP, which makes inter-operability across systems possible. A high level Petri-net, so-called XML-net is employed for modeling processes of interactions among resources. The operations on resources could be then translated into manipulations of XML documents. Finally, the unified RESTful services description in WADL (Web Application Description Language) and orchestration in extended BPEL are deduced from process models in XML-net. Thereby diversiform applications could be encapsulated in RESTful services and integrated into complete supply chain process. Finally, the software architecture for implementation and a case study are given to verify and illustrate the approach.

2 Overview of the Process Integration Approach Based on XML-Nets

From the resource-oriented viewpoint, business processes could be regarded as a set of activities, which change resource properties and transform their states. Therefore, resource, the interaction of resources with process, and IT implementation of process models are three crucial factors to realize process integration in a RESTful service enviorment. Hereby unified definition and representation of resource model are required firstly. In our approach, resource meta-model describes the structures of resources. Thereafter, a high-level Petri-net is employed to model and represent interactions among resources and their states transformation. Finally, for execution and deployment, RESTful web services description and orchestration are derived from process model in Petri-net, in this way the diversiform applications will be encapsulated in RESTful services and organized to complete execution of business processes.

Fig. 1 illustrates the main idea of our approach.

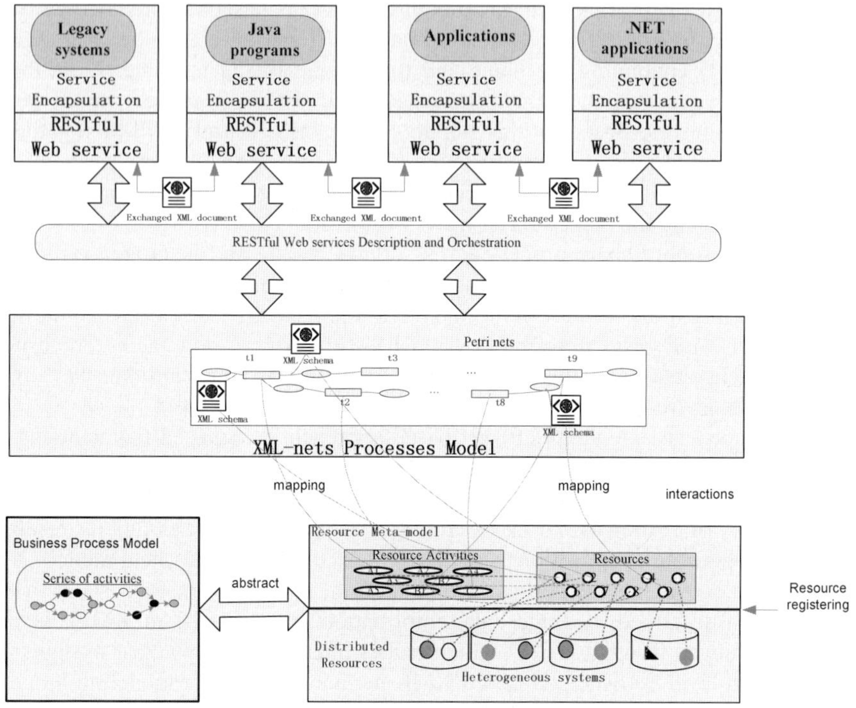

Fig. 1. The main idea of integrated framework based on XML-nets processes model

Firstly, the information of business entities is mapped as resources. To achieve supply chain process integration, all resources coming from existing distributed and heterogeneous systems have to be identified, analyzed and encapsulated, so as to hide the data differences among different systems. Therefore, it is very necessary to define an unified structure for resource modeling. Based on workflow analysis, a resource meta-model is built, which defines not only the internal attributes and properties of a specific resource, but also its external interfaces and behaviors. The resource meta-model contains four main components: Resource Properties, Operation Set, Activity State and Rule Model. The Resource Properties and Operation Set represent separately the internal attributes and operations of a resource, while Activity State and Rule Set define the external behaviors of a resource. A business process contains several activities, in which the correlative resources can be invoked by RESTful service. In section 3, the formal expression of resource meta-model will be presented.

After modeling resources with resource meta-model, XML schemas for corresponsive resources are derived. These XML schemas act as resource views in Petri nets during the process modeling. Likewise, concrete resource objects and their states are described in XML documents.

In a REST framework, the control flow of process is essentially regarded as resources states and their transformation. To describe the states transformation of resources and enact supply chain processes, a specific class of high-level Petri-net, so-called XML-net, is employed. XML-net is a formal, graphical modeling language which was developed by Oberweis and Lenz [4] [5]. It integrates the concept of complex data objects into Petri-net formalism [6]. In XML nets, the static components are typed by XML schema diagrams, each of them representing a XML schema. Places can thus be regarded as containers for XML documents that are valid for the corresponding XML schema. These XML documents conserve the resources states in runtime. The flow of XML documents is defined by occurrences of transitions, i.e., the activities taking place. So-called filter schemas label the adjacent arcs and thereby describe the documents that are related to the activities and the way these documents are manipulated. In our approach, places in XML-net could be regarded as containers for XML-documents and transitions in XML-net represent the operations on resources. With XML-net, control flow of supply chain processes and the states transferring of resources can be integrated well into process models.

However, a supply chain process model in XML-net could not be directly executed in a RESTful services system. Therefore it is necessary to interpret such processes with machine-processable specifications for HTTP based web applications [7]. In this paper, RESTful Web services are described with WADL, which is a resource-centric description language and can hereby combine resources representation with Web service description more appropriately. Briefly speaking, the resource definitions in WADL are derived from XML schemas. The operation definitions could be deduced from transitions, and operations set of resource builds up the port of correlative RESTful web service.

The Business Process Execution Language (BPEL) [8] is used as a standard service composition for modeling and executing business process. However it cannot be applied directly in RESTful service environment. To composite RESTful Web services into processes, a BPEL extension for RESTful is proposed in [9]. Based on process models in XML nets, the RESTful services description in WADL and services composition in extended BPEL could be automatically generated. In this way, RESTful web services provide a resource-oriented platform for interoperation among legacy systems, java programs, application interfaces, and .net applications.

The service descriptions in WADL provide general interfaces for accessing to resources, through which diversiform applications in heterogeneous systems are encapsulated into RESTful services. And exchanged data are packed in self-described XML documents, which facilitate the communication and cooperation among participants in supply chain process.

3 Resources and Process Modeling

In order to realize process integration for supply chain application in a resource-centric environment, resource model, interactive mechanism between resource and activities, and services description generation are main three theses of our approach. Therefore, resource model and its formal expression are given in section 3.1. Based on the XML-net, interactive mechanism is given in section 3.2. The last section is used to build WADL and BEPL expression for execution in the application.

3.1 Resource Modeling and Formal Expression

Based on information rather than process, the nature of business process is the executing and transforming process of resources. Generally speaking, a business process step includes some procedures, in which business information of resource models is carried out and properties of resource models are changed according to each certain operation, and the corresponding parameters along with the messages are transferred to the next business process step. To achieve the goal of business process integration, all resources in existing distributed and heterogeneous systems have to be firstly identified, analyzed and encapsulated. Then all encapsulated resources will be registered into the resource integration platform under the unified management, so as to hide the data differences among different systems. Finally the resources are able to achieve interoperability in the way of RESTful Web Services. Therefore, it is very necessary to define a unified structure for resource modeling.

Based on workflow analysis, a resource meta-model is proposed. With the aid of this method of resource modeling, the integrated platform can organize and manage the static and dynamic messages in the enterprise system, as well as establish an open and extensible framework for business application system so as to satisfy the integration requirements of distributed, dynamic, flexible, and typically heterogeneous systems.

Fig. 2 illustrates the resource meta-model. The resource meta-model contains four main components: Resource Properties, Operation Set, Activity State and Rule Model. The Resource Properties and Operation Set represent separately the internal attributes and operations of a resource, while Activity State and Rule Set define the external behaviors of a resource. A business process contains several activities, in which the correlative states transformation of resources can be implemented through RESTful service. All these components are defined accurately as following.

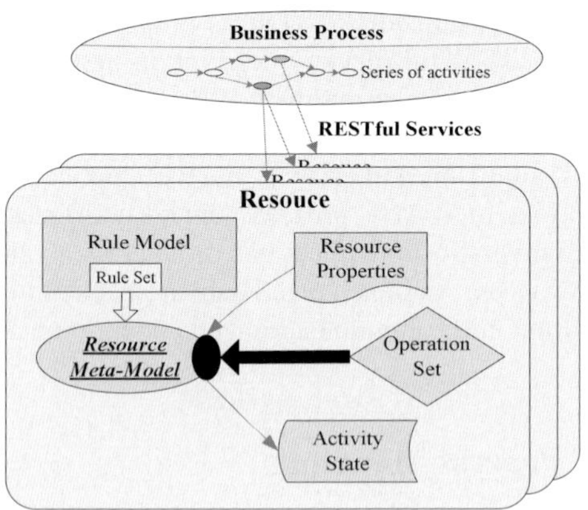

Fig. 2. Resource Meta-Model

Definition 1. Resource Meta-Model (RMM)
The Resource Meta-Model is defined as a set of Resource Property (RP), Operation Table (OT), Rule Set (RS) and Activity State (AS). Thus the Resource Meta-Model (RMM) is defined as the following tuple:

RMM = <RP, OT, RS, AS>

Definition 2. Property Set (PS)
The Property Set (PS) is defined as a following set of properties.

$PS = \{p_1, p_2, \ldots\}$, $|PS|>0$.

For each property $p_i \in PS$, p_i = <PropertyName, InitialValue>

Hereby p_i indicates the content of a single property, and the PS is a set of several properties. Every property p_i uses two main fields to keep its name and its initial value separately.

Definition 3. Resource Property (RP)
The Resource Property is defined by the following tuple:

RP = <ResID, ResType, SProps, DProps>

Where SProps = $\{ rp_i \mid rp_i \in PS \wedge Static(rp_i) = TRUE \}$,

DProps = $\{ rp_i \mid rp_i \in PS \wedge Static(rp_i) \neq TRUE \}$.

Hereby the ResID is the unique identifier of a resource entity. ResType is the underlying category, which describes the type of a resource with a specific static property and can be organized by using a hierarchical tree method. SProps includes the static properties of a resource, and here the "static" means that the values of these properties will never be changed once created, e.g. the manufacturing data of a product. Correspondingly DProps includes the dynamic properties, which are used to describe the real-time status, usage calendar and lifecycle status of resources and so on, e.g. the status (created, locked, outgoing, and archived) of orders in a sales process. Dynamic properties can also be used to describe the status of resources in business process steps, because that they are suitable for workflow modeling.

Definition 4. Operation Table (OT)
The Operation Table is defined as the following set of operations.

$OT = \{o_1, o_2, \ldots\}$, $|OT|>0$

where o_i = <OpName, OpBody, VariableTable>

Each operation o_i includes three fields: OpName indicates the name of an operation, OpBody defines the body of the operation and all the variables used by the operation are defined in the field VariableTable.

Definition 5. Activity State (AS)
The Activity State is defined by the following tuple:

AS = $<s_{start}, s_{end}, S_n>$

where $S_n = \{s_1, s_2, s_3, \ldots \}$, $|S_n| > 0$, $s_{start} \in S_n$, $s_{end} \in S_n$

s_i = <StateName>

s_{start} indicates the initial state of a resource in the state diagram, and s_{end} means the end state. Each s_i in the set S_n defines an intermediate state of the resource.

Definition 6. Rule Set (RS)
Rules Set is used to reflect the rules and restricts of a resource status transition in the process.

$$RS = \{r \mid r = \langle\langle s_i, RP,\rangle, s_j\rangle, (s_i \in \{\varnothing, s_{start}\} \cup S_n) \wedge (s_j \in \{s_{end}\} \cup S_n), c_k \in 2^{CS}\}$$

Here s_i indicates the current workflow state, which is generally defined in AS. If the s_i in the corresponding rule is null, the rule only needs to check the current resource property and the constraints defined by c_k. c_k defines a constraint combination. The status will be set to s_j if the constraint is satisfied with the current resource property. E.g., when the order's goods are available in the inventory, the status can be converted as activated. Business rules are organized based on resources, and the choreography of resource services can be more convenient through compounding with resource rules in the integration of processes.

3.2 XML-Net Component Derivation for Resource Process Interaction

In order to analyze the interactive mechanism of resources and resource-based activities, XML-net is introduced to act as a media. In a resource-based environment, all messages exchange between partners in supply chain via HTTP request/response interactions. Such interaction processes are modeled in XML-net. The formal definition of XML-net is presented as follows:

Definition 7. XML-Net (XN)
A XML net is a tuple= <P, T, F, SchemaXN , FuncP, FuncT, FuncURI> where

- P and T are disjoint sets of places and transitions
- $F \subseteq (P \times T) \cup (T \times P)$ is the flow relation
- SchemaXN are the XML schema related to places.
- A function FuncP assigns to each place an XML schema SchemaXN
- FuncT is a function assigning guard conditions to transitions, where a condition FuncT(t): ($\bullet t \rightarrow$ XML) specifies combination of input XML documents
- FuncURI is a function assigning URIs to tuples of communication transitions and combinations of input of input documents, i.e. FuncURI(t) : ($\bullet t \rightarrow$ XML) $\rightarrow$ URI
- A set of valid XML documents are assigned to each place of XML-net as initial marking.

Based on resource meta-model, process models in XML-net could be deduced. In a XML-net process model, XML schemas with unified definition are deduced from resources model. They prescribe the properties of a resource and possible value ranges. Thereafter these XML schemas are utilized for verifying the exchanged XML documents between transitions in XML-net process models. During runtime, XML documents are generated and contain the states of resources. And Places in XML-net could be regarded as containers for XML documents. Similarly, transitions in XML-net are derived from the operations set on resources. In Resource Meta-Model, each resource has its own operations set, in which operations are defined. These operations read or modify the properties values of the resource object, so that the state of resource as well as the state of process will also be changed. In XML-net, the operations which read or change states of resource will be interpreted as transitions. The rules in resource meta-model prescribe the states transferring. Combining with operations and rules set, the flow relation F in XML could be also deduced.

In process models, XML documents describing resource states are generated according to the XML schemas and the data fields will be evaluated according to resource model.

The mapping functions could be defined as follows:

Definition 8. XML schema Mapping $\phi_R(R) = \{(R_i, SchemaXN_i)\}$

The mapping function $\phi_R(R)$ defines the mapping relation between resource models and XML schemas in XML-net. For each resource, there is a corresponsive XML schema which describes it.

Definition 9. Transition Mapping $\phi_T(R, OT) = \{((R_i, o_k), t_j) \mid o_k \in OT_i, t_j \in T\}$

For operation $\mathbf{o_k}$ of resource model $\mathbf{R_i}$, there is a corresponding transition $\mathbf{t_j}$ in XML-net, which describes the transformation of the correlative resource states.

Definition 10. XML document Mapping $\phi_P(R, AS) = \{((R_i, rp_l, s_j), p_k) \mid rp_l \in PS_i, s_j \in AS_i, p_k \in P\}$

For the properties and states of resource, a XML document in XML net is used to describe resource runtime state. Places in XML net act as container for XML document. Therefore we could build up the function ϕP which maps from resource activity states with corresponding properties to places in XML net.

Definition 11. F relation Mapping $\phi_F(R, AS, RS, OT) = \{((R_i, s_j, r_k, o_l), f_m \mid s_j \in AS_i, r_k \in RS_i, o_l \in OT_i\}$

The relation F between transitions and places could be also mapped from the states, operations and rules of resources.

Fig.3 illustrates the course of generation of XML schemas and XML documents from resource meta-model.

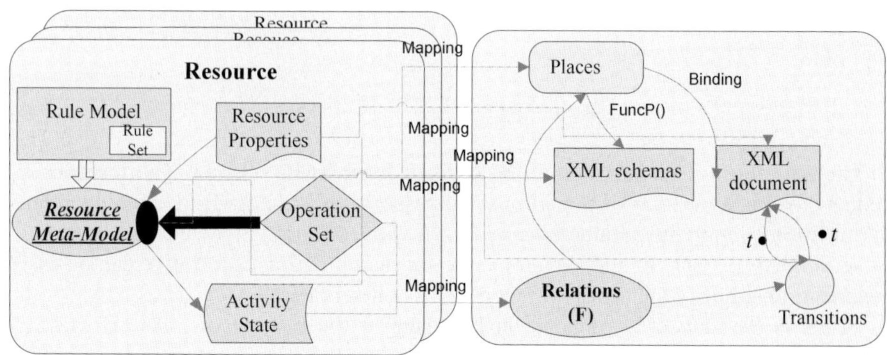

Fig. 3. Generation of XML-net Components for Process Model

3.3 RESTful Services Description and WADL Generation

As explained above, supply chain processes are modeled with XML-nets. The correlated resources and their states are described with XML documents. It implies that both data objects and control flow of processes are modeled in XML-nets completely. From such model, we could deduce RESTful services descriptions.

Currently there are two specifications used to describe REST services: WADL (Web Application Description Language) and WSDL 2.0 (Web Services Description Language) HTTP binding extension. WADL [10], proposed by Marc J. Hadley at Sun Microsystems, is a description language especially for HTTP-based web applications that follow REST architectural style.

The root of WADL is the application element, which consists of grammars and resources elements. The grammars element contains definition of the exchanged data format, e.g. in the language of XML schema or Relax NG. The resources element is a container for all resources and defines the base URI for all resources. Each resource is then represented as the resource element, which is identified with a path to the base URI. A resource element contains a list of HTTP method elements that are allowed to operate on this resource. Each method element contains a request element and a response element. The request element is composed by a few parameters. The response element contains the representation of the resource and a fault element if something goes wrong during the execution.

Fig.4. shows the mapping process from transition to method description in WADL. For simplicity we only focus on the main elements and ignore other less important elements in WADL such as doc, resource-type etc.

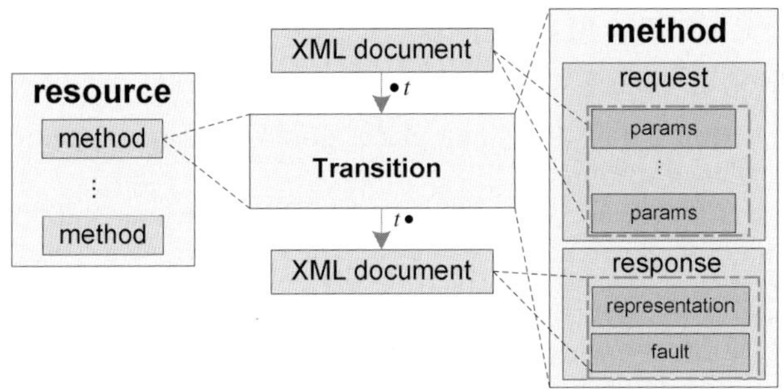

Fig. 4. Mapping WADL description

The definition of resource could be derived from XML schema. The methods are deduced from description of transitions. Corresponsively, the XML documents related to transition contain the parameters and return value of correlative deduced method.

With WADL, the stub and skeleton code for manipulation of resource can be easily generated to simplify the development of RESTful services.

BPEL is the standard composition language in the context of SOA. However it cannot be applied directly to REST. To composite RESTful Web services into processes, BPEL is extended for RESTful in [9]. And the new extension called BPEL for REST extends BPEL in the following two ways:

- Four activities <get>, <post>, <put> and <delete> are introduced to invoke RESTful web services, which corresponds the HTTP methods directly so that the client can use the familiar way to manipulate appropriate resources.
- The <resource> container element is introduced to declare resources involved in the process dynamically. The client can then access the published resources if they are available, that is, the process is being executed in their scope. Global resource is always available. Within the <resource> element, four elements, i.e. <onGet>, <onPost>, <onPut>, <onDelete>, are defined to declare the handlers that are triggered when the process receives the corresponding HTTP request.

With these two ways of extensions, BPEL is also appropriate to describe a business process implemented with RESTful web services. A. Koschmider proposed an approach to deduce BPEL components from XML nets [11]. Similarly, BPEL extension code could be also deduced from process model in XML nets. Especially, the <resources> container elements in BPEL extension are derived from XML schema. And the activities in BPEL extension are derived from transitions definitions in XML net.

4 Software Architecture and Case Study

4.1 Software Architecture

Based on resource model and process model in XML-net, Fig.5 illustrates the main framework supporting our idea, which involves five main modules: Heterogeneous Resources Adapter, Resource Model Management, Resource Object Management, XML-nets Processes Model Engine and RESTful Services Management.

The module Heterogeneous Resources Adapter supports heterogeneous systems to encapsulate operational resources, and then registers them to the Resources Registration Center on the integrated platform in order to make sure that the integrated platform can use the unified resource identifier to carry out operations on the heterogeneous resources.

The module Resource Model Management is responsible for the management of the resource model information from various heterogeneous information systems. The meta-information of the resource model and resource mapping strategies are stored in XML files and can be used by the resource object model. The module Resource Object Management manages the resource object information, which is obtained from different heterogeneous systems. Resources representation in XML documents, which is the result of the resources requests, is generated and sent to the XML-net model engine.

Based on resources model, the module XML-nets Process Model Engine is responsible for modeling processes. Thereafter, the RESTful services descriptions and orchestration descriptions are automatically generated according to process models.

The authority, transaction, discovery and matching of RESTful services are managed within this module.

4.2 A Case Study

In this section, a supply chain process with type "Make-to-Order" is proposed. This process could be seen as the states transformation of correlative resources. For simplicity, the payment processes and correlative resources are not considered in this

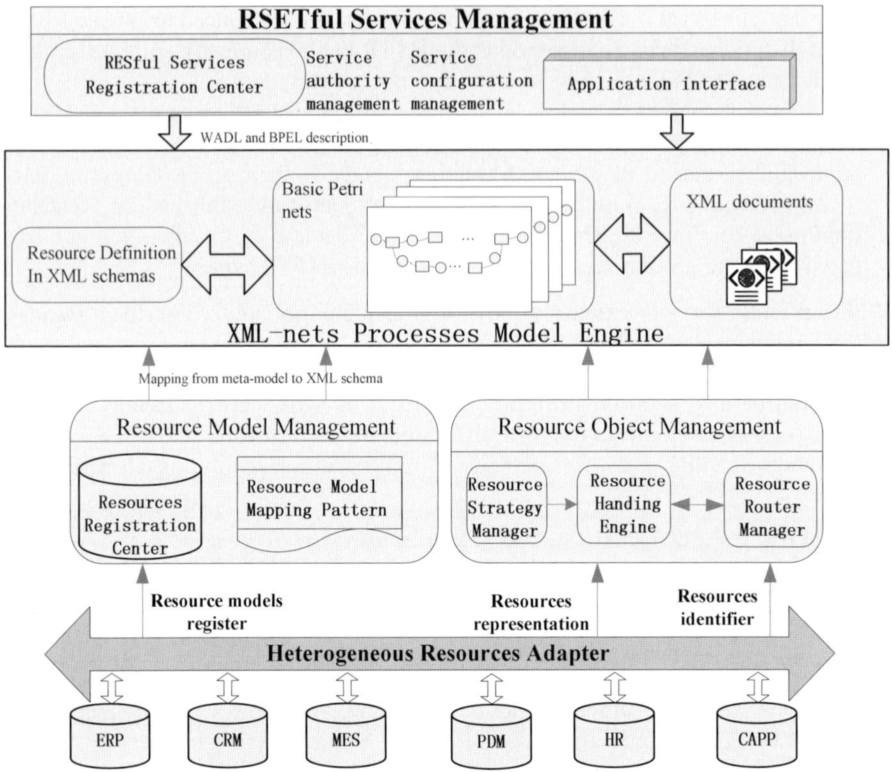

Fig. 5. The software architecture of the resource-based integration platform

paper. Several participants are involved in this scenario: at the beginning, a customer sends a purchase order request to manufacturer. The manufacturer receives the purchase order, checks inventory and organizes productions, including creating MRP (Material Request Planning). To prepare for production, the manufacturer authorizes a raw material supplier and sends raw materials order to the supplier. The material supplier deliveries raw materials to manufacturer and updates the state of his stock. The manufacturer checks the state of raw materials, when its state becomes to be "arrive", the product will be manufactured and then be delivered to the customer by the deliver. The corresponsive invoice is also created and sent to the customer.

Based on the process description above, there are 7 resources involved in the process. The Resource set is:

$ResS_t$ = {Purchase Order, Inventory, Material Order, Stock, Raw Material, Product, Invoice}

Each resource should be further defined based on resource meta-model using Definition 1, 2 und 3. For example, the meta-model of resource Purchase Order can be depicted as:

$R_{PURCHASEORDER} = \langle R_{URI}, OS_{PURCHASEORDER}, PS_{PURCHASEORDER}, AS_{PURCHASEORDER}, RS_{PURCHASEORDER} \rangle$

Which R_{URI} is the URI of resource Purchase Order, $OS_{PURCHASEORDER}$ is the operations set, here $OS_{PURCHASEORDER} = \langle OT_{CREATE}, OT_{UPDATE}, OT_{GET}, OT_{POST}, OT_{DELETE} \rangle$. To adapt to RESTful services environment, the operations of a resource degenerate to simplified HTTP verbs.

$PS_{PURCHASEORDER}$ describe the properties set of Purchase Order. In this example, it includes basic information of customer, product and vender.

$AS_{PURCHASEORDER}$ defines activity states of Purchase Order, in this example, $AS_{PURCHASEORDER} = \langle s_{START}, s_{CREATE}, s_{VERIFIED}, s_{FULFILED}, s_{END} \rangle$

$RS_{PURCHASEORDER}$ defines the rules set of resource states transferring. In this example $RS_{PURCHASEORDER} = \langle\langle s_{START}, s_{CREATE} \rangle, \langle s_{CREATE}, s_{VERIFIED} \rangle, \langle s_{FULFILED}, s_{END} \rangle\rangle$

From meta-model, the corresponsive XML schema is derived. In XML-net, XML schemas are represented in a graphical XML schema definition language (GXSL) [5]. Details about GXSL and its utility in XML-net could be found in [5] [6]. For example, the schema of resource Purchase Order in GXSL could be depicted as Fig.6:

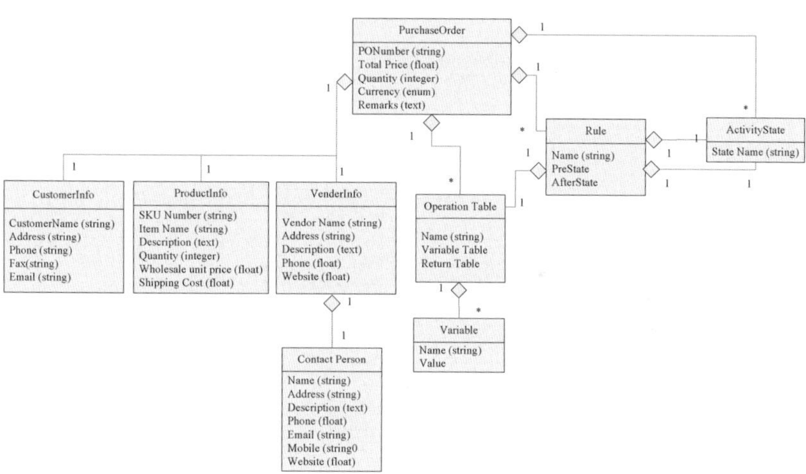

Fig. 6. The schema of resource Purchase Order in GXSL

XML documents in XML-net contain states of resources and are generated according to their corresponding XML schemas. Illustratively, a XML document describing verified Purchase Order is shown as Fig.7:

```
<?xml version="1.0" encoding="UTF-8" ?>
- <PurchaseOrder xmlns:xs="http://www.w3.org/2001/XMLSchema" xmlı
    <PONumber>2009047254744711</PONumber>
    <TotalPrice>302,452</TotalPrice>
    <ActivityState>Verified</ActivityState>
    <Quantity>240</Quantity>
    <Currency>Euro</Currency>
    <Remarks />
  - <CustomerInfo>
      <Customername>Autoankauf.Co</Customername>
      <Address>Feldberg str.22a Mannheim</Address>
      <Phone>+49 621 3547134</Phone>
      <Fax>+49 621 3547135</Fax>
      <EMail />
    </CustomerInfo>
  - <ProductInfo>
      <SKUNumber>223443562347634</SKUNumber>
      <Itemname>Zahnriehmen</Itemname>
      <Description>VW Zahnriehmen</Description>
      <Quantity>240</Quantity>
      <Wholesaleunitprice>1000</Wholesaleunitprice>
      <Shippingcost>62,452</Shippingcost>
    </ProductInfo>
  + <VenderInfo>
  - <OperationTable name="create">
      <Variable name="input" />
      <Variable name="input2" />
      <ReturnValue name="output" />
    </OperationTable>
  - <OperationTable name="get">
      <Variable name="input" />
      <Variable name="input2" />
      <ReturnValue name="output" />
    </OperationTable>
  + <OperationTable name="post">
  + <OperationTable name="update">
  + <OperationTable name="delete">
  + <Rule>
  </PurchaseOrder>
```


Fig. 7. XML documents describing verified Purchase Order

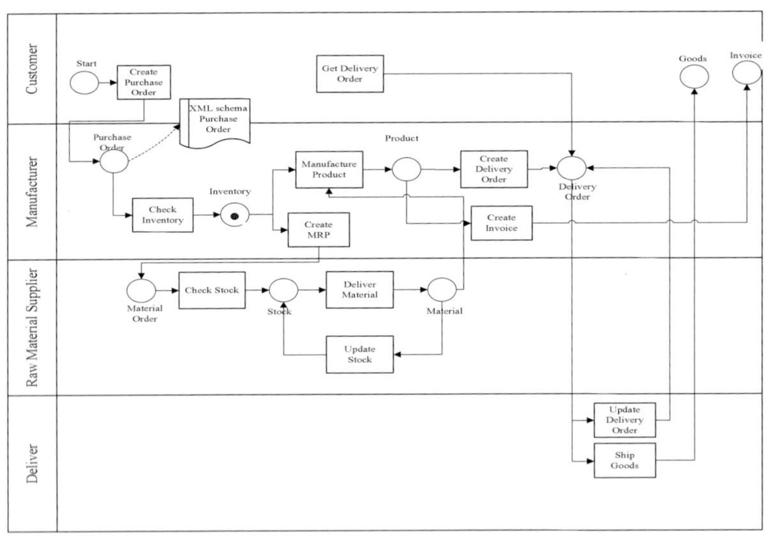

Fig. 8. Process model in XML-nets

The correlated places in XML-net is the set $P_{PURCHASEORDER}$=<$p_{PURCHASEORDER}$>, and the correlated transitions in XML-net is the set $T_{PURCHASEORDER}$ = <$t_{CREATEPURCHASEORDER}$, $t_{UPDATEPURCHASEORDER}$ >

Analogically, the other places, transitions and correlative XML schemas for resources could be derived. Hence, the complete XML-net which describes this supply chain process is shown as Fig.8.

From process model in XML-net, RESTful service descriptions in WADL and choreograph description in BPEL extension are deduced. Fig.9 shows the fragment of service description related to resource Purchase Order.

```
- <application xmlns:xsi="http://wss.w3.org/2001/XMLSchema-instance" xsi:schemaLocati
  - <resources base="http://example.org/customer">
    - <resource path="PurchaseOrder">
      - <methode name="CREATE">
        - <request>
            <param name="customerid" type="xsd:string" required="true" />
            <param mediaType="application/xml" element="order" />
          </request>
        - <response>
            <representation mediaType="application/xml" element="order" />
          </response>
        </methode>
      - <methode name="GET">
        - <request>
            <param name="porderid" type="xsd:string" required="true" />
          </request>
        + <response>
        </methode>
      - <methode name="UPDATE">
        + <request>
        + <response>
        </methode>
      + <methode name="DELETE">
      </resource>
    </resources>
  </application>
```

Fig. 9. Service description in WADL

```
  <?xml version="1.0" encoding="UTF-8" ?>
- <process name="replenishPrococess">
  - <resource uri="PurchaseOrder">
      <variable name="inventory" />
      <variable name="invoice" />
      <variable name="products" />
    - <onPut>
        <post uri="inventory/materialInventory" response="inventory" />
      - <if>
          <condition>$inventory.amount</condition>
        - <scope name="requestRawmaterial">
          - <resource name="MaterialOrder">
              <POST uri="http://www.supplier.com/MaterialOrder" />
              <resource uri="Stock" />
            - <flow>
                <PUT uri="http://www.supplier/stock" response="stock" />
                <PUT uri="http://www.manufacture.com/rawmaterial" />
              </flow>
            </resource>
          </scope>
        </if>
      + <scope name="production">
      + <flow name="sendGoods">
      </onPut>
    </resource>
  </process>
```

Fig. 10. Fragment of choreograph description in BPEL

With the extension of BPEL, process implemented with RESTful web services could be also described appropriately. Fig.10 shows the fragment of extended BPEL code that composites RESTful services into process.

5 Discussion

Compared with other approaches, which integrate business processes in resource-oriented architectures, our approach has the following features.

Firstly, resource meta-model provides a unified information framework to describe resource. By introducing resource adapter, resources from heterogeneous environments can be represented and managed in a uniform formats, which is the basis for resources manipulation across different systems.

Secondly, XML-net is employed both for representing data flow and modeling control flow. XML schemas derived from resource meta-model provide a unified view for resources with diversiform sources. XML documents encapsulate resources states, thereby the messages are self-descriptive and using HTTP protocol could be a standard protocol for data exchanging.

Last but not least, RESTful service description in WADL and services orchestration description can be automatically deduced from process model in XML-net. Hence, the heterogeneous systems are encapsulated in RESTful service, which makes it possible to integrate process cross organizations and systems.

6 Related Work

A high-level Petri-net, so-called XML-net, is used for integration and deployment supply chain process in a RESTful service environment. The development of XML nets has been done in [5]. Here, XML documents carry information and are transferred between transitions. The work of employing XML nets for cross organizational workflow modeling could be found in [6]. Similarly, in [12], XML net is used for process management in supply chains, but it lacks of consideration on IT implementations.

XML nets could be also used for Web services description and composition. In [11], Koschmider developed a XML nets based approach for deduction of Web service description in WSDL and orchestration description in BPEL. Nevertheless, these works are based on traditional WfMS (Workflow Management System) and towards SOA architecture.

Recently, resource-oriented architecture and RESTful Web service have received increasing attention for business process modeling and integration. A few papers that use REST principles to support the business process modeling and execution have been published [14] [15] [16]. Business process is then viewed as the evolution of the states of a collection of business entities. However, the unified resource representation is still missing.

To adapt to RESTful services environment, a few specifications for services description and composition appear, e.g. WADL [10] and WSDL 2.0 can be used to describe web application and also to describe RESTful services. Details about the comparison between WADL and WSDL 2.0 HTTP binding extension can be found in [7].

In [9] BPEL is extended to support the composition of RESTful services natively. The extension is achieved in the following two aspects. Firstly, a process can be published as resources dynamically by adding a new <resource> element. Secondly, the RESTful services with BPEL can be invoked by adding four activities according to the standard HTTP methods.

However, in these works, there is still lack of unified representation and management of resource from various service providers. Service description and orchestration that could be generally accepted by supply chain participants are also in absence. Such issues prevent the communication and coordination among supply chain participants. Hence further across-organizational integration of supply chain process could not be effectively achieved. It is therefore necessary to describe resources in heterogeneous systems formally and encapsulate of services so that the across-organizational integration of supply chain process becomes possible.

7 Conclusion

In this paper we propose a XML-net based approach for enactment of supply chain process in RESTful service environment. A resource meta-model is constructed to describe resources from various sources. XML-net is employed for resource representation, process modeling and generation of RESTful services description. Our approach is highlighted by its integration of unified representation for resources from heterogeneous systems. And by employing XML nets, both process modeling and the generation of RESTful services description can be enhanced, which facilitates the enactment of cross-organizational collaborative business process.

Acknowledgment

This paper is supported by the National High Technology Research and Development Program of China ("863" Program) under No.2008AA04Z126, the National Natural Science Foundation of China under Grant No.60603080, No.70871078, and the Aviation Science Fund of China under Grant No.2007ZG57012.

References

1. Cherbakov, L., Galambos, G., Harishankar, R., Kalyana, S., Rackham, G.: Impact of service orientation at the business level. IBM Systems Journal 44(4), 653–668 (2005)
2. Fielding, R.T.: Architecture Styles and the Design of Network-based Software Architectures. Doctoral dissertation, University of Califonia, Irvine (2000)
3. Kumaran, S., Liu, R., Dhoolia, P., Heath, T., Nandi, P., Pinel, F.: A RESTful Architecture for Service-Oriented Business Process Execution. In: Proceedings of IEEE International Conference on e-Business Engineering 2008, pp. 197–204 (2008)
4. Lenz, K., Mevius, M., Oberweis, A.: Process-oriented Business Performance Management with Petri Nets. In: Proceedings of the 2nd IEEE Conference on e-Technology, e-Commerce and e-Services, Hong Kong, pp. 89–92 (2005)

5. Lenz, K., Oberweis, A.: Modeling Interorganizational Workflows with XML Nets. In: Proceedings of the 34th Hawaii International Conference on System Sciences (2001)
6. Lenz, K., Oberweis, A.: Inter-organizational Business Process Management with XML Nets. In: Ehrig, H., Reisig, W., Rozenberg, G., Weber, H. (eds.) Advances in Petri Nets. LNCS, vol. 2472, pp. 243–263. Springer, Heidelberg (2003)
7. Takase, T., Makino, S., Kawanaka, S., Ueno, K., Ferris, C., Ryman, A.: Definition Languages for RESTful Web Services: WADL vs. WSDL 2.0. IBM Reasearch (2008)
8. Web Service Business Process Execution Language 2.0, OASIS, `http://docs.oasis-open.org/wsbpel/2.0/OS/wsbpel-v2.0-OS.pdf`
9. Pautasso, C.: BPEL for REST. In: Dumas, M., Reichert, M., Shan, M.-C. (eds.) BPM 2008. LNCS, vol. 5240, pp. 278–293. Springer, Heidelberg (2008)
10. Hadley, M.J.: Web Application Description Language (WADL). Sun Microsystems Inc. (2006), `https://wadl.dev.java.net/wadl20061109.pdf`
11. Koschmider, A., Mevius, M.: A Petri Net Based Approach for Process Model Driven Deduction of BPEL Code. In: Meersman, R., Tari, Z., Herrero, P. (eds.) OTM-WS 2005. LNCS, vol. 3762, pp. 495–505. Springer, Heidelberg (2005)
12. Mevius, M., Pibernik, P.: Process Management in Supply Chains – A New Petri-Net Based Approach. In: Procedings of the 37th Hawaii International Conference on System Sciences (2004)
13. Richardson, L., Ruby, S.: RESTful Web Services. O'Reilly, Cambridge (2007)
14. Xu, X., Zhu, L., Liu, Y., Staples, M.: Resource-Oriented Architecture for Business Processes. In: 15th Asia-Pacific Software Engineering Conference 2008, pp. 395–402 (2008)
15. Kumaran, S., Liu, R., Dhoolia, P., Heath, T., Nandi, P., Pinel, F.: A RESTful Architecture for Service-Oriented Business Process Execution. In: Proceedings of IEEE International Conference on e-Business Engineering 2008, pp. 197–204 (2008)
16. Muehlen, M., Nickerson, J.V., Swenson, K.D.: Developing web services choreography standards: the case of REST vs. SOAP. Decision Support Systems 40(1), 9–29 (2005)

Anonymity and Censorship Resistance in Unstructured Overlay Networks*

Michael Backes[4], Marek Hamerlik[1], Alessandro Linari[2], Matteo Maffei[1], Christos Tryfonopoulos[3], and Gerhard Weikum[3]

[1] Saarland University
{maffei,mhamerli}@cs.uni-sb.de
[2] Oxford Brookes University & Nominet UK
alessandro@nominet.org.uk
[3] Max Planck Institut für Informatik
{trifon,weikum}@mpi-inf.mpg.de
[4] Saarland University & MPI-SWS
backes@cs.uni-sb.de

Abstract. This paper presents Clouds, a peer-to-peer protocol that guarantees both anonymity and censorship resistance in semantic overlay networks. The design of such a protocol needs to meet a number of challenging goals: enabling the exchange of encrypted messages without assuming previously shared secrets, avoiding centralised infrastructures, like trusted servers or gateways, and guaranteeing efficiency without establishing direct connections between peers. Anonymity is achieved by cloaking the identity of protocol participants behind groups of semantically close peers. Censorship resistance is guaranteed by a cryptographic protocol securing the anonymous communication between the querying peer and the resource provider. Although we instantiate our technique on semantic overlay networks to exploit their retrieval capabilities, our framework is general and can be applied to any unstructured overlay network. Experimental results demonstrate the security properties of Clouds under different attacks and show the message overhead and retrieval effectiveness of the protocol.

1 Introduction

Over the last years unstructured overlays have evolved as a natural decentralised way to share data and services among a network of loosely connected peers. The popularity of systems like Gnutella and Freenet [5], has propelled research in this field, while lately the proliferation of social networking has added another interesting dimension to the problem of searching for content in such networks. In unstructured overlays, peers typically connect to a small set of other peers, and queries are propagated along connections in the overlay network, using some query forwarding strategy that aims at finding peers with resources matching the issued query.

* Work partially supported by the initiative for excellence and by Emmy Noether program of the German federal government and by Miur project SOFT: "*Security Oriented Formal Techniques*".

R. Meersman, T. Dillon, P. Herrero (Eds.): OTM 2009, Part I, LNCS 5870, pp. 147–164, 2009.

Semantic Overlay Networks (SONs) are an instance of unstructured networks that has lately been given considerable attention [6,18,2,20]. In a SON, peers that are semantically, thematically, or socially close (i.e., peers sharing similar interests or resources) are organised into groups to exploit similarities at query time. This flexible organisation improves query performance while maintaining high peer autonomy, and has proved a useful technology not only for distributed Information Retrieval (IR) applications, but also as a natural distributed alternative to Web 2.0 application domains such as decentralised social networking in the spirit of Flickr or del.icio.us. Query processing is achieved by identifying which region in the network is better suited to answer the query and routing the query towards a peer in that region. This peer is then responsible for forwarding the query to his neighbours in the region.

In such an information exchange, not all information providers are willing to reveal their true identity: for instance, publishers may want to present their opinions anonymously to avoid associations with their race, ethnic background or other sensitive characteristics. Furthermore, people seeking for sensitive information may want to remain anonymous so as to avoid being stigmatised or even to avoid physical, financial or social detriment by suppressors. The freedom of information exchange is another important issue that got increasing attention in the last years. Some organisations, such as governments or private companies, may regard a discussion topic or a report as inconvenient or even harmful. They may thus try to censor the exchange of undesired information by either suppressing resource providers or, if these are protected by anonymity, by taking control of strategic regions of the network, such as gateways and proxies, in order to filter the communication.

Despite the importance of SONs as a building block of data management and social networking applications, security issues have not been investigated in this setting. To the best of our knowledge, there exist no studies that try to solve the anonymity and censorship resistance problems arising in SONs. This, together with the observation that SONs are actually vulnerable to a number of different attacks, varying from surrounding to man-in-the-middle attacks, shows the importance of enforcing security in such an environment.

In this work we present *Clouds*, a P2P search infrastructure for providing *anonymous* and *censorship resistant* search functionality in a SON. We exploit the inherently high connectivity among similar peers for guaranteeing anonymity by relying on a self-organisation of peers into groups that we call *clouds*. Message routing is modified to take place among clouds instead of peers, thus hiding the identity of both the resource provider and the querying peer, while cloud size is a tunable parameter that affects anonymity and efficiency. Censorship resistance at communication level is achieved by a cryptographic protocol that guarantees the secrecy of the resource, thus avoiding censorship based on the inspection of the messages circulating in the network. This protocol achieves a number of challenging goals: enabling the exchange of encrypted messages without assuming previously shared secrets, avoiding centralised infrastructures, such as trusted servers or gateways, and guaranteeing efficiency without establishing direct connections between peers. The contribution of this paper is twofold:

- We present the first system to guarantee anonymity *and* censorship resistance in SONs. Although we instantiate our technique on SONs to leverage their retrieval capabilities and to support a rich data model and query language, our framework can be applied to any type of unstructured overlay network.
- We demonstrate the effectiveness of our architecture by analysing the anonymity and censorship-resistance properties provided by Clouds under different attack scenarios, namely surrounding, intersection, man-in-the-middle, and blocking attacks.

The rest of the paper is organised as follows. Section 2 discusses our data model and query language, and outlines the SON paradigm. Section 3 presents Clouds and its associated protocols. Section 4 introduces the attack scenarios addressed in this paper. Our experimental evaluation is given in Section 5, and related work is discussed in Section 6. Finally, Section 7 concludes the paper and gives directions for future research.

2 Background Information

This section outlines our data model and query language, and describes the construction and properties of SONs. For more detail, we refer the interested reader to [6,18,2,20].

2.1 Data Model and Query Language

We utilise the *Vector Space Model (VSM)* to represent documents, queries, peer and cloud descriptions. In our setting a *resource* is any piece of information that can be described by a set of *keywords*, such as a text document which is characterized by its terms, an image, which is often associated to a set of tags, or an mp3 file. We associate a *weight* to each keyword so as to represent the importance of the keyword as a description for the given resource. A query is a set of keywords, for which the weights are explicitly assigned by the user or implicitly by the system (e.g., through relevance feedback techniques [31]). A resource is characterized by a *resource description* r_i, that is a vector containing keywords and weights for these keywords. Similarly, the *profile* of a peer P, $profile(P)$, is computed using the descriptions $r_1,\ldots,r_m$ of the resources stored at this peer.

A standard technique to decide which resource description r best matches a given query q, is to utilise a *similarity function* $sim(q,r)$, which assigns a numerical score to each pair (q,r). The scores corresponding to different resource descriptions are then compared to derive a relevance ranking with respect to query q. A common similarity measure in IR is the cosine of the angle formed by the vector representations of q and r. Notice that, in practice, any function that models the similarity between a resource and a query can be used. For example, in the case of a social network the similarity function could also contain a social component, that also considers the strength of relations among users.

2.2 Semantic Overlay Networks

In a SON each peer P performs a variable number of *random meetings* with other peers in the network, during which they exchange their profiles and compute their similarity. Based on the similarity function $sim()$, a peer P establishes two types of links:

- *Short-distance links* towards the k most similar peers in the network discovered through the random meeting process. The number of short-distance links k is usually small (e.g., $O(\log N)$, where N is the number of peers in the network).
- *Long-distance links* towards k' (typically $k' < k$) peers chosen at random from the rest of the network.

To maintain short-distance links up-to-date and ensure the clustering property of the SON, a periodic *rewiring procedure* [6,18,2,20] is executed by all peers, aiming at discovering new more similar peers, or refreshing links that have become outdated due to network dynamics. Long-distance links, usually updated using random walks, are necessary to avoid creating tightly clustered groups of peers that are disconnected from the rest of the network. Query answering in SONs benefits from the fact that peers containing related information are directly linked or at a short-hop distance from each other. Thus, the task of finding a peer that can answer a query q reduces to locating the appropriate cluster of peers. Once a peer in the appropriate cluster is reached (i.e., if $sim(profile(P),q)) \geq \beta$, where β is called *broadcast threshold*), the query goes through a limited broadcast using short-distance links aiming at reaching all neighbours of P. Due to the SON properties, these peers are able to answer q with high probability.

3 The Clouds Protocols

This section describes in detail the protocols that regulate the interactions between peers and allow them to anonymously share and retrieve resources available in the network.

3.1 Protocol Overview

The key principle behind the anonymity mechanism of Clouds is to cloak both the querying peer and the resource provider behind a group of neighbouring peers, called cloud. Peers generate clouds at random, without necessarily using them, to minimise the correlation between the events of joining and using a cloud. Additionally, they non-deterministically decide to participate or not in clouds created by other peers. Clouds are created using the short-distance links of peers and are thus populated by peers in the neighbourhood of the cloud initiator. Communication takes place between clouds, and all peers in a cloud share the same probability of being involved in any communication which has this cloud as the start- or end-point (also known as *k-anonymity* [28]). To avoid correlation of roles in the protocol with specific actions, which would compromise anonymity, the protocol is designed so that the observable behaviour is the same for all peers, regardless of them being initiators of forwarders of a message.

The proposed cryptographic protocol aims at addressing the problem of censorship at the communication level, where a malicious party aims at filtering out any communication that contains unwanted content (either a query or a resource). The secrecy of the resource is protected by cryptography, making it hard for the attacker to censor the communication based on an inspection of the message content. The design of such a protocol is conceptually challenging since we do not assume previously shared secrets or centralised infrastructures.

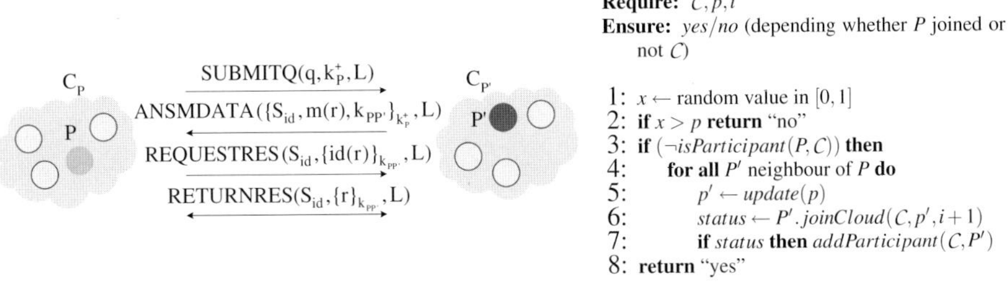

Fig. 1. Communication protocol and algorithm *joinCloud*(C, p, i)

The protocol is composed of four steps summarised in Figure 1. A querying peer chooses a cloud it participates in to issue a query. The query follows a random walk in the cloud to obscure the message initiator, leaves the cloud from multiple peers to ensure higher resistance to censorship and is routed towards a region in the network that possibly contains matching resources. A *footprint list* is used to collect the list of traversed clouds and facilitates the routing of the subsequent messages. A responder to the query encrypts the answer with a public key received with the query message and routes it towards the cloud of the querying peer, as specified in the footprint list. All subsequent messages between the querying peer and the responder will have a cloud as a destination and, when this cloud is reached, the message will be broadcasted to reach the intended recipient. Finally, notice that the query message does not contain any session identifier which would connect the query to the subsequent messages. In the last two protocol steps, however, a session identifier is used to avoid costly decryption checks, since the message content is encrypted.

3.2 Cloud Creation

The design of a cloud creation algorithm should satisfy some fundamental properties, such as randomness, tunability and locality, in order to reveal as little information as possible to potential attackers while maintaining the useful clustering properties of the underlying SON. According to our cloud creation algorithm, when a peer P generates a new cloud C, it selects the participants among its neighbours utilising short-distance links. This guarantees that clouds are populated by semantically close peers (cloud *locality*). As shown in Section 4.1, this property is crucial to guarantee that cloud intersections have high cardinality, thus preventing *intersection attacks* that aim at breaking anonymity. Peers that join a cloud in turn select other neighbours, and the protocol is executed in a recursive way with decreasing probability to join C.

The *joinCloud*() algorithm shown in Figure 1 shows this procedure from P's point of view, assuming that P has received a $joinCloud(C, p, i)$ message that may have been generated either by itself or by any other peer. This message specifies the cloud C, the probability p to join it, and the step i. With probability $1 - p$, P replies negatively (line 2) to the request, otherwise it accepts to join C (line 8). In this case, if P is not already in C, it triggers a recursive procedure in its neighbours (line 3) by sending a $joinCloud(C, p', i+1)$ message to each of them (lines 4-7) with join probability p' (line 5). Finally, the peers joining C are marked by P as neighbours in C (lines 6-7).

SUBMITQ

Require: SUBMITQ$(q, k^+, L), p_{rw}, \beta$

```
if L = [C] for some C then
    x ← random value in [0, 1]
    if x ≤ p_rw then
        forward SUBMITQ(q, k+, L) to one neighbor in C
    set TTL for SUBMITQ(q, k+, L)
if TTL ≥ 0 then
    if clouds(P) ∩ L = ∅ then
        select C_P ∈ clouds(P) with maximum sim(q, C_P)
        L ← L :: C_P
    if sim(q, profile(P)) ≤ β then
        forward SUBMITQ(q, k+, L) along a random subset of long-distance links.
    else
        forward SUBMITQ(q, k+, L) to all short-distance links.
```

ANSMDATA

Require: ANSMDATA$(\{S_{id}, m(r), k\}_{k+}, [C_1, \dots, C_n])$, peer P

```
if C_1 ∈ clouds(P) then
    if decryption of {S_id, m(r), k}_k+ succeeds then
        process m(r)
    forward ANSMDATA({S_id, m(r), k}_k+, [C_1, ..., C_n]) to all peers in C_1
else
    scan [C_1, ..., C_n] and find the left-most cloud C_i such that P has a neighbour in C_i
    forward ANSMDATA({S_id, m(r), k}_k+, [C_1, ..., C_n]) to a subset of the peers in C_i
```

REQUESTRES and RETURNRES

The last two messages are routed using the footprint list in the same way as ANSMDATA

Fig. 2. Behavior of peer P in the different protocol steps

The *update*(p) function used to obtain p' is the means to control cloud population and to offer tunability between anonymity and efficiency. For simplicity, in the experiments we consider the same function throughout the network, but in practice each peer may use its own *update*() function. Notice that peers only know the neighbours certainly belonging to their clouds (i.e., the neighbours that have sent or have positively answered to a *joinCloud* request). No information can be derived from a negative answer, since a peer already belonging to the cloud answers negatively with probability $1 - p$. This helps to avoid statistical attacks on cloud membership.

3.3 Query Routing

In this section, we present the routing algorithm used to route a query q from a peer P to a resource provider P'. Figure 2 gives the pseudocode for the query routing procedure followed by any peer P. Notice that the protocol is the same regardless of P being the initiator or the forwarder of the message, in order to avoid breaches in anonymity.

When P wants to issue a query q, it constructs a message $msg = \text{SUBMITQ}(q, k_P^+, L = [C_P])$, where k_P^+ is a public key generated by P especially for this session, and L is the *footprint list* that will be used to collect the list of clouds msg will traverse during the routing. P initialises this list with one of its clouds. The collection of clouds in the footprint list is performed as follows. A peer P_i that receives msg checks whether one of the clouds it participates in is already listed in L. If not, it chooses a cloud C_{P_i} and appends it to L (i.e., $L \leftarrow L :: C_{P_i}$). This information will be exploited by the successive phases of the communication protocol to optimise routing between P and the resource provider.

The routing algorithm for a SUBMITQ message is reported in Figure 2. The algorithm consists of two steps: an *intra-cloud* routing, during which the message msg performs a random walk in C, and an *inter-cloud* routing, during which msg is delivered to a peer P' not participating in C. Each peer receiving (or creating) msg, forwards it to a random peer participating in C with probability p_{rw} and also to a peer that subsequently enters the inter-cloud phase. The intra-cloud routing phase is necessary to avoid revealing the identity of the query initiator to a malicious long-distance neighbour that exploits the existence of a single cloud C in L. By adding the random walk phase within C, an attacker cannot know whether P is the initiator of *msg* or simply a forwarder entering into the inter-cloud phase.

Inter-cloud routing is based on the fireworks query routing algorithm [17]. A peer P receiving message *msg* computes the similarity $sim(q,profile(P))$ between q and its profile. If $sim(q,profile(P)) \leq \beta$, where β is the broadcast threshold, this means that neither P, nor P's neighbours are suitable to answer q. Thus, *msg* is forwarded to a (small) subset of P's long-distance links. If $sim(q,profile(P)) > \beta$, the query has reached a neighbourhood of peers that are likely to have relevant resources and *msg* goes through a limited broadcast using short-distance links.

In order to limit the network traffic, each message entering the inter-cloud phase is associated to a time-to-live (TTL), which is updated at every hop. Message forwarding is stopped when TTL reaches zero. Finally, all peers maintain a message history and use it to discard already processed messages.

3.4 Answer Collection

When a peer P' receives the message $\text{SUBMITQ}(q,k_P^+,L)$, it searches its local collection and retrieves the list $R = \{r_1, r_2, \ldots, r_n\}$ of resources matching q. Then P' constructs the reply message $msg = \text{ANSMDATA}(\{S_{id}, M, k_{PP'}\}_{k_P^+}, L)$, i.e., it encrypts with k_P^+ the list $M = \{m(r_1), m(r_2), \ldots, m(r_n)\}$ of metadata for the local result list R, a unique session identifier S_{id}, and a symmetric key $k_{PP'}$. S_{id} will be used by P' in the subsequent protocol steps to identify *msg* as an open transaction, while $k_{PP'}$ will be used to encrypt the remaining messages between P and P' and to avoid computationally expensive public key cryptography.

The routing algorithm for *msg* is based on the footprint list L, as described in Figure 2. A peer receiving *msg* and not belonging to the destination cloud C_P, forwards *msg* to a (small) subset of its neighbours that participate in the left-most cloud of L.

Finally, when a peer P participating in the destination cloud C_P receives *msg*, it broadcasts it to all its neighbours in C_P. Subsequently, it tries to decrypt the message using its session private key k_P^-, to discover if it is the intended recipient of *msg*. During the broadcast in C_P, the message history of each peer is used to discard already processed messages. Note that even the intended recipient P forwards *msg* in C_P, in order to avoid detection by malicious neighbours. After a predefined timeout or a large enough answer set, P chooses the resources to be retrieved and enters the last two phases of the protocol with the peers responsible for them.

Assume that P is interested in the resource corresponding to the metadata $m(r)$ and stored at P'. It creates the message $\text{REQUESTRES}(S_{id}, \{id(r)\}_{k_{PP'}}, L)$, where the footprint list L, the session identifier S_{id}, and the symmetric key $k_{PP'}$ have been delivered with ANSMDATA. Here and in the remainder of the protocol, S_{id} can be used in clear, since it cannot be associated with any of the previous transactions. Its usage reduces the amount of data that needs to be encrypted/decrypted at each step and allows the peers in the destination cloud to quickly discard messages that are not addressed to them.

Routing of REQUESTRES and RETURNRES messages is the same as ANSMDATA; it utilises the footprint list L to reach the destination cloud, and then a cloud broadcast to reach the intended recipient.

Note that the cloud-based communication protocol is largely independent on the underlying network: the only connection is given by the strategy utilised to route the query, while the rest of the messages is routed according to the footprint list. This makes it possible to apply our framework to any kind of unstructured overlay network, choosing an arbitrary routing strategy for the SUBMITQ message.

4 Attack Scenarios

In this section, we introduce the attacks that might in principle break anonymity and censorship-resistance in our framework. We qualitatively reason about the resistance of our framework against such attacks, referring to Section 5 for experimental evaluations.

4.1 Attacks on Anonymity

Surrounding Attack. Assume that a malicious peer P_{Adv} generates a fake cloud C_{Adv} and sends a message $JoinCloud(C_{Adv}, p, i)$ to P, and assume that P accepts to join cloud C_{Adv}. If i is the last step of the join algorithm, P does not have any other neighbours in C_{Adv}. The anonymity of P is thus compromised since P_{Adv} can monitor all P's activities in C_{Adv}. This attack can be generalized by considering a population of colluding malicious peers trying to surround P. They block all the *JoinCloud* messages received from honest peers and instead send P messages of the form $JoinCloud(C_{Adv}, p, i)$, where i is one of the last steps of the joining algorithm. Notice that the risk of incurring in surrounding not only depends on the topology of the network but also on its physical implementation. In a wireless scenario, for example, it is very unlikely that a malicious peer is able to gain exclusive control of the communication channel of another peer, which is constituted by its surrounding atmosphere.

To mitigate the threat of this attack, Section 5.1 presents an *update*() function that keeps the number of peers joining the cloud in the first steps small, letting the majority of the peers in the region join the cloud in the last steps. This guarantees that the number of peers joining the cloud after the invitation of the peer under attack is relatively high in the first steps of the joining algorithm, and this number represents the anonymity guarantee of P. In fact, each peer has a significant number of clouds which he joined in the first steps of the joining algorithm, and these clouds can be used for obtaining strong anonymity guarantees. The experimental evaluation in Section 5.2 shows that the surrounding attack is effective only if the adversary controls the majority (at least 50%) of the peers around the peer under attack.

Intersection Attack. Since peers participate in different clouds, a malicious peer might try to "guess" the identity of the querying peer based on the (even partial) information that it has available about the intersection of two clouds. Notice that computing cloud intersections is difficult because cloud topology is not known in general and a malicious peer only knows its neighbors.

Remember that clouds are generated using only short-distance links and that clouds are thus confined in a small region of the network (locality property). As confirmed by the experiments in Section 5.3, this guarantees that the intersection of the clouds that a peer participates in has high cardinality, demonstrating that our framework is resistant to intersection attacks.

4.2 Attacks on Censorship Resistance

Blocking Attack. The blocking attack aims at blocking (instead of monitoring) all SUBMITQ messages containing undesired queries and, tracking the cloud C of the querying peer, at subsequently blocking all ANSMDATA messages directed to C. Note that our query routing mechanism (see Section 3.3) guarantees that SUBMITQ is replicated and routed through different paths: this redundancy helps to overcome the blocking attack, since it requires the attackers to be located either in *all* the paths followed by SUBMITQ message or in one of the paths followed by SUBMITQ message and *all* the paths followed by ANSMDATA message. As confirmed by the experiments in Section 5.4, the blocking attack is effective only if the adversary has a pervasive control (at least 50%) of the region to surround.

In the following we present a theoretical characterisation of the resistance to blocking attacks. This is formalised as the *probability that a communication which an adversary is willing to censor will not be blocked.*

The probabilities that all SUBMITQ and all ANSMDATA messages are blocked is given by:

$$\Pr\{block(\text{SUBMITQ})\} = [\Pr\{A_1\}]^{p_1} \tag{1}$$

$$\Pr\{block(\text{ANSMDATA})\} = \Pr\{A_1\} \times [\Pr\{A_2\}]^{p_2} \tag{2}$$

where p_1 (resp. p_2) is the number of distinct paths followed by the first (resp. second) message and A_1 (resp. A_2) denotes the event that *at least one attacker resides in one of the paths of* SUBMITQ *(resp.* ANSMDATA*)*. For simplicity, we have assumed these events to be independent from each other and from the specific path followed by the messages. If the attackers are randomly distributed in the network, the probability associated to events A_1 and A_2 is:

$$\Pr\{A_1\} = \Pr\{A_2\} = 1 - \binom{N-k}{m} \Big/ \binom{N}{m} \tag{3}$$

where N is the number of peers in the network, k is the number of adversaries, and m is the average number of peers in a path connecting P to P' (where P and P' are not counted). The fraction in Equation 3 represents the number of *safe* (i.e., not including any attacker) paths divided by the total number of paths.

Finally, the degree of censorship resistance can be computed as

$$1 - \Pr\{block(\text{SUBMITQ}) \vee block(\text{ANSMDATA})\} \tag{4}$$

which can be obtained as the combination of the two non-independent events.

Man-in-the-middle Attack. In the man-in-the-middle attack (MITM), the attacker intercepts the SUBMITQ$(q, k_P^+, [C_P, \ldots])$ message sent by P and replaces it by a freshly generated message SUBMITQ$(q, k_{Adv}^+, [C_{Adv}])$, where k_{Adv}^+ and C_{Adv} are the adversary's public key and cloud, respectively. The adversary then runs two sessions of the protocol, one with the querying peer P and one with provider R, pretending to be the query responder and initiator respectively. This allows the attacker to filter and possibly interrupt the communication between P and R. Note that the attackers need to be located in *all* the paths followed by SUBMITQ, and hence the MITM is just a special case of the blocking attack.

5 Experimental Evaluation

In this section, we present our experimental evaluation, which is designed to demonstrate the anonymity and censorship resistance properties of our framework and the message overhead imposed by the protocol.

5.1 Experimental Setup

For our experiments, we used a subset of the OHSUMED medical corpus [16] with 32000 documents and 100 point queries. Documents were clustered using the incremental k-means algorithm and each cluster was assigned to a single peer to simulate different peer interests. This peer specialisation assumption is common in SONs [6,18,2], and does not restrict our setting since a peer with multiple interests might cluster its documents locally and create one identity for each interest in the spirit of [20]. These interests are then used to build a semantic overlay network according to standard techniques [6,18,2]. The resulting SON is composed of 2000 peers and the description of a peer's interest is represented by the centroid vector of its documents. Each peer maintains a routing index with 10 links to other peers, 20% of which are long-distance links. On top of the created SON we invoke the cloud creation process and implement the cryptographic cloud-based protocol presented in Section 3. The *update*() function used in our setting has values 0.25 for the first six steps of the cloud creation and 1.0 for the remaining 4 steps. Therefore the number of peers joining the cloud in the first steps is small, and the majority of the peers in the region joins the cloud in the last steps. As shown in Section 5.2, this choice of the *update*() function mitigates the threat of surrounding attacks. This *update*() function produces clouds with average size of about 70 peers, which is the baseline value for our experiments.

We have explored different values for the *update*() function, and the values considered in this section returned the best results in terms of anonymity and cloud locality. Due to space constraints, we do not further discuss other options.

Our simulator is written in Java, and our experimental results were averaged over 64 runs to eliminate fluctuations in the measurements. In the experiments presented below we consider the following system parameters that are varied appropriately to demonstrate Clouds' performance.

- κ: The average number of clouds created per peer. In our experiments the baseline value for κ is one, unless otherwise stated.
- μ: In our experimental setting, we consider colluding attackers that aim at attacking a region of the network (either for censoring or for breaking anonymity). Parameter μ represents the percentage of malicious peers in the region under attack, where the region size may vary depending on the attack type.
- β: The broadcast threshold is used to measure the similarity between a query and a peer profile, and assists a peer in deciding whether a query may be effectively answered by it and its neighbours. Decreasing β results in more broadcasts and thus higher message traffic and also higher recall. The baseline value used in our experiments is 0.6.

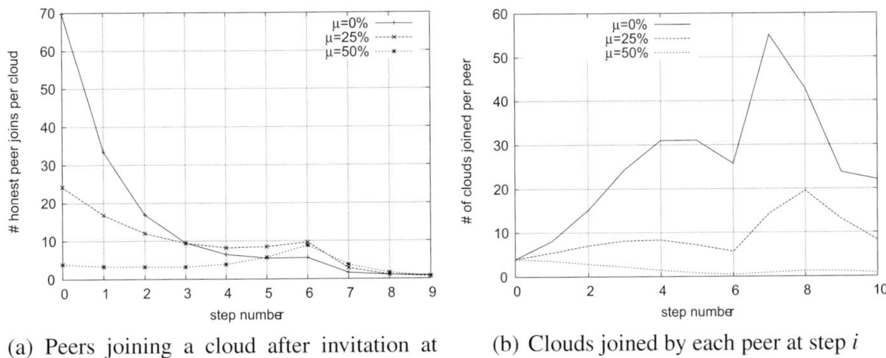

(a) Peers joining a cloud after invitation at step *i*

(b) Clouds joined by each peer at step *i*

Fig. 3. Surrounding attack when varying the percentage of malicious peers

5.2 Surrounding Attack

The experiments on the surrounding attack are conducted by selecting 100 peers with profiles closest to the peer under attack and, from this pool of peers, by selecting a percentage μ varying from 0% to 50% to be malicious. This surrounding scenario corresponds to the best strategy for the attackers to surround a peer, since in our system each peer randomly selects his neighbours among the peers with close profile. In other words, trying to surround a peer by positioning malicious peers as close as possible, so as to form a sort of barrier, is not the best strategy since that peer would not necessarily select them as neighbours. Notice that if the neighbours of a peer were selected to be the peers with the closest profiles, as in a typical SON, then the attackers might generate profiles extremely close to the peer under attack in order to occupy all its short-distance links and effectively isolate it from the rest of the network. We consider this architectural choice of paramount importance to enforce anonymity in SONs.

Figure 3(a) shows the average number of honest participants joining a cloud C after an invitation from peer P or from one of its descendants in C, depending on the step i in which P has joined C. This number represents the anonymity degree of P in C, assuming that the adversary knows i. In this experiment, malicious peers intercept and block all $JoinCloud(C, p', i')$ messages received from honest participants. The graph shows that under a scenario where the 25% of the peers in a region attack a single honest peer ($\mu = 25\%$), Clouds offers good anonymity guarantees, and at least 10 honest peers join the cloud after P in all the first six steps of the joining algorithm. Even under a scenario where $\mu = 50\%$, the peer under attack is not completely surrounded and in the first six steps a number between five and ten peers joins the cloud after P.

Observe that the number of peers joining the cloud decreases over the number of steps in the scenario without malicious peers, as expected, but tends to increase until the sixth step when half of the peers is set to be malicious. This positive and seemingly surprising result can be explained by the cloud shape induced by the *update* function. In the first phase of the joining algorithm (steps 0-5), the probability p that P's neighbours join the cloud is low and, since many of them are malicious, the blocking strategy of the adversary is effective and the number of peers joining the cloud after P is small. At the beginning of the second phase (steps 6-7), however, the joining probability p is high, the blocking strategy of the adversary is less effective, and so the number of P's neighbours joining the cloud is high. Finally, after the seventh step, most of the honest peers in the region have already joined cloud C, and so the number of peers joining after P in the last steps is very small. Notice that the numbers of peers joining in the last steps of the protocol is higher when under attack. This is easily explained since clouds are local and when all the peers in a region are honest, most of them have already joined the cloud in the first steps. Under attack, most of the honest peers in a region do not join the cloud in the first steps since malicious peers block most of the *JoinCloud* messages. In the last steps, however, the joining probability increases, the blocking activity of the attackers is less effective, and, since most of the honest peers have not joined the cloud yet, they do it in the last steps.

Figure 3(b) shows the number of clouds joined by each peer depending on the step number and the percentage of malicious peers. In this experiment, we consider a scenario where each peer generates four clouds on average. As we can see, even under attack, each peer has at disposal a number of clouds to join at each step, and in particular at the first steps which give in general better anonymity guarantees.

5.3 Intersection Attack

In Figure 4(a), we measured the average cardinality of all 2-wise and 3-wise intersections of the clouds which a peer participates in. Notice that the majority of the intersections has cardinality between 40 and 50 peers, and there exist very few cloud intersections with cardinality less than 30 peers. This result was expected due to the cloud creation process that exploits locality and creates clouds with high overlap, populated by semantically close peers. Figure 4(b) shows the probability for a peer to participate in clouds with 2-wise and 3-wise intersection cardinality lower than the value indicated in the x-axis. Notice that the probability to participate in clouds with intersection smaller than 40 is

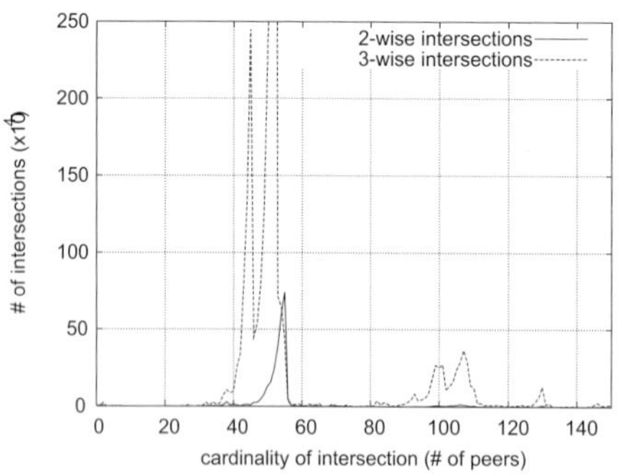

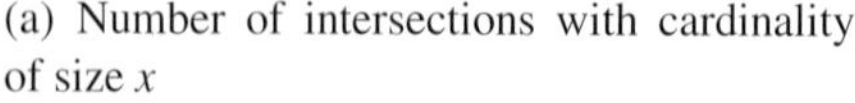
(a) Number of intersections with cardinality of size x

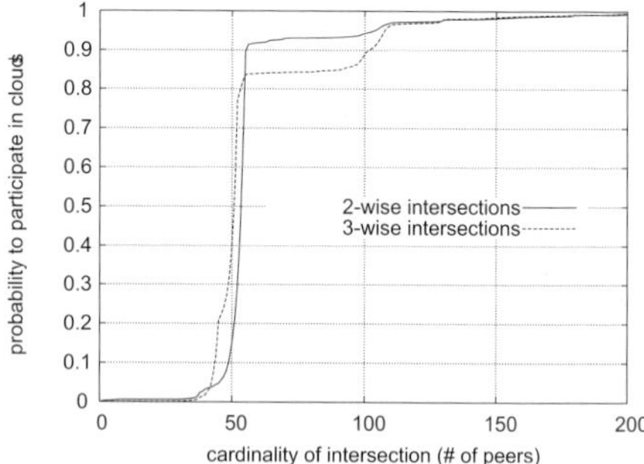

(b) Probability to participate in clouds with intersection cardinality lower than x

Fig. 4. Cardinality of intersections as a measure for intersection attacks negligible, and that this probability does not change when moving from 2-wise to 3-wise intersections, demonstrating the resistance of Clouds to intersection attacks.

5.4 Blocking Attack

This set of experiments shows the censorship resistance properties of our system under the strongest attack, i.e., the blocking attack. In this experiment we assume a set of colluding attackers that try to censor some specific topic; notice however that the attackers cannot have any precise information about which peer(s) store documents about this topic. Thus, the attackers aim at surrounding a region in the SON and blocking all non-qualifying information that is exchanged. To perform this attack we assumed a region of 200 peers that are the closest to the topic to be censored and varied the percentage μ of malicious peers in this region. Figure 5 shows the percentage of malicious peers in this region that is needed to censor a single topic. The most interesting observation emerging from this graph is that the attackers need to occupy at least 50% of the region (i.e., 100 peers or 5% of the network in our setting) to make the attack effective. Notice that this corresponds to censoring only a single topic, and an attacker should occupy a significant part of the network to mount an effective attack against multiple topics. Another interesting observation is that due to our fuzzy routing mechanism, some answers are not returned to the requesters even without malicious peers (leftmost point in the x-axis). Finally, observe that the number of clouds created per peer (κ) does not affect the censorship resistance properties of our approach. This is particularly interesting since, as stated before, anonymity and censorship resistance are often conflicting goals.

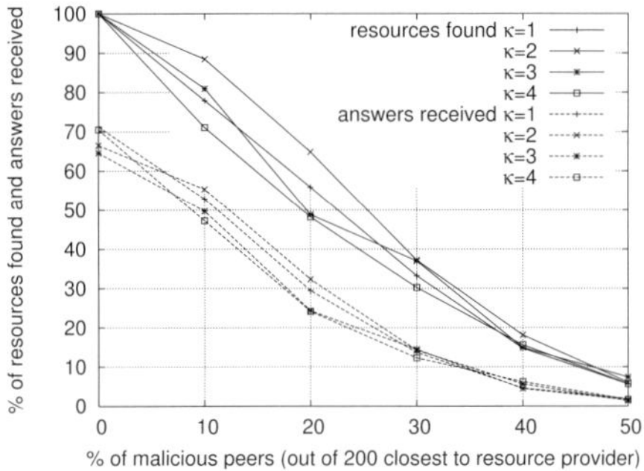

Fig. 5. Censorship resistance for colluding attackers surrounding a resource provider

5.5 Message Traffic

Figure 6(a) shows how the message traffic in Clouds is affected by the size and number of the clouds in the system. To control the former parameter, we modify the *update*() function described in Section 5.1, while for the latter we modify the average number of clouds created per peer through parameter κ. An important observation derived from this figure is that message traffic is highly sensitive to the cloud size. This was expected since our routing mechanism, that is based on the fireworks technique, depends

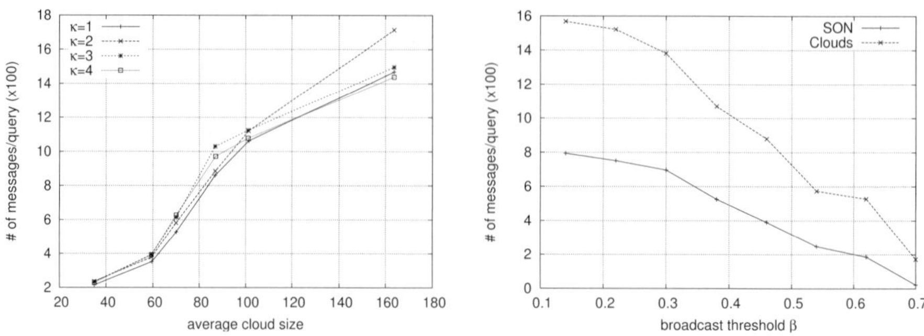

Fig. 6. Message overhead of Clouds

on message broadcasting inside clouds in all four phases of the protocol. Therefore, even a small increase in the average cloud size results in a significant increase in message traffic, since all cloud members have to receive all messages that have their cloud as a destination. Another interesting conclusion drawn from this experiment, is the insensitivity of message traffic to the number of clouds each peer creates. In fact, the increase in the number of clouds a peer participates in is dampened by the decrease of the probability to find one of these clouds in L (due to the increase in the domain of L). Finally, notice that the message traffic imposed on the network is heavily dominated by the broadcasting of messages within the clouds, rather than by the routing through them. This causes the routing of messages to contribute only for a small fraction to the observed traffic, thus resulting in a minimal effect in message overhead.

Figure 6(b) shows the overhead introduced by Clouds to achieve both anonymity and censorship resistance in a SON environment. The important observation in this graph is that both protocols are affected by the increase of the broadcast threshold in the same way, since both graphs present the same behaviour. Intuitively, increasing β results in less peers broadcasting the message to their neighbourhood, which eventually decreases retrieval effectiveness, but also heavily affects message traffic as also shown in the graph. The decrease in messages observed when increasing β is mainly due to less broadcasts taking place during a SUBMITQ message. The extra message traffic imposed by Clouds is mainly due to the broadcasting of messages inside clouds, while the main volume of traffic for the SON is the result of the resource discovery protocol. As already mentioned, this extra message traffic is the price we have to pay in order to avoid previous failure-prone and attack-prone solutions like static cloud gateways [22,26], or static paths between information providers and consumers [5,8].

5.6 Summary of Results

We have demonstrated the effectiveness of our architecture under different attack scenarios and we have shown that an attacker must control a large fraction of a network region to effectively cast any type of attack. Although adversaries can utilise the SON to place themselves near target peers, they need to control over 50% of a region to compromise a peer's anonymity or effectively censor a topic. Intersection attacks are also difficult to perform, even under the assumption that an adversary knows the cloud topology and all the participants in these clouds. Both 2-wise and 3-wise cloud intersections have a cardinality higher than 40, and the probability that a peer participates in two or three clouds with small intersection cardinality is below 10^{-4}.

Cloud size affects message traffic, due to the broadcasts performed at destination clouds to locate the intended recipient of a message. Contrary, our architecture is insensitive to the number of clouds generated by each peer, as shown in Figures 5 and 6(a), thus fitting dynamic scenarios where κ is neither predictable nor enforceable. Finally, as Figure 5 suggests, Clouds is able to retrieve about 70% of the answers created by the resource providers, even when not under attack. This is due to our cloud-based routing mechanism, and creates a decrease of about 30% in retrieval performance when compared to a SON without security features. Notice, however, that Clouds is not meant to be a full-fledged information retrieval system operating under ideal conditions, but is rather designed to deliver results even under challenging attack scenarios.

6 Related Work

In this section, we briefly discuss the papers that are related to our approach and overview proposals that support anonymity and censorship resistance in a P2P setting.

Systems for anonymity. In the recent past, several techniques have been presented for preserving anonymity in a P2P environment. One strand of research, pioneered by Freenet [5], tried to protect the communication channel, while other approaches tried to anonymize the communication parties by hiding them in groups.

Onion routing [14,15] belongs to the former family and is adopted by TOR [8], one of the most known systems for anonymity. An onion is a recursively layered data structure, where each layer is encrypted with the public-key of the peers it was routed through. Encrypting each layer with a different key preserves the secrecy of the path through which the onions is routed. Contrary to our approach, onion routing relies on static paths which makes the approach vulnerable to peer fails. A similar approach is followed by Tarzan [13] and MorphMix [23], where layered encryption and multi-hop routing is utilised. With the proliferation of DHTs, approaches exploiting the anonymity properties of onion routing in structured overlays have also been proposed [29].

Crowds [22] hides peers in groups, and all intra-group communication is orchestrated by a trusted server that knows all peers in its group, while Hordes [26] utilises multicast groups as reply addresses from the contacted server. Contrary to our approach, both systems rely on static groups and trusted servers, which introduce vulnerability (e.g., if servers are compromised) and central points of failure. The notion of groups (without a server component) as cloaks of peer identity was first introduced in Agyaat [1] for a DHT environment. We extend this approach to unstructured networks, and utilise a richer query language, provide cryptography-based censorship resistance, and support dynamic cloud creation and maintenance. Other approaches like DC-nets [4] and XOR-trees [9] guarantee anonymity by allowing at exactly one client to transmit a message at a given time slot, making it a restrictive model for large-scale applications. Finally, P^5 [25] provides an interesting architecture that enables tuning between efficiency and anonymity degree, a feature that we also support by the control of cloud population.

Systems for censorship resistance. Censorship at storage level consists in performing a selective filtering on the content of a peer and aims at blocking the resource as soon as its presence is detected. To avoid this type of attack, techniques such as replication of resources [5,11] and encryption of file chunks [19,30] have been proposed.

Censorship at communication level aims at corrupting the channel between communication participants. Approaches like [12,7,21] address the problem of an adversary which is able to take control of a large fraction of peers and to render unwanted resources unreachable. Similarly to our approach, these solutions are based on limited broadcasts of the query and the result, but can only be applied to structured overlays and do not address anonymity.

7 Conclusion and Future Work

We have presented Clouds, a novel P2P search infrastructure for proving anonymous and censorship-resistant search functionality in SONs. We have demonstrated that our system is resistant against a number of attacks. Although we instantiate our technique on SONs to leverage their retrieval capabilities and to support a rich data model and query language, the anonymity and censorship resistance mechanisms presented in this paper are general and can in principle be applied to any kind of unstructured overlay. The reason is that the cloud-based communication protocol is largely independent on the underlying network: the only connection is given by the strategy utilized to route the query, while the rest of the messages is routed according to the footprint list. This makes it possible to apply our framework to any kind of unstructured overlay network, by choosing an arbitrary routing strategy for the SUBMITQ message. For instance, our approach could be applied to Gnutella-style networks by adopting flooding as the routing for the SUBMITQ message, or small-world networks [24,20] by adopting interest-based query routing, while leaving the rest of the underlying protocols (e.g., rewiring in the case of small world networks) intact.

We emphasize that SONs represent a stress-test for our architecture, due to the number of attacks that adversaries can mount by exploiting the semantic correlation among peers. Applying our architecture to other types of unstructured overlay networks would enhance the security guarantees offered by our framework. For instance, a Gnutella-style overlay network, that would lack the semantic correlation between peers, would offer more security guarantees due to the randomness of peer connections. Additionally, in social networks peers create connections based upon trust relationships from real-life (e.g., do not connect to people that either you or your friends do not know). This trust mechanism would enhance the security properties of our architecture since mounting the surrounding and blocking attacks would be harder for adversaries.

As future work, we plan to extend our analysis to other sophisticated attack scenarios like the *Sybil attack* [10], where a malicious peer obtains multiple fake identities and pretends to be multiple distinct peers in the system and the more general *Eclipse attack* [3], where multiple identities are used to isolate a peer from the rest of the network. Syngh et al. [27] present a defense against Sybil attacks in structured overlay networks: Since a peer mounting an Eclipse attack must have a higher than average peer degree (i.e., incoming and outgoing links), the idea is to enforce a peer degree limit by letting peers anonymously *audit* each other's connectivity. This conceptually elegant solution needs the introduction of a central authority to certificate peer identifiers and the presence of a secure routing primitive using a constrained routing table [3]. It is also possible to exploit specific features of the network in order to prevent or mitigate the threat of Sybil attacks: for instance, Sybilguard [32] is a recently proposed protocol that exploits *trust relationships* among peers to limit the corruptive influences of Sybil attacks in social networks. It would be interesting to investigate a combination of these techniques in our system; in particular, trust relationships for choosing the bootstrap peer and for guiding the joining procedure and auditing-based Sybil identification for reducing the threat of Eclipse attacks against peers already present in the network.

Finally, given the importance of social networking in the Web 2.0 framework, an interesting extension of Clouds would be to exploit links among friends in a social network to determine the security and efficiency guarantees of our protocols against different types of attacks.

References

1. Singh, A., Gedik, B., Liu, L.: Agyaat: Mutual Anonymity over Structured P2P Networks. Emerald Internet Research Journal (2006)
2. Aberer, K., Cudré-Mauroux, P., Hauswirth, M., Van Pelt, T.: GridVine: Building Internet-Scale Semantic Overlay Networks. In: McIlraith, S.A., Plexousakis, D., van Harmelen, F. (eds.) ISWC 2004. LNCS, vol. 3298, pp. 107–121. Springer, Heidelberg (2004)
3. Castro, M., Druschel, P., Ganesh, A.J., Rowstron, A.I.T., Wallach, D.S.: Secure Routing for Structured Peer-to-Peer Overlay Networks. In: Proceedings of the USENIX Symposium on Operating Systems Design and Implementation, OSDI (2002)
4. Chaum, D.L.: The Dining Cryptographers Problem: Unconditional Sender and Recipient Untraceability. Journal of Cryptology (1988)
5. Clarke, I., Miller, S., Hong, T., Sandberg, O., Wiley, B.: Protecting Free Expression Online with Freenet. IEEE Internet Computing (2002)
6. Crespo, A., Garcia-Molina, H.: Semantic Overlay Networks for P2P Systems. In: Proceedings of the International Workshop on Agents and Peer-to-Peer Computing, AP2PC (2004)
7. Datar, M.: Butterflies and Peer-to-Peer Networks. In: Möhring, R.H., Raman, R. (eds.) ESA 2002. LNCS, vol. 2461, p. 310. Springer, Heidelberg (2002)
8. Dingledine, R., Mathewson, N., Syverson, P.: Tor: The Second-Generation Onion Router. In: Proceedings of the USENIX Security Symposium (2004)
9. Dolev, S., Ostrobsky, R.: Xor-Trees for Efficient Anonymous Multicast and Reception. ACM Transactions on Information and System Security (TISSEC) (2000)
10. Douceur, J.R.: The sybil attack. In: Druschel, P., Kaashoek, M.F., Rowstron, A. (eds.) IPTPS 2002. LNCS, vol. 2429, p. 251. Springer, Heidelberg (2002)
11. Endsuleit, R., Mie, T.: Censorship-Resistant and Anonymous P2P Filesharing. In: Proceedings of the International Conference on Availability, Reliability and Security (ARES) (2006)
12. Fiat, A., Saia, J.: Censorship Resistant Peer-to-Peer Content Addressable Networks. In: Proceedings of the ACM-SIAM Symposium on Discrete Algorithms (SODA) (2002)
13. Freedman, M.J., Sit, E., Cates, J., Morris, R.: Introducing Tarzan, a Peer-to-Peer Anonymizing Network Layer. In: Druschel, P., Kaashoek, M.F., Rowstron, A. (eds.) IPTPS 2002. LNCS, vol. 2429, p. 121. Springer, Heidelberg (2002)
14. Goldschlag, D., Reed, M., Syverson, P.: Onion Routing. Communications of the ACM (CACM) (1999)

15. Han, J., Liu, Y., Xiao, L., Ni, L.: A Mutual Anonymous Peer-to-peer Protocol Design. In: Proceedings of the IEEE International Symposium on Parallel and Distributed Processing (IPDPS) (2005)
16. Hersh, W., Buckley, C., Leone, T.J., Hickam, D.: OHSUMED: An interactive retrieval evaluation and new large test collection for research. In: Proceedings of the Annual International ACM SIGIR Conference (1994)
17. King, I., Ng, C.H., Sia, K.C.: Distributed content-based visual information retrieval system on peer-to-peer networks. ACM Transactions on Information Systems (2002)
18. Loser, A., Wolpers, M., Siberski, W., Nejdl, W.: Semantic Overlay Clusters within Super-Peer Networks. In: Proceedings of the International Workshop on Databases, Information Systems and Peer-to-Peer Computing (DBISP2P) (2003)
19. Waldman, A.R.M., Cranor, L.: Publius: A Robust, Tamper-Evident, Censorship-Resistant, Web Publishing System. In: Proceedings of the USENIX Security Symposium (2000)
20. Raftopoulou, P., Petrakis, E.G.M.: iCluster: A Self-organizing Overlay Network for P2P Information Retrieval. In: Macdonald, C., Ounis, I., Plachouras, V., Ruthven, I., White, R.W. (eds.) ECIR 2008. LNCS, vol. 4956, pp. 65–76. Springer, Heidelberg (2008)
21. Ratnasamy, S., Francis, P., Handley, M., Karp, R., Shenker, S.: A Scalable Content-Addressable Network. In: Proceedings of the ACM Special Interest Group on Data Communications (SIGCOMM) (2001)
22. Reiter, M., Rubin, A.: Crowds: Anonymity for web transactions. ACM Transactions on Information and System Security (TISSEC) (1998)
23. Rennhard, M., Plattner, B.: Introducing MorphMix: Peer-to-Peer Based Anonymous Internet Usage with Collusion Detection. In: Proceedings of the International Workshop on Privacy in the Electronic Society (WPES) (2002)
24. Schmitz, C.: Self-Organization of a Small World by Topic. In: Proceedings of the International Workshop on Peer-to-Peer Knowledge Management (P2PKM) (2004)
25. Sherwood, R., Bhattacharjee, B., Srinivasan, A.: P^5: A Protocol for Scalable Anonymous Communication. IEEE Security and Privacy (2002)
26. Shields, C., Levine, B.N.: A Protocol for Anonymous Communication over the Internet. In: Proceedings of the ACM Conference on Computer and Communications Security (CCS) (2000)
27. Singh, A., Ngan, T., Druschel, P., Wallach, D.S.: Eclipse attacks on overlay networks: Threats and defenses. In: Proceedings of the IEEE International Conference on Computer Communications (INFOCOM), pp. 1–12 (2006)
28. Sweeney, L.: k-Anonymity: A Model for Protecting Privacy. International Journal of Uncertainty, Fuzziness and Knowledge-Based Systems, IJUFKS (2002)
29. Tsai, H., Harwood, A.: A scalable anonymous server overlay network. In: International Conference on Advanced Information Networking and Applications (AINA) (2006)
30. Waldman, M., Mazières, D.: Tangler: a Censorship-Resistant Publishing System Based on Document Entanglements. In: Proceedings of the ACM Conference on Computer and Communications Security (CCS) (2001)
31. Wu, L., Faloutsos, C., Sycara, K.P., Payne, T.R.: FALCON: Feedback Adaptive Loop for Content-Based Retrieval. In: Proceedings of the VLDB Conference (2000)
32. Yu, H., Kaminsky, M., Gibbons, P.B., Flaxman, A.: Sybilguard: defending against sybil attacks via social networks. In: Proceedings of the ACM Special Interest Group on Data Communications (SIGCOMM), pp. 267–278 (2006)

An Information Brokering Service Provider (IBSP) for Virtual Clusters

Roberto Podesta', Victor Iniesta, Ala Rezmerita, and Franck Cappello

INRIA Saclay-Ile-de-France
Bat. 490 Université Paris-Sud
91405 Orsay, France
{podesta,Victor.Iniesta,rezmerit,fci}@lri.fr

Abstract. Virtual clusters spanning over resources belonging to different administration domains have to face with the computational resources brokering problem in order to allow the direct interaction among them. Typically, it requires some component collecting, elaborating and providing resources information. Currently, every virtual cluster project provides its own solution with a low degree of portability and isolation from the rest of the platform components. We propose the Information Brokering Service Provider (IBSP), a general approach which wants to represent an uniform solution adaptable to a wide set of existing platforms. Its pluggable and flexible design is based on a decentralized architecture where each of the cooperating nodes is composed by a resource interface exposing the brokering service, and a portion of a distributed information system. In this paper, we present the motivations, principles and implementation of IBSP as well as the performance of IBSP serving the Private Virtual Cluster system.

Keywords: Brokering, P2P, Virtual Clusters.

1 Introduction

The convergence between Peer to Peer (P2P) and Grid Computing [1] contributed to the advent of projects aiming to port the cluster abstraction of computational grids on top of common machines. Virtual cluster projects like VioCluster [2], Wide-area Overlay of virtual Workstations [3] (WOW), Private Virtual Cluster [4] (PVC), Cluster On Demand [5] (COD), etc. address this challenge. Typically, they rely on overlay networks of Internet-connected machines and aim to provide a transparent view of a single cluster where parallel applications may run unchanged.

Similar concepts are also adopted in the emerging cloud computing [6]. A transparent view of grouped machines is obtained by a cloud interface masking a distributed back-end where cluster virtualization techniques are adopted. The computational resources leveraged by clouds are physical clusters rather than common Internet-connected machines. Compared to P2P-based cluster virtualization, cloud computing gives a greater relevance to a clean user interface which exposes methods to configure the whished computational environment. Samples are the widely known Amazon

R. Meersman, T. Dillon, P. Herrero (Eds.): OTM 2009, Part I, LNCS 5870, pp. 165–182, 2009.

Elastic Compute Cloud [7] (EC2), based on a proprietary technology, and the open source Nimbus [8] and Eucalyptus [9].

P2P-based cluster virtualization presents not trivial issues related to the pooling of the computational resources. The arising problems may span from networks barriers to the lack of knowledge of resources location. These practical obstacles imply the adoption of components able to perform resources inter-mediation, information management and discovery. Such tasks involve a set of needs we define as brokering issues. Currently, every project provides its own customized solutions. It is difficult to individuate a reference model because the implemented mechanisms are often not completely isolated from other aspects of the platforms. Their level of reusability is consequently very low, even if the intrinsic nature of the handled problems is usually shared. A standardized and uniform solution might greatly simplify the deployment of current virtual cluster platforms and would free the development of future virtual cluster projects from a not negligible burden. In this paper, we present a possible solution to this challenge. Upon a categorization of brokering issues, we adopt the Service Oriented Architecture (SOA) [10] model to design a software entity providing a general resources brokering service, an Information Brokering Service Provider (IBSP) for virtual clusters. The widely known SOA pattern implies a single entity acting as the Service Broker, which exposes the information of services published by Service Providers. Service Consumers get that information from the Service Broker, and they can subsequently invoke the services on the Service Providers. In our case, the services become computational resources. The system architecture is built to maximize the flexibility in order to fit a wide range of possible requirements. Therefore, it leverages a distributed structure, which can grow and shrink depending on the optimization needs of the served platforms. The system design adopts a pluggable approach to allow the dynamic set-up of the modules supporting specific platforms.

The remainder of the paper is organized as follows: in section 2, we draw the categorization of brokering needs in a representative set of scenarios; section 3 describes the operational model, the software design and some implementation highlights of IBSP; section 4 presents the real applicative scenario we used to test the IBSP reference implementation, and the test results we obtained; and, finally, section 5 concludes with some remarks and future works.

2 Drawing a Categorization of Brokering Issues

The resources brokering problem is not only present in the domain of cluster virtualization. In fact, similar issues are tackled in several domains, and different tools solve them with approaches and architectural solutions which have to be considered. Furthermore, interesting issues are coming from the cloud computing adjacent field, where many cluster virtualization concepts are adopted. Table 1 lists a representative set of domains with some tool examples and the relative brokering issues. The last column shows the broker software architecture of those tools. The highlighted categorization of brokering issues has a double aim: first, to ease the idea of uniform brokering service provider by a clear list of needs in virtual cluster platforms; and, second, to explore possible intersections in other domains.

2.1 Virtual Cluster Platforms

VioCluster consists in a combination of the VIOLIN [11] network virtualization and XEN [12] machine virtualization and it is targeted to build a computational resources trade among computational clusters belonging to different administrative domains. The brokering issue in VIOLIN is bounded to the overlay namespace resolution which reflects a three layer hierarchy of entities (virtual routers, virtual switches and virtual machines). The WOW project aims to provide a communication infrastructure overlaying the Internet and routing messages among a set of world wide distributed virtual workstations. The brokering needs of WOW consist in the resource enrollment, detection and lookup. The adopted system consists in a distributed DHCP implementation. The DHCP client residing on each VM is redirected to the TAP virtual driver and encapsulated in a SOAP request, which accesses to a distributed database containing the configuration of the virtual ID (e.g. a class of private IP addresses). The PVC system builds an overlay network by the adoption of a virtual ID system corresponding to IP addresses belonging to the experimental class E. A central registry maps those virtual IPs to real IPs thus allowing the communication between resources. Moreover, the registry works also as resource connectivity enabler to allow the direct connection between peers. The COD [13] project uses a resource broker to manage the resource leases. Other aspects like resource identification, lookup and detection are managed by its overlay network. Virtuoso [14] is a virtual cluster system pooling dynamically virtual machines connected through an overlay network named VNET. The brokering issues in Virtuoso refer to virtual machines identification, lookup, detection and information. They are bounded to the overlay routing ruled by a

Table 1. Categorization of brokering issues

DOMAIN	TOOL	BROKERING ISSUE	SOFTWARE DESIGN
Web Services	UDDI	Yellow pages; White pages; Green pages	Centralized / UBR
Meta-computing	H2O	Service lookup	Centralized
Enterprise	JNDI	Remote Object Lookup	Centralized
Sensor Network	Data Fusion	Semantic based asynchronous object lookup	Centralized
Sensor Network	JAIN SLEE	Asynchronous object lookup	Centralized
Overlay Network	Narada	Tree-based routing	Hierarchically distributed
Virtual Cluster	Viocluster (VIOLIN)	Resource enrollment, identification, detection, lookup; network information	Hierarchically distributed
Virtual Cluster	PVC	Resource enrollment, identification, detection, lookup; network information	Centralized
Virtual Cluster	WOW	Resource enrollment, identification, detection, lookup; network information	Distributed DHCP
Virtual Cluster	COD	Resource enrollment, identification, detection, lookup; network information; resource information	LDAP
Virtual Cluster	Virtuoso (VNET)	Resource enrollment, identification, detection, lookup; network information	Hierarchically distributed
Virtual Cluster	ViNe	Resource enrollment, identification, detection, lookup	Distributed
Cloud	Nimbus	Hypervisor management; VM Images localization; resource information	Centralized registry
Cloud	Eucalyptus	VMs group management / configuration; VM Images localization; resource information	Hierarchically distributed
Cloud	Enomaly	Hypervisor management; VM Images localization; resource information	Centralized registry

statically configured two-level hierarchy of elements using an ID system based on MAC addresses tunneling. The ViNe[15] project is a communication infrastructure overlaying internet similar to WOW. It uses private IP addresses as ID systems which are configured through IP aliasing on the machine NIC interface. The resource lookup and detection is ruled by virtual routers onto which all the traffic among resources is redirected.

2.2 A Look at Other Domains

The Universal Description Discovery and Integration [16] (UDDI) standard plays the role of service broker in Web Services technology. Sites exposing Web Services register them on an UDDI broker. Web Services consumers looking for a certain type of service perform a request to an UDDI broker and obtain a network reference enabling the direct communication. The service provided by UDDI offers three types of possibilities: the white pages (e.g. it replays to a query wanting a specific business name), the yellow pages (e.g. a certain type of business) and the green pages (e.g. a certain type of technical information about a business). Therefore, it provides more than simple lookup by providing elaborated information in order to allow the direct communication between business entities. Typically, UDDI registries are isolated centralized server. The UDDI Business Registry (UBR) project was the only effort targeted to build a federation of UDDI registries. This project was also known as UDDI cloud, and federated a limited number of UDDI servers belonging to the project partners (i.e. IBM, Microsoft, SAP and HP). The basic mechanism behind the federation was a periodical propagation of the entries onto all the registries [17]. However, UBR ended in 2006 and the partners shut down the UDDI cloud [18]. Among the others, a motivation of the UDDI cloud end was the poor performance of the federation mechanism [19] based on a not optimized, pure data replication. The Narada Broker [20] is a distributed system targeted to route messages among remote nodes by supporting publish/subscribe communication and P2P interaction. Narada provides compatibility with already developed asynchronous communication standards like Java Message Service and JXTA [21], and has been adopted in several applicative scenarios. Narada works as overlay transport layer by taking care of all the communication among nodes composing the application using it. It is a middleware working as the glue connecting remote peaces of a distributed application. Narada adopts a hierarchical topology with master brokers and normal brokers. The resources brokering problems in Narada result in registration and lookup. They are solved by the hierarchical namespace ruling also the routing on the overlay transport layer. The table contains further samples of brokering needs of tools operating in various domains. The widely known JNDI registry serves the remote object look-up in enterprise application based on Java. The meta computing framework H2O [22] uses an extension of the Java RMI registry to lookup the deployed services. Some interesting solution comes from the sensor network world. The Semantic Data Fusion for sensor networks [23] introduces an ontology-based broker in a publish/subscribe system to get a smart integration of the data coming from sensors. Publishers and subscribers communicate via declarative grammar their respective notifications and interests to the semantic broker. Afterwards, its engine performs the matching thus enabling the smart information exchange. The Activity Context of the Java API for Intelligent Network Service Logical Execution

Environment (JAIN SLEE) [24], a Java standard targeted to event driven system, is also familiar to this approach. The Activity Context allows a smart event routing through declarative programming and a JNDI registry thus enabling the connection between asynchronous network events sources and Java components.

2.3 Cloud Computing

Cloud computing infrastructures like Nimbus [1], Eucalyptus [2] and Enomaly [3] coordinate distributed resources by the adoption of centralized registries working in cooperation with remote agents deployed on the hypervisor installed on each machine. Even if there are some slight differences, basically they use a master-worker information system where the master is the registry collecting information about the workers represented by the daemon running on the managed machines. Nimbus and Enomaly build this mechanism on top of a two-level hierarchy condensing the master registry in the cloud interface. Eucalyptus decouples the cloud interface from a set of super registries with a hierarchical 3-tier architecture. A cloud-level registry mediates the composition of virtual cluster instances based on physical resources collected by cluster-level registries. These architecture types reflect the normal target of resources set hidden by a cloud interface, namely a computational cluster. The handled information assumes the form of meta-data describing the groups of virtual machines fitting a specific request coming from the interface exposed to the cloud users.

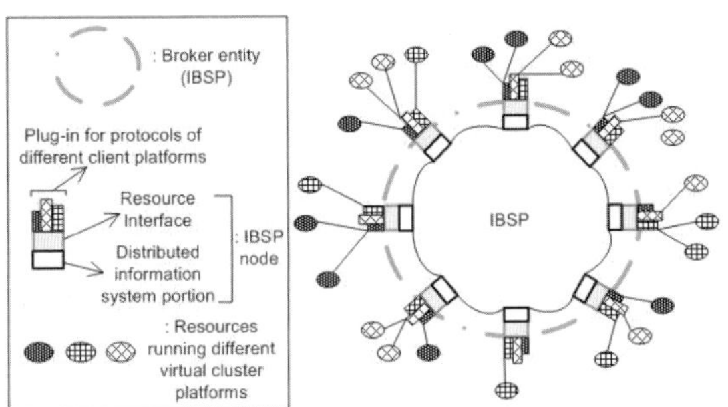

Fig. 1. IBSP Architecture

3 Information Brokering Service Provider (IBSP)

The operational model of IBSP derives from the categorization of the brokering issues shown in the previous section. We looked firstly at the P2P-based virtual cluster platforms which are the main target of our work. We extracted from that table the various brokering needs in order to compose a set of general functionalities. A look at the brokering issues in other fields is useful to reuse already developed concepts even if a completely re-usable solution does not exist. The table suggests also possible intersections with cloud platforms. However, the different kinds of used computational resources (i.e. computational clusters for cloud computing, common machines for P2P cluster virtualization) and the different targets (i.e. precise fulfillment of user requests for cloud computing, maximization of utilization of Internet-connected machines for P2P cluster virtualization) induced us to focus the broker functionalities definition on the needs of P2P-based virtual clusters we have shown in section 2.1 (i.e. VIOLIN, PVC, WOW, COD, VNET, ViNe).

3.1 IBSP Specification

The categorization shows several brokering issues shared among different platforms. Therefore, we extract the following list of functionalities: 1) resources enrollment, 2) resources identification system, 3) resources lookup, 4) resources presence detection, 5) resources fault detection, 6) resources connectivity enabler, and 7) resources information system.

Note that the virtual cluster platforms calling IBSP functionalities are referred as client platforms in the remainder of the paper. The rest of this sub-section examines the functionalities offered by IBSP. 1) The enrollment regards the registration of the resources into the broker. After this step, the broker is able to provide information about them. This functionality implies the exposition of an enrollment method for every type of resource required by a client platform. 2) The broker provides an ID system which is used to identify resources. An unique ID is assigned to a resource. An ID system has to be configured before the startup of the broker. The ID systems can be ranges of private IP addresses, namespaces or any other identification system required by a supported client platform. The assignment of IDs can be done by the broker (in a pseudo-DHCP way) or statically configured by the client platform. In both cases, the broker maintains the IDs consistence. Different ID systems can coexist within the broker in case of broker instance serving concurrently multiple platforms. This functionality is not bounded to the exposition of a specific method, while it is rather linked to the enrollment method. In fact, the invocation of the enrollment method ends with the memorization of an association between an unique ID and a resource representation, which includes at least its physical location. 3) The simple resource lookup is provided. Therefore, the broker exposes a simple lookup method to retrieve a target endpoint (e.g. physical location as an URL or an IP) when a resource is looking for another resource given a certain target ID. An extension of this method provides the additional information possibly associated to a given ID more than the mere location information. 4) The presence detection functionality serves to verify the actual participation of a given resource. A resource participates in a virtual cluster if it is enrolled into the broker. A resource presence detection method confirms or not if a resource is enrolled by a given ID. 5) The resource fault-detection may appear similar to the previous functionality. However, it refers to the notification of a resource failure. This functionality is offered though a subscription mechanism. A resource subscribes its interest in the aliveness of a target resource, and it is notified in case of fault of that resource. The subscription can be explicitly done by a resource or automatically performed by the broker which subscribes the interest of a resource in another one if the first one has previously looked up the second one. 6) The resources connectivity enabler refers to the actions performed by the broker to set-up the connectivity between two enrolled resources. It is an inter-mediation functionality, it supports the start-up of the connection between two resources, which otherwise would not be able to communicate. After this set-up phase, the resources communicate directly without any other broker intervention. The connectivity enabler functionality relies on schemas of messages relay between the requester resource and the requested resource through the broker. Those schemas may comprise a simple request/reply messages exchanges as well as complex inter-mediations composed by messages loops involving the requester, the callee and even third-part resources. Forwarded messages can be enriched and modified by the broker according to specified pattern depending on the client platform. This functionality can be useful for a variety of situation that can span from network information (e.g. bypass of routers and firewalls, overlay routing rules, etc.) to public security information exchange. This functionality is mapped to a set of connectivity methods depending on the client platform. Finally, 7) the broker has a flexible information system where resources information are stored. Basically, a resource is registered with an association between an unique ID and a representation of its actual location. However, typical virtual cluster platforms require more than just location (e.g. system information, leases, etc.). Therefore, it is possible to add additional information to an entry representing a resource. Such information can be registered at the enrollment stage as well as by an update method exposed to modify the entries content. Furthermore, the information system serves to provide a resource discovery not based on a precise ID. It can be useful for those cases requiring a resource search on the base of some feature without an a-priori knowledge of the resource ID. This kind of search can be considered rare compared to the ID-based search. However, it can be useful for a noticeable variety of cases (e.g. search for specific computational capabilities, specific software configuration, etc.). IBSP support this capability by a publish/subscribe based system. Every module supporting a client platform can implement an application grammar used by resources to publish topics which are useful for a not ID-based resource discovery. Resources can then subscribe to topics according to the module grammar. Such a grammar can be composed by a few keywords or based on some ontology. This approach resembles the semantic-based data fusion for sensor network mentioned in section 2.2.

IBSP structure wants to maximize its flexibility in order to be adaptable to a large set of client platforms. Mainly, it aims to: 1) avoid single points of failure, 2) provide multiple access points, 3) allow both loosely coupled and permanent connections between the broker and the resources. The first and the second constraints force a distributed structure where all the participating nodes have the same importance without any hierarchy. The third one represents an exception to the native SOA model. It is recommended for resources not able to receive incoming connections because they are hidden by NATs or firewalls.

3.2 Implementation Choices

The resulting architecture is a distributed set of cooperating nodes constituting the whole broker entity. Each node hosts a resource interface and a portion of a distributed information system. The first exposes the broker functionalities to resources, and different plug-ins implement the protocol syntaxes and the semantics required by the supported client platforms. Figure 1 shows a graphical representation of the architecture of IBSP. The number of composing nodes can dynamically grow and shrink. The assignment of this number can be done by some heuristic depending on the actual situation requiring IBSP. If it is serving just a single client platform, the number of IBSP nodes can be the optimal one calculated for that platform. If it is serving multiple client platforms, the maximum one can be used. However, the calculation of the optimal number of the IBSP instances required by a singe or more client platforms may depend on a wide set of factors. Just for a single client platform, reflections coming from the overlying parallel application as well as particular administrative situations can influence the number of the needed IBSP instances. Section 4 presents a set of experiments we ran in different case studies which contribute to ease the assignment of the number of IBSP instances. We focused on P2P data distribution models to design the distributed information system to avoid hierarchical layers bringing to critical points of failure. We selected the structured overlay network approach which relies on Distributed Hash Tables [28,29](DHTs). The meaning of a DHT is a hash table distributed among a set of cooperating nodes. As a normal hash table it allows to put and get items as key-value pairs. IBSP stores into a DHT the ID generated for an enrolled resource as key. The value associated to a key has to provide a certain information about the resource. That value can be a string reassuming directly all the required data depending on the client platform (e.g. resource actual location, resource information) or a network pointer (e.g. an URL) to a file containing that information. It depends on the amount and the nature of the data. The first solution is obviously faster and, therefore, preferable if applicable. We prefer the DHT adoption to the unstructured overlay network because DHTs provide better scalability. Moreover, the usage scenario of the broker should not affect the main known drawback of DHTs. Their main capability is to provide a scalable logarithmic keys lookup time. The scalability is not a trivial issue in unstructured overlay networks being based on queries broadcasting. The solution consists in the layering of nodes into a hierarchical classification. However, it introduces critical points of failure in the structure and does not nullify completely the lack of scalability. Unstructured overlay networks have actually good capabilities as the possibility to perform range queries which are not possible with DHTs. To overcome this limitation, DHT implementations provide basic APIs to perform multicast communication based on the publish/subscribe pattern. This capability is particularly useful for the functionalities exposing methods replaying to not ID-based requests (e.g. fault detection and some of the information system functionalities). Current IBSP reference implementation is written in Java. A wrapper hides an existing DHT API, namely FreePastry [30]. It provides access to the insert/lookup operations and the multicast communication based on the publish/subscribe pattern. FreePastry is an open source project and the main Java DHT solution with Bamboo [31]. Even if it has been shown that Bamboo may be preferable in some cases [32], the differences are not so remarkable. An evaluation of the Java DHT implementations is out of the scope of this paper. The default configuration exposes HTTP-REST interfaces and loads a resources ID system based on a class of private IP addresses. Such configuration is not bounded to any specific platform. Declarative programming allows specifying the plug-ins to be loaded to support specific platforms. Plug-ins are deployed even at run-time through a descriptor Java properties file. It serves to set-up the proper classes fitting a client platform. A set of interfaces are provided to implement the plug-ins. They let to customize the following aspects: 1) the ID system; 2) the grammar for topic publish/subscribe; 3) the message model; 4) the wire protocol; and 5) optional DHT interactions through the wrapper. While we already mentioned the first and the second aspects in the functional description, it is useful to spend a few words about the others. The third and fourth aspects determine the way to invoke the methods exposed by IBSP. More precisely, they define the semantic and the protocol syntax of a supported client platform. The fifth aspect serves to customize the format of the data to be stored and retrieved to and from the DHT. They can be simply represented as values of DHT entries thus requiring some rule to compose and parse the values strings. Alternatively, they can be network pointers to specific remote files containing all the information associated to a resource. In this case, this aspect allows specifying the way to access to remote sites and the handler to manage the adopted file format. A practical example showing the plug-in of an existing platform is included in the reference implementation. It regards the already mentioned Private Virtual Cluster (PVC) platform. The developed plug-in uses the PVC metric to indentify resources, namely the experimental class E of IP addresses. The messages are text based and exchanged on top of TCP. Apart the fault detection, all the available functionalities are used by PVC. The most remarkable implementation is a connectivity enabler method articulating a loop allowing the discovery of the traversing technique necessary for the direct connection with a resource. That loop involves different PVC entities, and implies a whole exchange of eleven messages. Given the small amount of data required by PVC, the wrapper stores PVC resources information directly as the DHT values instead of network pointers. A registered PVC resource is represented by a couple composed by its virtual class E IP address and an aggregate data including: peer real IP, resource nature, traversing technique required to access to itself, virtual hostname and optional data. Next session shortly provides some details about PVC to introduce the various set of case studies we tested with the IBSP plug-in for PVC.

4 Case Studies

PVC is a P2P middleware targeted to provide a computational cluster abstraction on top of machines connected to LAN or WAN networks through the TCP/IP suite. From the user point of view, the access to these resources is transparent, without changing the security policies of each participating domain. A PVC peer daemon runs on every participating machine. On top of those demons, a distributed application can run unmodified upon the virtual cluster. PVC defines a set of *interposition techniques* to intercept the network communications of the overlying application. Another set of techniques, called *connectivity techniques*, allows establishing direct connections between resources, which can be behind firewalls or NATs. The interposition techniques detects when a virtual resource attempts to establish a direct connection, triggering the suitable connectivity technique by means of the IBSP connectivity enabler functionality. Each interposition technique implies a constant overhead in the peers direct connection which is not under control of IBSP. Since our target is not the analysis of the PVC connectivity techniques impact, we used *Traversing TCP* in all the experiments we ran. The functionalities we measured are: the resource enrollment, the resource lookup and the resource connectivity enabler we describe before. The resource enrollment includes the interaction with the ID system. In the case of PVC, the connectivity enabler is called inter-connection establishment, and the term resource becomes synonym of PVC peer. They are the functionalities mainly used by virtual clusters as shown in table 1, and they produce the largest part of the overhead involved by the IBSP adoption. PVC does not use the presence detection and the fault detection methods, while the use of the information system is limited to a particular case which has not relevant effect on the performances. An implementation of the presence detection would behave similarly to the resource lookup because basically it would leverage the same primitives (as it is implemented in the neutral HTTP-REST interface). We ran three different stages of experiments to test IBSP serving PVC.

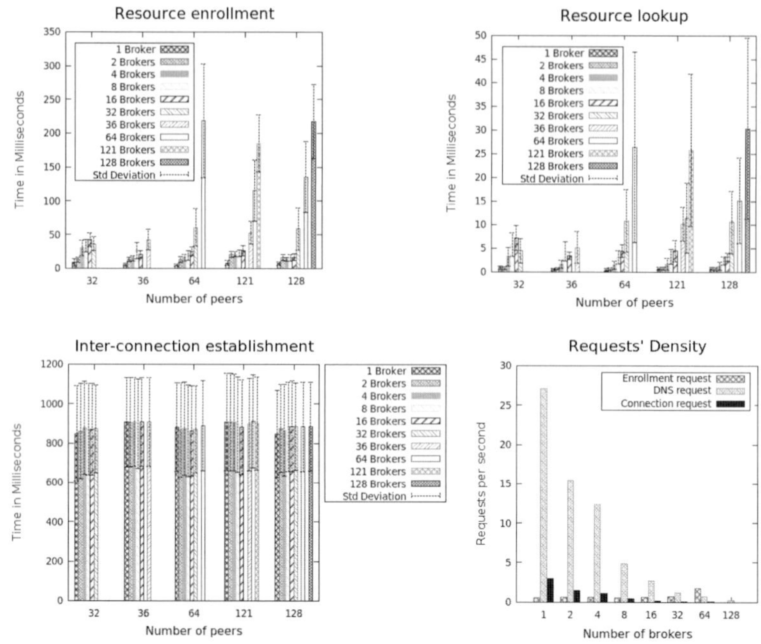

Fig. 2. NPB experiments results

We leveraged the Grid5000 [33] platform for the first two stages. The Grid5000 project offers an experimental test-bed for grid experiments. This platform is made up with computing resources coming from nine sites geographically distributed in France. Each site has its own hardware and network specifications, and clusters are inter-connected through a 10Gb/s link or at least 1 Gb/s, when 10Gb/s is not available. Depending on the size of the performed experiment, resources can belong to only one site, to a maximum of eight different sites. The third experiment deals with a heterogeneous environment we will describe later.

4.1 NAS Benchmarks

For the first group of experiments, we ran the NAS [34] benchmarks on top of PVC. The NAS Parallel Benchmarks (NPB) are used to evaluate the performance of parallel supercomputers. They provide a set of kernel and pseudo-applications to measure different aspects of parallel environments. We used the MPI-based implementation of the NPB. The CG, EP, LU, MG benchmarks were run for the following PVC virtual cluster sizes: 32, 64 and 128 nodes, while the BT and SP benchmarks for 36, 64 and 121 nodes. For a fixed virtual cluster size, each benchmark were run with different sets of IBSP nodes serving PVC: we increased the number of IBSP nodes starting from a single instance managing all the PVC nodes, to the match of the virtual cluster size (i.e. PVC peers) following the power of 2 (2, 4, 8, ... 256). An equitable number of PVC peers were assigned to each IBSP node. Note that the same experiments with a random assignment of PVC peers among the IBSP nodes gave similar results. In all the NPB experiments, we used the *Library interposition technique* on the PVC peers. This technique overloads certain standard system calls, by means of the Dynamic Library Linking feature. The different tested functionalities showed similar behaviors independently on which NPB it was run. Figure 2 shows the average time and standard deviation of each functionality for every configuration of IBSP nodes and PVC peers. The performance evolution of the resource enrollment and lookup operations for any virtual cluster size, in relation with the number of IBSP nodes, follows a similar trend. The larger overhead of the resource enrollment functionality is explained by the fact that the resource enrollment includes a resource lookup, in order to check if the chosen ID it is not already in use. The address generation itself and the creation of a new entry in the DHT also increase the execution time of this functionality. With only a single IBSP, the overhead is very limited, since these two operations remain local. When the number of IBSP nodes increases, the overhead due to coordination and network communication becomes higher: an IBSP node must interact with the nodes in its neighborhood and wait for their answers to complete the operation. From a cluster size of 64 nodes, the resource enrollment and lookup operations undergo a consistent decrease, even if they remain in a very low scale. This is explained by the influence of the Grid5000 inter-site communication arising when resources belong to more than one single site. It also explains a higher standard deviation. Finally, when the number of IBSP nodes is equivalent to the number of PVC peers, the information reaches its maximum distribution over the DHT. Experiments with more IBSP nodes than PVC peers would not provide any additional information, since the worst test case comprises each IBSP node managing a single PVC peer as maximum. These results might appear in contrast with the intuitive deduction that

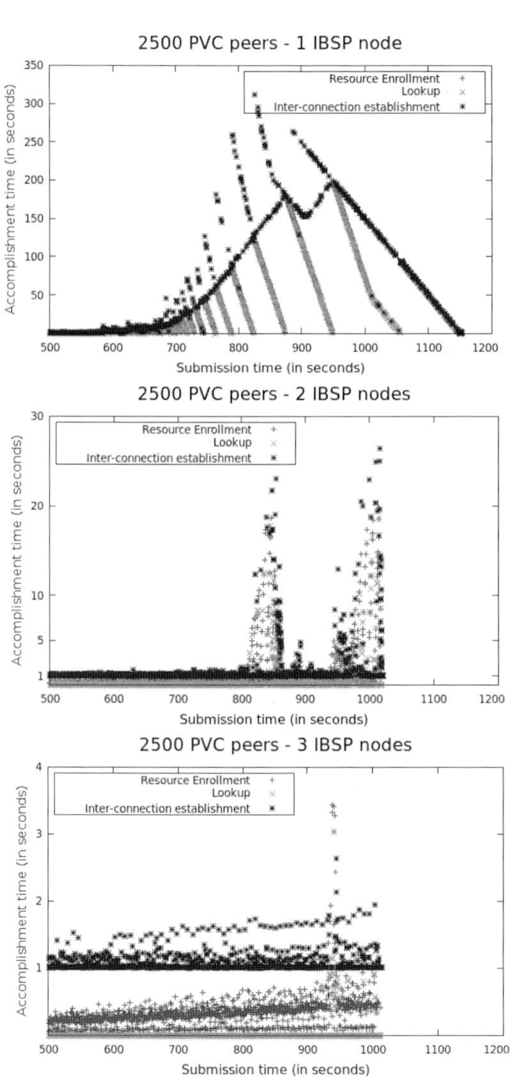

Fig. 3. Scalability experiments results

more brokers should be faster than a single one, because they are supposed to perform tasks in parallel. It may be true over a certain threshold causing a performance decrease because of the concurrent management of multiple peers. However, we did not reach this scalability threshold, even if the scale of real collected peers was not negligible. It means that the overhead due the connection broker-to-peer is lower than the overhead due to communication broker-to-broker. We did not get over that scale because of practical burdens hampering to set-up larger experiments on Grid5000. It is possible to say that a single instance of IBSP is enough up to manage 128 PVC peers. More instances can be used for any reason (e.g. administrative policies, reliability, high availability, etc.) with the shown overhead. In the inter-connection establishment graphic, the measures show a constant overhead in all the cases. Considering the scale, the behavior reflects the influence of the other measures plus a constant due to the PVC peers interposition and connectivity techniques, which are not under control of IBSP and take an active role during this process. Obviously, the overhead due to the Grid5000 inter-cluster communication has also an influence. Figure 2 shows also the requests density associated to a virtual cluster of 128 peers, the largest in the NPB experiments. The stress associated to the resources enrollment remains constant, since PVC peers were registered sequentially at the same rate, in all cases. For the other requests, we can easily notice a decrease in the stress the IBSP nodes are subjected to, when their number increases, since the workload is shared among them. It could suggest another reason to adopt multiple IBSP instances in order to reduce the load on the machines hosting IBSP.

Regarding the adoption of multiple IBSP instances and the mentioned scalability issue, we ran a further experiment with a simulated configuration. We implemented a simulated PVC peer which calls sequentially the enrollment, the lookup and the Interconnection establishment operations disregarding the NAS benchmarks. For each run, five Grid5000 machines were used to launch simultaneously 500 instances of simulated PVC peer per machine, at a rate of one instance every two seconds. Each IBSP node ran in a different machine. Figure 3 reports the interesting results we obtained. The graphics show for each run, the submission and accomplishment time of each operation request. It looks evident the performance decrease when a single broker has to manage the whole set of 2,500 simulated peers. The overlap of functionalities requests, the overhead associated to the ID generation process, the simultaneous access to the DHT, and the stress in the network management are added thus provoking

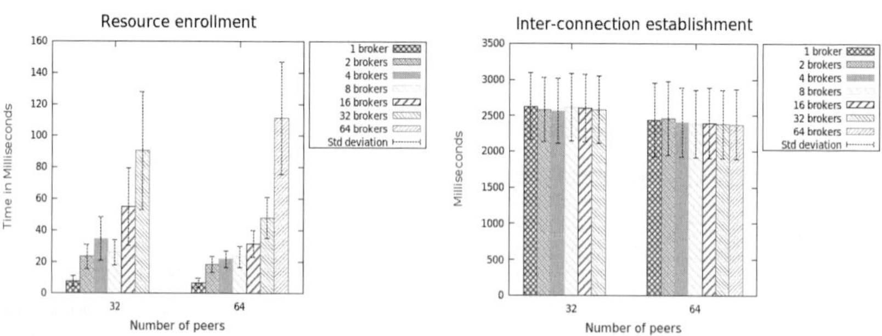

Fig. 4. CONDOR/BLAST experiment results

	Total Execution Time				
	cg	ep	lu	mg	condor
1 peer	64,6 s	3,9 s	99,4 s	6,9 s	7 s
2 peers	67,4 s	3,2 s	98,9 s	7,1 s	7 s
4 peers	32,1 s	3,9 s	97,8 s	4,9 s	6,9 s
8 peers	68,3 s	4,4 s	97,4 s	6,9 s	7 s
16 peers	70,6 s	4,4 s	95,3 s	6,4 s	6,7 s
32 peers	66,8 s	4,1 s	99,7 s	6,8 s	6,6 s
64 peers	32,2 s	3,9 s	81,6 s	3,3 s	6,9 s
128 peers	24,5 s	5,4 s	68 s	4,5 s	-

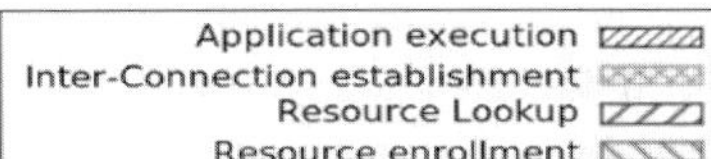

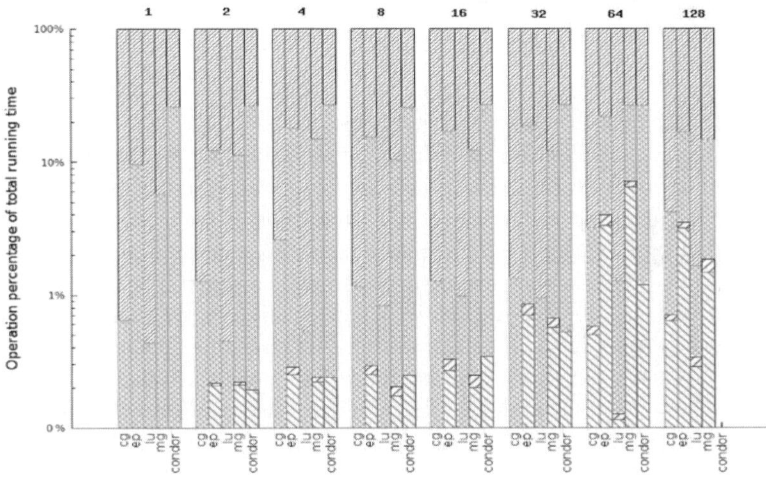

Fig. 5. IBSP/PVC Impact

a noticeable amount of peaks and a general increase of the measured accomplishment times of each functionality. As shown, the adoption of one instance more reduces greatly the general increase. Some peaks still remain when the structure is stressed by the accumulation of many concurrent requests. Finally, the third IBSP instance reduces significantly even the number of peaks arising in the most stressful moments. Obviously, these results are less meaningful than the ones obtained in a real execution environment running consolidated benchmarks. However, they provide a reasonable cue advising the use of multiple instances of IBSP for pure performances reasons, as it seems when the size of PVC virtual cluster is one order of magnitude larger than the largest we reached in the real environment.

4.2 Condor/BLAST

We set up a different PVC configuration to test the IBSP functionalities in another applicative scenario. We configured PVC peers to use the *Kernel Module interposition technique* instead of the Library interposition, used to run the NPB benchmarks.

Even if the communication protocol does not change, the different approach to manage the network communications brings PVC to behave like a different application compared to the previous case. We ran the job scheduler Condor on top of PVC. The jobs submitted to Condor belong to BLAST (Basic Local Alignment Search Tool) application [35]. This program compares nucleotide or protein sequences to sequence databases and calculates the statistical significance of matches. In the tests, different queries are performed against the Genome Survey Sequence (GSS) database, being managed each one of them as an independent job. Condor schedules these jobs, and assigns them to the suitable nodes. Condor has been set up without any special configuration and with a single resource pool, where the participating nodes are registered. We ran this test on virtual clusters composed respectively by 32 and 64 PVC nodes. The configuration of the PVC virtual cluster size follows the same pattern we used to run the NAS benchmarks. The number of BLAST queries submitted for each test is equal to the number of participating nodes.

Figure 4 reports the graphs of the average time obtained for the enrollment and the inter-connection establishment. The measures show trends similar to the previously seen ones. The different absolute values are a reflection of the different interposition technique used by PVC. Figure 5 provides a single view of the time percentage related to each IBSP operation compared to the whole execution time of each application we ran. We used the data coming from the experiments with the largest virtual cluster sizes, 128 nodes for NPB benchmarks and 64 nodes for Condor/BLAST. In the lowest total execution time, the IBSP overhead does not exceed the 20%. The Condor/Blast test over PVC configured with the Kernel Module interposition technique has the largest impact.

4.3 Heterogeneous Environment

The aim of the third type of test is to prove the capability of IBSP to manage PVC peers belonging to different domains. We collected PVC peers running over resources coming from: Amazon EC2 (5 PVC peers and 1 IBSP node), DSLlab [36] (1 PVC peer and 1 IBSP node), an experimental platform providing a set of nodes spread over

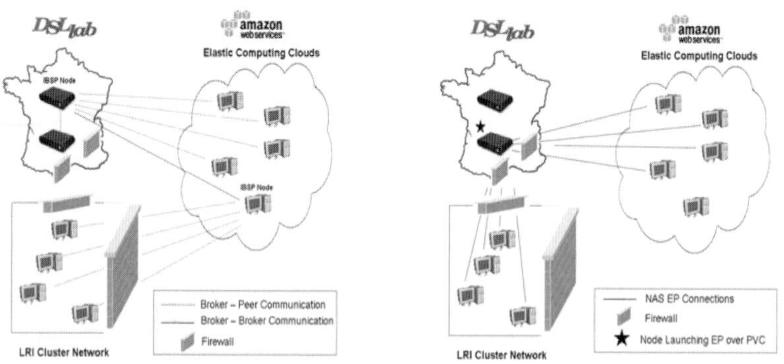

Fig. 6. Heterogeneous environment configuration

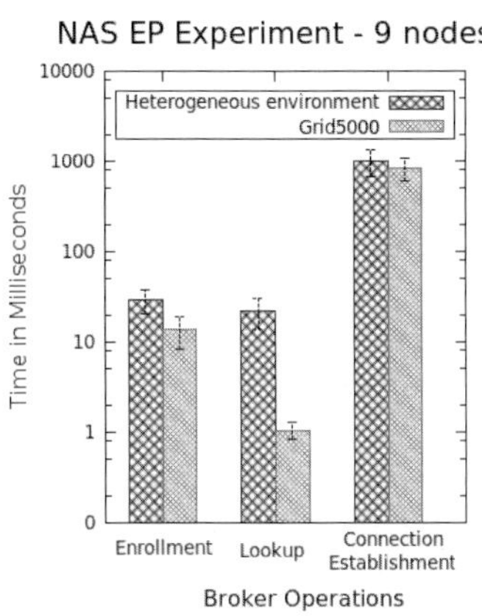

Fig. 7. Heterogeneous experiment results

the ADSL, and a private network (4 PVC peers) belonging to the Laboratoire de Recherche en Informatique (LRI), located at Orsay, France. We ran the EP NAS benchmark on top of PVC, using all the available resources (i.e. 9 PVC peers), with the same configuration used in the experiments shown in section 4.1.

Figure 6 shows the IBSP-PVC topology, and the connections established among PVC peers when the NAS EP benchmark is launched. Special attention is needed when the PVC-IBSP resources belong to different administration domains. PVC peers belonging to the same domain cannot connect to an IBSP node which is also running in the same domain: the IBSP node would register their private real IPs instead of the public real IPs associated to their administration domain. This would cause a conflict when an external PVC peer attempts to establish a connection because the IBSP node would provide the private addresses of the requested peers. This problem is solved by enrolling PVC peers to IBSP nodes running in different domains. This deployment operates on a real user environment instead of an isolated one as Grid5000. In fact, this experiment is able to collect PVC peers belonging from commercial providers to single Internet-connected machines (DSLlab), and goes a step further from isolated test-bed environments. We ran the same experiment on the Grid5000 platform to be able to compare the results coming from this heterogeneous environment. Figure 7 shows the measures we obtained. The resource enrollment and resource lookup operations present a longer execution time than their equivalents on Grid5000 because of the higher network latency due to the communication through the public Internet. There is not a substantial change in the Inter-connection procedure. In fact, the lightweight IBSP-PVC communication protocol is not affected by the lower bandwidth of the public Internet.

5 Conclusion and Future Works

In this paper, we have presented a distributed software architecture targeted to provide a brokering service seized to various needs of P2P-based virtual cluster platforms. Our work takes its origin from a classification of brokering needs arising in different platforms. The drawn categorization is the base of the Information Brokering Service Provider (IBSP), a SOA-compliant Resource Broker for virtual clusters. It is built on top of a flexible architecture which allows to be plugged-in in order to serve different platforms.

We tested the early IBSP reference implementation serving the Private Virtual Cluster (PVC) platform. As shown in section 4, the overhead added by IBSP to the considered applications and benchmarks is acceptable when compared to the total execution time. We performed our tests on an isolated test-bed as Grid5000 as well as on a real scenario involving resources distributed over Internet. The main contribution of the experiments on Grid5000 is a precise evaluation of IBSP capabilities without any disturbs possibly arising in real environments. The experiments showed that a single IBSP instance could be enough for a large number of PVC peers. However, as shown in the last experiment, there are real cases which require multiple instances because of practical burdens not permitting the optimal deployment. In fact, IBSP is built with a flexible architecture which allows administrators to tune its topology. Our evaluation quantifies the overhead involved by this possibility, and provides precise cues on the performance impact which can be brought by a particular topology choice. The encouraging results we obtained are the starting point for definition of new plug-ins in order to promote the IBSP adoption for the deployment of P2P-based virtual clusters. As immediate future works, we planned the development of the IBSP support for the emerging HIPCAL [37] cluster virtualization platform.

Acknowledgments. Experiments presented in this paper were carried out using the Grid'5000 experimental test-bed, being developed under the INRIA ALADDIN development action with support from CNRS, RENATER and several Universities as well as other funding bodies (see https://www.grid5000.fr).

References

1. Foster, I., Iamnitchi, A.: On Death, Taxes, and the Convergence of Peer-to-Peer and Grid Computing. In: Kaashoek, M.F., Stoica, I. (eds.) IPTPS 2003. LNCS, vol. 2735. Springer, Heidelberg (2003)
2. Ruth, P., McGachey, P., Xu, D.: VioCluster: Virtualization for Dynamic Computational Domains. In: Proceedings of the IEEE International Conference on Cluster Computing, Cluster 2005, Boston, MA, USA, September 26-30 (2005)
3. Ganguly, A., Wolinksky, D.I., Boykin, P.O., Figueiredo, R.: Improving Peer Connectivity in Wide-area Overlays of Virtual Workstations. In: Proceedings of the 7th International Symposium on High Performance Distributed Computing, HPDC 2008, Boston, MA, USA, June 23-27 (2008)
4. Rezmerita, A., Morlier, T., Néri, V., Cappello, F.: Private Virtual Cluster: Infrastructure and Protocol for Instant Grids. In: Nagel, W.E., Walter, W.V., Lehner, W. (eds.) Euro-Par 2006. LNCS, vol. 4128, pp. 393–404. Springer, Heidelberg (2006)
5. Cluster On Demand (COD), http://www.cs.duke.edu/nicl/cod/
6. Keahey, K., Figueiredo, R., Fortes, J., Freeman, T., Tsugawa, M.: Science Clouds: Early Experiences in Cloud Computing for Scientific Applications. In: Proceedings of the International Conference on Cloud Computing and its Application, CCA 2008, Chicago, IL, USA, October 22-23 (2008)
7. Amazon Elastic Compute Cloud (EC2), http://aws.amazon.com/ec2/
8. Nimbus, http://workspace.globus.org/
9. Eucalyptus, http://eucalyptus.cs.ucsb.edu/
10. Krafzig, D., Banke, K., Slam, D.: Enterprise SOA: Service-Oriented Architecture Best Practices. Prentice-Hall, Englewood Cliffs (2004)

11. Jiang, X., Xu, D.: VIOLIN: Virtual internetworking on overlay infrastructure. In: Cao, J., Yang, L.T., Guo, M., Lau, F. (eds.) ISPA 2004. LNCS, vol. 3358, pp. 937–946. Springer, Heidelberg (2004)
12. XEN, `http://www.xen.org/`
13. Irwin, D., Chase, J., Grit, L., Yumerefendi, A., Becker, D., Yocum, K.: Sharing Networked Resources with Brokered Leases. In: Proceedings of the annual USENIX Technical Conference, Boston, MA, USA, 30 May – 3 June (2006)
14. Sundararaj, A., Gupta, A., Dinda, P.: Dynamic Topology Adaptation of Virtual Networks of Virtual Machines. In: Proceedings of the Seventh Workshop on Languages, LCR 2004, Houston, TX, USA, October 22-23 (2004)
15. Tsugawa, M., Fortes, J.: A Virtual Network (ViNe) Architecture for Grid Computing. In: Proceedings of the 20th International Parallel and Distributed Processing Symposium, IPDPS 2006, Rodhes Island, Greece, April 25-29 (2006)
16. UDDI Specification, `http://www.uddi.org/pubs/uddi_v3.htm`
17. Sun, C., Lin, Y., Kemme, B.: Comparison of UDDI registry replication strategies. In: Proceedings of the International Conference on Web Services, San Diego, CA, USA, July 6-9 (2004)
18. UBR Shutdown FAQ, `http://uddi.microsoft.com/about/FAQshutdown.htm`
19. Banerjee, S., Basu, S., Garg, S., Lee, S.J., Mullan, P., Sharma, P.: Scalable Grid Service Discovery Based on UDDI. In: Proceedings of the 3rd International Workshop on Middleware for Grid Computing, Grenoble, France, November 28-29 (2005)
20. Pallickara, S., Fox, G.: NaradaBrokering: A Middleware Framework and Architecture for Enabling Durable Peer-to-Peer Grids. In: Endler, M., Schmidt, D.C. (eds.) Middleware 2003. LNCS, vol. 2672. Springer, Heidelberg (2003)
21. JXTA, `http://www.sun.com/software/jxta/`
22. H2O, `http://dcl.mathcs.emory.edu/h2o/`
23. Wun, A., Petrovi, M., Jacobsen, H.: A System for Semantic Data Fusion in Sensor Networks. In: Proceedings of the inaugural International Conference on Distributed Event-Based Systems, DEBS 2007, Toronto, Canada, June 20-22, 2007, pp. 20–22 (2007)
24. JAIN SLEE, `http://jainslee.org`
25. Freeman, T., Keahey, K.: Flying low: Simple leases with workspace pilot. In: Luque, E., Margalef, T., Benítez, D. (eds.) Euro-Par 2008. LNCS, vol. 5168, pp. 499–509. Springer, Heidelberg (2008)
26. Nurmi, D., Wolski, R., Grzegorczyk, C., Obertelli, G., Soman, S., Youseff, L., Zagorodnov, D.: The Eucalyptus Open-source Cloud-computing System. In: Proceeding of the International Conference on Cloud Computing and its Application, CCA 2008, Chicago, IL, USA, October 22-23 (2008)
27. Enomaly, `http://www.enomaly.com/`
28. Dabek, F., Brunskill, E., Kaashoek, M.F., Karger, D., Morris, R., Stoica, I., Balakrishnan, H.: Building Peer-to-Peer Systems With Chord, a Distributed Lookup Service. In: Proceedings of the 8th Workshop on Hot Topics in Operating Systems, HotOS-VIII, Elmau/Oberbayern, Germany, May 20-23 (2001)
29. Rowstron, A., Druschel, P.: Pastry: Scalable, decentralized object location, and routing for large-scale peer-to-peer systems. In: Guerraoui, R. (ed.) Middleware 2001. LNCS, vol. 2218, p. 329. Springer, Heidelberg (2001)
30. FreePastry DHT, `http://freepastry.org`

31. Bamboo DHT, http://bamboo-dht.org/
32. Kato, D., Kamiya, T.: Evaluating DHT Implementations in Complex Environments by Network Emulator. In: Proceedings of the 6th International Workshop on Peer To Peer Systems, IPTPS 2007, Bellewue, WA, USA, February 26-27 (2007)
33. Cappello, F., et al.: Grid'5000: a large scale, reconfigurable, controllable and monitorable Grid platform. In: Proceedings of the 6th IEEE/ACM International Workshop on Grid Computing, GRID 2005, Seattle, WA, USA, November 13-14 (2005)
34. NPB, http://www.nas.nasa.gov/Resources/Software/npb.html
35. BLAST, http://blast.ncbi.nlm.nih.gov/Blast.cgi
36. DSLlab, https://dsllab.lri.fr:4322
37. HIPCAL, http://hipcal.lri.fr/

Efficient Hierarchical Quorums in Unstructured Peer-to-Peer Networks

Kevin Henry, Colleen Swanson, Qi Xie, and Khuzaima Daudjee

David R. Cheriton School of Computer Science
University of Waterloo
Waterloo, Ontario, Canada
{k2henry,c2swanso,q7xie,kdaudjee}@uwaterloo.ca

Abstract. Managing updates in a peer-to-peer (P2P) network can be a challenging task, especially in the unstructured setting. If one peer reads or updates a data item, then it is desirable to read the most recent version or to have the update visible to all other peers. In practice, this should be accomplished by coordinating and writing to only a small number of peers. We propose two approaches, inspired by hierarchical quorums, to solve this problem in unstructured P2P networks. Our first proposal provides uniform load balancing, while the second sacrifices full load balancing for larger average quorum intersection, and hence greater tolerance to network churn. We demonstrate that applying a random logical tree structure to peers on a per-data item basis allows us to achieve near optimal quorum size, thus minimizing the number of peers that must be coordinated to perform a read or write operation. Unlike previous approaches, our random hierarchical quorums are always guaranteed to overlap at at least one peer when all peers are reachable and, as demonstrated through performance studies, prove to be more resilient to changing network conditions to maximize quorum intersection than previous approaches with a similar quorum size. Furthermore, our two quorum approaches are interchangeable within the same network, providing adaptivity by allowing one to be swapped for the other as network conditions change.

1 Introduction

Peer-to-peer (P2P) networks have become increasingly attractive over the last few years; systems such as Gnutella [1] and Kazaa/FastTrack [2] are used extensively for file-sharing over the Internet. The underlying idea of having a completely decentralized, dynamic configuration of heterogeneous peers promises several benefits over more traditional approaches, which are limited by the need for costly, high-maintenance central servers that often act as a bottleneck to the scalability of the system. P2P systems, on the other hand, can distribute workloads among all peers according to resource capabilities, thereby allowing the network size to grow. Moreover, with many nodes in the network, P2P systems can achieve high availability and redundancy at a lower cost.

R. Meersman, T. Dillon, P. Herrero (Eds.): OTM 2009, Part I, LNCS 5870, pp. 183–200, 2009.

An open question is how to efficiently use P2P networks in the distributed database setting. Several issues must be considered, including appropriate data placement and replication strategies, how best to achieve coordination among peers, and in particular, how to handle updates. It is this last issue that is the main focus of this paper; we seek to ensure consistency of updates in a completely decentralized, unstructured P2P setting. We devise a hierarchical quorum system for managing updates that i) requires no coordination among the nodes beyond the usual query messages and responses, ii) requires writes at only a small percentage of data copies (as low as 10% when replication levels are high), iii) guarantees that all requests will see an up-to-date version of the data in the absence of node failure, and iv) achieves a high probability of accessing up-to-date data even when the network suffers from a high degree of instability caused either by *network churn* (peers entering and leaving the network at a normal rate) or *peer failure* (peers leaving the network without new peers joining to maintain network connectivity).

Our work has been inspired by that of Del Vecchio and Son [3], who consider the problem of applying a traditional quorum consensus method to an unstructured P2P distributed database system. Our methods differ significantly from theirs, however, as they only give probabilistic estimates as to the likelihood of seeing up-to-date copies in a fully-functioning network. In addition, we achieve small quorums without sacrificing consistency, whereas Del Vecchio's scheme relies on a trade-off between quorum size and the probability of accessing stale data. Other than [3], little research has been done on quorum-based P2P update management, and none of the other schemes address the unstructured P2P setting. Instead, most relevant work relies heavily on structured P2P systems, such as those with distributed hash tables (DHTs). It is an interesting question whether or not there is a practical method of handling updates without the use of an overlay network such as a DHT, and it this problem that we explore in this paper.

We motivate our approach with a brief discussion of quorums and present Del Vecchio's flexible quorum system in Sect. 2, before giving a detailed description of our quorum selection algorithms in Sect. 3. We then analyze the performance of our system over a simulated Gnutella network in Sect. 4, and give a detailed comparison of our results with Del Vecchio's in Sect. 5. After presenting other relevant work in Sect. 7, we give concluding remarks and comment on future research directions in Sect. 8.

2 Quorums

Quorums are a well-known method of coordinating read and write operations on multiple copies of a data item. Each copy of the data item is assigned a certain number of votes, and to perform a read or write operation, a corresponding read or write quorum must be formed by assembling a given number of these votes, specified as the read/write quorum level (r and w, respectively). The underlying principle of a *correct* quorum system is that any two quorums intersect; in this way, an up-to-date version of the data copy will always be present in any quorum. We remark that the smaller the quorum size (i.e. the fewer required votes), the more efficient the quorum system; we always seek to minimize the amount of communication necessary to perform a read or write operation.

The traditional, or majority, quorum consensus method relies on the rules $r + w > n$ and $2w > n$, where n is the number of copies of the given data item. These rules guarantee the desired quorum intersection property, thereby ensuring data consistency. An obvious question is how we can effectively relax these consistency guarantees—for example, if we can tolerate reading stale data from time to time, perhaps we can make the read quorum more efficient by reducing the number of votes, while still keeping the probability of reading an up-to-date copy at an acceptable level for the given application.

Del Vecchio and Son [3] investigate this issue in the P2P setting using a probabilistic, flexible version of the traditional quorum consensus method. That is, they investigate the probability of accessing stale data when write quorum levels do not require a majority of the data copies, and experiment with different quorum levels in a simulated Gnutella network. Their results are promising, in that, if the data is replicated at a relatively large percentage of the peers (20% or greater), the probability of accessing stale data is low even with quorum levels as low as 20% and the presence of network churn. We refer to the quorum system of Del Vecchio and Son as the *flexible quorum* approach, since their basic idea is to allow peers to vary quorum levels; a more detailed look at the probabilistic guarantees their scheme provides is given in Sect. 5.

Given the unpredictable nature of P2P networks, some sacrifice of consistency is necessary: we will always be faced with the possibility of peer failure and network churn, so we cannot guarantee that even one up-to-date copy of a file is present at a given time. Whereas Del Vecchio chooses to sacrifice consistency in order to lower quorum size, we focus on achieving small quorums while relaxing consistency *only* in the presence of network churn or peer failure.

In this paper, we propose two *hierarchical quorum consensus* algorithms for managing updates in unstructured P2P systems. We propose a scheme that builds on a hierarchical scheme [4], which imposes a logical multi-level tree structure on the location of data copies and recursively applies the traditional quorum consensus method to each level. That is, data copies are viewed as leaves of an n-level tree, and for each level i ($i = 1, 2, \ldots, n$), a read quorum r_i and a write quorum w_i of nodes are assembled recursively that satisfy the following two properties: $r_i + w_i > \ell_i$ and $2w_i > \ell_i$, where ℓ_i is the number of nodes at level i of the tree. In particular, by recursive we mean that at each level i, we are assembling a quorum from the children of the nodes in the quorum at level $i - 1$. A relatively straightforward proof by induction shows the correctness of this quorum method; we shall not reproduce this result here. It is further shown in [4] that the optimal quorum size can be achieved using ternary, or degree three, trees, giving quorum sizes of $n^{0.63}$ copies as opposed to $\lceil \frac{n+1}{2} \rceil$ copies as in traditional quorum consensus. We describe our hierarchical P2P quorum system in the next section.

3 Hierarchical P2P Quorums

The benefits of the hierarchical quorum approach come at the cost of requiring a fixed tree structure. For this reason hierarchical quorums cannot be directly implemented in an unstructured P2P network. In this section two variants of the hierarchical quorum approach are developed that allow peers to randomly, but deterministically, select their position in a tree without any communication or coordination between themselves. Our approach is probabilistic in the sense that the tree structure is random; however, it has the property of failing gracefully toward the majority quorum in the worst case. Unlike the flexible quorum approach described in the previous section, our quorums are guaranteed to overlap when all peers are reachable.

Our first and simplest approach, the *random hierarchical quorum*, assigns peers to locations in the tree randomly and recursively builds quorum by selecting a random majority at each level. The second approach, the *hybrid hierarchical quorum*, utilizes a combination of random and fixed traversals of the tree in a manner that increases the expected intersection size, while still selecting most members of the quorum at random.

3.1 Random Hierarchical Quorums

Let M be an upper bound on the number of peers in the network; we assume this value is a public parameter. Let T be a complete ternary tree with depth d defined such that there are 3^d leaf nodes, where d is the smallest integer such that $3^d > M$, i.e., $d = \lceil \log_3 M \rceil$. The levels of the tree are labeled $0, \ldots, d$, with the root located at level 0 and the leaves at level d; the leaves of the tree are indexed $0, \ldots, 3^d - 1$. Peers will be located at level d of the tree. Since the network is unstructured, peers determine their own location in the tree by generating a random integer modulo 3^d for each data item. We recommend using a hash function, such as the SHA family [5], to compute location as a function of a peer's network address and the primary key of a given data item. In this manner, each tree is uniquely determined by the primary key of the data item, and each peer's location within the tree is determined by its network address, thereby allowing any peer to compute the location of any other peer without any communication. Given that each peer selects its position in the tree at random, it is expected that several peers will select overlapping locations, and also that several locations in the tree will be empty. Thus, instead of treating each leaf as a single peer, each leaf corresponds to a (possibly empty) set of peers. No peers are located at non-leaf positions in the tree, but one may conceptually visualize each non-root position in the tree as the set of all leaves/peers rooted at that position.

In our protocol, each data item is associated with a tree T as described above. Thus, each peer has a collection of tree locations: one for each data item the peer possesses. We comment that M may be chosen optimistically, as there is no penalty to over-estimating network size. Underestimating network size does not affect correctness, but will affect quorum size if the estimate is too low. In particular, choosing M, and hence the depth of the tree, to be significantly larger than the size of the network has only the drawback of increasing the maximum size of the integer associated with each peer's location in the tree. On the other hand, choosing a very small tree depth increases the size of quorums.

Figure 1a shows a sample tree assignment on 9 peers labeled $A, \ldots, H$. The rest of Fig. 1 shows how the quorum generation algorithm, given by **RandQuorum**, traverses the tree to build quorum; in each part of the figure, the nodes selected by **RandQuorum** are bold. A summary of the algorithm, together with the running example from Fig. 1, is as follows: the peer wishing to establish quorum sends a search query into the network to locate peers containing relevant data and their corresponding tree positions. (Alternatively, the locations could be computed by the peer receiving the responses, if location is computed using a hash of a peer's network address.)

Once all responses have been received, a quorum is generated by starting at the root of T and recursively building quorum at each level. There are two possible cases: 1) If there are two or more non-empty children (i.e. the number of the peers rooted at the child is non-zero), then two non-empty children are selected at random and the process is repeated at each selected child (lines 1-4). We see this in Fig. 1b; here, the leftmost and rightmost children on the first level are chosen. Similarly, we see that the leftmost node of the first level has three non-empty children, so we choose two of them at random, namely the middle and rightmost (shown in Fig. 1c). 2) If there are not at least two non-empty children, the union of all peers rooted at the current position in the tree is selected and a majority is returned (lines 6-9). We see this case in the rightmost node of the first level of Fig. 1b. Here, as shown in Fig. 1c, the set of all peers rooted at this point are collected into a single set and a random majority is returned.

The algorithm terminates when the bottom of the tree has been reached, and the set of all peers returned by the algorithm forms the quorum. We see this in Fig. 1d: since neither selected subtree (from Fig. 1c) has at least 2 non-empty children, a random majority of the peers rooted at this point are returned. The union of all peers returned by the algorithm $\{B, C, G, H\}$ forms the quorum.

Algorithm 3.1: **RandQuorum**

Input: A non-empty ternary tree T

Output: A random hierarchical quorum on T

1. **if** T has two or more non-empty children **then**
2.   Select two distinct non-empty children T_1 and T_2 at random
3. $S_1 \leftarrow$ **RandQuorum**(T_1) {Recursively build quorum on T_1}
4. $S_2 \leftarrow$ **RandQuorum**(T_2) {Recursively build quorum on T_2}
5.   **return** $S_1 \cup S_2$ {Combine quorums generated from T_1 and T_2}
6. **else**
7. $S \leftarrow$ **LEAVES**(T)
8.   **return MAJORITY**(S)
9. **end if**

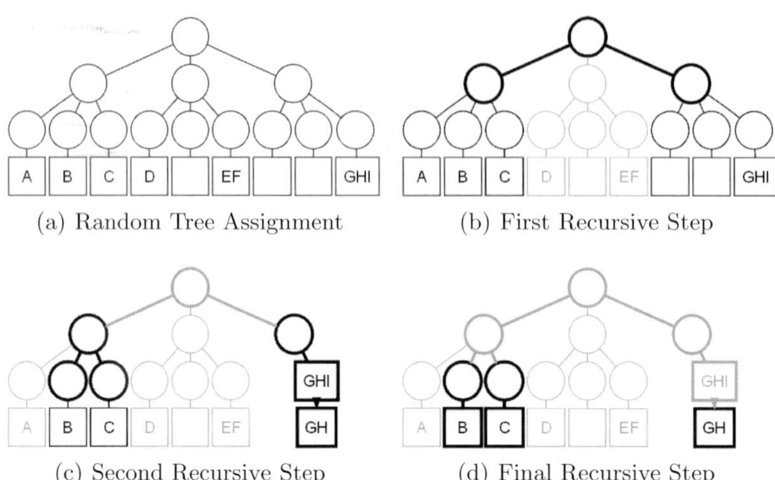

(a) Random Tree Assignment (b) First Recursive Step

(c) Second Recursive Step (d) Final Recursive Step

Fig. 1. A sample run of **RandQuorum** on the set of peers $\{A, \ldots, I\}$. The bold path demonstrates how the tree is traversed at each recursive step, with the bolded boxes representing those peers returned by the algorithm. The union of the sets $\{B\}$, $\{C\}$, and $\{G, H\}$ forms the quorum.

RandQuorum makes use of two sub-routines, namely, **LEAVES**(T) and **MAJORITY**(S). The former returns the set of leaf nodes of T, while the latter returns a random majority quorum from the elements of S.

Recall that each peer selects its position in the tree independently from all others. Thus, while it is unlikely that this process results in a completely balanced tree, it is also unlikely that the result is a significantly unbalanced tree. At level k of the tree there are 3^k nodes, each the root of a subtree with 3^{d-k} leaves. If n peers randomly join the tree, then the probability that a subtree at level k is empty is $(1 - 3^{-k})^n$. For example, if the network contains 1000 peers, then a tree of depth 7 will be chosen. This tree has capacity for 2187 peers, and thus will always contain empty nodes at level 7. The probability that level 6 contains empty subtrees is approximately 25%, whereas the probability that level 5 contains empty subtrees is approximately 1%. Because the probability of empty subtrees becomes exponentially unlikely as one moves closer to the root, it is reasonable to assume that such randomly constructed trees will always be fully populated at a reasonable depth relative to the number of peers in the network. For this reason, our quorums are comparable in size to optimal hierarchical quorums, despite the fact that our trees are generated randomly.

Figure 2a shows the average size of quorums generated by our approach in comparison with the theoretical best performance of hierarchical quorums. For reference, a line representing the size of the quorum obtained from the majority quorum consensus algorithm is also included. These results were generated by taking the average size of a quorum returned by **RandQuorum** on 1000 different randomly constructed trees for each network size.

In the absolute worst case, each peer may choose the exact same location in the tree. In this case, **RandQuorum** is unable to move lower in the tree and simply returns a majority of the entire set of peers. Thus, the worst case performance is equivalent to the majority quorum consensus approach. Clearly, the event that each peer selects the same location in the tree is exceptionally rare, as discussed earlier in this section.

As peers enter and leave the network, the structure of the tree changes. Despite this, because each peer generates its position in the tree independently, the distribution of peers across the leaves of the tree will remain uniform over time. Thus, incoming peers do not need to assume a vacancy left by a peer that has just exited. This approach also has the benefit of being robust against failures, such as network outages or partitions, in the sense that there is no correlation between a peer's physical location in the network and logical location in the tree. After a large scale failure, the remaining peers will still be uniformly distributed among the leaves of the tree, and near-optimal quorum sizes will still be expected. We investigate the problem of strengthening the random hierarchical quorum against peer failure in the next section.

3.2 An Improvement: Hybrid Hierarchical Quorums

The random hierarchical quorum approach developed in the previous section guarantees that any two quorums generated from the same tree overlap at at least one point. In a fixed network, this would be sufficient to guarantee correctness; however, in a P2P network this is not sufficient. Between the generation of any two quorums, multiple peers may enter or leave the system. Furthermore, it cannot be guaranteed that every peer in the network can reach every other peer. Thus, if the average intersection size of two quorums is small, then even a relatively small number of peers becoming unreachable could cause two quorums to not overlap with unacceptably high probability. For this reason, it is desirable to maximize the intersection of any two quorums.

One method of maximizing quorum intersection is to always choose the same set of peers to be in a given quorum. This could be accomplished in the random tree approach by always selecting the same majority of nodes at each level of the tree. This approach always generates the same quorum for a given data item, with the same average quorum size as a random traversal, but it also means that only a fixed, small percentage of peers will ever be chosen to receive reads or writes for a given data item. Thus, a high resilience to peers becoming unreachable comes at the cost of focusing the load on small set of peers, rather than distributing the workload randomly among the available replicas. As peers are distributed randomly at the leaves of the tree, however, fixing the traversal to always select the left and middle children, for example, will still result in a different random quorum for each data item, as each data item is associated with a unique tree.

To compromise between maximizing quorum intersection and balancing load, we combine the load balancing of a random traversal with the large intersection of a fixed traversal in the following manner: at the root, we traverse the leftmost subtree using a fixed majority at each level, and traverse one of the remaining two subtrees selecting a random majority at each level. This is summarized in the algorithms **HybridQuorum** and **FixedQuorum**.

FixedQuorum is identical to **RandQuorum** except that the algorithm proceeds recursively on the left and middle subtrees when both are non-empty, rather than on two random non-empty subtrees.

Algorithm 3.2: **HybridQuorum**

Input: A non-empty ternary tree T
Output: A hybrid hierarchical quorum on T

if left subtree and at least one remaining subtree are non-empty **then**
  $T_1 \leftarrow$ **LEFT**(T)
  $T_2 \leftarrow$ a random non-empty non-left subtree
  $S_1 \leftarrow$ **FixedQuorum**(T_1) {Build a fixed quorum from T_1}
  $S_2 \leftarrow$ **RandQuorum**(T_2) {Build a random quorum from T_2}
  return $S_1 \cup S_2$ {Combine quorums generated from T_1 and T_2}
else
  $S \leftarrow$ **LEAVES**(T)
  return **MAJORITY**(S)
end if

Figure 2b demonstrates the effectiveness of this approach. One can see that the intersection size is increased significantly compared to the randomly generated quorum, while still leaving plenty of room for load to be distributed among different peers in the tree. These results were generated by running either **RandQuorum** or **HybridQuorum** twice on the same tree and counting the number of common peers in each. This process was averaged over 1000 different trees.

One benefit of using the tree structure to select the fixed set of peers is that the intersection property is guaranteed between any two hierarchical based quorums, even if one is built using a random traversal, and the other a partially-fixed traversal. In a setting where peers are relatively stable, the random hierarchical quorum may be used, with the benefit of randomly distributing load among replicas of a data item. If network conditions change and a large number of failures or exiting peers are detected, peers can seamlessly switch to the hybrid hierarchical quorum, and all quorums generated from this point will be expected to intersect at a larger number of peers. Quorums generated prior to the switch will still be valid, although not as resilient to failure or changing network conditions.

Although the hybrid hierarchical quorum shifts extra load onto some fixed set of peers, it is important to recall that trees are built on a per-data item basis. Whether or not a given peer is a member of the fixed set for one data item does not have any bearing on whether or not it is a member of the fixed set for a different data item. The set of fixed peers for a given data item is chosen uniformly at random, and thus, although the load balance is skewed for any single data item, the load balance across all data items remains random.

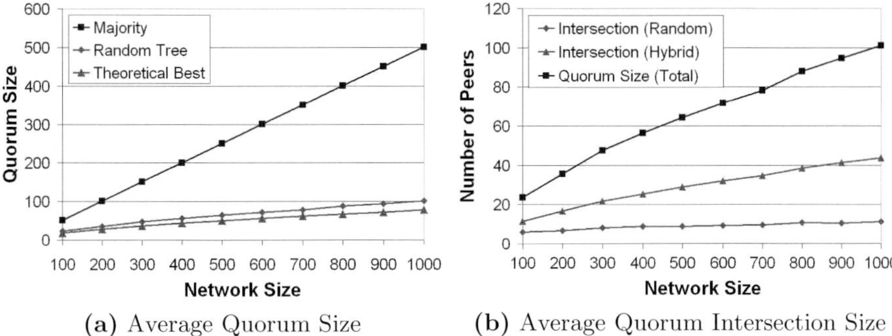

(a) Average Quorum Size (b) Average Quorum Intersection Size

Fig. 2. (a) Comparison of average random tree quorum size to the majority quorum and optimal hierarchical quorums. Widest 99% confidence interval: ±0.60. **(b)** Comparison of average quorum intersection size for a random traversal and hybrid traversal of the tree. Generated by randomly assigning peers to 1000 different trees and averaging the intersection size of two quorums from each. Widest 99% confidence interval: ±0.64.

4 Performance Evaluation

In order to evaluate the performance of our random and hybrid hierarchical quorum approaches, we performed several experiments using the Gnutella protocol [1] simulated under NeuroGrid [6]. We note that the Gnutella protocol is a rather basic P2P system that relies on flooding the network with messages in order to locate documents; as such, the number of times a message is forwarded is limited by a time-to-live (TTL) parameter. The Gnutella protocol is well documented, widely used, and is completely unstructured in its basic form. Unlike some unstructured P2P protocols, such as FastTrack, Gnutella benefits from being an open specification. These properties make Gnutella an ideal candidate for testing the effectiveness of our random hierarchical quorum approaches. An additional benefit of performing our experiments using Gnutella is the ability to directly compare our results to Del Vecchio and Son [3].

4.1 Experimental Environment

Although NeuroGrid is a stand-alone P2P protocol, the NeuroGrid simulator provides functionality for simulating several P2P protocols, including Gnutella. A Gnutella network is modeled as a set of peers, each possessing one or more documents, and each document is associated with one or more keywords. Random networks are generated with respect to a connectivity parameter, specifying the number of connections a peer establishes with others when it joins the network. At network creation time, each peer chooses k distinct peers (with a default value of $k = 2$) and establishes a connection with each. Thus, each node has, on average, approximately $2k$ connections to other peers in the system. When implementing our algorithms, we noticed that the simulator disregards the connectivity specified in the configuration file and instead uses a hard-coded value of 2 connections per node. For our simulations, we manually changed the connectivity of the network to $k = 3$. Our reasons for this choice are justified in Sect. 4.2

The main functionality provided by NeuroGrid is the ability to specify a distribution of documents across peers, perform a keyword search for a given document, henceforth referred to as a *query*, and generate result statistics, such as the number of peers with a match and number of messages sent between peers. By default, queries originate from a random peer and are for a single keyword. We set each node to contain a single document, with each document having a single keyword associated with it, and vary the number of keywords to model the replication rate of data items within the network. For example, 1 keyword implies this keyword is associated with every document (100% replication), whereas 5 keywords means each peer is assigned a given keyword with probability $\frac{1}{5}$, thus achieving 20% replication for each keyword. Because keywords are assigned randomly, we use a modified keyword distribution algorithm to guarantee exact replication rates.

For completeness, we also describe NeuroGrid's behavior with respect to the TTL of a message. The default TTL in NeuroGrid is 7, and the Gnutella specifications [7] state that the TTL is the "number of times the message will be forwarded by Gnutella servants before it is removed from the network." Each peer decrements the TTL by 1; if the TTL is non-zero and the message has not been seen before, the peer forwards the message to its neighbors. In NeuroGrid, the originator of the message decrements the TTL before forwarding it. Thus, a message with an initial TTL of 7 will only travel a maximum of 6 hops in NeuroGrid; we leave this functionality unchanged, as we find in Sect. 4.2 that 6 hops are sufficient for the network sizes we consider.

Unless otherwise specified, the results in this section are generated for a network size of 1000 peers, with replication rates of 1%, 5%, 20%, and 100%, generated as described earlier. When averaging our results, we sample 10 values for a given network, and then create a new network. This process is repeated 100 times for a total of 1000 random samples. The connectivity (number of connections each node establishes with others) is set to 3 and the default TTL is set to 7. For each data point a 99% confidence interval has been computed and a description of the largest interval accompanies each plot.

4.2 Network Reachability

In order to verify that our choice of network parameters (connectivity and TTL) is appropriate for a network of this size, we investigated the reachability of peers in the network as successively larger numbers of peers fail or leave the network. Reachability is determined by distributing a single keyword to all peers, terminating all connections for a fixed percentage of peers, and then counting the number of matches for this keyword when queried by a random non-failed peer. Our test included trials for connectivities of 2 and 3, respectively.

Our experiments showed that an initial connectivity of 3, or approximately 6 total connections per peer, is sufficient to retain access to virtually all peers in the network so long as the failure rate is less than 30%. As the number of failed peers increases, the network becomes partitioned and the average reachability declines quickly. If the connectivity is 2, then full reachability is unlikely, and the number of reachable peers drops to less than 50% once 20% of peers have failed. It should be noted that in our simulation, peers did not attempt to establish new connections when a neighbor failed. Applying such a technique could maintain consistent connectivity in the presence of failure, so long as a large number of peers do not fail simultaneously, thereby partitioning the network before connectivity can be restored. By choosing to model peers that do not seek out new connections when neighbors fail, we are modeling the worst case performance under failure.

4.3 Cost of Communicating with Quorum

We can now determine the average cost for communicating with a quorum once it has been established. The communication cost for establishing the quorum takes exactly one query, the cost of which depends on the number of peers and the number of connections between them. The average cost of a query in the networks we consider is roughly constant; we therefore omit the messages involved in establishing quorum from our results.

Figure 3a was generated by recording the TTL of the query message for each peer possessing a document with the target keyword that received the query. From this, we calculate the number of hops necessary to reach each peer in the quorum. In Gnutella, communication with each peer occurs along the path from which it was located, so this gives the total cost of sending individual messages to each peer in the quorum as a function of hops. Optimizations may be possible if a single message may be used to communicate with all peers along a given path. In our simulation we observed that the number of messages required to communicate with the quorum is approximately 4 times the size of the quorum itself.

4.4 Stale Access under Churn

As mentioned in Sect. 3.2, a key goal is to compromise between maximizing quorum intersection and balancing load among replicas. This is motivated by the fact that in a P2P environment, peers are expected to continually leave and join the system. As in [3], we model this with network churn. The churn rate of a network can be defined as the number of peers that are expected to leave and join the system between queries for a given data item. We assume that the network size is constant, i.e., peers enter and leave at the same rate, and also that the replication rate of a data item stays constant. This is modeled by selecting a fixed percentage of peers at random and setting their data to be stale; that is, we assume that all peers rejoining the system have stale data.

Intuitively, the average intersection size of two quorums provides a good indicator of how resistant to churn a given quorum system is. Thus, we expect a

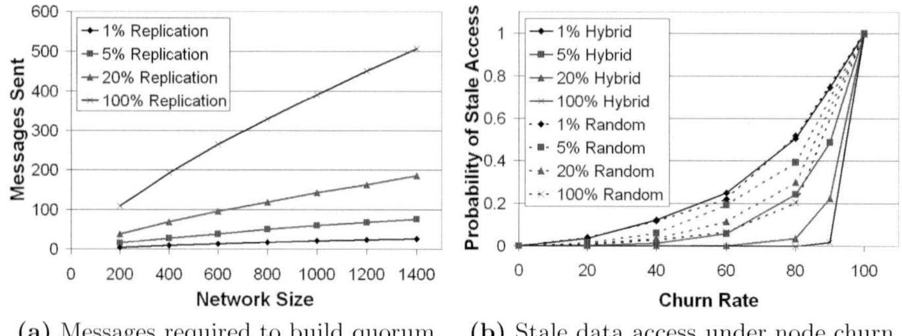

(a) Messages required to build quorum **(b)** Stale data access under node churn

Fig. 3. (a) The average cost of communicating with all peers in a quorum. Confidence intervals grow with the size of the quorum, ranging from ± 0.2 to ± 4.5 messages. **(b)** A comparison of the probability of stale data access under node churn for hybrid hierarchical quorums and random hierarchical quorums. Widest 99% confidence interval: ± 0.04.

hybrid hierarchical quorum to perform much better on average than a random hierarchical quorum. Figure 3b demonstrates the performance of the hybrid hierarchical quorum for several levels of data replication. At 100% replication, we can tolerate 90% node churn without any significant probability of a quorum containing only stale data. This result may seem counterintuitive, but our simulation shows that hybrid hierarchical quorums intersect at over 40 peers on average, and thus, the probability that some of these peers remain after 90% churn is quite high. Indeed, the probability that a random set of 40 peers and a random set of 100 peers (i.e., the non-churned peers) do not intersect is approximately 1.4%. Stale access for lower replication rates is far more probable, as expected, although most replication rates can still handle a reasonable amount of churn.

As we would expect, random hierarchical quorums provide much less resilience against network churn. Figure 3b also shows a direct comparison of hybrid hierarchical quorums and random hierarchical quorums. Even at 40% churn, stale access begins to become a problem at 100% replication. Recall that the hybrid hierarchical increases the expected quorum intersection, but at the cost of requiring a small subset of the peers to always be part of the quorum. Figure 3b suggests that if the level of churn is low, then random hierarchical quorums are sufficient and in this case, load can be distributed uniformly among replicas. Conversely, if the level of churn is high, then the problem of stale data access can be mitigated by biasing the load slightly towards some peers through the use of hybrid hierarchical quorums. Our two quorum approaches have the property that any hybrid quorum and random quorum on the same tree will intersect. Thus, peers can default to the fully load-balanced random hybrid quorums, but switch to hybrid quorums if a large amount of network churn is detected.

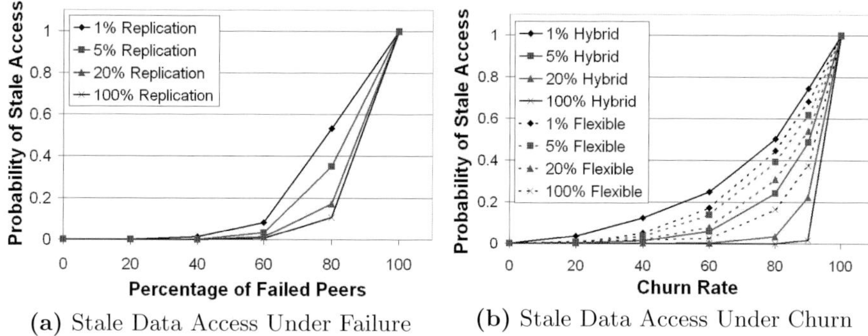

(a) Stale Data Access Under Failure (b) Stale Data Access Under Churn

Fig. 4. **(a)** The probability of accessing stale data after a fixed percentage of peers have left the network. Widest 99% confidence interval: ±0.04. **(b)** A comparison of the probability of stale data access under node churn for hybrid hierarchical quorums and flexible quorums. Widest 99% confidence interval: ±0.04.

4.5 Stale Access under Failure

The tests for stale access under network churn are based on the assumption that peers enter and leave the network at the same rate, thus keeping the average connectivity of the network constant. In reality, network failures may cause a large number of peers to simultaneously become unreachable. The main difference between the results in this section and the previous section is that as the failure rate increases, the average connectivity of the network decreases. Once a significant portion of peers have failed, the network may become partitioned and full reachability is no longer expected.

Figure 4a shows the probability of stale access as an increasing number of peers fail for the hybrid hierarchical quorum. As expected, the probability of stale access becomes noticeable for lower failure rates than the corresponding churn rate. At 40% failure, the reachability of the network quickly begins to drop and the probability of stale access rises accordingly. In practice, we expect that large scale failure is detectable by individual peers, and that read/write operations are suspended if less than 50% of the expected peers are reachable. Del Vecchio and Son assume that the replication rate for a given data item is public knowledge. If this is the case, then quorum generation can be aborted if the number of responses to a query is significantly lower than expected.

5 Comparison with Flexible Quorums

To our knowledge, the only other attempt to apply quorums to unstructured P2P networks is the flexible quorum approach of Del Vecchio and Son [3]. As previously discussed, we have chosen to evaluate our approach in many of the same situations so that we can directly compare the performance of the two approaches, although some considerations must be made. In the flexible quorum approach, the size of the quorum is a free parameter of the system and may be selected to provide the desired level of stale access. Because each flexible quorum consists of a fixed percentage of all replicas, this requires that the replication rate and total number of peers are known by each peer (or at least, that a reasonable estimate is known). In our hierarchical-based quorums, the size of the quorum is not a free parameter, but instead is a function of the replication rate; our approach constructs a quorum out of the entire set of peers that respond to a query. As seen in Fig. 4a, for 1000 peers with 100% replication, the probability of our biased tree quorums not intersecting at 50% failure is near 0. A flexible quorum with a quorum level of 60% would never succeed in establishing quorum at this failure rate, as fewer than 60% of the peers remain in the network. Additionally, the assumption that the replication rate is known allows peers to detect large failures, such as network partition, based on the number of responses to a query. In the presence of a large number of failures, read/write operations could be suspended until connectivity is restored.

A benefit of our hierarchical-based quorums over flexible quorums is that we can *guarantee* that two quorums always overlap if all peers are reachable. In order to guarantee intersection, a flexible quorum must contain at least 50% of the replicas. For 1000 peers with 100% replication, our tree quorums require approximately 10% of peers. It should be noted that, with high probability, flexible quorums do provide a low probability of stale access for quorum sizes smaller than 50%, but there is no guarantee. This behavior is shown in Fig. 4b, which directly compares our hybrid hierarchical quorum to flexible quorums in the presence of network churn.

Since quorum size is not a free parameter in our quorums, in order to compare our approach to flexible quorums, we first computed the expected quorum size of a hybrid hierarchical quorum for the given replication rate. To generate the probability of stale access for flexible quorums, we used quorums of the same average size as our tree quorums. For 1000 peers, the quorum sizes selected were 9.6%, 17.5%, 30%, and 60% for replication rates of 100%, 20%, 5%, and 1%, respectively. For all replication rates above 1%, biased tree based quorums provide a much lower probability of stale access. At the 1% replication level, there are too few peers to see any benefit from using a tree-based approach over the simple majority quorum. Comparing Fig. 3b and Fig. 4b, we see that random tree quorums are similar in performance to flexible quorums in general.

Another comparison point is the time necessary to build quorum, i.e., the query-response time, as a function of quorum size. Del Vecchio and Son observe that smaller quorums can be assembled faster than larger quorums, as it takes less time to hear a response from 10% of replicas than 90% of replicas. In comparison, our tree quorums require the full response set from the query before proceeding. In practice, taking the first set of responses to a query creates a bias towards peers located closer to the query originator. If two peers on opposite sides of a large network each assemble their closest 100 neighbors, then the probability of these two quorums overlapping in a large network is much less than if two nearby peers assemble quorums. To eliminate this bias towards nearer neighbors, it is necessary to wait for the entire set of responses from the query before choosing a random quorum. We note that it is unclear whether Del Vecchio's experiments take this local bias into account; in particular, the experimental section does not specify whether random peers are used for each query, or if the same peer is used for repeated queries. In generating the above flexible quorum results, we used a new peer chosen at random for each query.

We remark that in a real-word scenario peers are limited to a local view of the network and so must use estimates of current network size and document replication rates, which can be obtained via gossip protocols [8], in order to decide on an appropriate time to build quorum. In addition, given that both our hierarchical-based quorums and Del Vecchio's flexible quorums are built by first querying the network, followed by contacting a certain percentage of peers, we expect the number of messages required by each to be approximately equal, particularly if local bias is eliminated. For these reasons, we feel that comparing the two approaches based on quorum size is reasonable.

6 Discussion

Our hierarchical systems are quite flexible in that quorums generated from the random and hybrid algorithms are interchangeable; peers can switch dynamically from one system to the other. Both the random and hybrid hierarchical quorums use the same randomly constructed logical tree structure, which carries the benefit of distributing peers in a ternary tree such that their logical location in the tree is unrelated to their physical location in the network. This means that correlation between peers failing in the physical network does not translate into localized failure within the tree. As the network structure changes, the expected distribution of peers within the logical tree remains constant.

The key difference between the random and hybrid hierarchical quorums is in the traversal of the tree. The random approach chooses a random majority, resulting in a completely random quorum, which has the effect of distributing reads/writes equally across all replicas of a data item. The hybrid approach traverses one subtree with a fixed majority and another with a random majority at each level, and thus shifts some of the load to a fixed set of peers for that data item. The effect of this shift is an increase in the expected size of the quorum intersection, and hence a higher tolerance to network churn and failure. Despite the non-uniform load balance, the fact that a different random tree is used for each data item allows the hybrid hierarchical quorum to still retain load balancing across all data items in the network.

We emphasize that there is a tradeoff between increasing the expected intersection size of two quorums and load balancing when we use the hybrid hierarchical approach. In our current method, although we fix the traversal over part of the tree, thereby increasing the expected intersection size of two quorums, we still leave a large proportion of the peers in sections of the tree that are traversed randomly. We remark that by changing the degree to which the traversal is fixed, various average quorum intersection sizes can be achieved. Since fixing a subset of the peers has the affect of reducing the load balancing provided by our random tree assignment, it may be worth investigating the optimal intersection size for common sets of network parameters. For example, if network churn could be estimated ahead of time, a traversal strategy could be chosen that minimizes the fixed portion of peers, while still providing adequate stale access probabilities.

A drawback of our approach is that the entire network must be queried each time a quorum is constructed. However, the cost of sending messages is negligible compared to the cost of propagating and applying updates and our quorum approach minimizes the size of quorums, thus minimizing the number of peers an update must be sent to and thereby reducing the total cost of propagating and applying updates. A peer could choose to cache the results of a query, or more specifically the tree associated with a given data item. In most practical settings, however, it is unlikely that a peer could make use of a cached result before the instability of the network renders it invalid.

An interesting direction for future work is to consider the presence of Byzantine peers in the network. Such peers can misrepresent the freshness of their data or selectively choose not to forward messages in the network. Our hybrid hierarchical quorum could be used to guard against a small number of Byzantine peers, as the probability of stale access is low even if a large number of peers have failed or possess stale data. Because hybrid quorums are expected to intersect at a large number of peers, we could use a naive k-of-n voting protocol to combat Byzantine peers. In such an approach, a quorum is only accepted as valid if at least k of the n members agree on a specific version of the data.

7 Related Work

Much work has been done regarding the use of quorums in distributed database systems; a comprehensive survey and comparison of many quorum systems, including majority quorum consensus and weighted voting quorums [9], hierarchical quorum consensus [4], grid schemes [10,11,12], and tree quorum protocols [13] may be found in [14]. Wool [15] discusses general issues related to the usefulness of quorums for managing data replication and Naor and Wool [11], in addition to presenting the grid-based Paths quorum system, give a summary of the general quorum tradeoff between load balance and availability. We note that our choice of hierarchical quorum systems relies on its use of a logical structure in which responsibility is distributed symmetrically; as all copies of a data item are located at the leaves of the tree, there is no real hierarchy among the data copies themselves. This is in contrast to tree quorum protocols, for example, in which each node of the tree is assumed to have a copy of the data item and nodes higher in the tree are more important. Moreover, while grid protocols, in which nodes are organized into various grid patterns, appear promising at first glance, these protocols tend to suffer from poor availability in the presence of write operations.

Papers dealing with variations of the hierarchical quorum approach to the distributed database setting include [16], which analyzes multilevel, or hierarchical schemes, concentrating on the problem of determining the most suitable variant for a given application, and [17], which examines a layered multilevel quorum scheme that recursively assembles quorums by alternating between a read-one write-all (ROWA) and majority quorum consensus strategy.

Apart from the flexible quorum approach, research has focused on the application of quorum systems to *structured* P2P networks. Baldoni et al. [18] investigate variants of the hierarchical quorum scheme in Chord [19], a DHT-based P2P system. Their approach relies heavily on the infrastructure of the DHT itself, and so is not applicable to the unstructured setting we consider. In particular, we remark that, in order to make their generalization of the hierarchical quorum system more compatible with Chord's infrastructure, they sacrifice optimal quorum sizes by using degree four trees rather than degree three. Additionally, DHTs like Chord map a keyspace onto a set of peers, whereas in our setting a peer possesses a set of data upon entering the network and is only required to participate in reads and updates for those data items it already possesses. Other examples of DHT-based quorum systems include [20,21,22,23]; as these methods are both unrelated to the hierarchical quorum system and heavily dependent on the infrastructure of the P2P network, we do not discuss them here.

8 Concluding Remarks

We have presented two new approaches to establishing read/write quorums in an unstructured P2P network: the random hierarchical quorum and the hybrid hierarchical quorum. These quorum systems are interchangeable (i.e., a random quorum will always intersect a hybrid quorum), which allows for a seamless transition from one to the other. We have investigated the performance of our random and hybrid hierarchical quorums and demonstrated that our hybrid hierarchical quorum is highly resilient against network churn, allowing one to retrieve fresh data even at churn rates of 90% or when up to half of the network has failed. Our observations show superior performance over previous quorum approaches in unstructured P2P networks. Furthermore, under realistic network sizes, we achieve a lower probability of stale data access for quorums of similar size.

References

1. Kirk, P.: RFC-Gnutella 0.6, `http://rfc-gnutella.sourceforge.net/index.html`
2. Liang, J., Kumar, R., Ross, K.W.: The fasttrack overlay: a measurement study. Comput. Netw. 50(6), 842–858 (2006)
3. Del Vecchio, D., Son, S.H.: Flexible update management in peer-to-peer database systems. In: IDEAS 2005: Proceedings of the 9th International Database Engineering & Application Symposium, Washington, DC, USA, pp. 435–444. IEEE Computer Society Press, Los Alamitos (2005)
4. Kumar, A.: Hierarchical quorum consensus: A new algorithm for managing replicated data. IEEE Trans. Comput. 40(9), 996–1004 (1991)
5. National Institute of Standards and Technology. FIPS 180-2, secure hash standard, federal information processing standard (FIPS), publication 180-2. Technical report, Department of Commerce (August 2002)

6. Joseph, S.: Neurogrid simulation setup, `http://www.neurogrid.net/php/simulation.php`
7. Kirk, P.: Gnutella protocol development: Standard message architecture, `http://rfc-gnutella.sourceforge.net/developer/testing/message-Architecture.html`
8. Kostoulas, D., Psaltoulis, D., Gupta, I., Birman, K.P., Demers, A.J.: Active and passive techniques for group size estimation in large-scale and dynamic distributed systems. J. Syst. Softw. 80(10), 1639–1658 (2007)
9. Gifford, D.K.: Weighted voting for replicated data. In: SOSP 1979: Proceedings of the seventh ACM symposium on Operating systems principles, pp. 150–162. ACM, New York (1979)
10. Cheung, S.Y., Ammar, M.H., Ahamad, M.: The grid protocol: A high performance scheme for maintaining replicated data. IEEE Trans. on Knowl. and Data Eng. 4(6), 582–592 (1992)
11. Naor, M., Wool, A.: The load, capacity, and availability of quorum systems. SIAM J. Comput. 27(2), 423–447 (1998)
12. Kumar, A., Rabinovich, M., Sinha, R.K.: A performance study of general grid structures for replicated data. In: Proceedings the 13th International Conference on Distributed Computing Systems, May 1993, pp. 178–185 (1993)
13. Agrawal, D., El Abbadi, A.: The tree quorum protocol: an efficient approach for managing replicated data. In: Proceedings of the Sixteenth International Conference on Very Large Databases, pp. 243–254. Morgan Kaufmann Publishers Inc., San Francisco (1990)
14. Jiménez-Peris, R., Patino-Martínez, M., Alonso, G., Kemme, B.: Are quorums an alternative for data replication? ACM Trans. Database Syst. 28(3), 257–294 (2003)
15. Wool, A.: Quorum systems in replicated databases: science or fiction. Bull. IEEE Technical Committee on Data Engineering 21, 3–11 (1998)
16. Freisleben, B., Koch, H.-H., Theel, O.: Designing multi-level quorum schemes for highly replicated data. In: Proc. of the 1991 Pacific Rim International Symposium on Fault Tolerant Systems, pp. 154–159. IEEE Computer Society Press, Los Alamitos (1990)
17. Freisleben, B., Koch, H.-H., Theel, O.: The electoral district strategy for replicated data in distributed systems. In: Proc. of the 5th Intern. Conference of Fault-Tolerant Computing Systems, pp. 100–111 (1991)
18. Baldoni, R., Jiménez-Peris, R., Patino-Martínez, M., Querzoni, L., Virgillito, A.: Dynamic quorums for DHT-based enterprise infrastructures. J. Parallel Distrib. Comput. 68(9), 1235–1249 (2008)
19. Brunskill, E.: Building peer-to-peer systems with chord, a distributed lookup service. In: HOTOS 2001: Proceedings of the Eighth Workshop on Hot Topics in Operating Systems, Washington, DC, USA, p. 81. IEEE Computer Society, Los Alamitos (2001)
20. Zhang, Z.: The power of DHT as a logical space. In: IEEE International Workshop on Future Trends of Distributed Computing Systems, pp. 325–331 (2004)
21. Lin, S., Lian, Q., Zang, Z.: A practical distributed mutual exclusion protocol in dynamic peer-to-peer systems. In: Voelker, G.M., Shenker, S. (eds.) IPTPS 2004. LNCS, vol. 3279, pp. 11–21. Springer, Heidelberg (2005)
22. Naor, M., Wieder, U.: Scalable and dynamic quorum systems. Distrib. Comput. 17(4), 311–322 (2005)
23. Silaghi, B., Keleher, P., Bhattacharjee, B.: Multi-dimensional quorum sets for read-few write-many replica control protocols. In: Fourth International Workshop on Global and Peer-to-Peer Computing (2004)

Load-Aware Dynamic Replication Management in a Data Grid

Laura Cristiana Voicu and Heiko Schuldt

Department of Computer Science, University of Basel, Switzerland
{laura.voicu,heiko.schuldt}@unibas.ch

Abstract. Data Grids are increasingly popular in novel, demanding and data-intensive eScience applications. In these applications, vast amounts of data, generated by specialized instruments, need to be collaboratively accessed, processed and analyzed by a large number of users spread across several organizations. The nearly unlimited storage capabilities of Data Grids allow these data to be replicated at different sites in order to guarantee a high degree of availability. For updateable data objects, several replicas per object need to be maintained in an eager way. In addition, read-only copies serve users' needs of data with different levels of freshness. The number of updateable replicas has to be dynamically adapted to optimize the trade-off between synchronization overhead and the gain which can be achieved by balancing the load of update transactions. Due to the particular characteristics of the Grid, especially due to the absence of a global coordinator, replication management needs to be provided in a completely distributed way. This includes the synchronization of concurrent updates as well as the dynamic deployment and undeployment of replicas based on actual access characteristics which might change over time. In this paper we present the Re:GRIDiT approach to dynamic replica deployment and undeployment in the Grid. Based on a combination of local load statistics, proximity and data access patterns, Re:GRIDiT dynamically adds new replicas or removes existing ones without impacting global correctness. In addition, we provide a detailed evaluation of the overall performance of the dynamic Re:GRIDiT protocol which shows increased throughput with respect to the replication management protocol with a static number of replicas.

1 Introduction

Novel data-intensive applications are increasingly popular in eScience. In these applications, vast amounts of data are generated by specialized instruments and need to be collaboratively accessed, processed and analyzed by a large number of scientists around the world. Many good examples can be listed such as high energy physics, meteorology, computational genomics, or earth observation. Due to their enormous volumes, such data can no longer be stored in centralized systems but need to be distributed worldwide across data centers and dedicated storage servers. Motivated by the success of computational Grids in which distributed resources (CPU cycles) are globally shared in order to solve complex problems in a distributed way, a second generation of Grids, namely *Data Grids*, has emerged as a solution for distributed data management in data-intensive applications. The size of data required by these applications may be

R. Meersman, T. Dillon, P. Herrero (Eds.): OTM 2009, Part I, LNCS 5870, pp. 201–218, 2009.

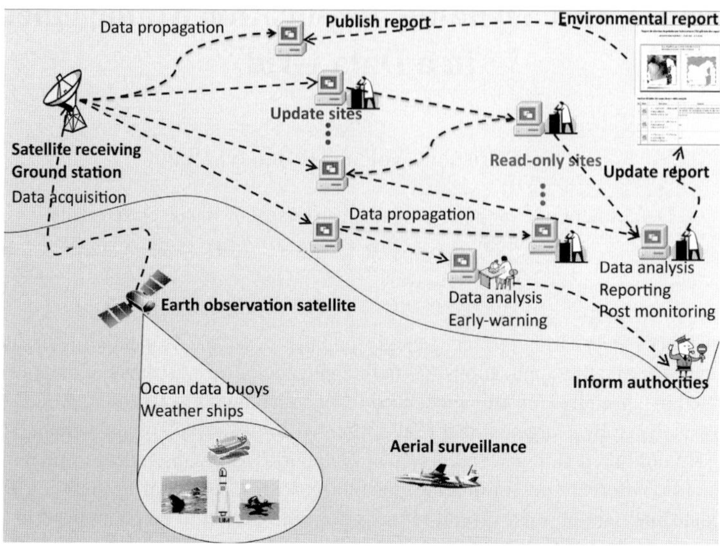

Fig. 1. Example Application Scenario: Data Grids for Earth Observation

upto petabytes. High energy physics data instruments like the large hadron collider at CERN, will produce vast amounts of data during their lifetime: "[...] will be the most data-intensive physics instrument on the planet, producing more than 1500 megabytes of data every second for over a decade" [7]. Similarly, in the earth observation domain, data archives around the world will grow to around 14.000 Tbytes by the year 2014 [14].

Consider the following earth observation scenario (depicted in Figure 1), where data are collected at one or more stations and maintained at multiple sites [26]. In order to improve availability, scalability and avoid single points of failure in the system as well as to best meet the users' requirements, copies of data are maintained at multiple geographically distributed sites. Assume that scientists are closely monitoring oil spills into the sea in a collaborative effort. The potential damage to the area at stake requires the readiness to detect, monitor and clean-up any large spill in a rapid manner. The occurrence of such an incident results in the acquisition and analysis of thousands of images summing up to Tbytes of data and generating multiple concurrent accesses to data. For this type of applications the most up-to-date data, provided in a timely manner is essential, even in the presence of concurrent updates. An efficient data replication protocol has to guarantee the success of such applications, by ensuring, to mention only a few of the advantages: availability of data, efficient access to data, freshness of data, consistency of replicas and reliability in accessing the data.

Replication techniques are able to reduce response times by moving and distributing content closer to the end user, speeding content searches, and reducing communication overhead. An efficient replication management protocol has to consider changes in the access patterns, i.e., should dynamically provide more replicas for data objects which are frequently accessed. However, increasing the number of updateable replicas per data object in an unlimited way may have significant drawbacks on the overall system performance. Therefore, the number of updateable replicas for data objects which are no longer of importance to a large group of users should be dynamically reduced, in order to reduce the overhead of replica maintenance. A number of existing approaches to replication management in parallel and distributed systems address data placement in regular network topologies such as hypercubes, rings or trees. These networks possess many attractive properties that enable the design of simple and robust placement algorithms. These algorithms, however, cannot be directly applied to Data Grid systems due to the lack of regular network structures and the data access patterns in Data Grid systems which are very dynamic and thus hard to predict.

In this paper, we propose the Re:GRIDiT approach to dynamic replica management in Data Grid systems, taking into account several important aspects described below. First, replicas are placed according to an algorithm that ensures load balancing on replica sites and minimizes response times. Nevertheless, since maintaining multiple updateable copies of data is expensive, the number of updateable replicas needs to be bounded. Clearly, optimizing access cost of data requests and reducing the cost of replication are two conflicting goals and finding a good balance between them is a challenging task. We propose efficient algorithms for selecting optimal locations for placing the replicas so that the load among these replicas is balanced. Given the data usage from each user site and the maximum load of each replica, our algorithm efficiently minimizes the number of replicas required, reducing the number of unnecessary replicas. This approach to dynamic replica management is embedded into the Re:GRIDiT replica update protocol which guarantees consistent interactions in a Data Grid and takes into account concurrent updates in the absence of any global component [25,26].

The remainder of the paper is organized as follows. Section 2 discusses related work. In Section 3, we introduce our system model. Section 4 gives details of the proposed protocol. Section 5 describes the implementation of our protocol and provides a detailed performance evaluation. Finally, Section 6 concludes.

2 Related Work

In contrast to its vital importance, dynamic replication management in Data Grids which takes into account its particular requirements of the Grid has not been the focus of previous work. In general, we can distinguish three related fields: i.) data replication in the Grid which mainly concentrates on read-only data and/or manual data placement; ii.) replication management in database clusters and peer-to-peer networks; iii.) dynamic replication of services in Grid environments.

In current Data Grids, replication is mostly managed manually and based on single files as the replication granularity [17]. In addition, these files are mainly considered as read-only [10,11]. There are also more comprehensive solutions as proposed in [20]. This solution, however, may suffer from consistency problems in case of multiple catalogs and while accessing more than one replica. Moreover, to the best of our knowledge, there is no replication solution for Data Grids which address freshness and versioning issues so far while also taking into account several updateable replicas.

In the context of replication management in database clusters, there are several well-established protocols in the database literature which deal with *eager* and *lazy* [13] replication management. Conventional eager replication protocols have significant drawbacks with respect to performance and scalability [13,15] due to the high communication overhead among the replicas and the high probability of deadlocks. In order to overcome these drawbacks, group communication primitives have been proposed in [16]. Lazy replication management, on the other hand, maintains the replicas by using decoupled propagation transactions which are invoked after the "original" transaction has committed. Older works on lazy replication focus only on performance and correctness [5,8]. The approach presented in [4] extends this focus by imposing much weaker requirement on data placement and recent works consider freshness issues as well [19,1]. The drawback of these approaches is that they assume a central coordinator which can become a bottleneck or a single point of failure. Dynamic replication in peer-to-peer networks has also known extensive development over the years [2,9,12,18]. Sophisticated solutions, which eliminate the need of a central coordinator, have been proposed, but they either do not take into account data freshness [2], or multiple data types ([18] for example, only considers images as data objects) or neglect consistency issues [12].

An early attempt to evaluate scalability properties of adaptive algorithms for replica placement services in dynamic, distributed systems was reported in [3]. Weissman and Lee [27] proposed a replica management system which dynamically allocates replicated resources on the basis of user demand, where resource requests by users can be transparently rerouted if individual replicas fail. More recently, other replication schemes have been proposed. Through experiments, most have demonstrated limited scalability and ability to operate under conditions of resource heterogeneity and dynamism. Valcarenghi [21] presented a replication approach in which replicas are located in proximity to each other to form service islands in a Grid network. These approaches have been developed at the service level and thus do not take into account data replication constraints or freshness and versioning issues.

3 The Re:GRIDiT System Model

Despite recent advances, Data Grid technologies often propose only very generic services for deploying large scale applications such as users authorization and authentication, fast and reliable data transfer, and transparent access to computing resources. In this context, we are facing important challenges such as the ones coming from the earth observation application presented in Section 1. We built therefore further functionality on top of the underlying Grid middleware, while taking into account particular requirements coming from both the Grid infrastructure and the Grid applications (dynamically distributed management of replicated data, different levels of freshness, the absence of a central coordinator). In this paper we present the Re:GRIDiT (Replication Management in Data GRid Infrastructures using Distributed Transactions) protocol that dynamically synchronizes updates to several replicas in the Grid in a distributed way. Re:GRIDiT was developed as a response to the need of a protocol that meets the challenges of replication management in the Grid, and that re-defines the Grid and re-discovers it (in other words, that "re:grids it"), bringing it to a level where it can satisfy the needs of a large variety of users from different communities.

Figure 2 sketches the Re:GRIDiT architecture. The bottom layer is built by the hardware: computing nodes, storage nodes and fast network connections. At the middleware level, Grid services provide homogeneous and transparent access to the underlying heterogeneous components while Re:GRIDiT services (which build on top of any Grid

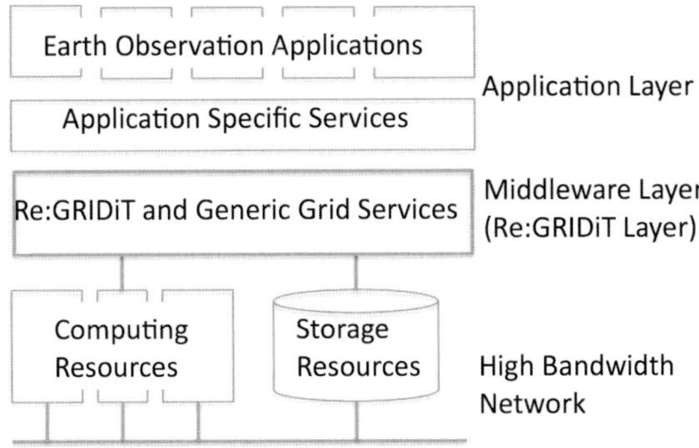

Fig. 2. A Layered Architecture, from Resources to Applications

middleware) provide dynamic and distributed replication of Grid-enabled applications with data sharing and distributed computation capabilities in a way that is transparent to the applications and the users. The application specific layer provides high level and domain specific services taking into account the data semantics, access to heterogeneous database formats, high level services, support for parallel and interactive applications, etc. The components in each layer build on new capabilities and behaviors provided by the lower layer. This model demonstrates flexibility and shows how Grid architectures can be extended and evolved upon.

At the middleware layer, we assume Re:GRIDiT as being part of the Grid middleware and present on each site. The sites hold data objects, replicated in the network, and operations through which the data objects can be accessed. A data object can be represented at the base level as a file stored on the local file system, or inside a database on the local database management system (similar to the model presented in [6]). Each data object is uniquely identified by an object identifier OID.

As illustrated in Figure 3 each site s_j offers a set of operations on data objects $Op^{s_j} = \{op_1, op_2, \ldots, op_n\}$. The operations can be invoked within transactions using the interface of the site. We consider the following update operations (shortly referred to as updates): insert object (denoted $insert(d, t, OID)$ for an insertion of a data object d into a storage type t (file or database), with a specified OID); delete object (denoted $delete(OID)$ for the deletion of a data object with a specified OID); replace object (denoted $replace(d, OID)$ for the replacement of an object with a specified OID). We consider the following read operations: read object (denoted $read(OID)$ for the reading of an object with a specified OID; read versioned object (denoted $read_v(vnr, OID)$ for the reading of a version of an object with a specified OID).

In addition to these operations, which we call *direct operations*, and which result from the translation of user operations, we assume two additional operations which are not available at the user level. These operations are used by the system in order to support replication (*indirect operations*). We consider the following indirect operations: copy object (denoted $copy(OID, vnr, dest)$ for the copying of a version of a data object to a specified destination site); remove object (denoted $remove(OID, vnr, dest)$ for the removal of a version of a data object from a specified destination site).

In our work the notion of conflict is defined based on the commutativity behavior of operation invocations (semantically rich operations [22]). Dependencies between transactions in a transaction's schedule occur when there is at least a pair of operation invocations that is in conflict. We model our system such that each site holds a copy of the conflict matrices, by means of which they can automatically detect conflicts. We assume conflicts to be rather infrequent, but they need to be handled properly in order to guarantee globally serializable executions.

In Re:GRIDiT one data object can reside only on one site, but could be replicated on several sites. We distinguish between two types of replica sites: *update* and *read-only* sites. Read-only sites allow read access to data only. Updates occur on the update sites and are propagated to other update sites and finally to the read-only sites. The nature of a site (i.e., update or read-only) is determined by the operations it provides [26]. We define an *n-to-m* relationship between the two types of sites; an update site can have any number of read-only children (to which updates are propagated). Read-only sites in turn can be shared with other update sites.

Re:GRIDiT assumes the presence of local logs on each site (see Figure 3), where operation invocations are recorded. These logs are used for conflict detection and recovery purposes. Each site uses a set of tools to obtain a (typically partial and sometimes even out-dated) information regarding the state of the system and take replication decisions. We introduce the following components to facilitate the scheduling of read-only transactions in the Grid and the replica management decision. These components are not centrally materialized in the system, and contain global information that is distributed and replicated to all the update sites in the system:

- *replica repository*: used to determine the currently available update replicas in the network. Moreover, each update replica is aware of its own read-only replicas, where it propagates update changes, in order to maintain a certain level of freshness/staleness in the network.
- *freshness repository*: used to collect freshness levels of the data objects periodically or whenever significant changes occur. In this paper, the term freshness level is used to emphasize the divergence of a replica from the up-to-date copy. Consequently update sites will always have the highest freshness level, while the freshness level of the read-only sites will measure the staleness of their data. Since update sites propagate changes to read-only sites, they are aware of the freshness levels of the read-only sites to which they propagate changes.
- *load repository*: used to determine an approximate load information regarding sites. This information can then be used to balance the load while routing read-only transactions to the sites or for the replica selection algorithm. Update sites periodically receive information regarding the load levels of other update sites and their readony children. Read-only sites are only aware of their own load levels.[1]

[1] In order to improve efficiency and not to increase the message overhead this information can be exchanged together with replica synchronization request. This information needs to be exchanged more frequently while there is a replica synchronization process in place, while the exchange is not needed during a site's idle time, when the load is unlikely to vary.

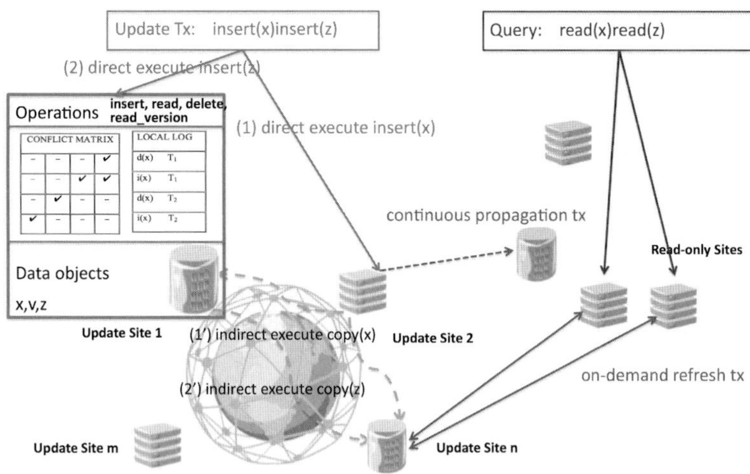

Fig. 3. System Model Architecture at the Middleware Level

Besides these local metadata each site maintains a set of global, pre-defined set of variables: δ_T, collection time interval (defined as the time between individual state information collection) and Δ_{load}. These values strongly depend on the nature of the supported application, but the protocol may allow these values to be self-defined on a per site basis, depending on the physical characteristics of the site.

With respect to transactions submitted by clients, i.e., client transactions, we distinguish between *read-only transactions* and *update transactions* (according to [1]). A read-only transaction only consists of read operations. An update transaction comprises at least one update operation next to arbitrarily many read operations. Decoupled *refresh transactions* propagate updates through the system, on-demand, in order to bring the read-only replicas to the freshness level specified by the read-only transaction. *Propagation transactions* are performed during the idle time of a site in order to propagate the changes present in the local propagation queues to the read-only replicas. Therefore, propagation transactions are continuously scheduled as long as there is no running read or refresh transaction. Re:GRIDiT exploits the sites' idle time by continuously scheduling propagation transactions as update transactions at the update sites commit. Copies at the read-only sites are kept as up-to-date as possible such that the work of refresh transactions (whenever needed) is reduced and the performance of the overall system is increased. Through our protocol a globally serializable schedule is produced (although no central scheduler exists and thus no schedule is materialized in the system) [24]. Moreover, the update transactions' serialization order is their commit order.

Note that a transaction may not always succeed due to several failure reasons. To satisfy the demand for an atomic execution, the transaction must compensate the effects of all the operations invoked prior to the failure [22]. This compensation is performed by invoking semantically inverse operations in reverse order.

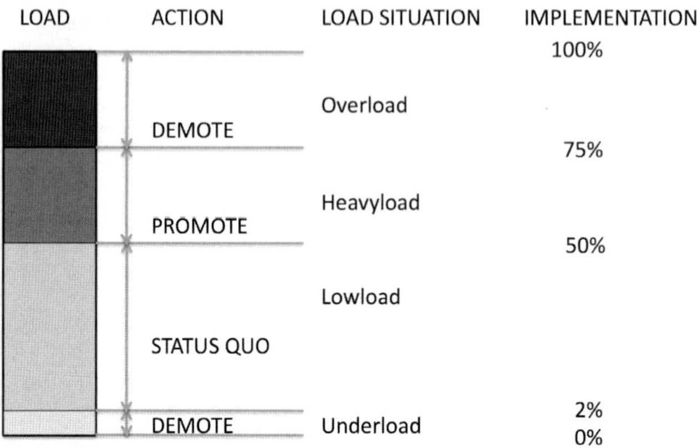

Fig. 4. Possible Load Situations

3.1 Load Metrics

In practical scenarios, the load metrics may reflect multiple components, such as CPU load, storage utilization, disk usage or bandwidth consumption. We abstract from the notion of load by defining a load function $load \in [0, 1]$, where the maximum, in the case of CPU load, for instance, corresponds to 100 % load.

We assume that an individual site can measure its total load. This can be done by keeping track of resource consumption (CPU time, IO operations, access count rate etc.) In order to ensure an accurate measurement, the load collection is averaged over a certain time window and consequently the decisions are made based on load trends rather than punctual values.

We define the mean load as $\overline{load(s)} = \frac{\sum_{i=0}^{n} load(s_i)}{n}$ where n represents the total number of load calculations within the given time window and classify the following load situations (for a site s):

- An *overload* situation can occur if $\overline{load(s)} \in (\alpha_{overload}, 1)$, where $\alpha_{overload} \in [0; 1]$.
- A *heavyload* situation can occur if $\overline{load(s)} \in (\alpha_{heavyload}, \alpha_{overload})$, where $\alpha_{overload} \in [0; 1]$ and $\alpha_{heavyload} \in [0; 1]$.
- A *lowload* situation can occur if $\overline{load(s)} \in (\alpha_{underload}, \alpha_{heavyload})$, where $\alpha_{heavyload} \in [0; 1]$ and $\alpha_{underload} \in [0; 1]$.
- An *underload* situation can occur if $\overline{load(s)} \in (0, \alpha_{underload})$, where $\alpha_{underload} \in [0; 1]$.

The parameters $\alpha_{underload}$, $\alpha_{heavyload}$ and $\alpha_{overload}$ are application dependent and obey the following relation: $0 \leq \alpha_{underload} \leq \alpha_{heavyload} \leq \alpha_{overload} \leq 1$. The possible load levels (together with an example of their implementation in a practical scenario) are presented in Figure 4. We extend the above classification to a system-wide level (i.e., a *system overload* can occur if $\frac{\sum_{j=0}^{N} \overline{load(s_j)}}{N} \geq \alpha_{overload}$, where

$\overline{load(s)}$ is the average load on a site s and N represents the total number of sites in the system) and define two additional situations that can be observed at a system level:

- A *majority overload* situation can occur if a majority of sites (of at least $\frac{N}{2}+1$) is in an overload situation. The remaining sites must be at least in a heavyload situation.
- A *minority overload* situation can occur if a minority of sites (of at most $\frac{N}{2}-1$) is in an overload situation. The remaining sites must be at least in a heavyload situation.

Note that the definition of the load levels in a practical scenario is of major importance since crossing the boundaries of a certain level will trigger a certain action in the dynamic replication protocol.

3.2 Best Replica

The efficiency of the replication system depends on the criteria used for the selection of replica sites. In our protocol, we choose the notion of *host proximity* as one of the criteria for replica selection. The proper definition of this metric impacts both the response time perceived by the user and the overhead involved in measuring the distance. In order to define the "closest" replica, the metrics may include response time, latency, ping round-trip time, network bandwidth, number of hops or geographic proximity. Since most of the above metrics are dynamic, replica selection algorithms such as [23] typically rely on estimating the current value of the metric using samples collected in the past. We abstract from the above notions by defining a *closeness function*, in which the notion of closeness can reflect any of the metrics above, and is defined as $close(s, c) \in [0, 1]$, where 0 is the absolute minimum value for this metric (for example, in the case of geographic proximity, 0 would correspond to the site s that is geographically farthest away from the client c and 1 to the site s that is geographically closest to the client c).

The second criterion for the replica selection is based on the notion of "freshness". We use in our approach a freshness function, $fresh(d, s) \in [0, 1]$, (as defined in [19]) that reflects how much a data object d on a site s has deviated from the up-to-date version. The most recent data has a freshness of one, while a freshness of zero represents infinitely outdated data (data object is not present on that site). For read-only sites, the freshest replica has the most up-to-date data (ideally as close to 1 as possible).

The third criterion is related to the concept of load (see Subsection 3.1) and defines the "least loaded" site as the one with the smallest value of the load within a given set.

The following situation may require a replica selection: a particular system load situation may require the acquisition of a new update replica from the read-only replicas available. For each replica s the replica selection algorithm keeps track of its load, $load(s)$ and the freshness of each data object d, $fresh(d, s)$. For read-only sites, the freshness is smaller or equal to one and represents a measure of the staleness of the data on a site. The algorithm begins by identifying replicas with the smallest load level, replicas with the highest freshness level and replicas that are the closest to the requester and chooses the best replica among them.

Definition 1. ***(Best replica function)*** *Let S be the set of all sites in the system and D the set of all data objects. Then, for a copy of a data object d, required by a given client site c, the function best(c,d) is defined as follows: $best : S \times D \rightarrow S$, such that $best(c, d) = s^* \in S$, where s^* corresponds to the greatest value of the sum: $\alpha close(s^*, c) + \beta fresh(d, s^*) + \gamma(1 - load(s^*))$, for given α, β, γ with $\alpha + \beta + \gamma = 1$, $\forall s^* \in S$.* □

Choosing replicas in the round-robin manner would neglect the proximity factor. On the other hand, always choosing the closest replicas could result in poor load distribution. Our protocol keeps track of load bounds on replicas, freshness levels and proximity factors, which enables the replica management algorithm to make autonomous selection decisions. We allow the decision to take into account the freshness level of a replica. The usage of freshness in this protocol is a trade off of consistency for performance. Furthermore, by using a weighted sum for the notion of best replica, we allow higher level applications to give preference to a certain parameter over the others.

4 The Re:GRIDiT Protocol

4.1 Replica Promote and Demote

In the following we define the conditions and consequences of dynamic replica acquisition or release. When a new updateable replica is created, it is updated with the content of existing update replicas (depending on the data objects that exist at that site). In order to ensure consistency the creation of a new update replica takes place in two phases: (i) *Phase 1*: sites are informed that a new replica will join the network, they finish currently executing transactions and start queueing any direct operations belonging to subsequent transactions; (ii) *Phase 2*: the replica has joined and is up-to-date, the sites start executing the queued direct operations taking into account the new replica when executing indirect operations.

However, a new replica promote may not always be beneficial. The decision whether to acquire or release a replica is made based on a combination of local (accurate) and global (partially accurate) information. We identify the following cases, depending on the local load of a site (see Figure 4):

- local *underload*; An *underload* situation intuitively implies that there are no updates and no queries at this site. In this case the site autonomously decides to self demote.
- local *lowload*; A *lowload* situation is considered to be a normal load situation. The site continues the normal execution.
- local *heavyload*; A *heavyload* situation is dealt with by the acquisition of additional replica(s). A site in *heavyload* will inform the other update sites of the intention to promote a new replica. If there is no other promote taking place at the same time and all the sites have acknowledged the promote, the site will proceed to choose the best replica, as described in Subsection 3.2, and promote it to update site.
- local *overload*; An *overload* situation is a special case in which the global state of the system needs to be taken into account, since a local situation may be the result of several influencing factors that do not necessarily reflect the state of the system. We distinguish two sub-cases:

Algorithm 1. Re:GRIDiT Site Protocol (Extension)

```
Main Execution Thread:
while true do
    Proceed normal Re:GRIDiT execution;
    wait for next message m;
    if m contains message: inform promote then
        finish direct operations of active transactions;
        update replica repository information;
        return ACK message;
        queue all incoming direct and indirect operations;
    else if m contains message: end inform promote then
        execute queued operations;
    else if m contains message: inform demote then
        update replica repository information;
    end if
end while
```

- In the case in which the system is in a majority overload situation the acquisition of a new replica is unlikely to improve the situation (since more replicas imply more synchronization). No action will be taken.
- In the case in which the system is in a minority overload situation, the overload can be assumed to be due to read operations (which are not replicated). The protocol will dictate the site to migrate the read request to other update sites in the system which are not in an overload situation.

Definition 2. ***(Replica promote).*** *The process by which a read-only site is transformed into an update site is called* **replica promote** *(replica acquisition).*
(Preconditions). *An update site s will begin the process of the acquisition of the best replica, from its own read-only children, which satisfies the criteria defined in Definition 1, if and only if the following statements hold: load(s) $= heavyload$ and there is no other promote initiated by a different site taking place at the same time and the already existing number of update replicas in the system has not reached a maximum.*
(Postconditions). *A read-only site r has been promoted to an update site and all the other update site are aware of the new replica. The information in the distributed replica repository is updated accordingly.* □

Definition 3. ***(Replica demote)*** *The process by which an update site is transformed into a read-only site is called* **replica demote** *(replica release).*
(Preconditions) *An update site s will begin the process of self release for a data object x, if and only if the following statements hold: load(s) $= underload$ or load(s) $= overload$ and no active transactions exist at s and the data are available elsewhere at a minimum number of replicas.*
(Postconditions) *Site s no longer exists as update site and all the other update sites are made aware of the disappearance of the replica site s. The information in the distributed replica repository is updated accordingly.* □

Algorithm 2. Re:GRIDiT Site Protocol (continued)

```
Background Thread:
while δ(T) do
    collect load
    if Promote Precondition then
        select best read-only replica according to Definition 1;
        inform update replicas of new promote;
        promote read-only replica;
        update replica repository information;
    else if Demote Precondition (in underload) then
        inform update replicas of self demote;
        update replica repository information;
    else if Demote Precondition (in overload) then
        if majority overload then
            inform update replicas of self demote;
            update replica repository information;
        else if minority overload then
            route read requests to sites ∉ overload
        end if
    end if
    if (load - previous load) > Δ_load then
        δ(T) - -;
        propagate load changes to update replicas
    end if
    if (previous load - load) > Δ_load then
        δ(T) + +;
        propagate load changes to update replicas
    end if
end while
```

Since our protocol is completely distributed the decision on what action to take and when is made mostly based on local information and the partially outdated information about the state of the system. Nevertheless, in order to maintain consistency, the updateable replicas are synchronously informed of any replica promote or demote. The extensions required by the Re:GRIDiT Protocol are illustrated in Algorithms 1 and 2.

4.2 Failure Handling

We identify the following types of failure that could occur in dynamic Data Grid environments. We suppose the sites fail-stop and that failures are detected by means of failure to acknowledge requests within a certain timeout. This timeout mechanism ensures that failure detection prevents processes from long delays but also, from suspecting sites too early which could incur unnecessary communication overhead. If a site crashes during an inform demote or inform promote phase, the failure is detected by means of the timeout mechanism. The site in charge of the inform will update the replica repository and reissue the inform request. If the site to be chosen promoted is failing to answer, the next best replica (according to Definition 1) is chosen to be promoted.

Since the number of copies of the object varies in time the dynamic replication and allocation algorithm is vulnerable to failures that may render the object inaccessible. To address this problem, we impose a reliability constraints of the following form: "The number of copies per data object cannot decrease below a minimum threshold". If such constraint is present, and the local site s fails to meet it, then any of sites informed of the demote p refuses to accept the exit of a data site, if such exit will downsize the replication scheme below the threshold. In other words, p informs s that the request to exit from the replication scheme is denied; subsequently, updates continue to be propagated to s. Site s continues to reissue the request whenever the preconditions for replica demote dictates to do so. The request may be granted later on, if the replication scheme is expanded in the meantime. Failure during other phases of the algorithm are considered in the semi-static replication protocol described in [26].

5 Implementation and Evaluation

The Re:GRIDiT protocol has been implemented using state-of-the-art Web service technologies which allows an easy and seamless deployment in any Grid environment. We have evaluated Re:GRIDiT with support for dynamic replica placement and deployment against Re:GRIDiT using a static replication scheme (presented in [25]), in a Java WS-Core Globus Toolkit container environment where the web services run. The experimental setup consists of multiple hosts equipped with a local Derby database and Java WS-Core. Both have been chosen due to their platform-independence, but any other existing off-the-shelf database can be used for this purpose.

The goal of these evaluations is to verify the potential of the protocol, in terms of scalability and performance, compared to a protocol allowing only a semi-static

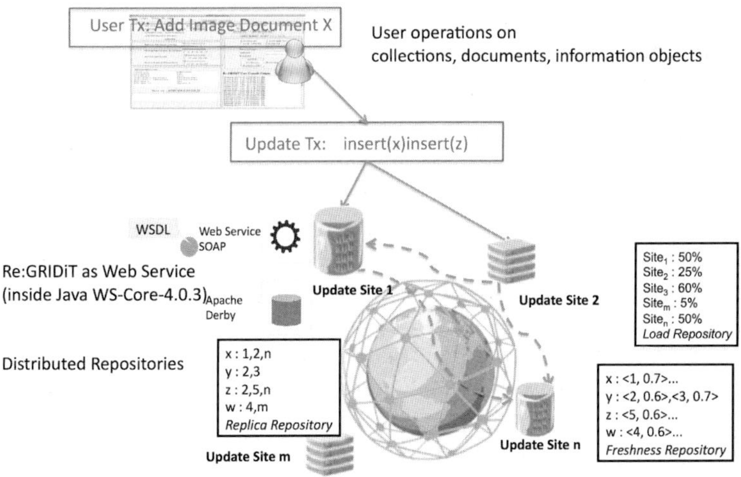

Fig. 5. System Model Implementation

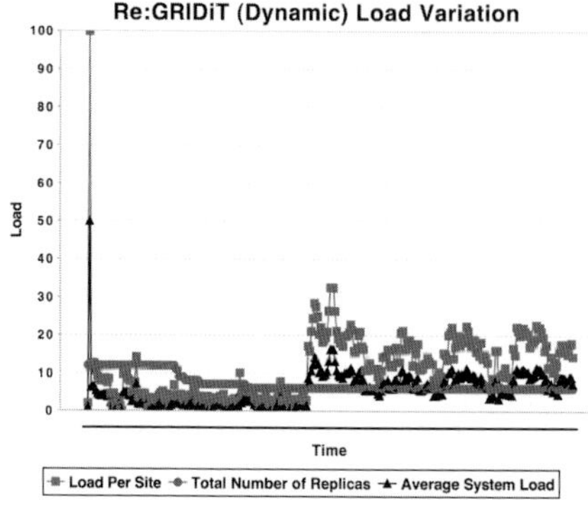

Fig. 6. Load Variation in Time. Dynamic Setup (12 Initial Update Sites).

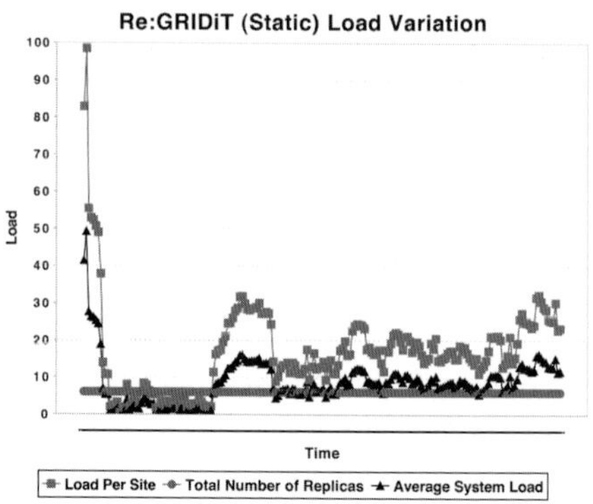

Fig. 7. Load Variation in Time. Static Setup (6 Initial Update Sites).

replication scheme. For the purpose of our evaluation we used up to 48 sites[2], equipped with a Dual Intel®CPU 3.20 GHz processor, 5 GB RAM and running Ubuntu Linux 8.0.4 as operating system. They are all running Java WS-Core version 4.0.3. The evaluation setup is schematically presented in Figure 5. For simplicity, we have used the site's CPU percentage as load measurement. The actual values used for the different load situations are depicted in Figure 4. These value have been chosen such that the intervals for a site promote or demote are comparable. Unless otherwise stated the measurements have consisted on runs of 100 transactions each. The conflict rate was set to 0.01 (since conflicts are assumed to be infrequent). Each transaction consists of 5 direct operations. The transactions were started sequentially, one after the other, with an update interval of milliseconds, such that as many transactions as possible are active at the same time.

5.1 Load Variation in the Presence of Active Transactions

In order to evaluate the performance of our protocol we have conducted several experiments that evaluate the load variation in the presence of active transactions at the sites. We have recorded the mean local load variations, the mean system loads and the evolution of the number of replicas in time (in a dynamic setting). As observed from Figure 6 for an initial number of 12 update sites, the average number of replicas in the system tends to remain stable at an minimum of 6. We have compared these results with the static replication protocol for which we used 6 update sites. The average system load shows more fluctuations in the dynamic case (due to the dynamic replica management overhead), and on average, the local site loads show a less than 5% increase with respect to the static case. In this case, the static setting of 6 update replicas has been chosen to match the optimal minimum which is reached by the dynamic setting; therefore an unfortunate choice of less than optimal number of replicas in the static setting produces non-negligible load differences with respect to the dynamic setting.

[2] This number refers to the maximum number of update sites in the system. In addition there may be hundreds of read-only sites present in the system.

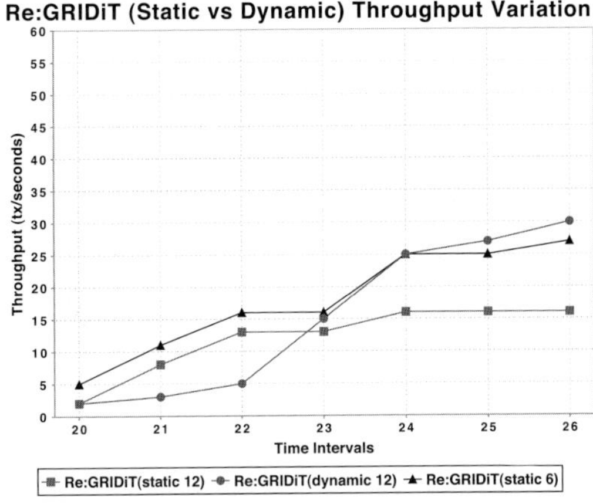

Fig. 8. Throughput Variation. 1 Run.

5.2 Evolution of Throughput

Another means of evaluating the performance of the dynamic and static Re:GRIDiT protocols is by comparing their throughput, calculated as the number of transactions committed within a time interval (in this example 20 seconds). In this measurement, both protocols with an initial setup of 12 sites and the static protocol with an initial setup of 6 sites have been compared. As in the previous cases, the dynamic replication protocol has stabilized the number of update sites at a minimum of 6. As it can be seen from Figure 8, the throughput in the dynamic setting is higher than in both static settings. The reason for this behavior is that transactions begin the commit much sooner in time in the dynamic case than in the static one. The rationale behind it is that the static Re:GRIDiT requires more time to synchronize the update to a higher (and constant) number of update sites in the case of 12 static sites, therefore the transaction duration is higher. It can also be observed that initially the throughput in the dynamic setting is smaller than in the case of the static setting with 6 update sites, due to the extra load imposed by the demote of the unnecessary update sites. Nevertheless, the throughput of the dynamic setting is increasing and outperforms the throughput of the static setting with 6 update sites, due to the selection of the best replica sites in the dynamic case.

5.3 Load Variation in the Presence of Active Transactions and Additional Load

As it can be seen from Figures 6 and 7, the distributed concurrency control of active running transactions and the replica management hardly drive the CPU load within the *heavyload* level and never in the *overload* level. In order to observe the system's behavior under heavier load stress we have tested the protocols in a setting using active transaction and artificial load variations (that mimic the behavior of additional read operations). These load variations introduce dynamic changes in the system which lead to promote and demote situations. In each of the cases the load has been maintained stable for several minutes in order to give the system enough time to react to the changes.

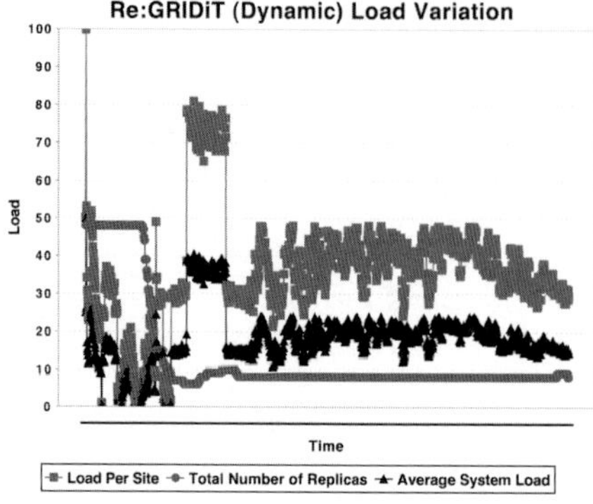

Fig. 9. Load Variation in Time. Dynamic Setup (48 Initial Update Sites).

Re:GRIDiT (Static) Load Variation

Load

Time

Load Per Site — Total Number of Replicas — Average System Load

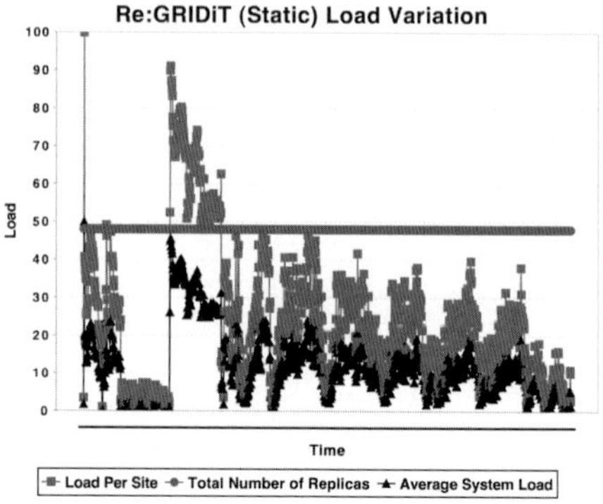

Fig. 10. Load Variation in Time. Static Setup (48 Initial Update Sites).

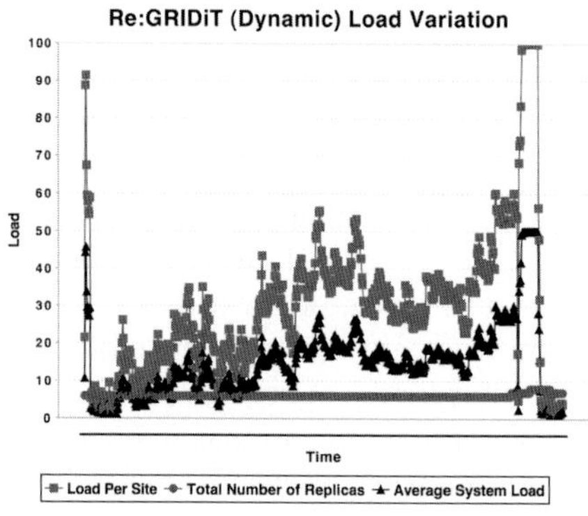

Fig. 11. Load Variation in Time. Dynamic Setup (6 Initial Update Sites).

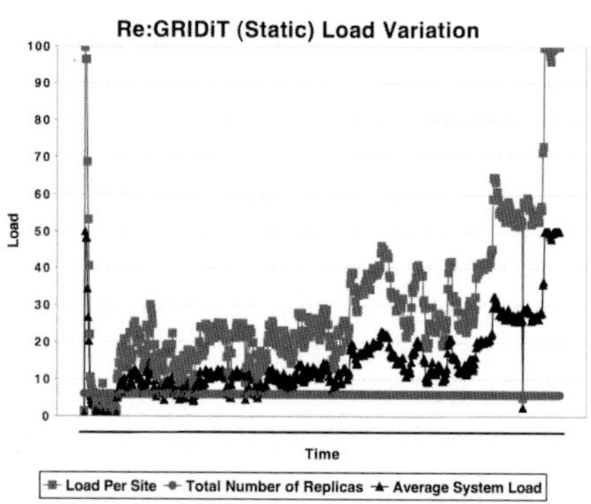

Fig. 12. Load Variation in Time. Static Setup (6 Initial Update Sites).

The results can be seen in Figures 9 and 10 where an initial setup of 48 update sites is compared and Figures 11 and 12 where an initial setup of 6 update sites is compared. It is clear that while both protocols are subject to the same load levels the dynamic Re:GRIDiT protocol is able to better cope with the varying load situations and consequently promote and demote sites as needed.

5.4 Dynamic Re:GRIDiT Protocol: Summary of Evaluation Results

Summarizing, the main achieved goals of our dynamic Re:GRIDiT protocol are:

- Dynamic: replicas can be created and deleted dynamically when the need arises. As it can be seen from Figures 9 and 11 dynamic changes in the load determines when new replicas need to be acquired or released.

- Efficient: replicas should be created in a timely manner and with a reasonable amount of resources. Figure 6 reflects the overhead in replica acquisition with respect to the static protocol presented in Figure 7 which is shown to be at approximately 5%.
- Flexible: replicas should be able to join and leave the Grid when needed. Our dynamic replication protocol allows replicas to join and leave the replication scheme as long as a minimum number of replica is present in the system. This limit is application dependent since different application scenarios might have different needs.
- Replica Consistency: in an environment where updates to a replica are needed, different degrees of consistency and update frequencies should be provided. In our measurements the Re:GRIDiT distributed and optimistic concurrency control protocol (presented in [25]) is enabled.
- Scalable: the replication system should be able to handle a large number of replicas and simultaneous replica creation. We have tested our protocol on a setup consisting of up to 48 update sites. In addition there may be hundreds of read-only sites.

6 Conclusions and Outlook

Data Grids are providing a cutting-edge technology of which scientists and engineers are trying to take advantage by pooling their resources in order to solve complex problems and have known intensive developments over the past years. Basic middlewares are available and it is now time for the developers to turn toward specific application problems. In this paper we propose an algorithm that aims at solving key problems of replication management in Data Grids. Re:GRIDiT hides the presence of replicas to the applications, provides replica consistency without any global component and allows an efficient, flexible and dynamic replication scheme where replicas can be created and deleted as the need arises. Furthermore we provide an evaluation of our dynamic replication protocol and compare it with a protocol which allows a static replication scheme, clearly proving the advantages of Re:GRIDiT.

Some open question remain for future research and certainly require future investigation: (i) how to respond in an efficient manner to read-only requests with different freshness levels or (ii) how to monitor access patterns (to improve access of heavy/frequent users to data).

References

1. Akal, F., Türker, C., Schek, H.-J., Breitbart, Y., Grabs, T., Veen, L.: Fine-Grained Replication and Scheduling with Freshness and Correctness Guarantees. In: VLDB, pp. 565–576 (2005)
2. Akbarinia, R., Pacitti, E., Valduriez, P.: Data currency in replicated DHTs. In: Proceedings of the ACM SIGMOD international conference on Management of data, pp. 211–222 (2007)
3. Andrzejak, A., Graupner, S., Kotov, V., Trinks, H.: Algorithms for Self-Organization and Adaptive Service Placement in Dynamic Distributed Systems. Technical report, HP (2002)
4. Breitbart, Y., Komondoor, R., Rastogi, R., Seshadri, S., Silberschatz, A.: Update propagation protocols for replicated databates. In: EDBT, pp. 97–108 (1999)
5. Breitbart, Y., Korth, H.F.: Replication and consistency: being lazy helps sometimes. In: PODS, pp. 173–184 (1997)

6. Candela, L., Akal, F., Avancini, H., Castelli, D., Fusco, L., Guidetti, V., Langguth, C., Manzi, A., Pagano, P., Schuldt, H., Simi, M., Springmann, M., Voicu, L.: DILIGENT: integrating digital library and Grid technologies for a new Earth observation research infrastructure. Int. J. Digit. Libr. 7(1), 59–80 (2007)
7. CERN. LHC Computing Centres Join Forces for Global Grid Challenge. CERN Press Release (2005), http://press.web.cern.ch/press/PressReleases/Releases2005/PR06.05E.html
8. Chundi, P., Rosenkrantz, D.J., Ravi, S.S.: Deferred Updates and Data Placement in Distributed Databases. In: ICDE, pp. 469–476 (1996)
9. Cohen, E., Shenker, S.: Replication strategies in unstructured peer-to-peer networks. In: SIGCOMM Comput. Commun. Rev, pp. 177–190 (2002)
10. EDG: The European DataGrid Project, http://eu-datagrid.web.cern.ch/eu-datagrid/
11. EGEE: The Enabling Grids for E-sciencE Project, http://www.eu-egee.org/
12. Gopalakrishnan, V., Silaghi, B., Bhattacharjee, B., Keleher, P.: Adaptive Replication in Peer-to-Peer Systems. In: ICDCS, pp. 360–369 (2003)
13. Gray, J., Helland, P., O'Neil, P., Shasha, D.: The Dangers of Replication and a Solution. In: International Conference on Management of Data, pp. 173–182 (1996)
14. Harris, R., Olby, N.: Archives for Earth observation data. Space Policy 16, 223–227 (2007)
15. Jiménez-Peris, R., Patiño-Martínez, M., Kemme, B.: Are Quorums an Alternative For Data Replication. ACM Transactions on Database Systems 28, 2003 (2003)
16. Kemme, B., Alonso, G.: A new approach to developing and implementing eager database replication protocols. ACM Transactions on Database Systems 25, 2000 (2000)
17. The Laser Interferometer Gravitational Wave Observatory, http://www.ligo.caltech.edu/
18. Rathore, K.A., Madria, S.K., Hara, T.: Adaptive searching and replication of images in mobile hierarchical peer-to-peer networks. Data Knowl. Eng. 63(3), 894–918 (2007)
19. Röhm, U., Böhm, K., Schek, H.-J., Schuldt, H.: FAS: a freshness-sensitive coordination middleware for a cluster of OLAP components. In: VLDB 2002, pp. 754–765 (2002)
20. SRB: The Storage Resource Broker, http://www.sdsc.edu/srb/
21. Valcarenghi, L., Castoldi, P.: QoS-Aware Connection Resilience for Network-Aware Grid Computing Fault Tolerance. In: Intl. Conf. on Transparent Optical Networks, vol. 1, pp. 417–422 (2005)
22. Vingralek, R., Hasse-Ye, H., Breitbart, Y., Schek, H.-J.: Unifying concurrency control and recovery of transactions with semantically rich operations. Theoretical Computer Science 190(2) (1998)
23. Vingralek, R., Sayal, M., Scheuermann, P., Breitbart, Y.: Web++: A system for fast and reliable web service. In: USENIX Annual Technical Conference, pp. 6–11 (1999)
24. Voicu, L., Schuldt, H.: The Re:GRIDiT Protocol: Correctness of Distributed Concurrency Control in the Data Grid in the Presence of Replication. Technical report, University of Basel, Department of Computer Science (2008)
25. Voicu, L.C., Schuldt, H., Akal, F., Breitbart, Y., Schek, H.-J.: Re:GRIDiT – Coordinating Distributed Update Transactions on Replicated Data in the Grid. In: Proceedings of the 10th IEEE/ACM Intl. Conference on Grid Computing (Grid 2009), Banff, Canada (October 2009)
26. Voicu, L.C., Schuldt, H., Breitbart, Y., Schek, H.-J.: Replicated Data Management in the Grid: the Re:GRIDiT Approach. In: ACM Workshop on Data Grids for eScience (May 2009)
27. Weissman, J., Lee, B.: The Virtual Service Grid: an Architecture for Delivering High-End Network Services. Concurrency And Computation 14, 287–319 (2002)

Resource Planning for Massive Number of Process Instances

Jiajie Xu, Chengfei Liu, and Xiaohui Zhao

Centre for Complex Software System and Services
Faculty of Information and Communication Technologies
Swinburne University of Technology
Melbourne, Australia
{jxu,cliu,xzhao}@groupwise.swin.edu.au

Abstract. Resource allocation has been recognised as an important topic for business process execution. In this paper, we focus on planning resources for a massive number of process instances to meet the process requirements and cater for rational utilisation of resources before execution. After a motivating example, we present a model for planning resources for process instances. Then we design a set of heuristic rules that take both optimised planning at build time and instance dependencies at run time into account. Based on these rules we propose two strategies, one is called holistic and the other is called batched, for resource planning. Both strategies target a lower cost, however, the holistic strategy can achieve an earlier deadline while the batched strategy aims at rational use of resources. We discuss how to find balance between them in the paper with a comprehensive experimental study on these two approaches.

1 Introduction

Nowadays, how to improve the business process performance and service quality becomes one of the key issues for organisations to survive and thrive in the market competitions. Works [4, 5, 7, 8] indicate that resource allocation is an important topic for business process management and service quality improvement. With limited resources, business processes are expected to execute with lower cost and in shorter period [3, 13]. In addition, resources should be properly allocated for business processes according to the natural characteristics and rationalities, such as periodical availability, fair use, and load balance.

The resource planning can be further complicated in the scenario where a massive number of process instances exist. The dependencies and conflicts between concurrent process instances and exclusively occupied resources have to be taken into account in the resource planning. For an organisation, resources must be allocated to guarantee the assigned volume of process instances can be finished before a negotiable deadline to customers at an acceptable lower cost to organisation. Due to those factors such as inventory capacity and resource availability patterns, the massive number of process instances have to be executed in batches for balancing the load.

Several works [1, 6, 13] have been done in investigating the process oriented resource planning. However, none of them considered all the above problems. Aiming

R. Meersman, T. Dillon, P. Herrero (Eds.): OTM 2009, Part I, LNCS 5870, pp. 219–236, 2009.

to address all these issues, we introduce a novel model to specify and analyse the relationships among roles, resources and process structures. For an optimised and rational resource planning for a massive number of process instances, we design a series of heuristic rules that consider a deadline posed by customers, and a lower cost, conflicts in concurrent use of resources, periodical resource availability and inventory capacity, etc. from the organisation point of view. Based on these rules, we propose two strategies, one is called holistic and the other is called batched, for resource planning. Both strategies target a lower cost, however, thc holistic strategy can achieve a shorter deadline while the batched strategy aims at rational use of resources. We discuss how to find balance between them in the paper. A comprehensive experimental study is also conducted to demonstrate how our approach can support organisations to plan and adjust resource allocations for a massive number of process instances. The contribution of this paper is listed as the following aspects:

- Take into account the batching of process instance execution, to enhance the rational scheduling and utilisation of resources;
- Plan the resource allocation at build time to guarantee that all requirements on time, cost, batch pattern, etc., can be satisfied;
- Consider the run time dependencies between process instance in resource planning, and therefore support the multi-instance parallel execution planning;

The remainder of this paper is organised as follows: Section 2 discusses our problem for requirement oriented build-time resource optimisation with a motivating example. Section 3 introduces a role based business process model, which defines the related notions for resource planning and their relationship. Based on this model the problem to be addressed in this paper is formalised. Two strategies for process oriented resource planning are presented and discussed in Section 4. In Section 5 we conduct an experimental study to evaluate and compare two strategies proposed in this paper. Section 6 reviews the related work and discusses the advantages of our approach. Lastly, concluding remarks are given in Section 7.

2 Motivating Example

In this section, we analyse the resource planning for massive number of process instances with motivating example.

Figure 1 shows a simplified business process, as well as a set of resources which belong to different roles. Here, each resource has a maximal work load of 24 hours, and different roles have different capabilities for performing certain tasks with given expense rates. For the organisation, it has to consider how to meet time and cost requirements and how to maintain the rationalities of resource utilisation when allocating resources to all process instances.

Given a massive number of process instances, their execution can be enforced in two strategies, i.e., the holistic strategy - handling all instances as a whole lot and the batched strategy - handling the instances in batches. For illustration purpose, the number of instances is set to 40. Different strategies result in different resource allocation plans. Figure 2 (a) shows the resource plan for a holistic solution, where instances may overlap with each other in execution, and therefore resources should be allocated in an interleaved way to avoid resource conflicts. According to this holistic

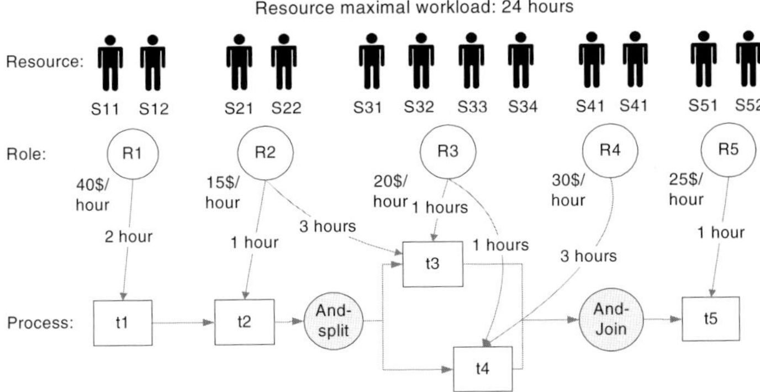

Fig. 1. Business process and environment

solution and the resource plan, 40 process instances can be finished in 120 hours at the cost of $1000. Using the batched strategy, a solution is shown in Figure 2 (b), where each batch contains 8 instances and can be finished in 30 hours, and 5 batches are executed in the sequential order. In this case, the total execution time becomes 150 hours and the cost is $980.

The second solution is intuitively preferable though it takes a bit more time for execution. This is because that it allows the resources to be used periodically, which adapts to the natural characteristics and rationalities of resources, such as periodical availability, fair use, and load balance. Figure 2 (c) shows the job schedule of s_{11} in the two discussed strategies. In the first solution, this resource has a rather skewed workload distribution, which may cause extra cost for the organisation to schedule the resource, e.g., double pay for overtime hours for human resources. While in the second solution the resource is evenly used.

Particular attention should be paid when evaluating a holistic solution and a batched solution, because their performance is sensitive to the given deadline, budget limit, number of instances, etc. In the example, if the deadline is set to 120 hours, then only the holistic solution is feasible. Yet, if the organisation is able to negotiate with customers to relax the deadline to 150 hours, the batched solution becomes feasible and more attractive, as it meets the deadline and follows a periodical resource utilisation. Alternatively, the organisation may negotiate with customers to change the number of instances to 36 if the deadline is strict. As shown in Figure 2 (d), we can have another batched solution that runs 4 batches at $900, with each batch running 9 instances within 30 hours, and thereby can meet the strict deadline of 120 hours.

From the above discussion, we can see that the alternative solutions for the batched strategy according to different input arguments are very useful for organisations to negotiate and determine resource allocations. Here, we summarise the main factors for tunning batched solutions and determining resource allocations:

- Cost difference relative to deadline relaxation;
- Time difference relative to cost increment;
- Time and cost difference relative to the number of assigned instances;
- Time and cost relative to the number of batches;
- Intention of using resources in a periodical manner.

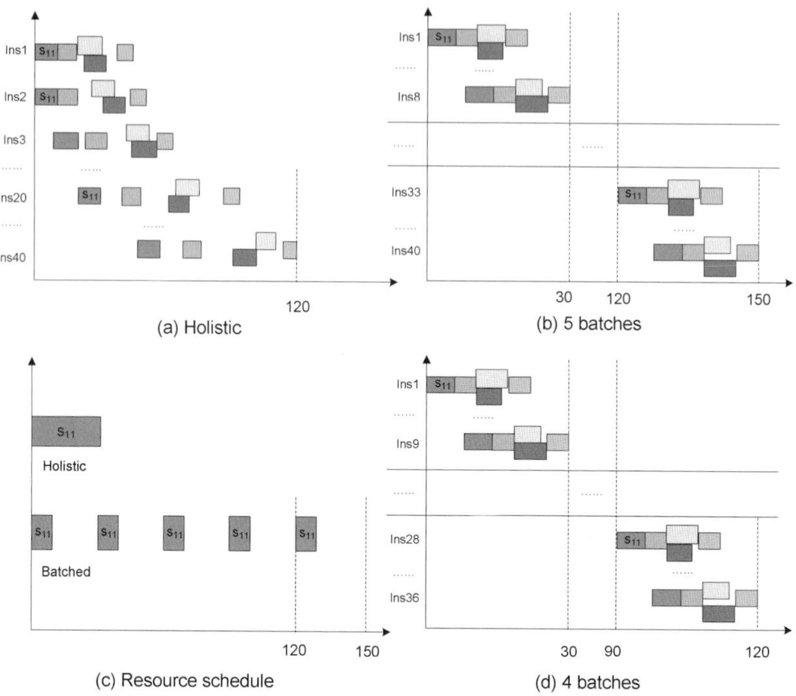

Fig. 2. Process execution

Aiming to assess the inter-influence between these factors, we propose a comprehensive resource planning approach to calculate these factors and determine the influences between them. Our approach is expected to support organisations to negotiate with customers and seek an optimal resource allocation plan.

3 Role Based Business Process Model with Problem Statement

In this section, a model comprising resources, roles, tasks, and business processes is presented to define those notions used in resource planning and describe their relationship. In this model, the scope of all definitions is on a single business process. A batch is a class of instances to be processed in parallel. Resource abundance is a concept to measure the capability and capacity relationship between the tasks in business process and the roles of available resources. Obligation indicates how much a role is required by the tasks, and support measures how much available workload can be used for a given task.

Definition 1 (Resource). A resource s denotes an available unit that is necessary for executing a task. In real cases, a resource can be a human, a machine or a computer. In this model, each resource has two attributes:

- *Role* indicates which group this resource belongs to according to its position. In this model a resource can have only one role to perform, yet a role may include multiple resources. For a resource *s*, function *r(s)* returns the role of this resource.
- *Maximal workload* denotes how much time this resource can provide service for the business process. Function *mload*(*s*) returns maximal workload of resource *s* for the business process.

Definition 2 (Role). A role *r* denotes a class of resources that own the same capability. Each role has an attribute of cost:

- *Cost* denotes the monetary cost a resource of role *r* needs to spend to perform a task. In this model it is valued by the hourly pay of the role. Given a role *r*, function *costRatio(r)* returns the cost of this role.

The resources belonging to role *r* may be capable to perform several tasks, where function *capable*(*r*, *t*) depicts such capability mapping. Function *capable*(*r*) returns the task set $\{t|capable(r, t)\}$ role *r* is able to perform.

Definition 3 (Task). A task *t* is a logical unit of work that is carried out by a resource to process one instance. A task has an attribute of *role*:

- *role* defines what kind of resources can perform this task. In other words, a task can be performed by resources that can match the role attribute of task. In this model, one task has at least one role, therefore has a none-empty role set R:$\{r\}$.

When a resource *s* is allocated to a task *t*, the execution time is determined by the role of *s*. It can be returned by function *time*(*t*, *r*), and *r*=*r*(*s*). The cost for *s* to execute *t* is returned by function $cost(t, r)= time(t, r)\times costRatio(r)$. Accordingly, workload of *time*(*t*, *r*) will be consumed from *s*.

Definition 4 (Business process). A business process represents a series of linked tasks, which collectively describe the procedure how a business goal is achieved. The structure of a business process *p* can be modelled as a directed acyclic graph in the form of $P(T, E, G, v_s, v_t, P)$, where

(1) $T = \{t_1,\ldots, t_n\}$, $t_i\in T\ (1 \leq i\leq n)$ represents a task in the business process;
(2) $G =\{g_1,\ldots, g_m\}$, $g_i\in G\ (1 \leq i\leq m)$ represents a gateway in the business process;
(3) E is a set of directed edges. Each edge $e = (v_1, v_2) \in E$ corresponds to the control dependency between *vertex* v_1 and v_2, where $v_1, v_2 \in T\cup G$;
(4) v_s is the starting node of the business process, which satisfies that $v_s\in T\cup G$;
(5) v_t is the terminating node of the business process, which satisfies that $v_t\in T\cup G$;
(6) $Pth =\{pth_1,\ldots, pth_k\}$, $pth_i\in P\ (1 \leq i\leq k)$ represents a path consisting a set of linked tasks from v_s to v_t. Total execution time of a path *pth* is *pth.tm*, which is calculated as the sum of processing time of each task on this path.

We analyse the environment to improve resource planning efficiency. Which resource is excessively required? Which task may have resource crisis? We use resource abundance to represent the capability and capacity between tasks and roles. Further, support of tasks and obligation of roles are defined based on resource abundance.

Definition 5 (Resource abundance). If a role r is capable of executing task t, resource abundance of r for t represents how much workload r can contribute to perform t. It is calculated as:

$$w(t, r) = \sum_{\forall s:r(s)=r} mload(s) / (n \times time(t, r))$$

where $\sum_{\forall s:r(s)=r} mload(s)$ is the total available workload of this role. n is the number of instances, and $n \times time(t, r)$ is the workload required to be assigned on t. $w(t, r)$ measures how much ratio of the total workload required by t can be contributed by r. For instance, task t_1 requires workload 18 hours on role r_1 to perform for all the instances, and resource of r_1 has workload 24 hours in sum, then $w(t_1, r_2) = 24/18=4/3$.

Definition 6 (Support). Support denotes the richness of resources available for a task t. Maximal support $supt_{max}$ is based on all resource used for t. Mean support $supt_{mean}$ is based on resource equally used for all capable tasks. They are calculated as:

$$supt_{max}[t] = \sum_{\forall r:capable(t,r)} w(t,r), \quad supt_{mean}[t] = \sum_{\forall r:capable(t,r)} w(t,r) / |capable(r)|$$

where $supt_{max}$ of t is the sum of value on resource abundance for all roles capable for t. Notice that if $supt_{max}[t]<1$, resource is insufficient to execute t and the process is not executable. In our example, task t_1 has two capable roles r_1 and r_2, $w(t_1, r_1) = (2\times 12)/(12\times 1.5)=4/3$ and $w(t_1, r_2) = 1$, then $supt_{max}[t_1] =7/3$. $supt_{mean}$ considers the situation resource workload is evenly used for capable tasks. In this example, we have $supt_{mean}[t_1] =4/3+1/3=5/3$.

Definition 7 (Obligation). Obligation denotes how much a role is obligated to the tasks. It reflects the tightness of this role in resource planning. Given the resource abundance relationship, obligation of a role is computed as:

$$o(r) = \sum_{\forall t:capable(t,r)} 1/ supt_{mean}[t]$$

In the pattern roles are equally used, for each task t executable by role r, the degree of t requires r is $1/supt_{mean}[t]$. Obligation of r is the sum for all tasks. For example, if both t_3 and t_4 can be performed by r_3, and $supt_{mean}[t_3]=4$, $supt_{mean}[t_4] \approx 4.42$, then $o(r_3)=(1/4)+(1/4.42) \approx 0.48$.

Definition 8 (Batch). Assume there are totally n instances to be processed, and a batch contains m instances ($m \leq n$). Those m instances in a batch are processed in parallel. We assume that there is no inference between instances belonging to different batches during execution.

Problem Statement

Suppose that a business process p together with its task set T, resource set S, role set R, the mappings from S to R, i.e., $S \rightarrow R$, and from R to T, i.e., $R \rightarrow T$, are given. The requirement is to process n instances of p in b batches before a deadline t_D.

Table 1 lists the involved variables in this problem statement, which will be used through out this paper.

Table 1. Variable definitions

n	The number of assigned instances
t_D	The given deadline
b	The number of batches
m	The size of batch
t_b	The execution time for one batch

Our goal is to find a batching scheme, i.e., the proper number of batches b, and a resource planning scheme ($S \rightarrow T$) such that:

(1) Time constraint can be satisfied ($t_b \times b \leq t_D$);
(2) Less total cost;
(3) The resource allocation is conflict free, i.e., no resource serves for multiple tasks or instances simultaneously.

Note, the holistic solution is a special case of batched processing, i.e., b=1.

4 Resource Planning

In this section, we discuss the strategies for finding resource planning solutions according to the problem statement discussed in the previous section. A set of heuristic rules for resource planning is first presented in Section 4.1. Based on the heuristic rules, Section 4.2 and Section 4.3 introduce the holistic and batched strategies, respectively. In Section 4.4 we discuss the balance of customer requirements and rationality of resource use.

4.1 Heuristic Rules

Because of the instance dependencies, resource allocation on one task may affect the execution of other tasks. Then, given a task in the business process, which resource to execute the task is optimal in the overall perspective? When multiple instances are executed in parallel, each of them needs to be allocated with a resource to execute this task. We use some heuristics in the allocation. The rules used in this paper are listed as follows:

Rule 1. Skewed tasks are allocated earlier than others in order to avoid crucial resources misused by other tasks.
Rule 2. If the obligation of role r is over a threshold ∂ , which means resources of this role are over-required, such resources should be preserved if possible in the resource allocation for later use.
Rule 3. Resource allocation should avoid, if possible, increasing the overall time of the longest path.
Rule 4. Resource allocation intends to use economical resource as task performer in order for minimal overall cost.

Rule 1 is about the allocation sequence, to improve the effectiveness of resource planning. *Rule 2* is intended to avoid workload problem and therefore improve efficiency. In the resource allocation, there is a balance between saving cost and saving time. *Rule 3* and *Rule 4* deal with time and cost saving respectively. The balance is, if resource allocation on a task is likely to increase overall time, *Rule 3* is used to avoid violating the deadline; otherwise, *Rule 4* is applied to reduce cost. Through the heuristic rules, many solutions can be applied to plan the resource and schedule instances. In the holistic planning strategy of this paper, we propose a two step solution. In the first step, a basic resource allocation is applied to plan the resources for all instances for build time optimisation. However, due to the process structural characteristic and instance dependency, run time conflicts may occur in the first step. The second step handles those conflicts. In the batch based resource planning strategy, instances are executed in batches. We will address the problem of how to partition instances into batches, and then optimise the use of resource according to the fixed batch pattern.

4.2 Holistic Based Resource Planning

In this strategy, resource planning is applied for all instances. Based on the heuristic rules discussed in Section 4.1, we develop a two step strategy in 4.2.1 and 4.2.2 to optimise the use of resources by considering the total time constraint, instance dependencies for run time execution and cost optimisation.

4.2.1 Basic Resource Allocation

We first plan resource for the business process without considering conflicts. Such planning is as n instances processed in parallel. According to *Rule 3*, we keep track of the longest path among all instances to avoid increasing of the time required for processing the batch t_b. Given n instances ($ins_1,\ldots,ins_n$) to allocate, every instance ins_i has k paths $Pth_i=\{pth_{i1},\ldots, pth_{ik}\}$, then among those $k\times n$ paths in the batch, the longest path is such a pth_{ij} ($1\le i\le n$, $1\le j\le k$) with the maximal value of processing time $pth_{ij}.tm=\sum_{t\in pth_{ij}} time(t)$, where we set the processing time of task $t\in pth_{ij}$ as minimal executing time of all capable resources $time(t)[ins_i]=min(\{time(t, s)|\ capable(t, s)\})$ by default, and after t is allocated with resource s processing time is updated to $time(t)[ins_i]=time(t, s)$. In this way we track the whole structure to target the longest path in overall perspective, and then manage resources reasonably to reduce the longest path and enable instances within a batch to be executed in balance.

According to the resource abundance, some tasks may have multiple allocable resources, while some tasks may have few. We call those tasks whose mean support is lower than a threshold σ as 'skewed task'. We have skewed task set $SkT=\{\ t\mid t\in T \wedge supt_{mean}(t) < \sigma\ \}$. According to *Rule 1*, tasks in *SkT* should be allocated before other tasks to avoid resource shortage on the skewed tasks in resource planning.

Therefore, skewed tasks in *SkT* are firstly allocated before others in the ascending order of their support value. As followed, the first unallocated task $t\in T$ on the longest path of the batch structure is selected for planning iteratively, until the whole batch of instances are allocated. When task t is selected, we apply Algorithm 1 to allocate suitable resources to this task for each instance in the batch. In Algorithm 1, the input *allocT* is a resource plan before task t of instances in the batch are allocated, while the output *allocT'* is the new plan after t is allocated.

Input:	n	-	number of instances
	S	-	set of resources that are available
	t	-	the task to be allocated for m instances
	allocT	-	resource planning before allocation on t
Output:	*allocT'*	-	resource planning after allocation on t

1	S_1={ $s \mid s \in S$, $r=r(s)$, $capable(t, r) \wedge mload(s) \geq time(t, r)$ };
2	S_2={ $s \mid s \in S \wedge o(r) < \partial$, $r = r(s)$ };
3	TS={ $t[ins_i] \mid 1 \leq i \leq m$ };
4	**if** $\exists t[ins_j] \in TS$: $t[ins_j] \in lpth(Pth_{11},\ldots,Pth_{mk})$ **then**
5	**if** $\mid S_2 \mid > m$ **then** $s=getMinTime(S_2)$;
6	**else** $s=getMinTime(S_1)$;
7	**end if**
8	$allocT'=alloc(t[ins_j], s)$; $TS = TS - \{t\}$;
9	**end if**
10	**for each** $t[ins_k] \in TS$ in descending order of $rlpth(t).tm$
11	S_3={s $\mid s \in S_1 \wedge rlpth(t[ins_k]).tm+time(t, s) \leq lpth(Pth_1,\ldots,Pth_m).tm$ };
12	**if** $S_2 \cap S_3 \neq \phi$ **then** $s=getMinCost(S_2 \cap S_3)$;
13	**else if** $S_2 \neq \phi$ **then** $s=getMinTime(S_2)$;
14	**else** $s=getMinOblig(S_1)$;
15	**end if**
16	$allocT'=alloc(t[ins_k], s)$; $TS = TS - \{t\}$;
17	**end for**
18	**return** *allocT'*;

Algorithm 1. Basic resource allocation

As task t in the business process is selected, there are m such tasks belonged to different instances in the batch need to be allocated, and they are included in the task set TS. Resource set S_1 includes all resources that can be allocated to t (Line 1). Among the resources in S_1, those under-required according to *Rule 2* are classified as set S_2 (Line 2). In Line 3, we check if there exists a task in set TS belonged to the current longest path of the batch, which is returned by function $lpth$ ($Pth_{11},\ldots,Pth_{mk}$), and Pth_i ($1 \leq i \leq m$) is the path set of the ins_i in batch. If it is true, the task on the longest path is allocated in such a way (Lines 4-9): if S_2 contains at least m resources with no less workload than t requires, S_2 is the candidate (Line 5). Otherwise, S_1 is the candidate (Line 6). This task is on the longest path, thus we select a resource from the candidate with minimal execution time to satisfy *Rule 3* by function *getMinTime*() (Lines 5-6). After that we apply function $alloc(t, s)$ to update *allocT* (to *allocT'*) and some necessary information (maximal workload, resource abundance, support and obligation) in Line 8. For the remaining tasks, they are processed in descending order of the overall time of relative longest path. In Line 10, relative longest path to task t is the longest path t is involved in, which is returned by function $rlpth(t)$. In Line 11, S_3 is a resource set selecting resource from S_1 that after using this resource does not increase time of the longest path. In Line 12, if $S_2 \cap S_3 \neq \phi$, then according to *Rule 4* a resource of minimal cost is selected by function $getMinCost(S_2 \cap S_3)$. Otherwise if $S_2 \neq \phi$, then the current longest path must be increased, and we select from S_2 with the one with minimal time to satisfy *Rule 3* by function *GetMinTime*(S_2) (Line 13). Otherwise, all resources are over-required, and then we select one resource of minimal obligation by function *GetMinOblig*(S_1) according to *Rule 2* (Line 14). Then selected resource is allocated to this task in Line 16.

This algorithm allocates suitable resource to execute m instances in a batch respectively. However, this algorithm does not consider run time features, thus there may exist allocation conflicts due to instance dependency caused by competing resources. Therefore, we handle the allocation conflict immediately after Algorithm 1.

4.2.2 Handling Allocation Conflict

The allocation conflict handling strategy ensures the run time correctness of resource planning. Algorithm 2 will not be applied if the outcome of Algorithm 1 is over the deadline. In the first place, this algorithm checks the allocation conflicts in the order that tasks in a batch are executed. Once conflicts are observed, the conflicts are handled directly. Each checked task may have several conflicts with other tasks, and we will firstly determine the sequence of the conflicts to be handled. For each allocation conflict to be handled, we can reallocate one task to another suitable resource, or make the conflicting tasks executed in a sequential order. In Algorithm 2, we analyse the solutions and their impact on overall performance to select a proper one based on the heuristic rules.

Task set *TS* includes all the tasks for allocation (Line 1).The overall structure *bp* is n instances executed in parallel, and each ins_i ($1 \le i \le n$) has a process structure p_i (Line 2). Firstly, we check all tasks in *TS* for the run time conflicts in the sequence they start. When a task t_0 is checked and there are conflicts in which t_0 is involved, those detected conflicts are handled in Lines 5-32. Task set *Tc* (Line 5) includes all tasks conflict with t_0. If a task in *Tc* can be reallocated in such a way that both time and cost are reduced, reallocation will be applied (Lines 9-12). Among the resources that can reduce both time and cost, the one of minimal time is selected for reallocation when the task is on longest path according to *Rule 3*, while the most economical one is chosen if not on longest path according to *Rule 4*. In Lines 15-34, the remaining conflicts are handled in the decreasing order of the relative longest path that tasks in *Tc* are involved in. Both reallocation and task/instance delay can be applied. The solution that is optimal in the overall perspective will be selected.

Lines 16-25 deal with conflicts in such case that one of the two conflicting tasks is on the longest path. Reallocation may reduce time but increase cost. However if we choose task delay time cannot be reduced. bp_1 and bp_2 are two operational choices of task delay in arbitrary order (Line 17), t_1 and t_2 are their corresponding overall time of all instances accordingly, and t_{min} denotes the minimal overal time if we choose task delay (Line 18). Our first option is reallocation to reduce the longest path of *bp* according to *Rule 3*. If tsk_l, which is the longer of the two conflicting tasks, is on the longest path, then our first option is to reallocate on tsk_l and hence reduce the critical path. In Line 20, function *TimeReduceRes*(tsk, t_{min}) returns the resource set that through such reallocation on *tsk* overall time is shorter than task delay. When the resource set S_1 returned by *TimeReduceRes*(tsk_l, t_{min}) is not empty, we select the resource in set that can mostly reduce the time (Line 21). When S_2 is empty, we reallocate the other task tsk_s to reduce time. If resource set S_3 returned by *TimeReduceRes*(tsk_s, t_{min}) is not empty, we choose the one of minimal cost and do not incur

Input:	n	-	number of instances
	S	-	set of resources that are available
	$allocT$	-	resource planning before allocation on t
Output:	$allocT'$	-	resource planning after allocation on t
	t_b	-	execution time of the batch

```
TS={T_1[ins_1]+…+T_m[ins_m]};
bp=parallel(p_1, p_2,…, p_m); allocT'=allocT;
while(TS ≠ ϕ)
  select task t_0: first task to start in TS in batch execution;
  Tc={t | t∈ TS has time overlap with t_0 and they use the same resource};
  for each t∈ Tc: (t, s)∈ allocT
    S_1={s' | time(t, r(s')<time(t, r(s) ∧ cost(t, r(s'))<cost(t, r(s))};
    if S ≠ ϕ
      if t∈ lpth(Pth_11,…,Pth_mk) then  s'=getMinTime(S_1);
        else  s'=getMinCost(S_1);
          end if
      allocT'=alloc(t, s'); Tc=Tc-{t};
    end if
  end for
  for each t∈ Tc in descending order of rlp(t)
    if rlpth(t_0)>rlpth(t) then tsk_l = t_0, tsk_s = t; else tsk_l = t; tsk_s = t_0; end if
    bp_1=delay(t→t_0, bp); bp_2=delay(t_0→t, bp);
    t_1 =lpth(bp_1).tm; t_2 =lpth(bp_2).tm; t_min=min(t_1, t_2);
   if tsk_l ∈ lpth(bp, allocT') then
     S_2=TimeReduceRes(tsk_l, t_min), S_3= TimeReduceRes(tsk_s, t_min);
    if S_2 ≠ ϕ then s=getMinTime(S_2); allocT'=alloc(t_l, s);
      else if S_3 ≠ ϕ then s=getMinCost(S_3); allocT' =alloc(t_s, s);
      else if t_1< t_2 then bp= bp_1 ;  break;
      else  bp= bp_2;
    end if
   else
    allocT_1=bestRealloc te(p', allocT'); t_3= lpth(bp, allocT_1).tm;
    if t_3=min(t_1, t_2, t_3) then  allocT'= allocT_1;
      else if t_1=min(t_1, t_2, t_3) then  bp= bp_1;  break;
      else  bp= bp_2;
    end if
   end if
   Tc=Tc-{t};
  end for
  if(Tc=ϕ) then TS=TS-{t_0};  end if
end while
output allocT', t_b =lpth(p',allocT').tm;
```

Algorithm 2. Conflict handling algorithm

new conflict with previous tasks (Line 22). However when both of tasks cannot be properly reallocated, we use task delay to handle the conflict (Lines 23-24). Two tasks can be executed in arbitrary order. We compare the two sequences of execution, and the pattern leading to less value of the longest path is chosen. Lines 28-31 deal with conflict in cases none of the tasks are on the longest path. Function *bestReallocate*() returns updated allocation table after proper reallocation. Such reallocation firstly ensures minimal overall time and secondly minimal cost if total time is same. According to *Rule 4*, when Reallocation results in shorter longest path than task delay, it is performed in most economical way (Line 28). Otherwise, we compare two task delay pattern, and the one of less value of longest path is applied (Lines 29-30). When all the conflicts related with t_0 is handled, we check and handle conflicts for the next one task of *TS*. This procedure continues until end node is reached. The output of Algorithm 2 is a new planning scheme for the n instances without run time execution error and the total execution time t_b (Line 37). If $t_b < t_D$, such allocation solution is valid because time constraint is satisfied.

In summary, in the first step we plan the resources for build-time optimisation based on a set of heuristic rules. The conflict handling guarantees it is run time conflict free in the second step.

4.3 Batch Based Resource Planning

In this section, we discuss the strategy to allocate resources for batches of instances, where the proper size of a batch is to be calculated first.

4.3.1 Pre-analysis

In this part, we conduct an analysis on available resources, business requirements and process structures to sketch out a preliminary batch pattern for the resource planning.

When each batch contains m instances, then the total batch number $b=n/m$, and each batch requires time t_b to execute. Given that overall time cannot exceed the deadline, we can conclude formula (1) for pre-allocation planning as listed below. As resources tend to be periodically available in practice, such as work shifts for staff, we choose to process instances in batches. To use resources rationally, we attempt to set more batches while each batch contains fewer instances. Therefore, resources can be scheduled easily with less conflict. If each task in the business process uses the most efficient resource and accordingly the longest path of the process structure is lp, then the processing time of a batch has a lower bound $t_{lowerb} = \sum_{t \in lp} min(\{time(t, s) \mid capable(t, s)\})$, because every single instance requires a duration no less than t_{lowerb} for execution. With formula (1) and the range of $t_{lowerb} \le t_b$, we can compute the minimal number of instance m_0, and conclude formula (2).

$$t_b \times n / m \le t_D \tag{1}$$

$$\lceil t_{lowerb} \times n / t_D \rceil = m_0 \le m \tag{2}$$

We use $m_0 = \lceil t_{lowerb} \times n / t_D \rceil$ and $b_0 = n/m_0$ as the initial batch pattern for resource planning in following sections. Previously, lower bound t_{lowerb} is obtained with the assumption that each batch contains only one instance. If each batch changes to contain m_0 instances, t_{lowerb} will be updated accordingly. Given task $t \in T$, different instances in a batch may perform t with different resources. Assume the resource set capable of executing t is $S=\{s_1, s_2, \ldots\}$, and for each resource s_i, it may serve for k instances on task t in the batch according to its workload limit. Then s_i can be expressed as $s_{i1}, s_{i2}, \ldots, s_{ik}$. Each of them is for one of k instances in the batch. Therefore, there is a corresponding resource set:

$$A(t) = \{s_{ij} \mid s_i \in S \wedge 1 \leq j \leq |m\_load(s_i) / (b_0 \times time(t, s_i))| \}$$

where s_{ij} is a resource with the exactly workload to execute task t of one instance in the batch, and $j \leq | m\_load(s_i) / (b_0 \times time(t, s_i))|$ because that is the maximal number of instances that s_i can serve for t in a batch. Then, we start to update lower bound of batch time t_b and update the batch pattern if formula (1) cannot be satisfied, until a suitable batch pattern is selected for resource planning. Details of this procedure are listed as below:

(1) Resource checking. For each $t \in T$, if $|A(t)| < m_0$, i.e., there is no sufficient resource to serve task t of all instances based on this batch pattern, the pattern has to be updated and m_0 is to be increased in (3).

(2) Lower bound update. Otherwise, we firstly refine the t_{lowerb} based on this batch pattern (m_0 instances). For each $t \in T$, there are m_0 instances to allocate, and different instances may use different resources to perform task t. From $A(t)$, we choose a subset $B(t)=\{ s_{ij} | s_{ij} \in A(t)$ is one of the top m_0 efficient resources $\}$, hence using resources in $B(t)$ can provide the best result in aspect of time. Based on $B(t)$, we will refine the lower bound of batch time t_{lowerb} based on the following computations:

(a) Let p (from v_s to v_t) be a path in the process and p include tasks $T=\{ t_1, t_2, \ldots\}$, for $\forall t \in T$ we use a resource from $B(t)$ to execute t. Assume the processing time of each instance $i\,(1 \leq i \leq m_0)$ the processing time is $pt(i)$, then the total time to execute m_0 instances is $\sum_{i=1}^{m_0} pt(i) = \sum_{t \in p} \sum_{s_{ij} \in B(t)} time(t, s_i)$. As we know, the time required for executing p for the whole batch is determined by the maximum path among m_0 instances, which is no less than the average of total time for executing p for all m_0 instances. Therefore we choose the lower bound for processing p as $TM(p) = \sum_{t \in p} \sum_{s_{ij} \in B(t)} time(t, s_i) / m_0$;

(b) According to the process structure, there may be multiple paths from v_s to v_t. Each path has its own lower bound according to (a). The lower bound of executing time for the whole process is the maximal value of $TM(p)$ for all paths. Therefore, given the paths $P=\{p_1, p_2, \ldots\}$ of the business process, the lower bound in this batch pattern is refined to:

$$t_{lowerb} = max(TM(p_1), TM(p_2), \ldots)$$

After the lower bound of batch time t_{lowerb} is updated, we check $t_{lowerb} \times \lceil n / m_0 \rceil < t_D$ again. If it is satisfied, this pattern is allocable and we move on to section 4.3 to do resource planning based on this pattern. Otherwise, we update the batch pattern in (3).

(3) Batch pattern update. In this pattern (m_0 instances), the total batch number is $b_0 = \lceil n / m_0 \rceil$. When we increase m_0, if b does not reduce accordingly, then the total time is definitely increased. Therefore in order to reduce total processing time, m_0 must be updated to such a value that b is reduced as a result. Hence, the new batch pattern is $m_0 = \lceil n/(b_0+1) \rceil$, and we go back to (1) with this new pattern.

The procedure from (1) to (3) continues until a batch pattern can be found. Then in the next step, we investigate how to plan the available resources for the given batch pattern to achieve the business goals.

4.3.2 Resource Allocation for Batches of Instances

Through the previous computation we selected a batch pattern. For each batch, we use the holistic algorithm discussed in Section 4.2 to plan resources for m instances within a batch. If the total time of returned resource plan exceeds the deadline, we adjust the batch pattern and replan as below:

In order to find a new batch pattern for reducing the overall time, we have to decrease the number of batches b by 1 to b', accordingly increase the number of instances within a batch from m to m' ($m' > m$), to make more instances processed in parallel. Consequently, the batch time will be increased from t_b to t_b'. With more instances to be processed in a batch, we know that $t_b' \geq t_b$. We know that the basic requirement is $t_b \times n/m \leq t_D$, and we can deduce that $m' > n \times t_b / t_D$. Hence based on the previous allocation result, we predict the new number of instances $m = \lceil n \times t_b / t_D \rceil$, and then based on this pattern we reuse the holistic strategy to plan for the new batch. This procedure continues until either we get the result that satisfies time constraint or b gets decreased to 1 which means that we cannot find an allocation plan for even a single batch.

4.4 Discussion

Both holistic and batched strategies can achieve a common goal, i.e., a lower cost by applying the set of heuristic rules. However, the holistic approach may suffer in a skewed resource allocation while the batched approach may not obtain a shorter deadline compared with the holistic approach. Due to the availability pattern of resources, an organisation always has a minimal batch number B for better arranging workload for resources. This implies that a batched solution with $b < B$ may not be an attractive solution to the organisation. However, a deadline may be required by customers. Therefore, it is necessary to balance the deadline and the number of batches.

If the batched approach fails to find a batched solution with $b \geq B$, the organisation may negotiate with customers to avoid losing the deal. Our strategy can provide suggestive information to help negotiation: (1) Extend the deadline. If the deadline is not strict, we calculate the total time t_a of the batched solution on $b = B$, and relax t_D to t_a if customers agree. (2) Reduce number of instances. Assume n_a ($n_a < n$) is the number of instances finished within the deadline when we apply the batched strategy on $b = B$. If the number of instance is negotiable, it can be relaxed to n_a. (3) Suggested batch number. We go on process batched strategy starting from $b_a = B\text{-}1$, and find a batch solution with batch number b_a. The organisation can judge if it is worthy to accept the order by using resources in this resource plan.

5 Experiment

In this section, we present an experimental study of the resource planning strategies proposed in this paper. Through the implemental data, we investigate and compare the performance of the two strategies with different volumes of instances, batch patterns and business process scales. Also, we also discuss how to apply the strategies to help organisations rationally use their resources.

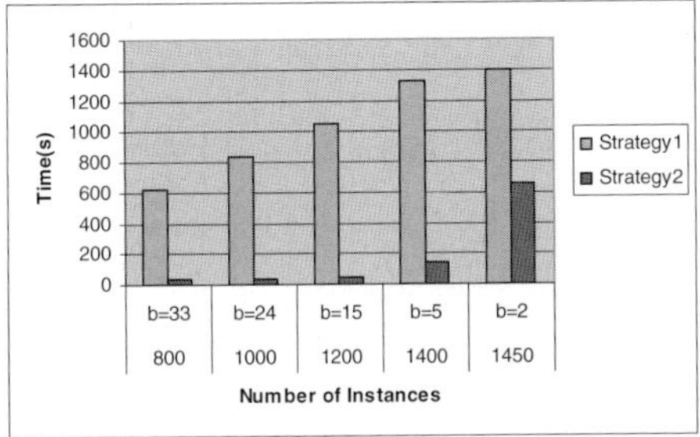

Fig. 3. Performance of two strategies

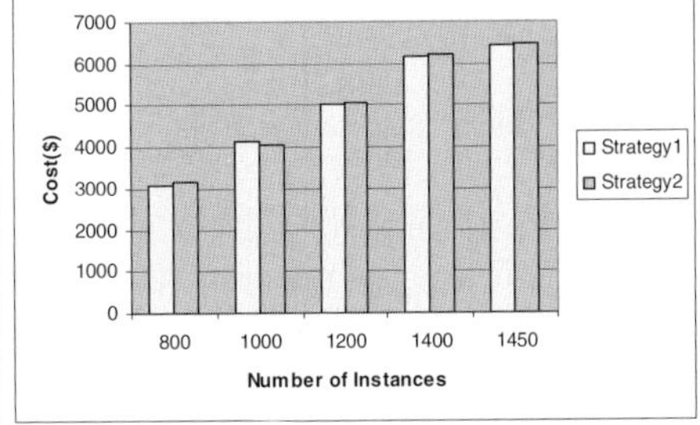

Fig. 4. Cost of two strategies

Firstly, we test the performance of the holistic planning strategy (as Strategy 1) and the batched planning strategy (as Strategy 2) with different amounts of instances. Extended from the motivating example discussed in Section 2, the test settings include a 30-task business process, a set of resources, and a requested completion time of 140 hours. In the same environment, we apply both strategies to plan resources for 800, 1000, 1200, 1400 and 1450 instances. For each test case, the schedule time of two strategies is compared in Figure 3, and the cost compared in Figure 4. Costs of two strategies are similar according to Figure 4, for example in producing 1000 instances the cost of strategy 1 is $4150, and the cost of strategy 2 is $4050. But strategy 2 has much shorter schedule time than that of holistic strategy according to Figure 3, for example, the schedule time on 1000 instances is 836s and 51s for two strategies respectively. We notice that the more batches we use, the more time can be saved by batched strategy. Suppose the organisation requires that the assigned 1450 instances to be executed in at least 5 batches, but the batched strategy can merely support 2 batches as shown in Figure 3. To cover the gap, the organisation may ask for either a relaxation on number of instances or an extension to the deadline. Figure 3 shows 5 batches can be supported with 1400 instances, and therefore we can recommend the organisation to negotiate with customer for relaxing number of instance to 1400.

Also, we compare the performance of planning with different batch patterns in a fixed environment. With 1450 assigned instances, Figure 5 lists the minimal time and overall cost to produce required instances based on 1, 2, 5, 10 and 20 batches. It shows that the total costs for different patterns are similar. However when the number of batches increases, the total time is likely to rise as compensation. When the deadline is 140 hours, only 2 batches can be supported according to Figure 5, although the organisation requires 5. Figure 5 shows it will take 145 hours to produce all instances if they are partitioned into 5 batches. Thus, we can recommend the organisation to negotiate with customers for extending the deadline to 145 hours.

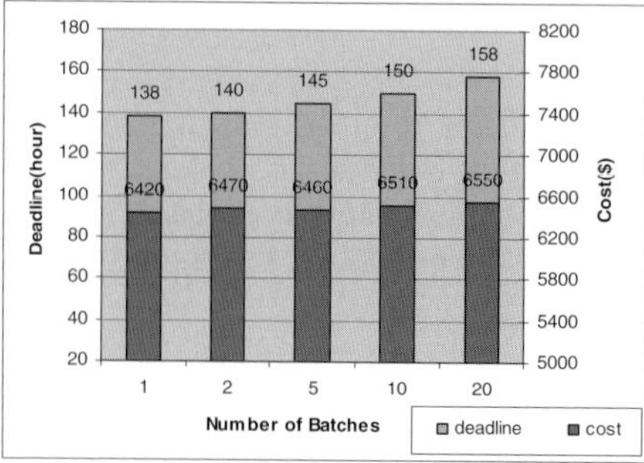

Fig. 5. Batch based resource planning

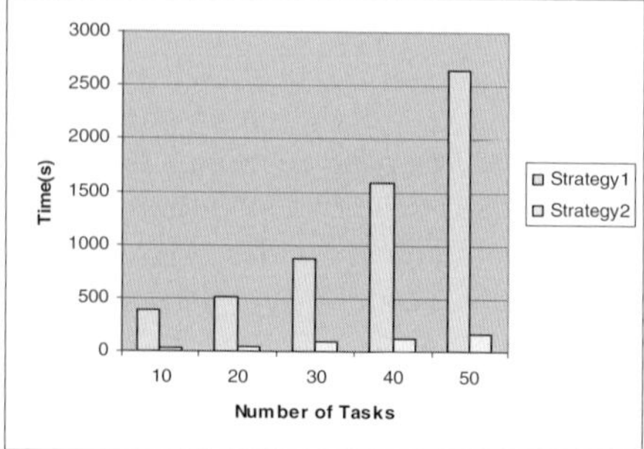

Fig. 6. Performance on different task scale

System performance is a crucial issue in some applications, especially when re-planning is frequently required because of the dynamic environment or requirements. Figure 6 shows the schedule time for 1000 instances under different task scales. As the figure shows, the schedule time increases dramatically when the number of tasks increases, and it is time consuming to plan resources for business processes containing a large number of tasks. While in comparison, the batched strategy is more efficient, and therefore is preferable and practical when the task scale rises.

According to these experiments, we see that the batch based strategy is of better performance and more rational in allocating resources while it sometimes may require more time as a compensation. By estimating the potential influences, our approach provides the organisation with the information on adjusting batch patterns for optimising resource utilisation.

6 Related Work and Discussion

Resource planning is a research topic related to task/workflow scheduling and resource allocation, which seeks an optimised execution sequence for a set of tasks to achieve certain goals. This procedure is dependent on which resources are allocated to execute the tasks. The previous related work can be generally classified into two categories: resource oriented scheduling and process oriented one. In the resource oriented scheduling such as [2, 12, 15], tasks are scheduled at the run time to achieve full utilisation of resources, and there is no dependency between these tasks. In contrast, the process oriented scheduling deals with tasks in the workflow where tasks have dependency with each other. In [14], Yu has proposed a genetic algorithm for scheduling scientific workflows for utility Grid applications by minimising the budget constraint while meeting user's deadline constraint. In his strategy, workflow is partitioned and deadline is assigned to the partitions for resource planning, but resource workload is not considered so resources are unlimitedly used. In [10], Topcuoglu, Hariri and Wu have proposed an Earliest-Finished-Time (HEFT) algorithm to reduce processing time, and HEFT has been demonstrated to be effective in the ASKALON project in [11]. While most work in this area of workflow scheduling focused on the temporal and causality constraints, [9] proposed a workflow scheduling framework under both temporal constraints and resource constraints, but did not discuss how to manage and allocate resource in workflow. Work [1] presented an adaptive scheduling algorithm that finds a suitable execution sequence for workflow tasks by additionally considering resource allocation constraint and dynamic topology changes. This work was applied in the embedded and wireless network systems. All the above approaches were proposed from the run time perspective, however sometimes we hope to manage resources at the build time to guarantee some requirements can be satisfied. In [6], Eden, Panagos and Rabinovich developed the technique to check time constraints at process build time and to enforce these constraints at run-time. However, this paper did not investigate the inter-instance dependency. In [13], we proposed a method to improve performance of business processes by integrating build-time resource allocation with business process structural improvement. However, the focus of this paper is to investigate the impact between resource allocation and process structural improvement, only intra-instance dependencies have been considered.

In contrast to the previous works, this paper focused on the planning of resources for massive number of instances to meet process requirements and rational utilisation of resources before execution. Particularly, our approach highlighted the influences caused by the process structure and the concurrency of process instances for resource planning, and our strategy can enable resources with balanced workload distribution to cater for their rationalities.

7 Conclusion

In this paper, we investigated how to plan resources for a business process optimally, in aspect of meeting process requirements and rational utilisation of resources. Also, our approach can provide information to organisation for decision making and negotiation with customer in order to better use resources. When a massive number of instances are given, both inter and intra instance dependencies are considered in the planning of each batch of instances to be executed in parallel. As the problem is computationally hard [14], a set of heuristic rules were designed, and two strategies based on these heuristic rules were proposed and compared.

In the future, we plan to improve our planning strategy for supporting complex resource patterns.

Acknowledgement

The research work reported in this paper is supported by Australian Research Council under Linkage Grant LP0669660.

References

1. Avanes, A., Freytag, J.C.: Adaptive Workflow Scheduling Under Resource Allocation Constraints and Network Dynamics. Journal of PVLDB 1, 1631–1637 (2008)
2. Braun, T.D., Siegel, H.J., Beck, N., Bölöni, L., Maheswaran, M., Reuther, A.I., Robertson, J.P., Theys, M.D., Yao, B., Hensgen, D.A., Freund, R.F.: A Comparison of Eleven Static Heuristics for Mapping a Class of Independent Tasks onto Heterogeneous Distributed Computing Systems. Journal of Parallel Distributed Computing 61, 810–837 (2001)

3. Cardoso, J., Sheth, A.P., Miller, J.A., Arnold, J., Kochut, K.: Quality of Service for Workflows and web Service Processes. Web Semantics 1, 281–308 (2004)
4. Etoundi, R.A., Ndjodo, M.F.: Feature-oriented Workflow Modelling Based On Enterprise Human Resource Planning. Business Process Management Journal 12, 608–621 (2006)
5. Huang, Y.-N., Shan, M.-C.: Policies In A Resource Manager of Workflow Systems: Modelling, Enforcement and Management. In: Proceedings of the 15th International Conference on Data Engineering, p. 104. IEEE Computer Society, Los Alamitos (1999)
6. Eden, J., Panagos, E., Rabinovich, M.: Time Constraints in Workflow Systems. In: Jarke, M., Oberweis, A. (eds.) CAiSE 1999. LNCS, vol. 1626, pp. 286–300. Springer, Heidelberg (1999)
7. Russell, N., van der Aalst, W.M.P., Hofstede, A.H.M.t., Edmond, D.: Workflow Resource Patterns: Identification, Representation and Tool Support. In: Pastor, Ó., Falcão e Cunha, J. (eds.) CAiSE 2005. LNCS, vol. 3520, pp. 216–232. Springer, Heidelberg (2005)
8. Russell, N., van der Aalst, W.M.P.: Work Distribution and Resource Management in BPEL4People: Capabilities and Opportunities. In: Bellahsène, Z., Léonard, M. (eds.) CAiSE 2008. LNCS, vol. 5074, pp. 94–108. Springer, Heidelberg (2008)
9. Senkul, P., Toroslu, I.H.: An Architecture for Workflow Scheduling Under Resource Allocation Constraints. Information System 30, 399–422 (2004)
10. Topcuoglu, H., Hariri, S., Wu, M.-Y.: Performance-Effective and Low-Complexity Task Scheduling for Heterogeneous Computing. IEEE Transactions on Parallel and Distributed Systems 13, 260–274 (2002)
11. Wieczorek, M., Prodan, R., Fahringer, T.: Scheduling of Scientific Workflows in The ASKALON Grid Environment. Journal of SIGMOD Record 34, 56–62 (2005)
12. Wu, A.S., Yu, H., Jin, S., Lin, K.-C., Schiavone, G.A.: An Incremental Genetic Algorithm Approach To Multiprocessor Scheduling. IEEE Transactions on Parallel and Distributed Systems 15, 824–834 (2004)
13. Xu, J., Liu, C., Zhao, X.: Resource Allocation vs. Business Process Improvement: How They Impact on Each Other. In: Dumas, M., Reichert, M., Shan, M.-C. (eds.) BPM 2008. LNCS, vol. 5240, pp. 228–243. Springer, Heidelberg (2008)
14. Yu, J., Buyya, R., Tham, C.-K.: Cost-Based Scheduling of Scientific Workflow Application on Utility Grids. In: International Conference on e-Science and Grid Technologies, Melbourne, Australia, pp. 140–147 (2005)
15. Zomaya, A.Y., Teh, Y.-H.: Observations on Using Genetic Algorithms For Dynamic Load-balancing. IEEE Transactions on Parallel and Distributed Systems 12, 899–911 (2001)

Assessment of Service Protocols Adaptability Using a Novel Path Computation Technique*

Zhangbing Zhou[1], Sami Bhiri[1], Armin Haller[1], Hai Zhuge[2], and Manfred Hauswirth[1]

[1] Digital Enterprise Research Institute, National University of Ireland at Galway
firstname.lastname@deri.org

[2] Institute of Computing Technology, Chinese Academy of Sciences
zhuge@ict.ac.cn

Abstract. In this paper we propose a new kind of adaptability assessment that determines whether service protocols of a requestor and a provider are adaptable, computes their adaptation degree, and identifies conditions that determine when they can be adapted. We also propose a technique that implements this adaptability assessment: (1) we construct a complete adaptation graph that captures all service interactions adaptable between these two service protocols. The emptiness or non-emptiness of this graph corresponds to the fact that whether or not they are adaptable; (2) we propose a novel path computation technique to generate all instance sub-protocols which reflect valid executions of a particular service protocol, and to derive all instance sub-protocol pairs captured by the complete adaptation graph. An adaptation degree is computed as a ratio between the number of instance sub-protocols captured by these instance sub-protocol pairs with respect to a service protocol and that of this service protocol; (3) and finally we identify a set of conditions based on these instance sub-protocol pairs. A condition is the conjunction of all conditions specified on the transitions of a given pair of instance sub-protocols. This assessment is a comprehensive means of selecting the suitable service protocol among functionally-equivalent candidates according to the requestor's business requirements.

1 Introduction

Given the inherent autonomy, heterogeneity, and continuous evolution of Web services, mediated service interactions are a common style of Web service interactions [9]. Hence, adaptability assessment is as important as compatibility analysis that targets direct service interactions. Following [3,4], by adaptation we mean the act of identifying, classifying, and reconciling mismatches between service behaviourial interfaces (the so-called service protocols) [5]. Adaptability assessment is further defined as the act of deciding whether or not service protocols of a requestor and a provider are adaptable without constructing an adapter,

* The work presented in this paper has been funded by Science Foundation Ireland under Grant No. SFI/08/CE/I1380 (DERI Lion-2).

R. Meersman, T. Dillon, P. Herrero (Eds.): OTM 2009, Part I, LNCS 5870, pp. 237–254, 2009.

computing their adaptation degree, and identifying conditions that determine when these two service protocols can be adapted. This assessment enables a requestor to identify and thus to select the most suitable service provider among functionally-equivalent candidates according to her business requirements.

As reviewed by Dumas et al. [5], previous works related to service interaction analysis mainly focus on either compatibility analysis [1,2] or adapter construction [3,4,7,9,12,13]. Adaptability can somehow be studied by techniques that construct adapters. However, these approaches are inadequate to provide the level of assessment we target. Indeed, being able to build an adapter merely implies that there are some situations where service protocols are possibly adaptable, while unsupported scenarios (due to un-reconcilable deadlocks [7] for instance) are excluded from the adapter protocol specification. An adapter does not differentiate the different possibilities of adaptability (i.e., the adaptation degree) between adaptable service protocols, and hence, these protocols are assumed to be the same to the requestor although they are actually different in the adaptation possibility. In addition, an adapter does not specify conditions that determine when two service protocols are adaptable. Consequently, these two service protocols may behave in such a way that their interaction fails, either because it is an unsupported scenario, or because some conditions are not satisfiable according to the exchanged message instances.

1.1 Motivating Example

Fig. 1 depicts, using guarded finite state automata (i.e., GFSA), the service protocols of two soft-drinks provider services (denoted *SP* and *SP_A*) and a possible interaction with a soft-drinks requestor service (denoted *SR*). A requestor, using *SR*, intends to buy soft-drinks online through one of these two provider services. Guards denoted Cd_i ($i \in [1, 5]$) correspond to conditions in the BPEL specification. Therefore we use guard and condition interchangeably in this paper.

The only difference between *SP* and *SP_A* is that *SP* allows canceling an interaction if some selected soft-drink is presently out of stock. *SP* and *SP_A* may apply a discount on the price depending on a *custInfo*. Hence, a *custInfo* is expected by *SP* (or *SP_A*) before the price is decided. Due to privacy concerns, *SR* only sends *custInfo* if the *price* is acceptable. This example shows a deadlock that occurs when *SR* directly interacts with *SP* (or *SP_A*). *SP* (or *SP_A*) is requesting a *custInfo* before sending a *price*, while *SR* is expecting a *price* before deciding upon continuation and sending *custInfo* or not. Adapters, such as the ones described in [13,7] can circumvent this problem if either (1) one of the receiving transitions is neither control nor mandatorily data dependent on one of its direct-succeeding-transitions [13], and hence, a *mock-up custInfo* message can be generated to resolve this deadlock, or (2) the adapter developer can provide a *custInfo* message using *evidences* [7], and thus, this deadlock is reconciled.

The *custInfo* is used for deciding whether or not a discount is to be applied. However, a *normalPrice* is always applied by default. For the requestor, both *SP* and *SP_A* are adaptable with *SR* (according to [13,7]) if the *price* is acceptable. Naturally, the requestor also wants to know whether the conclusion also holds

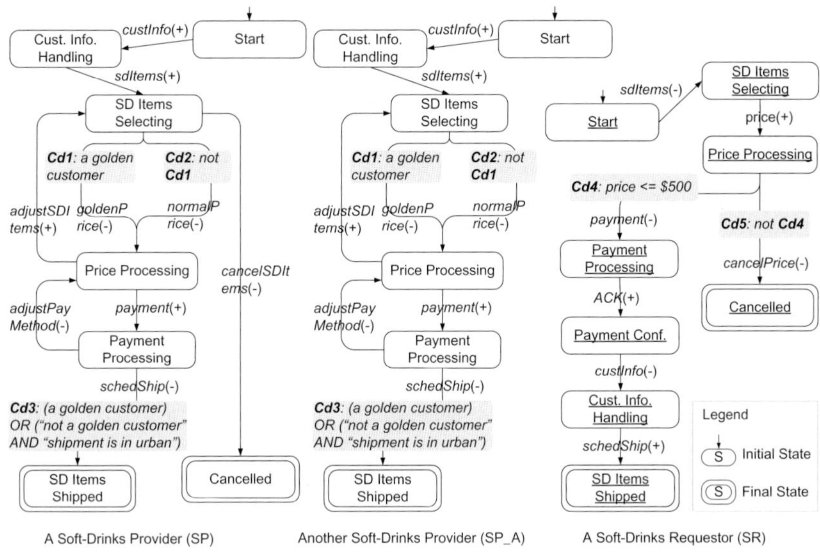

Fig. 1. Service protocols for soft-drink providers and requestor (*SP*, *SP_A* and *SR*)

when the *price* is unacceptable, so that the interaction can be cancelled if the price is unacceptable. *SP* can support this scenario according to [13,7], but, *SP_A* does not support this interaction because the interaction cannot be cancelled after the *Price Processing*.

Consequently, the requestor is given an initial impression that *SP* is more adaptable with *SR* than *SP_A*. If an expected interaction (for instance, the soft-drink items are to be shipped) can be supported by both *SP* and *SP_A*, *SP* is a more suitable candidate service provider because additional interactions (for instance, the interaction is to be cancelled) are also supportable by *SP* but not by *SP_A*. An adapter is ineligible to detect this difference since, in principle, an adapter can detect whether or not, but not how far, an adaptation is possible.

The reasoning so far only explores the legal message exchanges between *SP* (or *SP_A*) and *SR*. The success of an interaction in general and an adaptation in particular, however, depends also on conditions that guide which branches to follow and check whether transitions can be enabled. For instance, for an adaptation between *SP* (or *SP_A*) and *SR* leading to soft-drink items being shipped, a prerequisite is that Cd_2, Cd_3 and Cd_4 are able to be respected. Hence, $Cd_2 \bigwedge Cd_3 \bigwedge Cd_4$ is a condition that determines when *SP*, as well as *SP_A*, can be adapted with *SR*. For another adaptation between *SP* and *SR* leading to the interaction to be cancelled, a prerequisite is that Cd_2 and Cd_5 are able to be respected. However, this cancellation scenario is not supportable for *SP_A* and *SR*, and hence, $Cd_2 \bigwedge Cd_5$ is a condition for *SP* and *SR*, but is not a condition for *SP_A* and *SR*, that determines when they can be adapted.

To summarise, the conditions and the adaptation possibility (or the adaptation degree) are two complementary criteria to the requestor for selecting the suitable candidate service provider. Based on the conditions, the requestor first filters candidate service providers through checking whether the expected interactions can be supported. Thereafter, the requestor can choose a suitable service provider (possibly with the highest adaptation degree) among these functionally-equivalent candidates. The selected service provider is ensured to be interacted properly with the requestor service for achieving a certain goal.

The identification of these conditions, as well as the computation of the adaptation degree as mentioned above, is beyond existing adapter construction approaches, but is the concern of this paper. In the following sections, we show how they can be addressed in this particular example and in general.

1.2 Approach

We understand under adaptability assessment a combination of the following three perspectives: determining whether or not service protocols of a requestor and a provider are adaptable, computing their adaptation degree, and identifying conditions that determine when they can be properly adapted. In the following, we show how these three aspects are to be addressed in this paper.

1. We first construct a complete adaptation graph that captures all legal message exchanges between service protocols of a requestor and a provider (Section 3). A *complete* sequence of legal message exchanges reflects a mediated service interaction between a pair of instance sub-protocols of these two service protocols. An instance sub-protocol is a part of a service protocol that may be executed for a particular instance of this service protocol (Section 2). The emptiness or non-emptiness of the complete adaptation graph corresponds to the fact that whether these two service protocols are adaptable.
2. Assume m is the number of instance sub-protocols captured by the complete adaptation graph with respect to a certain service protocol p, and n is the number of instance sub-protocols specified by the specification of p. An adaptation degree is computed as a ratio between m and n (Section 5.1). Based on a novel path computation technique we propose in Section 4, we compute all instance sub-protocols for a service protocol specification (Section 4), and derive all instance sub-protocol pairs captured by the complete adaptation graph (Section 5.2).
3. Based on all instance sub-protocol pairs captured by the complete adaptation graph, we identify a set of conditions that determine when an adaptation is possible. For each pair, a condition is provided which corresponds to the conjunction of all conditions specified on the transitions of these two instance sub-protocols (Section 5.1).

Finally, we review the related work and conclude this paper in Section 6.

Our adaptability assessment builds upon our Space-based Process Mediator (SPM) that is detailed in our previous work [13]. It is important to notice that our technique is general and can be applied to other adapters as well.

This SPM is able to reconcile the deadlock in our motivating example in the *SD Items Selecting* state by providing a *mock-up custInfo* which is replaced whenever the *custInfo* arrives from *SR* to enable the *schedShip*(-) transition. Generally, the SPM can circumvent this kind of deadlock, if the dependency that exists between one of the receiving transitions with the state *enabled* and one of its direct- succeeding-transitions is neither control nor mandatorily data dependent. Then, the SPM can produce a *mock-up* message that acts as the data expected by this receiving transition. This *mock-up* message is replaced (through the space-based mechanism of the SPM) by a concrete message whenever it is produced by a peer protocol. More detail about the SPM is available at [13].

2 Preliminaries: Service Protocol, Control and Data Dependencies, and Instance Sub-protocol

Following [2], we adopt deterministic finite state automata for modeling service protocols, where transitions are triggered by message exchanges among partners. It should be noted that, a service protocol, or more generally a state automata, is sequentially executed. In other words, a state in a service protocol specification allows multiple transitions to follow, and hence, can lead to multiple states. However, at a certain point at runtime, only one transition can be enabled, and then be fired, and consequently, this state evolves to one of the following states.

In [2], the authors claim that different message and condition pairs are always possible to be mapped into new distinct message labels, and hence, conditions can be abstracted away. However, this simplification loses important information of conditions. For instance, the price in *SP* is mapped into *goldenPrice* and *normalPrice*. If Cd_1 and Cd_2 are not specified, the rationale for choosing *goldenPrice* or *normalPrice* is lost. Indeed, conditions are fundamental to retrieve prior known service protocols and to ensure possible interactions between these service protocols. Hence, we model service protocols in terms of GFSA. Note that the time property [17] is also an essential dimension of a service protocol specification, which is considered as our future work.

Formally, a service protocol is a tuple $p = (M, S, s, F, C, T)$, where M is a finite set of messages. Following [2], for each message $m \in M$, we use notations: $m(+)$ and $m(-)$, to denote the incoming or outgoing of m. S is a finite set of states, where s is the initial state and F is a finite set of final states. C is a finite set of conditions. $T \subseteq S^2 \times M \times C$ is a finite set of transitions. Each transition $\tau = (s_s, s_t, m(+/-), c \mid true) \in T$ defines a source state s_s, a target state s_t, an incoming or outgoing message m, and a condition $c \in C$ where *true* is the condition of default. Transitions are semantically described by specifying their *input*, *output*, *precondition*, and *effect* (i.e., IOPE in the OWL-S specification[1]).

As presented in [11], different kinds of dependencies can exist between transitions. These dependencies are often obfuscated by a service protocol specification that defines all possible execution sequencings of transitions. A sequencing constraint in a service protocol may originate from one or multiple kinds of dependencies [11]. We consider two kinds, namely control and data dependencies.

[1] See http://www.w3.org/Submission/OWL-S/

A transition τ_b is control dependent on another transition τ_a if the *completion* of τ_a (marked by its *effect*) is a necessary condition for the *enablement* of τ_b (guarded by its *precondition*). Data dependencies are classified as mandatory or optional. τ_b is mandatorily data dependent on τ_a if common data exist between the *output* of τ_a and the *input* of τ_b, whereas τ_b is optionally data dependent on τ_a if no common data exist between the *output* of τ_a and the *input* of τ_b, but, incoming conditions of τ_b use the data in the *output* of τ_a. We extract control and data dependencies from the semantic description of transitions [14].

Next, we introduce the concept of an instance sub-protocol. An instance sub-protocol represents a *valid* part of a service protocol that may be executed for a particular instance of this service protocol. *Valid* means free of dependency conflicts. Generally, if the destination transition of a dependency relation is in an instance sub-protocol, the source transition of this dependency relation must be in this instance sub-protocol as well. An instance sub-protocol itself is a *smaller* service protocol in size. For instance, *SP_A* is an instance sub-protocol of *SP*. However, *SP* itself is not an instance sub-protocol since no execution can lead a service protocol to two final states.

Formally, an instance sub-protocol of a service protocol $p = (M, S, s, F, C, T)$ is a tuple $ISP = (M_I, S_I, s_I, f_I, C_I, T_I)$ where: (1) $M_I \subseteq M$, $S_I \subseteq S$, $s_I = s$, $f_I \in F$, $C_I \subseteq C$, and $T_I \subseteq T$; (2) ISP is valid with respect to control and data dependencies of p; (3) there exists an execution instance of p with T_e as the set of transitions to be enabled and executed in this instance, then, $T_e = T_I$.

3 Service Protocols Adaptation Graph

This section addresses a joint analysis of two service protocols, that of a requestor and a provider, to decide whether they are able to be adapted. Intuitively, we need to examine all pairs of instance sub-protocols in these two service protocols to study whether each pair is adaptable. However, a service protocol may have many instance sub-protocols if it contains loop segments (for instance, *SP* has 17 instance sub-protocols, while *SR* has only 2). In addition, some instance sub-protocols in a service protocol may be much similar in their specifications, i.e., they share some transitions. Hence, an examination following this brute-force strategy is usually inefficient.

Inspired by the complete interaction tree proposed in [2], we introduce the notion of an adaptation graph that explores possible mediated service interactions between two service protocols. A mediated service interaction captures a legal message exchange sequence of two service protocols between a pair of their instance sub-protocols. As such, a mediated service interaction corresponds to a *complete* adaptation of these two service protocols leading from their initial states to a pair of their final states.

We also define an adaptation graph using GFSA, where a state is a combination of two states of participating service protocols. A transition is either a message exchange between these two service protocols, or a service protocol sending a message through an adapter (the SPM in our case), or an adapter forwarding a message (either a concrete message generated by a partner, or a *mock-up* message generated by the SPM) to a service protocol. Conditions associated with a transition in an adaptation graph are inherited from those of relevant transitions in service protocols. Note that conditions are not to be evaluated at the adaptation graph construction phase, since the evaluation of conditions depends on the exchanged message instances.

Definition 1 (Adaptation Graph). *Let* $p_1 = (M_1, S_1, s_1, F_1, C_1, T_1)$ *and* $p_2 = (M_2, S_2, s_2, F_2, C_2, T_2)$ *be two service protocols. An adaptation graph for* p_1 *and* p_2 *is a tuple* $adapt_{graph} = (M, S, s, F, C, T)$. $M \subseteq M_1 \cup M_2 \cup M_{SPM}$, *where* M_{SPM} *is a finite set of* mock-up *messages generated by SPM. The message polarity is defined as follows: messages outgoing in* M_1 *and* M_2 *are sent to SPM or a peer protocol, messages incoming in* M_1 *and* M_2 *are sent by SPM or a peer protocol, and messages in* M_{SPM} *are sent to* p_1 *or* p_2. $S \subseteq S_1 \times S_2$, *where* $s = (s_1, s_2)$ *and* $F \subseteq F_1 \times F_2$. $C \subseteq C_1 \cup C_2$, *and* $T \subseteq S^2 \times M \times C$.

An adaptation graph is complete if it includes all possible mediated service interactions between two service protocols. As an example, Fig. 6 in Section 5 illustrates the complete adaptation graph for *SP* and *SR* (denoted $adapt_{graph}$).

Due to space considerations we refer the interested reader to our technical report [16] for the algorithm to generate a complete adaptation graph, as well as its correctness proof. In a nutshell, the algorithm traverses from the initial states of two service protocols (i.e., p_1 and p_2) to their final states, and constructs the intermediate states through combining the intermediate states of p_1 and p_2 on the condition that there are legal message exchanges leading to them. A legal message exchange means either (1) a message exchange between p_1 and p_2, or (2) a message reordering and remembering [3,4,7,9,13] by means of SPM, or (3) a *mock-up* message generation by the SPM if one of receiving transitions (with the state *enabled*) in p_1 or p_2 is neither control nor mandatorily data dependent on one of its direct- succeeding-transitions [13].

The worst case time complexity of the algorithm is $O(k^3n^6)$ where k is the upper bound of transitions between a pair of source and target states, and n is the maximum number of states in two service protocols. Notice that as observed by [8], a service protocol tends to be a fairly simple model, because a service protocol, as well as a service in general, is designed by humans. This indicates that k is usually quite small. Moreover, the number of states in a service protocol is typically not big. Concretely, n is typically less than *100* [10], while n ranging from *50* to *100* is regarded as a large service protocol [1]. This indicates that our algorithm is able to construct a complete adaptation graph of practical relevance.

We also recognise that the size of a complete adaptation graph (i.e., the number of states and transitions) is not big. Since transitions in a service protocol are constrained by control and/or data dependencies, a state in a service protocol can combine with limited states (assume that l is the upper bound) in another service protocol which construct the states and transitions in this graph. Hence, the complete adaptation graph has $l \times n$ states in maximum, and there are k transitions at most between a pair of states.

We next explore how a complete adaptation graph contributes to the adaptability assessment. Given two service protocols p_1 and p_2, partial adaptability specifies that, some, but not all, instance sub-protocols in p_1 can be adapted with those in p_2, whereas full adaptability mandates that all instance sub-protocols in p_1 can be adapted with those in p_2. If this graph is not empty, which means that at least one pair of instance sub-protocols in p_1 and p_2 can be adapted, and hence, p_1 and p_2 are adaptable. However, to differentiate between partial and full adaptability requires checking whether or not all instance sub-protocols are reflected by the paths in this graph. To generate all instance sub-protocols of a service protocol is not a trivial task, which will be discussed in the next section. Hence, the emptiness or non-emptiness of the complete adaptation graph corresponds to the fact that whether or not p_1 and p_2 are adaptable.

4 Instance Sub-protocol Computation

Each path in a service protocol p, which leads from the initial state to one final state and is free of dependency conflicts, constitutes an instance sub-protocol of p. Hence, the problem of instance sub-protocol generation is reducible to that of path computation of a service protocol which is a directed cyclic graph. This path computation takes exponential time to the size of the graph in general.

As far as we know, *Path_BTC* [6] is the only promising technique that tackles this path computation problem. However, some assumptions made may not be satisfiable in our context. *Path_BTC* requires that two functions *concatenate* and *aggregate* are distributive. An example is given by [6] to explain this requirement as follows. Consider the case where there are two paths p_1 and p_2 from node j to k with associated labels l_1 and l_2 respectively. Let there also be a path p_3 from node i to j with label l. Then the path set from i to k (denoted P) is $\{p_3.p_1,\ p_3.p_2\}$. The symbol "." means path concatenation. This assumption may not be satisfied in our context since, in a service protocol, P may include another path that directly links i to k with label l_3. In addition, self-loops, which are excluded in *Path_BTC*, can exist in a service protocol.

In the following, we introduce a novel path computation technique to compute all paths in a service protocol. We use the service protocol SP_{Sib} depicted in Fig. 2 as a running example. SP_{Sib} is similar to SP apart from a self-loop specified on the state *SD Items Selecting* with a transition *moreSDItems*(+). This path computation procedure is composed of the following sequential steps:

1. Based on a service protocol we generate a FSA that (1) abstracts away conditions and self-loops, and then, (2) folds multiple transitions that share the source and target states into a single transition. Fig. 2 illustrates the FSA generated out of SP_{sib}, namely abstracted SP_{sib}. Thereafter, we identify back transitions in the generated FSA (detailed in Section 4.1).
2. We then compute all paths in the generated FSA. This path computation procedure is detailed in Section 4.2.

   By a path, we mean a sequence of alternating states and transitions from the initial state of this generated FSA to one of its final states. To align a path with an instance sub-protocol, we represent a path in terms of an FSA. Examples of paths are shown in Fig. 3. Note that multiple paths may share

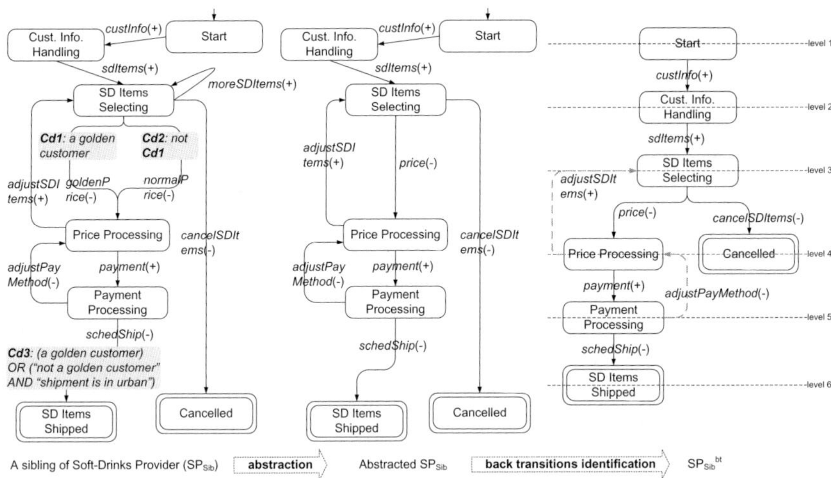

Fig. 2. Abstraction and back-transitions identification

the same FSA representation if (1) they share the same sets of states and transitions, and (2) they are different merely because of the number of times some states and/or transitions can be traversed.

3. Based on the paths computed in Step 2, we generate all paths of a service protocol by (1) unfolding transitions that have been folded in the first step, by (2) considering self-loops, and thereafter, by (3) integrating conditions that have been abstracted away previously. This procedure is detailed in Section 4.3. All paths free of dependency conflicts constitute the complete set of instance sub-protocols.

Except Step 2, other steps are computation light (with linear or polynomial time complexity). Regarding Step 2, as mentioned in Section 3, both a service protocol and a complete adaptation graph are typically not big in size. Hence, our technique is feasible to be used for the path computation of our context.

4.1 Back-Transitions Identification

A transition τ in the generated FSA is identified as a back-transition if (1) its target state is not a final state, and (2) the *distance* from the initial state of this FSA to the target state of τ is not longer than that to the source state of τ. We borrow the notion of *level* from tree automata to specify the level of a state in this generated FSA.

For instance, the level of the initial state *Start* in SP_{Sib} FSA is 1. The level of its immediately following state namely *Cust. Info Handling* is 2, and so on. The levels of states are computed using a breadth-first search, while ignoring all transitions whose level of the target state is not bigger than the level of the source state. These transitions ignored are in fact back-transitions. We refer by SP^{bt}_{Sib} in Fig. 2 to the generated FSA of SP_{Sib} where back transitions are identified and marked using dashed lines.

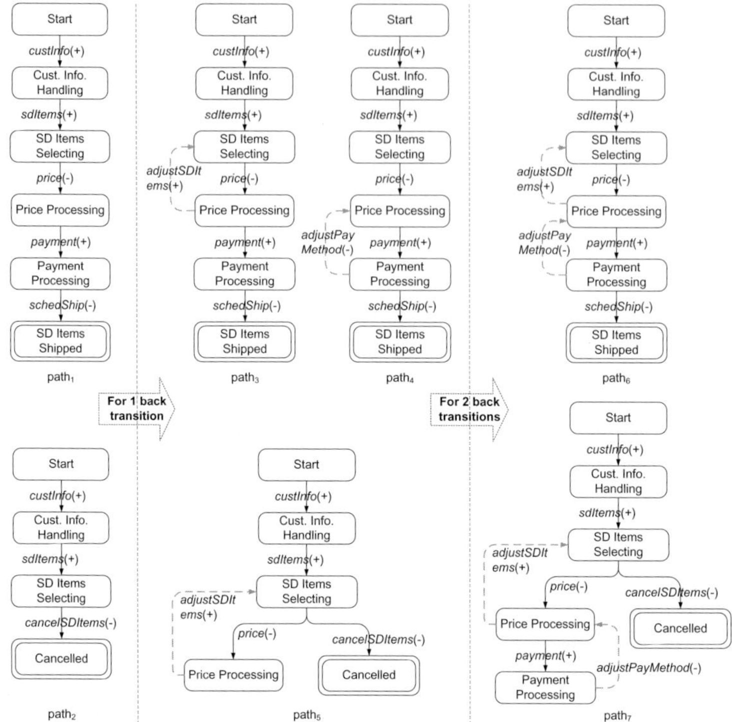

Fig. 3. Path computation

4.2 Path Computation of a Generated FSA

First, we generate all paths in the generated FSA while ignoring back-transitions using a breadth-first search. For instance, SP^{bt}_{Sib} has two paths $path_1$ and $path_2$, as shown on the left part of Fig. 3.

Based on the paths with i back-transitions (i = 0, 1, 2, ..., k-1, where k is the number of back-transitions in the generated FSA), we compute paths that contain i+1 back-transitions. In the following we show how, given a path (denoted p_{th}) with i back transitions and one of these k back-transition (denoted τ), we compute new paths with i+1 back-transitions. s_s and s_t denote the source and target states of τ respectively.

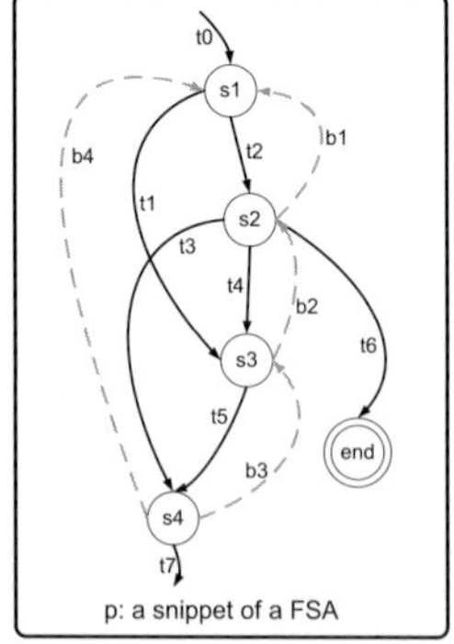

Fig. 4. A snippet of p

We distinguish between four situations for this path computation. The first situation corresponds to ***Case 1*** that (1) neither s_s nor s_t belongs to p_{th}, or (2) τ is a transition in p_{th} already. In this situation, no new path is to be generated.

The remaining three situations correspond to ***Case 2:*** both s_s and s_t belong to p_{th}, ***Case 3:*** s_t but not s_s belongs to p_{th}, and ***Case 4:*** s_s but not s_t belongs to p_{th}. For page limitation, we detail ***Case 2*** in this paper and refer the interested reader to our technical report [16] for the description about ***Case 3*** and ***4***.

Since SP^{bt}_{Sib} is simple and does not cover all possible situations, we use a more general FSA called p, and whose snippet is depicted in Fig. 4, to explain our computation steps. For ***Case 2***, we detail different steps to compute new paths. After each step, we refer to our example to illustrate how this step applies.

Case 2: Both s_s and s_t belong to p_{th}. This situation is illustrated in Fig. 5 where pa_{th} corresponds to p_{th} and $b2$ (see Fig. 5) corresponds to τ.

Step 1: We initialise two state sets ST_{upp} and ST_{low} to $\{s_t\}$ and $\{s_s\}$ respectively.

In our example, these correspond to $ST^{b2}_{upp} = \{s2\}$ and $ST^{b2}_{low} = \{s3\}$.

Step 2: We explore back-transitions in p_{th} to update ST_{upp} and ST_{low}. For each back-transition τ_1 in p_{th} (whose source state is s^1_s and whose target state is s^1_t), if $s^1_s \in ST_{upp}$ then $ST_{upp} = ST_{upp} \cup \{s^1_t\}$, and if $s^1_t \in ST_{low}$ then $ST_{low} = ST_{low} \cup \{s^1_s\}$. This procedure stops when no back-transition can be explored anymore.

In our example, this step leads to $ST^{b2}_{upp} = \{s1,\, s2\}$ and $ST^{b2}_{low} = \{s3,\, s4\}$.

Step 3: While ignoring back-transitions, we construct the set, denoted SEG, of all segments in p that start at one state in ST_{upp} and end at one state in ST_{low}.

In our example, SEG_{b2} = {s2-t4-s3, s2-t4-s3-t5-s4, s2-t3-s4, s1-t1-s3, s1-t1-s3-t5-s4, s1-t2-s2-t4-s3, s1-t2-s2-t4-s3-t5-s4}.

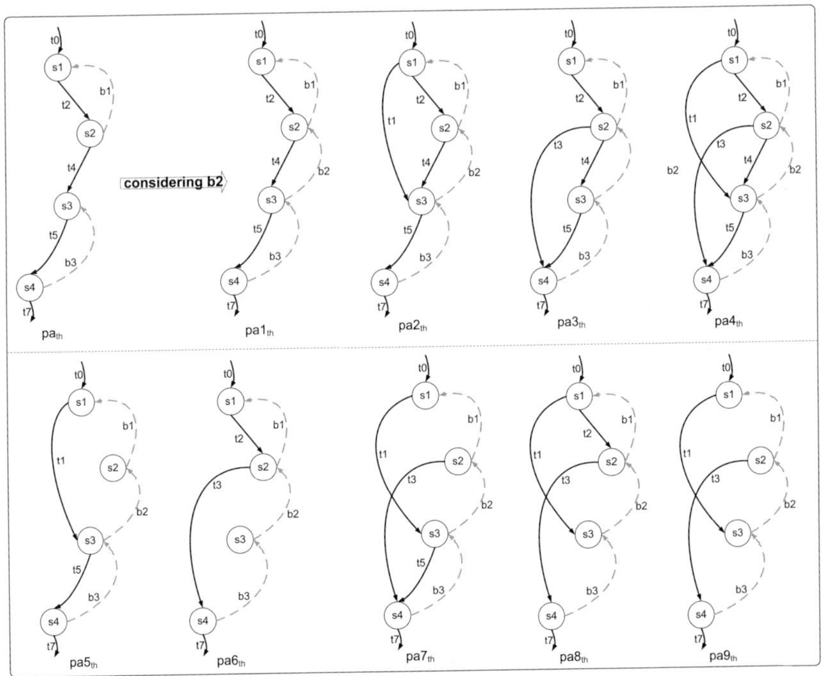

Fig. 5. Path computation for Case 2

Step 4: We remove each segment $seg \in SEG$ where all its states and transitions are contained in p_{th}. For SEG_{b2}, since some segments, such as s2-t4-s3, are already contained in pa_{th}, they are removed. This step leads to SEG_{b2} = {s2-t3-s4, s1-t1-s3, s1-t1-s3-t5-s4}.

Step 5: If two segments seg_1 and seg_2 in SEG are different merely because seg_2 includes more states and transitions than seg_1, and these additional states and transitions are already in p_{th}, seg_2 is removed. The reason for this removal is that, if two new paths (denoted p^1_{th} and p^2_{th}) are generated, p^1_{th} includes seg_1, and p^2_{th} includes seg_2, then p^2_{th} is actually a duplicate path with p^1_{th}.

Hence, SEG_{b2} = {s2-t3-s4, s1-t1-s3} since s1-t1-s3-t5-s4 is excluded. We use m to denote the number of segments in SEG. In our case, $m = |SEG_{b2}| = 2$.

Step 6: Then, 2^m new paths with i+1 back-transitions are generated. Each new path is made through cloning p_{th}, adding τ, and including either zero, or one, or multiple (even all) segments in SEG. The reason for this procedure is that, due to back-transitions in p_{th}, a new path with i+1 back-transitions possibly loops back from one state in ST_{low} to one state in ST_{upp} for traversing some or even all segments in SEG. In our case, $2^2 = 4$ new paths are generated as illustrated in Fig. 5, where $pa1_{th}$ is the new path that no segment in SEG_{b2} is included, $pa2_{th}$ and $pa3_{th}$ show the cases that one segment in SEG_{b2} is considered, and $pa4_{th}$ is the new path that includes both these two segments in SEG_{b2}.

Step 7: Each new path generated in Step 6 containing a segment seg in SEG, has another segment, alternative to seg, connecting the start state of seg to its final state. For instance, in $pa2_{th}$, besides the segment s1-t1-s3 which belongs to SEG_{b2}, there is another segment s1-t2-s2-t4-s3 starting at s1 and ending at s3. For each path of this kind, we generate an additional new path by removing the segment alternative to those in SEG. For instance, $pa5_{th}$ (see Fig. 5) is a new path generated from $pa2_{th}$ by removing s1-t2-s2-t4-s3, which is an alternative segment to s1-t1-s3 that belongs to SEG_{b2}. $pa6_{th}$ is another new path generated from $pa3_{th}$ following the same principal.

If a new path generated in Step 6 contains k segments in SEG, $2^k - 1$ additional new paths (some may be duplicate) are generated that correspond to all possible combinations of the k segments removal including the case where all of them are removed. As an example, $pa4_{th}$ in Fig. 5 contains 2 segments in SEG_{b2}. then $2^2 - 1 = 3$ additional new paths are generated, namely $pa7_{th}$, $pa8_{th}$ and $pa9_{th}$. $pa7_{th}$ is generated by removing s1-t2-s2-t4-s3, $pa8_{th}$ by removing s2-t4-s3-t5-s4, and $pa9_{th}$ by removing both s1-t2-s2-t4-s3 and s2-t4-s3-t5-s4.

A new path is discarded if it is duplicate with another path generated previously. By iteratively applying the steps above, all paths are generated. As shown in Fig. 3, there are 7 paths for SP^{bt}_{Sib} in total. Since there are finite paths with i back-transitions, this procedure stops with k times recursion, and each checks all k back-transitions with respect to all previously generated paths.

4.3 Path Computation of a Service Protocol

This section shows how we generate all instance sub-protocols of a service protocol based on the paths computed above. This procedure is achieved mainly by(1) unfolding transitions folded in the first step, and thereafter, by (2) taking self-loops and conditions into consideration that were abstracted away initially.

Given a transition τ in a path p_{th}, and assume that τ is folded from n transitions (i.e., they share the same source and target states), τ is to be unfolded as follows: (1) if τ is a part of a loop segment (including the case where τ is a back transition), then τ is unfolded to all possible combinations of these n transitions. Consequently, $2^n - 1$ additional paths are generated, and each path is made through cloning p_{th} where τ is replaced by a possible combination, otherwise, (2) τ is unfolded to these n transitions. Thereafter, n additional paths are generated, and each path is made through cloning p_{th} where τ is replaced by one of these n transitions.

A path that contains m folded transitions where each of them leads to t_j alternatives when unfolding it, is replaced by $\prod_{j=1}^{m} t_j$ new paths that correspond to all possible combination of unfolding these m folded transitions. For instance, assume that p_{th} contains m=2 folded transitions τ_a and τ_b. τ_a is folded from two transitions τ_1 and τ_2, and τ_a is a part of a loop segment. τ_b is folded from two transitions τ_3 and τ_4, and τ_b is not a part of a loop segment. Then, the alternatives to τ_a are τ_1, or τ_2, or both τ_1 and τ_2, i.e., $t_a = 3$. The alternatives to τ_b are τ_3 or τ_4, i.e., $t_b = 2$. Consequently, $t_a \times t_b = 6$ new paths are generated for p_{th}, τ_a and τ_b, where, as an example, one new path corresponds to the combination of (1) both τ_1 and τ_2 for τ_a, and (2) τ_3 for τ_b. In our case, there are 17 paths in total for SP_{Sib} after studying folded transitions in SP_{Sib}^{bt}.

After exploring folded transitions, we take self-loops into consideration. For a path p_{th} containing m self-loops, $2^m - 1$ new paths are generated where all possible combinations of these self-loops are considered. p_{th} represents the situation that no self-loop is included. For instance, there are 34 paths in total for SP_{Sib} after studying self-loops.

Finally, we reattach conditions with associated transitions (that have been abstracted away in Step 1). Paths free of dependency conflicts are instance sub-protocols. In our case, SP_{Sib} has 34 instance sub-protocols and SP has 17.

5 Service Protocols Adaptability Assessment

5.1 Computing the Adaptation Degree and the Condition Set

For two service protocols p_1 and p_2, and their complete adaptation graph $adapt_{graph}$, the adaptation degree is computed by means of Equation 1. The function $instSubProtocol(p_1, adapt_{graph})$ counts instance sub-protocols in p_1 that are captured by instance sub-protocol pairs of $adapt_{graph}$. The procedure of generating all instance sub-protocol pairs captured by $adapt_{graph}$ is to be detailed in Section 5.2. If the parameter $adapt_{graph}$ is set to $null$, the number of instance sub-protocols in p_1 is returned. The procedure of generating all instance sub-protocols in p_1 has been presented in Section 4.

$$adaptation(p_1, p_2) = \frac{instSubProtocol(p_1, adapt_{graph})}{instSubProtocol(p_1, null)} \tag{1}$$

For instance, $adaptation(SP, SR) = 8/17 \approx 0.471$, whereas $adaptation(SR, SP) = 2/2 = 1$. These show that the adaptability is an asymmetric relation between service protocols. In our motivating example, $adaptation(SR, SP\_A) = 1/2 = 0.5$. These mean that the adaptation degree informs the requestor about different adaptation possibilities between candidate service providers.

We next re-explore the problem of distinguishing between partial and full adaptability (pending in Section 3) by means of the adaptation degree. If $adaptation(p_1, p_2) = 1$, then p_1 is called fully adaptable with p_2, because each instance sub-protocol in p_1 can perform a mediated service interaction with at least one instance sub-protocol in p_2. If $adaptation(p_1, p_2) = 0$, p_1 is assumed not adaptable with p_2 since no instance sub-protocol in p_1 can have an adaptable instance sub-protocol in p_2. Otherwise, p_1 is regarded partially adaptable with p_2.

We recall that an adaptation graph reflects legal message exchanges between pairs of instance sub-protocols. For a particular pair of instance sub-protocols, the conjunction of conditions associated with their transitions constitutes another prerequisite for ensuring a proper adaptation. Such a *must-be-held* condition set is generated through studying all instance sub-protocol pairs. For instance for SP and SR, the condition set is expressed as follows: $\{Cd_2 \bigwedge Cd_3 \bigwedge Cd_4, Cd_1 \bigwedge Cd_2 \bigwedge Cd_3 \bigwedge Cd_4, Cd_2 \bigwedge Cd_4, Cd_2 \bigwedge Cd_5, Cd_1 \bigwedge Cd_2 \bigwedge Cd_4\}$. It is important to note that some conditions contain both Cd_1 and Cd_2. This reflects the fact that, in some adaptation scenarios, when SP loops back to the state *SD Items Selecting*, the transition *custInfo*(-) in SR has been executed, and the message *custInfo* can make Cd_1 satisfiable.

5.2 Instance Sub-protocols Pairs in an Adaptation Graph

Based on the complete adaptation graph and all instance sub-protocols of a service protocol generated previously, in this section, we explore how to generate all pairs of instance sub-protocols that can conduct mediated service interactions between service protocols of a requestor and a provider. This procedure includes the following two sequential steps: we first generate all paths captured by this complete adaptation graph, and then, we project each path to a pair of instance sub-protocols and thereafter cluster these instance sub-protocol pairs.

Path Computation for a Complete Adaptation Graph. As the first step, we generate all paths in the complete adaptation graph using the technique presented in Section 4. For instance for $adapt_{graph}$ as shown in Fig. 6, since it has 11 back-transitions and this graph is complex, thousands of paths are to be generated. Generally, computing all paths for a complete adaptation graph directly is usually inefficient.

Indeed, different from transitions in a service protocol where all transitions are necessary for achieving a particular business objective, some back-transitions in a complete adaptation graph may have no contribution to static analysis, and hence, can be ignored. Recall Definition 1, a transition in an adaptation graph specifies a legal message exchange either (1) between two service protocols (like $SP \rightarrow SR$: *normalPrice* in Fig. 6 that is between SP and SR through the SPM) or (2) between a service protocol and the SPM (like $SP \rightarrow SPM$:

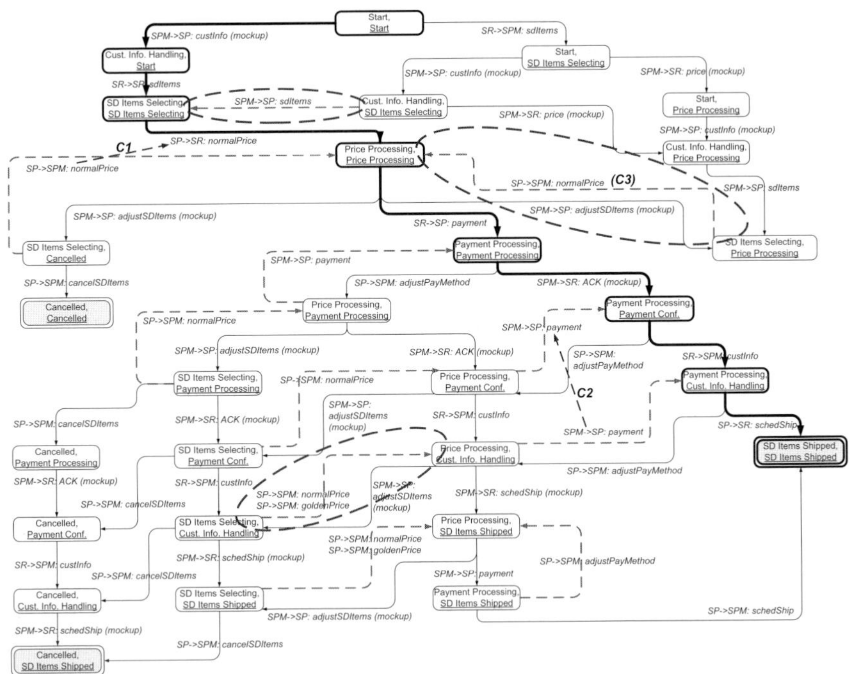

Fig. 6. $adapt_{graph}$: the complete adaptation graph for SP and SR service protocols

normalPrice in Fig. 6 that is between SP and SPM). Given a back transition, if its message exchange has been covered by all traces from the initial state of the complete adaptation graph to the source state of this back transition (i.e. the back transition repeats a message exchange that has been performed before), this back transition can be ignored for static analysis purposes.

A back-transition τ (the source state s_s and target state s_t) in a complete adaptation graph can be ignored if the following two conditions are satisfied:

Condition 1: For any trace from the initial state of the complete adaptation graph to s_s, another (back) transition exists in this graph which (1) is *before* τ, and (2) covers the message exchange by τ. An example is the exchange for message *normalPrice* by the back transition $SP \rightarrow SPM$: *normalPrice* and by the transition $SP \rightarrow SR$: *normalPrice* as shown in Fig. 6 by a dashed line with a mark *C1*. The transition $SP \rightarrow SR$: *normalPrice* implicitly specifies the following two message exchanges through the SPM: $SP \rightarrow SPM$: *normalPrice* and $SPM \rightarrow SR$: *normalPrice*. Hence, the message exchange specified by the back transition $SP \rightarrow SPM$: *normalPrice* has been covered by that of the transition $SP \rightarrow SR$: *normalPrice*. Another example is two back transitions $SPM \rightarrow SP$: *payment* as shown in Fig. 6 by a dashed line with a mark *C2*. The back transition, which is the source of this dashed line, meets this condition.

Condition 2: Ignoring τ does not cause that s_s is unreachable to any final state of this graph. A counterexample is the back transition $SP \rightarrow SPM$: *normalPrice* in Fig. 6 with a mark *C3*. If it is ignored, its source state (*SD Items Selecting, Price Processing*) is unreachable to any final state of this complete adaptation graph.

The first condition can be checked by a reversed breadth-first search on the graph starting at s_s, while the second condition can be verified through checking whether s_s has other outgoing transitions which are not back-transitions.

After examining all back-transitions in $adapt_{graph}$, only three out of these eleven back-transitions, which are enclosed by means of dotted ellipses as shown in Fig. 6, are necessary. Hence, there are 432 paths to be computed in total.

Instance Sub-protocol Pairs Generation. From the perspective of legal message exchange, each path in the complete adaptation graph leading from the initial state to one final state respects a mediated service interaction between a pair of instance sub-protocols. For instance, in Fig. 6, a path leading from the initial state (*Start*, *Start*) to one final state (*SD Items Shipped*, *SD Items Shipped*) (denoted p_{th}) is marked by means of thick lines on the states and transitions. The instance sub-protocol of SP involved in this adaptation (denoted ISP_{SP}) is $path_1$ in Fig. 3 where the transition *price*(-) is to be unfolded to *normalPrice*(-). The instance sub-protocol of SR involved is the one that leads to the final state *SD Items Shipped*.

On the other hand, for a certain instance sub-protocol pair, it possibly corresponds to more than one mediated service interaction since some messages can be exchanged in different orders. For instance, transitions *adjustPayMethod*(+) in SP and *custInfo*(-) in SR can be enabled in any order. Thereafter, multiple paths may reflect the mediated service interactions of the same instance sub-protocol pair. For instance, another path (denoted pd_{th}) shares the same instance sub-protocol pair as that of p_{th}, if p_{th} and pd_{th} have a common segment as specified in p_{th} from the intermediate state (*SD Items Selecting*, *SD Items Selecting*) to the final state (*SD Items Shipped*, *SD Items Shipped*).

To make the instance sub-protocol pairs unique, we cluster paths if they are captured by the same pair of instance sub-protocols. This requires a technique to identify the pair of instance sub-protocols that a path reflects. We project a path to a service protocol. The result is a complete execution path [2] leading from the initial state to one final state. The projection is an operator [2] that identifies transitions in a path associated with a service protocol, and restores their polarity according to the service protocol specification. For instance, the transition $SP \rightarrow SR$: *normalPrice* as indicated in Fig. 6 by a dashed line with a mark *C1* is projected into one transition in SP (i.e., *normalPrice*(-)) and another transition in SR (i.e., *price*(+)), while $SP \rightarrow SPM$: *normalPrice* is projected into a transition in SP (i.e., *normalPrice*(-)).

We then identify which instance sub-protocol this projected path belongs. This is achieved by comparing the transition set in this projected path to that in an instance sub-protocol. For instance, after projecting p_{th} to SP, the transition set is the same as that of ISP_{SP}. Hence, the instance sub-protocol pair is provided. We study other paths in such fashion. In our case, these 432 paths in $adapt_{graph}$ are clustered into 8 instance sub-protocol pairs of SP and SR.

6 Related Work and Conclusion

Adaptability analysis. A work similar to ours is the *adapter compatibility* analysis in the *Y-S* model [12] which checks whether or not two component protocols are adaptable with a particular adapter. The criteria are (1) no *unspecified receptions*, and (2) *deadlock free*. No *unspecified receptions* is restrictive, since message production and consumption in mediated service interactions are time-decoupled, and extra messages are often allowed. Our approach does not have this limitation. We give an adaptation degree which is more accurate than a binary answer. Since conditions in protocols are not explored, this work does not specify conditions that determine when two protocols are adaptable. Our approach specifies such necessary conditions. In addition, this work depends on the synchronous semantics. This assumption simplifies the problem, but it fails to capture most Web service interactions, since they are normally asynchronous. Our approach does not depend on such an assumption.

Adapter construction. Adapters are important for supporting interactions in the context of both software components [12] and Web services [3,4,7,9,13].

As shown in [12], an adapter is automatically constructed for two incompatible protocols. The adapter tackles order mismatches with *unspecified receptions*, but considers any deadlock as unresolvable. The same limitation exists in the adaptation mechanisms of mediation-aided service composition approaches [9].

In [3,4], possible mismatches are categorised into several mismatch patterns, and adaptation templates [3] or composable adaptation operators [4] are proposed for handling these mismatch patterns. However, mismatches between two protocols are identified by a developer and an adapter is constructed manually.

Besides the mismatches covered by adapters above, [7] handles a deadlock through *evidences*. The choice of an *evidence* for reconciling a given deadlock is decided by adapter developers, and hence, this method is not generic. This technique presumes that recommended business data is consistent with a certain interaction context, but data recommended by some *evidences*, (e.g., *enumeration with default* and *log based value/type interface*) may not satisfy this assumption, since enumeration may not be the default value, and some business data may differ in different interactions. In contrast, a *mock-up* message, generated by the SPM [13], is consistent with a certain interaction context.

In a nutshell, adapter construction means that service protocols are adaptable in some case. The possibility and the conditions are not specified. Adapter building constitutes a starting point, but is inadequate for assessing the adaptability. This paper builds upon the adapter construction presented in our previous work [13] and provides adaptability assessment, whose result is a key criterion to a requestor for identifying and selecting a suitable service provider from functionally-equivalent candidates according to her specific business requirements.

In conclusion, the two major contributions of this paper are as follows: first, our adaptability assessment, as reflected by the motivating example, is a comprehensive means of selecting the appropriate service provider among functionally-equivalent candidates. To the best of our knowledge, this assessment is new. Besides the technique presented in this paper, we have proposed another approach in [15] that assesses the adaptability using protocol reduction and graph-search with backtracking techniques. Compared with [15], this paper's second major contribution is a novel path computation technique which computes all instance sub-protocols in a service protocol as well as their pairs in a complete adaptation graph. This computation is general and can be applied, besides adaptability assessment, also to compatibility assessment where a numerical compatibility degree and a set of conditions are still unaccounted.

References

1. Backer, M.D., Snoeck, M., Monsieur, G., Lemahieu, W., Dedene, G.: A scenario-based verification technique to assess the compatibility of collaborative business processes. Data & Knowledge Engineering 68(6), 531–551 (2009)
2. Benatallah, B., Casati, F., Toumani, F.: Representing, analysing and managing web service protocols. Data & Knowledge Engineering 58(3), 327–357 (2006)
3. Benatallah, B., Casati, F., Grigori, D., Nezhad, H.R.M., Toumani, F.: Developing Adapters for Web Services Integration. In: Pastor, Ó., Falcão e Cunha, J. (eds.) CAiSE 2005. LNCS, vol. 3520, pp. 415–429. Springer, Heidelberg (2005)
4. Dumas, M., Spork, M., Wang, K.: Adapt or perish: Algebra and visual notation for service interface adaptation. In: Dustdar, S., Fiadeiro, J.L., Sheth, A.P. (eds.) BPM 2006. LNCS, vol. 4102, pp. 65–80. Springer, Heidelberg (2006)
5. Dumas, M., Benatallah, B., Nezhad, H.R.M.: Web Service Protocols: Compatibility and Adaptation. IEEE Data Engineering Bulletin 31(3), 40–44 (2008)
6. Ioannidis, Y., Ramakrishnan, R., Winger, L.: Transitive closure algorithms based on graph traversal. ACM Trans. on Database Systems 18(3), 512–576 (1993)
7. Nezhad, H.R.M., Benatallah, B., Martens, A., Curbera, F., Casati, F.: Semi-Automated Adaptation of Service Interactions. In: Proc. of WWW, pp. 993–1002 (2007)
8. Nezhad, H.R.M., Saint-Paul, R., Benatallah, B., Casati, F.: Deriving Protocol Models from Imperfect Service Conversation Logs. IEEE Trans. on Knowledge and Data Engineering 12, 1683–1698 (2008)
9. Tan, W., Fan, Y., Zhou, M.: A Petri Net-Based Method for Compatibility Analysis and Composition of Web Services in Business Process Execution Language. IEEE Trans. on Automation Science and Engineering 6(1), 94–106 (2009)
10. van der Aalst, W.M.P., Weijters, T., Maruster, L.: Workflow Mining: Discovering Process Models from Event Logs. IEEE Trans. on Knowledge and Data Engineering 16(9), 1128–1142 (2004)
11. Wu, Q., Pul, C., Sahai, A., Barga, R.: Categorization and Optimization of Synchronization Dependencies in Business Processes. In: Proc. of ICDE, pp. 306–315 (2007)
12. Yellin, D.M., Strom, R.E.: Protocol Specifications and Component Adaptors. ACM Trans. on Programming Languages and Systems 19(2), 292–333 (1997)
13. Zhou, Z., Bhiri, S., Gaaloul, W., Hauswirth, M.: Developing Process Mediator for Supporting Mediated Service Interactions. In: Proc. of ECOWS, pp. 155–164 (2008)
14. Zhou, Z., Bhiri, S., Hauswirth, M.: Control and Data Dependencies in Business Processes Based on Semantic Business Activities. In: Proc. of iiWAS, pp. 257–263 (2008)
15. Zhou, Z., Bhiri, S., Zhuge, H., Hauswirth, M.: Assessing Service Protocols Adaptability Using Protocol Reduction and Graph-Search with Backtracking Techniques. In: Proc. of SKG (2009)
16. Zhou, Z., Bhiri, S.: Assessment of Service Protocols Adaptability (2009) DERI technical report, `http://www.deri.ie/fileadmin/documents/DERI-TR-2009-06-10.pdf`
17. Zhuge, H., Cheung, T.Y., Pung, H.P.: A timed workflow process model. Journal of Systems and Software 55(3), 231–243 (2001)

Enhancing Business Process Automation by Integrating RFID Data and Events

Xiaohui Zhao[1], Chengfei Liu[1], and Tao Lin[2]

[1] Centre for Complex Software Systems and Services
Faculty of Information and Communication Technologies
Swinburne University of Technology
Melbourne, Australia
{xzhao,cliu}@groupwise.swin.edu.au

[2] Amitive Inc.
1400 Fashion Island Blvd
San Mateo, CA 94404, USA
tao.lin@amitive.com

Abstract. Business process automation is one of the major benefits for utilising Radio Frequency Identification (RFID) technology. Through readers to RFID middleware systems, the information and the movements of tagged objects can be used to trigger business transactions. These features change the way of business applications for dealing with the physical world from mostly quantity-based to object-based. Aiming to facilitate business process automation, this paper introduces a new method to model and incorporate business logics into RFID edge systems from an object-oriented perspective with emphasises on RFID's event-driven characteristics. A framework covering business rule modelling, event handling and system operation invocations is presented on the basis of the event calculus. In regard to the identified delayed effects in RFID-enabled applications, a two-block buffering mechanism is proposed to improve RFID query efficiency within the framework. The performance improvements are analysed with related experiments.

1 Introduction

RFID is a re-emerging technology intended to identify, track, and trace items automatically. Nowadays, the adoption of RFID is increasing significantly [1-3] in sectors of retailing, manufacturing, supply chain, military use, health care, etc. A bright forecast from IDTechEx [4] expects that the total RFID market value will rise from $4.96 billion in 2007 to $26.88 billion in 2017.

Attempts to apply RFID to facilitating the business running of real enterprises are often made by monitoring the material flows through reading events or collecting data from physical world. However, there is a distinct lack of business process management (BPM) support, and this lack barriers RFID-enabled applications from the benefits of on-site responses according to business logics, system agility against changing requirements, seamless integration with enterprise application systems, etc. Business process management blends business logics and related resources into reusable models for efficient transaction management. Once such business logics are incorporated into

R. Meersman, T. Dillon, P. Herrero (Eds.): OTM 2009, Part I, LNCS 5870, pp. 255–272, 2009.

RFID systems, it can provide real-time information in a network consisting of different enterprises, applications, and business partners in collaboration scenarios, and enable automatic reactions and execution based on the captured real-time information while eliminating manual inputting, processing, and checking of information. As a milestone towards such integration, Boeing has already embarked on a new assembly paradigm that is focused on using RFID and collaborative processes for its 787 (formerly 7E7 Dreamliner) aircraft to reduce the assembly time [5]. The integration of RFID and business logics seeks methodological advances and facilitating architectures to merge the wire level deployment and business level applications under one umbrella.

With advancement of RFID technologies and the wider adoption of RFID applications, both research communities and industry companies are drawing attention to the fusion of RFID edge systems and application systems. By integrating business processes into RFID data management, we can effectively facilitate business process automation, and thereby improve the agility and efficiency of current business operations in the end. Towards this ultimate goal, this paper focuses on modelling and integrating RFID data and events according to existing business processes. This integration is expected to enable the awareness of RFID edge systems to both real-time context and business logic, and in turn the automation and orchestration of business processes can be well supported.

In this paper, we propose a framework for modelling business logics from an object-oriented and event-oriented perspective. This framework makes the following contributions to current RFID data management:

- Propose a formalised object-oriented and event-driven RFID data management model;
- Establish an event calculus based mechanism for designing event-based rules to capture the dynamics in the event-rich environment of RFID-enabled applications;
- Discuss the deployment of business logics and semantics to RFID edge system;
- Present a buffering mechanism for efficient query executions in conformity with the proposed data/event integration framework.

The remainder of this paper is organised as follows: Section 2 discusses the requirements and challenges for incorporating business logics into RFID edge systems by analysing an RFID-enabled application. Section 3 introduces the object-oriented modelling perspective and the event-driven mechanism, and then presents the event calculus based RFID data management model, and finally demonstrates the deployment of business logics to RFID edge systems with an example. Section 4 analyses the delayed effects in RFID query executions, and proposes a 2-block buffering mechanism with related experiment results. Section 5 reviews the related work and discusses the advantages of the proposed framework. Finally, we conclude the paper with an indication on future work.

2 Looking into RFID-Enabled Applications

In this section, we take a distribution centre for assembling shipments as an example of RFID-enabled application. As a typical chain of logistics processes, a distribution centre is often selected to discuss RFID deployment [6], since a distribution centre assembles a large volume of shipments every day using pallets. Here we assume this distribution centre only packs pallets of two types of products and this distribution centre handles a large volume of shipments every day. We use this case to illustrate the nature of an RFID-applied environment.

Figure 1 shows the packing process at this distribution centre with a Business Process Modelling Notation (BPMN) [7] diagram. This packing process starts when receiving an order from customers, and thereafter pallets are transferred to the distribution centre, while the pickers are picking the ordered goods from the inventory and transferring them to the distribution centre. When receiving these goods, the packing station will periodically pack the goods at two packing lines, where line A can only do full pallet packing with the goods of the same type, and line B can do partial packing and mixed packing. After all ordered goods are packed, they will be sent for shipment. The details of sub process "Send Goods to Proper Lines" will be discussed in Section 3.4.

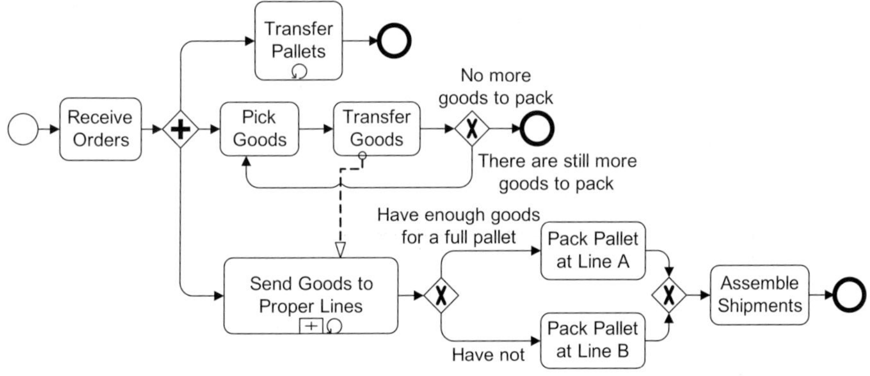

Fig. 1. Simplified business process diagram for the goods packing process

This distribution centre is an RFID "rich" environment. Not only the products and pallets, the equipments and tools are also tagged with RFID. The deployed readers in inventories, packing lines, transporting vehicles, etc., construct a network that monitors the movements of products, pallets, vehicles, etc. In this scenario, using RFID with the goods packing process in a distribution centre can bring the following benefits:

- Improve the visibility of the status of a business transaction. With RFID technology, which packing processes have started and which have not can be seen clearly. Therefore, early warnings for potential late shipments can be easily discovered.
- Improve the content accuracy for shipments.
- Record association hierarchy (association relationship of serialised cases with the corresponding pallet, pallets with the corresponding shipment). This information can be used for validating shipments.
- Recover recall or counterfeit goods.

In an RFID-applied environment, edge systems are responsible for most on-site operations, while monitoring the object movements [8]. If edge systems own the knowledge of business logics, they can handle business operations intelligently, and as such the business efficiency and effectiveness can be improved. To this end, we pose the following questions to initiate the thinking on the integration of business logics in an RFID-applied environment:

- How to design event patterns and use them to elicit meaningful business information from low level tag events?
- How to blend the RFID events and event patterns into business logic modelling, so that the defined business rules are naturally sensitive to RFID events?
- How to characterise the real-time status of objects in an event-dominated context?
- How to track the real-time status of objects by sorting out related events, and can that be achieved in an efficient way?
- How to deploy the modelled business logics to RFID edge systems, and enable them to intelligently operate in response to run time dynamics of a real application environment?

Our work targets at incorporating business logics into RFID edge systems with a novel event calculus based data management framework. To cater for the event-driven and data-intensive execution mechanism of RFID applications, this framework defines rules ad RFID queries to elicit the business meaning from RFID tag read events. In addition, this model encapsulates the related classes, rules, event patterns into RFID contextual scenarios. Once such RFID contextual scenarios are deployed to the data-driven middleware systems, the business logics can be pushed down to RFID edge systems. Therefore, the RFID edge systems will become aware to both real-time object information and business logics. By modifying the deployed RFID application contexts, the same edge systems and infrastructures can change to support different applications. This feature enhances the flexibility and customisability of RFID application systems.

3 Object-Oriented RFID Data Management Framework

3.1 Object-Oriented Perspective and Event-Driven Mechanism

Classic business process management approaches specify the handling procedures and the involved business logic in form of graph-based workflow models. Such models emphasise the control flow between activities, and these activities execute along the pre-defined sequence, i.e., an activity's start-up is triggered by the completion of preceding activities. However, in the RFID-applied environment, RFID tag read events are the raw data to trigger operations and activities, and thereby control the business process execution. To effectively utilise such real-time object level information, we claim that the RFID data management framework should be fully object-oriented and event-driven. Further, the control of the event processing in the corresponding edge systems needs to be managed by a business process management system, yet the execution of the business logic will be conducted by the edge systems.

At technical level, we choose the event calculus as the tool for event and rule modelling. The event calculus is a logic-based formalism that infers what is true when given what happens, when and what actions do, and it is based on the supposition that "all change must be due to a cause, while spontaneous changes do not occur" [9]. The event calculus particularly fits into the event based rule design and analysis in the event-rich environment of RFID-enabled applications. Compared with other event/state modelling approaches, such as UML state diagram and Event Process Chain (EPC) [10], event calculus owns advantages in modelling actions with indirect or non-deterministic effects, concurrent actions, and continuous changes, in a much concise representation.

3.2 Preliminaries of Event Calculus

The main components of event calculus include *events* (or action), *fluents* and *time points*. A *fluent* is anything whose numerical or boolean value is subject to change over time [9]. In this paper, we confine our attention to propositional fluents, i.e., Boolean fluents for simplification (note, this can be easily extended to situational fluents, i.e., the range of the fluents can be extended to be situations [9]). A scenario modelled by the event calculus constitutes predicates and axioms, which may refer to fluents, events, and time points as parameters. Table 1 lists the primary event calculus predicates.

In addition to these event calculus predicates, some *domain-specific predicates* may be used to describe special relations in a given scenario. For example, in the pallet packing scenario, we may use predicates *near_by*($pallet_1$, $pallet_2$), *contain*(*product*, *pallet*), etc., to specify the spatial relations between pallets and goods.

Table 1. Event calculus predicates

Predicates	*Explanation*
$Initiates(e, f, t)$	Fluent f starts to hold after event e at time t
$Terminates(e, f, t)$	Fluent f ceases to hold after event e at time t
$Initially_P(f)$	Fluent f holds from time 0, i.e., the initial time point
$Initially_N(f)$	Fluent f does not hold from time 0
$t_1<t_2$	Time point t_1 is before time point t_2
$Happens(e, t)$	Event e occurs at time t
$HoldsAt(f, t)$	Fluent f holds at time t
$Clipped(t_1, f, t_2)$	Fluent f is terminated between time points t_1 and t_2
$Declipped(t_1, f, t_2)$	Fluent f is initiated between time points t_1 and t_2

To guarantee its applicability, the event calculus introduces particular axioms to constrain the relationships between predicates, and solve the classic logic frame problem (please check reference [9] for details about the frame problem).

Domain-independent Event Calculus (EC) Axioms

(*EC*1) $Clipped(t_1, f, t_4) \leftrightarrow \exists e, t_2, t_3\ [Happens(e, t_2, t_3) \wedge t_1<t_3 \wedge t_2<t_4 \wedge Terminates(e, f, t_2)]$;

(*EC*2) $Declipped(t1, f, t_4) \leftrightarrow \exists e, t_2, t_3\ [Happens(e, t_2, t3) \wedge t_1<t_3 \wedge t_2< t_4 \wedge Initiates(e, f, t_2)]$;

(*EC*3) $HoldsAt(f, t_3) \leftarrow Happens(e, t_1, t_2) \wedge Initiates(e, f, t_1) \wedge t_2<t_3 \wedge \neg Clipped(t_1, f, t_3)$;

(*EC*4) $\neg HoldsAt(f, t) \leftarrow Happens(e, t_2, t_3) \wedge Terminates(e, f, t_1) \wedge t_2<t_3 \wedge \neg Declipped(t_1, f, t_3)$;

(*EC*5) $t_1 \leq t_2 \leftarrow Happens(e, t_1, t_2)$.

(*EC*6) $HoldsAt(f, t) \leftarrow Initially_P(f) \wedge Clipped(0, f, t)$;

(*EC*7) $\neg HoldsAt(f, t) \leftarrow Initially_N(f) \wedge \neg Declipped(0, f, t)$.

The first five axioms capture the behaviours of fluents once initiated or terminated by an event. To describe fluents' behaviours before the occurrence of any action which affects them, we axiomatise a general principle of persistence for fluents using (*EC*6) and (*EC*7), i.e., fluents change their values only via the occurrence of initiating and terminating actions. Here, we use symbol *EC* to represent the conjunction of these seven axioms, i.e., $EC = \bigwedge_{i=1}^{7} EC_i$.

Uniqueness-of-names (UNA) Axioms

To explicitly guarantee that there are no overlapping effects between events, it is needed to specify that the involved events are not identical. For example, UNA[*loaded*, *Alive*, *Dead*] indicates events *Loaded* ≠ *Alive*, *Alive* ≠ *Dead* and *Loaded* ≠ *Dead*. Here, we use symbol Ω to represent the conjunction of these UNA axioms.

3.3 RFID Data Management Model

Based on event calculus, we define an RFID data management model to characterise the behaviours of RFID objects.

Basic Definitions

Definition 1 (*RFID Class*)**.** An RFID class denotes a type of objects, which abstracts the common properties of these objects. Formally, an RFID class *c* is defined as tuple (*n*, *A*, *Q*), where

- *n* is the name of *c*;
- *A* is a set of attributes that *c* owns;
- *Q* is a set of fluent names, which can characterise the status of *c*'s instances.

Definition 2 (*Event Observation*)**.** The action of observing of an event raised by a reader or the system can be characterised as *Happens*(*e*, *t*), which denotes that event *e* occurs at time *t*.

In practice, events flow as a series, which consists of several event observations, such as, …, $Happens(e_1, t_1)$, $Happens(e_2, t_2)$, …, where $t_1<t_2$. We call such a flow of event observations as an *event series*.

Definition 3 (*Domain-dependent Rule*)**.** The domain-dependent logic is represented as the rules constituting the fluents and predicates introduced in the preliminary of event calculus. Syntactically, a domain-dependent rule *r* can be defined as $P \leftarrow \bigvee_{i=0}^{n} [(\bigwedge_{j=0}^{m} exp_{ij}) \vee exp_i]$, where

- $P \in \{Initiates(e, f, t), Terminiates(e, f, t), HoldsAt(f, t)\} \cup \{$domain-specific predicates$\}$;
- exp_{ij} and $exp_i \in \{Happens(e, t), HoldsAt(f, t)\} \cup \{$domain-specific predicates$\}$.

Definition 4 (*RFID Class Schema*)**.** An RFID class schema is a finite set Γ of RFID classes with distinct names such that every class referenced in Γ also occurs in Γ. i.e.,

$\forall$class $a \in \Gamma$, if a refers to class b, then $\exists b \in \Gamma$.

Definition 5 (*RFID Scenario*)**.** An RFID scenario S for an RFID schema Γ denotes an event calculus scenario with the RFID classes defined in Γ. Syntactically, S can be defined as tuple (Γ, R, EC, Ω)

- Γ is the RFID class schema;
- R is the set of rules, which are defined on the classes and fluents in Γ;
- EC and Ω represent the conjunctions of EC axioms and UNA axioms, respectively.

Definition 6 (*RFID Environment*)**.** An RFID environment *env* denotes the execution environment with actual events for an RFID scenario. An RFID environment *env* can be defined as tuple (I, Δ), where

- I denotes *env*'s initial settings, which are described using predicates $Initially_N$ and $Initially_P$;
- Δ denotes the received event series.

Definition 7 (*RFID Query*)**.** A query q_t over an RFID scenario S in a given RFID environment *env* can retrieve the real-time value of the fluents that are defined in S at a given time point t. Syntactically, a query q_t can be defined as $\leftarrow_{\Delta(t_0,t) \wedge I_{t0}} \rho_t$, where

- ρ_t denotes the target statement for the query in form of a conjunction of several $HoldsAt(f, t)$ predicates, i.e., $\rho_t = \bigwedge_i HoldsAt(f_i, t)$;
- $\Delta(t_0, t)$ denotes the set of events that are occurred from t_0 to t in environment *env*;
- I_{t0} denotes the initial settings at time t_0, i.e., the values of fluents at time t_0 in *env*;

In this paper, we confine that queries are all boolean ones, i.e., whether statement ρ_t is true or not at a given time point t according to RFID scenario S in environment *env*.

Definition 8 (*Operation Trigger*)**.** Driven by query results, an operation trigger *ot* can invoke corresponding operations of RFID edge systems in response to the dynamics of the RFID scenario. Syntactically, *ot* can be defined as

$| \leftarrow_{\Delta(t_0,t) \wedge I_{t0}} \rho_t | \Rightarrow invoke(op)$, where

- symbol "| |" denotes the boolean result of the query;
- *invoke*(*op*) denotes the action of invoking operation *op* of the edge system, if the query returns true.

Figure 2 illustrates the relationship between the aforementioned notions. An RFID schema consists of a set of RFID classes, which specify the fluents and the attributes to be used in describing domain-specific rules in the RFID scenario. Such an RFID scenario represents a self-contained system at build time, while an RFID environment represents the run time dynamics including the fluent values at the beginning time and the continuous event flow. Queries are used to retrieve the real-time fluent values, and the edge system can invoke proper operations in response to the query results via operation triggers.

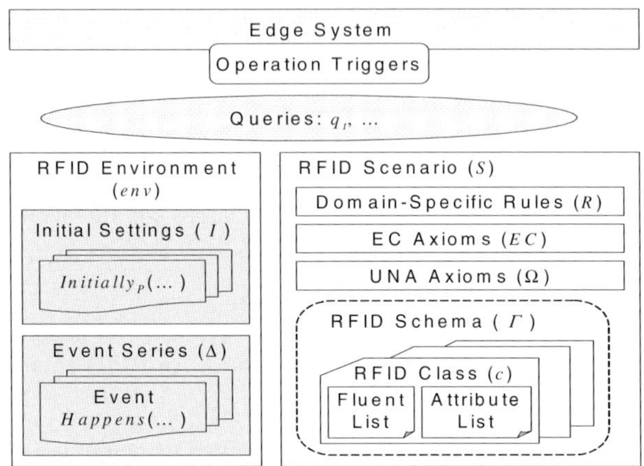

Fig. 2. RFID data management model architecture

The core part of our event calculus based model mainly relies on temporal predicates and rules, but this does not mean that our model is limited to temporal logics. Users can define domain-specific predicates, expressions and rules to represent complex business logics, as mentioned in Section 3.2.

3.4 Deployment of RFID Scenarios

We have conducted a pilot implementation of the proposed framework on a packing station at a distribution centre, which runs the same business process as shown in Figure 1. Here, we take sub process of "Send Goods to Proper Lines" as an example to evaluate if and how our approach can support business process automation in practice.

Figure 3 shows the details of this sub process in a BPMN diagram. In this scenario, the packing station has a temporary repository to store the received products. Once a product of type 1 comes, an event of "g1Arr" will be sent out, and be captured by task "Add the Number of Product Type 1", which is responsible for counting the products of type 1 in the repository. Similarly, task "Add the Number of Product Type 2" is responsible for counting the products of type 2 in the repository.

Event "stateCheck" is a periodical event, which triggers the execution of task "Check Repository". This task will check the number and types of the received products in the repository, and determines to send the products to which assemble line for packing. Here, line A can only pack full pallets of products of the same type, and line B can do partial or mixed pallets.

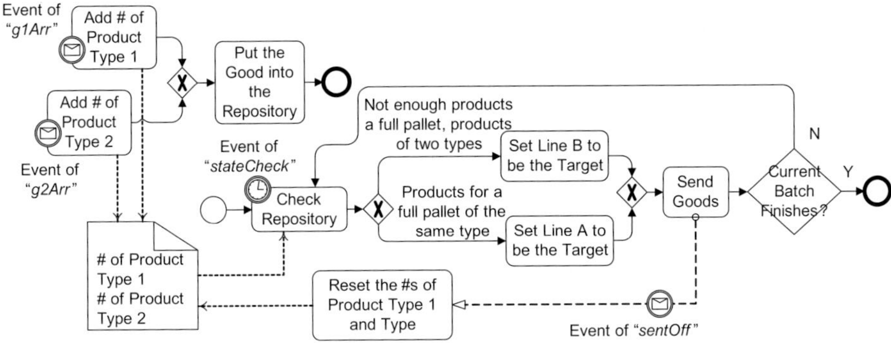

Fig. 3. Sub process "Send Goods to Proper Lines"

The current batch of products is considered to be over, if no more than a pallet of goods arrives in 4 seconds, and then the packing station will empty the repository and wait for the next batch. The main process covered by RFID technology was the tracking of the movement of goods from the band conveyer to the dispatcher. Tagging was done on individual product level.

According to the proposed model, we identify RFID classes *Dispatcher*, *Product Type* 1 and *Product Type* 2 in this scenario. With the classes, the following events are involved:

g1Arr –	a product of type 1 arrives to the dispatcher;
g2Arr –	a product of type 2 arrives to the dispatcher;
sentOff –	the deposited products are sent to a packing line;
stateCheck –	a periodical event to initiate the state checking of the dispatcher.

For class "Dispatcher", the following fluents are used to characterise a dispatcher's run time status.

Mixed –	the deposited products are of two types;
Full –	the repository has enough products for a full pallet;
Finish –	the current batch has been handled;
NotEmpty –	the repository is occupied;
Idle –	the dispatcher is standing by, rather than working.

In addition to these fluents, two variables, num_1 and num_2, are used to record the numbers of deposited products of type 1 and type 2, respectively.

The corresponding RFID scenario constitutes the RFID class schema including the three aforementioned classes, axioms *EC* and Ω, and the domain-dependent rule set *R* which comprises the following rules:

(R1) $[num_1$++, num_2++$]\leftarrow Happens([g1Arr, g2Arr], t)$;
(R2) $Initiates([g1Arr, g2Arr], NotEmpty, t)\leftarrow Happens([g1Arr, g2Arr], t)\wedge\neg HoldsAt(NotEmpty, t)$;
(R3) $Initiates(Mixed, t)\leftarrow Happens([g1Arr, g2Arr], t)\wedge(num_1\neq 0)\wedge(num_2\neq 0)\wedge\neg HoldsAt(Mixed, t)$;
(R4) $Initiates(Full, t)\leftarrow Happens([g1Arr, g2Arr], t)\wedge(num_1+num_2=MAX)\wedge\neg HoldsAt(Full, t)$;
(R5) $Terminates(Idle, t)\leftarrow Happens([g1Arr, g2Arr], t)\wedge HoldsAt(Idle, t)$;
(R6) $Terminates([Mixed, Full, NotEmpty], t)\wedge num_1=0\wedge num_2=0\leftarrow Happens(sentOff, t)\wedge HoldsAt([Mixed, Full, NotEmpty], t)$;
(R7) $Initiates(Idle, stateCheck, t)\leftarrow Happens(stateCheck, t)\wedge\neg HoldsAt(Idle, t)\wedge NoSentOff(t\text{-}4, t)$;
(R8) $Initiates(Finish, stateCheck, t)\wedge Terminates(Idle, stateCheck, t)\leftarrow Happens(stateCheck, t)\wedge\neg HoldsAt(Finish, t)\ )\wedge HoldsAt(Idle, t)\wedge KeepsIdle(t\text{-}4, t)$;

Here, (R1) uses num_1 and num_2 to record the numbers of arrived products of *Product Type* 1 and *Product Type* 2, respectively. The square brackets denote a selective relation. (R2-5) adjust the values of fluents *NotEmpty*, *Mixed*, *Full* and *Idle* when a product arrives. (R6) resets the values of the mentioned fluents to be false once a "sentOff" event occurs. (R7-8) work on "*stateCheck*" events, were (R7) turns the dispatcher into "Idle" mode if no "sentOff" events occurred in last 4 seconds before the latest "stateCheck" event, and (R8) turns the dispatcher into "Finish" mode if the "Idle" mode has been lasting in last 4 seconds before a "stateCheck" event. The referenced predicates *NoSentOff* and *KeepsIdle* are defined as follows,

$$NoSentOff(t_1, t_2) = \bigwedge_{t_i\in[t_1,t_2]} \neg Happens(sentOff, t_i);$$

$$KeepsIdle(t_1, t_2) = \bigwedge_{t_i\in[t_1,t_2]} HoldsAt(Idle, t_i).$$

These two predicates in (R8-9) are subject to the event or fluent values prior to present time, and therefore they result in typical delayed effects in RFID queries. We will dedicatedly discuss about their influence to query execution in next section.

Once the RFID scenario is defined, it can be inputted into the edge system, i.e., the dispatcher. Thus, the dispatcher is empowered with the awareness about the products' arrivals and the business logics on where to send products for packing. Further, the following queries and operation triggers can be deployed to the dispatcher, and enable the dispatcher to intelligently react to real-time dynamics.

Query $q_{1t}: \leftarrow_{\Delta(t_0,t)\wedge I_{t0}} HoldsAt(Full, t)\wedge\neg HoldsAt(Mixed, t)$;

Query $q_{2t}: \leftarrow_{\Delta(t_0,t)\wedge I_{t0}} HoldsAt(Full, t)\wedge HoldsAt(Mixed, t)$;

Query $q_{3t}: \leftarrow_{\Delta(t_0,t)\wedge I_{t0}} HoldsAt(Idle, t)\wedge\neg HoldsAt(Full, t)\wedge HoldsAt(NotEmpty, t)$.

Trigger 1: $|q_{1t}|\Rightarrow$ invoke operation "send to Line A";
Trigger 2: $|q_{2t}|\vee|q_{3t}|\Rightarrow$ invoke operation "send to Line B".

An RFID environment contains the initial settings and the real-time event series of a concrete execution environment for an RFID scenario. In this example, the RFID environment may consist of the following content:

I={ $Initially_N(NotEmpty)$; $Initally_N(Full)$; $Initally_N(Mixed)$; $Initally_N(Idle)$; $Initally_N(Finish)$; MAX=4; num_1= num_2=0. };

Δ={ $Happens(g1Arr, t_1)$, $Happens(g2Arr, t_2)$, $Happens(g2Arr, t_3)$, $Happens(g1Arr, t_4)$, … ($t_1 < t_2 < t_3 < t_4 < \ldots$)}.

Please note that MAX=4 denotes that a full pallet contains four products.

4 Efficient RFID Querying in Event Calculus Context

According to the proposed model, RFID queries are running to monitor the changes of the RFID scenario, and thereby invoke proper operations of edge systems to enable the business process automation. From our deployment practice, we find that the event and fluent dependencies influence a lot on the query execution performance. In this section, we are to investigate such dependencies.

4.1 Event/Fluent Delayed Effects in Queries

From Definition 7, we can see that for each query execution it needs to calculate the events occurred from t_0 up to the current time point. As time goes, the number of events increases towards infinite, and in turn this will reluctantly increase the query execution time towards infinite. To optimise the RFID query execution, we intend to cut off the event series for each query from the endless to a limited scope by eliciting useful information from previous query results. This intention is based on the fact that RFID-enabled applications always run queries continuously to monitor the real-time variations.

To optimise the event series, we have to consider the delayed effects caused by the rules referring to different time points. Take (R7) in last section as an example, fluent *Idle* changes to be true only if there are no "*sentOff*" event occurred in recent 4 seconds. We classify such delayed effects caused by events as *event delayed effects*, and the ones caused by fluent as *fluent delayed effects*, for example, fluent *Finish* in R(8) is subject to the values of fluent *Idle* of recent 4 seconds.

As shown in Figure 4, a series of events occur along the timeline. At time t_i, query q_{ti} is executed, and we can save the retrieved fluent values as a *fluent snapshot* for time t_i. When another query q_{tj} is to execute at time t_j, we intend to calculate the events from t_i rather than t_0 with the knowledge of the fluent snapshot at t_i. However, due to the delayed effects, some events occurred before t_i and some fluent values before t_i are also needed to when calculating for q_{tj}. Once q_{tj} is worked out, the fluent snapshot at t_j will be stored for the execution of later queries.

To specify the event delayed effects and fluent delay effects in RFID queries, we have regulated the form for general RFID queries, and proved this form can cover all possible cases.

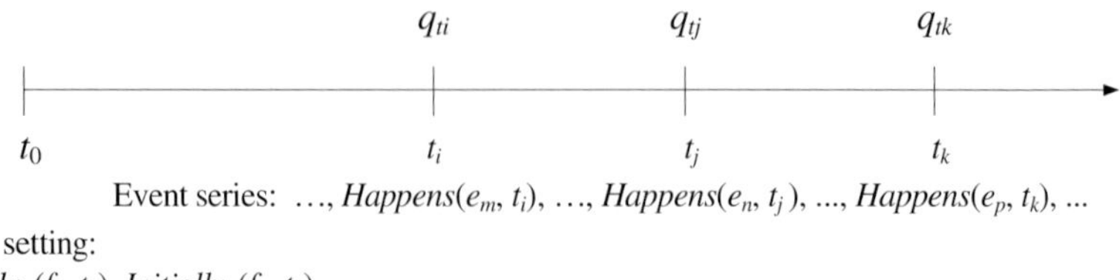

Fig. 4. Historical queries and subsequent queries

Equivalent Query Transformation

Suppose RFID scenario S has rule set $R=\{P\leftarrow \bigvee_{i=0}^{n} \{[\bigwedge_{j=0}^{m} Happens(e_{ij}, t\text{-}\Delta t_{ij})]\wedge[\bigwedge_{k=0}^{h} HoldsAt(f_{ik}, t\text{-}\Delta t_{ik})]\wedge exp_i\}\vee exp^{\circ}\}$, where variables i, j and k start from zero to cover all possible combinations. In S, Query $\leftarrow_{\Delta(t_0,t)\wedge I_{t0}} \rho_t$ is equivalent to $\leftarrow_{(\Delta(t_1,t)+\Delta')\wedge(I_{t0}+\Delta I)} \rho_t$, where

- $\Delta'=\{Happens(e_{ij}, t') \mid Happens(e_{ij}, t')\in\Delta(t_0, t_1), t_1\text{-}\Delta t_{ij}<t'<t_1$, where e_{ij} and Δt_{ij} are referred to by rules in $R\}$ (1)
- $\Delta I=\{HoldsAt(f_{ik}, t_1\text{-}\Delta t_{ik}) \mid f_{kj}$ and Δt_{ik} are referred to by rules in R $\}$ (2)

Details of the proof can be referred to report [11].

4.2 Two-Block Buffering Mechanism

Technically, the introduced query transformation deploys a specific space to buffer the previous events and fluent snapshots that may own delayed effects to later queries. This event buffering is similar to the slide window control in data stream processing, yet we only keep the events having delayed effects for future queries. The fluent snapshot buffering keeps the fluent values at historical time points by running extra queries. However, we need to note that such extra queries may refer to the fluent snapshots of earlier time points, which will require more queries. As such, an update to the snapshot buffer may result in a series of recursive queries, which will seriously damage the query execution performance. To avoid such recursive querying, we propose a two-block buffering mechanism. First, we define the data structure for a buffer.

For a given RFID scenario S and a given RFID environment *env*, the data structure of a buffer b can be defined as tuple $(t, \Delta', I, \Delta I)$, where

- t, the time point for buffer b;
- Δ', the buffered events that occurred before t, i.e., Δ' of formula (1) for S and *env*;
- I, the fluent snapshots at time point t;
- ΔI, the buffered fluent snapshots that occurred before t, i.e., ΔI of formula (2) for S and *env*.

The two-block buffering mechanism runs two such buffers to record the data with delayed effects for later queries. The one with an earlier time point is called *back buffer* (*bb*), while the other is called *fore buffer* (*fb*). As Figure 5 shows, each buffer reserves the related past events and fluent snapshots, which are indicated by the shadowed rectangle left to the buffer. The buffer size, which is measured by the time period that the buffered data span over, is subject to the rule set *R* of RFID scenario *S*, and is therefore bounded. In addition, the two buffers must be guaranteed to be non-overlapped with each other. The initial content of *bb* can be obtained by calculating from time t_0, while *fb*'s content can be done by calculating from *bb.t* with *bb*'s buffered content.

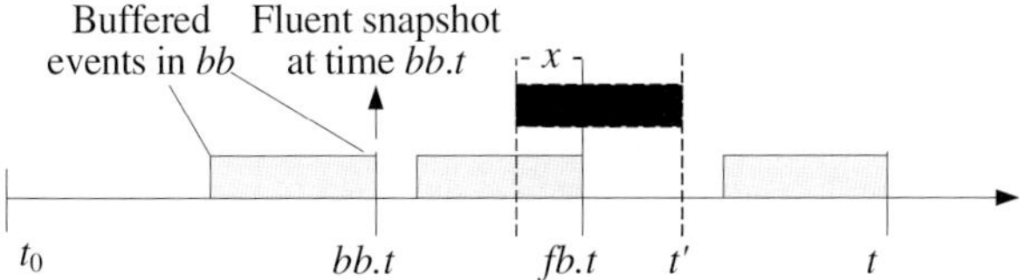

Fig. 5. Two-block buffering

If a query q_t runs at time *t* (*t*>*fb.t*) and all the required historical data occurred after *fb.t*, these historical data can be obtained by calculating from *fb.t*, with *fb*'s buffered content. If a query q_t' runs at time *t*′ (*t*′>*fb.t*), where this query requires some historical data of the time earlier than *fb.t*, it indicates that some past events and fluent snapshots occurred in time period *x*, as shown in Figure 5, have delayed effects to query q_t'. For the fluent snapshots taken during period *x*, they cannot be calculated out from *fb*'s buffered content, and will require more queries over earlier historical data, which may cause query recursions. We prevent such recursions by using *bb*'s buffered content instead of *fb*'s to calculate out q_t'.

This buffer updating mechanism guarantees that each query only needs to calculate the events occurred from *bb.t* or *fb.t* to current time plus the buffered events, rather than the whole event series from t_0. This mechanism is independent with queries, and therefore can support multiple heterogeneous queries to run over the same RFID scenario in an RFID environment.

The details of this 2-block buffering mechanism is given in report [12].

4.3 Experiment Results

To test the efficiency of the buffering mechanism, we have implemented a simulation environment on Cygwin (Linux Emulation for Windows) with IBM Discrete Event Calculus (DEC) Reasoner [13], relsat 2.02 [14] as the propositional satisfiability solver. The event input stream is simulated by a large event log file which can be randomly generated by programs. The simulation was performed on a personal computer with a Core 2 6300 CPU at 1.86 Ghz, 1 GB memory, and Windows XP OS.

The rule set for testing is based on the one discussed in the deployment section with slight modifications to adapt to DEC reasoner's encoding format. Queries are executed every 5 seconds, and we take the execution time as the main performance indicator.

To demonstrate the performance improvement, we execute the same queries under three mechanisms.

- Naive mechanism. This mechanism calculates all occurred events from time point 0s without any buffering.
- Periodical buffering mechanism. This mechanism periodically stores the fluent snapshots and discarding some old events. In this test case, to prevent the delayed effects, we set the interval to be 8 seconds (4 seconds for fluent delayed effect and 4 seconds for event delayed effect as indicated by (R7-8)), i.e., every 8 seconds it will record the current fluent snapshot and discard the events occurred 8 seconds before. This mechanism follows the idea of the fixed period partition mechanism proposed by Siemens Lab in [15].
- The proposed 2-block buffering mechanism. Here, we set the length of fore buffer and back buffer to be 4 seconds according to (R7-8).

Table 2. Comparison of query execution times

Query time point (seconds)	Naive (s)	Periodical (s)	2-Block (s)	Periodical * (s)
5	1.6	1.6	1.6	3.1
10	3.4	3.4	2.9	4.9
15	5.4	5.4	3.6	5.8
20	7.7	3.5	2.4	5.0
25	11.2	2.6	2.7	4.1
30	13.0	4.1	3.0	5.6
35	16.0	3.4	3.3	4.9
40	19.6	4.8	2.5	6.3

Table 2 lists the query execution time under these three mechanisms. Due to the limit of the simulation platform, the simulation results are only used to show the performance difference between buffering mechanisms, yet not indicate the practical execution time on real systems.

Table 3. Extra time cost for updating fluent snapshots in the periodical buffering mechanism

Query time points (s)	8	16	24	32	40
Execution time	2.8	2.4	2.4	2.5	2.4

Compared to the other two buffering mechanisms, the periodical buffering has to run extra queries to update the fluent snapshots on certain time points, i.e., time points $t \times 8$ (t=1, 2, 3, …). Table 3 lists the time cost for these queries, and from these data we can work out the mean extra time cost for each query is $(2.8+2.4+2.4+2.5+2.4)/8 \approx 1.5$s. Take into account the extra query cost, we change the query execution time by adding the mean time cost 1.5s, and put the new data in the column titled "Periodical *" in Table 2.

The query execution time of the periodical buffering mechanism fluctuates as the query time point changes. This indicates the execution time is influenced by the distance between the query time point and the latest buffering point. As the time for updating fluent snapshots follows the pattern of $t\times 8$ (t=1, 2, 3, …), at time point 15s, the latest updating is at time point 8s, and events from 0s to 8s are stored in the buffer. Therefore the calculation starts from 0s to 15s (range of 15s) with the fluent snapshot at time point 8s. At time point 25s, the latest updating is at time point 24s, and the calculation starts from 16s to 25s (range of 9s) with the fluent snapshot at time point 16s. The different event calculation ranges result in the variation of query execution time.

In contrast, the query execution time for the 2-block buffering mechanism keeps very smooth around 2.7s. This is because the 2-block buffering mechanism only updates fluent snapshots at the time of executing users' queries. For example, at time point 15s, the back buffer (*bb*) and the fore buffer (*fb*) were created at 5s and 10s, respectively, where *bb* buffered the events from 1s to 5s with the fluent snapshots at time point 5s, and *fb* did the events from 6s to 10s and with fluent snapshot at time point 10s, respectively. When a users' query is run at time point 15s, it calculates with the events occurred from 6s to 10s and from 11s to 15s, with fluent snapshot at time point 10s. The query result will be used to update the fluent snapshot at time point 15s. In addition, this updating mechanism does not involve extra queries cost like the periodical one does. This experiment clearly shows that our 2-block mechanism stands out with higher performance.

5 Related Work and Discussions

Major software vendors, like IBM (WebSphere RFID), Sun (Java RFID System), BEA (BEAWebLogic RFID), SAP (AII), etc., attempt to create RFID integration platform and deploy business process management to RFID-enabled applications. For instance, SAP's Auto-ID Infrastructure (AII) project aims to facilitate the connection between RFID devices, middleware systems, and business application systems [16, 17]. Components of the association data management (ADM) and the action & process management (APM) are responsible for managing the contextual information and the received events, and navigating the activity handling, respectively.

To better describe business logics in data (event)-intensive business scenarios, the object-oriented (or artifact-oriented) perspective has been proposed recently as a new modelling method [18-20]. Compared with traditional business process modelling approaches, the object-oriented modelling approach focuses on the business contexture and behaviours with declarative logics, rather than the sequencing of activities. In this way, the object-oriented modelling enables business actors to be aware of what can be done instead of what should be done.

A lot of research efforts have been put to tackle RFID complex event processing, yet most of them mainly focus on data cleansing and filtering. SASE [21, 22] has defined a SQL like complex event language to aggregate RFID events. The implemented SASE system uses a persistence storage component to support querying over historical data and to allow query results from the stream processor to be joined with stored data. In addition, the extended sliding window control and indexing techniques have been adopted by work [23] and [24] to improve the performance of continuous query processing over RFID event flows. Hu, Misra and Shorey have addressed the query issue from the perspective of energy efficiency in [25]. Wang, Liu, and Bai have investigated the temporal management of RFID data [15]. They have adapted traditional database query techniques to the temporal relationships of RFID data, and thereby defined a set of temporal complex event constructors in [26]. Two partitioning mechanisms have also been proposed in their work to support efficient queries. However, none of the mentioned works have provided an explicit solution on how to handle the delayed effects in event management, or how to integrate business process automation into RFID event management.

Our work aimed to integrate the business rules and RFID event data management together, and thereby empowered the RFID edge systems with the awareness to both business logics and real-time object-level information. A formal RFID data management framework was presented to model the involved business logics, RFID scenarios, queries and so on, on the mathematical basis of the event calculus. Particularly, the following features distinguished our RFID data management framework from others:

- Concise representation

With time dependent fluents, our RFID data management model is more elegant in representation and more powerful in expressivity, in comparison with traditional event/status modelling approaches, like UML state diagram and EPC, which has to awkwardly employ a large number of intermediate states to represent all possible status and delayed effects.

- Integration of business rules to RFID data management

The proposed model well composes the business rules, operation invocations, into the RFID scenario modelling to realise the design of self-contained and autonomous RFID systems. The customisable rules also enable the re-configurability of RFID edge systems.

- Edge system level business logic deployment

Our approach focused on deploying the modelled business scenarios to RFID edge systems. With the injected business logics, RFID edge systems can respond to the received events on site according to the defined rules and operation triggers. This feature is practically important for real-time handling applications.

- RFID query optimisation

In regard to the delayed effects in RFID queries over continuous event flows, our proposed 2-block buffering mechanism reused historical data to shorten the event series for calculus. The experiment results proved our mechanism is with better performance than others.

6 Conclusion

This paper looked into the business logic modelling for RFID-enabled applications. By establishing a formal RFID data management model, we proposed a novel method to specify the RFID classes, domain-dependent rules, real-time event series, queries and operation triggers, as well as inter-relations among these notions. This method fully catered for the features such as event-rich communications, large event volumes, event and fluent delayed effects, etc., in the RFID-applied environment. In compliance with this method, a query optimisation scheme with a two-block buffering mechanism was discussed for improving RFID query performance.

Our follow-up work is to further refine the proposed RFID data management model to support more complex business transactions, and enrich the optimisation mechanism with more techniques for different query cases.

References

1. Banks, J., Pachano, M., Thompson, L., Hanny, D.: RFID Applied. John Wiley & Sons, Chichester (2007)
2. Koh, S.C.L.: RFID in Supply Chain Management: A Review of Applications. In: Kumar, S. (ed.) Connective Technologies in the Supply Chain, pp. 17–40. Taylor & Francis, Abington (2007)
3. Liu, C., Li, Q., Zhao, X.: Challenges and Opportunities in Collaborative Business Process Management. Information System Frontiers 11, 201–209 (2009)
4. IDTechEx (2007), http://www.IDTechEx.com
5. Gillette, W.: Boeing Presentation at the Aerospace & Defense Summit (2004)
6. Bottani, E.: Reengineering, Simulation and Data Analysis of an RFID System. Journal of Theoretical and Applied Electronic Commerce Research 3, 12–29 (2008)
7. OMG: Business Process Modeling Notation (2009), http://www.omg.org/spec/BPMN/1.2/
8. Glover, B., Bhatt, H.: RFID Essentials. O'Reilly, Sebastopol (2006)
9. Shanahan, M.: The Event Calculus Explained. Artificial Intelligence Today, pp. 409–430. Springer, Heidelberg (1999)
10. Scheer, W.A.: Business Process Engineering, ARIS-Navigator for Reference Models for Industrial Enterprises. Springer, Berlin (1994)
11. Zhao, X.: Report: Proof of RFID Query Equivalent Transformations (2009), http://www.ict.swin.edu.au/personal/xzhao/proof.pdf
12. Zhao, X.: Report: 2-Block Buffering Mechanism for RFID Query Optimisation (2009), http://www.ict.swin.edu.au/personal/xzhao/algorithm.pdf
13. IBM Discrete Event Calculus Reasoner, http://decreasoner.sourceforge.net/
14. relsat 2.02, http://www.bayardo.org/resources.html
15. Wang, F., Liu, P.: Temporal Management of RFID Data. In: The 31st International Conference on Very Large Data Bases, Trondheim, Norway, pp. 1128–1139 (2005)
16. Bornhövd, C., Lin, T., Haller, S., Schaper, J.: Integrating Automatic Data Acquisition with Business Processes - Experiences with SAP's Auto-ID Infrastructure. In: The 13th International Conference on Very Large Data Bases, Toronto, Canada, pp. 1182–1188 (2004)
17. Götz, T., Safai, S., Beer, P.: Efficient Supply Chain Management with SAP Solutions for RFID. Galileo Press (2006)
18. Nigam, A., Caswell, N.S.: Business Artifacts: An Approach to Operational Specification. IBM Systems Journal 42, 428–445 (2003)
19. Liu, R., Bhattacharya, K., Wu, F.Y.: Modeling Business Contexture and Behavior Using Business Artifacts. In: The 19th International Conference on Advanced Information Systems Engineering, Trondheim, Norway, pp. 324–339 (2007)

20. Hull, R.: Artifact-Centric Business Process Models: Brief Survey of Research Results and Challenges. In: Meersman, R., Tari, Z. (eds.) OTM 2008, Part II. LNCS, vol. 5332, pp. 1152–1163. Springer, Heidelberg (2008)
21. Gyllstrom, D., Wu, E., Chae, H.-J., Diao, Y., Stahlberg, P., Anderson, G.: SASE: Complex Event Processing over Streams. In: The 3rd Biennial Conference on Innovative Data Systems Research, Asilomar, CA, USA, pp. 407–411 (2007)
22. Wu, E., Diao, Y., Rizvi, S.: High-performance Complex Event Processing over Streams. In: The ACM SIGMOD International Conference on Management of Data, Chicago, Illinois, USA, pp. 407–418 (2006)
23. Bai, Y., Wang, F., Liu, P., Zaniolo, C., Liu, S.: RFID Data Processing with a Data Stream Query Language. In: The 23rd International Conference on Data Engineering, Istanbul, Turkey, pp. 1184–1193 (2007)
24. Park, J., Hong, B., Ban, C.: A Continuous Query Index for Processing Queries on RFID Data Stream. In: 13th IEEE International Conference on Embedded and Real-Time Computing Systems and Applications, Daegu, Korea, pp. 138–145 (2007)
25. Hu, W., Misra, A., Shorey, R.: CAPS: Energy-Efficient Processing of Continuous Aggregate Queries in Sensor Networks. In: The 4th Annual IEEE International Conference on Pervasive Computing and Communications, Pisa, Italy, pp. 190–199 (2006)
26. Wang, F., Liu, S., Liu, P., Bai, Y.: Bridging Physical and Virtual Worlds: Complex Event Processing for RFID Data Streams. In: The 10th International Conference on Extending Database Technology, Munich, Germany, pp. 588–607 (2006)

An Integrated Approach to Managing Business Process Risk Using Rich Organizational Models

M.M. Zahidul Islam, Moshiur Bhuiyan, Aneesh Krishna, and Aditya Ghose

Decision Systems Laboratory, School of Computer Science and Software Engineering
University of Wollongong, NSW 2522, Australia
{mmzi44,mmrb95,aneesh,aditya}@uow.edu.au

Abstract. Business processes represent the operational capabilities of an organization. In order to ensure process continuity, the effective management of risks becomes an area of key concern. In this paper we propose an approach for supporting risk identification with the use of higher-level organizational models. We provide some intuitive metrics for extracting measures of actor criticality and vulnerability from organizational models. This helps direct risk management to areas of critical importance within organization models. Additionally, the information can be used to assess alternative organizational structures in domains where risk mitigation is crucial. At the process level, these measures can be used to help direct improvements to the robustness and failsafe capabilities of critical or vulnerable processes. We believe our novel approach, will provide added benefits when used with other approaches to risk management during business process management, that do not reference the greater organizational context during risk assessment.

1 Introduction

A Business Process can be described as a set of dynamically coordinated activities that are controlled by a number of socially dependant participants and aimed towards the achievement of a specific operational objective [7] [11]. Business Process Management is a re-emerging discipline, aimed towards supporting the effective and automated [11] management of business processes within an organization via specialized tools and methods. Business Process Management promotes that a clear understanding through the explicit modeling of the processes underlying an organization is required to support effective organizational management / improvement practices [4].

An effective means to represent and manage operational risk is one of the most important capabilities within an enterprise. Some of the most prominent applications of risk management techniques include financial / operational management and modeling of organization. Risk management techniques have also been extensively studied and applied within software process management, requirements engineering and project management disciplines [16][17][18][19]. More recently, risk management has been applied to the business process management and modeling domain that as a whole, aims to bridge the gap between organizational and I.T. level conceptual / management concerns [14] [15]. These approaches provide a more direct association between organizational risks at an activity level.

R. Meersman, T. Dillon, P. Herrero (Eds.): OTM 2009, Part I, LNCS 5870, pp. 273–285, 2009.

There are difficulties associated to addressing risk at process level. We believe that by taking actor-level considerations of vulnerability and criticality (at organizational level) is important for process-level risk management. We provide an approach to support risk management by supporting the identification of risk factors (in terms of vulnerability and criticality) at organizational level prior to their propagation and reflection at a process level. We believe that such an approach will provide a higher-level scope for risk that may span numerous processes within an organization. Business process risk analysis should be based on higher-level organizational models. A high-level approach to iterative risk assessment should be integrated throughout the business process lifecycle. Therefore, risks may be identified and managed at an organizational level prior to their delegation to actual business processes. We provide an enhanced capability to relate risk at an organizational level by looking at the strategic relationships between functional units and process participants. We define risk at organizational model level on the basis of vulnerability and criticality. For organizational models we use the agent-oriented organizational modeling notation – *i** [13] that describes the organizational relationships among various actors and their rationales. For business process model representation we use a standardized, operational and executable process modeling notation – BPMN [12]. The authors consider that the majority of risks identified lie in mismatch with the methods employed within the various phases of the process lifecycle, a lack of clarity who is responsible for the individual phases or their results and a mismatch of process design, automation and evaluation objects. We believe that risk can be better viewed by using a combined notation proposed in [12].

The following section starts with a discussion of risk and risk management and our chosen notations. We then describe our approach to identify risk factors including our proposed measurement for vulnerability and criticality of actors at organizational level. Next section we then illustrated the integration of risk factors in process model with examples. Finally we provided a case study and then some concluding remarks.

2 Background

2.1 Agent Oriented Conceptual Modeling

The agent metaphor is powerful in modelling organizational contexts. Agent-Oriented Conceptual Modelling (AOCM) in notations such as the *i** framework [13] (see: figure 1) have gained considerable currency in the recent past. Such notations model rich organizational contexts and offer high-level social/anthropomorphic abstractions (such as goals, tasks, soft goals and dependencies) as modelling constructs.

It has been argued that notations such as *i** help answer questions such as what goals exist, how key actors depend on each other and what alternatives must be considered. Furthermore, *i** has been acknowledged as illustrating the key social/strategic inter-relationships between actors [6] [13] required for effective business process redesign. This is achieved via support for reasoning about organizational activities and their assignment to various organizational agents [13] in respect to: the ability, workability, viability, and believability of their routines; and, level of commitment [13].

The central concept in i^* is that of intentional actor. These can be seen in the below Emergency Service Provider model as nodes representing the intentional/social relationships between six (6) actors required to schedule a meeting: an Emergency Coordination Center Coordinator (ECCC); Field Control Center Coordinator (FCCC); Volunteer/Emergency Workers; Community; Weather Bureau and Call taking supervisor/system.

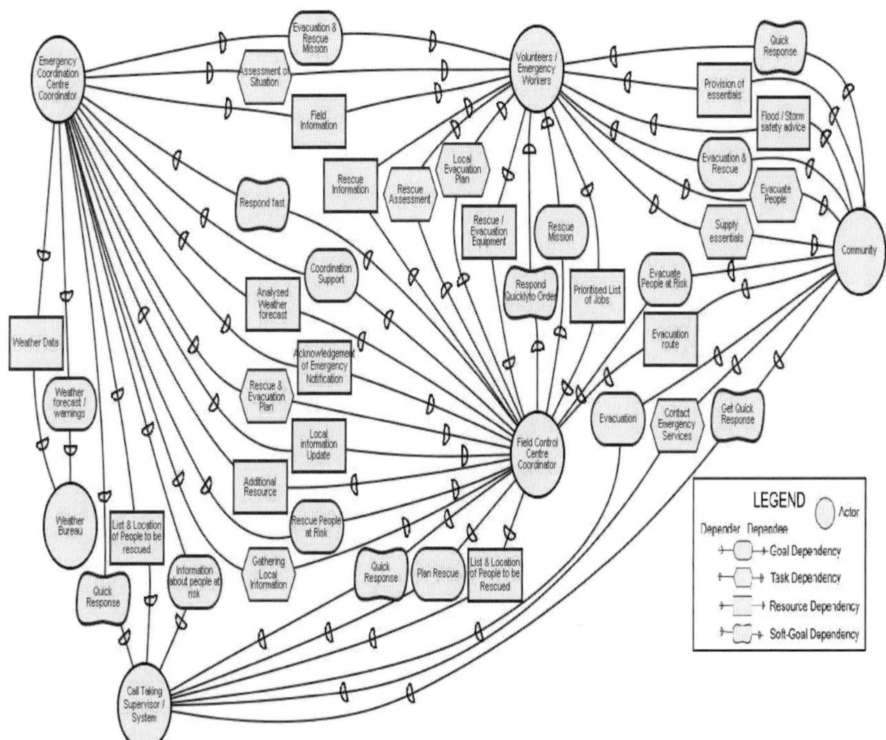

Fig. 1. Strategic Dependency Model of an Emergency Service Provider

The i^* framework consists of two modelling components: Strategic Dependency (SD) model and Strategic Rationale (SR) model [13]. The SD model consists of a set of nodes and links. Each node represents an actor, and each link between the two actors indicates that one actor depends on the other for something (i.e. goals, task, resource, and softgoal) in order that the former may attain some goal. The depending actor is known as depender, while the actor depended upon is known as the dependee. The object around which the dependency relationship centres is called the dependum. The SR mode further represents internal motivations and capabilities (i.e. processes or routines) accessible to specific actors that ensure dependencies can be met.

2.2 Business Process Modelling with BPMN

Many existing BPM notations primarily focus on technical process aspects including the flow of activity execution/information and/or resource usage/consumption [13]. This perspective is aimed at describing the sequence of activities, events and decisions that are made during process execution, however social and intentional components lack representation. The technical focus of these notations is especially suited for applications in the description, execution and simulation of business processes but is lacking in support for process redesign and improvement [13].

One such notation is the Business Process Modelling Notation (BPMN), developed by the Business Process Management Initiative (BPMI.org). BPMN can be seen as primarily a technically-oriented notation that is augmented with an ability to assign activity execution control to entities (e.g. roles) within an organization with

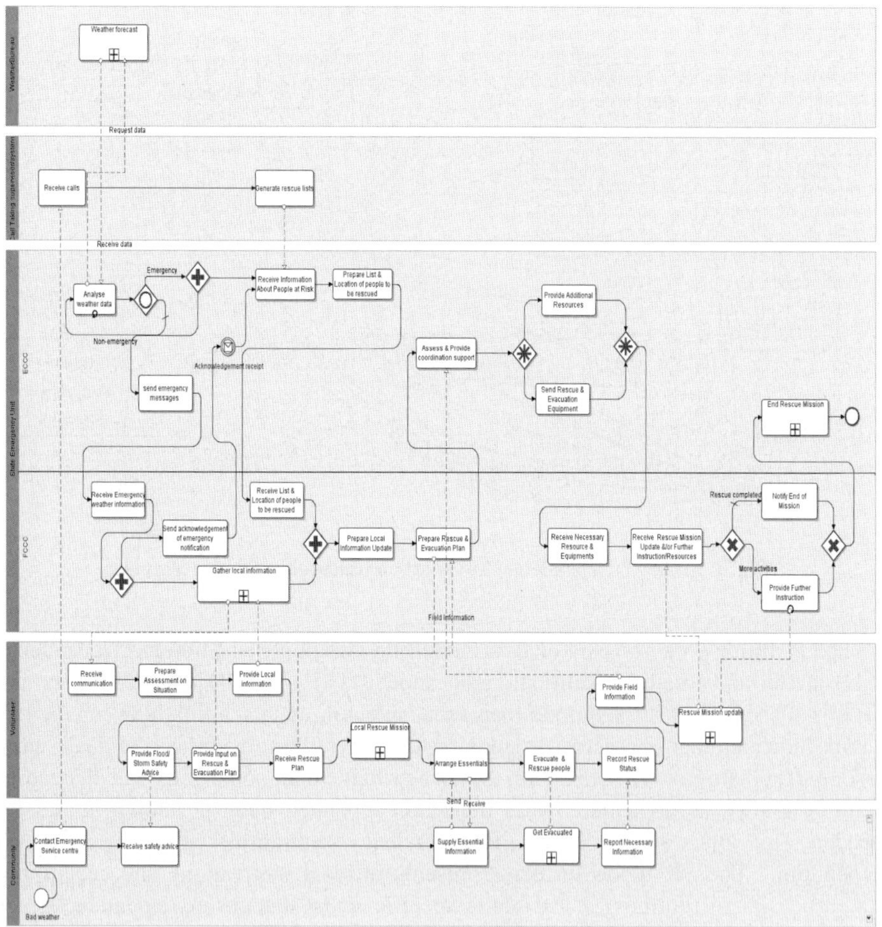

Fig. 2. BPMN model of the Emergency Service Provider

'swim-lanes'. This effectively provides a view of the responsibilities and required communications between classes of process participants, but does not provide a view of other social and intentional characteristics including the goals of participants and their inter-dependencies.

Processes are represented in BPMN using flow nodes: events (circles), activities (rounded boxes), and decisions (diamonds); connecting objects: control flow links (un-broken directed lines), and message flow links (broken directed lines); and swim-lanes: pools (high-level rectangular container), and lanes partitioning pools. These concepts are further discussed within [12]. Since its initial publication [12], BPMN has been accepted by the greater Business Process Management community [1] [11], due to its expressiveness and ability to map directly to executable process languages including XPDL [4] and BPEL [10] [12].

3 Identifying Risk within Organizational Models

In this section we will describe our intuitive approach to analysis and design with regards to organizational risk. In order to achieve this task, we propose an analysis of strategic dependencies between actors in order to measure and identify each actor's vulnerability and criticality. Once determined, the design task will be focused towards the area of process modelling that requires most attention.

3.1 Vulnerability

The vulnerability of an actor plays a vital role for identifying and measuring risk. The *i** model provides an intentional description of a process in terms of a network of dependency relationships among actors [13]. We believe because of its richer modeling concepts, the model provides a better basis for an analyst to explore the broader risk implications of alternative organizational structure. It can help analyze opportunities and vulnerabilities and recognize patterns of relationship. A depender actor's intention is to have the dependency goal achieved, task performed, or resource available. Failure to obtain the dependum can affect the process by making it more vulnerable and hence increasing the likelihood of risk occurrence. In our work we propose a way of measuring vulnerability of actors at organizational model. The analyst can then take necessary steps to mitigate these vulnerabilities in process models. A stronger degree of vulnerability implies that a stronger initiative to mitigate vulnerability is necessary. Such initiative can be taken by increasing the monitoring process of dependee actor's activities.

We propose a metric for actor vulnerability. This metric is effectively divides the number of outgoing dependencies by the number of dependee actors. A depender actor with more outgoing dependencies implies a greater degree of vulnerability. We consider outgoing dependencies for vulnerability measurement as we believe that outgoing dependencies indicate delegation of tasks and activities. If the tasks are delegated to other actors the depender actor becomes vulnerable. In case of the failure of dependee actor to satisfy the dependency, the corresponding task/goal might not be satisfied (a considerable risk). The vulnerability of actors thus is related to the likelihood of a risk occurring. We believe if an actor is vulnerable, an increase in the overall likelihood of risk occurrence is apparent. Intuitively, if the likelihood increases risk will increase as well.

The formula we use to assess the vulnerability measurement (VM) of actors at organizational level is as follows:

VM_{org} =No of Outgoing Dependencies / No of Dependee Actors
For example, for actor EmergencyCoordinationCentreCoordinator,
No of Outgoing dependencies = 12 and No of Dependee Actors = 4
So, Vulnerability at Organizational Model, VM_{org}= (12/4) = 3

Table 1. Vulnerability Measurement of Actors at Organizational Model

Name of the Actor	No of Outgoing Dependecies	No of Dependee Actors	VM_{org}
Emergency Coordination Centre Coordinator	12	4	3
Weather Bureau	0	0	MinimalVulnerability
Call Taking Supervisor/ System	0	0	MinimalVulnerability
Volunteer/Emergency Workers	4	2	2
Flood Control Centre Coordinator	7	3	2.33
Community	8	3	2.66

In a softgoal dependency, a depender depends on the dependee to perform certain goals or task that would enhance the performance. The notion of a softgoal derives from the Non-Functional Requirements (NFR) framework [2] and is commonly used to represent optimization objectives, preferences or specifications of desirable (but not necessarily essential) states of affairs. So, softgoals are non-functional requirements of the system, which have positive or negative contribution toward achieving a goal, task, or resource. While measuring the vulnerability of actors we do not include the softgoal dependencies. We believe these non-functional requirements of the system have minimal impact on risk either in the organizational level or on the process level. When we calculate the outgoing dependencies of actors we exclude the softgoal dependency.

If any actor has no outgoing dependency with other actors, we consider that the actor has minimal vulnerability as we believe it can not affect the likelihood of occurrence in a greater extent. From figure-1 we find that the actors WeatherBureau and CallTakingSuperviosr/System do not have any outgoing dependencies. It means they have not delegated their responsibilities or tasks to other actors. But, actor with no vulnerability does not necessarily mean that it is not critical enough to affect the consequences if it fails. In this case criticality of the actor is considered to measure the risk. Now we need to refine the vulnerability calculation by relating it at process level. The formula we use to calculate vulnerability measurement (VM) at process level is as follows:

VM_{bp} = Organizational Level Vulnerability (VM_{org}) * Number of Incoming Flows (control flow and message flow)

For example, for actor EmergencyCoordinationCentreCoordinator,
Organizational Level Vulnerability, VM_{org} = 3 and No of Incoming Flows = 6
So, Vulnerability at Process Level, VM_{bp}= 18

Table 2. Vulnerability Measurement at Process Level

Name of the Actor	VM_{org}	Incoming Flow	VM_{bp}
Emergency Coordination Centre Coordinator	3	6	18
Weather Bureau	Minimal Vulnerability	1	Minimal Vulnerability
Call Taking Supervisor/ System	Minimal Vulnerability	1	Minimal Vulnerability
Volunteer/Emergency Workers	2	5	10
Flood Control Centre Coordinator	2.33	6	13.98
Community	2.66	2	5.32

3.2 Criticality

Criticality is the consequence factor that is measured from the impact of an actor's performance where the actor is assigned to satisfy responsibilities/incoming dependencies. The more critical an actor is, the more ability it carries to impact other actors and the organizational context. Incoming dependencies towards an actor are taken into consideration to measure the criticality of an actor. The incoming dependencies describe responsibilities are assigned to an actor from other actor. By receiving dependencies from other actor makes the dependency receiving actor crucial. If it fails to satisfy the incoming dependencies the depender actors are widely affected which possibly affect the context as a whole. In order to mitigate the risks associated with the system the criticality measurement of actors should be taken into consideration. Measuring critical factors of actors helps the analysts to analyze and construct alternative options to achieve the aim of the system. This will alleviate the risk management and increase the robustness of the system.

Criticality of actors at Organizational Model is measured by multiplying number of incoming dependencies and number of depender actors. The formula we use to assess the criticality measurement (CM) of actors is as follows:

CM_{org} = No of Incoming Dependencies * No of Depender Actors
For example, for actor Volunteer,
No of Incoming Dependencies = 10 and No of Depender Actors = 3
So, Criticality at Organizational Model, CM_{org} = 10*3 = 30

Table 3. Criticality Measurement of Actors at Organizational Model

Actors	No of Incoming Dependencies	No of *Depender* Actors	CM_{org}
Emergency Coordination Centre Coordinator	4	1	4
Call Taking Supervisor/System	9	3	27
Volunteer/Emergency Workers	10	3	30
Flood Control Centre Coordinator	9	3	27
Weather Bureau	2	1	2
Community	2	1	2

According to the result from Criticality Metrics, Volunteer actor is more critical than other actor in the model. Volunteer has ten incoming dependencies from three other actors and its existence is more crucial because if it fails to satisfy any of the incoming dependencies received from other three actors it will have greater impact on other actors and to system as a whole. We have not considered the softgoal dependencies while calculating the criticality of the actors for the same reasons of vulnerability measurement.

If an actor does not have any incoming dependencies from another actor of the model then it portrays that the actor has distributed his dependencies to other actor but no other actor has delegated any tasks, resources and goals into this actor. So the actor will have minimal impact on the consequences of the performance of other actors in the strategic context of the model. For this reason an actor with no incoming dependencies will be positioned with minimal criticality fact towards it but the vulnerability factor of that actor will take it into the consideration of the risk measurement in the

Table 4. Criticality Measurement at Process Level

Actors	CM_{org}	No of Outgoing Flows	CM_{bp}
Emergency Coordination Centre Coordinator	4	4	16
Call Taking Supervisor/System	27	1	27
Volunteer/Emergency Workers	30	6	180
Flood Control Centre Coordinator	27	6	162
Weather Bureau	2	1	2
Community	2	3	6

strategic framework. Now we need to refine the criticality calculation by relating it at process level. The formula we use to calculate criticality measurement (CM_{bp}) at process level is as follows:

CM_{bp} = Organizational Level Criticality (CM_{org}) * No of Outgoing Flows
For example, for actor Volunteer,
Organizational Level Criticality = 30 and No of Outgoing Flows = 6
So, Criticality, CM_{bp}= 30*6 = 180

4 Integrating Risks in Business Process Models

4.1 Treating Vulnerable Actors

We believe that every actor in the business processes should be given a relative level of effort to mitigate vulnerability via robustness and efficiency. We suggest for the more vulnerable actors more monitoring of the tasks/ sub-tasks is necessary. Monitoring of the business process means tracking the individual tasks or subtask in a process so that information on their state can be easily made visible. It is done to measure the satisfactory performance of a business process. Business process tasks of the vulnerable actors need more monitoring so that we can continually refine them based on feedback that comes directly from operational level.

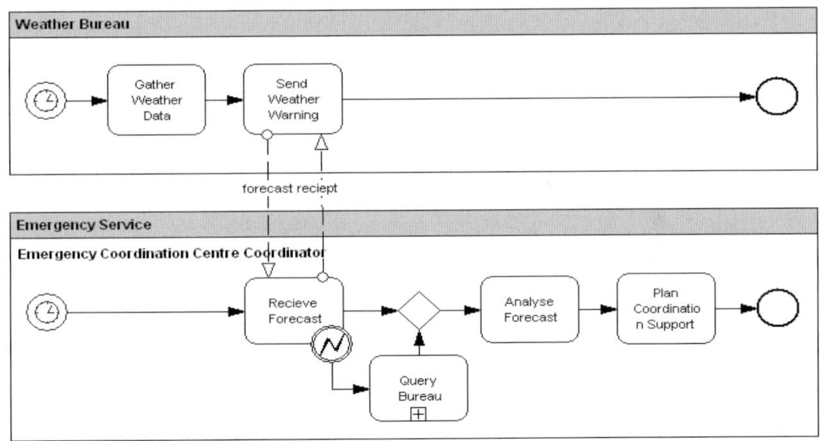

Fig. 3. Business Process Model in BPMN

This process model has two actors WeatherBureau and ECCC with few tasks and subtask. The model also represents exception handling procedure for RecieveForecast task. From table-1, we find that ECCC is the most vulnerable actor which implies more monitoring of the tasks and subtask inside this process is required.

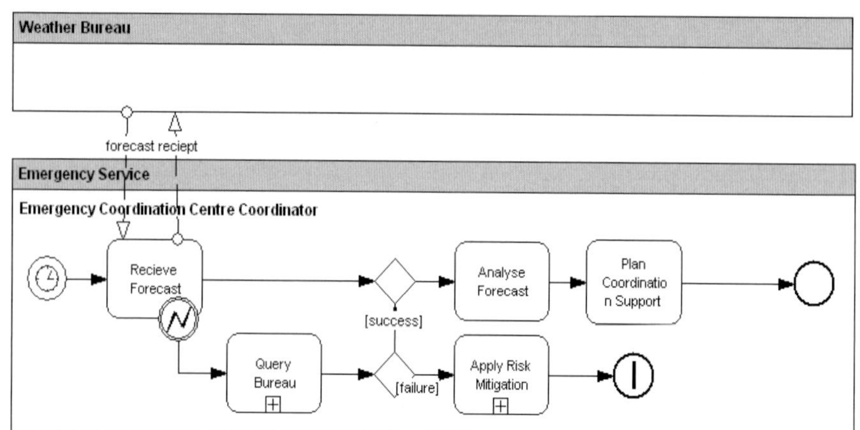

Fig. 4. Extended Process Model Reflecting the Vulnerable Actor

The process model in figure 3 is improved in figure 4 by using our notion of vulner-ability. The exception for RecieveForecast task is handled by QueryBureau sub-process. We extend this model by integrating ApplyRiskMitigation sub-process. This sub-process includes the risk mitigation procedures which takes place in case of the failure of QueryBureau sub-process.

The analyst should design the organizational model or process model carefully while delegating the dependencies from one actor to other actors. Actor with dependencies over only one actor is more vulnerable than actor with dependencies with multiple actors. For example in figure 5, the vulnerability level of actor A1 is 4 and actor B1 is 1. Actor A1 has four dependencies over A2. If actor A2 fails then all the dependencies will remain unsatisfied. On the other hand actor B1 has delegated its dependencies over four actors. If any of the four dependee actor fails one dependency will remain unsatisfied, but the others might be satisfied. Thus actor A1 is more vulnerable than actor B1.

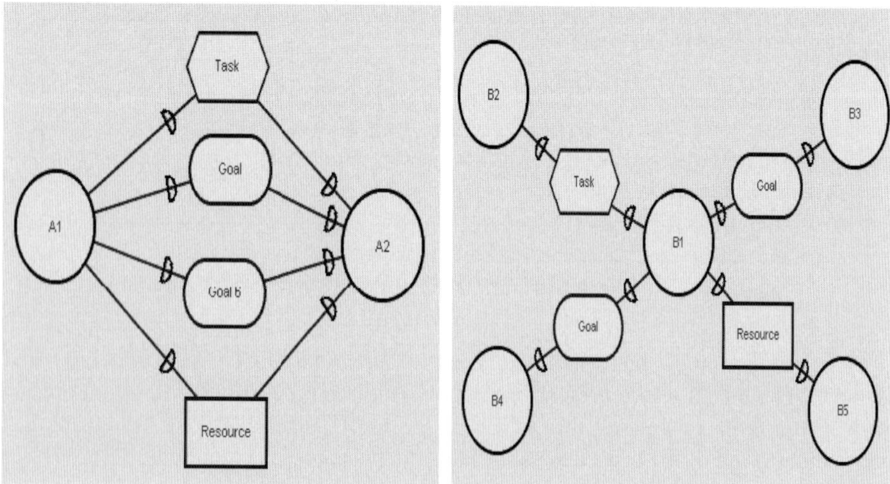

Fig. 5. Delegation of Dependencies among Actors

4.2 Treating Critical Actors

Volunteer actor is the most critical actor according to the matrix. In this case the three actors ECCC, FieldControlCentreCoordinator(FCCC) and Community are dependant on Volunteer actor to accomplish their certain objectives. Failure to satisfy these objectives/incoming dependencies will have a big impact on the performance of the depender actors and to the system as a whole. To minimize the criticality levels of actors, the analyst needs to have pragmatic and profound process delegation strategy.

The tasks and sub-processes of the most critical actors should be robustly planned to make the whole process successful. To make the process robust the analysts need to identify what is the overall objective of the process. This should describe problems to be solved, issues to be addressed, key participants, whether all the tasks are well integrated within the process and how the processes add values and quality to the system.

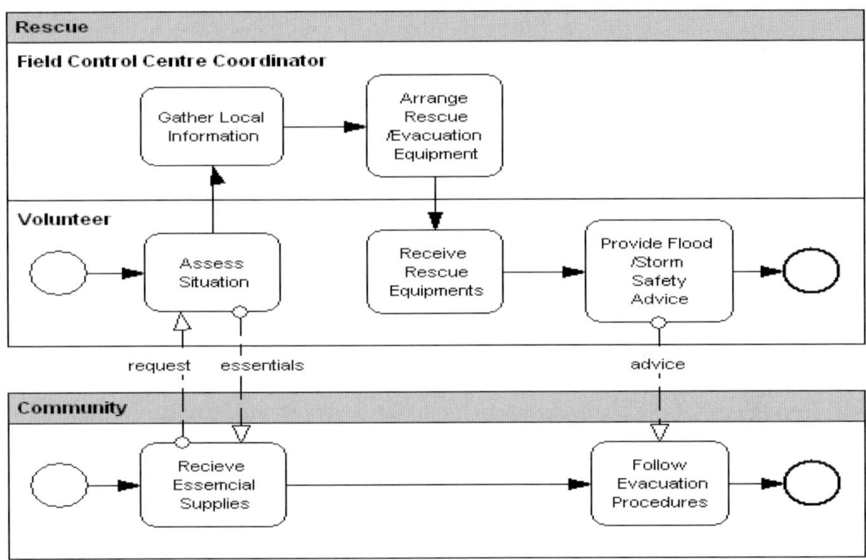

Fig. 6. Business Process Model in BPMN

The objective of the process in figure 6 is to provide a Flood/Storm Safety advice to the Community. Volunteer provides the safety advice to the Community. For the well completion of the process Volunteer needs to have local information and rescue equipments which are done by FieldControlCenterCoordinator by accomplishing two tasks GatherLocalInformation and ArrangeRescue/EvacuationEquipment. Upon successful completion of the task ReceiveRescueEquipments the Community receives the message ProvideFlood/StormSafetyAdvice from Volunteer in the FollowEvacuation-Procedures tasks which add values to the process of evacuation.

The below process model is extended from figure 6 by introducing an exception handling technique in Volunteer's RecieveRescueEquipments task to manage its satisfactory performance. If the Volunteer does not receive the rescue equipments from FCCC the process will throw an exception which sends query to FCCC. To handle the

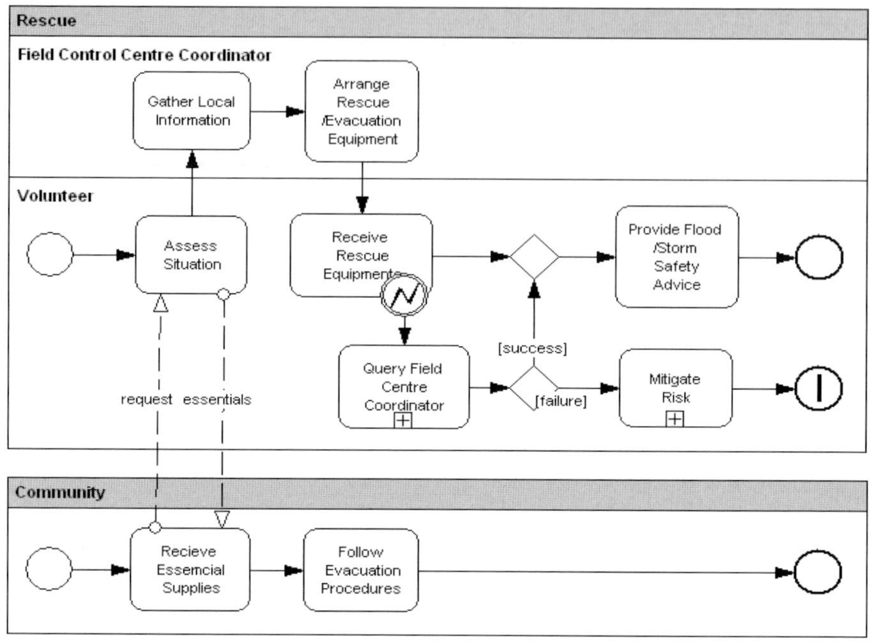

Fig. 7. Extended Process Model Reflecting the Critical Actor

risks from negative response from the FCCC a MitigateRisk sub-process is introduced. Exception handling should be taken into thoughtful consideration by the analyst as exceptions may arise in any stages of the process.

The processes of the critical actors should have mutual consistency to reduce criticality and increases process performance. If a process is allocated to an actor, which the actor may not be capable of performing, it is likely to delay the process which could lead to a disaster. Clearly specified activities for the actors should be one of the most important priorities to the analyst. It makes easy to comprehend and allocate resourceful process design to ease the modification of processes.

5 Conclusions

In this work we have presented a discussion on how we can identify risk in terms of vulnerability and criticality in organizational models. We have also provided a way to integrate risks within the process model. We believe it helps the analyst while to design organizational models, delegate dependencies among various actors, choose alternatives, decompose tasks, maintain consistency among organizational and process models, handle exceptions etc. However, we have considered the concept of vulnerability and criticality of actors only. We have not considered our notions on the activities and sub-process for assessing risks. Our future work will deal with the combination of actors and their tasks and sub-processes. Our proposal is based on a combined notation (*i**-BPMN) which might not be suitable for organizations using different notations. However, we wish apply our proposal to different notations in the future.

References

1. Becker, J., Indulska, M., Rosemann, M., Green, P.: Do Process Modelling Techniques Get Better? A Comparative Ontological Analysis of BPMN. In: Campbell, B., Underwood, J., Bunker, D. (eds.) Proceedings 16th Australasian Conference on Information Systems, Sydney, Australia (2005)
2. Chung, L.: Representing and Using Non - Functional Requirements for Information System Development. A Process-Oriented Approach. PhD Thesis, Graduate De-partment of Computer Science, Toronto, University of Toronto (1993)
3. Fischer, L.: Workflow Handbook, Workflow Management Coalition, WfMC (2005)
4. Hall, C., Harmon, P.: The 2004 Enterprise Architecture, Process Modelling & Simulation Tools Report, Technical Report (2005), `http://bptends.com`
5. Hammer, M., Champy, J.: Reengineering the Corporation: A Manifesto for Business Revolution, Harper Business (1993)
6. Katzenstein, G., Lerch, F.J.: Beneath the surface of organizational processes: a social representation framework for business process redesign. ACM Transactions on Information Systems (TOIS) 18(4), 383–422 (2000)
7. Koliadis, G., Vranesevic, A., Bhuiyan, M., Krishna, A., Ghose, G.K.: A combined approach for sup-porting the business process model lifecycle. In: Proceedings of the Asia-Pacific Conference on Information System (2006)
8. Loucopoulos, P., Kavakli, E.: Enterprise Modelling and the Teleological Approach to Requirements Engineering. International Journal of Intelligent and Coop-erative Information Systems 4(1), 45–79 (1995)
9. Miers, D.: The Split Personality of BPM, Business Process Trends (2004), `http://bptrends.com`
10. Ouyang, C., van der Aalst, W.M.P., Dumas, M., ter Hofstede, A.H.M.: Translating BPMN to BPEL, BPM Center Report BPM-06-02 (2006), `http://BPMcenter.org`
11. Smith, H., Fingar, P.: Business Process Management – The Third Wave. Meghan-Kiffer Press, Tampa (2003)
12. White, S.: Business Process Modelling Notation (BPMN), Version 1.0, Business Process Management Initiative (2004), `http://BPMI.org`
13. Yu, E.: Modelling Strategic Relationships for Process Reengineering. PhD Thesis, Graduate Department of Computer Science, University of Toronto, Toronto, Canada, p. 124 (1995)
14. Muehlen zur, M., Ho, D.T.-Y.: Risk management in the BPM lifecycle. In: Bussler, C.J., Haller, A. (eds.) BPM 2005. LNCS, vol. 3812, pp. 454–466. Springer, Heidelberg (2006)
15. Muehlen zur, M., Rosemann, M.: Integrating Risks in Business Process Models. In: Proceedings of Australasian Conference on Information Systems (ACIS 2005), Manly, Sydney, Australia (2005b).
16. Silveira, C.: A Knowledge-Based Risk Management for the Utility Business Service Model, Informing science and Information Technology Education Joint Conference, Pori, Finland (2003)
17. Nogueira, J.C., Luqi, B.S.: A Risk Assessment Model for Software Prototyping Projects. In: Proc International Workshop on Rapid System Prototyping, pp. 28–33 (2000)
18. Schmitt, M., Grégoire, B., Dubois, E.: A risk based guide to business process design in inter-organizational business collaboration. In: International Workshop on Requirements Engineering for Business Need and IT Alignment (REBNITA 2005), Paris (2005)
19. Sumner, M.: Risk Factor in Enterprise-wide/ERP projects. Journal of Information Technology 15, 317–327 (2000)

Revisiting the Behavior of Fault and Compensation Handlers in WS-BPEL

Rania Khalaf[1], Dieter Roller[2], and Frank Leymann[2]

[1] IBM TJ Watson Research Center, 1 Rogers St, Cambridge MA 02142, USA
rhkalaf@us.ibm.com

[2] Institute of Architecture of Application Systems, University of Stuttgart, Germany
Universitätsstraße 38, 70569 Stuttgart, Germany
{Dieter.H.Roller,Frank.Leymann}@iaas.uni-stuttgart.de

Abstract. When automating work, it is often desirable to compensate completed work by undoing the work done by one or more activities. In the context of workflow, where compensation actions are defined on nested 'scopes' that group activities, this requires a model of nested compensation–based transactions. The model must enable the automatic determination of compensation order by considering not only the nesting of scopes but also the control dependencies between them. The current standard for Web services workflows, Business Process Execution Language for Web Services (WS-BPEL), has such compensation capabilities. In this paper, we show that the current mechanism in WS-BPEL shows compensation processing anomalies, such as neglecting control link dependencies between nested non-isolated scopes. We then propose an alternate approach that through elimination of default handlers as well as the complete elimination of termination handlers not only removes those anomalies but also relaxes current WS-BPEL restrictions on control links. The result is a new and deterministic model for handling default compensation for scopes in structures where: (1)both fault handling and compensation handling are present and (2)the relationships between scopes include both structured nesting and graph–based links.

Keywords: WS-BPEL, Compensation, Transactions, Error handling, Workflow.

1 Introduction

WS-BPEL [18] is the business process modeling language for Web services. The language is rich in functionality, providing capabilities for fault and compensation handling. Fault handling provides a process fragment that runs if a fault is raised by any of a set of running activities. Compensation handing provides a process fragment that runs to reverse the cumulative work of a set of successfully completed activities. Compensation is usually run due to a fault in another part of the process. A set of WS-BPEL activities is defined by grouping them in a 'scope'. Scopes in WS-BPEL are also used to provide additional capabilities such as scoping of variables, event handlers that can run multiple times as long as the scope is active, and termination handlers that are run if the scope is running when its parent scope is trying to exit. For fault, compensation, and termination handlers, WS-BPEL provides default behavior for each that is attached

R. Meersman, T. Dillon, P. Herrero (Eds.): OTM 2009, Part I, LNCS 5870, pp. 286–303, 2009.

on every single scope that does not explicitly provide one of them. This behavior is complicated by WS-BPEL's mixed control model which allows control dependencies to be defined using both/either structured activities and explicit control links. Structured activities like 'sequence' or 'flow' impose control logic on activities nested within them, while a control link from a source activity to a target activity specifies that the target activity must not start until the source activity has completed. WS-BPEL allows control links to cross the boundaries of structured activities, for example going from an activity inside one 'flow' to another activity nested inside another 'flow'. In this paper, we argue that WS-BPEL's default handling behavior leads to high complexity and a violation of control link reversal expected in workflow rollback. We show how using only explicit handlers simplifies the model making it more usable, leads to less surprise behavior, makes calculating compensation order more efficient, and allows one to relax certain restrictions that WS-BPEL places on allowed constructs.

The over-loaded functionality of scopes leads to complexity in handling compensation. For example, a scope added just to scope a variable will by default affect the fault and compensation handling behavior of the process. A scope will always run its default compensation handler if: (1)a fault is thrown inside it and a matching fault handler is on one of its ancestor scopes, or (2)if it needs to terminate its nested scopes due to a fault thrown not within it but in one of its ancestor scopes. Several scopes can be compensated by a single handler by running their compensation handlers in some order. This order may be either an explicit order defined by the process designer or the 'default order' calculated according to the WS-BPEL standard. The latter is the prescribed order for a scope's default compensation handler. The default order is conceptually aimed at occurring in the reverse order of scope completion. However, calculating the reverse order in the presence of both scope nesting and links that cross scope boundaries is not straightforward. In WS-BPEL 2.0, the manner in which this order is calculated at runtime is made complex due to the interleaving of compensation, termination and fault handling as a fault is being propagated up the scope hierarchy to a matching fault handler. While the standard provides enough information for the resulting compensation order to be calculated at runtime, we doubt a process designer can take all the possibilities into account whenever adding a scope to a process. Examples of matters that complicate this order include that (1)a fault always propagates one level of scope nesting at a time and is re-thrown by the default fault handler which first terminates and compensates immediately nested scopes and (2)one must handle child scopes that were still running when the fault was caught before compensating completed ones.

A key goal of this work is to provide a more deterministic, simplified model for determining the default compensation order without sacrificing key functionality for workflow languages such as WS-BPEL that support a mix of graph and structured control constructs. We aim to reduce the large number of possible orderings created by the current behavior, which is highly sensitive to runtime state. Our approach is based on the concept of considering only the view of the world from the scope where the fault can be handled. This entails removing default fault and compensation handlers from WS-BPEL, as well as default and explicit termination handlers. Note that we propose removing 'default compensation handlers' which is not to be confused with 'default compensation order'. We keep the behavior that an explicit fault handler first terminates all running activities inside its scope. Default compensation *order*, however, is kept only using the <compensate/> activity. In the absence of these default handlers, we propose a new compensation and fault handling scheme that is closer to what the designer expects and is still rooted in the same abstractions as WS-BPEL's.

The scenarios used in this paper are based on those brought forward by the WS-BPEL Technical Committee in OASIS (WS-BPEL TC) [21] as drivers for the current compensation order behavior in WS-BPEL. We compare the current behavior with the behavior that would result using our approach. In fact, the current WS-BPEL behavior is too restrictive because it does not allow a process to have 'peer-scope cycles', which are themselves included in some of the committee's own motivating scenarios. A 'peer-scope-cycle' is a cycle created by links between two scopes nested in the same parent scope. We will show how this restriction can be eased in our approach so that such scenarios can be constructed by workflow designers.

The rest of this paper is structured as follows: related work is followed by a presentation and illustration of WS-BPEL's current compensation behavior and the problems it presents. Then, an alternative approach is presented leading to the conclusion and future work.

2 Related Work on Recovery in Workflow

WS-BPEL combines both structured nested activities with explicit control dependencies in the form of links, combining the block structured (calculus based) and graph-based approaches of earlier workflow languages [13]. This has presented new challenges in adapting known compensation models to suit this change.

The concept of compensation scopes (known at the time as 'compensation spheres') was introduced for graph–based flow systems such as the model defined in [14,15]. A compensation sphere is a view on the process, grouping together a set of constructs with corresponding properties. Default compensation order was determined by reversing the control edges between the process's activities. However unlike WS-BPEL, such graph-based flow models did not include scope-level fault handlers and the scopes were not first class constructs with a place in the workflow's navigation. That is, spheres could not be the sources or targets of control links. Du et. al [7] presented a mechanism for automatically determining compensation scope boundaries at runtime based on dependency analysis, state, and user hints to provide more flexibility beyond requiring statically defining a scope at design time. In contrast, WS-BPEL's scopes are always defined at design time because they have a place in the navigation.

The need to rollback long running transactions, for which ACID is impractical, led to the creation of Sagas [9]. A Saga is made of several tasks, each being an ACID transaction and having an optional compensator which defines undo actions for that task. In the case the Saga needs to be aborted, then already running tasks are aborted and those that had completed have their compensators run in reverse order of completion. Nested Sagas [10] extended the model, so a task in a Saga could be either an ACID transaction or another Saga. Considering the scopes in a WS-BPEL process as representing a nested Saga, one can see a parallel between Sagas and WS-BPEL's default compensation and termination behavior today. However, the tasks in a Saga do not have the complex relationships between them beyond nesting that WS-BPEL scopes do and neither do they combine fault handling as a different and complimentary capability to rollback.

It was shown in [1] that while the model presented by Sagas is too limited for a workflow environment, most advanced transaction models could be modeled directly into a workflow using patterns of existing workflow constructs. This was used in the Exotica project to provide a pre-processor that took a high level specification of a transaction model and generated the corresponding workflow artifacts to implement it. YAWL [22] is one example of a flow language which does not support compensation directly but in which one could model compensation behavior using patterns of YAWL constructs. [2] defines such a mapping from WS-BPELs compensation model to a YAWL 'compensation pattern'. They cover only explicit compensation, calling default WS-BPEL compensation 'troublesome'.

We conclude from the above that the combination of nested scopes as first class constructs in a workflow with the merging of calculus and graph based processes required additional work beyond existing rollback models. The result was today's WS-BPEL default compensation and termination behavior that combines the concepts from nested Sagas with those from compensation spheres in [15].

There is a large body of work on the use of transaction and coordination in the context of workflows and long running activities. We focus on the part most relevant to WS-BPEL itself and that would provide the context as to what the new problems are that it introduced and how this leads to the conclusion just presented. We do refer the interested reader to [6] for a more in-depth look at these models.

New research on the WS-BPEL compensation model itself has focused on requesting new features and on the explicit compensation case as opposed to default order and hence would not be adversely affected by our approach. In [3], the author describes WS-BPEL's compensation mechanism and requests a richer capability enabling a compensation handler access to current process state as well as the ability to call a compensation handler from outside of a fault handler. The former has already been done in WS-BPEL 2.0, while the latter is still not possible. In [11], the authors provide a look at workflow and transaction models, present a classification of the types of cancellation and recovery that are common in workflows and a statement of which of these can be supported by WS-BPEL directly. Some of the critiques of WS-BPEL's compensation capabilities have been addressed since then in WS-BPEL 2.0. They present an argument for having more flexibility in defining compensation and present a plan for creating a consistency language and tools to help with richer rollback capabilities. It would be interesting to see whether these would layer on top of the existing WS-BPEL compensation capabilities or perhaps make use of the extraction of compensation to an external coordinator done in [19].

Transaction models that go beyond compensation are overlaid on WS-BPEL processes to govern sets of interactions with external partners in [8] and [20]. Where [8] directly added transaction management capabilities using WS-BPEL language extensions, Tai et. al [20] addressed the problem by using policies, on existing WS-BPEL constructs, that are related to WS-BusinessActivity [17] and WS-AtomicTransaction [16]. A more detailed look into the relationship between WS-BPEL compensation and WS-BusinessActivity is presented in [19], where the main focus is to externalize the coordination taking place between a scope and its child scopes by using an extended version of the WS-BusinessActivity protocol.

We had examined the intricacies of WS-BPEL's fault handling due to its combination of structured and graph–based flow models in [4]. We now do the same for compensation handling while also proposing an approach (based on section 3.6 of [12]) to simplify the default compensation order.

3 Current WS-BPEL 2.0 Compensation Processing

In this section, we describe how compensation handling occurs in the WS-BPEL standard. Compensation processing in WS-BPEL is based on the notion of scopes. A WS-BPEL scope is an activity that itself contains a group of activities, which themselves may also be scopes. Thus, scopes may be nested. The scope assigns certain characteristics to the activities nested within it. The entire process, by default, is a scope.

WS-BPEL provides mechanisms for trying to recover from faulty situations. Central to these mechanisms are 'fault handlers' that are attached to a scope and can catch and deal with faults in that scope. There are several ways in which a fault may be thrown in a WS-BPEL process, e.g.: A Web service called from a process instance may respond with a fault message or a process itself might detect erroneous situations that result in internal faults. When a fault occurs, the regular processing in the scope in which the fault occurred is interrupted and the fault is passed to a fault handler of that scope. The fault handler aims to correct the situation such that regular processing can continue outside the scope or alternate ways to complete the process can be taken. All of this might require undoing actions that have already been completed within the scope. Actions required to undo already completed activities are defined via 'compensation handlers'. Thus, a fault handler of a scope may start the compensation handlers of its nested completed scopes to undo their actions. It does so via a <compensate> activity. The compensate activity may optionally name a particular scope. If it names a scope, only the compensation handler associated with the specified scope is called; otherwise, it performs compensation on all enclosed completed scopes according to the default compensation order. A 'termination handler' aborts a running scope. A handler (fault, compensation, or termination)of a scope may contain any type of activity, including the empty activity. If any of the compensation or termination handlers is not specified for a scope, a corresponding *default handler* is provided. For fault handlers, a default fault handler is always provided on every scope to catch all faults except those for which an explicit fault handler is defined.

Figure 1 helps illustrate the details of the WS-BPEL fault handling and compensation mechanism. All the figures in this paper use the visual cues shown in figure 1's key. As shown, each scope is associated with three handlers: a fault handler, a compensation handler, and a termination handler. When navigation through the process enters a scope, the fault handler and the termination handler are installed. When the scope has completed processing, the termination handler is de-installed and the compensation handler is installed.

Figure 1 shows the following situation: the outer scope G is running with its fault and termination handlers active. Enclosed are scope F, which is still running with the

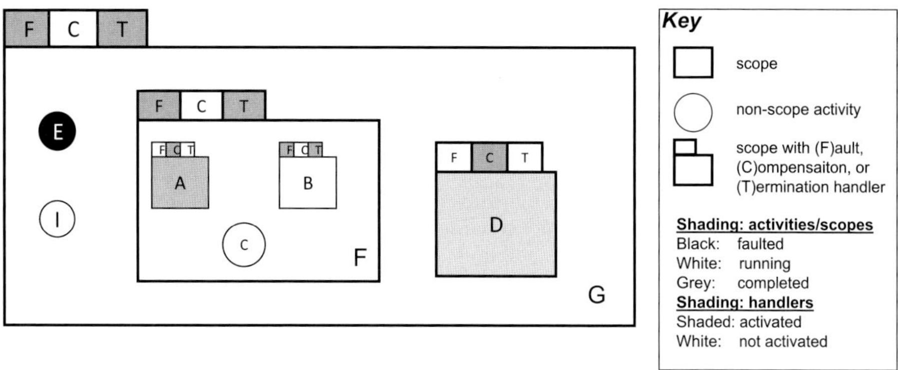

Fig. 1. Scopes With Associated Handlers

fault handler and the termination handler active, scope D which has completed, so its compensation handler is active, and the non-scope activity E, which has just faulted. Scope F has enclosed two scopes: scope A which has completed, scope B which is running. Scope F also encloses non-scope activity C which is running.

Fault handling is always initiated through an error raised in the flow. Each fault is associated with some fault code. When a fault occurs, all running activities within the scope in which the fault occurs are terminated. If the activity is not a scope, it is terminated according to the WS-BPEL specifications (wait for it to complete or simply abort it). If the activity is a scope, the associated fault handler is de-activated and the termination handler is carried out. This causes the following, shown in figure 1 : when activity E in figure 1 faults, first the running activity I is aborted. Next, the fault handler of scope F is de-installed and its termination handler is invoked.

The termination handler performs the following on its nested activities, in this order:

1. Enclosed non-scope activities: stop them if they are running. Otherwise, do nothing. This is done in arbitrary order.
2. Enclosed running scopes: deactivate their fault handlers and invoke their termination handlers. This is done in arbitrary order.
3. Enclosed completed scopes: invoke their compensation handlers. This is done according to the default compensation order. This is equivalent to performing a single <compensate> activity in the termination handler.

Thus, when the termination handler associated with scope F is invoked: First it terminates the non-scope activity C. Next, it processes the running scope B by de-activating the fault handler and invoking the termination handler. Last, it invokes the compensation handler of scope A.

Following termination, the matching fault handler is invoked. The default fault handler is used if an explicit matching one is not found on the current scope. In the case of figure 1, this means invoking the fault handler of G which in this case is a default fault handler. In WS-BPEL, a *default fault handler* runs a compensate activity and then re-throws the fault up to its parent (i.e. enclosing) scope. Thus, the fault handler of scope G

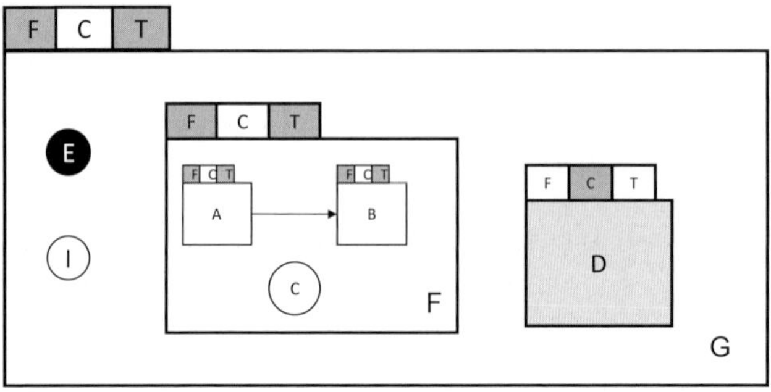

Fig. 2. Arbitrary Termination Sequence

therefore first invokes the compensation handlers of its nested completed scopes: here, scope D.

A scope's compensation handler usually contains the set of activities that undo the effects of the completed scope. For example, a scope that performs a payment commonly has a compensation handler that causes a refund. The *default compensation handler* runs a compensate activity, which causes the compensation handlers of all completed enclosed scopes to be invoked in default compensation order.

It is important to highlight that the described processing of fault, compensation, and termination handling causes first the running scopes (and thus all their enclosed scopes) to be processed before processing the already completed scopes. The processing of *running* scopes is in arbitrary order.

If, for example, as in Figure 2, two peer scopes are still running, then WS-BPEL does not prescribe any sequence in which the two scopes are to be terminated, even if they are connected via a control link. Thus, the WS-BPEL engine may first invoke the termination handler of scope A before invoking the termination handler of scope B. This may introduce some non-determinism; however, we can safely ignore this because termination in itself is somewhat non-deterministic (dependent on the state of the activity it may or may not be terminated).

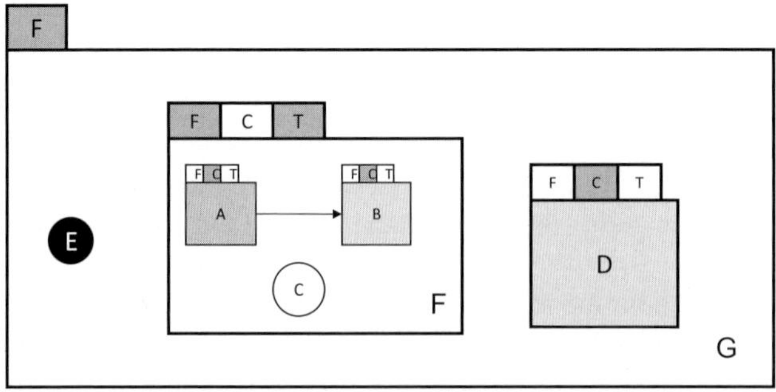

Fig. 3. Honoring Control Links during Compensation

Figure 3 shows the situation where the two peer scopes A and B have already completed. In this case, compensation honors the control link dependency between scopes A and B by first invoking the compensation handler of scope B before that of scope A.

The WS-BPEL specification mandates that the control links are honored when compensating peer scopes that have completed and that have a control link between them. This is the backbone of the concept of default compensation order. However, it violates this for non-peer scopes.

4 Problems with the Current Compensation Mechanism

The fault and compensation mechanisms suffer two main problems:

1. In the presence of control links that cross scope boundaries, the compensation order can violate the control link dependencies. An illustration of this problem is provided in section 4.1, showing that even for a simple case the default handlers of a scope may have devastating effects on the business logic and that WS-BPEL does not providing guarantees on compensation order between non-peer scopes.
2. A *zigzag* behavior of compensation is introduced by the default handling mechanism when scopes on different levels are being compensated, because compensation, termination and fault handling act on only one level of scope nesting at a time. An illustration of this behavior is provided in section 4.2.

The complexity of compensation and in particular the zigzag behavior is a problem because it is hard for process modelers, testers, and business analysts to comprehend the behavior a process will have due to compensation: they would have to always keep in mind all the current states in all the different scopes and their dependencies. The problem manifests itself in the modeling and particularly testing of workflows: one must really dive deeply into the process to understand the subtleties. One aim of our approach is to simplify the design and analysis of processes for our target audience of process modelers, testers, and business analysts. We do so by only performing compensation when explicitly requested by the designer and providing one possible default compensation order for each scope in the process. As a result, one can easily view and reason about that order to determine how best to design, refactor, or analyze one's processes.

4.1 Illustrating the Anomalies

We illustrate the example using the scenario in figure 4. This example is similar to the one in [5] and is an extension to figure 1. A control link has been added from scope B to scope D. It should be noted that this is only possible if the scope F is a non-isolated scope, otherwise the link would be blocked until scope F has completed. Furthermore all fault and termination handlers are default handlers. This case exhibits a compensation order which violates reversal of the control prescribed by the control link.

First, we provide a high-level view of part of this sample to show a concrete case of the problem and then perform a step by step walk through for the entire sample. Assume that the sample is for a service to special order an item and uses compensation for an error that requires a recall. Assume scope B processes the payment and D ships

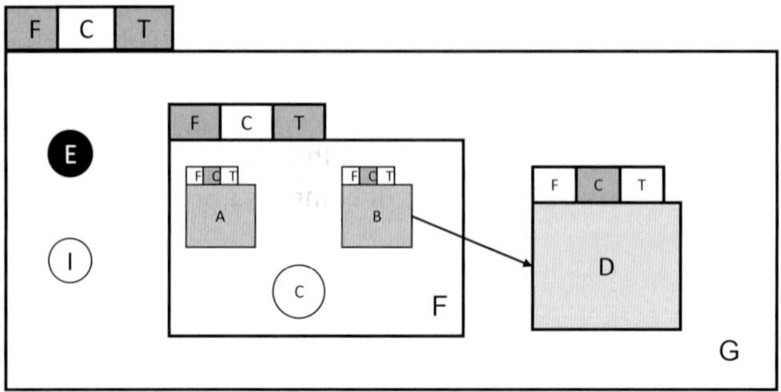

Fig. 4. Wrong Compensation Sequence

the product and waits for the customer to acknowledge receipt. Both B and D have explicit compensation handlers defined. The compensation handler of B refunds the customer. The compensation handler of D requests that the customer send the item back and wait until the merchant receives and checks the returned item. Assume that scope F is added to supply a shared, local variable ‘creditCard’ between customer-specific payment activities A, B, and C that shadows another process-wide ‘creditCard’ variable that stores corporate billing information. We stated that compensation is expected to occur so as to reverse control dependencies and thus the process would be expected to compensate D then B: first return the item and have it checked by the merchant then refund the customer. The next paragraph will show that for some (and not all) executions of even this simple process, WS-BPEL will compensate B before D which would cause the customer to be refunded before the item is returned by the customer and checked by the merchant. Some may argue that perhaps a solution is to redesign the process. However, not only would the reasons and options be confusing for the designer but even this simple example with a handful of scopes and two levels of nesting illustrates key problems: default handlers of a scope may have devastating effects on the business logic and there are no guarantees by the engine on compensation order between non-peer scopes such as B and D.

Now we consider the step by step runtime behavior of the sample in figure 4 when activity E faults. At that time, termination processing is initiated as follows. First the non-scope activity I is terminated. Next the fault handler of scope F is deactivated and the termination handler is invoked. The termination handler first terminates the running non-scope activity C, then it invokes the compensation handlers of scope A and B in default compensation order. Since A and B are not joined by a control dependency, then the default order is simply any order. When the two compensation handlers have finished, control goes to the fault handler of scope G. The fault handler, since it is the default fault handler, invokes a compensate activity which causes the compensation handler of scope D to be invoked.

This sequence of events shows that WS-BPEL carries out compensation (A then B or B then A, followed by D) violating the basic concept of reverse control order as determined by the control links between completed scopes.

4.2 Complexity of WS-BPEL2.0 Compensation Behavior

We have claimed that the current default handler behavior causes high complexity in the default compensation order making it difficult for a designer to anticipate the resulting behaviors when making process design decisions. In this section, we back this claim.

The following example shows the complexity involved due to the existence of default handlers. Notice in particular the 'zigzag' behavior as scopes from different levels in the hierarchy get compensated.

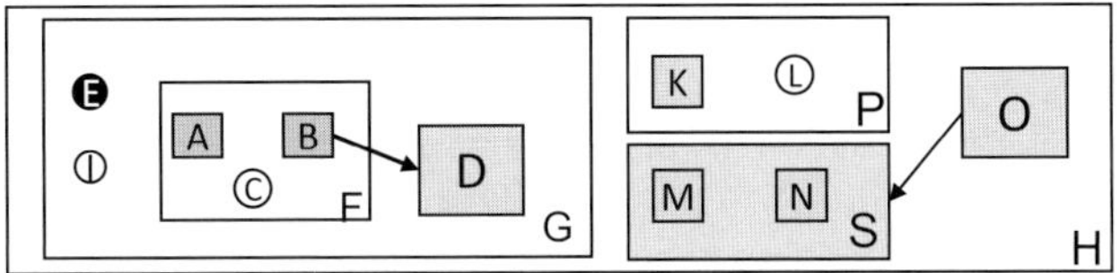

Fig. 5. Example of Compensation Complexity. Assume all scope have default FCT handlers.

In the example in Figure 5, we nest scope G from figure 4 within scope H along with three other scopes P, S, and O, which themselves contain scopes K, M, and N. The same behavior as described in the previous section above happens first, ending with the compensation of D. This is shown as steps 1 through 5 in figure 6 where the fault, compensation and termination of scopes is shown.

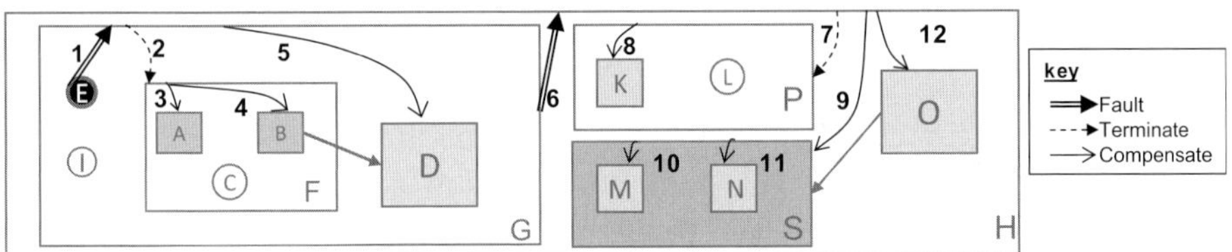

Fig. 6. Runtime FCT handling of scopes

To illustrate the behavior of what happens next, the rest of the description will refer in parenthesis to the step numbers on the arrows in figure 6: (6)the default fault handler of scope G rethrows the fault up to scope H. Recall that upon receiving a fault, a scope without a matching fault handler first terminates its running children and then runs its default fault handler. The default fault handler compensates completed immediate child scopes and finally rethrows the fault to its parent scope.

This is illustrated for scope H: it first terminates its running child P (7). P's default termination handler terminates L and runs the compensation handler of K (8).

Termination of P completes at this point. The default fault handler of scope H then compensates the immediate children of scope H in default order. This means reversing control order between peer scopes, i.e. S then O: First, the default compensation handler of scope S is invoked (9). This invokes the compensation handlers of scopes M (10) and N (11). Finally the compensation handler of scope O is called (12). When this handler

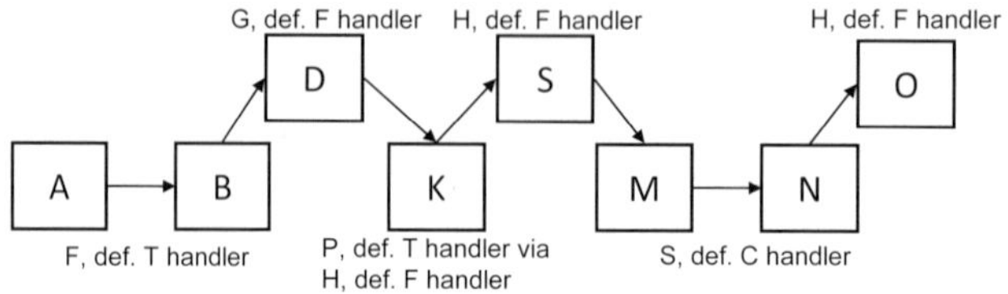

Fig. 7. WS-BPEL 2.0 Compensation order and triggers for scope H

completes, the default fault handler of scope H rethrows the fault to its parent scope (not shown).

As can be seen, compensation happens by first going down each of the scopes, compensating the innermost scopes first. Figure 7 illustrates the resulting order in which the compensation handlers of the shown scopes are called along with the respective handler that triggered the compensation. The zigzag behavior of the control leading to compensation is illustrated in Figures 7 and 6: from G to F, down to A and B, up to G, down to D, up to H, down to P then to K, back up to H down to S, down to M and N, and finally back up to O. In our approach, we avoid the zigzag behavior by immediately going to the scope with the matching explicit fault handler and performing the fault handling and optionally compensation as well on all nested scopes directly from that one matching scope, avoiding all these levels of indirection. Even though we remove default handlers, process designers will still be able to create more advanced compensation and fault handling patterns through the use of explicit handlers.

5 An Approach That Modifies Handler Behavior

We propose reducing the complexity of compensation processing and removing the anomalies presented in the previous sections by (1)Eliminating default fault handlers while specifying that if a fault reaches the top most scope (i.e. the process) without being caught then the process is terminated without any compensation, (2)Eliminating default compensation handlers as well as both default and explicit termination handlers, and (3)Modifying the reachability of compensation handlers such that a compensation handler of a scope may be called from any ancestor scope instead of only from the immediate parent scope.

The corresponding runtime behavior for our approach is as follows:

1. If a fault occurs in a scope that has no explicit fault handler for the particular fault, the fault is automatically propagated to outer scopes until an explicit matching fault handler is found. If no fault handler is found, the process is terminated. Otherwise, all running activities within the fault handler's scope are terminated (with no special termination handling).
2. Then, the fault handler is run.
3. If the fault handler calls default compensation via the <compensate> activity, then nested scopes are compensated according to the default compensation order. This order is dictated by the scope's 'Compensation Order Graph' (COG), constructed according to the detailed algorithm in section 5.1.

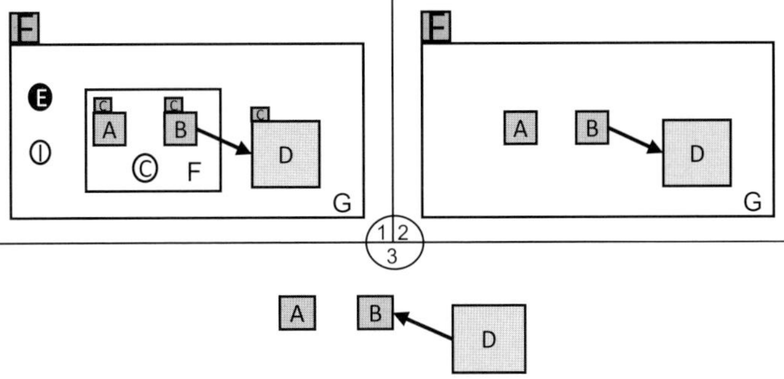

Fig. 8. Compensation of G: (1)scope G, (2)the view of relevant scopes, (3)the default compensation order in a COG

Note that the approach does not change the WS-BPEL behavior in which a fault not explicitly caught at any level in the process results in terminating the process without any requirement of notification. While this is reminiscent of silent termination, it is expected that such cases are reported to the process owner or administrator by implementation specific means and are thus left out of the standard (and therefore the modifications presented in this paper). WS-BPEL by design does not enforce management or administration behavior, leaving those up to WS-BPEL engine vendors.

An important part of the COG's construction is its treatment of scopes according to the presence of explicit compensation handlers: if a scope has no explicit compensation handler, the scope does not exist from a compensation perspective, it is *transparent*; if a scope has an explicit compensation handler, then the inner structure is not known to the compensation mechanism, that means the scope is *opaque*. Note that this meaning of opaque is unrelated to the use of opaque in Abstract WS-BPEL Processes.

In addition to respecting linking behavior, COGs are calculated at design time and always dictate the order for default compensation, thereby allowing a designer to study the (small set of) possible execution sequences of the process. In contrast, the behavior in WS-BPEL today results in sequences that will differ at runtime based not only on scope nesting and linking but also on scope completion times and termination handler behavior as shown in section 4. In WS-BPEL today, one can still create graphs at design time to encode compensation order (section 5.5.3 in [12]); however, the default order is not always run according to these graphs because (1)they represent one level of scope nesting at a time and (2)WS-BPEL dictates that active immediate child scopes are terminated (and their children possibly compensated) before compensation of completed immediate child scopes.

In order to illustrate our approach, consider that the fault handler in figure 4 has been defined explicitly and just carries out compensation through running a compensate activity. Since our approach centers on dropping default handlers, explicit compensation handlers on relevant scopes as shown in figure 8(1). Then the compensation processing would see the situation as shown in figure 8(1). Since F is transparent, it is obvious that the control links are honored for compensation.

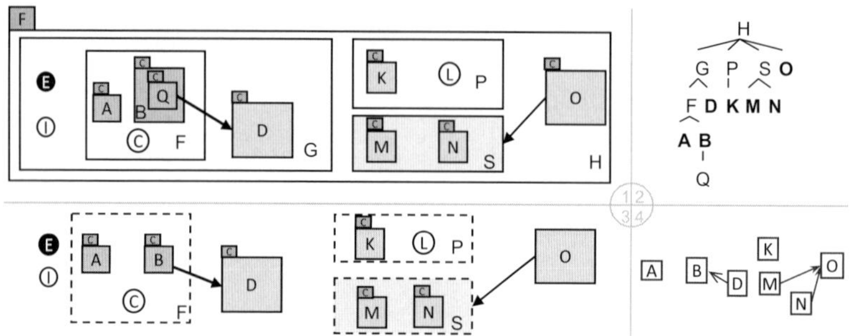

Fig. 9. COG of H: (1)scope H, (2)the scope-and-loop hierarchy subtree rooted at H, (3)determining COG edges, (4)the COG of H

Figure 9(1) shows a sample scope H and the steps, described in the next section, for determining its compensation order and encoding it in a Compensation Order Graph (figure 9(4)). Scope H is similar to the scope in figure 7 but has explicit compensation handlers, no default handlers, and changes the source of the link from scope B to D to be from scope Q in B. An interesting point is the necessity in the COG to draw control links between scopes M and N and scope O as will be described in section 5.1.

5.1 Building the Compensation Order Graph

The Compensation Order Graph (COG) of a scope is used to calculate the default compensation order for that scope; that is, the order in which explicit compensation handlers of nested scopes must be run. In our approach the only way that compensation in default order gets triggered is by a compensate activity in a fault handler. Thus, the COG need only be calculated for scopes that have at least one explicit fault handler containing a compensate activity. In contrast, any scope in a standard WS-BPEL process may have to perform default compensation.

Conceptually, the COG of a scope is the graph formed when looking down from that particular scope at scopes nested within it at any level, such that opaque scopes are collapsed to a single node and transparent scopes are ignored without obscuring nested opaque scopes from view. The opaque scopes thus found form the COG's nodes. The edges between the nodes reverse the control dependencies present in the process between the corresponding scopes. In reality, it is slightly more complicated because one must also handle scopes nested in loops and links that cross scope boundaries as will be shown in the rest of this section. We first present background on the treatment of control dependencies and then detail the algorithm for constructing the COG of a scope, starting with determining COG nodes and proceeding to COG edges.

The edges of a COG are based on the control dependencies in the process between the corresponding scopes, including both: dependencies specified by explicit control links and the implicit dependencies imposed by structured activities like ‘flow’, ‘sequence’, and ‘while’. WS-BPEL does not allow control links to cross the boundary of the looping structured activity ‘while’, nor does it allow explicit control links to create cycles.

Having presented the intuition behind COGs, we now present the algorithm for creating them. Consider a loop in the process to be a 'compensation relevant' loop of scope s if it (1)is nested in s, (2)not nested in any child scope of s, and (3)contains, at any level of nesting, at least one immediate child scope of s that has an explicit compensation handler and is not in a fault handler of s.

Consider a 'scope-and-loop hierarchy tree' whose nodes are the scopes and loops of the process. Each loop or scope has an edge to its immediately enclosing loop or scope. For example, if a scope s_2 is in loop l_1, l_1 is in another loop l_2, and l_2 is in a second scope s_1, then the tree's edges would be $(s_2, l_1), (l_1, l_2), (l_2, s_1)$. Figure 9(2) shows the scope and loop hierarchy subtree rooted at scope H. The COG nodes (in bold in the figure) are found by navigating this tree as follows:

Consider $N_{COG}(s)$ to be the set of nodes of the COG of s. Then, $N_{COG}(s) = S_{COG}(s) \cup L_{COG}(s)$, where $S_{COG}(s)$ is a subset of the scopes nested in s and $L_{COG}(s)$ is a subset of the compensation relevant loops of s. The nodes are determined by walking the scope-and-loop hierarchy sub-tree rooted at s in a depth first manner. For each visited node, if it (1)is a scope, (2)has an explicit compensation handler, and (3)is not in a fault handler of s, then it is added to $S_{COG}(s)$. Alternatively, if it is a compensation relevant loop of s, then it is added to $L_{COG}(s)$. Otherwise, it is ignored. During the depth first traversal, the children of a node added to $S_{COG}(s)$ or $L_{COG}(s)$ are *not* visited. Thus, $N_{COG}(H) = S_{COG}(H) = \{A, B, D, K, M, N, O\}$. Notice that Q is not visited.

For each loop l in $L_{COG}(s)$, create a corresponding 'loopCOG'. A loopCOG's nodes and edges are constructed the same way as for a normal COG except starting from the scope-and-loop hierarchy sub-tree rooted at l.

Determining the edges for the COG of a scope s requires determining the control dependencies between the COG's nodes. To do so, perform the following on the activities in the process nested in s. (Any changes to process structure described in this section are for the purposes of COG construction only and must not be reflected in the workflow designer's actual process.)

1. Collapse each loop l that belongs to $N_{COG}(s)$ into one node by hiding activities nested within it. Recall that no links may cross the boundary of a WS-BPEL loop so no dangling links will result from collapsing l.
2. Collapse each 'opaque scope' s_o that belongs to $N_{COG}(s)$ into one node by hiding activities nested within it. Handle any dangling links that result by moving the target/source of each link crossing the boundary of s_o to have s_o as its target/source. More precisely, if the source activity of a link is nested in s_o but its target is not, then treat s_o itself as the link's source; if the target of a link is nested in s_o and its source is not, then treat s_o itself as the link's target.

   The result of this step is shown in figure 9(3): scope B is made the source of the link from scope Q (nested in B) to scope D. In the figure, the transparent scopes P, F, G, and S are shown with dashed lines to illustrate that they are ignored in the COG since they are not among its nodes. Notice that activities such as M and N that are nested in transparent scopes, however, may appear in the COG.
3. Create the edges: if after these modifications there exists a control path in the process between a node n_1 in the COG (or any activity in n_1) to a node n_2 in the COG

(or any activity in n_2), such that the path does not contain any other COG node n_3 (or activity in n_3), then an edge (n_2, n_1) is added to the set of edges of the COG.

The only relevance of transparent scopes to the COG is that they may impose implicit control dependencies that form segments of paths between COG nodes, such as the control paths from scope node O to M and to N via the transparent scope S. This path is due to the nesting of M and N in S, in which case WS-BPEL semantics dictate that there is an implicit control dependency such that M and N may only start if S has started. One way to model both implicit and explicit dependencies in order to determine paths is to flatten the process into a graph-based model using the algorithms in ([12], Chapter 3.4), whose details are out of scope for this paper. Figure 9(4) shows the resulting COG of scope H, including the edges, while figure 8(3) shows the COG of G.

The third step above reverses the dependencies in the process between COG nodes. For example, the COG of G contains the edge (D, B) between the COG nodes representing scopes B and D, whereas the process itself had a link from Q in B to D.

One must also create, for each loop node l in the COG, the edges of the corresponding loopCOG. This is done in the same manner as for the edges of a COG, but on the activities in the process nested in l and resulting in edges between the nodes of the loopCOG.

No cycles are allowed in any of the generated compensation graphs. This is a relaxation of the current WS-BPEL restriction, mentioned in section 1, disallowing peer-scope-cycles regardless of whether and where explicit compensation handlers are used. The main reasons why this is a relaxation are that COGs are not created for every scope and not every scope appears as a node in the COG of another scope. Thus some peer scopes cycles are allowed when using our approach, such as scopes that do not have compensation handlers or those that do but are not in the COG of another scope.

5.2 Running Compensation According to the Default Order

Having described how the COGs are created to encode the default compensation order, we now explain how compensation in default order can be executed by navigating the COGs at runtime.

When compensation of a scope in the default order is triggered, the process engine navigates the scope's COG. For the purposes of navigation, each COG can in fact be treated as a process itself where (1)the transition and join conditions, which are WS-BPEL constructs that enable conditional branching and joining, are set to true, (2)the implementation of each node created from collapsing a scope results in a call to the compensation handler of the corresponding scope, and (3)the implementation of each node created from collapsing a loop corresponds in navigating (possibly more than once) the loopCOG of that node.

The number of times to run the COG of a loop depends on the number of times the loop ran in the instance of the scope being compensated. This number of iterations, along with information for keeping track of scope instances and whether or not they can be compensated, requires book-keeping. The details of this book-keeping are left out of this paper because it is also required in standard WS-BPEL (without our modifications). The interested reader is referred to Section 3.6.3 in [12] for details of a structure for this purpose named the 'ScopeLoopQ'.

6 Advantages and Drawbacks

In this section, we summarize how the new approach meets our goals of simplifying designing, refactoring and analyzing processes with regards to default compensation order as well as respecting control order. We lay out the gains and highlight a few drawbacks. We have shown the following advantages of this approach:

- Simpler for designers, testers, and business analysts to design and analyze their processes. The compensation order is easy to understand and visualize, consequently simplifying the design process. It is also only triggered explicitly, making it more predictable. This simplification could lead to richer process semantics due to providing the designer with more control over what will take place at runtime. Whereas the default compensation order in BPEL today is a function of which activity raised a fault, which scope it was caught in, and the levels of scope nesting in between, we have shown a default compensation order that is simply a function of the process structure and can be calculated at design time and presented to the designer.
- Enables separation of concerns between different capabilities of a scope. For example, a designer who adds a scope in order to simply scope a variable will not see a change in resulting compensation order at runtime. If, however, the designer does want such a change, she may simply add explicit fault and compensation handlers.
- Provides more efficient execution due to the ability to calculate the COGs before runtime and avoid the zigzag behavior. In WS-BPEL today, the order can only be determined at runtime and involves several levels of scope nesting. Additionally, we know apriori which subset of the process's scopes may be compensated and calculate the default compensation order only for that subset.
- Respects the order specified by the explicit process control flow and performs repairs at the level where the fault is caught. WS-BPEL today exhibits some cases where default compensation order is in contradiction with order specified by reversing control edges.
- Loosens the WS-BPEL restriction on cycles between peer scopes.
- Easily injectable into WS-BPEL engines because it is a variation on the existing compensation mechanism.

The drawbacks include the removal of custom termination handling and the deviation from the WS-BPEL standard so this behavior cannot be added as a WS-BPEL extension. However, termination handlers have not been a common feature in workflow systems, so it is unclear that this will be a major problem for process designers. On the other hand, it would be possible for our approach to allow termination handlers as long as they are explicit and disallowed from triggering compensation.

7 Conclusion and Future Work

In this paper, we investigated how the advent in process models of combining structured and graph based process modeling techniques, as well as a combination of fault and compensation handling, affects the roll-back and recovery capabilities of a workflow language. We framed the discussion in context of existing approaches to workflow recovery and prior art in nested transactions. Focusing on WS-BPEL, we provided an in–depth look into WS-BPEL's compensation mechanisms with a focus on default compensation, showed where it causes complications and possibly unexpected results, and provided an alternate approach for calculating the default compensation order that is based on the view from the scope where a fault will be explicitly handled. We also advocated for the removal of default fault and compensation handlers as well as termination handlers. The proposed approach presents several advantages that we illustrated, including a deterministic order for default compensation of scopes and decoupling the compensation and fault handling behaviors of scopes from each other and from other scope functions such as variable scoping. This leads to less complexity and less surprise for the user. It also enables one to relax the peer-scope dependency restriction currently in the standard.

We continue to investigate several open areas such as allowing an opaque scope to be treated as transparent if it did not successfully complete at the time it became a candidate for compensation and the effect on event handlers. A longer term research goal is to explore the ramifications of this approach beyond workflow and WS-BPEL; for example, to determine whether it could augment existing transaction models such as nested Sagas or provide a new transaction model generalized beyond WS-BPEL.

References

1. Alonso, G., Agrawal, D., El Abbadi, A., Kamath, M., Gunthor, R., Mohan, C.: Advanced transaction models in workflow contexts. In: Int'l Conference on Data Engineering (1996)
2. Brogi, A., Popescu, R.: From BPEL Processes to YAWL Workflows. In: Bravetti, M., Núñez, M., Zavattaro, G. (eds.) WS-FM 2006. LNCS, vol. 4184, pp. 107–122. Springer, Heidelberg (2006)
3. Coleman, J.: Examining BPEL's compensation construct. In: Workshop on Rigorous Engineering of Fault-Tolerant Systems, REFT (2005)
4. Curbera, F., Khalaf, R., Leymann, F., Weerawarana, S.: Exception handling in the BPEL4WS language. In: van der Aalst, W.M.P., ter Hofstede, A.H.M., Weske, M. (eds.) BPM 2003. LNCS, vol. 2678, pp. 276–290. Springer, Heidelberg (2003)
5. König, D.: R26: Default Compensation Order Conflict (2006), `http://www.oasis-open.org/committees/download.php/21303/WS_BPEL_review_issues_list.html#IssueR26`, `http://www.oasis-open.org/committees/download.php/21199/Issue%20R26.ppt`
6. Dayal, U., Hsu, M., Ladin, R.: Business process coordination: State of the art, trends, and open issues. In: Very Large Databases Conference, VLDB 2001 (2001)
7. Du, W., Davis, J., Shan, M.-C.: Flexible specification of workflow compensation scopes. In: GROUP 1997: Proceedings of the international ACM SIGGROUP conference on Supporting group work. ACM, New York (1997)
8. Fletcher, T., Furniss, P., Green, A., Haugen, R.: BPEL and business transaction management (2003), `http://www.oasis-open.org/committees/download.php/3263/BPEL.and.Business.Transaction.Management.Choreology.Submission.html`
9. Garcia-Molina, H., Salem, K.: Sagas. Proc. ACM Sigmod (1987)
10. Garcia-Molina, H., Gawlick, D., Klein, J., Kleissner, K., Salem, K.: Modeling long-running activities as nested sagas. IEEE Data Eng. Bull. 14(1) (1991)

11. Greenfield, P., Fekete, A., Jang, J., Kuo, D.: Compensation is not enough. In: International Conference on Enterprise Distributed Object Computing Conference (2003)
12. Khalaf, R.: Supporting Business Process Fragmentation While Maintaining Operational Semantics: A BPEL Perspective. PhD thesis, University of Stuttgart (2008), http://elib.uni-stuttgart.de/opus/volltexte/2008/3514/ ISBN 978-3-86624-344-6, dissertation.de
13. Kopp, O., Martin, D., Wutke, D., Leymann, F.: The Difference Between Graph-Based and Block-Structured Business Process Modelling Languages. Enterprise Modelling and Information Systems 4(1), 3–13 (2009)
14. Leymann, F.: Supporting Business Transactions via Partial Backward Recovery in Workflow Management Systems. In: Proc. BTW 1995. Springer, Berlin (1995)
15. Leymann, F., Roller, D.: Production Workflow: Concepts and Techniques. Prentice-Hall, Upper Saddle River (2000)
16. OASIS. Web Services Atomic Transaction (WS-AtomicTransaction) version 1.1 (2007), http://docs.oasis-open.org/ws-tx/wstx-wsat-1.1-spec.pdf
17. OASIS. Web Services Business Activity (WS-BusinessActivity) version 1.1. (2007), http://docs.oasis-open.org/ws-tx/wstx-wsba-1.1-spec.pdf
18. OASIS. Web Services Business Process Execution Language (WS-BPEL) Version 2.0 (2007), http://docs.oasis-open.org/wsbpel/2.0/wsbpel-v2.0.html
19. Pottinger, S., Mietzner, R., Leymann, F.: Coordinate BPEL Scopes and Processes by Extending the WS-Business Activity Framework. In: Meersman, R., Tari, Z. (eds.) CoopIS 2007. LNCS, vol. 4803, pp. 336–352. Springer, Heidelberg (2007)
20. Tai, S., Khalaf, R., Mikalsen, T.A.: Composition of coordinated web services. In: Jacobsen, H.-A. (ed.) Middleware 2004. LNCS, vol. 3231, pp. 294–310. Springer, Heidelberg (2004)
21. Thatte, S., Roller, D.: Default compensation order (2003), http://www.oasis-open.org/committees/download.php/4449/Default%20Compensation%20Order.ppt
22. van der Aalst, W.M.P., ter Hofstede, A.H.M.: Yawl: yet another workflow language. Inf. Syst. 30(4) (2005)

Understanding User Preferences and Awareness: Privacy Mechanisms in Location-Based Services

Thorben Burghardt, Erik Buchmann, Jens Müller, and Klemens Böhm

Universität Karlsruhe (TH), 76131 Karlsruhe, Germany
{burgthor,buchmann,muellerj,boehm}@ipd.uka.de

Abstract. Location based services (LBS) let people retrieve and share information related to their current position. Examples are Google Latitude or Panoramio. Since LBS share user-related content, location information etc., they put user privacy at risk. Literature has proposed various privacy mechanisms for LBS. However, it is unclear which mechanisms humans really find useful, and how they make use of them. We present a user study that addresses these issues. To obtain realistic results, we have implemented a geotagging application on the web and on GPS cellphones, and our study participants use this application in their daily lives. We test five privacy mechanisms that differ in the awareness, mental effort and degree of informedness required from the users. Among other findings, we have observed that in situations where a single simple mechanism does not meet all privacy needs, people want to use simple and sophisticated mechanisms in combination. Further, individuals are concerned about the privacy of others, even when they do not value privacy for themselves.

1 Introduction

Location based services (LBS) are important in many application areas, e.g., platforms for socializing like Google Latitude or geotagging applications like Panoramio. LBS require their users to disclose private information. Think of a person who is using a geotagging service with her GPS cellphone to annotate locations. This allows other users to find out (1) where the person has been at a certain time, (2) which data she finds interesting, and (3) her itinerary, by constructing coarse tracks from subsequent annotations. LBS that manage high-resolution tracks reveal further details, e.g., the fitness of a hiker. Almost any LBS application comprises similar privacy threats. Research has proposed various privacy enhancing technologies (PETs) for LBS. They vary in complexity and require different levels of awareness, mental effort and understanding from their users. However, as people adapt to the functionality available, it is unclear which PET is indeed suitable for LBS users. In this paper, we describe the results of a study that addresses the following research questions:

Q1: Which information do people disclose in the LBS context? To deploy PETs it is important to know (1) which data people disclose in LBS, and (2) what they generate for personal use only. We distinguish between information

R. Meersman, T. Dillon, P. Herrero (Eds.): OTM 2009, Part I, LNCS 5870, pp. 304–321, 2009.

on locations, tracks, content (e.g., annotations of a location) and metadata (e.g., creation date). We want to find out which, how much and what kind of data people reveal, and if the locations disclosed refer to hot spots of the daily life.

Q2: Which social groups are allowed to see private information? LBS are frequently used to share personal locations, e.g., to make appointments. Thus, we want to observe which information people disclose to the general public and which one to their social relationships.

Q3: How do people use privacy mechanisms? If people are unable to use a PET properly, it fails in practice. We find out how people cope with PETs ranging from simple to complex ones, and how they use mechanisms that require awareness and mental effort.

Q4: Which kind of privacy mechanisms do people prefer? PETs have to meet the expectations and desires of their users. We want people to rank PETs and PET combinations to find out which ones they would like to use.

These questions can be answered only by means of a user study. This is challenging: First, privacy preferences obtained from offline surveys or artificial lab experiments are not necessarily the natural ones [1]. Instead, people tend to overestimate their privacy needs. Second, as mobile devices with integrated GPS, cellphone and broadband Internet connection (XDAs) are just about to enter the mass market, people do not yet possess in-depth knowledge of privacy threats of LBS applications. Furthermore, we cannot assume that users know PETs proposed in the scientific literature [2,3,4,5]. Third, it requires high effort and financial expenses to equip study participants with equipment, to ensure that technical issues do not bias the study results and to keep participants motivated over several weeks. Fourth, many privacy threats in LBS originate from sharing information and from social interactions. Thus, meaningful results require a group of participants who are used to interact with each other.

We have conducted an extensive user study that considers these issues. Our primary interest is on privacy relationships between LBS users, i.e., we assume the service provider to be a trusted third party and leave aside privacy issues between service provider and consumers. We have developed a mobile geotagging application, which is a popular example of LBS that share user-generated content among many individuals. Thus, our study results will be relevant for other LBS as well. We are first to study LBS privacy in a real setup: We let individuals use our geotagging application in their everyday life for two weeks. Our participants access the geotagging application from XDAs and via a web application. Without revealing our interest in privacy right away, we ask the participants to assign tags, e.g., 'best coffee ever' or 'house of friend', to locations of interest. To raise privacy threats, we invite family members, teachers, friends, classmates and acquaintances to participate in our study as well. They are able to browse and search our LBS for any tags, tracks, metadata and time information, unless the participant who has provided this information has classified it as private. In order to observe privacy needs related to these threats and the usage of privacy mechanisms, we have implemented a number of well-known PETs, ranging from a straightforward on/off switch to (de-)activate track recording as used in Google Latitude, to sophisticated mechanisms like anonymized requests for tags.

Our analysis shows that participants do not like to introduce differentiations between individuals with regard to privacy when it comes to making information available. For more than 83% of all information generated, they either make it visible to anybody or to nobody. From the opposite perspective, participants wanted to use fine-grained privacy mechanisms for 17% of the information generated, i.e., our study participants are privacy-aware. We also find that they want to combine PETs of different complexity. Finally, people are concerned about the privacy of their friends and acquaintances, even if they do not mind disclosing information on themselves.

Paper structure: We discuss related work in Section 2, privacy threats and PETs in Section 3 and our study methodology in Section 4. In Section 5 we present our results, Section 6 concludes.

2 Related Work

Today virtually everybody is aware of the potential of LBS as well as of their privacy threats. The EU directive (2002/58/EC) [6] requests the consent of a user before her exact location may be processed. However, following the well known definitions of privacy of Alan Westin [7] – "Privacy is the claim of an individual to determine what information about himself or herself should be known to others." – such undifferentiated, binary decisions to always disclose everything when having given consent once are inappropriate. Research offers PETs to protect privacy in a fine grained way. However, there is a well-known gap between the social requirements and the technical mechanisms [8], and it is important to consider how people perceive and use PETs. In the following we give an overview of relevant PETs and of existing user studies.

PETs for LBS range from simple switches to (de-)activate GPS [9] over mechanisms where the user can differ between social groups and vary the accuracy for each piece of information disclosed [10] to mechanisms that work without any user interaction. Automatic mechanisms include [2] where the authors adapt the idea of k-anonymity to LBS, i.e., a request is indistinguishable from k-1 requests of other users. [3] introduces algorithms to create false dummies to conceal the real location. CliqueCloak [4] allows for a variable k and combines spatial and temporal aspects. Casper [5] provides a location anonymizer extending k-anonymity with a minimal area and offers a privacy aware query processor. [11,12] describe PET tailored to continuous LBS and trajectories. We adapt the idea of k-anonymity in one PET evaluated in our study.

Studies on LBS: Studies, e.g., [9], examine the privacy implications of several imaginary location-based services. They asked 16 participants questions referring to usefulness, how often they would use the service, and which privacy implications they see. The main result concerning privacy is that continuous location tracking causes more concerns than services that simply are location-aware. [13]

presents a questionnaire survey with 130 participants. Participants stated their privacy preferences for different situations (e.g., work lunch and leisure activity) and for location requests from different persons. They found that the requester is paid more attention to than the situation. Two further studies address a similar scenario: In [10], 16 participants had to give the names of persons from different social relationships. Over two weeks, participants received randomly generated hypothetical requests from these persons and replied with their current occupation, and what they would disclose about their position to the requester. The results are that the main factors to disclose information is the requesting person, the reason for the request, and if the information is deemed useful for the requester. However, requests have been hypothetical, and participants knew this. In our study, requests are real, and decisions to disclose information take immediate effect. In [14], the authors used a real application (running on a mobile phone) involving location requests and automatic notifications that are triggered when participants approached places defined previously. The location is the cell tower reachable. The study was performed with 8 participants (developers of the application and their spouses) over a period of five days. As a result, most of the requests granted came from spouses, but the small volume of data did not allow for any further conclusions. The application used is based on SMS communication. In our study, participants are permanently connected via broadband Internet, exchange their location and, in addition, content. We offer much higher accuracy by using GPS, i.e., threats are more obvious, and the areas defined are more precise.

To our knowledge, no study has yet explored the usage of different PETs under real life conditions, with a real application and realistic privacy threats.

3 Privacy and PETs in LBS

In this section, we describe our study scenario, respective privacy threats and the PETs we consider.

3.1 Mobile Geotagging

A popular variant of LBS is mobile geotagging where people assign various kinds of content to locations. For example, Panoramio.com allows to browse pictures, tags, comments, location, camera information and the photographer on a Google Maps mashup. Panoramio shows that the locations managed by LBS are related to places where the user has been or currently is, and the content provided is informative regarding user interests and attitudes.

To observe how people deal with privacy threats in an LBS context, we have implemented a mobile geotagging application based on the Streamspin framework [15] and a spatial database. With our application, a user could assign the position of his house the tags 'my home'. Our LBS stores the real name of the person tagging, a user pseudonym[1], creation times, tags, geo-coordinates and

[1] We distinguish between real name and pseudonym, because we want to observe if individuals differentiate between these when disclosing personal identifiers.

tracks of consecutive locations. The users of our LBS can browse and search for all tags and the corresponding metadata if the person who has generated the information has made it visible. Our application can be used via a XDA (a HTC Trinity), or with a web application. The XDA allows to assign tags to the current position of its user, i.e., it connects to the LBS, and it stores tracks and tuples <*user, timestamp, location, tag list*> in a database. Furthermore, the XDA allows to search for tags assigned to locations next to the current position. The XDA is updated in real time, if new tags appear in the vicinity. The web application displays the tags, positions and tracks with a Google Maps mashup. It lets each user view and edit all information provided. To fill the database with examples, we imported 200.000 geo-referenced keywords from Wikipedia.

3.2 Privacy Threats

When considering information accessible to other LBS users, privacy problems can arise from four sources:

Content. Content is the information people generate explicitly [16]. For example, the content of a geotagging application is the tags or photos provided, and the content of a people-finder service can be the social network of the participants. Clearly, content can be sensitive information.

Location. Mobile LBS obtain the current position of the user whenever she provides content or issues a query [5]. Other LBS might also manage locations the user has visited before. This is problematic, as location information can be private, e.g., the house of a friend, a hospital or a religious building.

Tracks. Tracks containing user itineraries can be the result of a LBS that records the movement of an XDA. Alternatively, LBS users can generate tracks by sorting locations publicly visible chronologically. Tracks can reveal where a person lives, buildings visited, relationships, daily routine etc [17].

Metadata. In LBS, metadata relate to the content, locations and tracks provided. Examples are the user name or the creation date. Metadata of content depend on the application area. For example, photos could contain EXIF information on the camera model. This information may be specific to individuals, and its dissemination may be a privacy leak.

3.3 Privacy Enhancing Technologies for LBS

A huge number of PETs is available for LBS. They range from simple, intuitive mechanisms to sophisticated ones that require a fundamental understanding of privacy threats and the technical background. To investigate a wide range of approaches, we have decided to implement and make available five PETs as representatives of PET classes from literature and real applications. Since we want to observe which social groups are allowed to see which personal information, our mechanisms include approaches where a user can say when information is disclosed to certain groups. More specifically, the study participants had to name persons belonging to each of the social groups *teacher, parents, classmates, friends* and *acquaintances.*

$PET_{checkbox}$ *(Public/Private Checkbox):* To observe how our study participants make use of simple and intuitive privacy mechanisms, we have implemented a straightforward checkbox on the XDA that allows or prohibits disclosing content, locations and metadata to anybody. The default is 'private'.

PET_{fine} *(Fine-grained Control):* This PET allows to specify privacy preferences for each piece of information. It is part of our web application, i.e., a user can define her privacy settings after she has generated the tag. Checkboxes let the participants define for each location, tag, real name, user pseudonym and time which social group is allowed to see the information. For example, a participant can specify 'Make the tags *my school* public for my friends, but hide my real name and the creation time.'. We ensured that it requires exactly the same effort to make a piece of information public or private.

In order to observe the privacy preferences and the privacy mechanisms preferred for tracks, we have implemented three mechanisms. Furthermore, we have adapted PET_{fine} so that it can be used on fragments of a track.

PET_{areas} *(Private Areas):* This PET lets the user specify areas where a track will not be disclosed to others. Thus, when a user enters a private area, the part of her track visible for others ends at the border of the area. The users can configure this for each social group by using an intuitive Google Maps mashup, integrated in our web application. The mashup lets the participants draw closed polygons, each representing a private area. Examples of such areas span from small rectangles around the user's home to polygons that cover the entire world. Private areas are specified in advance, and PET_{areas} takes effect at the moment of data acquisition on the XDA.

PET_{switch} *(GPS Switch):* This mechanism is motivated by [9] that request a simple option to turn off location tracking. It is similar to the PET implemented in Google Latitude. PET_{switch} lets the participants (de-)activate the GPS receiver, i.e., it provides immediate control over the disclosure of track information. However, it also requires permanent attention to deactivate the GPS whenever entering a sensible region.

PET_{anon} *(Anonymizer):* This PET, motivated by [5], deploys the notion of k-anonymity so that queries for tags at a location nearby cannot be distinguished from queries of k-1 other users. The anonymizer works as follows: Before issuing a query, the user specifies a k, i.e., a number of users from which she wants to be indistinguishable. Now suppose a user issues a query for tags in her vicinity. The anonymizer extends the query region so that at least k other persons could have sent the same query. The LBS answers with the tags for the extended area, which the XDA filters for the original query region. It is known that this algorithm does not prevent the construction of precise tracks for (i) frequent (continuous) requests (therefore participants had to refresh tags shown manually) and (ii) in populated regions where the extended areas are small. Note that literature, e.g., [5], provides a minimum region size and an adaptive location optimizer to avoid these problems. Others group 'similar' tracks. However, we only consider k to not burden the participants with too many parameters. Using this PET properly requires a fundamental understanding of its operation principles and of the privacy threats of LBS.

4 Study Design

In the following, we describe our key design decisions and how we have structured and conducted our user study.

4.1 Key Design Decisions

The design of the study is based on five key decisions:

Intensive Study. We have decided to run a study requiring a high degree of participant involvement, under real conditions, with a relatively small number of individuals. The rationale is as follows: (1) Competence regarding the technology is required to obtain meaningful study results [18]. Thus, we have to *intensively train* our participants on GPS, the XDAs, and our applications. (2) Participants of offline surveys tend to overestimate their privacy needs [1]. We expect more realistic results if our participants use an application in their *daily lives* in a natural way; and we observe their behavior.

Two Phases of Evaluation. We have tested PETs in two phases: (i) A tagging phase where we focus on locations, content and metadata disclosed, and (ii) a track-recording phase where we analyze the behavior of the participants for each PET in isolation. We do so because some PETs interfere with others, and we do not want to overstrain the participants with several complex PETs active at the same time.

Web Application. We have designed and implemented parts of our setup as a web application, for two reasons. First, some PETs need a detailed parameterization. To avoid that unintuitive user interfaces bias our results, we have implemented the preference dialogs of PET_{fine} and PET_{areas} as a web application. Second, we have invited persons from the social environment of our participants to observe the web application in order to generate real privacy threats.

Participants. Typically, tagging services are used to facilitate retrieval, for opinion expression, sharing, play and competition etc. [19]. Thus, our study requires technology-affine individuals who are part of the same social network, i.e., who live in a nearby area to see the tags of friends and interact with each other frequently. Furthermore, our study participants should be a relevant target group for LBS providers. We decided for a German upper grade high-school class of 25 students aged between 16 and 17 years. We equipped 10 of them (eight female and two male) with our XDAs. In order to come up with real privacy threats, we provided the parents, teachers and other pupils who did not receive a XDA with access to our web application. We have equipped parents and teachers with accounts. Thus, many persons from the social environment of our participants could see content, metadata, locations and tracks which participants have made public. For contacts with individual accounts in turn, participants could differentiate in more detail what to reveal.

Incentives. To encourage participants to use our LBS, we let them send one SMS for free for any 5 different locations tagged and for every 30 minutes the XDA was connected to the LBS. The payment was independent from the position or the kind of tag provided. Our participants have used the XDA and our LBS very frequently.

4.2 Study Procedure

We have divided our study into the four phases *Introduction*, *Tagging*, *Track Recording* and *Completion*. Each phase starts and ends with a meeting where we instruct the participants and ask control questions to ensure proper usage of the mechanisms. In the following, we will describe these phases. During the entire experiment, we record any keystroke and any GPS information from the XDA. Thus, our logs contain complete track information even if a participant has required that some tracks shall remain private[2]. More details on our study, e.g., the questionnaires used and screenshots of our web application, can be found on our complementary web page[3].

Introduction. We start with a questionnaire (Q.1) to obtain information on (i) demographic data, (ii) the usage habits regarding cellphones and Internet services, and (iii) general privacy attitudes. In order to provide a plausible motivation for our application without revealing our interest in privacy, we describe popular tagging applications and LBS, e.g., del.icio.us and mobile city guides. Furthermore, we introduce the XDA with the mobile location tagging application. We explain our application and guide participants to tag training objects at our university campus. After that, we demonstrate how those objects appear in our web application. Finally, we tell our participants that mobile tagging applications usually contain features to decide who may access which content. Therefore, we explain how to use the privacy mechanisms for tags ($PET_{checkbox}$, PET_{fine}). To this end, we let our participants name the members of their social groups and configure their buddy lists accordingly.

Tagging. The Tagging Phase lasts one week. Without yet revealing our interest in privacy, we encourage our participants to use the XDA in their everyday life

[2] Note that this has been the only chance to really measure the gap between actual privacy preferences and stated ones. We have informed their teacher on the privacy focus of our study in advance, and we did not make any track accessible to others before activating the PETs. Further, all parents gave their consent that their children participate in the location tagging experiment.

[3] http://privacy.ipd.uka.de

and to tag different objects. We give examples like sights, meeting points, own house and house of friends. We secretly record all movements of the XDAs, i.e., we log participant tracks. At the end of the week, we reveal our research interest in location privacy to the study participants. First, we hand out a questionnaire (Q2.A) on the usability of our tagging application, technical problems that might have biased results up to this point and on the incentives. Afterwards, we explain that privacy threats do not only arise from the tags and locations provided, but also from tracks that a mobile tagging application can record.

Track Recording. At the beginning of this phase we hand out a questionnaire (Q2.B) asking for all situations and locations where our participants do not want their tracks to be visible for others. We then activate the privacy mechanisms that allow the participants to hide their tracks. Note that we have trained our participants on the PETs but have not introduced them to different privacy threats regarding tracks. As some of the PETs are not orthogonal, e.g., defining private areas (PET_{areas}) and the GPS switch (PET_{switch}), we decided to evaluate each individual PET in isolation.

We start with PET_{areas} and ask the participants to specify private areas where their track may not be disclosed. Afterwards, to analyze if users have defined areas that meet their preferences, they are allowed to look at the tracks we have recorded during the tagging phase, and we let them refine their private areas if needed. In the following two days, our participants use PET_{areas} in their everyday life to hide track information that is supposed to be private. For another two days, we provide the GPS switch that activates and stops track recording (PET_{switch}). For the last two days, we let the participants use the anonymizer (PET_{anon}).

At the end of both the Tagging and the Track Recording phase, participants can use the fine grained control (PET_{fine}). With PET_{fine} users can specify who can see what for tracks recorded during the Tagging phase, and tracks that are not hidden by one of the other mechanisms during the Track Recording phase.

Completion. We finish our study with a questionnaire (Q.3) on the study and on the privacy mechanisms offered.

5 Study

In this section we will answer the research questions form the introduction.

5.1 Location, Content and Metadata

Our participants assigned 1042 tags to 442 different locations, i.e., they have provided 442 <*location, tag list, metadata*>-tuples that make locations of their everyday life explicit. Most locations have been tagged with 1 to 3 tags. This is in line with other tagging applications [20]. 41% of all locations were tagged with two or more tags.

Which Information Do People Disclose? (Q1). To find out which information people would like to share with others, we evaluate which tags and metadata our participants have made publicly visible by $PET_{checkbox}$ and PET_{fine}. Our expectation is that our participants want to disclose most of the information.

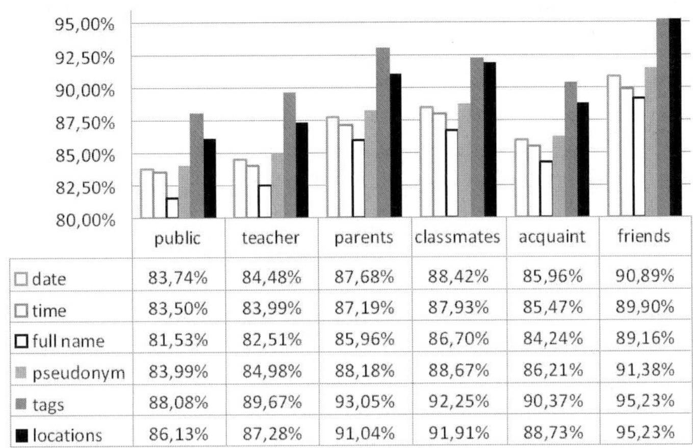

	public	teacher	parents	classmates	acquaint	friends
date	83,74%	84,48%	87,68%	88,42%	85,96%	90,89%
time	83,50%	83,99%	87,19%	87,93%	85,47%	89,90%
full name	81,53%	82,51%	85,96%	86,70%	84,24%	89,16%
pseudonym	83,99%	84,98%	88,18%	88,67%	86,21%	91,38%
tags	88,08%	89,67%	93,05%	92,25%	90,37%	95,23%
locations	86,13%	87,28%	91,04%	91,91%	88,73%	95,23%

Fig. 1. Content and metadata disclosed

The Column 'public' in Figure 1 shows how often information in the various categories has been disclosed. The figure distinguishes between the date and time tags have been created, the name and the pseudonym, the tag per location, and the location itself. The figure shows that our participants wanted 88% of all tags generated and 86% of all locations tagged to be visible for anybody. 81% of all <*location, tag list, metadata*>-tuples are marked public, and 2% are specified private in their entirety.

The participants have made fine distinctions regarding the disclosure of information from the remaining tuples, i.e., they disclosed only some attributes from a <*location, tag list, metadata*> tuple. In 10% of all tuples, only some tags from the tag list were made public to anybody. In 4% of the tuples, only the location and the tags are disclosed, but all metadata is kept private. 3% of the tuples reveal only some of the metadata. Creation date, creation time and user pseudonym have been marked public nearly as often as the tags. Our participants disclosed their real name least often. None of our participants has considered that linking data from multiple tags with different privacy settings might allow to infer data marked private only for some tuples. For example, if a participant has ever revealed her pseudonym together with her real name, the real name can be reconstructed for tuples where only the pseudonym was public.

Leaking information on important locations of everyday life can be a severe privacy threat. In order to find out if participants let others know their centers of life, we have measured the time a person stays in the vicinity of a location she has made public to someone else. Figure 2 shows the logarithmic cumulative

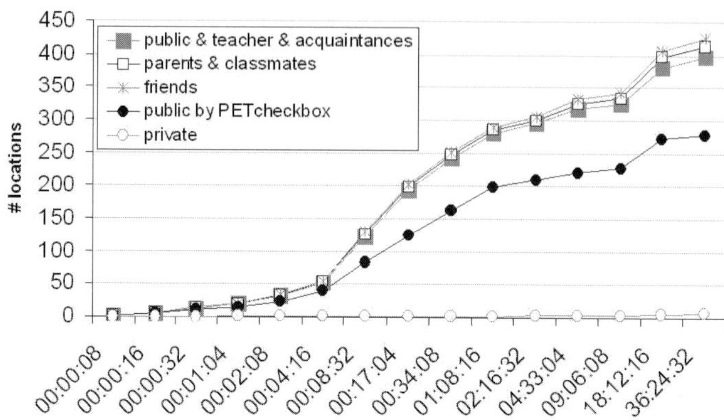

Fig. 2. Time users spent in the vicinity of their tags – time.log(hh:mm:ss)

distribution function of how long participants stayed in a distance of less than 200m from a location they have tagged. 20% of all tuples refer to locations where the user has stayed for at least 20 hours during 14 days of our study. Such locations include the school building or the home of the participant. Remember that our participants made 2% of all tuples private. 50% of these private tuples refer to locations where its creator spent at least 20 hours.

These findings lead to the following conclusions: (1) Instead of disclosing a minimum of information, people tend to disclose everything to everybody. Only very sensitive information is kept private. (2) People do not think that joining data might lead to privacy leaks. This calls for automated PETs which warn the user if she is about to disclose privacy-threatening combinations of information. (3) Even if many tuples were disclosed in an all-or-nothing manner, there is a need for PETs that specify in a fine-grained way which particular information is publicly visible. (4) The time spent in the vicinity of tags disclosed indicates that people frequently disclose important locations of their everyday life.

Which Social Groups Are Allowed to See Private Information? (Q2). PET_{fine} lets the participants distinguish not only between kinds of data, but also between social groups, namely teacher, parents, classmates, acquaintances and friends. Figure 1 shows which groups are frequently allowed to see which information. Remember that the participants made 81% of all data visible to anybody and kept 2% entirely private. Thus, each social group was allowed to see at least 81% of all tuples (public). Our participants had disclosed most to the group friends and least to the group teacher. When comparing our initial questionnaire with Figure 1 we found some inconsistencies. In the questionnaire, 6 of our 10 participants did not want to reveal their whereabouts to their parents, 2 were undecided, and the remaining 2 were inclined to tell their locations. However, our study results show that parents are allowed to see between 86% and 93% of all information, depending on the kind of data. For all social groups the correlation coefficient between disclosed metadata and disclosed locations is 99%, for tags 97%.

We have provided our participants with a list of example locations which they could tag, including their home and houses of friends. As expected, all participants have tagged their home but even four of them have made this tag public. 8 of 10 participants tagged at least one house of a friend. However, only one participant made these tuples public, and one disclosed it to friends. This is in line with our preceding questionnaire, where 8 of the participants stated to care about the privacy of friends.

Figure 2 shows the time the participants have spent close to the locations they have tagged, and which social groups were allowed to see this location. Thus, the figure tells who might observe important locations of the daily life. To avoid clutter in the figure, there is one curve representing several groups with very similar values. 16% of all *<location, tag list, metadata>*-tuples describe where the creator of the tuple stayed for at least 20 hours during our study. Friends are allowed to see 95% of all locations. 19% of these locations are close to places where the issuer of the tuple stayed at least 20 hours.

Summing up, (1) our participants differentiate between social groups for 17% of all information provided, and (2) behave similarly regarding the different social groups and metadata disclosed. Furthermore, (3) they care for privacy of certain others more than for their own privacy.

How Do People Use PETs, and What Kind of PETs Do They Prefer? (Q3, Q4). Our participants had three options to keep their privacy: (i) They could use a checkbox on the device that makes an entire tuple public ($PET_{checkbox}$). (ii) They could use our web application to specify in detail which group is allowed to see which information from a tuple (PET_{fine}). (iii) Our participants could abstain from providing information, if they deem it too cumbersome to use the PETs provided. We have observed that the participants have made 63% of all tuples publicly visible using $PET_{checkbox}$. The privacy settings of 37% of the tuples were specified with PET_{fine}. The participants used PET_{fine} to make an additional 23% of all locations public. Thus, though our participants prefer simple mechanisms, they readily use sophisticated PETs. Furthermore, in many cases, e.g., for their homes or for houses of friends, they spent effort in providing detailed privacy settings, instead of simply not providing information. Thus, there clearly is a need for sophisticated PETs.

5.2 Location Tracking

In the following, we will compare the usage of PET_{areas}, PET_{switch} and PET_{anon}, followed by an analysis of the tracks that have been made visible.

How Do People Use PETs, and What Do They Disclose? (Q1, Q3). To find out how our participants use the PETs, we look at the track information we have recorded.

PET_{areas} (Private Areas). This PET lets the participants define areas where tracks are not shown to others, e.g., not to 'anybody'. We have obtained private area definitions at the beginning and at the end of the track recording phase.

At the beginning, we handed out a questionnaire where the participants specified private areas in plain text. Furthermore, our participants had to draw these areas in a Google Maps mashup. Our participants defined 26 private areas, which are related to leisure activities (11), home (6), school (5), (boy)friends (2), work (1) and relatives (1). Table 1 shows how many areas mask a track, the average area size and its standard deviation. In order not to bias our statistics, we have excluded two outlier areas that cover the country and the city. Note that our participants have removed those areas in the real usage phase. To measure if the areas cover hot spots of the daily lives, we have also calculated for how long our participants stayed in one of the private areas. The initial areas cover regions where the participants stay for 25% of the time they have used our system.

We obtained refined areas at the end of the track recording phase, i.e., after the participants have seen the tracks recorded in the tagging phase, and after two days of using PET_{areas}. The refinements include 6 new areas that prevent the disclosure of tracks to anybody. On average, our participants more than doubled the size of their initial areas. One area was increased by factor 16 after a participant realized that an area of the size of a house allows to see when entering the building. Overall, we found that the area specification includes 29 areas which the participants actually pass from time to time. When comparing the time the participants have spent in private areas to the time spent in the initial areas (for the entire experiment time recorded), it increased from 24.5% to 33.2%.

Table 1. PET_{areas}: private areas initially and refined

Initial Areas	anybody	teacher	parents	friends	acquaint.	classmates
Number of Areas	12	23	16	16	18	19
Avg. size (km^2)	0.712	1.121	0.578	1.576	0.504	1.339
Stddv. size	1.191	3.496	1.051	4.146	1.007	3.826
Time in areas (%)	24.5%	26.9%	24.8%	26.9%	25.0%	26.5%
Refined Areas	anybody	teacher	parents	friends	acquaint.	classmates
Number of Areas	18	28	23	20	22	25
Avg. size (km^2)	0.984	0.720	0.823	0.966	0.832	0.806
Stddv. size	1.494	1.258	1.356	1.418	1.387	1.307
Time in areas (%)	33.2%	40.2%	33.9%	34.7%	34.0%	34.3%

Our results show that people wish to adapt their privacy settings frequently. Furthermore, people cannot imagine location based privacy threats without seeing their tracks. The fact that our participants significantly increased their areas after we confronted them with track recordings supports this.

PET_{switch} (GPS Switch). The second mechanism we evaluated is the manual on/off switch for the GPS receiver. We have logged when and where the GPS has been turned off, i.e., when the participants wanted to keep their movements

private. We expected our participants to use the switch when entering one of the private areas they have defined for PET_{areas}. But the GPS switch was rarely used. During two days, the participants used the switch 6 times on average, and each of them used it less than 12 times.

To evaluate PET_{switch}, we assume that the area definitions from PET_{areas} are an exhaustive specification of the private areas. As Table 2 shows, our participants have made 44% of their tracks visible on average while beeing in a private area, i.e., when they wanted to be unobserved.This indicates that PET_{switch} fails in practice. Thus, PETs that require continuous attention might lead to unintended disclosure of personal information, even if they are as simple as a switch for a GPS receiver. Our final questionnaire supports this finding.

Table 2. Comparing PET_{switch} to PET_{areas}

GPS / Position	in private area	out of area	sum
active (%)	43.56%	31.71%	75.27%
inactive (%)	16.45%	8.27 %	24.73%
sum (%)	60.01%	39.99%	**100.00%**

PET_{anon} (Anonymizer). This PET adapts k-anonymity to avoid that sequences of positions can be combined to a track. But the choice of an appropriate k requires technical understanding. Thus, we have explained PET_{anon} in detail. Note that the participants expected that others would use the application, i.e., values greater than 10 have been possible.

Table 3. PET_{anon}: Average distance to the k-nearest neighbor

k	2	3	4	5	6
distance (km)	1.06	1.67	2.00	2.35	2.72
k	7	8	9	10	–
distance (km)	3.17	3.83	5.79	13.43	–

Our participants have chosen a k that varies between 1, i.e., no anonymity, and 30. The average k was 7.33 with standard deviation 8.64. In order to find out which value of k is appropriate, we calculate the average distance between each of the 14,015 positions recorded and the k nearest neighbor. For example, Table 3 shows that the average distance to the 3rd neighbor is 1.67km. Thus, a position anonymized with $k = 3$ means that the user can be located anywhere in a circular area[4] of $8.87km^2$ on average. The average k corresponds to a circle of $38.5km^2$, which is much more than the average area size of $0.7km^2$ specified

[4] We have used a circular area just to provide an intuition. [5] uses rectangular grid cells interleaved to a pyramid scheme.

with PET_{areas}. Thus, we conclude that our participants are unable to use the anonymizer in line with their preferences.

Which Social Groups Are Allowed to See Private Track Information? (Q2). PET_{fine} and PET_{areas} allow to restrict the visibility of parts of a track for certain social groups. At the end of the study, 5 of our participants have used PET_{fine} to assign privacy preferences to 270 tracks. The other half has not disclosed any track information. Only 28% of all tracks have been made visible for others. Table 4 shows which social groups these 28% of the tracks have been disclosed to. Only 2% of all tracks have been made visible for anybody, 4% for teachers; parents could see 21%, friends 51%.

Table 4. Tracks disclosed using PET_{fine}

Phase	pub	teacher	parents	friends	acqu.	classm.
Tagging (%)	2%	4%	13%	28%	7%	10%
Track Rec. (%)	0%	0%	8%	23%	2%	4%
Sum (%)	2%	4%	21%	51%	9%	14%

PET_{areas} lets the participants configure who is not allowed to see a track recorded in a private area. Table 1 compares the number of areas defined, their size and the time the users have spent in it, for each social group. Our participants have defined most areas to hide from teachers (28) and fewest to hide from friends (20). This is in line with our findings from PET_{fine}. When considering the time spent in a private area, our participants do not want to be seen from their teachers for 40% of the time. For all other social groups, the participants want not to be seen for less than 35% of the time.

Our evaluation supports two findings. First, people seem to be more concerned about revealing tracks than disclosing locations, content and metadata. While the participants make 81% of the latter public, they disclose hardly any tracks to anybody. Second, there is a difference in the track information disclosed with PET_{fine} and PET_{areas}. As the disclosure rate is larger for the PET that lets the users see the tracks they are about to reveal, we conclude that people are not aware of track based privacy threats. This is in line with our findings on the usage of the GPS switch.

Which Kind of Privacy Mechanisms Do People Prefer? (Q4). To answer this question, we evaluate a final questionnaire. We ask our participants to assign a score of 1 to 5 in categories safety, ease of usage, complexity to understand, and effort to use to PET_{areas}, PET_{switch} and PET_{anon}. The average scores are displayed in Figure 3. Furthermore, we ask our participants to rank PET or PET combinations they would like to use in the future. Figure 4 shows the average rank.

Among all PETs in isolation, our participants prefer the GPS switch. They assigned it the highest grades in all categories. PET_{areas} obtained the worst grades in all categories. Since our participants noticed that the GPS switch needs permanent attention, they prefer combining it with other PETs. Here, our participants preferred the GPS switch together with the anonymizer. We found this surprising, because our participants neither used the GPS switch nor the anonymizer in line with their privacy preferences. A possible explanation is that the participants wanted to combine the most intuitive PET with a PET that needs only one parameter.

5.3 Discussion

It has been our design decision to execute an elaborate study under real conditions with a limited number of well-prepared participants. Nevertheless, we deem our study results representative, for three reasons:

Relevant study group. Our study included 25 individuals, their parents and one teacher. We equipped 10 of them with mobile devices. Our participants were interested in new technologies, all are used to cellphones and Internet applications, and they are mobile in the sense of visiting many different locations instead of having a fixed schedule each day. Furthermore, our participants are a relevant target group for commercial mobile applications.

Real application. Our results have been obtained with a geotagging application similar to a wide class of services. This includes all applications where users share location-based content related to a position that has been visited before or is visited currently. The study participants displayed usage patterns comparable to those described in the literature [19]. At the end of our study, some participants asked if they could continue using our application.

Real privacy threats. The participants integrated our application into their everyday life. The study period included working days, holidays and weekends. The average usage time of our application was 6 hours per day, i.e., the usage intensity was very high. During the study, the XDAs logged spatial resolutions of the GPS receiver that were better than 100m. This is sufficient to distinguish the individual buildings the participants have entered.

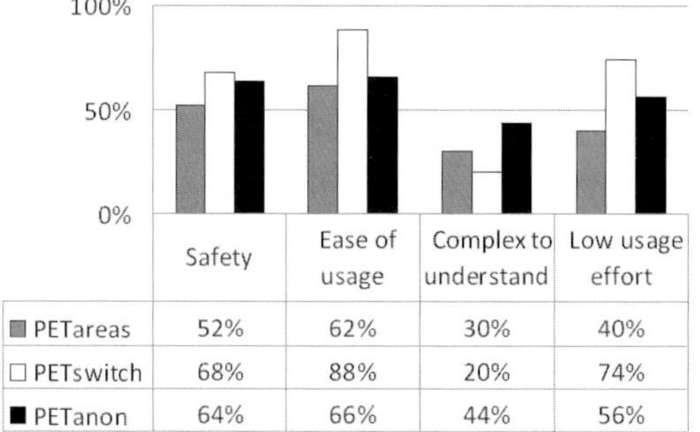

	Safety	Ease of usage	Complex to understand	Low usage effort
PETareas	52%	62%	30%	40%
PETswitch	68%	88%	20%	74%
PETanon	64%	66%	44%	56%

Fig. 3. Comparison of Mechanisms

PET Ranking	∅
GPS switch & Anonymizer	2,9
All mechanisms combined	3,1
Private Areas & GPS switch	3,4
Private Areas & Anonymizer	3,9
GPS switch	4,2
Anonymizer	4,8
Private Areas	5,4

Fig. 4. PET Ranking

We summarize the lessons learned from our study as follows:

Social groups. Participants frequently wanted to disclose information to certain groups, e.g., to friends or classmates. Thus, we recommend PETs that allow a differentiation between social groups.

Different information. LBS do not only process locations and tracks, but other kinds of data as well, e.g., content or metadata. PETs for LBS have to consider such information, because privacy threats arise from any data, and people frequently want to disclose a subset of information, e.g., a position without the date of creation.

Complex PETs. For transparency reasons, it is often assumed that users prefer PETs which are easy to understand. However, we have shown that people favor complex, sophisticated PETs over simple ones – if the sophisticated PETs reduce the effort of the users to have their privacy preferences enforced.

6 Conclusions

Location based services (LBS) are an important recent development. However, LBS put user privacy at risk. In this paper, we have investigated which privacy mechanisms individuals want to use, and how. To this end, we have implemented a fully operational geotagging service, and we let our study participants use this application in their daily lives.

Among other results we found that people tend to use mechanisms that are easy to understand. However, when mechanisms require constant awareness, they fail in practice, i.e., cannot serve all privacy needs. Thus, our participants wanted to combine them with automated approaches that allow a fine-grained control over the data disclosed. Although people tend to disclose most information, we have observed for 17% of the data provided that our participants wanted to specify who is allowed to see it in a very detailed manner. Finally, we found that people care for the privacy of their friends even if they disclose everything about themselves.

Acknowledgements. This work was partly funded by DFG BO2129/8-1. We thank Ursula Kotzur and the authors of [15] for supporting our study.

References

1. Spiekermann, S., Grossklags, J., Berendt, B.: E-privacy in 2nd generation e-commerce: privacy preferences versus actual behavior. In: EC. ACM, New York (2001)
2. Gruteser, M., Grunwald, D.: Anonymous usage of location-based services through spatial and temporal cloaking. In: MobiSys. ACM, New York (2003)
3. Kido, H., Yanagisawa, Y., Satoh, T.: Protection of location privacy using dummies for location-based services. In: ICDEW. IEEE, Los Alamitos (2005)
4. Gedik, B., Liu, L.: A customizable k-anonymity model for protecting location privacy. Technical report, Georgia Institute of Technology (2004)

5. Mokbel, M.F., Chow, C.Y., Aref, W.G.: The new casper: query processing for location services without compromising privacy. In: VLDB (2006)
6. European Parliament: Directive 2002/58/ec (2002)
7. Westin, A.F.: Privacy and Freedom. Bodley Head (April 1970)
8. Ackerman, M.S.: The intellectual challenge of cscw: The gap between social requirements and technical feasibility. In: HCI. ACM, New York (2001)
9. Barkhuus, L., Dey, A.: Location-based services for mobile telephony: A study of users' privacy concerns. In: CHI INTERACT (2003)
10. Consolvo, S., Smith, I.E., et al.: Location disclosure to social relations: Why, when, & what people want to share. In: SIGCHI. ACM, New York (2005)
11. Xu, T., Cai, Y.: Location anonymity in continuous location-based services. In: GIS, pp. 1–8. ACM, New York (2007)
12. Terrovitis, M., Mamoulis, N.: Privacy preservation in the publication of trajectories. In: MDM. IEEE, Los Alamitos (2008)
13. Lederer, S., Mankoff, J., Dey, A.K.: Who wants to know what when? privacy preference determinants in ubiquitous computing. In: CHI. ACM, New York (2003)
14. Smith, I., Consolvo, S., et al.: Social disclosure of place: From location technology to communication practice. In: Gellersen, H.-W., Want, R., Schmidt, A. (eds.) PERVASIVE 2005. LNCS, vol. 3468, pp. 134–151. Springer, Heidelberg (2005)
15. Wind, R., et al.: A testbed for the exploration of novel concepts in mobile service delivery. In: MDM (2007)
16. Ames, M., Naaman, M.: Why we tag: motivations for annotation in mobile and online media. In: SIGCHI. ACM, New York (2007)
17. Gonzalez, M.C., Hidalgo, C.A., Barabasi, A.L.: Understanding individual human mobility patterns. Nature 453 (2008)
18. Babbie, E.R.: The Practice of Social Research, 10th edn. Academic Internet Publ. (2007)
19. Marlow, C., et al.: Ht06, tagging paper, taxonomy, flickr, academic article, to read. In: HYPERTEXT. ACM, New York (2006)
20. Millen, D.R., Feinberg, J., Kerr, B.: Dogear: Social bookmarking in the enterprise. In: SIGCHI. ACM, New York (2006)

Information Sharing Modalities for Mobile Ad-Hoc Networks

Alexandre de Spindler, Michael Grossniklaus, Christoph Lins, and Moira C. Norrie

Institute for Information Systems, ETH Zurich
CH-8092 Zurich, Switzerland
{despindler,grossniklaus,lins,norrie}@inf.ethz.ch

Abstract. Current mobile phone technologies have fostered the emergence of a new generation of mobile applications. Such applications allow users to interact and share information opportunistically when their mobile devices are in physical proximity or close to fixed installations. It has been shown how mobile applications such as collaborative filtering and location-based services can take advantage of ad-hoc connectivity to use physical proximity as a filter mechanism inherent to the application logic. We discuss the different modes of information sharing that arise in such settings based on the models of persistence and synchronisation. We present a platform that supports the development of applications that can exploit these modes of ad-hoc information sharing and, by means of an example, show how such an application can be realised based on the supported event model.

1 Introduction

Recent advances in mobile phone technologies have promoted the development of a new generation of mobile applications that allow users to interact and share information opportunistically based on ad-hoc network connections. Information may either be shared between mobile devices in a peer-to-peer (P2P) manner or between mobile devices and fixed installations. Both forms of information sharing have been used in different ways in a variety of applications. For example, ad-hoc network connectivity between peers has been used to detect the physical copresence of users in social settings as a basis for exchanging rating data in a P2P manner in recommender systems [1]. However, P2P connections between mobile devices have also been used as a means for storing data associated with a location based on the movement of data between peers as they pass through that location [2]. The sharing of data between mobile devices and fixed installations has been used to provide location-based services such as allowing mobile users to access local data from a server when in proximity to particular wireless hotspots or Bluetooth stations [3,4].

An analysis of the different forms of opportunistic information sharing based on ad-hoc connectivity reveals two main characteristics that determine the form of sharing, namely *persistence* and *synchronisation*. If information is shared in

R. Meersman, T. Dillon, P. Herrero (Eds.): OTM 2009, Part I, LNCS 5870, pp. 322–339, 2009.

a persistent manner, it means that the data will be stored on the client after disconnection. Otherwise, information is only shared when the devices are connected and hence data is transient on the client. If information is shared in a synchronised manner, it means that copies of data should eventually merge so that the effects of any updates are reflected in all copies. These characteristics are orthogonal to each other, meaning that they can be combined in different ways to offer four basic modes of information sharing.

In this paper, we examine these different modalities of opportunistic information sharing in detail and present a framework that allows application developers to select and combine these in flexible ways according to the requirements of particular applications. By using a specific application scenario that exhibits all four forms of information sharing, we explain how the framework supports application development.

We start in Sect. 2 with an overview of related work on data sharing in mobile settings. Section 3 then examines the different forms of information sharing in detail using our particular application scenario to show how these can be used to meet different requirements in terms of persistence and synchronisation. Section 4 presents a platform and abstract model for the realisation of these modes based on shared data collections, while Sect. 5 shows how an application can be developed using the platform. Section 6 describes how we implemented the platform based on an extension of an existing P2P collection framework together with a unified event model for distributed object databases. We compare the resulting framework with existing platforms and frameworks for the development of mobile applications in Sect. 7. Concluding remarks are given in Sect. 8.

2 Background

Although mobile phone vendors now offer tools for the development of mobile applications, these tend to offer only basic features for data persistence and sharing. For example, the platform independent Java Wireless Toolkit (Java WTK) uses a simple key-value store for data persistence which means that developers have to define and implement a mapping between Java application objects and key-value pairs for each application. With Android[1], Google has taken another path by integrating SQLite[2] and offering the developer methods that take SQL statements as string-typed arguments similar to JDBC[3]. This results in the classic impedance mismatch problem between the application model and the storage model of data along with a lack of compile-time safety. Alternative methods are available where the single components of SQL statements can be provided individually, but this results in methods with many arguments, some of which are not used in most cases and therefore must be set to null. This still results in Java code not being checked during compilation as the table and column names are provided as string values. In contrast, development platforms for PCs such as

[1] http://code.google.com/android

[2] http://www.sqlite.org

[3] http://java.sun.com/products/jdbc

db4o[4] support Java object persistence and therefore avoid both the impedance mismatch problem and the lack of compile-time safety checks.

Support for information sharing is also limited in these platforms and data exchange must be implemented based on sockets able to send and receive binary data. Short-range connectivity such as Bluetooth or WiFi can be used to react to peers appearing in the physical vicinity, but there is no high-level support for vicinity awareness and data exchange. For example, using Java WTK, the developer has to implement two listeners, one registered for the discovery of a device and another which is notified about application-specific services discovered on a particular device.

Within the research community, the ad-hoc nature of connectivity leading to frequent and unpredictable disconnections is usually regarded as problematic and there is a lot of work on providing distributed persistence and synchronicity despite the dynamics of mobile environments [5,6,7,8]. However, a few researchers have investigated ways in which ad-hoc connectivity can be exploited as a means of detecting the proximity of users to locations, devices or other users. For example, several projects have developed location-based services that use short-range connection technologies such as WiFi or Bluetooth to provide mobile users with access to information about a location based on their proximity to that location [9,3,10,11,4]. When a user moves into the vicinity of the server fixed at that location, a connection is established and information relevant to the location can be viewed on the user's mobile device. When the user moves out of the connection range, typically that information is no longer accessible.

In previous work [1], we have shown how ad-hoc connectivity can be used to adapt collaborative filtering (CF) algorithms to mobile settings. Typical CF algorithms such as user-based filtering proceed in two steps, first selecting a set of similar users and then aggregating the ratings made by these users. In our approach, users share ratings when they are in physical proximity by means of ad-hoc connectivity between their mobile devices and recommendations are based on the aggregation of ratings stored locally on their device. The underlying assumption is that users who are physically copresent in the same social context will be similar and the more often such encounters happen, the greater the similarity. Thus, the very way in which data is shared filters that data according to user similarity and, therefore, the first step of user-based filtering to compute similar users is not required. In [1], we also report on studies carried out to validate the underlying assumption. Note that if users have exchanged ratings and then edit them afterwards when they are no longer connected, then the updates should be propagated if and when the users encounter each other again.

Both of the above examples, show how mobile applications may take advantage of the presence or absence of connectivity to share information opportunistically. The underlying physical proximity serves as a filter mechanism inherent to the application logic.

Some frameworks have been developed specifically for P2P connectivity in mobile settings including Mobile Web Services [12] and JXTA [13]. However,

[4] http://www.db4o.com

these tend to focus on lower-level forms of data exchange rather than information sharing. For example, in JXTA, the application development consists of specifying message formats and how they are processed in terms of request and response handling similar to that of service-oriented architectures. This results in a blending of the application logic typically embedded in an object-oriented data model and the collaboration logic specified based on a request-response scheme. Efforts to provide higher level abstractions of P2P networks have either focussed on the allocation and retrieval of identifiers to resources in fixed networks without considering any notion of handling [14] or they offer only a few limited collaboration primitives and lack support for vicinity awareness [15,16].

Within the database research community, a number of P2P database systems, overlay networks and middlewares have been developed, including Pastry [17], Piazza [18], PeerDB [19], Hyperion [20], P-Grid [21] and GridVine [22]. However, research efforts have tended to focus on issues of object identity, schema matching and query reformulation, distributed retrieval, indexing and synchronisation as well as transaction management. To date, there has been little effort on supporting developers of mobile applications that utilise P2P connectivity to share information opportunistically with other users in the vicinity.

3 Modes of Opportunistic Information Sharing

In this section, we examine the different modes of opportunistic information sharing in mobile ad-hoc networks by looking at an example application that features all four basic modes. The application scenario is graphically depicted in Fig. 1. The numbers in the figure are used in the following description to refer to the individual steps of a use case.

Assume a user intends to go to a movie theatre where they are a regular customer. Further, they have installed an application on their mobile phone that allows them to take advantage of the new technical facilities that this particular theatre offers to registered customers. Equipped with their mobile phone, our customer heads to the theatre (1). As soon as the user enters the building, the phone connects to a theatre server and the current movie schedule is displayed on their device (2). The programme contains all movies that will start playing within the next three hours along with the number of available seats. While browsing the movies, the number of available seats is updated in real time. The programme is stored persistently on the device and kept up to date by the theatre server as long as the user is connected.

While our customer is trying to decide on a movie, an advertisement suddenly pops up informing them that they have won a 50% reduction voucher for a particular movie (3). Such pop-ups are only available on the device while it is connected to the server and may be updated by the server, for example in order to upgrade from a 50% reduction voucher to a free ticket as the start time approaches. Having decided to accept the voucher, our customer orders a ticket for the proposed movie and the ticket is sent to their phone (4). The ticket data will be stored on the phone persistently, and is handled as a stand-alone piece of data not kept in sync with any data residing on the server.

Now our customer realises that they have left their spectacles at home and decides to return home to collect it before the start of the movie. When they leave the building, their application indicates that it is no longer connected with the server (5). As a consequence, the previously received advertisements vanish. In contrast, the ticket along with the list of movies remains reminding them of the start time. They later return and watch the movie (6). Their ticket gets marked as expired (7) and therefore will be removed from their mobile phone when leaving the movie theatre (8).

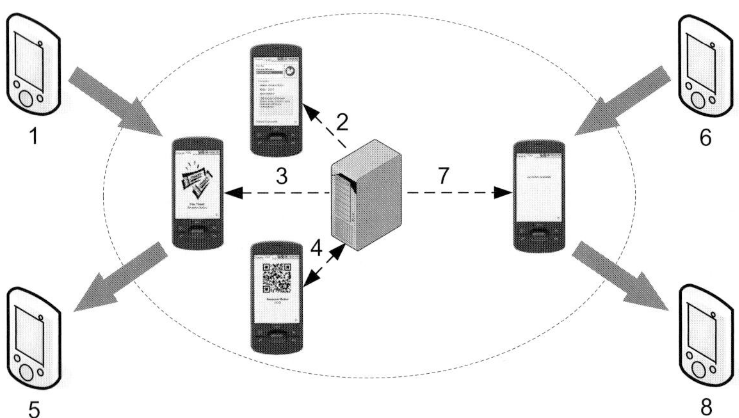

Fig. 1. Application

The short-range connections between personal mobile devices or between mobile devices and fixed installations described in this scenario allow for information to be shared based on physical proximity. In a mobile environment, connections are likely to be short-lived as devices move in and out of range. Therefore, one consideration is whether shared data should still be available when the connection is lost or if it should only be available during the connection. For example, the movie schedule that users receive as soon as they enter the theatre is stored persistently so that they can still access it after leaving the theatre. In contrast, pop-up advertisements are only available on the mobile device while the users are in the theatre. Once they have left, these advertisements are automatically removed from the mobile device.

Furthermore, in some cases, shared information should be kept synchronised to whatever extent is possible after devices are disconnected, while in other cases the copies of data should be decoupled. In our scenario, the movie schedule along with the number of available seats displayed on the mobile device is continuously updated by the theatre server. Changes carried out on the server are automatically propagated to the mobile devices. When a customer leaves the theatre, their mobile device is no longer connected and hence the copy of the movie programme stored on their device will not reflect the most recent updates. However, the next time that the customer enters the theatre and connection is re-established that data will be updated immediately. In contrast, a movie ticket is copied to the mobile device in such a way that it is decoupled from the original object residing on the server.

Taking these cases into consideration, two orthogonal concepts can be identified. Data *continuity* refers to the question of whether the lifetime of an object on a client is limited to the lifetime of the connection. Data *coupling* addresses the question of whether objects on different devices are kept synchronised. The composition of these concepts defines four modalities of information sharing as shown in Fig. 2. The horizontal axis covers the data coupling concept and the vertical axis represents the concept of data continuity. In this figure, we assume two devices are connected and take on the perspective of information being sent from a local to a remote device.

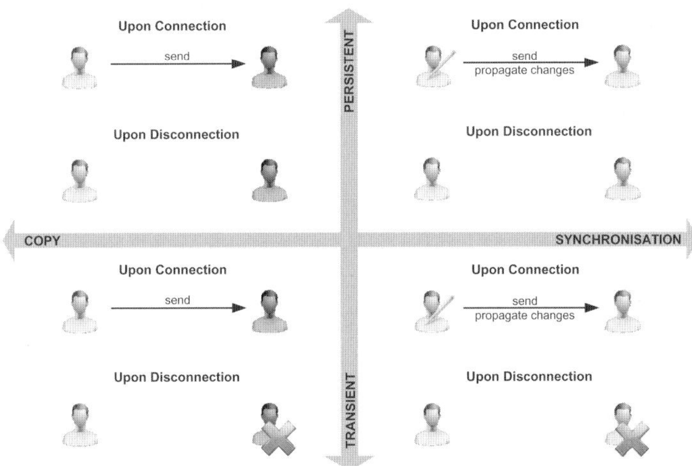

Fig. 2. Sharing Modes

The *persistent copy* mode is regarded as the default mode because it is the simplest one and is likely to be the one used most often. Upon connection, objects are copied to the remote device. The copied objects are not linked to the original ones. Instead, the original and the remote data will be treated as two distinct objects that can be manipulated independently. After disconnection, the received data remains visible allowing further offline access.

In *transient copy* mode, received objects are deleted as soon as the devices disconnect. Access to information on the receiver's device is therefore determined by the state of connectivity. However, the received objects are still treated as independent and, therefore, while the devices are connected and the data is visible, updates are not propagated.

In *persistent synchronisation* mode, when devices connect, data is transferred in such a way that it remains available after disconnection. As long as the connection is established, updates are propagated. Clearly, object synchronisation cannot take place after disconnection. However, data will be synchronised as soon as the connection is re-established. The access to local data is always possible, hence allowing offline operation.

Finally, the *transient synchronisation* mode covers the case when data on the receiver device is only visible during connection and updates are propagated. Manipulations of the data do not need to be tracked after disconnection because, upon disconnection, the data on the remote device no longer exists.

Note that the mode in which objects are shared may be altered after they have been shared. For example, objects shared in transient mode can be set to be persistent by the receiver. As a result, these objects will not be deleted upon disconnection. Persistent objects may be set to be transient in which case they will be deleted. Similarly, synchronised objects can be decoupled and objects that have been shared in copy mode may be set to be synchronised.

4 Platform for Opportunistic Information Sharing

To support the development of mobile applications that take advantage of ad-hoc network connections for opportunistic information sharing, we have developed a general platform based on an abstract model of information sharing that offers the four basic modes of information sharing presented in the previous section. The model is based on the classification of objects into collections that determine the basic sharing mode. Further, an event model allows handlers to be registered for particular events such as the addition/removal of an object to/from a collection or updates to attribute values of objects within a collection. Upon such events, the registered handlers are notified and, consequently, the handlers execute their actions which implement the specific sharing logic.

All sharing modes are implemented based on the collections shown in Fig. 3. The root collection named `Objects` contains all existing objects and the `Shared Objects` collection is a subcollection that contains all objects that have either been sent or received. In the previous section, we identified two orthogonal sharing mode dimensions, namely transient—persistent and copy—synchronisation. Each of these dimensions is represented as a single subcollection of the `Shared Objects` collection. Thus, the `Transient` collection represents those objects which are shared and not persistent while the `Synchronised` collection represents those objects which are shared and synchronised i.e. not decoupled as in the case of copy. Objects can belong to neither, one or both of these subcollections, thereby representing the four different modes of information sharing.

Handlers registered to be notified about the arrival or departure of peers to and from physical proximity, as well as for changes to object attribute values, drive the sharing process in terms of actions. The actions consist of executing queries, propagating updates, sending objects and adding or removing sent and received objects to and from collections. The sequence of actions specific to particular sharing modes are summarised in Tab. 1.

We now outline how the collections are used in order to realise the four different sharing modes. For the sake of simplicity, we assume two peers sharing

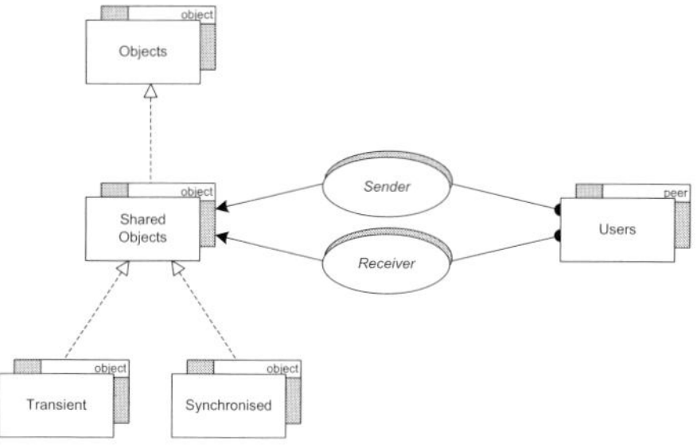

Fig. 3. Collection model of the sharing modes

Table 1. Operations associated with sharing modes

	Sender	**Receiver**
Persistent Copy	**Upon connection**	
	Insert into Shared Objects	Insert into Shared Objects
	Insert into Receiver	Insert into Sender
	Upon disconnection	
	—	—
Transient Copy	**Upon connection**	
	Insert into Shared Objects	Insert into Shared Objects
	Insert into Receiver	Insert into Sender
		Insert into Transient
	Upon disconnection	
	—	Delete
Persistent Synchronisation	**Upon connection**	
	Insert into Shared Objects	Insert into Shared Objects
	Insert into Receiver	Insert into Sender
	Insert into Synchronised	Insert into Synchronised
	Handler on synchronised objects	Handler on synchronised objects
	Propagate/Apply changes	Propagate/Apply changes
	Upon disconnection	
	Track changes	Track changes
Transient Synchronisation	**Upon connection**	
	Insert into Shared Objects	Insert into Shared Objects
	Insert into Receiver	Insert into Sender
	Insert into Synchronised	Insert into Transient
	Handler on synchronised objects	Insert into Synchronised
	Propagate/Apply changes	Handler on synchronised objects
		Propagate/Apply changes
	Upon disconnection	
	Remove from Synchronised	Delete

a single object under a particular sharing mode, where one of them acts as a sender and the other one as a receiver. One-to-many and many-to-one object sharing which typically occur in mobile settings can always be represented by multiple and directed one-to-one object transfers. Note that, for a particular object to be shared, the sharing mode is defined and known to the system on both the sender and receiver side. The way in which objects are associated with a particular sharing mode will be explained in the following section when we show how applications are developed.

Figure 4 highlights those collections and associations to which a shared object or other objects are added and how they are associated. In the four sharing modes, all sent and received objects are inserted into the `Shared Objects` collection. Additionally, received objects are associated with the sending peer using the `Sender` association on the receiver side. Thus, we can avoid exchanging objects that have already been sent in persistent sharing modes. Conversely, sent objects are associated with the receiving peer using the `Receiver` association on the sender side. This information is exploited for the propagation of updates.

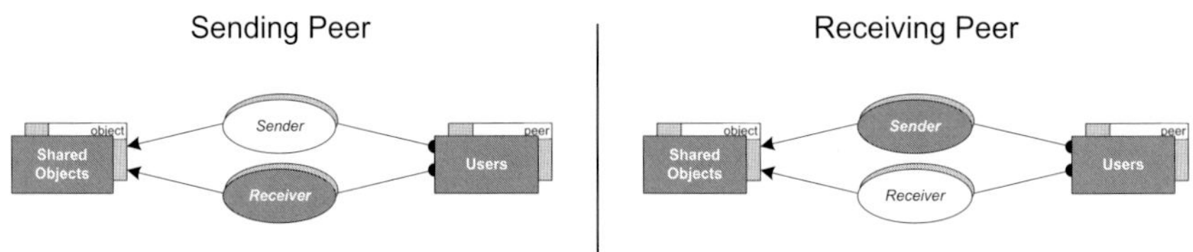

Fig. 4. Association of shared objects with sending and receiving peer

Figure 5 shows the collections to which an object being shared is added. The set of collections to which it is added can be seen as a configuration resulting in the object being shared in a configuration-specific sharing mode. The configuration is a classification and therefore it is the classification that determines the mode. Since objects can be dynamically classified at runtime, the sharing mode can also be managed dynamically at runtime.

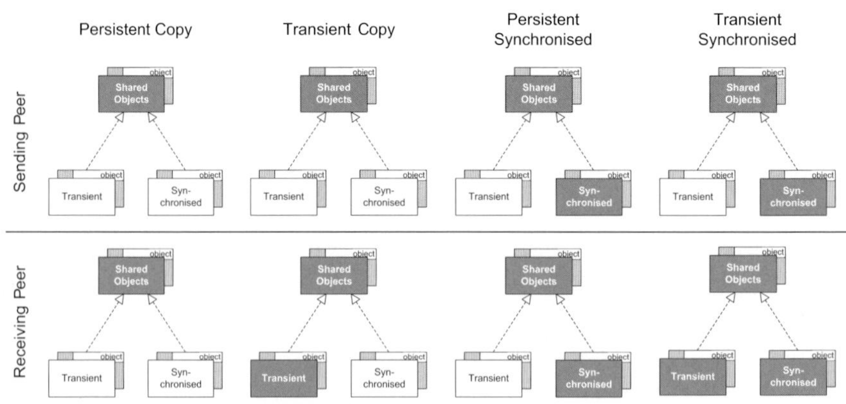

Fig. 5. Sharing mode-specific classifications of shared objects

If an object is to be shared in persistent copy mode, it is not classified any further. As a result, the object will simply be sent or received, persist despite disconnections and not be kept synchronised. If an object is shared in transient mode, it will additionally be put into the `Transient` collection on the receiver side, when it is received. Upon disappearance of the sending peer, a handler retrieves all members of this collection which are associated with the disappearing peer by means of a query. These objects are deleted in order to guarantee that they are no longer visible nor accessible.

In the case that an object should be shared in synchronised mode, it is put into the `Synchronised` collection on the sender and receiver sides, before being sent and after having been received, respectively. On both sides, a handler is registered to observe changes on all members of that collection. When updates are performed, the handler action will propagate the changes to the peers associated with the updated objects. If such peers are not in the vicinity, the handler retains its action so that it will be re-executed as soon as a connection has been re-established. Finally, if an object is to be shared in transient synchronised mode, it will be put into the `Synchronised` collection on both the sender and receiver sides as well as into the `Transient` collection on the receiver side. The effect will be the combination of the effects of having the object in one of the collections as described for the persistent synchronised and transient copy mode.

The sharing mode of an object is changed by adapting its classification. An object configured to be shared in transient mode can be removed from the `Transient` collection during connection. As a result, it will not be deleted when the connection is lost. Conversely, persistent objects may be added to this collection in order to make them transient. The synchronisation mode may be altered by adding or removing objects from the `Synchronised` collection.

5 Application Development

Having presented our model of the sharing modes based on collections, we now describe the implementation of the application scenario presented in the previous section. For the sake of a comprehensive overview, we start at the beginning by presenting how a database is created and opened using the following code.

```
DatabaseManager dbms = new DatabaseManager();
dbms.createDatabase("MovieTheatreApplication");
Database db = dbms.open("MovieTheatreApplication");
```

The application domain is modeled by four different types which are depicted in Fig. 6 by means of UML classes. Since these types are implemented as regular Java classes, we do not show the program code. The `Movie` class consists of a name, a start time, number of available seats and a description. A `Ticket` has attributes referencing the customer peer and the movie. A peer object contains all necessary information to identify and connect to a peer. An `Order` contains a peer and a movie object. In terms of advertisements, an image together with a string phrase can be specified when an object of type `Advertisement` is created.

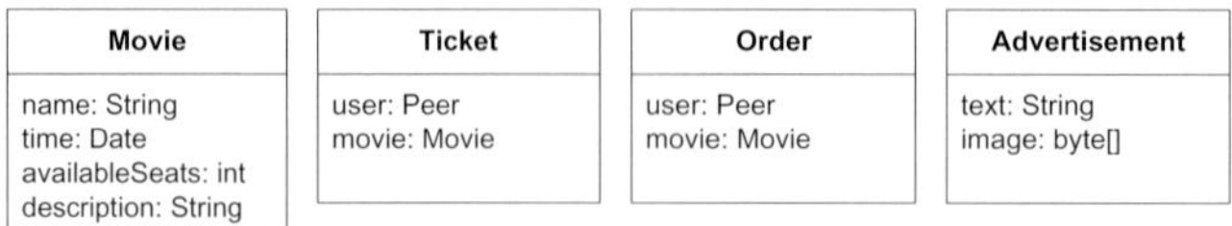

Fig. 6. UML diagrams of domain types

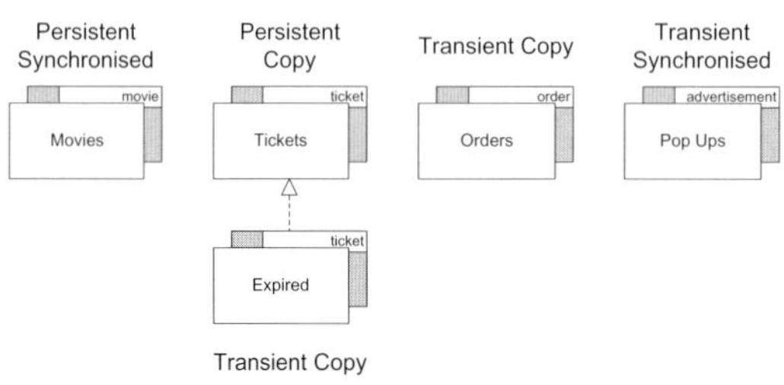

Fig. 7. Collection model

As shown in Fig. 7, a collection for every type is created to manage the extents of the types as well as one subcollection to classify tickets as expired. These collections are all set to be persistent collections with the effect that all members are stored in the database. A collection is created by providing its name and membertype and made persistent as shown by the following statements.

```
Collection<Movie> movies = db.createCollection("Movies", Movie.class);
movies.setPersistent();
```

When the user in our scenario enters the building for the first time, their mobile device automatically connects to the server provided by the movie theatre. The current movie programme is presented to the user based on the following query to select movies showing in the next 3 hours:

```
QueryNode<Movie> selection = db.queries().
  select(movies, Movie.time, System.currentTimeMillis()
    + 3 * 3600000, BinaryOperator.SMALLER);
```

The application running on the handheld device makes the `Movies` collection available and therefore it is able to receive members. Available collections contain those objects that are shared and are made available in a particular sharing mode. Therefore, the sharing mode for a particular object is defined by the collection through which it is shared. In the case of the `Movies` collection, whenever changes to the movie items are carried out on the server, the updates are forwarded to all visitors. This is realised by making the movie collection available in persistent synchronised mode as shown in the following code:

```
movies.makeAvailable(selection, Synchronised.TRUE && Persistent.TRUE);
```

The sharing is initiated by the appearance of a peer in physical proximity to the movie theatre which is translated to a new member added to the `Vicinity` collection. Consequently, a handler must be created which executes the query specified above and sends the query result to the new peer in the vicinity. This handler is then registered with the vicinity collection for addition events.

```
class VicinityHandler implements Handler {
  private Collection collection;
  public VicinityHandler(Collection collection) {
    this.collection = collection;
  }
  public void action(Event event) {
    Peer peer = ((AddEvent)event).getMember();
    Collection<Movie> result = this.collection.query().execute();
    Connectivity.send(peer, result);
  }
};
Handler handler = new VicinityHandler(movies);
db.queries().collection("Vicinity").addAdditionHandler(handler);
```

The other collections are created and made available similarly. The `Advertisements` collection is made available on the device of the user as well as on the server. As long as the user resides within the building and stays connected, advertisements can be distributed and updated at any time as defined for the transient synchronised mode.

The functionality to order a ticket is also implemented based on the notion of shared collections. The `Orders` collection is made available on both the mobile devices and the server. The server application acts as a client and receives order items. Since these items are only required to exist while the user is at the cashpoint, the availability is set to transient. The receipt of an order item will be treated by a handler registered for addition events on the `Orders` collection. The handler action is executed resulting in the creation of a ticket object that is added to the `Tickets` collection which will be transfered directly to the requesting device. Orders are handled immediately which leaves no time for updates. Therefore this collection is not set to be in synchronisation mode.

A ticket is a unique item that cannot be changed by the client nor by the server application. It stays visible independently of whether a connection is established or not. Both of these requirements are covered by the persistent copy mode. However, as soon as a movie starts, the server application sets the user's ticket to be expired. This is achieved by adding the ticket object to the subcollection of the `Tickets` collection named `Expired`. This subcollection has been made available with a transient copy sharing mode which overrides the sharing mode of the supercollection. Therefore, the ticket object residing on the mobile device will be removed from the device as soon as the user leaves the theatre.

6 Platform Implementation

Our platform is composed of three components as shown in Fig. 8. A collection framework allows objects to be classified, shared and stored persistently. An event system offers the means to define events, to register handlers to be notified

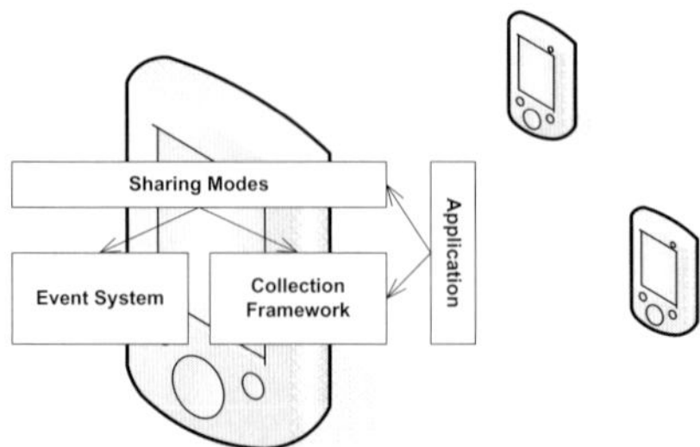

Fig. 8. Architecture

about events and to associate actions with handlers. Finally, a sharing mode component implements the modalities introduced in Sect. 3 based on the other two components.

An application programming interface (API) provides access to a database management system that manages collections, events, handlers and query representations. Through the API, database instances can be created, opened, closed and deleted, as well as allowing objects, collections, events, handlers and queries to be created, retrieved and deleted, defined and executed with a database instance. The platform is implemented in Java allowing mobile applications to be implemented using the regular Java platform.

6.1 Collections

Our platform is based on an extension of a P2P collection framework that we developed previously [23]. We therefore first give an overview of the concept of a P2P collection. Fig. 9 presents the concept as a UML diagram. A P2P collection has a name with which it can be retrieved within a query and membertype defining the type of its members. It allows members to be accessed and dynamically added and removed at runtime. A P2P collection may optionally store its members persistently which can be configured at design and runtime. Furthermore, a query system allows P2P collections to be retrieved, unified, intersected and objects to be selected, mapped and accessed.

As shown in Sect. 5, sharing is implemented by making collections available or unavailable. An available P2P collection can send and receive members to and from compatible collections residing on remote peers. Two P2P collections are compatible if they have the same name and compatible membertypes. Members may be sent upon explicit request or automatically as a result of a peer appearing in physical proximity to another one. Members that are sent from a source collection to a target collection will appear as members of the target collection. In some cases where members are shared automatically as soon as users are in each other's proximity, there must be a way of selecting which members of a collection to send. For this purpose, a query can be defined and provided when the P2P collection is made available to return the required members.

Fig. 9. Collections

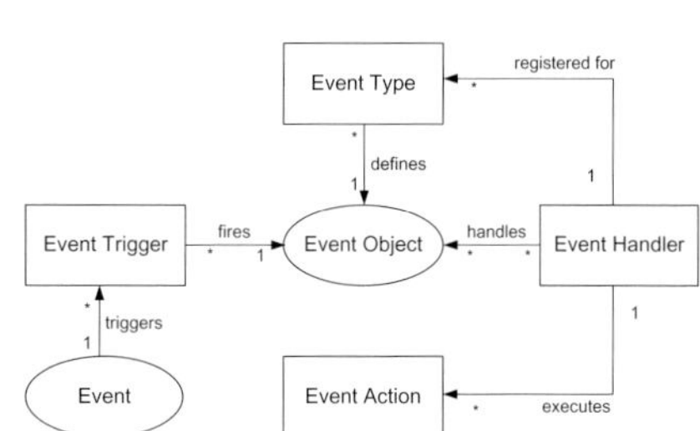

Fig. 10. Event concepts

Having P2P collections as a central point for information sharing affects the manner in which applications are developed. In a model-driven development scenario, this results in one additional step. First, the application domain is modeled in terms of types which map to Java classes. In a second step, P2P collections are defined. By default, a P2P collection is created for each domain type to represent its extent. Collections and subcollections as well as associations may be added at any time in order to further classify domain objects.

6.2 Events

The second component of our platform for information sharing in mobile ad-hoc networks is an event system based on a unified event model for object databases [24,25]. We describe here those concepts required in the scope of this work. The original motivation of our event system was to unify concepts for event handling found in object-oriented programming languages with trigger mechanisms from database systems in the setting of object databases. Current object-oriented databases offer limited support for event handling and often delegate this issue to the programming language. We believe that object databases need a well-defined event model that offers the generality and flexibility supported in active databases and modern event processing systems.

We base our event model on the four concepts of *event triggers*, *event types*, *event handlers* and *event actions* shown in Fig. 10. Event triggers are an abstraction of low-level operations that occur in the database such as an object being created, changed or added to a collection. Apart from being bound to such database operations, event triggers can also be bound to a schedule that determines when they are fired. Finally, they can also be fired explicitly by the user or the application code. Each event trigger is associated with an event type that defines the type of the event objects that are fired by the trigger. In addition to an attribute identifying the source of the event, event objects can also carry other information depending on the context in which they were fired. Whenever an event object is created by a trigger, our system matches it to event handlers that are registered to process events of a given type. All matching handlers are notified and the event object is passed to them. A handler can define a conditional expression that guards the execution of the event handling logic. If this expression evaluates to true, the handler invokes its associated event action and passes the required parameters.

By introducing separate concepts for event triggers, types, handlers and actions, event processing functionality can be made orthogonal to persistence. The model supports user-defined as well as system-defined event types with a registration service that allows one or more handlers to be associated with the same event type. Further, the model supports reuse of event types and event distribution which is an important requirement in the context of this work.

6.3 Sharing Modes

Once a P2P collection has been made available, it can either share members by explicit request or automatically upon the arrival of a peer in the physical vicinity. In the latter case, the collection is notified about the new peer, executes the query and sends the query result with the sharing mode specified by the available collection. As part of the P2P collection framework, a system collection named `Vicinity` contains objects representing peers. This collection is a direct representation of a peer's physical proximity and contains all those peers that are in the vicinity. A new peer appearing in the vicinity will trigger the creation of a new peer object which will be added to this collection. Conversely, the disappearance of a peer will trigger the removal of the respective object from the vicinity collection. Available collections register handlers with this vicinity collection to be notified about the addition or removal of peer objects. Upon notification, the mode-specific sharing process is executed.

Our platform for opportunistic information sharing integrates the P2P collection framework with the event system and also introduces new functionality to implement the sharing modes. By default, an available collection shares its members in persistent copy mode. If another mode is favoured, a particular mode can be provided as a second argument when the collection is made available. As a result, the sharing mode for a particular object is defined by the available collection of which it is a member. A sharing mode may be changed at any time by making a collection available once more with the new mode. Note that sharing modes of subcollections override the modes of the supercollections, only for those objects that are members of the subcollections. As a result, not only can the sharing mode of an available collection be changed dynamically at runtime but, using subcollections, it can also be adapted for dynamic subsets of objects.

7 Discussion

In [23], we compare the effort required to implement mobile applications using our P2P collection framework with that required if using mobile phone SDKs such as Java WTK. Figure 11 compares the components needed and the amount of interaction required to implement data persistence, vicinity awareness and

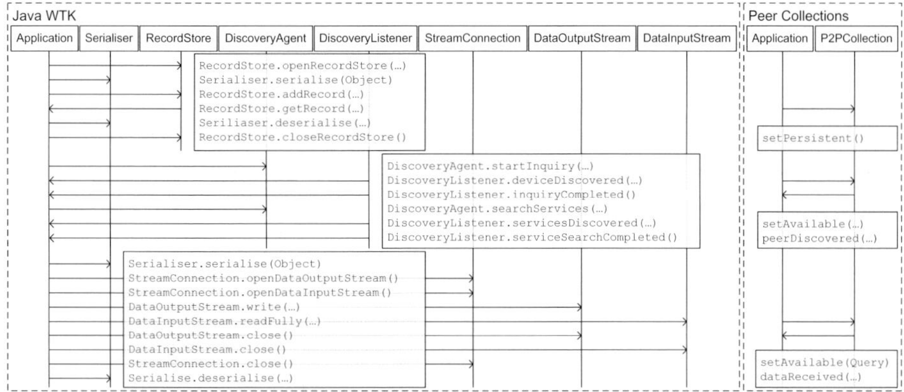

Fig. 11. Comparison of using Java WTK (left) and our platform (right)

data sharing using Java WTK, on the left hand side, and our P2P collection framework, on the right hand side. The ability to be able to simply set collections to store their members persistently is a great improvement compared to Java WTK where application objects have to be serialised manually and stored using key-value records. In order to store objects of a particular type based on key-value pairs, an application developer has to program a database-like component and put a lot of effort into overcoming the impedance mismatch between objects and key-value pairs. As opposed to the simple vicinity awareness mechanism provided by our platform, the developer of a Java WTK application needs to implement the scanning of the environment based on low-level connection technologies such as Bluetooth or WiFi. Moreover, using our framework, the application developer does not have to bother with low-level socket-based connectivity and data transmission in terms of serialisation and deserialisation. The fact that the developer can work at the level of the application model by deciding how data collections should be shared, and in which mode, presents a significant contribution to the development of mobile applications.

If a developer manages to implement data persistence, vicinity awareness and data sharing, the resulting sharing mode will be persistent copy. If another mode is required, the developer needs to additionally implement the mode-specific behaviour as described in Tab. 1. For transient mode, shared objects must be deleted from the device as soon as a peer is no longer in the vicinity, while the synchronisation of objects requires objects to be observed and updates to be conveyed and applied. The fact that our platform does all of this greatly simplifies the developer's work as they are simply required to select sharing modes rather than having to implement them.

8 Conclusions

Mobile applications with functionality to share information in ad-hoc peer-to-peer networks are becoming increasingly popular. These applications target domains such as collaborative filtering and location-based services which exploit the ad-hoc connectivity as a means of sharing information opportunitistically, thereby filtering information based on proximity. Therefore, in this novel class of applications, connections and disconnections throughout the life-time of a program are an integral part of the application logic rather than something to be hidden from the user and developer alike.

Due to the limited support for opportunistic information sharing in existing development platforms for mobile applications, we have designed and implemented a platform that provides high-level support to application developers. Central to the platform is support for the four basic modes of opportunistic information sharing that we have identified. These modalities are based on the observation that ad-hoc information sharing is mostly governed by two main factors, namely persistence and synchronisation. This platform builds on a P2P collection framework for sharing and persistence, an event system for the specification of event-driven application behaviour and a component that provides support for the four different sharing modes. We have shown how applications can make use of the functionality offered by the platform and compared it to the effort required using an off-the-shelf development kit.

References

1. de Spindler, A., Norrie, M.C., Grossniklaus, M.: Recommendation based on Opportunistic Information Sharing between Tourists. Information Technology & Tourism 10(4) (2009)
2. Xu, B., Ouksel, A., Wolfson, O.: Opportunistic Resource Exchange in Inter-Vehicle Ad-Hoc Networks. In: Proc. Intl. Conf. on Mobile Data Management, MDM (2004)
3. Burrell, J., Gay, G.K.: E-graffiti: Evaluating Real-World Use of a Context-Aware System. Interacting with Computers 14(4) (2002)
4. Fernandez, C., Escudero, C., Iglesia, D., Rodriguez, M., Marcote, E.: MOVIL-TOOTH: a Bluetooth Context-Aware System with Push Technology. In: Proc. IEEE Mediterranean Electrotechnical Conference, MELECON (2006)
5. Park, I., Hyun, S.J.: A Dynamic Mobile Transaction Management Strategy for Data-intensive Applications based on the Behaviors of Mobile Hosts. In: Information Systems and Databases (ISDB), pp. 92–97 (2002)
6. Repantis, T., Kalogeraki, V.: Data Dissemination in Mobile Peer-to-Peer Networks. In: Proc. Intl. Conf. on Mobile Data Management, MDM (2005)
7. Lee, M., Helal, S.: HiCoMo: High Commit Mobile Transactions. Distrib. Parallel Databases 11(1) (2002)
8. Padmanabhan, P., Gruenwald, L., Vallur, A., Atiquzzaman, M.: A Survey of Data Replication Techniques for Mobile Ad Hoc Network Databases. The VLDB Journal 17(5) (2008)
9. Leonhardi, A., Kubach, U., Rothermel, K., Fritz, A.: Virtual Information Towers – a Metaphor for Intuitive, Location-Aware Information Access in a Mobile Environment. In: Proc. IEEE Intl. Symp. on Wearable Computers, ISWC (1999)
10. Cano, J.-P., Manzoni, P., Toh, C.K.: First Experiences with Bluetooth and Java in Ubiquitous Computing. In: Proc. IEEE Symposium on Computers and Communications, ISCC (2005)

11. D'Souza, M.J., Postula, A.J., Bergmann, N.W., Ros, M.: Mobile Locality-Aware Multimedia on Mobile Computing Devices; A Bluetooth Wireless Network Infrastructure for a Multimedia Guidebook. In: Proc. Intl. Conf. on E-Business and Telecommunication Networks, ICETE (2005)
12. Srirama, S.N., Jarke, M., Prinz, W.: Mobile Web Services Mediation Framework. In: Proc. Workshop on Middleware for Service Oriented Computing, MW4SOC (2007)
13. Traversat, B., Arora, A., Abdelaziz, M., Duigou, M., Haywood, C., Hugly, J.C., Pouyoul, E., Yeager, B.: Project JXTA 2.0 Super-Peer Virtual Network. Technical report, Sun Microsystems, Inc. (2003)
14. Aberer, K., Alima, L.O., Ghodsi, A., Girdzijauskas, S., Haridi, S., Hauswirth, M.: The Essence of P2P: A Reference Architecture for Overlay Networks. In: Proc. IEEE Intl. Conf. on Peer-to-Peer Computing, P2P (2005)
15. Wang, A.I., Bjornsgard, T., Saxlund, K.: Peer2Me – Rapid Application Framework for Mobile Peer-to-Peer Applications. In: Intl. Symp. on Collaborative Technologies and Systems, CTS (2007)
16. Kortuem, G., Schneider, J., Preuitt, D., Thompson, T.G., Fickas, S., Segall, Z.: When Peer-to-Peer comes Face-to-Face: Collaborative Peer-to-Peer Computing in Mobile Ad hoc Networks. In: Proc. Intl. Conf. on Peer-to-Peer Computing, P2P (2001)
17. Rowstron, A., Druschel, P.: Pastry: Scalable, decentralized object location, and routing for large-scale peer-to-peer systems. In: Guerraoui, R. (ed.) Middleware 2001. LNCS, vol. 2218, p. 329. Springer, Heidelberg (2001)
18. Tatarinov, I., Ives, Z., Madhavan, J., Halevy, A., Suciu, D., Dalvi, N., Dong, X.L., Kadiyska, Y., Miklau, G., Mork, P.: The Piazza Peer Data Management Project. SIGMOD Rec. 32(3) (2003)
19. Ooi, B.C., Tan, K.L., Zhou, A., Goh, C.H., Li, Y., Liau, C.Y., Ling, B., Ng, W.S., Shu, Y., Wang, X., Zhang, M.: PeerDB: Peering into Personal Databases. In: Proc. ACM SIGMOD Intl. Conf. on Management of Data, SIGMOD (2003)
20. Rodríguez-Gianolli, P., Kementsietsidis, A., Garzetti, M., Kiringa, I., Jiang, L., Masud, M., Miller, R.J., Mylopoulos, J.: Data Sharing in the Hyperion Peer Database System. In: Proc. Intl. Conf. on Very Large Databases, VLDB (2005)
21. Aberer, K., Datta, A., Hauswirth, M., Schmidt, R.: Indexing Data-Oriented Overlay Networks. In: Proc. Intl. Conf. on Very Large Databases, VLDB (2005)
22. Cudré-Mauroux, P., Agarwal, S., Budura, A., Haghani, P., Aberer, K.: Self-Organizing Schema Mappings in the GridVine Peer Data Management System. In: Proc. Intl. Conf. on Very Large Databases, VLDB (2007)
23. de Spindler, A., Grossniklaus, M., Norrie, M.C.: Development Framework for Mobile Social Applications. In: Proc. Intl. Conf. on Advanced Information Systems Engineering, CAiSE (2009)
24. Grossniklaus, M., Norrie, M.C., Sgier, J.: Realising Proactive Behaviour in Mobile Data-Centric Applications. In: Proc. Intl. Workshop on Ubiquitous Mobile Information and Collaboration Systems, UMICS (2007)
25. Grossniklaus, M., Leone, S., de Spindler, A., Norrie, M.C.: Unified Event Model for Object Databases. In: Proc. Intl. Conf. on Object Databases, ICOODB (2009)

Unveiling Hidden Unstructured Regions in Process Models

Artem Polyvyanyy[1], Luciano García-Bañuelos[2,*], and Mathias Weske[1]

[1] Hasso Plattner Institute at the University of Potsdam
Prof.-Dr.-Helmert-Str. 2–3, D-14482 Potsdam, Germany
{Artem.Polyvyanyy,Mathias.Weske}@hpi.uni-potsdam.de
[2] Institute of Computer Science, University of Tartu
J. Liivi 2, Tartu 50409, Estonia
luciano.garcia@ut.ee

Abstract. Process models define allowed process execution scenarios. The models are usually depicted as directed graphs, with gateway nodes regulating the control flow routing logic and with edges specifying the execution order constraints between tasks. While arbitrarily structured control flow patterns in process models complicate model analysis, they also permit creativity and full expressiveness when capturing non-trivial process scenarios. This paper gives a classification of arbitrarily structured process models based on the hierarchical process model decomposition technique. We identify a structural class of models consisting of block structured patterns which, when combined, define complex execution scenarios spanning across the individual patterns. We show that complex behavior can be localized by examining structural relations of loops in hidden unstructured regions of control flow. The correctness of the behavior of process models within these regions can be validated in linear time. These observations allow us to suggest techniques for transforming hidden unstructured regions into block-structured ones.

Keywords: Process structure tree, process model analysis, process model correctness, process model transformation.

1 Introduction

Software engineers employ principles of conceptual modeling to encapsulate all the information about real world entities in formal models, e.g., specification of behavior. The research field of business process management [1] investigates the problem of capturing behavioral aspects of real world entities in process models. Process models are widely used to design, analyze, and improve how companies organize operational processes. Furthermore, process models are used in the design of distributed software systems and to provide a blueprint for systems that realize the processes.

* Supported by the Estonian Science Foundation and the European Regional Development Fund through the Estonian Centre of Excellence in Computer Science.

R. Meersman, T. Dillon, P. Herrero (Eds.): OTM 2009, Part I, LNCS 5870, pp. 340–356, 2009.

Process modeling languages, e.g., BPMN [2], formalize process models as directed graphs, where edges specify execution order constraints, task nodes represent business activities, and gateway nodes define the control flow routing logic of a model. Graph-based process modeling languages allow a great level of expressiveness for business analysts capturing process scenarios in models. In order to fulfill business goals, analysts can come up with arbitrarily structured process models. However, the degree of freedom which analysts gain is the primary source of errors [3,4,5]. One can easily end up with a model which encodes undesired scenarios, e.g., ones that never reach the goal state of a process or ones that result in the uncontrolled concurrent execution of business activities. Additionally, process models with complex structures aggravate the computational complexity of model analysis tasks, e.g., the task of checking the correctness of the process model's behavior—the (behavioral) soundness [6]. One widely accepted solution to address identified problems is to restrict the freedom, i.e., to restrain the structural principles of the composing models. This, however, limits the creativity during the business process model design phase. Furthermore, the restrictions on allowed structural patterns often result in models with replicated task nodes as well as replicated structural patterns, which are heavily introduced in order to capture envisioned scenarios.

In this paper, we use the SPQR-tree process model decomposition technique, known from compiler theory [7] and introduced to the business process management field in [8], to derive a structural classification of process models. We identify a structural class of process models which are composed solely of block-structured fragments. However, when these fragments are combined in a model they form regions of unstructured process behavior. In these regions control flow can enter and afterwards leave an individual fragment through the same node in a looping pattern. We refer to such regions as "hidden" unstructured regions.

Hidden unstructured regions have been identified before [7,8]. They are referred to as *non-prime subprograms* in [7] and as *directed bond fragments* in [8]. However, we take a step forward as we provide a characterization of their structural properties and describe a linear time method for verifying the soundness of the underlying process model. Additionally, we show that existing flow graph restructuring techniques can be adapted to transform hidden unstructured regions. Restructured process models become suitable for translation to block-structured languages such as BPEL [9,10], e.g., for execution.

The rest of the paper is organized as follows: The next section discusses the related work. In section 3, the SPQR-tree decomposition technique is presented. First, we exemplify the technique with the help of undirected graphs. Afterwards, we discuss the implications of the decomposition if performed on process models. In section 4, we present the notion of a behavioral correctness for process models. Section 5 presents structural process model classes and discusses the identification and behavioral analysis of hidden unstructured regions in process models. Section 6 shows that existing techniques can be adapted to transform hidden unstructured regions into block-structured process models. The paper closes with ideas on future steps and conclusions that summarize our findings.

2 Related Work

Kiepuszewski, Hofstede, and Bussler [4] identify the classes of unstructured process graphs that can be transformed to equivalent structured versions. Their classification relies on two parameters: the presence of parallel control flow (i.e., *and* gateways) and the presence of loops. Three categories are identified. The first category is restricted to *xor* logic only and allows loops. As the underlying logic is simple, one can use techniques developed for flowchart restructuring to transform a model in this category to an equivalent structured form (e.g., [11]). They also showed that restructuring can be achieved either by node duplication or by the use of auxiliary variables, but some cases can only be restructured with the use of the aforementioned variables. The second category comprises acyclic models which allow internal *and* logic. As the authors point out, the presence of *and* gateways may induce structural problems such as deadlocks or lack of synchronization. They further analyze this category and conclude that most unstructured models carrying internal *and* logic cannot be restructured except for a subset of the models with an overlapping structure [12]. Finally, the authors describe a category consisting of models with loops and with internal *and* logic. They concluded that, in general, well-behaved cyclic process can be restructured. But, the authors also presented a well-behaved cyclic process that cannot be restructured that uses variables to synchronize some of the parallel paths. An open question remains, whether there exist well-behaved arbitrary processes without the variables that cannot be restructured.

Liu and Kumar [5] extended the work by Kiepuszewski et al. Their aim was to identify the source of unstructuredness in process models and to investigate the scenarios and configurations giving rise to structural conflicts. Their taxonomy is built on top of three dimensions. First, the notion of *corresponding control elements* or *corresponding pair*, i.e., a split and a join gateways which have at least two distinct control flow paths going from the split gateway and reaching the join gateway. Corresponding pairs are further classified in *proper* ones if both gateways use the same logic and, otherwise, in *mismatched* ones. Second, the notion of *nesting of corresponding pairs*. Nesting is further divided in proper and improper, and quantified according to the number of intermediate improperly nested gateways: first-order improper nested, if only one intermediate gateway exists, and so on. The third parameter being considered is the presence of loops in the process model. The authors proceed by describing families of process models which are variations on the three dimensions. The first family corresponds to first-order improper nested graphs. To illustrate the characteristics of this family, they present a process model topology with four gateways and enumerate all variants (changing the gateway logic) and identify the potential structural conflicts. The second family corresponds to second-order improper nested graphs, for which a similar analysis is performed. In this context the authors identify the overlapping structure as the only sound configuration allowing mismatched corresponding pairs. The last family analyzed is that of first-order improper nested graphs with loops. The analysis of higher level improper nesting has been left open.

In this paper, we do some first steps in investigating how a process structure tree, in particular the SPQR-tree, can be employed for the efficient analysis of process model behavior. Similar to [4], we identify and investigate different structural process model classes. However, their focus is on restructuring suitability. In contrast, we want to investigate the structural fragment types discovered by the SPQR-tree decomposition. The fragments are finer-grained, and allow us to analyze both their structural and their behavioral properties. In this way, we identify classes of fragments for which correctness of process model behavior can be validated in linear time. This also differs from the approach described in [5]. Their study is based on the analysis of a single case for each family of process models with an arbitrary number of gateways, for which the set of possible variants on gateway logic has been enumerated. However, we consider that such approach lacks of generality. In our approach, the analysis is simplified by the fact that the SPQR-tree decomposition by itself gives the hints about the structural properties. We base our reasoning on this properties and derive the restrictions to be observed by sound process models.

3 The SPQR-Tree Decomposition

This section explains the technique of structural process model decomposition. First, the technique is exemplified by performing the decomposition of undirected graphs. Afterwards, we discuss the implications of the decomposition technique if performed on process models.

3.1 Graph Decomposition

The SPQR-tree decomposition is a decomposition of an undirected biconnected multigraph aimed at identifying its triconnected fragments. A graph is k-connected if there exists no set of $k-1$ elements, each a vertex or an edge, whose removal makes the graph disconnected. Such a set is called a separating $(k-1)$-set. Separating 1- and 2-sets of graph vertices are called *cutvertices* and *separation pairs*. 1-, 2-, and 3-connected graphs are referred to as *connected*, *biconnected*, and *triconnected*, respectively.

The algorithm for the discovery of triconnected fragments of a graph was first proposed in [13]. Afterwards, in [7], the algorithm was applied to sequential program parsing. It was proposed to decompose the directed program graph into the *parse tree* (or *the tree of the triconnected components*). The parse tree is a hierarchical representation of graph fragments induced by its *split pairs*, where a split pair is either a separation pair, or a pair of adjacent vertices. The parse tree was studied as SPQR-tree in [14,15]. [13,16,17] show the path towards a linear time complexity algorithm implementation of the SPQR-tree decomposition. The decomposition results in process fragments of four structural types: S-, P-, Q-, and R-type fragments.

- *Series case.* A split pair is a pair of graph vertices giving a maximal sequence of vertices and consists of k nodes and k edges ($k \geq 3$)—the S-type fragment.

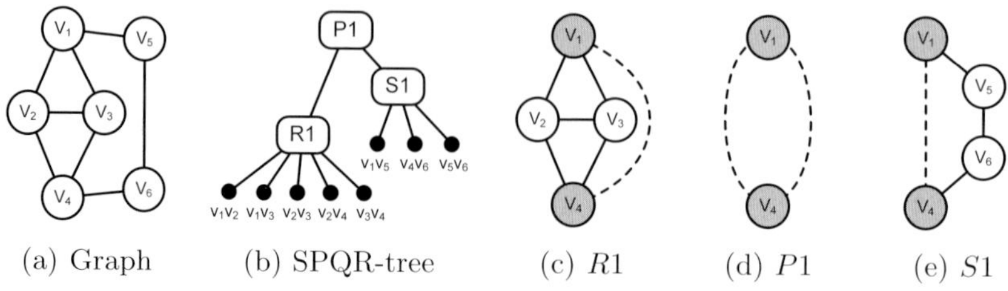

(a) Graph (b) SPQR-tree (c) $R1$ (d) $P1$ (e) $S1$

Fig. 1. SPQR-tree decomposition of a graph

- *Parallel case.* A split pair is a pair of adjacent graph vertices in k distinct edges ($k \geq 2$)—the P-type fragment.
- *Trivial case.* A split pair is a pair of adjacent graph vertices—a fragment consists of one edge—the Q-type fragment.
- *Rigid case.* If none of the above cases applies, a fragment is a triconnected fragment—the R-type fragment.

Figure 1 exemplifies the SPQR-tree decomposition. Figure 1(a) shows an undirected graph, whereas Figure 1(b) gives its SPQR-tree decomposition. The names of SPQR-tree nodes hint at the structural types of their underlying fragments, e.g., $S1$ is the series case fragment, $P1$ is the parallel case fragment, and $R1$ is the rigid case fragment. The v_iv_j nodes represent Q-type structural fragments—graph edges.

Each SPQR-tree node represents a *fragment skeleton*, i.e., the basic structure of a fragment and its relations with other fragments. Figures 1(c), 1(d), and 1(e) show the fragment skeletons of the SPQR-tree from Figure 1(b). Each fragment skeleton consists of the original graph edges (drawn with solid lines) and *virtual edges* (drawn with dashed lines). Each virtual edge is shared between two fragment skeletons and hints at a structural relation between skeletons in the SPQR-tree, whereas each original edge is contained in one skeleton. The nodes that form the separation pairs of the graph are highlighted with a grey background, e.g., nodes v_1 and v_4 in fragment skeleton $R1$. These nodes, when removed, disconnect fragment $R1$ from the rest of the graph. Observe that separation pairs are only discovered in a set of graph nodes with the number of coincident edges higher than two. This aligns with the definition of the series case fragment. In order to obtain a maximal sequence, any adjacent S-type fragments within the SPQR-tree must be combined. Otherwise, one can discover a combinatorial set of separation pairs within the series case fragments, e.g., v_1v_4, v_1v_6, and v_4v_5 within the $S1$ fragment skeleton (see Figure 1(e)). Similarly, the structural relation of a parallel case fragment within a parallel case fragment should be recognized as a single P-type fragment.

In order to obtain the original graph, one must "glue" all the fragment skeletons pair-wise in any order, along the virtual edges, i.e., merge adjacent vertices of the shared virtual edge. Once fragment skeletons are combined, the shared virtual edge is removed.

3.2 Process Model Decomposition

In this section, we examine fragments obtained after the SPQR-tree decomposition of a process model. We start with the definition of a process model adopted from [1], which is also the generalization of the definition proposed in [18]—gateways that are both split and join are allowed in a process graph.

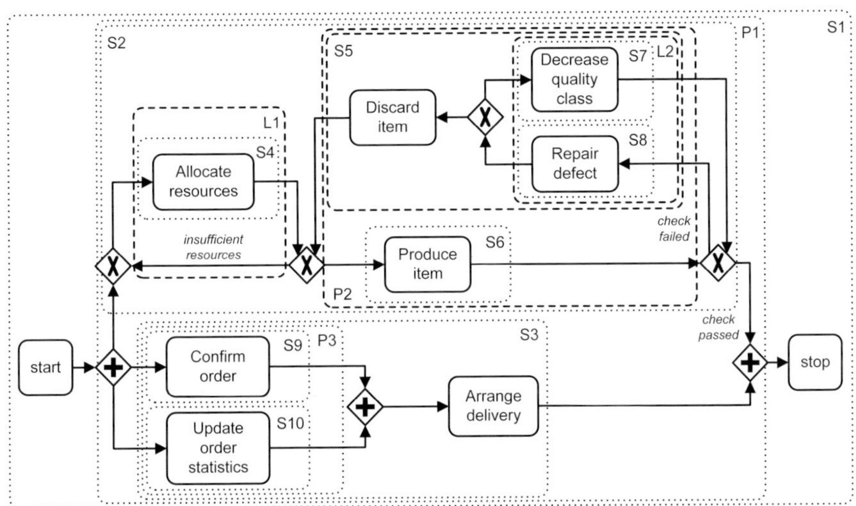

Fig. 2. A process model and its SPQR-tree node fragments

Definition 1 (Process Model). *A* process model *is a tuple* $P = (N, E, type)$, *where:*

- $N = N_T \cup N_G$ *is a set of nodes, where* N_T *is a nonempty set of tasks and* N_G *is a set of gateways; the sets are disjoint,*
- $E \subseteq N \times N$ *is a set of directed edges between nodes defining control flow,*
- $type : N_G \rightarrow \{and, xor\}$ *is a function that assigns a control flow construct to each gateway.*

Moreover, (N, E) is a connected graph—a *process graph*. Each task $t \in N_T$ can have at most one incoming and at most one outgoing edge ($|\bullet t| \leq 1 \wedge |t\bullet| \leq 1$), where $\bullet t$ stands for a set of immediate predecessor nodes ($\bullet t = \{n \in N | (n,t) \in E\}$) and $t\bullet$ stands for a set of immediate successor nodes ($t\bullet = \{n \in N | (t,n) \in E\}$) of task t. A task $t \in N_T$ is a *process entry* if $|\bullet t| = 0$. A task $t \in N_T$ is a *process exit* if $|t\bullet| = 0$. There is at least one process entry task and at least one process exit task. Each gateway $g \in N_G$ has either more than one incoming edge, or more than one outgoing edge. A gateway $g \in N_G$ is a *split* if ($|\bullet g| = 1 \wedge |g\bullet| > 1$). A gateway $g \in N_G$ is a *join* if ($|\bullet g| > 1 \wedge |g\bullet| = 1$). A gateway $g \in N_G$ is a *mixed* gateway if ($|\bullet g| > 1 \wedge |g\bullet| > 1$).

We require process models to be structurally correct, i.e., structurally sound. Figure 2 gives an example of a structurally sound process model.

Definition 2 (Structural Soundness). *A process model is* structurally sound *if there is exactly one process entry, exactly one process exit, and each process model node is on a path from the process entry to the process exit.*

Similar to the case with graphs, one can construct an SPQR-tree of a process graph and obtain the decomposition of a process model's control flow edges. In general, an SPQR-tree can be rooted to any node. However, in the context of process models it makes sense to root the tree to the node which represents an S-type fragment that contains a process entry and a process exit. In this case, one obtains a structural process model hierarchy [8,18,7], i.e., a containment hierarchy of sets of process model edges. The hierarchy shows a refinement of structural patterns that collectively build up a process model, starting with a top level series case fragment.

In order to properly address parts of a process model, we define a process fragment as a connected subgraph of a process model.

Definition 3 (Process Fragment). *A* process fragment *of a process model* $P = (N, E, type)$ *is a tuple* $F = (N_F, E_F, type_F)$*, where* $N_G \subset N$ *is a set of gateways of* P*, which consists of a connected subgraph* (N_F, E_F) *of the process graph* (N, E) *of* P *and function* $type_F$*, which is a restriction of function type of* P *to the set* $N_F \cap N_G$*.*

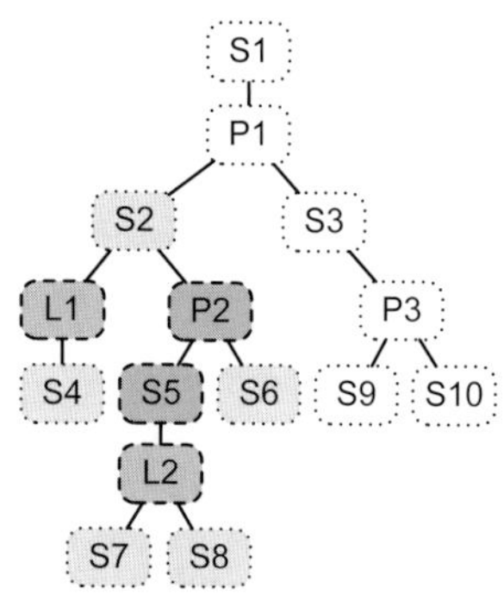

Fig. 3. SPQR-tree of the process model from Figure 2

The fact that the SPQR-tree decomposition of a process model delivers a concrete hierarchical containment of edges (the root node is always fixed) allows to uniquely identify process fragments that are represented by fragment skeletons. A process fragment that corresponds to a certain node, or a fragment skeleton, of the SPQR-tree is obtained by gluing its fragment skeleton with all its descendent skeletons in the SPQR-tree hierarchy. We refer to such a process fragment as an *SPQR-tree node fragment.* For instance, the $S1$ fragment in Figure 3 is a series case fragment that corresponds to the whole process model shown in Figure 2. In general, the SPQR-tree root fragment corresponds to the whole process model. Figure 3 shows the SPQR-tree decomposition of the process model from Figure 2. Observe that the Q-type fragments, i.e., the control flow edges, are not visualized for simplicity reasons. In Figure 2, each SPQR-tree node fragment is enclosed in the region with a dashed or dotted borderline, i.e., a fragment is formed by control flow edges enclosed in or intersecting the region and nodes adjacent in a set of edges that form the fragment.

In the following, we give a classification of nodes contained in a process fragment.

Definition 4 (Boundary, Internal, Entry, Exit nodes). *A node is either a boundary or an internal node of a process fragment F in a process model P:*

- *A node $n \in N_F$ is a* boundary *node of F if n is a process entry or a process exit of P, or there exist edges $e_i \in E_F$ and $e_j \in E \backslash E_F$ adjacent through n. A boundary node can be a fragment entry or a fragment exit:*
  - *A node $n \in N_F$ is a fragment* entry *if all the incoming edges of n are outside of F ($\bullet n \subseteq N \backslash N_F$) or all the outgoing edges of n are inside of F ($n \bullet \subseteq N_F$)*
  - *A node $n \in N_F$ is a fragment* exit *if all the outgoing edges of n are outside of F ($n \bullet \subseteq N \backslash N_F$) or all the incoming edges of n are inside of F ($\bullet n \subseteq N_F$)*
- *A non-boundary node is an* internal *node of a process fragment.*

We also employ the directed property of the process graph edges and the definition of entry and exit nodes to recognize a special type of process fragments—process components.

Definition 5 (Process Component). *A* process component *of a process model is a process fragment $C = (N_C, E_C, type_C)$ with exactly two boundary nodes: one fragment entry and one fragment exit.*

This notion of a component was first introduced in [7] as a concept of a *proper subprogram.* Process fragments that are not components are *non-components.* The importance of process fragments with a single entry and a single exit logic was identified in [19,8]. In [18], we showed how process components can be used for the task of process model abstraction, i.e., the discovery of reducible process fragments. The implication of the technique is fragmentation of structural as well as behavioral process model analysis tasks. In [18], we proved that for a process model which forbids mixed gateways, all SPQR-tree node fragments are process components. The observation is captured in Theorem 1.

Theorem 1. *Any SPQR-tree node fragment of a structurally sound process model, which forbids mixed gateways, is a process component.*

However, for the generalized definition of a process model (see Definition 1), it does not necessarily hold that all SPQR-tree node fragments are components. In the process model from Figure 2, fragments $S5$, $L1$, $L2$, and $P2$ (the corresponding SPQR-tree nodes are highlighted with dark grey background in Figure 3) are non-components. For each of the fragments, one of the boundary nodes is neither an entry, nor an exit. In Figure 2, non-components are enclosed in the regions with a dashed borderline, whereas components are enclosed in the regions with a dotted borderline.

In the following part of the paper, we will investigate the structural particularities of the parallel case process fragments, which are non-components, and their influence on the behavioral analysis of process models.

4 Correctness of Process Models

Section 3 stated a structural requirement of a process model correctness—structural soundness. In this section, we discuss the correctness property relevant to the dynamics of processes—(behavioral) soundness.

Before one can judge the correctness of process behavior, the execution semantics of process models has to be specified. We specify execution semantics of process models by proposing a mapping to Petri nets [20]. The mapping procedure is adopted from [21] and is visualized in Figure 4. Figure 4(a) shows all possible patterns of a process model edge that connects tasks or gateways of either *xor* type or *and* type. During the mapping procedure, each process model edge is mapped onto the corresponding Petri net pattern proposed in Figure 4(b). In addition to the mapping rules proposed in Figure 4, one has to add a source place i which enables a transition corresponding to the process entry, and a sink place o which is the only output place of the transition corresponding to the process exit. Similar to [21], we name the described mapping—the *petrify* mapping. The result of the *petrify* mapping is a workflow net [6]. Also, similar to [21], one can show that the mapping of a process model onto a Petri net results in a free choice Petri net [22]. Places with multiple outgoing arcs can only result when mapping process model edges that have a common *xor* type gateway as a source node. However, each of these outgoing arcs always has only one transition as a target; this transition models the choice decision.

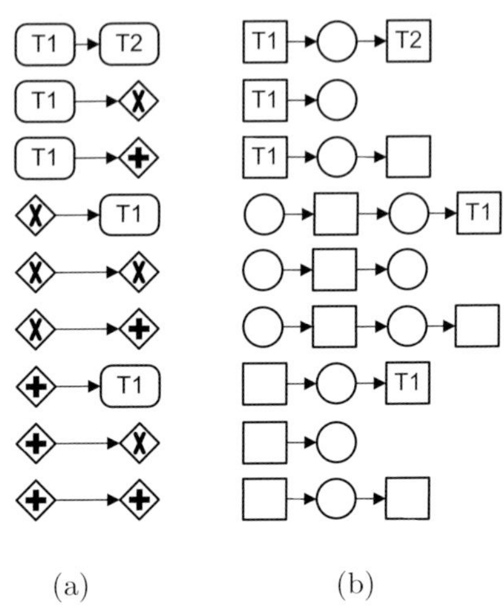

Fig. 4. Mapping process models to Petri nets

The mapping specifies execution semantics of process models as follows: A *xor* split forwards control flow along one of the outgoing edges. A *xor* join merges multiple alternative threads of control flow without synchronization. An *and* split concurrently forwards control flow along all the outgoing edges. An *and* join synchronizes multiple alternative threads of control flow. A mixed gateway first behaves as a join and then as a split of the corresponding gateway type.

At this point, we are ready to define the (behavioral) soundness property of a process model.

Definition 6 (Soundness). *A process model is* sound *if the* petrify *mapping of the process model results in a sound workflow net.*

Wil van der Aalst showed in [6] how the soundness property of workflow nets relates to the properties of liveness and boundedness, i.e., a workflow net is sound

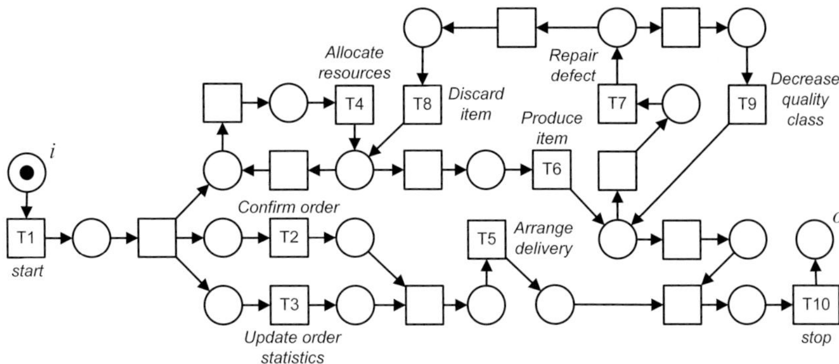

Fig. 5. A workflow net, the *petrify* mapping of the process model from Figure 2

if and only if the extended net is live and bounded, where the extended net is obtained by adding an extra transition $t*$ to the workflow net which connects the sink place o with the source place i. Liveness and boundedness of free choice Petri nets can be checked in polynomial time [22]. Therefore, the soundness of free choice Petri nets as well as the soundness of process models can be validated in polynomial time.

Alternatively, the soundness property of a process model can be deduced from the absence of structural conflicts in the process model.

Theorem 2. *If a workflow net is a result of the petrify mapping of a process model which is free of dead-locks and free of lack of synchronization, it is sound.*

Taking into consideration that any workflow graph obtained as a mapping of a process model is a free choice Petri net, the proof of Theorem 2 is analogous to the proof of Theorem 2 in [21].

Figure 5 shows a workflow net, which is the result of the *petrify* mapping of the process model from Figure 2. The workflow net is sound; the soundness can be validated in polynomial time. Therefore, according to Definition 6, the process model from Figure 2 is sound.

5 Hidden Unstructured Regions in Process Models

In this section, we identify regions in process models that hide unstructured process logic and discuss structural constraints of these regions implied by model correctness properties. We start the discourse by identifying structural classes of process models based on the SPQR-tree decomposition for which soundness can be decided in linear time. One can define structural process model classes based on the presence or absence of certain structural case fragments in process models and the notion of a process component. For instance:

Definition 7 (Block-structured Process Model). *A process model is* block-structured *if the SPQR-tree decomposition of the process graph contains no R-type fragments and all SPQR-tree node fragments are process components.*

In the general case, SPQR-tree node fragments can also be non-components:

Definition 8 (Quasi Block-structured Process Model). *A process model is* quasi block-structured *if the SPQR-tree decomposition of the process graph contains no R-type fragments.*

To conclude, a process model is *graph-structured* if the SPQR-tree decomposition of the process graph contains at least one R-type fragment. In the following, we examine in detail the block-structured classes of process models. Before we continue with the discussion, we identify loop case fragments which are the special types of parallel case fragments.

Definition 9 (Directed Fragment). *An SPQR-tree node fragment is* directed *if one of its boundary nodes has only outgoing incident edges among fragment edges, and the other has only incoming incident edges among fragment edges.*

The directed property assures that once control flow enters the fragment, it does not reach the fragment entry before the fragment exit. Also, if control flow reaches the fragment exit it passes control outside the fragment. Process fragments that are not directed are *non-directed* fragments. Fragment $S3$ from Figure 2 is directed, whereas fragment $S5$ is not. A directed process fragment is clearly a process component.

Process components are useful for behavioral analysis. One can expect that once control flow enters a component through the fragment entry, it also leaves the component through the fragment exit exactly once. This observation was developed in [18] to propose the process model abstraction technique which aggregates process fragments, per process component base, into tasks of a higher abstraction level. Therefore, if one is assured of the correctness of a process component, the component can be seen as a Q-type fragment which passes control flow from its entry to its exit.

Definition 10 (Loop Case Fragment). *An L-*type *(or* loop case*) fragment is a parallel case fragment for which: (i) the entry of the fragment is also the exit for at least one of its child fragments, and (ii) each child skeleton specifies a directed SPQR-tree node fragment.*

Fragment $L2$ in Figure 2 is an L-type fragment. $L2$ contains two directed fragments $S7$ and $S8$ and the entry of $L2$ is the exit of $S8$; there exists a cyclic path that goes through the boundary nodes of $L2$. Contrary, fragment $P2$ in Figure 2 is a P-type fragment, but not an L-type fragment; $P2$ contains a non-directed fragment $S5$.

It is a straightforward task to check whether a block-structured process model is sound. One must check if the gateway types of the boundary nodes for each P-type fragment which is also a component match, meaning both gateways are either of *xor* type or of *and* type. To ensure the soundness of a block-structured process model, any L-type component of the process model must be structured by boundary gateways of *xor* type. If the fragment entry of an L-type component is an *and* type gateway, there is a dead-lock situation, while if the fragment exit of an L-type component is an *and* type gateway, there is a live-lock situation.

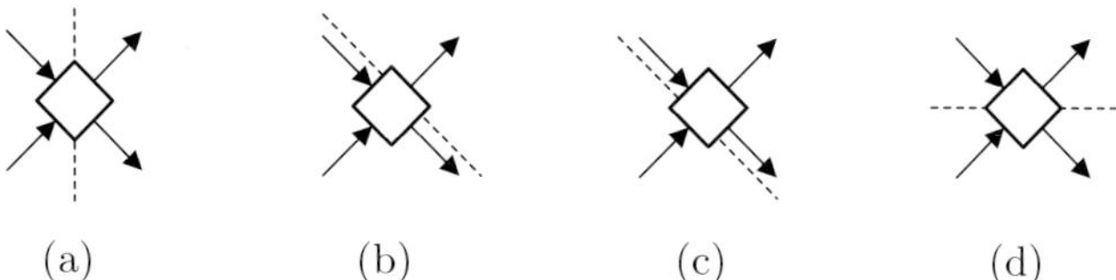

Fig. 6. All unique combinations for edge separation on internal and external fragment edges for a boundary node of a fragment which is a mixed gateway

This also implies a lack of synchronization at the fragment exit. We summarize the conditions in the following theorem:

Theorem 3. *A block-structured process model is sound if and only if:*

- *for each P-type component the boundary gateways match in type,*
- *for each L-type component the boundary gateways have xor type.*

In the case of a quasi block-structured process model, it is not obvious what checks should be applied to non-components in order to validate the soundness of the process model. To address the challenge, we investigate structural particularities of non-components. The first characteristic of non-components follows from Theorem 1:

Corollary 1. If an SPQR-tree node fragment of a process model is a non-component, then at least one of its boundary nodes is a mixed gateway.

By examining the structure of a boundary fragment node which is a mixed gateway, one can make a stronger statement, which is captured in Lemma 1.

Lemma 1. *An SPQR-tree node fragment is a non-component if and only if it has a boundary node that is a mixed gateway that has at least one incoming and at least one outgoing edge both among internal and external fragment edges.*

Proof. Figure 6 shows combinations of incident edge settings with a boundary node which is a mixed gateway. The dashed lines divide the edges onto internal and external fragment edges. The combinations are obtained by allowing only incoming, only outgoing, or both incoming and outgoing edges at each side of a dashed line. Each combination represents a collection of edge settings, where every edge stands for an arbitrary number of edges (but at least one) that have the same structural relation (incoming or outgoing, internal or external) with the boundary node. Out of the total of nine possible combinations, two interfere with the requirement of structural soundness: all incident edges are incoming or all incident edges are outgoing. Moreover, such nodes are not even gateways. Three combinations are the mirror copies of patterns shown in Figures 6(a), 6(b), and 6(c). Hence, the four unique combinations are visualized in Figure 6. Observe that out of all possible edge settings, only the one given in Figure 6(d) allows paths in both directions across the fragment's boundary and, therefore, is neither an entry nor an exit of the fragment (see Definition 4). Finally, any other edge setting allows paths only in one direction and, hence, a node is either an entry or an exit of the fragment (see Definition 4). □

A fragment with at least one boundary node like the node from Figure 6(d) is a non-directed fragment. For quasi block-structured process models there is a strong relation between directed fragments and L-type fragments.

Lemma 2. *An SPQR-tree node fragment of a quasi block-structured process model is a non-directed fragment if and only if it is an L-type fragment or it contains an L-type fragment that shares a boundary node with it.*

Proof. A non-directed fragment contains incoming and outgoing incident edges with one of its boundary nodes. In a quasi block-structured process model this node is the boundary node of some P-type fragment. The P-type fragment must have Q-type and/or S-type fragments as child fragments (see [15,18]). Every Q-type fragment is directed. If all the S-type fragments which are the children of the P-type fragment are directed, then the fragment is an L-type fragment. If there exists an S-type fragment which is a non-directed fragment, then it must share a boundary node with a P-type fragment. Hence, the above described logic can be recursively applied to this P-type fragment. Eventually, we will reach an L-type fragment which also shares a boundary node with the initially investigated non-directed fragment. Therefore, if a fragment is a non-directed fragment, then it is either an L-type fragment or it contains an L-type fragment that shares a boundary node with it.

The reverse direction of the proposition is trivial to show. At least one of the boundary nodes of the SPQR-tree node fragment is incident with incoming and outgoing edges contained in the fragment. This node is also the boundary node of the L-type fragment. Hence, the fragment is a non-directed fragment. □

Finally, we are ready to make the concluding statement which characterizes the nature of non-components in quasi block-structured process models.

Lemma 3. *If an SPQR-tree node fragment of a quasi block-structured process model is a non-component, then either it is an L-type fragment, or it shares a boundary node with an L-type fragment.*

Proof. If an SPQR-tree node fragment of a quasi block-structured process model is a non-component, then it has a boundary node which is a mixed gateway that has incoming and outgoing edges among internal and external fragment edges (Lemma 1). Every fragment with a boundary node as in Figure 6(d) is clearly a non-directed fragment. In a quasi block-structured process model, a non-directed fragment is an L-type fragment or it contains an L-type fragment that shares a boundary node with it (Lemma 2). □

The fact that each non-component of a quasi block-structured process model shares a node with an L-type fragment allows us to define criteria for checking the soundness of a quasi block-structured process model.

Theorem 4. *A quasi block-structured process model is sound if and only if:*

- *for each P-type component the boundary gateways match in type,*
- *for each L-type component the boundary gateways have xor type,*
- *for each non-component the boundary gateways have xor type.*

Proof. In addition to the criteria proposed in Theorem 3, one needs to show that non-components cause structural conflicts, either a dead-lock, or a lack of synchronization, when they have a boundary node of the *and* type. Every non-component is either an L-type fragment, or it shares a boundary node with an L-type fragment (Lemma 3). Moreover, a boundary node of the non-component which is neither its entry nor its exit is a boundary node of an L-type fragment contained in the non-component and has incoming and outgoing incident edges outside the non-component. If this node is an *and* type gateway, it introduces the uncontrolled concurrency (a lack of synchronization) conflict to the process model. If the process model is mapped to the workflow net using the function *petrify*, one can always observe an unbounded place, which is an output place of the transition that corresponds to the *and* type gateway of the L-type fragment and is on a path that leads outside of the fragment. □

Theorems 3 and 4 conclude that the soundness of quasi block-structured process models can be checked in linear time. The SPQR-tree of a process graph can be constructed in linear time [17]. The directed property of a fragment as well as an L-type fragment structure can be checked locally. Finally, model soundness can be checked by performing a post-order traversal to the SPQR-tree of the process graph while verifying if the structural constraints defined in Theorem 4 hold for every fragment.

6 Transformation of Hidden Unstructured Regions

The problem of program control flow graph restructuring has largely attracted the attention of the compiler construction research community. In contrast to flow-graphs, process models can be structured with advanced constructs such as *and* type gateways that may preclude the use of those general approaches. However, our focus is on non-component fragments which, according to Theorem 4, have *xor* type gateways as boundary nodes. This fact allows us to conclude that there exist techniques that can be applied for transforming quasi block-structured models into block-structured ones.

Please note that restructuring techniques do not consider mixed gateways. Hence, prior to applying a restructuring, conflicting mixed gateways must be split as follows: A first gateway is used for collecting all incoming edges, whereas a second one is used for collecting outgoing edges. Finally, an edge is added to connect the first gateway with the other one. To illustrate this procedure, consider Figure 7, which presents fragment $S2$ in the running example after splitting the two conflicting gateways (i.e., shaded gateways). $S2$ corresponds to the non-component fragment highlighted in the SPQR-tree in Figure 3.

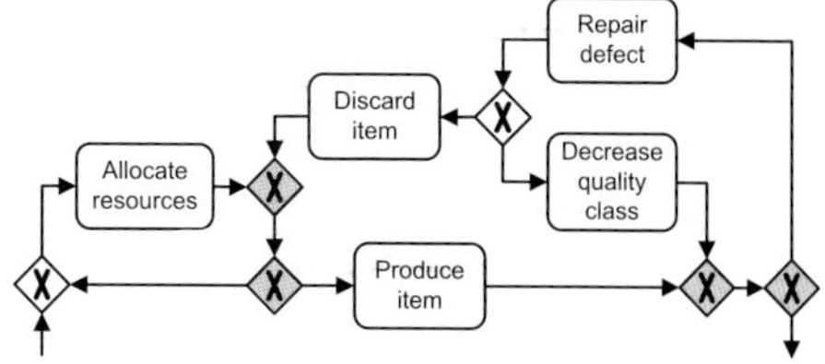

Fig. 7. Splitting conflicting gateways on the running example process

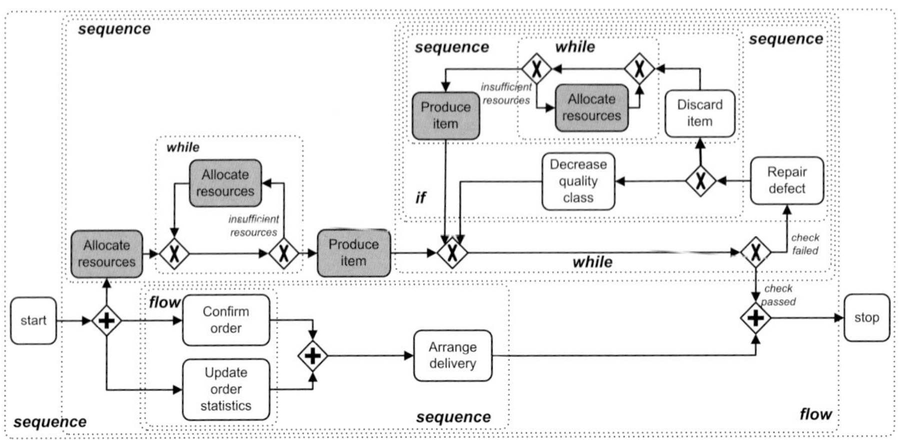

Fig. 8. The restructured process model for the process model from Figure 2 and its mapping to BPEL

Splitting conflicting gateways unveils the unstructured logic which is hidden by non-component fragments. For instance, a closer look at Figure 7 allows us to identify two unstructured loops, one with two entry points (left-hand side) and the other with two exit points. It is worth noting that the resulting region would be enclosed in a R-type fragment if the corresponding SPQR-tree was updated.

Figure 8 presents the restructured process model of the running example (the transformation was done using the technique proposed in [11]). Please note that some nodes have been replicated, i.e., tasks with dark grey background. In general, restructuring methods rely on node replication and/or on encoding the control flow via additional variables and gateways. The translation of the resulting process model to a block-structured language is straightforward (see [8]). Although the process model still contains mixed gateways, the SPQR-tree decomposition provides enough structural information to support the translation. To illustrate this, Figure 8 presents the mapping of the control flow of the running example to BPEL constructs.

7 Conclusions

In this paper, we describe a structural classification for process models based on the SPQR-tree decomposition. Three classes are identified: block-structured, quasi block-structured, and graph-structured process models. We focus on quasi block-structured models which contain regions with hidden unstructured behavior. Although this kind of regions was described before [7,8], we take a step forward as we characterize their structural properties and show that their soundness can be verified in linear time. We also show that existing techniques for flow-graph restructuring can be applied to transform hidden unstructured regions into block-structured equivalent process fragments. It is worth noting that in the cases where a quasi block-structured process graph contains *and* type gateways, all of them are part of block-structured fragments. As a consequence, hidden unstructured fragments rely only on *xor* logic, which enables the use of restructuring techniques. The evoked transformation may be required if a quasi block-structured process model must be translated into a block-structured language such as BPEL, e.g., for execution.

In the future work, we want to extend our approach to address the analysis and transformation of unstructured process models, i.e., process models whose SPQR-tree decomposition contains R-type fragments. We believe that the structural constraints imposed by L-type fragments to their enclosing regions can be exploited to speed up the verification of unstructured process models, for instance, if R-type fragments contain internal loops. We are also interested in further investigating the analysis and the transformation of process models which use *or* type gateways.

References

1. Weske, M.: Business Process Management: Concepts, Languages, Architectures. Springer, Heidelberg (2007)
2. OMG: Business Process Modeling Notation, Version 1.2. (January 2009)
3. Laue, R., Mendling, J.: The Impact of Structuredness on Error Probability of Process Models. In: Kaschek, R., Kop, C., Steinberger, C., Fliedl, G. (eds.) UNISCON. LNBIP, vol. 5. Springer, Heidelberg (2008)
4. Kiepuszewski, B., ter Hofstede, A.H.M., Bussler, C.: On Structured Workflow Modelling. In: Wangler, B., Bergman, L.D. (eds.) CAiSE 2000. LNCS, vol. 1789, p. 431. Springer, Heidelberg (2000)
5. Liu, R., Kumar, A.: An Analysis and Taxonomy of Unstructured Workflows. In: van der Aalst, W.M.P., Benatallah, B., Casati, F., Curbera, F. (eds.) BPM 2005. LNCS, vol. 3649, pp. 268–284. Springer, Heidelberg (2005)
6. Aalst, W.: Verification of Workflow Nets. In: Azéma, P., Balbo, G. (eds.) Application and Theory of Petri Nets, pp. 407–426. Springer, Berlin (1997)
7. Tarjan, R.E., Valdes, J.: Prime Subprogram Parsing of a Program. In: Proceedings of the 7th Symposium on Principles of Programming Languages (POPL), pp. 95–105. ACM, New York (1980)
8. Vanhatalo, J., Völzer, H., Koehler, J.: The Refined Process Structure Tree. In: Dumas, M., Reichert, M., Shan, M.-C. (eds.) BPM 2008. LNCS, vol. 5240, pp. 100–115. Springer, Heidelberg (2008)
9. Alves, A., Arkin, A., Askary, S., Barreto, C., Bloch, B., Curbera, F., Ford, M., Goland, Y., Guízar, A., Kartha, N., Liu, C.K., Khalaf, R., König, D., Marin, M., Mehta, V., Thatte, S., van der Rijn, D., Yendluri, P., Yiu, A.: Web Services Business Process Execution Language Version 2.0. OASIS Standard (April 2007)
10. Margolis, B.: SOA for the Business Developer: Concepts, BPEL, and SCA (Business Developers series). Mc Press (2007)
11. Oulsnam, G.: Unravelling Unstructured Programs. The Computer Journal 25(3), 379–387 (1982)
12. Lin, H., Zhao, Z., Li, H., Chen, Z.: A Novel Graph Reduction Algorithm to Identify Structural Conflicts. In: Proceedings of the 35th Annual Hawaii International Conference on System Sciences (HICSS), Washington, DC, USA, vol. 9, p. 289. IEEE Computer Society, Los Alamitos (2002)

13. Hopcroft, J.E., Tarjan, R.E.: Dividing a Graph into Triconnected Components. SIAM Journal on Computing 2(3), 135–158 (1973)
14. Battista, G.D., Tamassia, R.: Incremental Planarity Testing. In: Proceedings of the 30th Annual Symposium on Foundations of Computer Science, FOCS (1989)
15. Battista, G.D., Tamassia, R.: On-Line Maintenance of Triconnected Components with SPQR-Trees. Algorithmica 15(4), 302–318 (1996)
16. Fussell, D., Ramachandran, V., Thurimella, R.: Finding Triconnected Components by Local Replacement. SIAM Journal on Computing 22(3), 587–616 (1993)
17. Gutwenger, C., Mutzel, P.: A Linear Time Implementation of SPQR-Trees. In: Marks, J. (ed.) GD 2000. LNCS, vol. 1984, pp. 77–90. Springer, Heidelberg (2001)
18. Polyvyanyy, A., Smirnov, S., Weske, M.: The Triconnected Abstraction of Process Models. In: Proceedings of the 7th International Conference on Business Process Management (BPM), Ulm, Germany (September 2009)
19. Vanhatalo, J., Völzer, H., Leymann, F.: Faster and More Focused Control-Flow Analysis for Business Process Models Through SESE Decomposition. In: Krämer, B.J., Lin, K.-J., Narasimhan, P. (eds.) ICSOC 2007. LNCS, vol. 4749, pp. 43–55. Springer, Heidelberg (2007)
20. Petri, C.: Kommunikation mit Automaten. PhD thesis, Institut für instrumentelle Mathematik, Bonn, Germany (1962)
21. Aalst, W., Hirnschall, A., Verbeek, H.: An Alternative Way to Analyze Workflow Graphs. In: Pidduck, A.B., Mylopoulos, J., Woo, C.C., Ozsu, M.T. (eds.) CAiSE 2002. LNCS, vol. 2348, pp. 535–552. Springer, Heidelberg (2002)
22. Desel, J., Esparza, J.: Free Choice Petri Nets. Cambridge University Press, New York (1995)

Cafe: A Generic Configurable Customizable Composite Cloud Application Framework*

Ralph Mietzner, Tobias Unger, and Frank Leymann

Institute of Architecture of Application Systems University of Stuttgart
Universitaetsstr. 38 70569 Stuttgart, Germany
firstname.lastname@iaas.uni-stuttgart.de
http://www.iaas.uni-stuttgart.de

Abstract. In this paper we present Cafe (Composite Application Framework) an approach to describe configurable composite service-oriented applications and to automatically provision them across different providers. Cafe enables independent software vendors to describe their composite service-oriented applications and the components that are used to assemble them. Components can be internal to the application or external and can be deployed in any of the delivery models present in the cloud. The components are annotated with requirements for the infrastructure they later need to be run on. Providers on the other hand advertise their infrastructure services by describing them as infrastructure capabilities. The separation of software vendors and providers enables end users and providers to follow a best-of-breed strategy by combining arbitrary applications with arbitrary providers. We show how such applications can be automatically provisioned and present an architecture and a prototype that implements the concepts.

1 Introduction

Driven by the need to outsource (part of) their IT-infrastructure, companies have shifted considerable amounts of their IT from their own premises to external companies. Therefore new delivery models for software have been emerging that allow the outsourcing of different aspects of an application. The software as a service (SaaS) model is a prominent example of such an outsourcing model. Other delivery models such as Infrastructure as a Service (IaaS) and Platform as a Service (PaaS) aim at providing (parts of) the necessary infrastructure and platform support to easily built and host applications at a provider.

The advent of these delivery models and new architectural styles such as service-oriented architecture (SOA) and emerging technologies such as Web services, virtualization and the abundance of fast internet connections has lead to new ways to deliver software. Service-oriented applications can now be assembled out of services that are running in one's own infrastructure as well as at

* This work is partially funded by the EU 7th Framework Project ALLOW (http://www.allow-project.eu/) (contract no. FP7-213339).

R. Meersman, T. Dillon, P. Herrero (Eds.): OTM 2009, Part I, LNCS 5870, pp. 357–364, 2009.

third parties and can be offered "as a service" to multiple customers. However, current approaches for modeling service-based applications often neglect the aspects concerning the modeling of requirements of an application on the runtime infrastructure or modeling of the runtime infrastructure itself. This is due to the fact that traditionally, applications are often developed with certain knowledge about the runtime infrastructure on which they are later deployed. In the new delivery models, the runtime infrastructure is (partially) outsourced to "the cloud" and the application must be automatically deployed at one or several providers. Therefore applications must be developed independently of the provider infrastructure they are later run on. This enables their automated provisioning at and across different providers. In this paper we introduce Cafe (composite application framework) a model and provisioning infrastructure to describe and provision composite, service-oriented applications and their required infrastructure. We start the paper with a motivating scenario in Section 2. In Section 3 we then briefly recapitulate how customizable composite applications and their requirements for the infrastructure can be modeled using so-called component and deployment graphs as well as variability descriptors. We show how different delivery models can be integrated for different components. We then show in Section 4 how the model can be used to automatically provision applications across different providers. We finish the paper by comparing our approach to related work and by giving hints for future work.

2 Periodic Inspection Reminder Scenario

In Germany, passenger cars must be safety-inspected every two years. As customers often forget that an inspection is required, the car dealer "Perfect Cars" wants to offer a new service to its customers, which reminds them via a SMS message or E-Mail when an inspection is required.

The inspection reminder software from "Perfect Software" (cf. Figure 1) provides a Web interface to the car dealer for triggering inspection reminders. Customer data is stored in a customer relationship management system (CRM service) that is accessed by the application. The CRM service can either be the one internal to the application or the Salesforce[1] Web service. In case the Salesforce Web service is used, a Salesforce wrapper component is to access the Salesforce API. Another component of the application is an E-Mail service that sends inspection reminders, this internal E-Mail service can be substituted for an external SMS service. The CRM service and SMS/E-Mail services are orchestrated by a BPEL workflow that implements the core application logic (reminder workflow in Figure 1). "Perfect hosting" is an application hosting company that offers the inspection reminder application they bought from "Perfect Software" to customers "as a service". New tenants (such as "Perfect Cars") can subscribe to the application via the application portal of perfect hosting.

[1] http://www.salesforce.com

3 Application Model

In this section we briefly introduce the formal concepts that are necessary to understand the algorithms in Section 4. A full formalization can be found in [9].

Applications $\mathcal{A}$ are assembled out of a set of *components* $\mathcal{C}$. Components can be application components (such as UIs, services and workflows) and infrastructure components (such as servers or middleware). These components are wired together via a set of *wires* $\mathcal{W}$. A wire between two components denotes a directed usage link. In addition to wires, the model allows to describe a set of *deployment relationships* $\mathcal{D}$. A deployment relationship between two components c_1 and c_2 denotes that the first component (c_1) is deployed on the second component (c_2).

Definition 1 (Applications). *The set of applications is defined as:*

$$\begin{aligned} \mathcal{A} &= \{a_1, \ldots, a_n\} \\ &= \{(C, W, D) | C \subseteq \mathcal{C} \wedge W \subseteq C \times C \subseteq \mathcal{W} \\ &\quad \wedge D \subseteq C \times C \subseteq \mathcal{D}\} \end{aligned}$$

We call the components and their wires the *component graph*. This is a directed, possibly cyclic graph. Besides communicating with each other (expressed in our model through wires) components may have a second relationship: the deployment relationship. We call the components and their deployment relations the *deployment graph*. The deployment graph is a directed, acyclic graph.

Components may be associated with certain properties. In the following we introduce the properties of a component that we need for the algorithms below, for additional properties see [9]: The function *implType* : $\mathcal{C} \rightarrow \mathcal{I}$ assigns an *implementation type* to a component. A member of the set of implementation types $\mathcal{I}$ can be any programming language. For example a workflow component could be implemented in BPEL. It can also be "providerSupplied" to denote that the provider must somehow provide the component.

Each component has an assigned *multi-tenancy pattern*. The multi-tenancy pattern can either be: (i) *single instance* - i.e. the component is shared by multiple

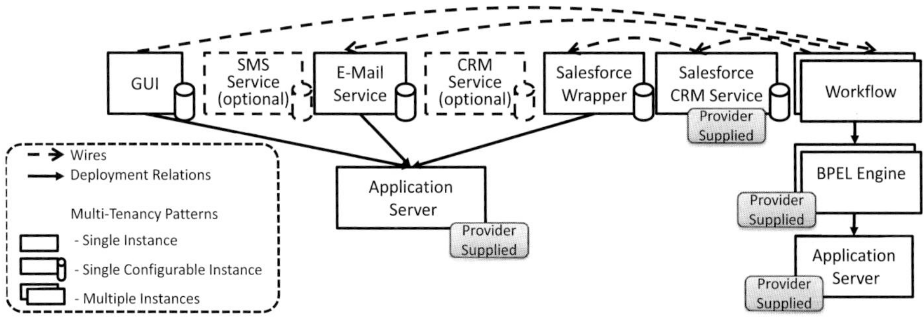

Fig. 1. Application Model

tenants, (ii) *single configurable instance* - i.e. the component is shared by multiple tenants and must be configured for every tenant, or (iii)*multiple instances* - i.e. the component must be deployed separately for each new tenant. The function $pattern : C \rightarrow \{singleInstance, singleConfInstance, multiInstances\}$ assigns a multi-tenancy pattern to a component.

Components can have requirements on other components (i.e. an application component can have a requirement on a middleware component it needs to be deployed on. A requirement has a type (e.g. availability or response time). For a discussion and formal representation of requirements and their types see [9].

4 Application Provisioning

During the deployment of a new tenant for an application four main steps are taken. In the first step the user selects a template application from a directory. In the second step the tenant customizes the application through a so-called *customization workflow* (for a discussion of these steps see [4]). In the third step the components required to run the new tenant are mapped to the already available environment and in the fourth step those application components that are not already available are provisioned. During customization, components might be added or removed from the template component and deployment graphs of the application. Additionally, alternatives are bound. For example the tenant selects that he wants to use the E-Mail service and the Salesforce CRM application in our sample application shown in Figure 1. Therefore the SMS service and the internal CRM service are removed from the component graph and deployment graph. Thus, customization can be a modification of the component and deployment graphs.

4.1 Matching Step

Once an application is ready to be provisioned for a tenant the first step is to examine which components of the application are already deployed for other tenants and can be reused. Therefore the deployment graph of the application is traversed top down. I.e. for each *root component* of the application it is checked whether a suitable match can be found in the environment. A root component is a component that has no incoming deployment relationship. We can restrict this step to the matching of components in either the single instance or single configurable instance patterns. Multiple instance components must be provisioned anyways so no matching is needed. Algorithm 1 shows the whole algorithm.

A suitable match is found if at least one component of the same type as the component in the application exists in the environment that has capabilities that match the requirement annotations of the component in the application.

The set of matching components is determined using the *match* function. The set $matchingComponents(c) = \{d \in \mathcal{CA} : match((c, d)) = true\}$ is defined as the set of components available at any provider $\mathcal{CA}$ that match the component

c. The function $match : \mathcal{C} \times \mathcal{CA} \rightarrow \{true, false\}$ is used in order to determine whether a component c in an application can be matched to a component or sub-graph represented through it's root component d at a provider. I.e., the application can be configured to use that component d.

In [6] we describe how policies expressed in WS-Policy can be matched to properties of resources in an environment. These algorithms can be used as the capabilities as well as the requirements for components can be described as policies. Other matching techniques such as semantic matching of services can be also employed to find suitable available components that can be assigned to components in an application.

Having found several possible matches, one of these matches must be selected. This can for example be done based on performance or cost. We will present different optimization algorithms that we have investigated in future work. For now we introduce a *selectBest* function that encapsulates the selection. In our prototype we simply take the first component out of the set of matching components. Formally the *selectBest* function is defined as: $selectBest : 2^{\mathcal{C}} \rightarrow \mathcal{C}$ The function $matchedComponent : \mathcal{C} \rightarrow \mathcal{CA}$ assigns exactly one component out of the set of available components to a component.

In case no matching is found for a component all the "child" components of this component are examined and matched. We define the set of "child" components $children(c)$ of a component $c \in C_a$ as the set of those components of application $a = C, W, D$ on which c has a deployment relationship. $children(c) = \{c_1 \in C_a | \exists d \in D : \pi_1(d) = c \wedge \pi_2(d) = c_1\}$.

In case a component is a child of multiple other components, such as the application server in our example, this component is matched the first time one of it's parent components cannot be matched. The next time the previous matching is reused.

Once the children of a component have been matched (or a suitable provisioning service has been found) a suitable *provisioning service* for the component needs to be found. A provisioning service [5] is a service that can provision a certain set of components. The available provisioning services at all providers is described through the set PS. We do not go into details here, on how to find such services or how a registry for different providers of such services can be built. In [6] we describe how WS-Policies can be used to find suitable Web services that describe their non-functional properties with WS-Policies. This work can be reused here. The function $findProvisioningService : \mathcal{C} \rightarrow PS$ searches for a suitable provisioning service for a certain component c A suitable provisioning service is a service that can provision the component or the virtual leaf graph with the necessary quality of service. The function $provisioningService : \mathcal{C} \rightarrow PS$ assigns a provisioning service to a component. Implicitly we assume that a component that can be matched will never be redeployed. However, there might be situations where it is cheaper to redeploy such a component (for example because a new provider offers a cheaper component than the one used before). Such a component can be treated as a component that could not be matched.

Algorithm 1. Matching Algorithm

{Algorithm is started with the set of root components RC}
function findMatch ($Components$)
{examine all components in the set}
for all $c \in Components$ **do**
  { check if the component is in the single instance or single configurable instance pattern and there exists a match in any of the available infrastructures }
  if $(pattern(c) \in \{singleInst, singleConfInst\}) \wedge ((matchingComponents(c) != \emptyset)$ **then**
    {select the best matching component for c}
    $matchedComponent(c) = selectBest(matchingComponents(c))$
  else
    {no matching component found, examine the underlying components }
    findMatch($children(c)$)
    {after all children have been treated - find a suitable provisioning service}
    $provisioningService(c) = findProvisioningService(c)$
  end if
end for
end function

4.2 Configuring the Properties of Components

After the matching of components to actual providers, components that use other components (i.e., components that are the source of a wire) must be configured with the properties (e.g. the EPR) of the component the wire points to. Two situations can arise now: (i) The property value is already available because the component is already deployed or the property is static. (ii) The property becomes available after provisioning. In case (i) the source component can be configured and then provisioned as described in the next section. In case (ii) the source component must be provisioned after the target component has been provisioned. Therefore the deployment step must take this ordering into account. This is done by reordering sibling components in the deployment graph. Reordering is not always possible, for example if two components have a cyclic wire dependency. In this case two solutions can be employed: (i) One of the components can be configured after deployment, then, this one is provisioned first. (ii) A mediator component (such as a virtual endpoint at an ESB for endpoint resolution) can be put between those components and can be configured with the concrete properties later. The components are then configured to use the mediator component.

4.3 Deployment of Components

Having obtained the properties of all components, the components that are not already matched can be provisioned, and those that are already matched can be configured. In our prototype we use a BPEL provisioning flow that traverses the deployment graph depth-first (taking the ordering of siblings into account that has been determined based on the wires) and calls the necessary provisioning services (that can be at different providers).

Using a full-blown workflow system such as a BPEL engine has the advantage over other scripting languages that the features of BPEL such as transactionality, human tasks (via BPEL4People), Web service invocations directly out of the process can be reused. In [5] we describe the architecture for a unified provisioning infrastructure that spans across multiple providers and that can be used to deploy the applications described in this paper.

In short the algorithm calls the provisioning service for each component that is not already matched and provisions this component. Components that are already matched and are deployed in the single configurable instance pattern need to be configured for the new tenant. In this case the "addTenant" method of the component is called that allows to deploy new configuration data for a new tenant. Therefore all components that can be deployed in single configurable instance mode must provide such an "addTenant" method. In [5] we describe how such methods can be provided on top of existing provisioning engines.

5 Related Work

The authors of [1] describe a framework for the configuration, distribution and deployment of Web services. They describe how Web services based applications can be configured and deployed. Our work differs from their approach as it allows to explicitly model multi-tenancy in the application. Additionally our model allows to model requirements on the necessary infrastructure for components. In [8] the provisioning and adaptation of composite Web service based applications is discussed with regards to hybrid environments. This approach nicely complements our approach as it allows providers to optimize the allocation of services to resources. Therefore this approach can be seen as one strategy for providers to provision an application in our scenario. However, the approach must be extended by the notion of multi-tenancy to be fully usable in our scenarios which also holds true for the SLA-driven provisioning approach presented in [3]. In [7] the Vienna component framework is presented that allows to combine components from different component models into a new application, however it does not include multi-tenancy patterns and different delivery models as well as the modeling of the required infrastructure. In [2] a framework for the deployment modeling of (SOA) applications is presented that is similar to our approach. While the approach presented in [2] allows to explicitly model the deployment relationships between infrastructure components down to the bare metal level (although you do not have to do this) we delegate this to the provider. Thus our approach is more focused on cross-provider based applications. Also the variability mechanisms in [2] differ from ours as they rely on templates that describe components or deployments that can then be customized. We focus on the variability in the application and the component graph and our orthogonal model allows to combine both. The topology definition described in [2] could be used as an input for a highly configurable provisioning service.

6 Conclusions and Future Work

In this paper we presented Cafe a model and provisioning architecture on how to specify composite applications that can be offered on demand by different providers. We introduced a component graph as a graph-based structure to capture the dependencies between components . A so-called deployment graph has been introduced to capture deployment relationships between components. We used these structures to show how components can be matched against the infrastructures offered by different providers. We showed how components that are not available can be automatically provisioned and how various delivery models can be integrated in the approach. In future work we will show how the distribution of components at one provider or across different providers can be optimized regarding costs for the running of the application. We presented the necessary formalisms based on the deployment and component graphs to describe how the matching and provisioning algorithms work in detail. We also showed how variability is captured in the model and how it affects the deployment of components. We have implemented a prototype that serves as a proof of concept of the conceptual work presented before.

References

1. Anzböck, R., Dustdar, S., Gall, H.: Software Configuration, Distribution, and Deployment of Web-Services. In: SEKE (2002)
2. Arnold, W., Eilam, T., Kalantar, M.H., Konstantinou, A.V., Totok, A.: Pattern based soa deployment. In: Krämer, B.J., Lin, K.-J., Narasimhan, P. (eds.) ICSOC 2007. LNCS, vol. 4749, pp. 1–12. Springer, Heidelberg (2007)
3. Ludwig, H., Gimpel, H., Dan, A., Kearney, B.: Template-Based Automated Service Provisioning-Supporting the Agreement-Driven Service Life-Cycle. In: Benatallah, B., Casati, F., Traverso, P. (eds.) ICSOC 2005. LNCS, vol. 3826, pp. 283–295. Springer, Heidelberg (2005)
4. Mietzner, R., Leymann, F.: Generation of BPEL Customization Processes for SaaS Applications from Variability Descriptors. In: SCC 2008 (2008)
5. Mietzner, R., Leymann, F.: Towards Provisioning the Cloud: On the Usage of Multi-Granularity Flows and Services to Realize a Unified Provisioning Infrastructure for SaaS Applications. In: SERVICES 2008 (2008)
6. Mietzner, R., van Lessen, T., Wiese, A., Wieland, M., Karastoyanova, D., Leymann, F.: Virtualizing Services and Resources with ProBus: The WS-Policy-Aware Service and Resource Bus. In: Proceedings of the 7th International Conference on Web Services, ICWS 2009 (2009)
7. Oberleitner, J., Gschwind, T., Jazayeri, M.: The Vienna Component Framework enabling composition across component models. In: Proceedings of 25th International Conference on Software Engineering, pp. 25–35 (2003)
8. Sheng, Q.Z., Benatallah, B., Maamar, Z., Dumas, M., Ngu, A.H.H.: Enabling Personalized Composition and Adaptive Provisioning of Web Services. In: Persson, A., Stirna, J. (eds.) CAiSE 2004. LNCS, vol. 3084, pp. 322–337. Springer, Heidelberg (2004)
9. Unger, T., Mietzner, R., Leymann, F.: Customer-defined Service Level Agreements for Composite Applications. Enterprise Information Systems 3(3) (2009)

Implementing Isolation for Service-Based Applications

Wei Chen[1], Alan Fekete[1], Paul Greenfield[2], and Julian Jang[3]

[1] School of Information Technologies, University of Sydney, NSW 2006 Australia
[2] CSIRO Mathematical and Information Sciences
Locked Bag 17, North Ryde NSW 1670 Australia
[3] CSIRO ICT Centre, Locked Bag 17, North Ryde NSW 1670 Australia
{weichen,fekete}@it.usyd.edu.au,
{paul.greenfield,julian.jang}@csiro.au

Abstract. Loosely-coupled distributed systems can be difficult to design and implement correctly, with time-of-check-to-time-of-use flaws arising from the lack of isolation being of particular concern. It is not feasible to use traditional distributed ACID transactions to solve such problems because the business activities being integrated are typically long-running and the interacting participants have incomplete mutual trust. 'Promises' were recently proposed as a solution to this problem. This paper discusses how promise-based isolation can be implemented when resources are described by predicates over properties, rather than being identified explicitly.

1 Introduction

Many distributed applications are based on long-running business processes that call each other back-and-forth in extended patterns of interaction. The designers of these systems have to correctly handle all of the different possible sequences of interaction, a task which can be hard for the normal ('happy') case, and much harder still when all possible error paths and all possible application states are considered.

One particular difficulty for application developers comes when services may fail if they are called while they are in an unsuitable internal state. For example, in an accommodation service, the operation used to book a hotel room will return an error if no appropriate room is available, and the calling application must include code to do something sensible on receiving such an 'operation unavailable in this state' error.

The traditional way of solving this problem is for such services to provide operations that clients can use to check that this internal state is appropriate before going on to call the state-sensitive operations. If this preliminary check succeeds the client can call subsequent operations that should not fail because the service is in an inappropriate state, and if it fails the client can take an early remedial/alternative processing path. For example, an accommodation service could have operations that report on room availability, allowing the client to check that a room was actually available before trying to book one. In an environment with concurrent threads or concurrent clients, the correctness of this 'check-then-use' programming style depends on the platform providing 'isolation' support to ensure that the relevant internal state does not change from being 'suitable' to 'unsuitable' between the check and the subsequent use.

R. Meersman, T. Dillon, P. Herrero (Eds.): OTM 2009, Part I, LNCS 5870, pp. 365–372, 2009.

Traditional 'ACID' transactional technologies implement isolation through locking, logging and two-phase commit [1] but these mechanisms are not appropriate for loosely-coupled, inter-organisational systems because they are based on unviable assumptions of trust and timeliness. As a result, the designers of these kinds of systems cannot assume that a state-sensitive operation will succeed just because they had earlier checked that it ought to succeed, and not considering this failure scenario, where a precondition value changes between time-of-check and time-of-use, can be a common source of programming errors.

We recently proposed an approach called 'Promises' [2, 3] which generalises coding patterns such as 'soft locks' and 'reservations' into a framework that brings the benefits of isolation to loosely-coupled systems. Promise-based applications can obtain 'promises' of later resource availability and then invoke state-dependent operations with confidence that they will succeed.

A promise is a time-limited commitment from a service that a requested resource will be available for later use. In this paper we go beyond our earlier work and explain how to implement promise-management for resources which are described implicitly and non-deterministically by their properties. An additional contribution of this paper is showing how the work of promise-management can be divided among several subsystems, separating for the handling of promises over one type of resource (e.g. hotel rooms) with its own syntax and semantics for promises, from the handling of promises for another type of resource (e.g. bank balances). We also describe a proof-of-concept implementation that includes these novel features and shows that acceptable performance can be obtained in promise-based systems.

2 Background and Related Work

As defined in our previous papers on Promises [2, 3], a 'promise' is a resource availability commitment given by a service to a client application. By accepting a promise request, a service guarantees that a client-specified set of conditions ('predicates') will be maintained over a set of resources for an agreed period of time.

Promises provide an isolation mechanism for loosely-coupled business processes. Client applications have to accumulate a set of promises that guarantee the availability of the resources required for the successful execution of their proposed actions. Then, the client application can send 'action' service requests to the corresponding application services to complete its business processes, knowing that these requests cannot fail as a result of conflicts over promised resources.

Promises are made over the availability of abstract resources and [2] defined three important ways that clients can view these resources. *Anonymous View:* Clients see a pool of identical and indistinguishable resource instances, such as the copies of a title in a bookstore. *Named View:* Clients see each resource instance as unique, with its own unique identifier that can be used in promise predicates. Specific airline seats (seat 24G on QF1 departing on 8/10/2009) are an example. *Property View:* Clients see abstract resources that are identified by their properties. For example, a hotel room may or may not have an ocean view, and may contain twin, double or king beds. These particular properties could be used in promise predicates that ask for any room with a view and a double bed.

The need for mechanisms that provide isolation across the different steps of an activity has attracted many researchers. Traditional ACID transactions can be used within a trust boundary, and can be implemented with a range of techniques including shared and exclusive multi-granularity locking [1]. For numeric values with increment and decrement operations, there are special high-performance mechanisms such as escrow locking [4]. A technique for locking predicates has been proposed in [5], but this requires determining the NP-hard solution to satisfiability of an arbitrary Boolean expression. We are not aware of any techniques in the database domain that correspond to our proposal for consistency checking with property-based resources. Rather than using locks in the DBMS, some applications implement isolation through application-level soft locks. Promises over property-based resources give the application more flexibility as soft locks can only be taken on named resources.

The needs of long-running applications, such as cooperative design tools, led to the study of extended transaction models. The ConTract model [6] specifies preconditions needed for operations to be successful. In ConTract there are defined language mechanisms to indicate these conditions, whereas Promises allows diversity in the way that properties for different types of resources are defined and supported by Promise Handlers. Another extended model is Sagas [7], where a business activity is treated as a sequence of separate steps with a compensator defined for each. Sagas provided all-or-nothing outcomes but do not give Isolation.

Several protocols have been published for managing activities in loosely-coupled distributed systems and B2Bi domains. Tentative Hold Protocol [8] does not itself ensure consistency of granted Holds and so does not prohibit several clients getting a Hold on a single resource at the same time. The reservation-coordination protocol [9] restricts the application to a two-phase structure, first obtaining reservations on a number of resources and then making use of those resources. In contrast, promise allows the application to have complex sequences of interactions, using some promises and then obtaining others. None of these protocols include consistency-checking algorithms for property-based resources which we describe for Promises.

Much other research has been focused on ensuring common all-or-nothing outcomes for a long-running activity spread over different sites, rather than isolation support. WS-BA is a standard for this; and the BTP protocol was proposed as a standard. Our ideas on achieving outcome consistency are described in [10].

3 Conceptual System Architecture

The conceptual architecture of our current (second-generation) prototype is shown in Fig. 1. The Promise Server is an intermediary along the path from the client to the service application and its role is to process promise-related elements found in messages sent from the client to the application server. These promise elements are passed to Promise Handlers, each of which deals with resources of a particular kind only, and is closely associated with a corresponding Resource Manager.

The Promise Server is implemented as a SOAP intermediary that intercepts messages sent from the client to application services. These messages may contain promise-related header elements and application service requests ('actions'). The Promise Server parses incoming messages, split them into promise parts and actions, and forwards the promise parts to the appropriate Promise Handlers and the actions to the Application Servers.

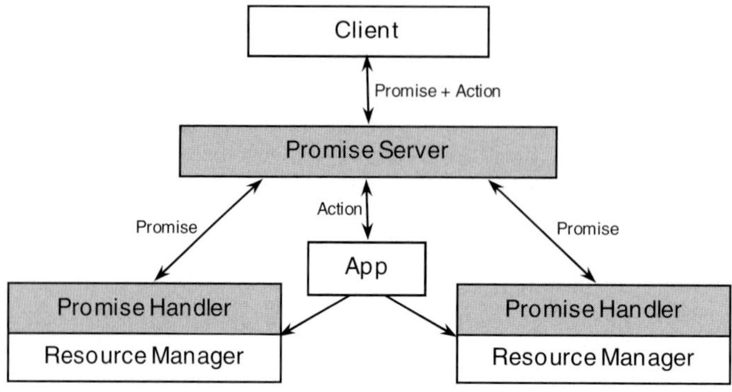

Fig. 1. Architecture of promise-based system

Promise operations have to support atomicity and consistency across internal promise state and resource allocations, even when errant or non-conforming application code allocate resources that could result in promises being violated. Our current prototype uses conventional ACID transactions as part of the implementation of this functionality. Transactions are started by the Promise Server and propagated to (multiple) Promise Handlers and Application Servers. Transactional isolation is used to ensure that concurrency problems cannot arise while promises and resources are being checked for consistency. Transactional atomicity is used to ensure that we can back out changes to promises or resource availability if promise requests cannot be granted or are violated. This use of ACID transactions is regarded as 'trusted' because client applications are excluded from the transaction scope.

Each class of resources supported by a Promise system has a corresponding Promise Handler that processes promise parts and maintains sets of promises over its resources. Promise Handlers are responsible for implementing promise operations, such as granting and releasing promises, and for consistency checking between promises and actual resource availability. Promise Handlers are resource-specific and so have close links to their related Resource Manager, including understanding the schema and semantics of the data used to define resource availability.

Promise Handlers maintain internally-consistent sets of promises that are also consistent with the resource availability state maintained by the Resource Manager. The code that is used to ensure this consistency has to have fast access to the availability state held within Resource Managers if the performance of promise-based systems is to be acceptable. Our previous papers suggested that Promise Handler functionality could be moved into Resource Managers, letting them manage both promised-based resource constraints as well as more conventional resource allocation state based on data values held in database tables. Our current prototype maintains the conceptual separation between Promise Handlers and Resource Managers, but links the two together closely through the use of a shared and unified data structure that describes all allocatable resources. This shared data structure is accessed and updated by both the Promise Handler and the Resource Manager.

4 Promises over Property-Based Resources

The actual work that has to be performed by a Promise Handler to make and keep promises depends on the nature of the resources involved. We have previously defined [2] three ways that resources can be viewed: anonymous view, named view and view via properties. The implementation issues for the first two views have been discussed in [3] and in this paper we focus on novel aspects in the implementation of promises over property-based resources.

4.1 Representing Resource Properties

Promises are made over resources that are modelled by one or more database tables held by Resource Managers. Columns of these resource tables provide the values for each property of a resource, including its current availability status. Promise predicates specify a set of properties and values that define needed resources in terms of these property columns. For example, an accommodation service could define information on hotel rooms, such as the number and size of beds, whether or not there is a sea-view, and which floor it is on. These properties can then be used in predicates to request that double-bed room with a sea view be kept available on a given date. The syntax used to specify these predicates is resource-specific as it has to reflect the semantics of the resource tables and their property columns.

Promise Handlers receive promise requests containing these predicates and turn them into searches for sets of satisfiable resources. If satisfiable resources are available, the new requests are checked to ensure that they are consistent with any already-granted promises before returning a 'promise granted' response to the caller.

4.2 Consistency Checking

As described in [3], checking the consistency of a set of promises is an essential step that takes place on almost every promise operation. The Promise Handler must ensure that the set of resources that are available are always sufficient to meet every promise that has been granted. This was easier for named and anonymous resources [3], but, where promises are based on predicates that implicitly identify a set of possible resources viewed via properties, it is not straightforward to see whether or not a group of competing promise requests based on different sets of properties can all be granted.

This problem can be resolved by using a bipartite graph matching [11] algorithm. A set of available resources form one partition of nodes, and promises form another. An edge links a resource node to a promise node whenever the promise can be satisfied by the resource (that is, the resource has the properties required by the predicate of the promise). The consistency check is the existence of a match in this bipartite graph which covers every promise, that is, each promise is satisfied by a resource, and each resource is only matched at most once.

The standard bipartite graph matching algorithm either finds a match or decides that there is no match. The algorithm is incremental, extending a partial match or else re-arranging the matches along an *augmenting path* of edges in the graph which are alternately edges which are used in the matching, and edges which are not used. We exploit this property for performance reasons by maintaining a match at all times, and incrementally updating it as new promises are made or resource availability changes.

4.3 Resource Suggestion and Resource Query Process

The applications discussed in [3] only dealt with named and anonymous resource views and were completely unaware of promises and promised resources. We have found that this simple approach is not possible with property-based views of resources because the correct choice of which resource to use in fulfilling an action depends on both the set of granted promises and availability data as recorded in database tables. Applications coded without knowledge of promises can only be aware of the database-resident availability state, and can mistakenly assign resources that are really needed to meet promises granted to other clients. This conflict will be picked up by the consistency check that follows the completion of the action. The problematic resource allocations (and other application actions) will then be undone to ensure the integrity of the already-granted promises, even if the application was executing under the protection of promises that guaranteed that suitable resources would be available. This is the very situation that promises were intended to prevent from happening.

To illustrate this problem, suppose a Promise Handler first receives a promise request for a room with sea-view. It finds that rooms 15 and 17 could be used to satisfy this request so the promise is granted. A promise request then arrives for a non-smoking room. This can be satisfied by room 15 and so this promise is granted as well. An action then arrives that is covered by the first promise, such as a request to book a room with sea-view. In the previous architecture of [3], the hotel booking service knows nothing about promises, so when its database queries on resource availability say that both rooms 15 and 17 are available and have sea-views, it can validly decide to use room 15 to meet the booking request. Unfortunately, this will result in the second promise being broken, and the completed action will be aborted by the post-execution consistency check, even though it was promised a suitable resource and such a resource is available. This would not have happened if the service had chosen room 17, and we need some way of conveying this hint to the application.

In our new architecture, application services find out about allocatable resources by sending a Query message to the Resource Manager responsible for the resource class. The Resource Manager then forwards this query on to the Promise Handlers responsible for the promises held over this resource class. These Promise Handlers then examine both their sets of promises and the available resources, and make a suggestion as to which resources can be safely used by the service. These suggestions are passed back to the Resource Manager and thence back to the application. In many cases, there will be multiple promise-consistent resources that could be used by the application and our design lets Promise Handlers suggest lists of allocatable resources, not merely one suitable resource. When the application service code receives the suggestion response, it chooses one resource from the suggested set, and uses it in completing the action.

5 Prototype Implementation and Performance

The second-generation prototype described in this section was built to explore issues arising from providing support for promises over property-based resource views. This new prototype is built on Windows Communication Foundation (WCF), a framework provided by Microsoft for service-oriented applications. Microsoft SQL Server is used as the Resource Manager and the Microsoft Distributed Transaction Coordinator (MSDTC) is used as the coordinator for the two-phase commit protocol. The Promise Server is implemented as a routing service [12] that forwards client messages (or parts of them) to Promise Handlers and application services. This routing architecture is extensible, and additional resources and Promise Handlers can be added by simply adding new Promise Server configuration entries that specify the resource name and the endpoint of the corresponding Promise Handler.

The prototype maintains an in-memory data structure that represents the current bipartite graph of promises and all resources that can satisfy each promise. We also store a current matching within this graph. This data structure represents a unified view of allocatable resources and is (conceptually) shared by both a Promise Handler and its related Resource Manager. This data structure is contained within a Common Language Runtime (CLR) library hosted by the database engine and is built from the database tables holding resource availability state and granted promises whenever the system starts up. It is kept up-to-date by stored procedures and triggers that are invoked whenever resource availability or promise status is changed. Keeping a current matching allows incremental calculation of the consistency check whenever the graph changes through searching for an augmenting path.

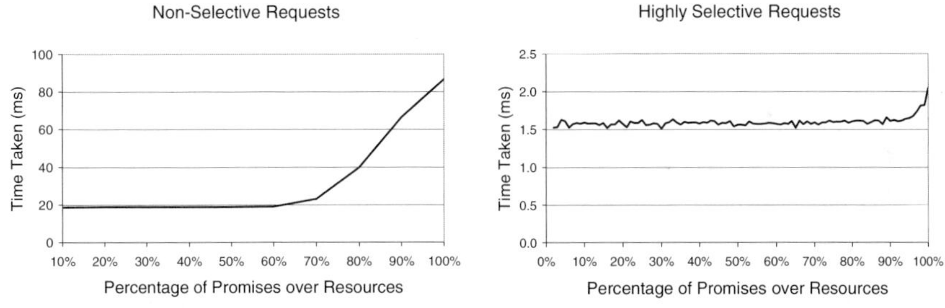

Fig. 2. Time taken to find path

Two application scenarios are used to investigate how the performance of promise-related operations changed as the number of satisfiable resources is scaled up. One scenario is *Non-selective Requests*, where each promise request can be satisfied by about 25% of all resources. The other is *Highly Selective Request*, where only 0.1% of resources can be used to satisfy any given promise request and these are distributed widely across all the available resources. The effects of these scaling characteristics can be seen in the time taken to process a new promise request (Fig. 2). The overhead caused by the consistency check is quite reasonable in both scenarios.

The time spent on allocating a resource and updating the allocatable resources graph is shown as 'Allocate + Update' in Fig. 3. It represents the whole cost of using a promised resource to fulfil an action. Compared with a simple SQL update, the overhead cost in this part of the Promises system is also quite acceptable.

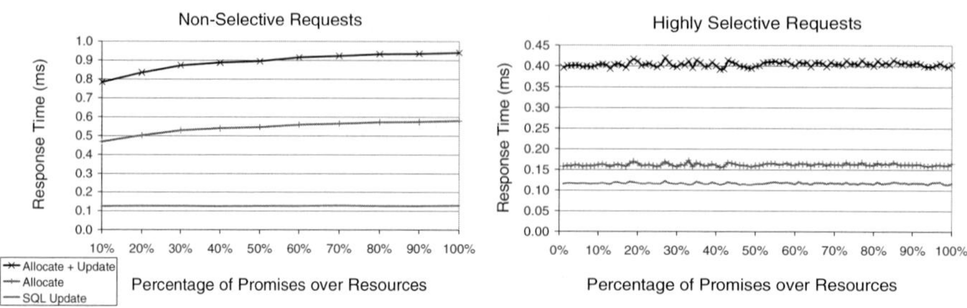

Fig. 3. Time taken to allocate & update

References

1. Gray, J., Reuter, A.: Transaction processing: concepts and techniques. Morgan Kaufmann, San Francisco (1993)
2. Greenfield, P., Fekete, A., Jang, J., Kuo, D., Nepal, S.: Isolation Support for Service-based Applications: A Position Paper. In: Proc. of CIDR, pp. 314–323 (2007)
3. Jang, J., Fekete, A., Greenfield, P.: Delivering Promises for Web Services Applications. In: Proc. of IEEE ICWS, pp. 599–606 (2007)
4. O'Neil, P.: The Escrow Transactional Methods. ACM TODS 11(4), 405–430 (1986)
5. Eswaran, K., Gray, J., Lorie, R., Traiger, I.: The Notions of Consistency and Predicate Locks in a Database System. Comm. ACM 19(11), 624–633 (1976)
6. Wachter, H., Reuter, A.: The ConTract Model. In: Elmagarmid, A. (ed.) Database Transaction Models for Advanced Applications, pp. 219–263 (1992)
7. Garcia-Molina, H., Salem, K.: Sagas. In: Proc. of ACM SIGMOD, pp. 249–259 (1987)
8. Srinivasan, K., Malu, P., Moakley, G.: Automatic Multibusiness Transactions. IEEE Internet Computing 7(3), 66–73 (2003)
9. Zhao, W., Moser, L.E., Melliar-Smith, P.M.: A Reservation-Based Coordination Protocol for Web Services. In: Proc. of IEEE ICWS, pp. 49–56 (2005)
10. Greenfield, P., Kuo, D., Nepal, S., Fekete, A.: Consistency for Web Services Applications. In: Proc. of VLDB, pp. 1199–1203 (2005)
11. Kleinberg, J., Tardos, E.: Algorithm Design. Addison-Wesley, Reading (2006)
12. Bustamante, M.: Building a WCF Router. In Service Station, MSDN Magazine (April 2008), `http://msdn.microsoft.com/en-us/magazine`

An Executable Calculus for Service Choreography

Paolo Besana[1] and Adam Barker[2]

[1] University of Edinburgh
[2] University of Melbourne

Abstract. The Lightweight Coordination Calculus (LCC) is a compact choreography language based on process calculus. LCC is a directly executable specification and can therefore be dynamically distributed to a group of peers for enactment at run-time; this offers flexibility and allows peers to coordinate in open systems without prior knowledge of an interaction. This paper contributes to the body of choreography research by proposing two extensions to LCC covering parallel composition and choreography abstraction. These language extensions are evaluated against a subset of the Service Interaction Patterns, a benchmark in the process modelling community.

1 Introduction

A core challenge in today's ever connected world is in combining distributed resources on-demand to perform coordinated tasks. Coordination of distributed resources can be achieved through the use of workflow technologies. Workflow specifications defined using standard orchestration languages such as the Business Process Execution Language[1] typically facilitate statically defined workflows to be enacted by a centralised workflow engine. The resulting workflow specifications can be brittle due to the highly dynamic nature of distributed resources and scalable only to a point [1]. Dynamic composition of resources around tasks can address these issues.

The OpenKnowledge project[2] has produced a framework providing the middleware that assorted peers can use to interact within an open system. Peers interact following protocols, named *interaction models*, that define their externally observable behaviours. These protocols provide a global view of the interactions, and therefore fit into the definition of choreography, as opposed to the one of orchestration, considered to be a description of coordination from the perspective of a single process. The protocols are written in the Lightweight Coordination Calculus (LCC), a compact choreography language based on process calculus.

This paper introduces the LCC syntax, corresponding OpenKnowledge framework (Section 2) and discusses its current limitations and possible workarounds (Section 3). We propose two extensions to the LCC language (Section 4) and evaluate these changes by analysing the representations of two interaction patterns (Section 5). Related work is discussed (Section 6) in the context of our language extensions. Finally our contributions are summarised (Section 7).

[1] `www.oasis-open.org/committees/wsbpel/`
[2] `www.openk.org/`

R. Meersman, T. Dillon, P. Herrero (Eds.): OTM 2009, Part I, LNCS 5870, pp. 373–380, 2009.

2 OpenKnowledge and LCC

The OpenKnowledge kernel [7] has been designed with the goals of light weightness and compactness. The core concept is the set of shared *interaction models*, enacted by participants, called *peers*, that play *roles* within them. Interaction models are written in the *Lightweight Coordination Calculus* (LCC) [5], a compact, executable choreography language based on process calculus.

An interaction model in LCC is a set of role clauses. Clauses can refer to *entry-roles*, which participants initially assume, and *auxiliary-roles*, that can be reached only from other roles. Participants in an interaction take their *entry-role* and follow the unfolding of the clause specified using a combination of the sequence operator ('*then*'), or choice operator ('*or*') to connect messages and changes of role. A participant can take several roles during an interaction and can recursively take the same role (for example when processing a list). Messages are either outgoing to ('⇒') or incoming from ('⇐') another participant in a given role. Message input/output or change of role is controlled by constraint satisfaction.

Figure 1 shows an interaction model for an auction. The interaction starts with the `auctioneer` role that receives the product to sell and the list of bidders as an input parameters. It immediately changes its role to `caller` passing in the list of bidders and the product to sell. The `caller` recurses over the list of bidders in `Bidders`. If the list of peers is empty, it returns to the calling role;

```
a(auctioneer(Product, Bidders), A) ::
  a(caller(Product, Bidders), A)
  then a(waiter(Bidders, curwinner(nul, 0), Winner)
  then sold(Product, Price) ⇒ a(bidder, WB) ← curwinner(WB, Price) = Winner

a(caller(Product, Bidders), A) ::
  null ← Bidders = [] % no bidders left
  or ( invite_bid(Product) ⇒ a(bidder, BH) ← Bidders = [BH|BT]
       then a(caller(Product, BT), A) % recursion )

a(waiter(Bidders, Bids, curwinner(WinBidder, WinBid), Winner) ::
  null ← allarrived(Bids, Bidders) and Winner = curwinner(WinBidder, WinBid)
  or null ← timeout() and Winner = curwinner(WinBidder, WinBid)
  or ( bid(Offer) ⇐ a(bidder, B) then
       ( a(waiter([B|Bidders], curwinner(B, Offer), Winner) ← Offer > WinBid
         or a(waiter([B|Bidders], curwinner(WinBidder, WinBid), Winner) ) )
  or a(waiter(Bidders, curwinner(WinBidder, WinBid), Winner) ← sleep(1000)

a(bidder, B) ::
  invite_bid(Product) ⇐ a(caller, A)
  then bid(Product, Offer) ⇒ a(caller, A) ← bid_at(Product, Offer)
  then sold(Product, Price) ⇐ a(auctioneer, A)
```

Fig. 1. Auction protocol, % represents a comment

otherwise, it sends the `invite_bid` message to the peer at the head of the list and recurses over the remaining peers. Once all the messages are sent, the peer takes the `waiter` role, passing the list of bidders. The parameter `Winner` is an output parameter, and its value is set when the role `waiter` ends. The `waiter` role first checks if all the replies have arrived or if the period has timed out: if one of these two conditions is true, then it assigns the current winner as the final winner. Otherwise, it checks if there is a message in the incoming queue. If there is an offer and it is higher than the current highest offer, it recurses making the current bidder the current winner, otherwise it simply recurses. If there is no offer in the queue, then it waits for a second and recurses.

Symmetrically, the `bidder` receives the request to bid, and sends the offer. It may then receive the `sold` message if the offer was successful. If unsuccessful, the framework will signal the end of interaction. Through this pattern (an implementation of the *synchronisation* pattern [6]) asynchronous message reception is possible: messages are received in any order, and bidders act independently.

3 Limitations

While developing interactions for the various scenarios, we encountered limitations in the current version of OpenKnowledge. It was often possible to find workarounds but these ad-hoc solutions lacked generality and clarity. The limitations can be divided in two categories: design and execution. We will describe two design-time limitations: the impossibility of representing different levels of abstraction in a clean way and the lack of a parallel operator.

In LCC, a single interaction model includes all activities and messages at all levels of abstraction. The only abstraction available is provided by roles, that have to belong to the same interaction model. One possible work-around, applied in various cases in the testbeds used for evaluating OpenKnowledge [8], is for a peer to start a new interaction from within a constraint. However, this solution has two drawbacks: starting a new interaction is the action of single peer, of which other peers are not aware. This makes it hard to include participants involved in the first interaction into the sub-interaction. It is also a brittle solution, as it is not possible to specify how constraints are solved by peers.

Another important limitation is the lack of a parallel operator. Parallel operations, as described in the *parallel split* pattern [6], can be obtained by sending a sequence of messages to a set of roles waiting for them. Sending a message is a non-blocking operation (we saw before that it requires only to insert the message in a queue) so, from the perspective of the `auctioneer`, the operations in the bidders are started nearly simultaneously. In the example the replies from the bidders are merged back by the `waiter` role.

However, in this specific case we know the number of parallel operations (that is, the number of bidders) before the start of the interaction. In the general case it is not possible to know in advance how many parallel operations need to be performed: if peers have to be bound to these roles in advance it is not possible to increase their number once the interaction has started. Moreover, it may not be clear who should perform these roles.

4 Proposed Design Extensions

4.1 Scene Operator

To address the lack of an abstraction mechanism and to maintain at the same time clarity at design-time, we introduce the concept of *scenes*, which are abstractions of interaction models. In turn interaction models implement scenes. We also introduce the new operator `scene(scenename, role)`, that defines the execution of a role in another scene.

```
a(auctioneer(Product), A) ::
  a(caller(Product, Bidders, Winner), A) ← getPeers("bidder", Bidders)
  then sold(Product, Price) ⇒ a(bidder, WB) ← curwinner(WB, Price) = Winner
  then scene(payment, a(payee(Price), A))

a(bidder, B) ::
  invite_bid(Product) ⇐ a(caller, A)
  then bid(Product, Offer) ⇒ a(caller, A) ← bid_at(Product, Offer)
  then sold(Product, Price) ⇐ a(auctioneer, A)
  then scene(payment, a(payer(Price), B))
```

Fig. 2. Extending the auction with scenes

An interaction models makes no assumption of how the scene will be performed: the operation has to be matched to another interaction model implementing the scene. This has to be performed by the enactment framework. A scene can succeed or fail, like a standard interaction model. Figure 2 shows how the auction protocol could be modified to include scenes. Once the winning bidder has been alerted, both the auctioneer and the bidder go into the *payment* scene, respectively in the `payee` and `payer` roles. The requirement is that all the peers in the calling interaction model that have encountered the same scene invocation are registered to participate in the run of the matched interaction model. The new scene can also include other roles and other participants. The addition of scenes does not influence the correspondence between LCC and π-calculus: at run-time the operation is equivalent to a normal role change. What changes is how the role is located and matched.

The use of scenes allows the creation of hierarchies of scenes at different levels of abstraction, in which the root is itself a scene. At each level, scenes are implemented by interaction models, that can contain other scenes, implemented by further interaction models and so forth.

4.2 Parallel Operator

The parallel operator we introduce here focuses around the operation of role change. While in the current version of LCC the role change operator inside a clause is always a sequential operation, we distinguish between two different role calls, one for blocking and one for non-blocking execution of a role clause: `b:a(type,ID)` and `nb:a(type,ID)`.

The current role call is blocking: the execution of the calling role is halted, the called role is executed, and when it terminates the caller resumes. The non-blocking call corresponds to spawning a new role in a parallel process: a role can spawn a new role, executed by the same peer. The spawned process has its own process identifier and its own incoming and outgoing message queues, like a normal participant.

```
a(auctioneer(Product, Time), A) ::
 b : a(caller(Product, Bidders), A) ← getPeers("bidder", Bidders)
 then nb : a(timer(Time), G) then
 then b : a(waiter(Bidders, curwinner(nul, 0), Winner)
 then sold(Product, Price) ⇒ a(bidder, WB) ← curwinner(WB, Price) = Winner

a(timer(Time), T) ::
 timeout ⇒ a(waiter, A) ← wait(Time)

a(waiter(Bidders, Bids, curwinner(WinBidder, WinBid), Winner) ::
 null ← allarrived(Bids, Bidders) and Winner = curwinner(WinBidder, WinBid)
 or (bid(Offer) ⇐ a(bidder, B) then ...)
 or timeout ⇐ a(timer, T)
```

Fig. 3. Auction interaction model with `timer` role

While in a blocking call, it is possible to have both input and output parameters; in non-blocking calls, to avoid concurrency issues in accessing the parameters, the called role can only have input parameters. In the blocking role call the variable containing the process identifier contains the caller process, while in the non-blocking role call it is instantiated with the process identifier of the newly created process. To avoid zombie processes, the spawned role maintains a link with the spawner: if the spawner terminates, the spawned also terminates.

An interesting use of the parallel role is for timers. Using a parallel operator the auctioneer can start a parallel timer role that waits a finite amount of time and then sends a message. For instance, in Figure 3 the auction protocol is modified to include a timer, that after a fixed amount of time sends a message to the `waiter`. The `waiter` receives the message if some bidder has not replied before the deadline.

5 Evaluation

In order to evaluate our proposed extensions, we discuss how a subset of the interaction patterns described in [3] can be represented in the extended version of LCC. While all patterns that use time-frames can benefit from the introduction of the non-blocking role change, we will analyse in detail two of these patterns: the *one-from-many receive* and the *one-to-many send/receive*, as their representation most benefits from its introduction.

5.1 One-from-Many Receive

A party receives several related messages from autonomous events at different parties. Correlation of messages should occur within a time-frame. The number of messages may not be known at design or run-time [3].

```
a(receiver(Gs), R) ::
  ( msg(X) ⇐ a(customer, C)
    then b : a(findgroup(Gs, X, G)
    then
    ( null ← isStopCondition(G)
      then nb : a(groupHandler(G), GH)
      then b : a(receiver(NewGs), R)
           ← subtract(G, Gs, NewGs) ) )
  or
  ( timeout(G) ⇐ a(timer, T)
    then b : a(groupTimeout(G), R)
    then b : a(receiver(NewGs), R)
         ← subtract(G, Gs, NewGs) )

a(timer(Group, Rx), T) ::
  null ← wait(Time)
  then timeout(Group) ⇒ a(receiver, Rx)

a(findgroup(Gs, X, G), R) ::
  ( null ← Gs = [] and newGroup(X, G)
    then
    nb : a(timer(G, R), T) )
  or
  ( null ← Gs = [G|GT]
    then
    ( null ← inGroup(X, G)
      then
      null ← addToGroup(X, G) )
    or
    b : a(checkgroup(GT, X, G) )
```

Fig. 4. One-from-many receive implementation in LCC

The solution is to create separate processes for each time-frame, for handling the proceeding after the stop condition, and using the central process for managing the reception and the dispatch of messages in groups. Figure 4 shows the implementation of the pattern using LCC. The `receiver` role is the recipient of the message about an event `X` from a `customer` or of a `timeout` message from a timer. When the message arrives, the peer in the `receiver` role first finds the group corresponding to the event `X` using the auxiliary role `findgroup`. This role recurses through the list of groups: at each recursion the membership of `X` to the current group is checked. If the corresponding group is found, it is returned in the output parameter `G`. If no group is found, a new group is created and a new parallel role `timer` is spawned. Back to the main role receiver, if the group is in `stopCondition` (sufficient messages have been received), a new role is spawned for handling the group. This role (not listed here) could contain the reference to a new scene that all the senders should perform. If a timeout message is received, the receiver takes the `groupTimeout` role. The role is not listed, but we assume a message is sent to every sender in the group to inform them of the timeout.

5.2 One to Many Send/Receive

A party sends a request to several parties. Responses are expected within a time-frame and some parties may not respond. The number of parties may not be known at design time and the responses need to be correlated to their request [3].

Sending the requests is handled easily using the original LCC and a pattern similar to the one used in the auction protocol by the `caller` role. The second part is difficult and clumsy to represent using the original LCC syntax. Using the non-blocking call we can write compact code, as it is shown in Figure 5.

```
a(caller(Rs), D) ::
  null ← Rs = []
 or
 ( (nb : a(invoker(R), C) ← Rs = [R|RT] )
 (  then b : a(caller(RT), D)            )

a(invoker(R), C) ::
  msg(X) ⇒ a(remote, R)
  then nb : a(timer(T, C), Z)
  then
  ( reply(Y) ⇐ a(remote, R)       )
  ( or timeout() ⇐ a(timer, Z)    )
```

Fig. 5. One to many send/receive implementation in LCC

The peer in the `caller` role recurses over a list of recipients and spawns a new role `invoker` for each message to be sent. The `invoker` role, initialised with the identifier of the remote peer to contact, sends the message to the peer, spawn a `timer` role for the timeout and then waits for the reception of either the reply from the remote peer or a timeout from the timer process.

6 Related Work

There are relatively few languages targeted specifically at service choreography, for a survey refer to [2][1]. The most widely known are:

- WS-CDL [9] or the Web Services Choreography Description Language is the proposed W3C standard for service choreography. WS-CDL has native constructs supporting the functionalities provided by the extensions described in this paper, that is a mechanism for dealing with complex interactions and support for parallel operations. A package in WS-CDL can contain more than a single choreography. The `perform` activity can be used to launch other choreographies sharing variables between the caller and the called choreography. It also has a specific `<parallel>` construct.
- Let's Dance [10] is a language that supports service interaction modelling both from a global and local viewpoint. In a global (or choreography) model, interactions are described from the viewpoint of an ideal observer who oversees all interactions between a set of services. Local models, on the other hand focus on the perspective of a particular service, capturing only those interactions that directly involve it. Let's Dance supports parallel operations with a specific operator for repeated interactions; it does not provide a mechanism for abstraction.
- BPEL4Chor [4] is a proposal for adding an additional layer to BPEL to shift its emphasis from an orchestration language to a complete choreography language. BPEL4Chor is a collection of three artifact types: participant behaviour descriptions, participant topology and participant groundings. In BPEL4Chor it is possible, using an attribute, to define repeated interactions as parallel. It does not provide mechanisms for abstraction.

7 Conclusions

OpenKnowledge and LCC have been deployed in various scenarios, such as bioinformatics, emergency response simulation and health informatics. The extensive use (more than 200 different interaction models) has highlighted some of the limitations of the language and of the framework. In this paper we have analysed two limitations encountered in designing interactions: the lack of an abstraction mechanism to separate different levels of detail and the lack of a parallel operator.

In order to address these limitations, this paper proposed two extensions to LCC. The first is the introduction of the concept of scenes: a scene is an abstraction of an interaction model. The second is the introduction of a non-blocking role invocation, that allows the creation at run-time of new processes performing roles. The evaluation has shown that the new extensions allow a cleaner representation of service interaction patterns.

References

1. Barker, A., Besana, P., Robertson, D., Weissman, J.B.: The Benefits of Service Choreography for Data-Intensive Computing. In: Proceedings of CLADE 2009, pp. 1–10. ACM, New York (2009)
2. Barker, A., van Hemert, J.: Scientific Workflow: A Survey and Research Directions. In: Wyrzykowski, R., Dongarra, J., Karczewski, K., Wasniewski, J. (eds.) PPAM 2007. LNCS, vol. 4967, pp. 746–753. Springer, Heidelberg (2008)
3. Barros, A., Dumas, M., ter Hofstede, A.H.M.: Service Interaction Patterns. In: van der Aalst, W.M.P., Benatallah, B., Casati, F., Curbera, F. (eds.) BPM 2005. LNCS, vol. 3649, pp. 302–318. Springer, Heidelberg (2005)
4. Decker, G., Kopp, O., Leymann, F., Weske, M.: BPEL4Chor: Extending BPEL for Modeling Choreographies. In: Proceedings of IEEE ICWS 2007, pp. 296–303. IEEE Computer Society, Los Alamitos (2007)
5. Robertson, D., Walton, C., Barker, A., Besana, P., et al.: Models of Interaction as a Grounding for Peer to Peer Knowledge Sharing. In: Dillon, T.S., Chang, E., Meersman, R., Sycara, K. (eds.) Advances in Web Semantics I. LNCS, vol. 4891, pp. 81–129. Springer, Heidelberg (2009)
6. Russell, N., ter Hofstede, A., van der Aalst, W., Mulyar, N.: Workflow Control-Flow Patterns: A Revised View. Technical Report BPM-06-22, BPM Center (2006)
7. Siebes, R., Dupplaw, D., Kotoulas, S., de Pinninck, A.P., van Harmelen, F., Robertson, D.: The OpenKnowledge System: An Interaction-Centered Approach to Knowledge Sharing. In: Proceedings of CoopIS (2007)
8. Trecarichi, G., Rizzi, V., Vaccari, L., Marchese, M., Besana, P.: OpenKnowledge at Work: Exploring Centralized and Decentralized Information Gathering in Emergency Contexts. In: ISCRAM 2009 (2009)
9. W3C. Web Services Choreography Description Language Version 1.0 (November 2005), `http://www.w3.org/TR/2005/CR-ws-cdl-10-20051109/`
10. Zaha, J.M., Barros, A., Dumas, M., ter Hofstede, A.: Let's Dance: A Language for Service Behaviour Modeling. In: Meersman, R., Tari, Z. (eds.) OTM 2006. LNCS, vol. 4275, pp. 145–162. Springer, Heidelberg (2006)

The Influence of an External Transaction on a BPEL Scope

Oliver Kopp, Ralph Mietzner, and Frank Leymann

Institute of Architecture of Application Systems, University of Stuttgart, Germany
lastname@iaas.uni-stuttgart.de

Abstract. Business processes constitute an integral part of today's IT applications. They contain transactions as essential building blocks to ensure integrity and all-or-nothing behavior. The Business Process Execution Language is the dominant standard for modeling and execution of business processes in a Web service environment. BPEL itself contains a transaction model based on compensation, that describes the (local) transactions in a business process. The WS-Coordination framework deals with (external) transactions between Web services and is used to define the transaction behavior between a BPEL process and its partners. In this paper, we investigate how external transactions between Web services interrelate with local transactions of BPEL.

1 Introduction

The Web Services Business Process Execution [1] language (BPEL for short) is the dominant language used for defining business processes based on Web services. BPEL is designed to support business processes that are long-running, possibly running for years. Since a human may use other programs than data bases, the traditional ACID principle cannot be used. Therefore, a compensation based principle has been introduced and is used in BPEL. Activities belonging together are grouped by a BPEL 'scope'. A scope offers the possibility to attach a compensation handler to the grouped activities. This compensation handler is run if the scope should be compensated.

Transactions internal to business processes are well understood. BPEL, however, is agnostic to transaction behavior offered by calling services: Current BPEL engines and workflow languages do not support a scope being compensated from outside. This paper presents the first concept, where a local transaction (marked as a scope) can be part of two transactions: the local transaction and an external transaction.

Consequently, the paper is structured as follows: Section 2 provides an overview of related work in the field. Section 3 deals with the case, where a scope is part of a partner's transactions and presents necessary extensions to current coordination protocols. Finally, Section 4 concludes and provides an outlook on future work.

R. Meersman, T. Dillon, P. Herrero (Eds.): OTM 2009, Part I, LNCS 5870, pp. 381–388, 2009.

2 Background and Related Work

A history of transactions in workflows is presented in [2,3]. The only work allowing overlapping spheres of transaction is presented in [4]. That work specifies semantics for completed spheres, but neither specifies consequences of a fault in a sphere nor specifies coordination protocols.

In general, there are two different types of transaction styles available in the business area: compensation-based transactions and ACID style transactions [3]. In the field of Web services, WS-Coordination [5] and its protocol specifications are the de jure standard for coordinating transactions [6]. WS-C offers WS-Atomic Transactions (WS-AT, [7]) for 2PC transactions and WS-Business Activity (WS-BA, [8]) for compensation based transactions. WS-Coordination itself specifies an extensible framework for service coordination. It offers three types of services: activation service, registration service and protocol service. The initiator of a coordination requests a coordination context from the activation service. This context is used to distinguish the concrete coordination from other coordinations running in parallel and contains an endpoint reference for the registration service. The context is sent to the participants, which register themselves at the registration service, which returns an endpoint for a protocol service used for the concrete coordination. WS-Coordination is not restricted to a fixed set of possible coordination protocols, since it allows the use of arbitrary coordination protocols. This extensibility has been used to offer coordination for split BPEL process [9], coordinating auctions [10] as well as externalizing the coordination of BPEL scopes as a whole [11].

It is shown in [12] how WS-Coordination (WS-C, [5]) can be used to include services in local transactions. In the case of a BPEL process being completely under control of a caller, [13] shows the necessary coordination protocols.

A comparison of WS-BA and the transaction model of BPEL is provided in [14]. In summary, the transaction model of BPEL and WS-BA follow the Saga principle, but they differ in the treatment of external parties: BPEL (without the extensions presented in [11,12]) only sees sub-scopes as participants of a transaction and does not forward the transaction context to other services nor allows for inclusion of a scope into an external transaction.

A general classification of workflow and transaction languages is presented in [15]. BPEL and the related WS-Coordination specifications belong into the class *Tx+WF*, where “workflow and transaction models exist at the same level” [15].

[16] shows how a coordinator can use completion rules to automatically decide whether a WS-BA transaction should be completed or compensated. This work can be integrated in our work to gain a more sophisticated coordinator for deciding on the completion of the transaction.

Criticism on BPEL’s transaction model is presented in [17,18]. The main point of criticism is that a compensation handler may only be called at a fault handler. Even if this issue was solved, the work did not present solutions for the interplay of local and external transactions.

Besides WS-Coordination, the business transaction protocol (BTP, [19]) and the Web Services Composite Application Framework (WS-CAF, [20]) are relevant. While BTP extends the 2PC protocol towards long-running transactions, WS-CAF offers a true compensation-based protocol. A comparison between WS-Coordination, BTP and WS-CAF is presented in [20]. It is shown in [21] that BTP and WS-CAF can be mapped to WS-Coordination. In the context of BPEL, WS-Coordination is the de facto standard and we show how WS-Coordination can be used to coordinate internal and external transactions.

UN/CEFACT's Modeling Methodology (UMM, [22]) offers a way to model business transactions. A business transaction is a collaboration between two parties and does not include compensation and ACID transaction concepts. It is shown in [23] how UMM Business Transactions can be transformed into abstract BPEL processes. Faults of BPEL scopes are communicated to the partner using explicit messages and not by an underlying transaction protocol.

Interactions between processes are captured in process choreographies [24]. A choreography description consists of a list of all participants, the messages exchanged and the ordering of the message exchange. There are several choreography modeling languages available [25]. The most prominent are WS-CDL [26] and BPMN [27]. WS-CDL mentions "transaction protocols" but does not state how these can be applied in the choreography. BPMN only regards local transactions and does not regard cross-process transactions. Currently, BPEL4Chor [25] is the only language extending BPEL towards choreography modeling. Cross-process transactions are added to BPEL4Chor in [28] by introducing the concept of a choreography sphere, where activities of arbitrary participants are marked to belong to one transaction. The work regards choreography spheres as the only inter-process transaction dependency and does not regard influences of a scope to activities nested in a partner's scope.

3 Process Partially Controlled by an External Partner

In this section we present the case, where a BPEL scope is "infected" by a transaction of an external partner. In this case, the first messaging activity in the scope is an activity receiving a message from a partner, which includes a transaction context. The last activity is the only sending a message to the partner in the transaction context. The scope of the partner has to start with the message sending activity and end with the message receipt. Thus, the transaction boundaries are marked by a scope at both partners.

Figure 1 presents an airline booking process and its interaction with a travel agency. To illustrate the examples, we use the Business Process Modeling Notation (BPMN, [27]) as graphical notation for BPEL. If the semantics of BPMN and BPEL are different, we explicitly point that out. Otherwise, the semantics of the BPMN diagram itself and the rendered BPEL process coincidence. This approach is described in detail in [29].

The travel agency is responsible for booking a flight and organizing a visa. In case the visa application is rejected, the activity "Visa Application" fails

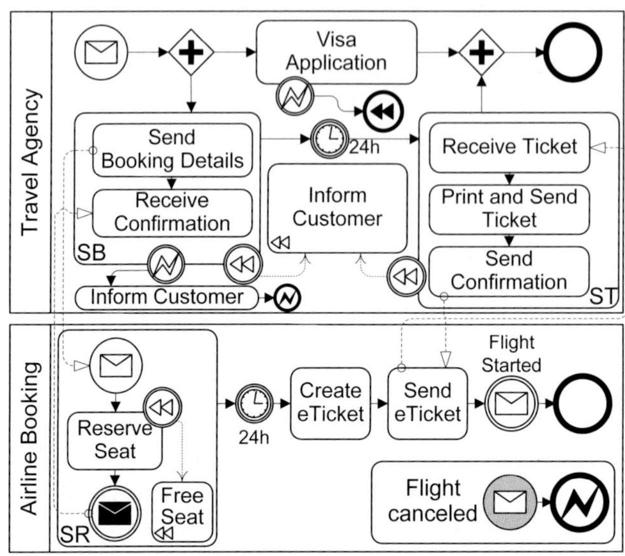

Fig. 1. Airline Booking Process

and throws an error. This error causes the compensation of already completed activities. The airline booking process first receives the booking details, reserves a seat and sends a confirmation to the travel agency. After 24 hours passed, the airline booking creates an eTicket, sends that ticket to the customer and waits for a confirmation. As soon as the flight starts, the process receives a message by the flight management system and the airline booking process ends.

A flight can always be canceled by the airline deposit process. In this case, the airline deposit process sends a flight canceled message to the airline booking process. This message is received by an non-interrupting event. Non-interrupting events are modeled as gray message start events and have BPEL's `onEvent` semantics. The reception of the message is enabled as long as the process is active. The reaction to the flight canceled message must be sending information to the travel agency. This can be achieved in two ways: (a) explicitly sending a message to the travel agency with explicit handling at the travel agency or (b) using the mechanisms of the underlying transaction protocol.

To illustrate the interplay between local transactions and external transactions, we focus on option (b). In this setting, the flight canceled message leads to a fault being thrown. This fault is handled at the process level and leads to the termination of all nested activities and the compensation of all completed activities in reverse order of their completion. The scope SR may be in two states if the fault is raised: (i) running or (ii) completed. In the first case, the BPEL engine needs to terminate the scope which leads to a termination of all nested activities. In addition, the coordination message `Fail` has to be sent to the travel agency. In the second case, the scope completed and sent the coordination message `Completed` to the travel agency. Now, the scope is in the state

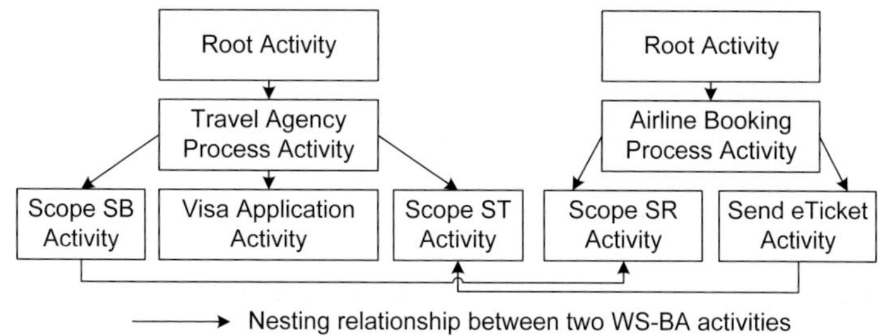

Fig. 2. WS-BA activities and their nesting in the travel agency scenario

"Completed" which enables the travel agency to compensate the scope SB. Due to the fault raised by the flight cancellation, the scope is compensated by the airline booking process. The current standard for compensation-based transactions, WS-Business Activity (WS-BA, [8]), does not allow a participant to compensate if it is in the state "Completed". In our setting, a compensation in the state "Completed" has to lead to a compensation of the other involved transactions, since the effect of one participant has been undone.

By using WS-BA, a transaction context is sent from the "Send Booking Details" activity to the airline booking process [12]. The scope SR is now part of two transactions: The scope is a sub-transaction of scope SR and also a sub-transaction of the airline booking process. Figure 2 presents the nesting graphically. As described in [11], each BPEL process, BPEL scope and invoke activity is a WS-C activity. These WS-C activities are nested in the case that the respective BPEL activities are nested in the process. As shown in [12], a WS-C activity a gets nested into an WS-C activity b if b calls a with a transaction context. Thus, the scope SR is nested in *two* WS-C activities: The WS-C activity of the scope SR and the WS-C activity of the airline booking process.

To enable a compensation in the state "Completed" by a participant, we extend WS-BA with Participant Completion to WS-BA with Participant-Triggered Compensation (WS-BA w/ PTC) as presented in Fig. 3. In WS-BA w/ PTC, the state "Closing" is replaced with a subgraph containing the states "Preparing Closing", "Closing Prepared" and "Closed". These three states are similar

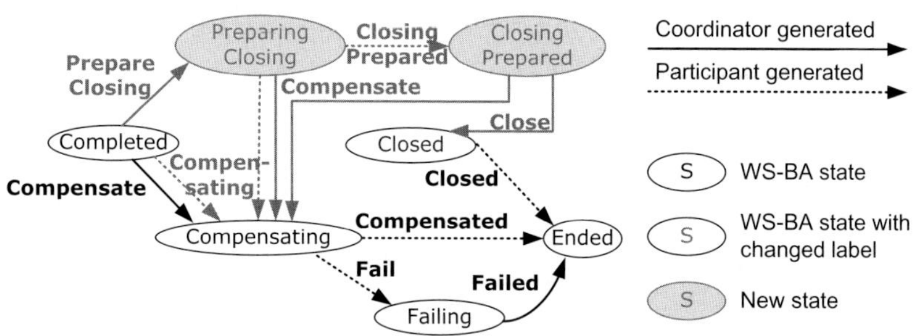

Fig. 3. WS-BA with participant-triggered compensating

to the 2PC protocol, where the participants first vote for the decision and the coordinator finally commits the transaction if all participants voted for commit. Otherwise, a rollback is made at each transaction [30]. In the case of WS-BA w/ PTC, all participants have to be in the state "Completed" to start the closing phase. In the case that not all participants are in the state "Completed" and one participant sends "Compensating" out of the "Completed" state, the transactions of the other participants have to be canceled or compensated: A "Cancel" message is sent to all participants in the state "Active" and a "Compensate" message is sent to all participants in the state "Completed". As soon as all participants are in the state "Completed", the message "Prepare Closing" is sent to all participants. The participants now change to the state "Preparing Closing". From herein, they can either sent "Closing Prepared" or "Compensating". For instance, the latter occurs in the case of the airline booking process: As soon as the scope SR is complete, the flight is canceled and thus the scope SR gets compensated. In case a "Compensating" message is received by the coordinator, it sends the message "Compensate" to all other participants. Otherwise, all participants sent "Closing Prepared" to the coordinator. A participant sends this message in case the transaction cannot be compensated by the participant any more. In case of the airline booking, this happens as soon as the message "flight started" is received and the process reaches its final state. In case all participants are in the state "Closing Prepared", no participant can send a compensation message to the coordinator. Thus, the coordinator can close the transaction by the message "Close". The participant is now in the state "Closed", which equals the state "Closing" of WS-BA. The participant acknowledges the closing by the message "Closed".

In case a fault happens at the activity "reserve seat" (scope SR), this fault has to be communicated to the scope SB. Since the scope SR does not contain a fault handler, the coordination message `Fail` can be directly sent to the coordinator. That message leads to a fault in the scope SB, which in turn leads to informing the customer and a process termination. In case the scope SR had a fault handler, two different messages may be sent to the travel agency process: `CannotComplete` or `Fail`. `CannotComplete` indicates that a fault has occurred and that the fault can be successfully handled by a fault handler. `Fail` indicates that the fault handler did run, but rethrew the fault. The coordinator, however, will be notified about the outcome after the fault handling. It is shown in [11] how WS-BA can be extended to support a notification before the fault handler runs.

The activity "Send eTicket" also starts a transaction. The message containing the eTicket also contains a transaction context. The travel agency receives the eTicket, generates a hardcopy of the ticket and sends it to the customer. Afterwards, the airline booking process receives a conformation that the ticket was handled successfully. In case an error occurs during the printing and sending of the ticket at the travel agency, the transaction protocol ensures that the activity "Send eTicket" raises a fault: The protocol sends `Fault` to the coordinator which in turn sends `Cancel` to the BPEL engine, which in turn raises the fault leading to a compensation of the SR scope and especially a compensation of the reserve seat activity. By using WS-BA as the underlying coordination protocol, the activity "Send Confirmation" is not necessary any more: The scope ST is a WS-C sub-activity of the WS-C activity for "send eTicket". Thus, the activity "send eTicket" may only complete if the WS-C sub-activity for ST completes.

4 Conclusion and Outlook

This paper presented the situation, where a BPEL scope is part of two transactions. We showed that the coordination protocol WS-Business Activity has to be extended to deal with that case.

In general, the boundary of the transaction is not be equal to the scope boundary. Thus, our ongoing work focuses on defining transaction boundaries using pairs of receive/reply activities, activities accessing the same data item and user defined boundaries. Our future work includes a formal description of the problem as well as a treatment of other coordination protocols such as WS-Atomic Transaction.

Acknowledgment. This work is supported by the BMBF funded project Tools4BPEL (01ISE08B).

References

1. OASIS: Web Services Business Process Execution Language Version 2.0 – OASIS Standard (2007), `http://docs.oasis-open.org/wsbpel/2.0/wsbpel-v2.0.html`
2. Dayal, U., Hsu, M., Ladin, R.: Business Process Coordination: State of the Art, Trends, and Open Issues. In: VLDB 2001, 27th International Conference on Very Large Data Bases. Morgan Kaufmann, San Francisco (2001)
3. Wang, T., Vonk, J., Kratz, B., Grefen, P.: A survey on the history of transaction management: from flat to grid transactions. Distributed and Parallel Databases 23(3), 235–270 (2008)
4. Leymann, F.: Supporting Business Transactions Via Partial Backward Recovery In Workflow Management Systems. In: Datenbanksysteme in Büro, Technik und Wissenschaft, BTW 1995 (1995)
5. OASIS: Web Services Coordination (WS-Coordination) Version 1.2 (2009), `http://docs.oasis-open.org/ws-tx/wscoor/2006/06`
6. Curbera, F., Leymann, F., Storey, T., Ferguson, D., Weerawarana, S.: Web Services Platform Architecture: SOAP, WSDL, WS-Policy, WS-Addressing, WS-BPEL, WS-Reliable Messaging and More. Prentice Hall PTR, Englewood Cliffs (2005)
7. OASIS: Web Services Atomic Transaction (WS-AtomicTransaction) Version 1.2 (2009), `http://docs.oasis-open.org/ws-tx/wsat/2006/06`
8. OASIS: Web Services Business Activity (WS-BusinessActivity) Version 1.2 (2009), `http://docs.oasis-open.org/ws-tx/wsba/2006/06`
9. Khalaf, R., Leymann, F.: Coordination Protocols for Split BPEL Loops and Scopes. Technical Report Computer Science 2007/01, University of Stuttgart (2007)
10. Leymann, F., Pottinger, S.: Rethinking the Coordination Models of WS-Coordination and WS-CF. In: ECOWS 2005 (2005)

11. Pottinger, S., Mietzner, R., Leymann, F.: Coordinate BPEL Scopes and Processes by Extending the WS-Business Activity Framework. In: Meersman, R., Tari, Z. (eds.) CoopIS 2007, Part I. LNCS, vol. 4803, pp. 336–352. Springer, Heidelberg (2007)
12. Tai, S., Khalaf, R., Mikalsen, T.A.: Composition of Coordinated Web Services. In: Jacobsen, H.-A. (ed.) Middleware 2004. LNCS, vol. 3231, pp. 294–310. Springer, Heidelberg (2004)
13. IBM, SAP: WS-BPEL Extension for Sub-processes – BPEL-SPE (2005), http://www.ibm.com/developerworks/library/specification/ws-bpelsubproc/
14. Sauter, P., Melzer, I.: A Comparison of WS-BusinessActivity and BPEL4WS Long-Running Transaction. In: Kommunikation in Verteilten Systemen (KiVS). Informatik aktuell, pp. 115–125. Springer, Heidelberg (2005)
15. Grefen, P.W.P.J., Vonk, J.: A Taxonomy of Transactional Workflow Support. Int. J. Cooperative Inf. Syst. 15(1), 87–118 (2006)
16. Riegen, M.V., Husemann, M., Ritter, N.: Providing Decision Capabilities to Coordinators in Distributed Processes. In: ICIW 2008: Third International Conference on Internet and Web Applications and Services, pp. 500–505 (2008)
17. Coleman, J.: Examining BPEL's Compensation Construct. In: Workshop on Rigorous Engineering of Fault-Tolerant Systems, REFT 2005 (2005)
18. Greenfield, P., Fekete, A., Jang, J., Kuo, D.: Compensation is Not Enough. In: EDOC 2003: 7th International Conference on Enterprise Distributed Object Computing, Washington, DC, USA, p. 232 (2003)
19. Dalal, S., et al.: Coordinating Business Transactions on the Web. IEEE Internet Computing 7(1), 30–39 (2003)
20. Little, M., Webber, J.: Introducing WS-CAF—more than just transactions. Web Serv. J. 3(12), 52–55 (2003)
21. Vetter, T.: Anpassung und Implementierung verschiedener Transaktionsprotokolle auf WS-Coordination. Diploma thesis, University of Stuttgart, IAAS (2006) (in German)
22. UN/CEFACT: UN/CEFACT's Modeling Methodology (UMM), UMM Meta Model – Foundation Module. Technical Specification V1.0 (2006)
23. Hofreiter, B., et al.: Deriving executable BPEL from UMM Business Transactions. In: SCC 2007: IEEE International Conference on Services Computing (2007)
24. Peltz, C.: Web Services Orchestration and Choreography. IEEE Computer 36(10), 46–52 (2003)
25. Decker, G., et al.: Interacting services: from specification to execution. Data & Knowledge Engineering 68(10), 946–972 (2009)
26. Kavantzas, N., et al.: Web Services Choreography Description Language Version 1.0 (2005), http://www.w3.org/TR/ws-cdl-10
27. Object Management Group: Business Process Modeling Notation, V1.2. (2009), http://www.omg.org/spec/BPMN/1.2/PDF
28. Kopp, O., Wieland, M., Leymann, F.: Towards Choreography Transactions. In: 1st Central-European Workshop on Services and their Composition, ZEUS 2009 (2009)
29. Schumm, D., et al.: On Visualizing and Modelling BPEL with BPMN. In: 4th International Workshop on Workflow Management, ICWM 2009 (2009)
30. Gray, J., Reuter, A.: Transaction Processing: concepts and techniques. Morgan Kaufmann, San Francisco (1993)

All links were last followed on June 22, 2009.

Cooperating SQL Dataflow Processes for In-DB Analytics

Qiming Chen and Meichun Hsu

HP Labs
Palo Alto, California, USA
Hewlett Packard Co.
{qiming.chen,meichun.hsu}@hp.com

Abstract. Pushing data-intensive analytics down to database engines is the key to high-performance and secured execution; however, the existent SQL framework is unable to express general graph-based dataflow processes, and unable to orchestrate multiple dataflow processes with inter-operation data dependencies.

In this work we extend SQL to Functional Form-SQL (FF-SQL) based on a calculus of queries, to *declaratively* express complex dataflow graphs. A FF-SQL query is constructed from conventional queries using Function Forms (FFs). While a conventional SQL query represents a dataflow tree, a FF-SQL query represents a more general dataflow graph. Further, with FF-SQL, a group of SQL dataflow processes with data dependency among their operations can be specified as a single, integrated FF-SQL definition, and executed cooperatively inside the database engine without repeated data retrieval, duplicated computation and unnecessary data copying. A novel extension to the PostgreSQL query engine is made to support FF-SQL dataflow processes.

1 Introduction

Executing data-intensive BI analytics inside the database engine can provide benefits in scalability, performance and security [3,5]. However, this approach is not yet generally applicable since the SQL language and SQL engine lack the capability to express a general graph structured dataflow process (beyond a query tree), and to orchestrate multiple intra-process and inter-process operations for sharing database access and query evaluation results. With the existent SQL framework, a single SQL query can only express tree-structured operations with coincident dataflows and control-flows. The result of a (sub)query can only be delivered to a single parent operation; in case it is requested by multiple operations, the query, must be evaluated multiple times. These limit the SQL framework in supporting graph-structured dataflows at both language and implementation levels. In this paper we present our solutions to these problems.

Let us consider 3 applications that use star-joins of a fact table and multiple dimension tables, as shown in Fig. 1.

With the existing SQL framework, the fork of the star-join result to multiple destination operations is not possible; instead, multiple separate queries must perform the star-join repeatedly; as for analytical computation, if we were to push *clustering* into the database engine, SQL is unable to express the iteration and the cache of the *customer feature vectors* across iterations without repeated data loading or derivation.

R. Meersman, T. Dillon, P. Herrero (Eds.): OTM 2009, Part I, LNCS 5870, pp. 389–397, 2009.

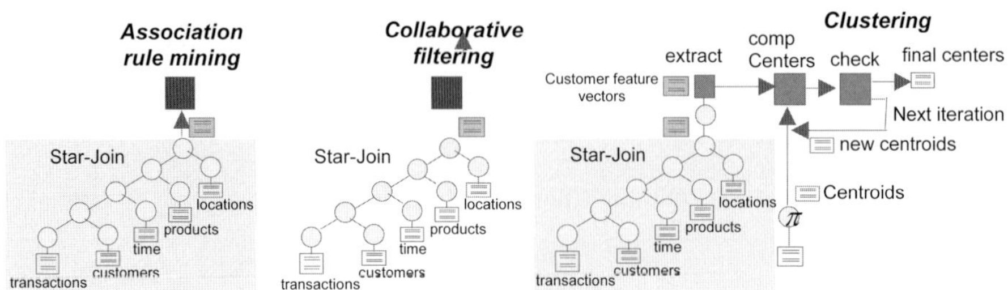

Fig. 1. SQL is unable to express the "share" of sub-query result, leading to duplicate evaluation

Other approaches to dealing with such problems exist. Cooperative file scan among multiple queries [2,7] focuses on attaching queries started later to already active scans. However, as indicated in [7], such a strategy is not effective when queries scan different ranges rather than the full content of a table. Common Subquery Optimization [6] focuses on sharing query plans rather than evaluation results. In contrast to the general business process management, our work is characterized by in-DB SQL dataflow process management.

In this work, we introduce "functional forms" to specify multiple cooperative queries with user-defined functions using a single declarative expression. We propose to (a) extend SQL to express general graph structured dataflow processes beyond query trees; and (b) extend query engine to orchestrate multiple dataflow processes for sharing database retrieval and query evaluation results, without repeated data retrieval, duplicated computation and unnecessary data copying.

Viewing queries as functions applied to relation objects, we introduce a set of meta-operators called Functional Forms (FFs) for constructing new functions from existing ones. We then provide an algebraic system made of queries, user-defined Relation Valued Functions (RVFs)[4], constructive primitive functions and FFs, referred to as the FF-SQL system. Applying a FF to query functions denotes a combined function, and applying that function to relations denotes a FF-SQL query. While a regular SQL query represents tree-structured dataflow, a FF-SQL query represents general graph-structured dataflow. We use FF-SQL to specify a dataflow process made of multiple correlated queries, which may invoke RVFs, for a particular application goal. We also use FF-SQL to specify a group of correlated dataflow processes with common data sources and data dependencies among their operations.

We have extended the PostgreSQL query engine to execute FF-SQL dataflow processes cooperatively without duplicate data access and query evaluation. A specific in-DB cooperative layer is provided to control application dataflows, schedule queries, and interact with query processing. It isolates much of the complexity of data streaming into a well-understood system abstraction. Our implementation will be reported separately.

The rest of this paper is organized as follows: Section 2 introduces the FF-SQL framework; Section 3 shows the FF-SQL specification of cooperative dataflow processes; Section 4 concludes the paper.

2 FF-SQL – An Algebraic Framework for Queries

A FF-SQL system is used to combine queries and user defined RVFs for representing application dataflow graphs. In the following we describe the operators – functional forms, and operands – query functions and RVFs, of a FF-SQL system.

2.1 Relation Valued Functions

As mentioned above, to extend the action capability of the database engine, we have generalized the table valued functions with scalar input to RVFs with relation input [3,4]. In this way RVFs can be treated as a relational operator or data source and integrated to SQL queries.

2.2 Query Variables vs. Query Functions

We distinguish the notion of *Query Function* from the notion of *Query Variable*. A query variable is just a query such as

```
SELECT * FROM Sales, Customers WHERE Sales.customer_id = Customers.id;
```

A query variable can be viewed as a relation data object, say v_Q, denoting the query result.

A query function, however, is a function applied to a sequence of parameter relations. For instance, the query function corresponding to the above query can be expressed as

f_Q := SELECT * FROM $1, $2 WHERE $1.customer_id = $2.id;

Then applying f_Q to a sequence of relations < Sales, Customers> with matched schemas, is expressed by

f_Q : < Sales, Customers> → v_Q (bind Sales to $1 and Customers to $2)

The major constraint of query functions is *schema-preserving*, i.e. the schemas of the parameter relations must match the query function. It is obvious that the above query function is not applicable to arbitrary relations.

2.3 Functional Forms

A functional form (or function combining form), FF, is an expression denoting a function; that function depends on the functions which are the parameters of the expression. Thus, for example, if *f* and *g* are RVFs, then *f*•*g* is a functional form denoting a new function such that, for a relation r, (*f*•*g*):r = *f*:(*g*:r) provided that r matches the input schema of *g*, and *g*:r matches the input schema of *f*.

2.4 FF-SQL Framework

A FF-SQL system is founded on the use of a fixed set of FFs for combining query functions (a query can invoke RVFs). These, plus simple definitions, provide the simple means of building new functions from existing ones; they use no variables or substitution rules, and they become the operations of an associated algebra of queries. All the functions of a FF-SQL system are of one type: they map relations into relations; and they are the operands of FFs. In general, a FF-SQL system comprises the following.

- A set O of objects. An object is either a relation, a query variable, a sequence $<r_1, \ldots, r_n>$ whose elements r_i are objects, or Δ (undefined), T (true), F (false), ø (empty).
- A set F of functions which are query functions, RVFs, construct primitives and their combinations that map objects into objects. Queries (and RVFs) are schema-aware.
- A meta-operator, apply; applying a function f to an object r is expressed as $f{:}r$, and in case the object is a sequence $<r_1, \ldots, r_n>$, is expressed by $f : <r_1, \ldots, r_n>$;
- A set C of functional forms for combining existing functions to new functions in F;
- A set D of definitions that define some functions in F and assign a name to each.

A set of constructive-primitives are provided but we only list the ones used in this report, such as

- **Selector** $\$i : <r_1, \ldots, r_n> = r_i$
- **Identity** $id : r = r$
- **Constant** !y : x = y (can also be treated as a FF but we opt to treat it as a function)
- **Union** union: $< r_1, r_2 > = r_1 \cup r_2$

A FF is an expression denoting a function. A FF is primarily used to combine queries with RVFs into higher level ones for expressing dataflow graphs, in a style not expressible by the conventional SQL. A subset of FF primitives used later in this paper is listed below (schema preserving is implied):

- **Composition** $(f \bullet g) : r = f(g{:}r)$
- **Construction** $[f_1 \ldots\ldots f_n] : r = <f_1{:}r \ldots\ldots f_n{:}r>$
- **Condition** $(p \rightarrow f, g) : r = ((p{:}r)\text{=T} \rightarrow f{:}r;\ g{:}r$
- **Map** (Apply to all) $\alpha f : <r_1 \ldots\ldots r_n> = <f{:}r_1 \ldots\ldots f{:}r_n>$
- **Reduce** $/f : r = r{=}{=}<r_1> \rightarrow r_1;$
  $r{=}{=}<r_1 \ldots\ldots r_n>\ \&\ n{>}{=}2 \rightarrow f{:}<r_1, /f{:}<r_2 \ldots\ldots r_n>>$

Map represents data-parallelism, construction represents task-parallelism. Reduce can be generally used for stepwise merge and aggregate purposes, such as

$/\$2 : <r_1 \ldots\ldots r_n> = r_n$ (last)

$/\text{union} : <r_1 \ldots\ldots r_n> = r_1 \cup r_2 \ldots \cup r_n$

To be specific to relational data manipulation, in the FF-SQL system we do not introduce computation primitives other than queries and user-defined RVFs; and we do not introduce constant values such as a number or a string. To have a non-relation constant passed in an RVF, the "constant" primitive can be used. For instance, given the RVF g_{rvf} with argument list (R_1, R_2, k) where k is an integer and R_1, R_2 are relations, then, for applying g_{rvf} to a sequence of relations $<R_1, R_2>$, a composite function G can be defined by (denoted as :=)

$$G := g_{rvf} \bullet [\$1, \$2, !k]$$

Applying G to $<R_1, R_2>$ can be expressed as

$$G : <R_1, R_2> = g_{rvf} \bullet [\$1, \$2, !k] : <R_1, R_2> = g_{rvf} : <R_1, R_2, k> = g_{rvf}(R_1, R_2, k)$$

In the FF-SQL system, functions can be defined level by level from query functions/RVFs in terms of FFs. Apply a FF to functions denotes a new function; and apply that function to relations denotes a *FF-SQL query*. A FF-SQL query has the expressive power for specifying a dataflow graph, in the way not possible by a regular query. In the other words, a regular SQL query represents tree-structured dataflows, a FF-SQL query can further represent graph structured dataflows.

The notion of FF is analogous to the function combining form found in FP system [2], however, the FF-SQL system is a declarative system rather than a functional programming system. Besides queries and RVFs, we do not introduce any computational primitives (such as +, - *, /); instead, we only introduce several constructive primitive functions. Note that FFs are also constructive rather than computational. FF-SQL further differs from FP in taking strongly typed (i.e. schema-aware) relations as data objects, where a query is viewed as a relation data consumer and producer, i.e. a data source.

3 FF-SQL Specification of Cooperative Dataflow Processes

In this section we shall illustrate how to use the proposed FF-SQL to express some typical dataflow schemes in the way not expressible by using regular SQL queries.

As shown in Fig 2, the modulated version of Fig. 1, we assume a table, *Sales*, holding shopping transaction data is retrieved, filtered and aggregated by a star-join query Q_1, resulting a relation, *Txs* with the following schema.

[TxDetailID, Customer, Item, DayOfWeek, Amount, Subtotal]

The attribute *Location* is left for further dimensioning the analysis by state, which is not referred to here for simplicity.

In addition, customer shopping behavior is described by a "feature vector" expressing the average daily spending in a week. A feature vector has 7 values (since a week has 7 days) thus can be viewed as a 7 dimension point in the feature space. For customer segmentation, they are clustered based on such feature vectors representing one aspect of their shopping behaviors. We also assume the existence of initial clusters, each identified by a *cid* and having a centroid feature vector, *cv*.

3.1 The High-Level Function

The star-join result of query Q_1 is delivered to RVF *CF* for generating collaborative filtering matrix, RVF *AR* for generating association rules, and a composite function *CL* for clustering the customers based on their shopping behavior feature vectors. The feature vectors are extracted from the result of Q_1 by query functions Q_f followed by RVF "*extract*". The initial and updated clusters are kept in relation Centroids [*cid*, *cv*].

[Query Variables]

```
Q1 = SELECT txDetail-id, item, customer, dayOfWeek, amount, subtotal
     FROM Sales, Customers, Product, Time  WHERE /* star-join conditions */;

Q2 = SELECT cid, cv FROM Centroids;
```

[RVFs]

$AR(Q_1)$	- mining cross-selling association rules
$CF(Q_1)$	- collaborative filtering of customer shopping preference

[Composite Function]

CL(Q_1, Q_2) - cluster customers based on the "feature vectors" representing their average daily spending in a week

Then multiple cooperative dataflow processes are expressed by the following single FF-SQL function.

[FF-SQL main function]

[*AR* • \$1, *CF* • \$1, *CL*] : <Q_1, Q_2>

Applying it to data objects <Q_1, Q_2> forms the following FF-SQL query

[*AR* • \$1, *CF* • \$1, *CL*] : <Q_1, Q_2>

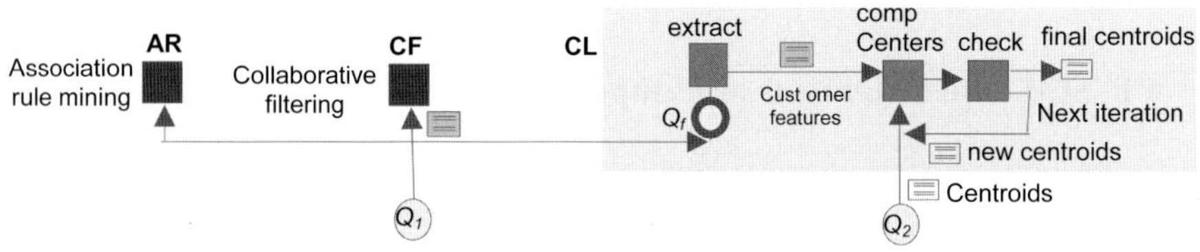

Fig. 2. Cooperative analytics applications with query result sharing

3.2 The Cluster Function

Now let us refine the clustering function CL. It is based on the *k*-means algorithm to cluster *n* objects based on attributes into *k* partitions, $k < n$. It is similar to the expectation-maximization algorithm for mixtures of Gaussians in that they both attempt to find the centroids of natural clusters in the data. It assumes that the object attributes form a vector space. The objective it tries to achieve is to minimize total intra-cluster variance. In our example, customers are clustered by the average daily spending in a week which is represented as a 7-valued vector corresponding to the 7 days in a week, or a 7 dimension point in the feature space.

In this K-Means clustering example, the customers are kept in a derived relation

CustomerFeatures [*customer*, *pv*]

where *pv* stands for the feature vector of a customer. The clusters are stored in relation

Centroids [*cid*, *cv*])

where *cid* stands for the ID of a given centroid, and *cv* stands for its feature vector.

For simplicity, let us abbreviate relation CustomerFeatures by P (since a vector can be considered as a 7-dimension Point), and Centroids by C.

[Query Function]

```
Qf = SELECT customer, dayOfWeek, SUM(subtotal) AS spending FROM $1
     GROUP BY customer, dayOfWeek;
```

[RVFs]

The RVF

extract($Q_f(Q_1)$)

is used to build feature vectors; it returns relation P (CustomerFeatures). The RVF

compCenters: <P, C> → C'

is used to derive a new set of centroids, i.e. a new instance of relation C, from relations P and C in a single iteration; which has the following two steps:

- the first step is for each customer in relation P to compute its distances to all centroids in relation C and assign its membership to the closest one, resulting an intermediate relation Nearest_centroids [pv, cid];
- the second step is to re-compute the set of new centroids based on the average location of member vector points.

After each iteration, the newly derived centroids, C', are compared to the old ones, C, by another RVF

check : <C', C> → {T; F}

for checking the convergence of the sets of new and old centroids to determine whether to terminate the K-Means computation or to launch the next iteration, using the current centroids, C', as well as the original points, P, as input data.

Our goal is to define a FF-SQL query

CL : <P, C>

that derives, in multiple iterations, the centroids of the clusters with minimal total intra-cluster variance, from the initial C relations towards the final instance of relation C.

During the *CL* computation, the relation C is updated in each iteration, but the relation P remains the same. A key requirement is to avoid repeated retrieval/derivation of either relation C or relation P from the database, which should be explicitly expressible at the language presentation level.

The function *CL* is defined by the following.

comp := [$1, $2, *compCenters*];

renew := (*check* • [$2, $3] → $3; *renew* • *comp* • [$1, $3]) ;

CL := *renew* • *comp* • [*extract* • Q_f • $1, $2];

Applying function *CL* to the points, P, and the initial centroids, C, for generating the converged centroids is expressed by the FF-SQL query

CL : <P, C>

The execution of FF-SQL query *CL* : <P, C> is explained as below.

- *comp*, i.e. [$1, $2, *compCenters*], maps relations P and C to a list of relations P, C and C'.

  [$1, $2, *compCenters*] : <P, C> → <P, C, C'>

  where C' is derived by *compCenters* : <P, C>
- then <P, C, C'> becomes the input of *renew*, where *check* • [$2, $3] is applied to C and C', i.e.

  check • [$2, $3] : <P, C, C'> = *check* : <C, C'> → {T; F}

If T is returned the *CL* function terminates with C' (the 3rd element in the above sequence) as its result; otherwise goes to the next iteration;

else if the above check fails, *renew* • *comp* • [$1, $3] is applied for the next iteration for re-generating a set of centroids, say C", as

renew • *comp* • [$1, $3] : <P, C, C'> = *renew* • *comp* : <P, C'>
= *renew* • [$1, $2, *compCenters*] : <P, C'> = *renew* : <P, C', C">

With the above CL definition, the refined main function for these three cooperative applications is specified in the following way.

[FF-SQL main function]

main := [*AR* • $1, *CF* • $1, *CL*]
:= [*AR* • $1, *CF* • $1, *renew* • *comp* • [*extract* • Q_f • $1, $2]]

comp := [$1, $2, *compCenters*];

renew := (*check* • [$2, $3] → $3; *renew* • *comp* • [$1, $3]) ;

The FF-SQL query for applying *main* to the data objects, i.e. the results returned from queries Q_1, Q_2, is expressed by the following.

main : <Q_1, Q_2> = [*AR* • $1, *CF* • $1, *CL*] : <$Q_1$, Q_2> = <*AR*:Q_1, *CF*:Q_1, *CL*:< Q_1, Q_2>>

As described later, the different versions of the instances of relation C output from *compCenters* in multiple iterations, actually occupy the same memory space.

4 Conclusions

In this work we proposed an approach for pushing data-intensive analytics down to database engines for high-performance and secured execution: integrating general analytic operations into SQL queries, handling general graph structured dataflows, and executing multiple dataflow processes with common subqueries or data sources.

To allow general analytic operations to be naturally and efficiently integrated with SQL queries, we support *RVF* at SQL language level. To *declaratively* express graph based dataflow we extend SQL to FF-SQL. To execute a group of correlated dataflow processes cooperatively without repeated data retrieval, duplicated computation and unnecessary buffering, we extend the query engine with a memory-based, embedded middleware layer.

The major advantage of FF-SQL lies in its expressive power for specifying complex dataflows. Specifically, a group of correlated SQL dataflow processes with common data sources or subqueries can be specified by a single, integrated FF-SQL dataflow definition, which provides the basis for their cooperation inside the database engine. The advantage of FF-SQL also lies in its simplicity, as it uses only the most elementary fixed naming system (naming a query) with a simple fixed rule of substituting a query for its name. Most importantly, they treat names as functions that can be combined with other functions without special treatment.

In this paper we described FF-SQL informally; the detailed formalisms, including the algebraic laws on FFs, will be documented separately. We are also developing support for FF-SQL by building a middleware layer inside the PostgreSQL database engine that deals with data buffering, dataflows, control flows and function scheduling.

References

1. Backus, J.: Can Programming Be Liberated from the von Neumann Style? A Functional Style and Its Algebra of Programs. ACM Turing award lecture (1977)
2. Cao, Y., Das, G.C., Chan, C.-Y., Tan, K.-L.: Optimizing Complex Queries with Multiple Relation Instances. In: ACM SIGMOD 2008 (2008)
3. Chen, Q., Hsu, M.: Data-Continuous SQL Process Model. In: Proc. 16th International Conference on Cooperative Information Systems, CoopIS 2008 (2008)
4. Chen, Q., Hsu, M., Liu, R.: Extend UDF Technology for Integrated Analytics. In: Proc. 10th Int. Conf. on Data Warehousing and Knowledge Discovery, DaWaK 2009 (2009)
5. DeWitt, D.J., Paulson, E., Robinson, E., Naughton, J., Royalty, J., Shankar, S., Krioukov, A.: Clustera: An Integrated Computation And Data Management System. In: VLDB 2008 (2008)
6. Tao, Y., Zhu, Q., Zuzarte, C.: Exploring Common Subqueries for Complex Query Optimization. In: IBM Centre for Advanced Studies Conference (2002)
7. Zukowski, M., Héman, S., Nes, N., Boncz, P.: Cooperative Scans: Dynamic Bandwidth Sharing in a DBMS. In: VLDB (2007)

Process Fragments*

Hanna Eberle, Tobias Unger, and Frank Leymann

University of Stuttgart, Institute of Architecture of Application Systems
Universitätsstraße 38, 70569 Stuttgart, Germany
lastname@iaas.uni-stuttgart.de

Abstract. The concepts presented in this paper are motivated by the assumption that process knowledge is distributed knowledge and not completely known just by one person. Driven by this assumption we deal in this paper with the following questions: How can partial process knowledge be represented? How can this partial knowledge be used to define something more complete? To use higher level artefacts as building blocks to new applications has a long tradition in software engineering to increase flexibility and reduce modeling costs. In this paper we take a first step in applying this concept to processes, by defining process building blocks and operations which compose process building blocks. The building blocks will be referred to as process fragments in the following. The process fragment composition may take place either at design or runtime of the process. The design time approach reduces design costs by reusing artefacts. However the runtime fragment composition approach realizes high flexibility due to the possibility in the dynamic selection of the fragments to be composed. The contribution of this work lies in a fragment definition that enables the fragment modeler to represent his 'local' and fragmentary knowledge in a formal way and which allows fragment models to be composed.

1 Introduction

Making business processes more flexible is a broad research area in BPM today. Flexibility in business processes means that the business process is able to react to varying situations and requirements. Flexibility in business processes can be realized in different ways. One way to realize flexibility is to model all alternatives possible as process. Often this approach is not possible, because the relevant knowledge becomes available e.g. not until the process is running. Flexibility is required especially in highly dynamic environments. Environments where not everything can be known at one time, e.g. design time, where one person has not a complete view on the process, and where things change very often. In these highly dynamic environments we consider process knowledge as something local and fragmentary. Where 'local' means that the knowledge is bound e.g. to a certain

* This work is partially funded by the ALLOW project. ALLOW (http://www.allow-project.eu/) is part of the EU 7th Framework Programme (contract no. FP7-213339).

R. Meersman, T. Dillon, P. Herrero (Eds.): OTM 2009, Part I, LNCS 5870, pp. 398–405, 2009.

location, to a certain situation or context or time or person. The term fragment is derived from the Latin verb *frangere* which stands for the english word *to break*. Therefore a fragment is defined as a part broken off something whole, like a shard is a part of a broken vase. A shard becomes a fragment, if it can be identified and glued to the vase it belongs to. The whole thing has some properties which are not evident in a fragment, e.g. a shard cannot be used as a container for something to store, whereas a vase can used as one. Fragments of a vase can not be glued together arbitrarily. The fitting fragments must be found and glued together in a suitable way. Having this assumption in mind, that knowledge is something 'local' and fragmentary, which needs to be build together depending on time or location, to receive a more complete view on the current knowledge for a certain situation, we adopt this idea for process modeling and execution. In our approach presented in this paper we model local process knowledge as process fragments, where process fragments can be glued together to a complete process. Process fragments are reflecting the partial and intermittent knowledge one modeler has at a certain time about a specific situation. Process fragments to be executed are chosen depending on data like the current location, and modeling data, like business case or other annotations. The work done in this paper in summarized in short in the following. We implement flexibility as 'local' process fragments, which can be glued together depending on situation, location or other context information. In this paper we motivate and draw an overall picture of the development and execution of fragment-based processes in Sections 2 and 3. We develop a fragment model definition, which leaves the modeler enough freedom in modeling 'local' knowledge and provides enough information to be able to compose fragments using syntactical correct operations in Section 5. We close this paper by relating our work to other work done in this area in Section 6 and a conclusion and a outlook for further work in 7.

2 Motivation and Application Scenario

As scenario we regard the local knowledge bank employees and other involved people have about the loan approval process [6]. Local knowledge about this domain is available from following persons: The bank front desk officer, the approver, the assessor, and the applicant. In the following we regard the local knowledge of the involved roles, which usually know at least the activities they have to perform, but the activities are not necessarily performed by them. The applicant knows that an application form has to be filled in to apply for a loan and that a notification is sent out by the bank to inform the applicant about either the reject or the accept of the application. The bank's front desk officer knows that credit information must be collected using a certain form. Afterwards the application is routed depending loan amount. The assessment officer knows about two assessment activities which are sequenced one after another. Depending on the outcome of the first assessment step either the other assessment step is performed or routed another way the risk is too high. The loan approval officer knows that a approval activity has to be done. This approval step is necessarily done by the loan approval officer, because only he has the right and the duty to carry the responsibility of his decision. The loan approval officer knows that the assessment activity is performed in two cases; either the amount of the loan exceeds 10000 or the the risk is too high and someone has to take responsibility. After his work is done the loan approval is either rejected or approved. The local knowledge that the bank employees apply depends not only on the business case to be executed but other differentiations to be made are necessary. E.g. an assessment officer has differing checks to perform depending on the economic situation. If the economic situation is stable fewer checks have to be performed than if the economic situation is an instable one. Therefore the local knowledge affiliated with the a business case might also be context depended. Other scenarios e.g. in real world scenarios where actors and executors move through space are conceivable, where local knowledge gets successively available depending on the executors location.

3 Fragment-Based Process Modeling and Execution

All people having fragmentary knowledge about the process model shall be able to model their knowledge as process fragment. Therefore lots of people have access to modeling process. The new designed process fragments get later annotated with local information, e.g. at what situation this fragment is used. Because there might be different behaviors a bank exhibits depending on the overall current economical situation. The process fragments get stored in a fragment repository. Having process fragments acquired the fragment definitions are ready to get composed by the fragment composer. The fragment composer implements a set of operations on how fragment definitions can be glued together. One operation is presented in detail in the remaining parts of this work. Two different approaches for composition times are possible. Either the 'local', situation-depended knowledge gets integrated to a complete process before the

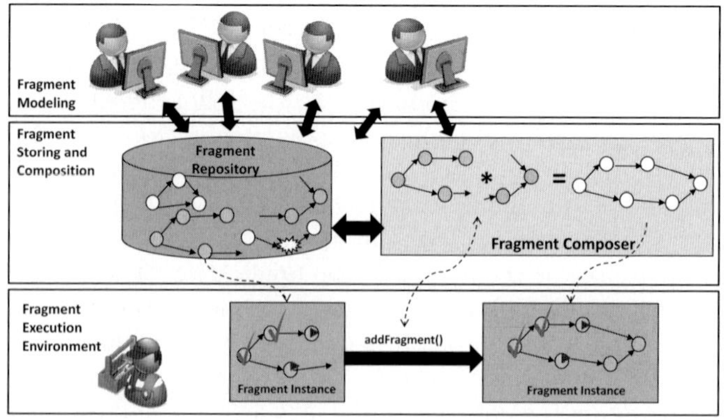

Fig. 1. Architectural Overview

process starts or it gets composed over a period of time, while known parts of the composed fragments can be executed in the meantime. The build-time approach can be used for either to build ad-hoc processes, which get executed only once, because it is strongly depending on a certain situation at composition time, or to as process which gets executed over and over again, as long as e.g. certain situation constraints hold. This build-time requires all fragments to be composed to be known before execution, which is a hard requirement for long running business processes, which run over years. For processes which have very high dependencies on current situations, e.g. like location, because they live in highly dynamic environments, the build-time approaches do not fit. Fragments and situation information becomes available not until runtime. This approach requires the running fragment instances to be migrated to the newly composed fragment definition. A overview of the architecture is shown in Figure 1.

4 Process Fragments

We consider a process fragment a fragmentary 'local' building block, representing incomplete 'local' knowledge, which can be composed with other fragments to build a complete and correct process model. Therefore a fragment model definition must serve following needs: 1. Fragment models must be able reflect the 'local' fragmentary process knowledge. 2. Fragment models must be composable to a complete process model. A fragment model definition must support and enable the fragment modeler to pour all knowledge he has into the formal form of a process based representation, a process fragment. Therefore we must provide modeling elements, which allow the modeler to specify, what is known, and to specify, what he knows he doesn't know. Therefore it must be possible to leave gaps in the definition, saying "'I know there happens something, but I cannot exactly tell what."'. All things that the modeler has concrete knowledge about can be modeled using established process modeling elements. Orderings of sets of activities can be specified by defining control connectors. Hyper-edges can be defined upon the set of activities and annotated with some alternative execution paths in case of faults appear within the contained activity set. Additionally to the established process modeling elements we provide modeling elements, which enable to model, where the exact business logic is not known. E.g. if it is known that activity A must be before B, but what exactly happens between A's completion and B's execution, is not known. Therefore we introduce a new modeling element called *region*. A region stands for business logic, which is not further defined. It can be connected with activities by defining normal control connectors (Figure 2). In the case that it is known, that three distinguishable proceedings are possible after activity C, but it's not known what activities are to be done afterwards, we leave the modeler the freedom to model this as activity C with three outgoing control connectors with no target activities defined. We leave the modeler even the possibility not to model control flow relation between activities at all. As a second step we need to define, what makes a fragment correct. We base our considerations herein on the assumption, that the composition of fragments should result in a correct process model. Therefore we derive correctness

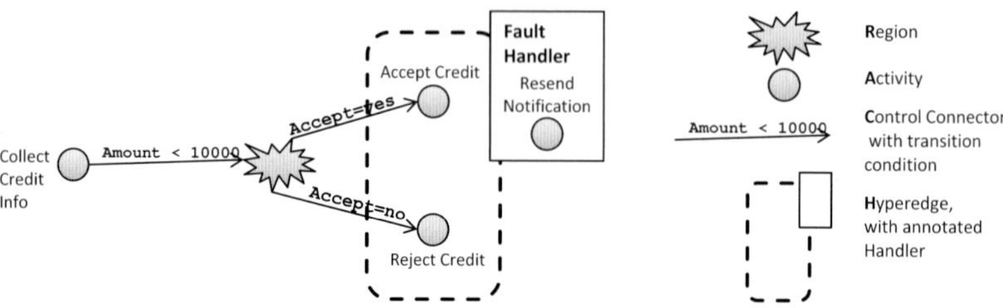

Fig. 2. Fragment Example

criteria directly from correctness criteria for process models. Fragment correctness criteria are defined in the following. A correct process model is required to have **no circles**. So does the fragment model. Because if a fragment building block contains a circle, the process completed of this fragment contains a circle. A correct process model is required to have no two control connectors to have the same target and source activities. The same holds for fragments. The hyperedges concerning transactional behavior of a correct process model are required to be properly nested. Again, each fragment must comply to this constraint. In order to restrict the numbers of dangling control connectors we define following constraint, where dangling means, that either the source or the target of a control connector is not defined. If there are dangling control connectors within a **connected fragment partition** with an equal transition condition one missing a target activity and the other one lacking in the source activity definition and if these two dangling control connectors can be glued together without defining a circle in the fragments control paths, than these two connectors must be glued together. We call this constraint *compactness constraint*, where compactness stands for the least fan out possible. A example of a non-compact fragment and a compact fragment is shown in Figure 3.

Fig. 3. Non-Compact Fragment vs. Compact Fragment

5 Fragment Composition Using Redundancy

Due to the 'local' knowledge modeling approach it is possible that some information is modeled redundantly. Two modeler might have partially overlapping knowledge i.e. some activities are contained in several fragments as well as some links connected to such activities. This redundancy can be used to define a first fragment composition operation by using the redundant information as gluing points between these two fragments. Based on the two fragment definitions we have given a set of the redundant activities, mapping exactly one activity onto one activity. There are lots of conceivable criteria [e.g. [8]], where the activity equality definition can be based on. The definition of these criteria is out of scope of this work. Furthermore, we assume, that no fragment contains more than one activity of the same activity definition. Note, that the fragments to be composed must not be contradicting, what means that two fragments are contradicting with respect to a composition operation, if the composed fragment is not compliant to the correctness criteria defined above, e.g. if the composed fragment contains circles and not properly nested hyperedges. The composition operation consists of following steps also presented in Figure 4. Firstly fragments activity, region, hyperedge and control connector sets, get united, where two equal activities get united to one activity. Now, the result may contain two control connectors connecting the same activities. To comply to the fragment correctness constraints, we merge these control connectors as follows: If the two control connectors have the same transition condition, we simply delete one connector, if the two control connectors have differing transition conditions the control connectors get merged by joining the transition condition with a OR-operator. After this step, we carry on with further refinement steps to make it compliant to further fragment constraints. Step two we are trying to substitute parts of

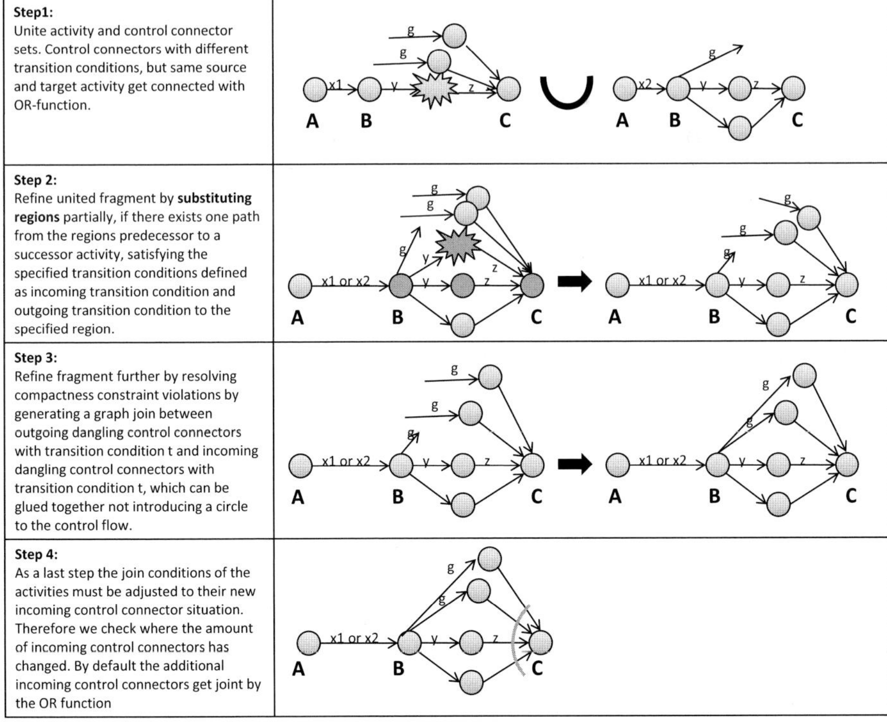

Fig. 4. Fragment Composition Operation

regions, with concrete parts. Afterwards we connect all dangling control connectors with the same transition condition to comply to the compactness constraint and not introducing circles by connecting the wrong control connectors. The last step deals with the adjustment of join conditions. Join conditions must be adjusted, where the amount of incoming control connectors has changed during the operation. The composition operation can be also applied to non-overlapping fragment knowledge. If this is the case the result fragment of the operation is a fragment with two unconnected parts. The composition operation has following algebraic properties. It is closed, commutative and has a neural element. It is not associative. The proof of these properties is out of scope of this paper.

6 Related Work

Composition operations are well known in the domain of formal languages, automaton theory [4], whereas the composition operation approach is new to the workflow domain. Since we use composition to realize flexibility in workflows we focus in the following on further flexibility approaches in the workflow domain. One possibility to deal with flexibility in workflows is to model all alternatives into the workflow model. The result of the modeling process is a very complex and very probably an unreadable workflow model. All information has to be available at design-time. A more dynamic approach in dealing with flexibility is to leave some gaps in the workflow specification, by e.g. defining some abstract activities with no fix activity implementation. The gaps are to be stuffed at runtime, when additional information gets available. There are many possible realizations of this gap-flexibility approach. In [1] a concept is introduced, on how subprocesses can be found and bound at runtime as implementation to a abstract activity. The subprocesses are annotated with contextual information, which enables a context aware selection of suitable subprocesses at runtime according to the current context configuration. These subprocesses are called 'worklets'. Other approaches dealing with dynamic service selection can be found here [5]. There exist also aspect-oriented approaches in workflow management to constitute frlxibility. Activities get annotated with aspects. If the aspect is hit by process context information the normal execution of a workflow can be interrupted and inter-weaved with additional activities [2]. In [9] process fragments are inserted into a process model in order to realize flexibility both at design- and runtime. Process fragments are also discussed in [7] as special reuse approach. Process variants as described in [3] also used for process flexibility could be mapped onto process fragments as representation. If we compare these approaches with the fragment-based flexibility approach, it strikes that all approaches base on one main process definition, which can be slightly changed and adapted during runtime, which is not the case in our fragment-based approach.

7 Conclusion and Outlook

In this paper we presented a new flexibility approach for processes. We motivated our approach with a scenario where the process knowledge is local and fragmentary and needs to be composed to receive a more complete view on the whole process. We presented a fragment definition including some correctness criteria to be able to represent 'local' knowledge. Based on fragment definition we defined a composition operation, which composes two fragments based on knowledge redundancies of both fragments. Lots of work towards a realization of our vision remains to be done, e.g. more composition operations, their formal specification and a formal fragment metamodel needs to be defined, building together a process fragment algebra. Tools supporting the fragment concept and an execution environment need to be built to compose fragments and execute fragments. The question on how fragments can be annotated to enable a automatic fragment selection for the next composition step will be part of the ongoing work.

References

1. Adams, M., ter Hofstede, A.H.M., Edmond, D., van der Aalst, W.M.P.: Worklets: A Service-Oriented Implementation of Dynamic Flexibility in Workflows. In: Meersman, R., Tari, Z. (eds.) OTM 2006. LNCS, vol. 4275, pp. 291–308. Springer, Heidelberg (2006)
2. Courbis, C., Finkelstein, A.: Towards an Aspect Weaving BPEL Engine. In: Coady, Y., Lorenz, D.H. (eds.) Third AOSD Workshop on Aspects, Components, and Patterns for Infrastructure Software 2004, Lancaster, United Kingdom (March 2004)
3. Hallerbach, A., Bauer, T., Reichert, M.: Issues in Modeling Process Variants with Provop. In: BPD 2008, pp. 54–65 (2009)
4. Hopcroft, J.E., Motwani, R., Ullman, J.D.: Introduction to Automata Theory, Languages, and Computation. Addison Wesley, Reading (2000)
5. Karastoyanova, D., Leymann, F., Nitzsche, J., Wetzstein, B., Wutke, D.: Parameterized BPEL Processes: Concepts and Implementation. In: Dustdar, S., Fiadeiro, J.L., Sheth, A.P. (eds.) BPM 2006. LNCS, vol. 4102, pp. 471–476. Springer, Heidelberg (2006)
6. Leymann, F., Roller, D.: Production Workflow - Concepts and Techniques. PTR Prentice Hall, Englewood Cliffs (2000)
7. Ma, Z., Leymann, F.: BPEL Fragments for Modularized Reuse in Modeling BPEL Processes. In: ICNS 2009, pp. 1–6. IEEE Computer Society, Los Alamitos (2009)
8. Martin, D., et al.: OWL-S: Semantic Markup for Web Services (2004), `http://www.w3.org/Submission/OWL-S/`
9. Reichert, M., Rinderle-Ma, S., Dadam, P.: Flexibility in Process-Aware Information Systems. In: Petri Nets 2009, vol. 2, pp. 115–135 (2009)

Complex Schema Match Discovery and Validation through Collaboration

Khalid Saleem and Zohra Bellahsene

LIRMM - UMR 5506 CNRS University Montpellier 2,
161 Rue Ada, F-34392 Montpellier
{saleem,bella}@lirmm.fr

Abstract. In this paper, we demonstrate an approach for the discovery and validation of n:m schema match in the hierarchical structures like the XML schemata. Basic idea is to propose an n:m node match between children (leaf nodes) of two matching non-leaf nodes of the two schemata. The similarity computation of the two non-leaf nodes is based upon the syntactic and linguistic similarity of the node labels supported by the similarity among the ancestral paths from nodes to the root. The n:m matching proposition is then validated with the help of the mini-taxonomies: hierarchical structures extracted from a large set of schema trees belonging to the same domain. The technique intuitively supports the collective intelligence of the domain users, indirectly collaborating for the validation of the complex match propositions.

Keywords: Complex Schema Matching, Mini-taxonomies, Collaboration, Tree Mining, Large scale.

1 Introduction

Schema matching relies on discovering correspondences between similar elements of two schemata. Several types of schema matching techniques [8] have been studied, demonstrating their benefit in different scenarios. Schema matching is categorized as *simple element level matching* and *complex structural level matching* [8]. Simple matching comprises of 1:1, 1:n and n:1 match cardinality, whereas n:m match cardinality is considered to be complex. Figure 1 demonstrates the 1:1 (author : writer) and n:m ({FName,LName} : {FirstName,MI,LastName}) match scenarios. Most of the existing approaches and tools give good 1:1 local and global match cardinality but lack the capabilities for discovering n:m matches among the schemata. In this paper, we present an idea for complex match proposition and its validation through an indirect user collaboration technique. We consider the schemata to be rooted, labeled trees, with the tree nodes representing the schema elements. This supports the computation of contextual semantics of the elements in the tree hierarchy. Research literature supports the assumption that there exists a hierarchy among the schema elements of most of the data models. For example, He et al. [5] extracted the hierarchical structures out of the web interface forms and Lee et al. [6] converted the relational database into XML structures, which are inherently trees.

R. Meersman, T. Dillon, P. Herrero (Eds.): OTM 2009, Part I, LNCS 5870, pp. 406–413, 2009.

Our approach is XML/ web interface forms schemata centric, where data lies at the leaf nodes. If a 1:1 correspondence is found between two non-leaf schema elements, we make the proposition that a complex match may exist between the descendant leaf nodes of the respective non-leaf nodes. Next, we utilize an already available repository (automatically generated) of mini-taxonomies [9] to validate the complex match proposition.

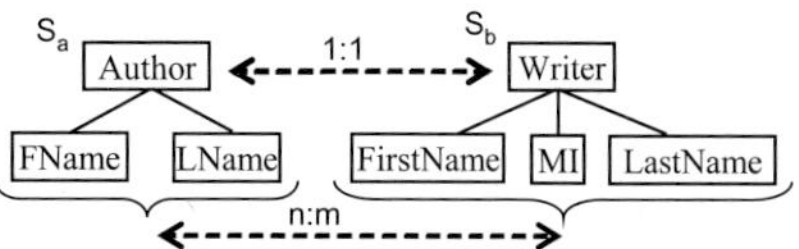

Fig. 1. 1:1 and n:m mappings between two schemata

Mini-taxonomy is basically a small hierarchical structure representing a concept, used in the input schemata. For example the two element hierarchies given in Figure 1 are mini-taxonomies representing the same concept *Author Name*. Mini-taxonomies repository is built with the help of a tree mining technique [9]. Tree mining extracts similar and frequent sub-trees from a large set of schema trees. Therefore the mini-taxonomies repository is in fact the collection of representations of the domain concepts, most frequently utilized in the respective domain. The validation process thus presents a collaborative support of the domain users, in the matching process.

The approach provides an automated technique with an approximate complex schema match quality [2], supporting a large scale scenario (schemata have large number of elements). The individual semantics of the node labels within the schema trees, have their own importance. We utilize linguistic matching algorithms, based on tokenisation, and synonym and abbreviation tables, to extract the concepts hidden within them.

Our Contributions

In this paper, we focus particularly, though not exclusively, on n:m complex mappings. We present a methodology for matching two schemata, covering the element level and structural level matchings discovery. The main features of our approach are as follows:

1. The approach is based on a tree mining technique. It utilizes the ancestor nodes details of the schema nodes, within the tree mining framework, to enable fast calculation of the contextual (hierarchical) similarity between the schemata.
2. It suggests the possible n:m matches between the two schemata and then validates these propositions using already available mini-taxonomies [9], representing the domain concepts.
3. The mini-taxonomies are automatically extracted from a large set of domain specific schemata using the tree mining technique. Mini-taxonomies represent the domain users perspective about the concepts' structural representation within the schemata. Therefore the technique demonstrates the use of collective intelligence in schema matching, in an indirect collaborative manner.
4. Overall, the approach is automatic and hybrid in nature to support the large scale scenarios.

Outline

The remainder of the paper is organized as follows. Section 2 gives the related work in schema matching, specifically the complex schema matching. In Section 3 we describe the overall architecture of our approach. Section 4 discusses the core of the paper, complex match proposition validation appproach with the help of an example. Section 5 presents the lessons learned from the evaluation of the approach. Section 6 outlines the future perspective and concludes.

2 Related Work

Simple matching with acceptable quality has been successfully demonstrated in [1,4,7] by utilizing the element level and structural level schemata knowledge. There has been very limited work on the complex schema matching.

Formally speaking the complex schema matching is n:m match cardinality problem [8]. But most of the work done in the complex schema matching research domain, revolves around the 1:n match discovery. In n:m match discovery, SCIA [11] is one such work but it is dependent on the manual input during the matching process. Another research, by Embley et al. [3] uses manually created mini-ontologies for the domain specific concepts, to generate n:m correspondences between the two schemata. Work in our paper is similar to [3] but the ontological structures, i.e., mini-taxonomies, are generated automatically [9] and we follow the n:m notion of the complex matching.

3 Complex Match Discovery - Our Approach

The architecture of the approach for complex match discovery is shown in Figure 2. It is composed of four modules: (i) *Pre-Phase*, (ii) *Mini-Taxonomies Generation*, (iii) *Simple Schema Matching*, and (iv) *Complex Match Proposition Validation*, supported by a repository which houses the synonym and abbreviation lists, mini-taxonomies, schemata and match results for future reuse.

The *Pre-Phase* module processes the input schemata as trees, calculating the depth-first node number and the scope (number of nodes in the schema tree rooted at that node), for each of the nodes in the input schema trees [10]. At the same time, for each schema tree a listing of the nodes is constructed, sorted in depth-first traversal order. As the trees are being processed, a sorted global list of labels over the whole set of schemata is created by the *Terms Extraction* sub-module.

In the *Similar Terms Computation and Clustering* sub-module, label concepts are computed using the linguistic techniques. The labels are tokenized and the abbreviated tokens are expanded using an abbreviation oracle. Currently, a domain specific user defined abbreviation table is being utilized. Label comparison is based on the similar synonym token sets, supported by a manually defined domain specific synonym table. The architecture is flexible enough to employ additional abbreviation, synonym oracles or arbitrary string matching algorithms. Similar labels are clustered together, intuitively clustering nodes with similar labels [10].

First task performed by the *Mini-Taxonomies Generation* module is to compute the frequency of each term in the forest of trees. Next, within each labels cluster, the term with the highest frequency in the forest of schema trees is taken as the symbol representing the cluster. The frequency of the cluster symbol is computed by adding the frequencies of all the terms in the cluster. From here on the algorithm executes similar to the frequent sub-tree mining algorithm given in [12]. The cluster representative symbols act as the starting labels for the data structure storing frequent sub-tree patterns. The output of the process is a list of the sets of mini-taxonomies. Each list representing a set of mini-taxonomies of same size. Next, these mini-taxonomies lists are replicated by replacing the cluster level similar labels in the list, thus producing all the possible mini-taxonomies which can be considered as the concept representation, frequently utilized by the domain users.

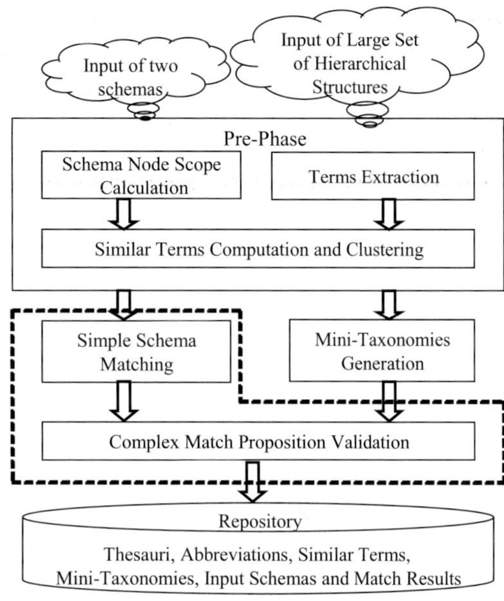

Fig. 2. Architecture for complex matching discovery and validation using automatically generated mini-taxonomies

The *Simple Schema Matching* module tries to compute a correspondence for every node from source schema tree to target schema tree. For each input node a set of possible matching nodes in the target schema is created, producing the target search space, based on node label similarity. For contextual similarity, *ancestor node match* is checked for each possible target matching node, to confirm that there exists a match between an ancestor node of the current source node and some ancestor node of the target node, except for the root node. This contextual proximity is calculated as α:

$$\alpha = 1/(ddif_s + ddif_t)$$

where $ddif_s$ is the depth difference between the current source node and the ancestor node for which a match exists and $ddif_t$ is the depth difference between the candidate target node and the ancestor node in target schema to which the source ancestor node is matched. The candidate target node with the highest α value is selected as the mappable target node.

Next, the method proposes the type of mapping (β). The initial matches are marked as 1:1 and extended to 1:n, n:1 or n:m categories, depending upon the leaf or non-leaf status of the the matched nodes. The complex match propositions discovered between the leaf nodes of the similar non-leaf nodes are validated in the complex match proposition validation part.

The *Complex Match Proposition Validation* (CMPV) module forms the main core of this research work. We use a novel method to validate the proposed complex matches with the help of automatically generated domain specific mini-taxonomies.

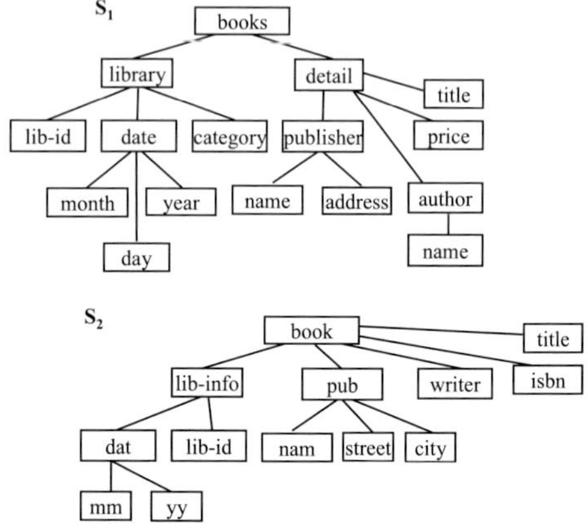

Fig. 3. Two schema trees S_1 and S_2 for complex match discovery

4 Complex Match Proposition Validation Algorithm

The Complex Match Proposition Validation (CMPV) module validates the n:m match propositions discovered in the simple schema matching module. The technique utilizes small conceptual taxonomies already generated from a large number of schemata within the specific domain.

4.1 Complex Match Validation Example

Figure 3 shows two schema trees from the books domain. A list of correspondences is shown in Figure 4, after the execution of simple matching module. The result is discovery of one to one matches along with complex match propositions (in brackets). The scenario presents one n:1 and four n:m complex match situations. The size of the sub-tree rooted at the non-leaf nodes is verified by scanning the scope of the nodes[1]. Large size sub-trees tend to be collection of concepts rather than a single concept.

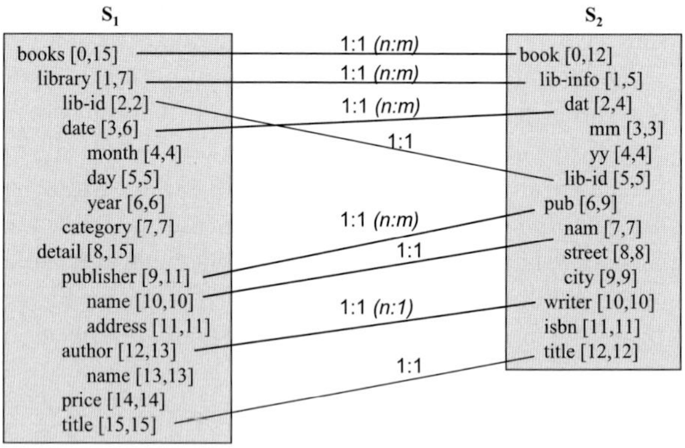

Fig. 4. Element level matches between schemata S_1 and S_2 after execution of simple schema matching

[1] For any node : label[X,Y]; X is depth-first order number and scope is given as (Y-X) i.e, the number of nodes in the sub-tree rooted at that node.

Next, the validation of the proposed complex matchings is done by our Complex Match Validation algorithm with the help of a set of already acquired mini-taxonomies. There are two mini-taxonomies, shown in Figure 5, representing (a) the *date* and (b) the *publisher* concepts respectively. The *date* concept can be represented by *date*, *dat* or some other similar string, as the root node for the concept. The leaf nodes collection of *month*, *day* , *year*, *mm*, *yy* represent attributes describing the concept. Within one instance of the mini-taxonomy, the presence of synonymous leaf nodes is not possible e.g. if *month* and *mm* are synonymous then the two nodes will not exist together in a mini-taxonomy with root node *date* or *dat* [5]. Similarly, Figure 5b presents the mini-taxonomy instances of the concept *publisher address*.

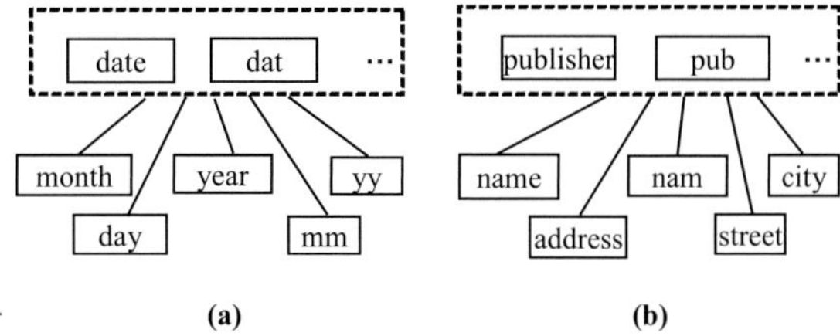

Fig. 5. Mini-taxonomies extracted from large input of books domain schemata

The execution of the algorithm occurs for each of the four n:m propositions. In case of S_1.books[0,15] $\leftrightarrow$ S_2.book [0,12] and S_1.library [1,7] $\leftrightarrow$ S_2.lib-info[1,5], no mini-taxonomy is found with root element similar to *books* or *book* and *library* or *lib-info*. As a result, the two propositions are discarded and only the 1:1 simple match is considered. In the other two cases the algorithm finds mini- taxonomies in the form of date-month/day/year, dat-mm/yy, publisher-name/address and pub-nam/street/city[2]. The algorithm substantially authenticates the propositions and creates two more complex matches as S_1.(month[4,4], day[5,5], year[6,6])$\leftrightarrow$$S_2$.(mm[3,3],yy[4,4]) and S_1.(name[10,10], address[11,11])$\leftrightarrow$$S_2$.(nam[7,7], street[8,8],city[9,9]). The final correspondences are shown in Figure 6.

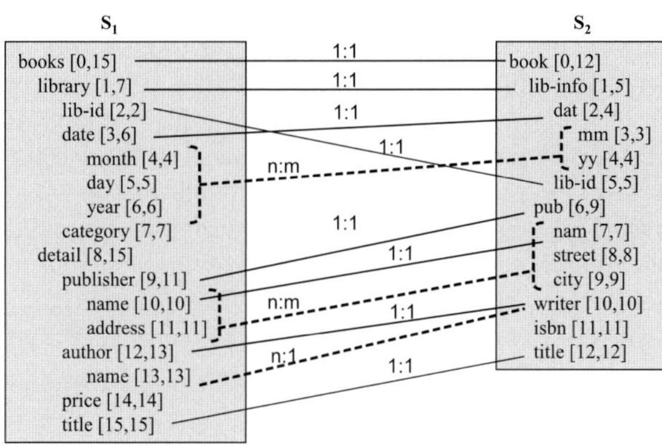

Fig. 6. Mappings between schemata S_1 and S_2 after execution of complex match validation algorithm

5 Lessons Learned

Several data sets[3] were selected as the input for the experiments. Characteristics of these sets of schemata are given in Table 1.

[2] - and / delimiters denote downward and upward traversal, respectively.

[3] http://metaquerier.cs.uiuc.edu/repository

Table 1. Characteristics of domain schema trees used in the CMPV experiments

Domain	BOOKS1	BOOK SEARCH	JOBS	AUTO	AIR TRAVEL	REAL ESTATE	COURSES
Type	Synthetic	Real	Real	Real	Real	Real	Real
Number of schemata	176	19	20	14	20	14	42
Average nodes per schema	8	6	5	5	14	9	8
Largest schema size	14	12	8	10	21	20	17
Smallest schema size	5	3	4	3	6	4	2
Schema Tree Depth	3	2	2	2	2	2	4
n:m match propositions	2	2	1	1	3	2	2

CONCEPT	COMPLEX MAPPINGS
date/depart/return	month,day,year↔mm,yy↔month,day,time
address/location	name,address↔nam,street,city↔street1,street2,city↔ address1,address2↔AreaCode,Country↔city,state
telephone	tel_res,tel_off↔morn_tel, even_tel,night_tel↔tel_mobile,tel_fix
name	firstName,lastName↔f_name,mi,l_name
passengers	adult,child,infant↔adult,senior,child↔adult_12-65,adult_65,child_2-4, child_5-12,infant
return/depart	month,day↔month,day,year,time
schedule	days,time,room↔DayTime,room↔Times,Place↔TimeBegin,TimeEnd, Room,Building↔time,building
car model	vfrom,vto↔fyear,tyear

An analysis of these domain schemata showed that only very small sets of elements (sets of leaf nodes representing some concept), could participate in a complex mapping. The idea of using mini-taxonomies works well, if their leaf nodes can represent some real complex match with a contextual map at ancestor level. Whereas, the ancestor level mapping is highly dependent on the label matching. Working with real world schemata of the web interface forms showed that the possibility of a complex match is very limited in this domain. The reason being the depth of such schema trees is very less, due to which the ancestor level matching requirements are restricted. Secondly, there existed a very vast variation of concept name for a certain concept. For example, there were 13 different labels for passenger concept in the travel domain.

6 Conclusion and Future Work

In this paper, we have presented an approach for discovering and validating the complex matches. Our approach is based on the leaf or non-leaf status of the node, putting forward the match proposition that when a non-leaf node is matched to a non-leaf node, there is the probability of an n:m match between the leaf nodes of the two non-leaf nodes. Next, the method validates this proposition, indirectly utilizing the collective intelligence of the domain users. This is achieved by using mini-taxonomies, extracted using frequent tree mining technique, from a large number of input schemata used over the specific domain. The technique is in fact collaboration of the domain users for complex match validation.

In the future, we plan to extend the label level matching techniques, utilizing state of the art lexical matchers and linguistic dictionaries. Secondly, we intend to exploit other application domains over the semantic web with our approach of indirect collaboration of the domain users.

Acknowledgements

K.S. is funded by the Higher Education Commission of Pakistan.

References

1. Do, H.-H., Rahm, E.: Matching large schemas: Approaches and evaluation. Information Systems 32(6), 857–885 (2007)
2. Doan, A., Madhavan, J., Dhamankar, R., Domingos, P., Halevy, A.Y.: Learning to match ontologies on the Semantic Web. VLDB J. 12(4), 303–319 (2003)
3. Embley, D.W., Xu, L., Ding, Y.: Automatic Direct and Indirect Schema Mapping: Experiences and Lessons Learned. ACM SIGMOD Record 33(4), 14–19 (2004)
4. Giunchiglia, F., Shvaiko, P., Yatskevich, M.: S-Match: an Algorithm and an Implementation of Semantic Matching. In: Bussler, C.J., Davies, J., Fensel, D., Studer, R. (eds.) ESWS 2004. LNCS, vol. 3053, pp. 61–75. Springer, Heidelberg (2004)
5. He, B., Chang, K.C.-C., Han, J.: Discovering complex matchings across web query interfaces: a correlation mining approach. In: KDD, pp. 148–157 (2004)
6. Lee, D., Mani, M., Chiu, F., Chu, W.W.: Net Cot: Translating relational schemas to XML schemas using semantic constraints. In: CIKM (2002)
7. Melnik, S., Rahm, E., Bernstein, P.A.: RONDO: A Programming Platform for Generic Model Management. In: SIGMOD, pp. 193–204 (2003)
8. Rahm, E., Bernstein, P.A.: A survey of approaches to automatic schema matching. VLDB J. 10(4), 334–350 (2001)
9. Saleem, K., Bellahsene, Z.: Automatic extraction of structurally coherent mini-taxonomies. In: ER (2008)
10. Saleem, K., Bellahsene, Z., Hunt, E.: PORSCHE: Performance ORiented SCHEma mediation. Information Systems - Elsevier 33(7-8), 637–657 (2008)
11. Wang, G., Zavesov, V., Rifaieh, R., Rajasekar, A., Goguen, J., Miller, M.: Towards User Centric Schema Mapping Platform. In: VLDB Workshop Semantic Data and Semantic Integration (2007)
12. Zaki, M.J.: Efficiently Mining Frequent Embedded Unordered Trees. Fundamenta Informaticae 66(1-2), 33–52 (2005)

Trust- and Location-Based Recommendations for Tourism

Annika Hinze and Qiu Quan

University of Waikato, New Zealand
a.hinze@cs.waikato.ac.nz, quan.qiu.yz@gmail.com

Abstract. Recommender systems in a travel guide suggest touristic sites a user may like. Typically, people are more willing to trust recommendations from people they know. We present a trust-based recommendation service for a mobile tourist guide that uses the notion of directly and indirectly trusted peers. The recommendations combine information about the peers ratings on sights, interpersonal trust and geographical constraints. We created two trust propagation models to spread trust information throughout the traveller peer group. Our prototype supports six trust-based, location-aware recommendation algorithms.

1 Introduction

Travellers using a mobile tourist guide like to receive recommendations about interesting sights. Typical recommender algorithms were developed for product recommendations in e-commerce, which do not consider location context. In addition, the classical techniques (collaborative filtering and content-based filtering) are not very transparent to the users so that they wonder: *why did I get this recommendation?* Long-term usage of recommenders is often related to the user's confidence and trust in the system. We found that people typically turn to travel agents for recommendations but they have the greatest confidence in recommendations by friends and others travellers they meet on a journey. We strongly believe that the information domain also plays an important role – data used for tourist recommendations has a location context and is thus different compared to commercial products. A tourist's preferences change depending on their location: e.g., interest in museums while visiting Paris and interest in outdoor activities in New Zealand. We identified five factors that should be taken into account: location, personal interest, travel history, feedback from peers, and trust in peers. This paper introduces recommendation algorithms that consider these factors.

Our goal is to increase the user loyalty, understanding and trust in a tourist recommender system. We therefore first analysed who users trust for travel recommendations. Our user study comprised a user study on three aspects: the typical sources of travel recommendations, the perceived reliability of recommendations based on their source, and the propagation of trust within a community. We performed four questionnaire-based surveys. Here we use selected results of the first survey focussing on perceived reliability of travel recommendation sources. The other surveys explored trust and similarity, trust in indirect friends, and design issues for trust-based recommenders. The questionnaires and a complete analysis is provided in [12].

R. Meersman, T. Dillon, P. Herrero (Eds.): OTM 2009, Part I, LNCS 5870, pp. 414–422, 2009.

The results of the first survey confirms that people prefer recommendations from friends. The most trustable friends are those who are most similar but people are also interested in recommendations from friends with different interests. With no recommendations from their friends, people are willing to consult tourist experts or geographically close travellers, but require a clear explanation of the results. Particularly interesting is that in addition to similar interests, also similar locations (current or in travel history) build trust. To support recommendations from tourist peers, we need to present users with transparent recommendation from geographically trusted users.

2 Trust-Based Recommendations in Tourism

2.1 Terms and Definitions

Following the results of our user study, we use the notions of *local trust* and *global trust* (=reputation) and additionally introduce the concepts of *geographic trust* and *location-aware reputation*. Two local trust metrics are *Personal trust* and *person-group confidence*. Personal trust refers to the subjective trust from one user to the other. Person-group confidence is a set of trust statements from a user to a direct peer a sight group (i.e., a set of sights) A *peer* p refers to a user. A *peer group* $P = \{p_1, p_2, \ldots, p_n\}$ represents a tourist community in the system. The corresponding peer group for a peer p_i ($p_i \in P$) is referred to as P_i, with $P_i \subseteq P$. Peers issue *personal trust* T_p to direct acquaintances: $T_p : P \times P \rightarrow [0,1] \subset \mathbb{R}$. $T_p(p_a, p_b) = 1.0$ with $p_a, p_b \in P$ means the source peer p_a completely trusts the target peer p_b. If $T_p(p_a, p_b) = 0$, all information coming from this peer will will be ignored. Peers outside the peer group have a $null$ trust value. Note that personal trust is not symmetrical. Different to our approach, the Advogato [9] trust metric only allows *Boolean* trust; this directly classifies the peer groups into entrusted or distrusted ones. The reputation of peer p constitutes a global *reputational trust* T_r, it is calculated by averaging the received personal trust values from the other users: $T_r : P \rightarrow [0,1] \subset \mathbb{R}$.

If the distance between the source peer and the target peer is less or equal to a given threshold λ, the target peer is given a *Geographic trust* T_g by the source peer. $T_g : P \times P \rightarrow [0,1]$ with $T_{g_{(p_i,p_j)}} = 1 - \frac{distance(p_i,p_j)}{\lambda}$ and $i,j \in [1,n]; p_i, p_j \in P; distance(p_i,p_j) \leq \lambda$ where p_i is the source peer, and p_j is the target peer. Target peers who are closer to the source peer receive a higher geographic trust since their domain knowledge may benefit the source peer more. Users with high *location-aware reputation* T_{lr} have visited a given location the peer is interested in: $T_{lr} : P \times P \rightarrow [0,1]$. The user-defined *Trust threshold* μ is used to identify individual trustable peers; this restricts the peer group when propagating trust on the trust network.

A *Trust path* graph ρ consists of two finite sets (see Fig. 1): a vertex set $V(\rho) \subseteq P$ for the peers and a directed edge set $E(\rho)$ representing the trust relationships [5]. The personal trust value is the weight of the edge issued by a to b. A trust path is a finite sequence of adjacent edges connected via vertices: $\rho[p_i, p_j] = p_i - p_k - \ldots - p_j$; $p_i, p_k, \ldots p_j \in P$ where p_i represents the source peer and p_j represents the target peer. All vertices (peers) on one path have to be unique.

Personal trust can only be propagated along the trust path if the peer has at least one (trusted) friend. Direct friends of the source peer are all adjacent peers on the trust paths which are connected by exactly one directed edge. The remaining peers on the same trust path are indirect peers of the source peer. The length of a trust path is calculated in the number of the edges between two peers; e.g., $\rho[p_a, p_e]$=$p_a - p_b - p_e$ is $s = 2$. As the trust path of the source peer to the target peer becomes longer, the trust decreases gradually. This trust decay D is defined as: $D : P \times P \rightarrow [0, 1] \subset \mathbb{R}$ The trust decay to each direct friend is 1. The value $d \in [0, 1]$ is the *decay constant*. Close friends have more influence on the recommendations. The community trust network graph N is formed by connecting direct trust paths of all personal trust network graphs N_i.

The *sight set* $S = \{s_1, s_2, \ldots, s_n\}$ contains all uniquely identifiable sights in the system. Sight group set $G = \{g_1, g_2, \ldots, g_k\}$ contains all classifications of sights. Each sight group $g_i \in G$ represents a category to which a sight $s_i \in S$ might belong. A nearby-sight group S_λ, $S_\lambda \subseteq S$, contains sights that are geographically close to the location of the user. λ is the distance threshold utilized for constraining nearby sights. The user feedback f is a numeric value between 0 and 10 regarding a sight. Trust-based recommendation R_{p_i} for a particular user p_i is an ordered list of the top N recommended nearby sights that have been associated with computed scores and a list of trustable peers: $R_{p_i} = \{(s_{\lambda_1}, f_1, P_{i_1}), \ldots, (s_{\lambda n}, f_n, P_{i_n})\}$, where $s_{\lambda_n} \in S_\lambda$ is one of the nearby sights, f_n is the computed score given to this sight. $P_{i_n} \subseteq P_i$ is the set of peers that recommended this sight.

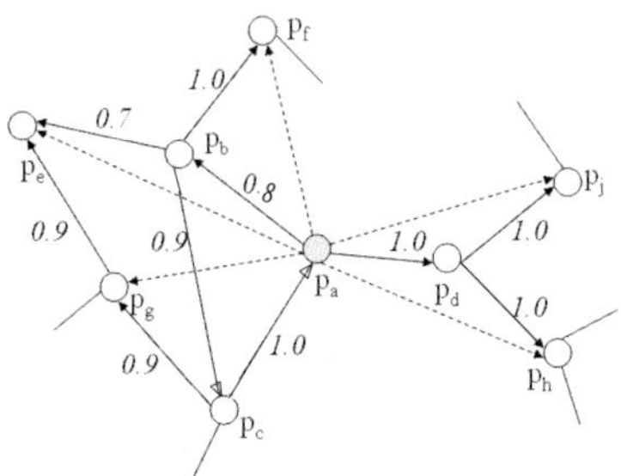

Fig. 1. Trust path graph

2.2 Trust Propagation Model

We designed three trust propagation models for the trust-based recommendation in TIP.

Model 1: Propagating Boolean trust. This trust model is the implementation of the most well-known local group trust metric, the Advogato metric [14]. Every user issues a Boolean decision (0 or 1) about the trustworthiness of direct peers. Trust will only be propagated if the personal trust of the peer is 1.

Model 2: Propagating numeric trust. Each user issues trust values to direct friends. Propagation of trust allows for predicting trust in indirect peers in two steps:

1. Calculation of trust on a given path. Three steps to calculate the trust value T between source and target on path ρ: (1) The propagated personal trust of source peer p_s to to target peer p_t is the product of all personal trust values on a path between them: $T_p(p_s, p_t) = T_p(p_s, p_i) * T_p(p_i, p_j) * \ldots * T_p(p_n, p_t)$, where $T_p(p_i, p_j)$ is the personal trust value issued by p_i to p_j. (2) The trust decay from the source peer to the target peer is computed from the number of edges on the path between two peers. (3) The trust value T from source peer to target peer is gained by multiplying the propagated personal trust T_p with the trust decay D. $T(p_s, p_t) = T_p(p_s, p_t) * D[p_s, p_t]$.

2. Trust path selection. If there are several paths existing between the source peer and the target peer, one needs to be selected to propagate the trust. By definition a trust path does not contain cycles. Trust path selection chooses the trust path with the overall maximum trust flow. First we need to identify all trust paths between source and target. Then the trust on each path is calculated. Finally, the path with the maximum trust value is selected: $T(p_i, p_j) = Max_{\forall\rho[p_i,p_j]}\{\prod_{k=1}^{w} T_p(p_i, p_j)(1-d)^{s-1}\}$, where w is the number of trust paths that connect the source peer and the target peer, $\forall\rho[p_i, p_j]$ includes all w paths between p_i and p_j, $p_i, p_j \in P$, and $p_j \in P_i$.

Model 3: Propagating trust person-group confidence. This is a modification of the previous model: Personal trust information as well as person-group confidence is propagated. This will generate more precise information of the user's preferences.

1. Person-group confidence definition. We identify the personal confidence regarding the sight groups ($\theta = \{g_1, g_2, \ldots, g_m\}$, the immediate sight categories of sights) of the friend as the *person-group confidence* vector.. $C_\theta^{p_i}(p_i, p_j) = \{c_{g_1}^{p_i}(p_i, p_j), \ldots, c_{g_m}^{p_i}(p_i, p_j)\}$ where $p_i, p_j \in P$, $p_j \in P_i$ and $g_m \in \theta$. Each element in the vector represents a confidence value on a particular sight group. Similar to personal trust, the person-group confidence needs to be explicitly issued to direct acquaintances: $C_\theta^P : P \times P \times \theta \rightarrow [0, 1]$ User p_i may hold a person-group confidence matrix $C_{Matrix_{p_i}}$ of friends: $C_{Matrix_{p_i}} : P \times \theta \rightarrow [0, 1]$. By using the person-group confidence, users can precisely express their preferences.

2. Propagating the person-group confidence on the trust network. The person-group confidence in an unknown peer is calculated similar to propagating trust: (1) select the trust path with the maximum trust flow between source peer p_1 to target peer p_n; (2) obtain person-group confidence vector issued by p_{n-1} to p_n; (3) update the person-group confidence vector using the maximum trust flow. The resulting vector is the computed *trust person-group confidence vector* of the source to the target. The trust person-group confidence is the combination of the person-group confidence and trust: $C_\theta : P \times P \times \theta \rightarrow [0, 1]$ with $C_\theta(p_i, p_j) = C_\theta^{p_i}(p_i, p_j)T(p_i, p_j)$, $p_i, p_j \in P$.

2.3 Recommendation Generation

We present four base approaches of trust-based recommendations. The fourth approach uses three different options of how to mitigate missing feedback scores.

Approach 1: Boolean trust propagation. This approach involves three steps. The first step is trust propagation in which user p_1 can find a trust peer group P_1. The second step is to collect the history data H_{p_1} of feedback of the peer group regarding all nearby sights. In Step 3, recommendations are generated based on the peers' historic data only. The data in H_{p_1} is aggregated by using the average of the feedback issued by all friends for the same sight. The recommendation generation occurs after completing the data collection. In this approach, the recommended sights from direct and indirect recommenders are treated equally.

Approach 2: Numeric trust propagation. This approach integrates the level of trust into recommendation generation; it involves four steps. The first step is the trust propagation to identify the peer group of the user – similar to approach 1. The second step is to collect the historic data set H_{p_i} from all peers as described above. The third step is to integrate the trust into H_{p_i} to form a trust-based rating data set $H_{p_i}^t$. The trust matrix is used to update the data set H_{p_i} by multiplying the corresponding trust values with the ratings from peers. In step 4, the recommended sights are generated based on $H^t_{p_i}$. Recommendations are generated from most trustable peers.

Approach 3: Propagation of person-group confidence. Often a person may trust someone only in some aspects; the trust person-group confidence vector is used to express this. Four steps are needed for the recommendation process. Firstly, the peer group P_i is created for the user; secondly, the trust person-group confidence matrix $C_{Matrix_{p_i}}$ is computed; thirdly, the trust person-group confidence matrix is used to update the extracted historic data set from H_{p_i} to $H^c_{p_i}$; finally recommendations are generated from the confidence-matrix-based historic data set. The updated historic data set contains the information about user preferred recommenders and the user preferred recommended subjects; which better models the user's interests and more easily accepted by the user. The computational cost in the recommending process will be higher.

Approach 4: Trust and location-aware collaborative filtering. Pure collaborative filtering (CF) is based on the similarity between the source peer and the target peer. In our case, we assume that if A trusts B, a similarity between A and B exists. The recommendation of trust-enhanced collaborative filtering is a list of the top N appreciated near sights generated from most trustable and similar peers. This approach needs four steps to generate recommendations. Firstly the trust propagation; then the user similarity computation; thirdly the prediction of ratings for the active user; and in the final step the use of the trust matrix to weight predicted ratings of recommended sights. For the user similarity calculation (Step 2), the rating statements of each peer are taken from the current user's peer group P_i as input from the user similarity metric, and the output is the similarity value of current user p_i against the peer. We use the *Pearson Correlation coefficient*, which is between $+1$ and -1.

Handling of missing feedback scores. Since each user only rates some sights, the resulting rating matrix is sparse. We replace no-feedback values by either a high score or a neutral score based on three strategies using:

(F1) *information from the user profile:* A score of 10 is allocated to a sight that belongs to a sight group preferred by the user; a neutral score of 5 otherwise.

(F2) *information from travel history:* If the user has visited a sight at least twice, the no-feedback value replaced by a score of 10; or with 5 otherwise.

(F3) *information from both profile and travel history:* a disjunction of F1) and (F2).

We expect the computational cost to be much lower than pure CF. Moreover, users know the information source and are able to influence the process. Hence this solution is more transparent and controllable than pure collaborative filtering.

3 Evaluation

We implemented our 6 recommender strategies in the context of the TIP project [6]. We refer to the implementation of Approaches 1–3 as TL-1–3, respectively, and to the variations of Approach 4 as TCR-1–3. We examined the strategies in a quantitative

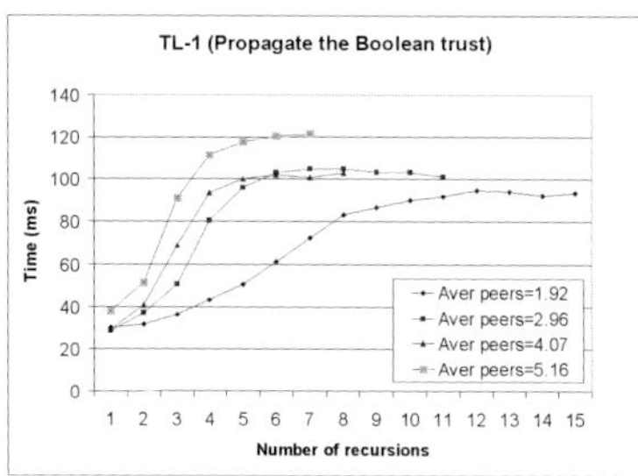

Fig. 2. Performance TL-1

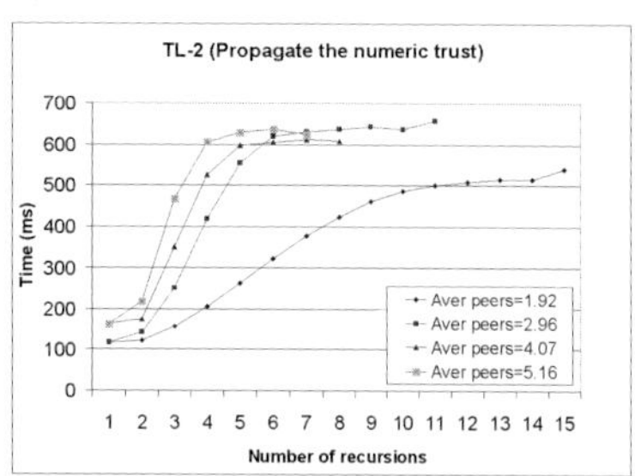

Fig. 3. Performance TL-2

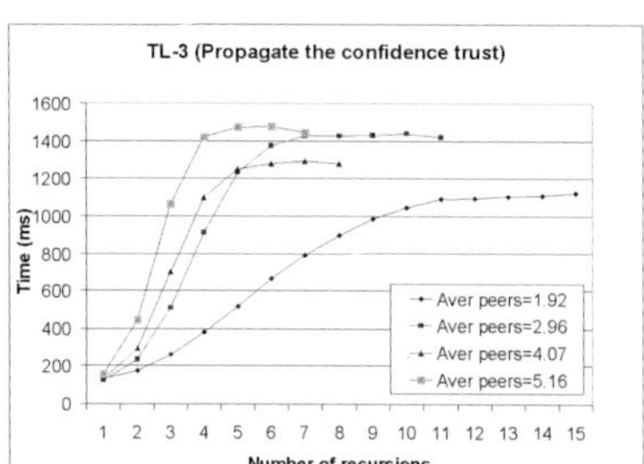

Fig. 4. Performance TL-3

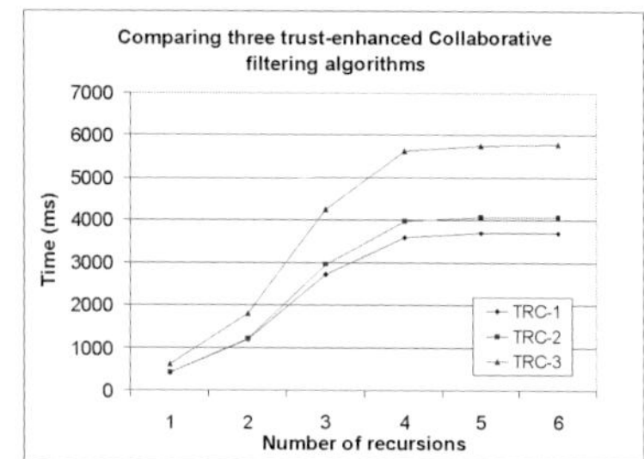

Fig. 5. TLC1–TLC3

performance evaluation as well as in a qualitative exploratory study on system functionalities to investigate the *transparency* of trust-based recommendations and *controllability* of the recommending process. Here we can present only a few results of our performance evaluation; for an extensive discussion see [12].

We varied the scale of the direct peer group to analyze time efficiency. All experiments were running on an Intel Pentium CPU 2.80GHz with 1.00GB RAM under Windows XP. In the test data sets, 100 users were created with each user having a direct peer group. The maximum number of peers in each direct peer group was 3, 5, 7 and 10. 500 unique sights were considered, equally classified into 10 sight groups. The maximum number of ratings issued by each user was 10. The direct trust value issued was a random real number value between 0 and 1, and the rating value was a random integer value between 1 and 10. Figs. 2 to 4 show that the response time for the three trust-based filtering algorithms TL-1 to 3 increases linearly with the number of peers in the peer group. The locations of recommended sights were not restricted. We see that performance depends heavily on the trust propagation model. The Boolean model (in TL-1) is a very simple and fast strategy. In TL-2, the trust model distributes the numeric trust on the network. This model needs to predict the trust of every indirectly trusted peer. In addition, trust values are integrated into peers' ratings from which recommendations are generated. As a result, the second model involves additional computation and is about five times slower. The trust model used in TL-3 specifies the trust on each particular sight group. In addition to propagating trust, more computation is needed for computing trust on different sight groups. The number of sight groups is an important factor that increases the time of trust propagation.

Fig. 5 shows the response times for the three trust-enhanced collaborative filtering algorithms TLC-1–3. The cost of computing user similarities and ratings prediction are similar as they use the same trust propagation model and sight rating matrixes ($users \times sights$). The main difference is the strategies for filling the no-feedback value. Response time for TLC-1 is influenced by the time to check the number of sight groups in the user's profile. TLC-2 needs to check the user's historic data, and then counts the number of times that each sight has been visited by the user. TLC-3 needs to check both the user profiles and user histories.

4 Related Work

Context-sensitive electronic tour guides. We analysed a number of tourist information systems [2,4,11,7,13] and found that none of them supports a rating component or collection of user feedback. Secondly, the displayed information follows a scheme that was pre-defined in the system or in users' profiles; other potentially interesting kinds of information would not be presented to users. Thirdly, users did not have access to information about other travellers' behaviours on similar travels nor to their comments.

Trust-based recommender systems. The trust concept has been evaluated to improve e-commerce recommenders, but most projects remain theoretical. The only running trust-based tourist recommender we encountered is the Moleskiing project [3] for skiing recommendations. Their trust metric MoleTrust [10] computes the user's trust in an unknown peer by tracing their personal social network. Users of Moleskiing need to manually record their skiing routes and experiences after finishing their trip. This prototype is not suitable for supporting ongoing travels. Ziegler and Lausen propose decentralized social filtering that uses trust network structures [14] for recommending books. The project combines the Appleseed trust propagation model (an adapted Advogado model) with a taxonomy-driven filtering technique to deal with data sparseness based on trust. The Appleseed model integrates the numeric trust weight, trust decay and trust normalization into the trust propagation, which make rankings feasible. Appleseed operates on a trust graph where a predefined trust threshold is used to detect entrusted peers in the process of exploring the social network. However, recommending sights is different to recommending books. In travelling, tourists' interests might change frequently according to the location, weather, season. Context-dependent changes of preferences have not been considered. The Recommend-Feedback-Re-recommend (RFR) conceptual framework [8] learns the users' preference from an FOAF-based environment. The system keeps updating distributed users' profiles to find a reliable user group. This framework provides a solution to group similar users in the distributed environment and recommendations automatically transfer to the other users starting from an initial recommender. Recommendations can only be transmitted to users through similar users who have rated the same item and where the given ratings satisfy the threshold. A distributed trust model [1] proposes a conditional transitive trust propagation model for a distributed environment and a protocol for recommendations: Trust is only allowed to be propagated if some trust conditions have been met.

5 Summary

Our initial research demonstrated that travellers prefer recommendations either from their friends or from people whose travel habits are similar to theirs. However, in practice these sources of information are not typically those most used. People are also willing to consult tourism experts or travellers nearby, but they then require a clear explanation of recommendations.

The goal of the research presented in this paper was to undertake a systematic interrogation of trust models in tourist recommender systems.

We proposed two trust concepts specifically designed for the tourist information domain: location-aware reputation and geographic trust. Geographic trust is higher the shorter the the distance between the active user and another user. Location-aware reputation is the reputation of users who have visited a certain location. These concepts give alternatives to users who have not defined their friends to the system or who do not agree with recommendations of their friends.

We created two trust propagation models and implemented six trust-based recommender algorithms. Our first trust propagation model uses direct numeric trust between users. In the second model a user's trust expresses confidence in users for certain sight groups. For comparison, we also implemented the well known Boolean trust model (Levien's Advogato local trust metric).

We evaluated three trust-based filtering algorithms and three trust-enhanced and location-aware collaborative filtering algorithms based on the second trust model. The efficiency of the trust-based filtering algorithms depends on the trust propagation model, while trust-enhanced collaborative filtering depends on the method to replace no-feedback values in the rating matrix. Due to its complexity, the trust-enhanced collaborative filtering is slower than the trust-based filtering.

The focus of this paper was local trust. In future work, we can compare the local trust metric against the global trust metric (reputation), and weigh trust-based recommendations constructed based on the local trust against recommendations created on reputation.

References

1. Abdul-Rahman, A., Hailes, S.: A distributed trust model. In: New Security Paradigms Workshop, pp. 48–60. ACM Press, New York (1998)
2. Abowd, G., Atkeson, C., Hong, J., Long, S., Kooper, R., Pinkerton, M.: Cy-berguide: A mobile context-aware tour guide. In: ACM Wireless Networks (1997)
3. Avesani, P., Massa, P., Tiella, R.: Atrust-enhanced recommender system application: Moleskiing. In: ACM Symposium on Applied Computing (2004)
4. Cheverst, K., Davies, N., Mitchell, K., Friday, A.: Experiences of developing and deploying a context-aware tourist guide: The guide project. In: Proc. of MOBICOM 2000 (2000)
5. Epp, S.: Discrete Mathematics With Applications. Brooks Publishing Company (1996)
6. Hinze, A., Voisard, A., Buchanan, G.: Tip: Personalizing information delivery in a tourist information system. Journal on Information Technology and Tourism 11(4) (2009)
7. Hristova, N., O'Hare, G.: Ad-me: A context-sensitive advertising system within a mobile tourist guide. In: Conf. on Artificial Intelligence and Cognitive Science, AICS (2002)

8. Kim, H., Jung, J.J., Jo, G.: Conceptual framwork for recommendation system based on distributed user ratings. In: Li, M., Sun, X.-H., Deng, Q.-n., Ni, J. (eds.) GCC 2003. LNCS, vol. 3032, pp. 115–122. Springer, Heidelberg (2004)
9. Levien, R., Aiken, A.: Attack-resistant trust metrics for public key certification. In: USENIX Security Symposium, pp. 229–242. USENIX Assoc. (1998)
10. Massa, P., Avesani, P.: Controversial users demand local trust metrics: An experimental study on epinions.com community. In: Nat. Conf. on Artificial Intelligence (2005)
11. O'Grady, M., O'Rafferty, R., O'Hare, G.: A tourist-centric mechanism for interacting with the environment. In: Proc. of MANSE 1999 (1999)
12. Qiu, Q., Hinze, A.: Trust-based recommendations in for mobile tourists in TIP. Technical report, University of Waikato (June 2007)
13. Youll, J., Morris, J., Krikorian, R., Maes, P.: Impulse: Location-based agent assistance. In: International Conference on Autonomous Agents (2000)
14. Ziegler, C., Lausen, G.: Paradigms for decentralized social filtering exploiting trust network structure. In: Meersman, R., Tari, Z. (eds.) OTM 2004. LNCS, vol. 3291, pp. 840–858. Springer, Heidelberg (2004)

Collaborative Ad-Hoc Information Sharing in Cross-Media Information Environments

Beat Signer[1], Alexandre de Spindler[2], and Moira C. Norrie[2]

[1] Vrije Universiteit Brussel
Pleinlaan 2
1050 Brussels, Belgium
bsigner@vub.ac.be
[2] Institute for Information Systems, ETH Zurich
CH-8092 Zurich, Switzerland
{despindler,norrie}@inf.ethz.ch

Abstract. Due to the ever increasing number of different digital media types that we use in our daily work, it is no longer sufficient to manage them in an isolated way but desirable to define associations across the media boundaries. While cross-media information systems allow different forms of media objects to be linked, there is often a lack of support for the flexible authoring and sharing of these links. We present a solution for exchanging cross-media link information in an ad-hoc way among communities of users based on a notion of a cooperative cross-media link server. We describe how the system has been implemented based on an existing cross-media link server (iServer) and peer-to-peer technologies.

1 Introduction

It is common nowadays for user communities to want to create, access and share heterogeneous collections of associated multimedia objects. While hypermedia systems, and specifically cross-media link servers, allow different types of media objects to be linked together in flexible ways, often there is a lack of support for flexible means of authoring and sharing these links.

Our interest was to support dynamic and temporary forms of user communities where users come together in physical or virtual space to support each other in a specific task. For example, a group of designers attending a meeting and a group of students attending a course can both be thought of as dynamic and temporary user communities. Typically, these user communities will want to create ad-hoc and transient forms of collaborative information spaces that support a particular activity such as planning a project or learning for an exam. It should therefore be possible to share link data in an ad-hoc way to support the community-based authoring of links between resources that define such transient collaborative information spaces.

We note that allowing users to exchange information in a peer-to-peer (P2P) manner rather than via a central server offers the potential for more dynamic

R. Meersman, T. Dillon, P. Herrero (Eds.): OTM 2009, Part I, LNCS 5870, pp. 423–430, 2009.

forms of information sharing. For example, in the case of a design meeting, information could be exchanged directly between personal devices such as laptops and also between these and fixed devices such as an interactive table.

Our solution was to develop a cooperative cross-media link server for exchanging link information in an ad-hoc way among communities of users. It is important to note that, in our solution, link services are discovered in a completely decentralised manner. Users can dynamically join the community and start annotating or adding links to arbitrary third-party resources as well as access link metadata from other users. To avoid issues of information overload and link fraud, we introduce a collaborative filtering mechanism based on a combined ranking of users and link resources.

In this paper, we describe how the system was implemented based on an existing cross-media link server, called iServer [1], and P2P technologies. We start in Sect. 2 with a more detailed look at the motivation for collaborative cross-media information spaces and a discussion of related work. In Sect. 3, we briefly introduce the existing iServer platform and its underlying cross-media link model before going on to describe the functionality and operation of the cooperative iServer version. Different approaches for user and link rating are discussed in Sect. 4. Concluding remarks are given in Sect. 5.

2 Motivation

The original hypertext and hypermedia information models are based on the concept of connected document spaces in which additional meaning is usually associated with the links between documents. Early models have been extended over the years, both to broaden the scope of the model and to improve the functionality and maintenance of systems through more powerful and flexible link management. With the emergence of ubiquitous and pervasive computing, physical hypermedia systems have been proposed that enable real-world objects to be linked to digital media, and vice versa, by allowing physical resources to also be included as nodes in the connected information space [2].

Underlying all of these developments, the basic information model remains the same and an information space is defined by a connected graph where the nodes are resources and the links are represented by edges. The anchor and target of a link can either be an entire resource or an element within a resource. We have realised a general framework to support a range of hypermedia tools and services by adopting a database approach to the problem that first involved developing a generic link metamodel and then implementing it using an object database management framework.

A benefit of managing links separately from resources is that new links can easily be created by arbitrary users. In these *open link authoring* systems the information space evolves over time based on the users' current interest. For example, in a teaching and learning environment, students can not only consume the material prepared and published by a teacher, but also add their own links between different resources (e.g. slide handouts, web pages and PDF documents).

Links can not only be created between existing resources but also associations to content that has been created during a lecture can be defined. Examples of solutions for capturing notes during a lecture or meeting and sharing them online include the Pulse Smartpen application from Livescribe[1] or the PaperPoint [3] interactive paper presentation manager.

When it comes to link sharing, students will no doubt find that certain students provide more useful links than others. A course will evolve over time and information of fellow students might be more reliable than information added by former students. On the other hand, information of previous students who obtained excellent grades might also be considered as especially valuable. The introduction of a notion of link quality based on user trust helps to deal with these issues. Our open link authoring approach combines the traditional publisher and consumer model with a democratic authoring process based on an open link platform that supports existing and emerging types of resources

The teaching environment is only one application scenario where people want to organise and share information with other local or remote users. Our iServer solution is not limited to a specific application, but rather provides a general platform for cross-media information management that can be applied to a large variety of application domains. The community-based link sharing, as provided by the presented P2P iServer framework, is related to proposals in distributed hypermedia [4]. A benefit of our approach lies mainly in the means to realise the goal. By building on advanced database technologies and exploiting practices of metamodel-driven system engineering to the full, we have succeeded in designing a flexible link server platform that supports digital and physical resources and is open to new communication paradigms including P2P technology.

3 Cooperative Cross-Media Link Server

iServer is a cross-media link server capable of supporting an extensible set of digital and physical media. The framework is based on the resource-selector-link (RSL) metamodel and supports the integration of new media types based on a resource plug-in mechanism. It is beyond the scope of this paper to describe all details of the RSL model but the complete model can be found in [5].

The RSL model consists of three core *entity* types represented by *resources*, *selectors* and *links* as highlighted in Fig. 1. A link is associated with one or more source and target entities where each of these entities might be either a resource, a selector or another link. While a resource represents an entire information unit or service (e.g. a web page or a PDF document), a selector allows us to address parts of a resource. Note that a selector always has to be associated with a single resource but a resource may have multiple selectors. By modelling links as an entity subtype, we gain the flexibility to define links with other links as source or target entities.

The user model defines the access rights associated with an entity and is essential for information sharing issues discussed later in this paper. Users may

[1] http://www.livescribe.com

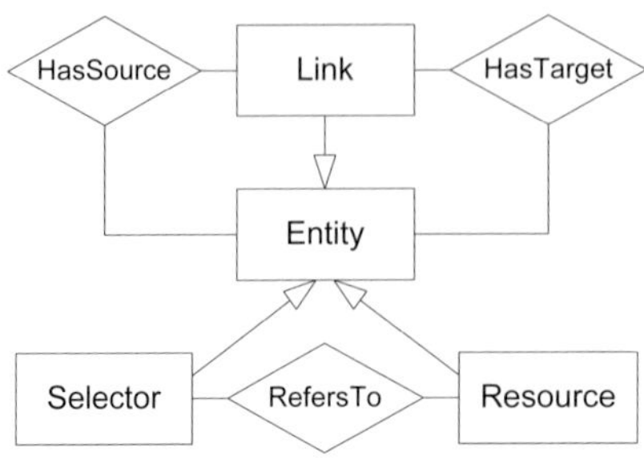

Fig. 1. Resource, selector and link

be either *individuals* or *groups* and each entity has exactly one individual creator who defines the user access rights. When it comes to sharing links and resources in the P2P version, it is up to the creator of an entity to specify whether the information is private or should be publicly available.

While the RSL model defines the abstract concepts of resources and selectors, the system can be extended to support new media types by providing concrete plug-in implementations for the resource and selector types. In the past, various iServer plug-ins have been developed to support different media types. This includes the iWeb plug-in for HTML resources as well as resource plug-ins for images, sound and movies or an RFID plug-in for tagging physical objects. Furthermore, a plug-in for linking paper documents has been realised based on interactive paper (iPaper) technologies [1]. A number of applications have been implemented based on the iServer framework, including generic cross-media browsers, a paper-based mobile tourist information system [6] as well as educational applications and information services. Based on our experience, the authoring of links in such highly-connected cross-media information spaces can be time-consuming and tedious. By making the cross-media authoring tools available to the user, we move from the publisher and consumer model to an open authoring system where every user becomes a potential publisher and may share their link information with other users.

In our distributed iServer solution, a client always accesses a resource from its original location and, only in a second step, is additional link metadata acquired over the dynamic P2P iServer network as illustrated in Fig. 2. The separation of content and metadata implies that a resource should always be available, provided that the server on which the resource is hosted is up and running, whereas any additional link information may change dynamically over time based on the set of iServer peers currently available in the network.

We have to distinguish between *persistent link metadata* in the form of link information that is stored in a personal iServer instance and the *transient link metadata* received from the set of remote iServer peers. While the quality of the persistent link metadata can be ensured by controlling the users who have access to a personal iServer instance, we do not have any direct control mechanism to guarantee the integrity of information provided by remote peers. Therefore, an

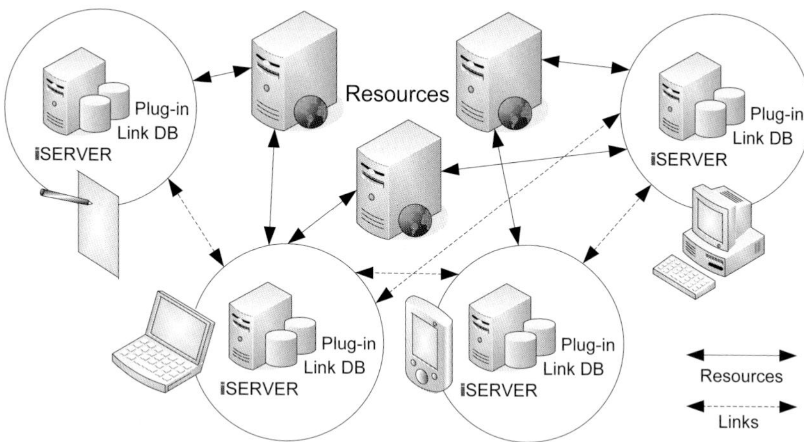

Fig. 2. Cooperative iServer architecture

essential part of the cooperative iServer framework is a collaborative filtering mechanism based on the rating of remote users and links.

Of course, we are not alone in this view since distributed link services, where link information is stored separately from the resources, have investigated associative linking of resources. More recently, P2P technologies have been considered for building new forms of open hypermedia systems where central link databases are no longer required [7]. Our approach differs in several aspects from other P2P-based open hypermedia systems. First, we do not use the P2P network functionality to replicate any hypermedia documents and perform searches in the distributed network of peers as proposed by Larsen and Bouvin [8]. In our solution, the original hypermedia documents are clearly separated from any annotations or external links and it is only the link or annotation metadata that is shared in an ad-hoc manner. In contrast to other systems, our approach is not based on finding resources which are similar to the resource at hand, but rather on associations between resources that users themselves have found to be meaningful which is closely related to the idea of associative linking originally proposed by Bush [9].

In designing a distributed architecture for iServer, it was one of our goals to ensure that the framework kept to its principle of being as general as possible, with a clean separation of concerns. Even if iServer is primarily an extension of an object database system, it provides well-defined Java, XML and Web Service APIs for accessing and updating information. The general interaction between peers consists of sending single API calls from one peer to another, the execution of that request on the remote site and the transfer of the result back to the initiating peer. The information returned by the remote peers has to be combined and integrated with information that is available from the local iServer instance. The functionality of a remote iServer system is offered to a local iServer instance via a peer service. More details about the P2P-based distribution architecture can be found in [10].

4 User and Link Rating

In a cooperative community of publishers and consumers who are equally responsible for the available content, a variety of issues must be addressed. Communities where participants consume without contributing produce little content and result in a heavy load on a few nodes. A community may also be a target for the distribution of unsolicited or malicious content. Some systems try to avoid the lack of contribution by keeping designated publishers to ensure the information supply. The content in our information spaces consists of resources and links and we primarily envisage shifting the authoring of links between resources to the user community. The effort required from the user should be minimal to encourage them to contribute.

The rapid and considerable emergence of technologies enabling democratic publishing has led to the development of rating and filtering systems addressing the issues of unsolicited or malicious content. Existing filtering techniques such as content-based or collaborative filtering have in common the fact that the rating inference is based on some notion of similarity [11,12]. However, the extraction of relevant features used for comparing resources, the computation of similarities and the cold start problem still form a bottleneck in current systems.

In real world situations, the filtering of resources is often based on the trustworthiness of the provider in addition to the relevance and quality of the resource. Electronic communities such as eBay[2] have successfully implemented trust and reputation techniques supporting users in taking decisions. Research has given rise to a variety of techniques for the rating of users and the interpretation of rating values [13,14]. In all these cases, a set of user ratings can be represented by a *rating graph* where a directed weighted edge connects a rating user with a rated user. If two users are not connected by an edge, there possibly exists one or multiple paths connecting them transitively. Therefore, a user who knows a relatively small number of other users and is able to rate them can infer ratings from a vast number of other users by a transitive propagation of ratings.

Our framework supports individual users in deciding on their own levels of trust in other users. We provide a rating implementation where a rating value is interpreted as the amount of trust flowing from a rating user to a rated user. The transitive propagation of trust consists of searching the minimum trust along a path connecting the two users. If multiple such paths exist, all minima are summed up. This can be achieved by running a maximum flow algorithm on the rating graph with the rating user as a source and the rated user as a sink.

The main interaction of iServer peers consists of exchanging link metadata. A common request is a query for in- and outgoing links of a particular resource. As a result, the user receives a set of links from multiple users. Filtering can be achieved by deciding, for each item received from a particular user, whether to accept or reject it. The information available for this decision consists of the currently received resource item, the sending user and the set of items previously received from other users. The currently received item can be rated in the context

[2] http://www.ebay.com

of the previously received ones, for example by computing its frequency. Therefore, we combine user ratings with response ratings to filter responses returned from remote peers. This may, not only help to improve the quality of information presented to the user, but also reduce the quantity, thereby preventing the information overload that could result from a large, highly-connected information space.

We encode user ratings as tuples containing the rating user, the rated user and the rating value. Such a tuple is created by the rating user and propagated to all other members of the peer group. A user rating manager ensures that all peers have the same set of tuples stored locally. The tuple synchronisation is achieved as follows:

1. On startup, the peer reads a file containing tuples stored in previous sessions. If this file does not exist, a new empty one is created.
2. The peer creates a tuple set S_{local} containing all tuples in the file. Whenever S_{local} changes, the file is updated.
3. When a peer joins the group, it requests the tuple set S^i_{remote} from all other members i of the group.
4. Every incoming tuple set S^i_{remote} is compared with the local set S_{local} as follows:
   - If S_{local} does not contain all tuples in S^i_{remote} then the local set is updated.
   - If S^i_{remote} does not contain all tuples in S_{local} then the local set is broadcast to all other members.
5. When no more sets are broadcast, all tuple sets contain the same tuples.
6. Whenever a new rating is set locally, the local set of tuples is broadcast to all other members of the group.

Every iServer P2P API request is broadcast to all members of the groups. As a result, a requesting peer possibly receives multiple responses. A response rating manager filters incoming responses before they are made accessible to the user. The selection is based on rating values that are computed by response raters for every response. The rating values of multiple response raters are combined by an aggregator function. We implemented two response raters, one returning the rating of the responding user as described above and the other returning the frequency of the response within the collection of previous responses. Further details about the link rating and the visualisation of shared links as well as applications that are based on the presented framework can be found in [10].

5 Conclusion

We have presented a notion of collaborative cross-media information environments based on community-based link authoring between arbitrary resources and elements within these resources. Cooperation is supported on two levels—the user management integrated into the core RSL model and a P2P architecture for distributed information sharing. In order to avoid information overload and ensure link quality, we introduced a collaborative filtering mechanism that is based on a graph model. A rater has been presented that implements filtering using a maximum flow algorithm. We are currently evaluating the similarity measure in the filtering process described in this paper which involves the implementation of a third response rater and its deployment within our iServer P2P framework.

References

1. Signer, B.: Fundamental Concepts for Interactive Paper and Cross-Media Information Spaces. PhD thesis, ETH Zurich, Dissertation ETH No. 16218 (May 2006)
2. Hansen, F.A., Bouvin, N.O., Christensen, B.G., Grønbæk, K., Pedersen, T.B., Gagach, J.: Integrating the Web and the World: Contextual Trails on the Move. In: Proc. of Hypertext 2004, Santa Cruz, USA (August 2004)
3. Signer, B., Norrie, M.C.: PaperPoint: A Paper-Based Presentation and Interactive Paper Prototyping Tool. In: Proc. of TEI 2007, Baton Rouge, USA (February 2007)
4. DeRoure, D., Carr, L., Hall, W., Hill, G.: A Distributed Hypermedia Link Service. In: Proc. of SDNE 1996, Washington, USA (1996)
5. Signer, B., Norrie, M.C.: As We May Link: A General Metamodel for Hypermedia Systems. In: Parent, C., Schewe, K.-D., Storey, V.C., Thalheim, B. (eds.) ER 2007. LNCS, vol. 4801, pp. 359–374. Springer, Heidelberg (2007)
6. Belotti, R., Decurtins, C., Norrie, M.C., Signer, B., Vukelja, L.: Experimental Platform for Mobile Information Systems. In: Proc. of MobiCom 2005, Cologne, Germany (August 2005)
7. Bouvin, N.O.: Open Hypermedia in a Peer-to-Peer Context. In: Proc. of Hypertext 2002, College Park, USA (June 2002)
8. Larsen, R.D., Bouvin, N.O.: HyperPeer: Searching for Resemblance in a P2P Network. In: Proc. of Hypertext 2004, Santa Cruz, USA (August 2004)
9. Bush, V.: As We May Think. Atlantic Monthly 176(1) (July 1945)
10. Signer, B., de Spindler, A., Norrie, M.C.: A Peer-to-Peer-based Distributed Link Service Architecture. Technical Report TR636, ETH Zurich (August 2009)
11. Claypool, M., Gokhale, A., Miranda, T., Mumikov, P., Netes, D., Sartin, M.: Combining Content-Based and Collaborative Filters in an Online Newspaper. In: Proc. of the ACM SIGIR Workshop on Recommender Systems, Berkeley, USA (August 1999)
12. Sarwar, B.M., Karypis, G., Konstan, J.A., Riedl, J.T.: Item-based Collaborative Filtering Recommendation Algorithms. In: Proc. of WWW10, Hong Kong (May 2001)
13. Jøsang, A., Ismail, R., Boyd, C.: A Survey of Trust and Reputation Systems for Online Service Provision. Decision Support Systems (2005)
14. Gambetta, D.: Can We Trust Trust? In: Trust: Making and Breaking Cooperative Relations, Department of Sociology, University of Oxford (2000)

DOA 2009 – PC Co-chairs' Message

On behalf of all those involved, we would like to welcome you to the proceedings of the 11th International Symposium on Distributed Objects, Middleware and Applications (DOA 2009), held in the Algarve, Portugal, November 2009.

Over the years, the DOA conferences have been an excellent venue to showcase cutting-edge research and development that is happening in both academia and industry. This nice blending of short-term and long-term work has allowed us to see presentations on Web services, cloud computing, virtualization, SOA and other topics that have often foretold the evolutions in these important areas. In this regard, DOA has always managed get the right combination of the theoretical and practical aspects of a wide range of topics, laying the groundwork for researchers to address new challenges in the years to come. In fact it is interesting to consider earlier DOA conference papers and their direct impact on those submissions that came afterwards.

Previous DOA conferences have stood at the forefront of showcasing real-world experiences of researchers, whether academic or industrial.

This year is certainly no different, with papers covering topics as diverse as dependency injection and Grid computing. Contributions covering experience, practice and research have been particularly encouraged, with student papers and those discussing early results also accepted where they were considered to be sufficiently important to a wider audience.

The quality of submissions this year has been extremely high, making the selection process more difficult than usual. All of the papers passed through a rigorous selection process, with at least three reviewers per paper and much discussion on the relative merits of accepting each paper throughout the process. As such we must thank the diligent work of all of the authors, whether or not your paper was accepted: the quality of the conference is down to you so please keep up the great work. In fact this year we had so many excellent submissions that we decided to accept 21 of the original 57 submissions.

Finally, we are grateful to the dedicated work of the experts in the field from all over the world who served on the Program Committee and whose names appear in the proceedings. Special thanks go to the external organizers for the event. The proceedings you hold in your hand are testament to the hard work that everyone has put into DOA. We hope you enjoy them and consider submitting something in the future.

August 2009

Mark Little
Fabio Panzieri
Jean-Jacques Dubray

R. Meersman, T. Dillon, P. Herrero (Eds.): OTM 2009, Part I, LNCS 5870, p. 431, 2009.

Aspect-Oriented Space Containers for Efficient Publish/Subscribe Scenarios in Intelligent Transportation Systems

Eva Kühn[1], Richard Mordinyi[1], Laszlo Keszthelyi[1], Christian Schreiber[1], Sandford Bessler[2], and Slobodanka Tomic[2]

[1] Space-based Computing Group
Vienna University of Technology
1040 Vienna, Austria
{eva,rm,lk,cs}@complang.tuwien.ac.at
[2] Telecommunications Research Centre Vienna
Donau-City 1, A-1210 Vienna, Austria
{bessler,tomic}@ftw.at

Abstract. The publish/subscribe paradigm is a common concept for delivering events from information producers to consumers in a decoupled manner. Some approaches allow the transportation of events even to mobile subscribers in a dynamic network infrastructure. Additionally, durable subscriptions are guaranteed exactly-once message delivery, despite periods of disconnection from the system.

However, in some application areas, like in the safety-critical telematics, durable delivery of events is not sufficient enough. Short network connectivity time and small bandwidth limit the number and size of events to be transmitted hence relevant information needed for safety-critical decision making may not be timely delivered.

In this paper we propose the integration of publish/ subscribe systems and Aspect-oriented Space Containers (ASC) distributed via Distributed Hash Tables (DHT) in the network. The approach allows storage, manipulation, pre-processing, and prioritization of messages sent to mobile peers during bursts of connectivity.

The benefits of the proposed approach are a) less complex application logic due to the processing capabilities of Space Containers, and b) increased efficiency due to delivery of essential messages only aggregated and processed while mobile peers are not connected.

We describe the architecture of the proposed approach, explain its benefits by means of an industry use case, and show preliminary evaluation results.

1 Introduction

The publish/subscribe (pub/sub) paradigm [1] is a common and largely recognized concept for delivering messages (events) in an anonymous decoupled fashion from publishers to peers subscribed for a topic or for the content of a message.

R. Meersman, T. Dillon, P. Herrero (Eds.): OTM 2009, Part I, LNCS 5870, pp. 432–448, 2009.

Current implementations of and research in notification systems are mostly focusing on an effective and large-scale dissemination [2], [3], [4] of huge quantity of information from publishers to subscribers in a fault-tolerant manner, how to improve the semantical quality or the expressiveness of subscriptions [5], [6], how to ensure durability or the correct order of messages [7], [8]. Other pub/sub approaches deal with these issues as well but assume additionally that peers are mobile [9] or the entire network is completely dynamic [10], [11].

In some application areas the durable delivery (in other words the guaranteed delivery with "exactly once" semantics) of subscribed messages is essential. However, there are application domains, like safety-critical telematics, in which this kind of reliability for subscribed events may be considered a precondition for operation, due to jurisdictional reasons, but is not adequate at all. Among others, a durable notification service has to store any events a peer has subscribed for while the subscriber is off-line. Once the peer is reachable again, the saved events have to be delivered to the associated subscriber. This means that the peer would receive a large amount of data that it has to process locally in order to extract relevant information. However, in scenarios from Intelligent Transportation Systems (ITS), mobile peers (vehicles) have only a few seconds of connectivity and very limited bandwidth [12], [13]. This may cause several problems: the reconnecting peer should receive all stored events which may have very different importance for the user or be even stale, but due to the limited bandwidth and connectivity window only a very few messages can be forwarded to the peer creating a kind of back-pressure in the system. Furthermore, due to the small connectivity window, there is a possibility that essential information, such as safety-critical ghost driver warnings, cannot be transmitted to the peer. If such messages are not forwarded to the peer on time humans lives may be jeopardized. Therefore, the safety risk grows with the amount of irrelevant or even incorrect information delivered instead of important life-saving information.

In [14] we described a customizable storage component, called Space Container, for efficient storage and retrieval of structured data. In [15], [16], [17], [18] we presented the SABRON approach on how to distribute and replicate such Space Containers by means of Distributed Hash Tables (DHT) [19] to efficiently store and retrieve structured, spatial-temporal data in a fault-tolerant manner. In this paper we propose the concept of Aspect-oriented Space Container (ASC), an extension for event processing of the original capabilities of Space Containers, for linking pub/sub systems and mobile peers with short connectivity time. Extending the ideas in [20], this paper reviews existing related work, describes the concept in more detail, and presents first performance evaluation results of the implemented system.

A Space Container allows to store entries in a customizable structured way by means of so called Coordinators. DHTs are used to place such Space Containers in the network in a fault-tolerant and scalable manner. Aspects are components that are triggered whenever a Space Container is accessed. It is a customizable application logic executed either before or after the operation on the Space Container for events processing. An Aspect can be used to check security policies, to persist, filter or manipulate incoming events, or to alter the content of a Space Container based on the received event. The combination of a pub/sub medium and ASC allow the mobile peer to inset a Space Container in the network which then acts as an intermediate-subscriber for events in the pub/sub medium, and processes the delivered events on behalf of that mobile peer.

The benefits of integrating pub/sub systems and the ASC approach in an ITS scenario are a) less complex application implementation since the processing logic has been moved to the customizable Aspects, b) efficient delivery of events to mobile peers, since relevant information have been extracted while the peer was off-line, thus minimizing the number of messages to be transmitted and avoiding any additional events processing at the peer's site, and c) releases resources of the pub/sub medium since message can be delivered to the intermediate subscriber, the Space Container.

The remainder of this paper is structured as the following: section 2 pictures the use case, section 3 defines the research questions, section 4 summarizes related work, section 5 describes the concept and the architecture of Space Containers and Aspects, while section 6 discusses preliminary evaluation results. Finally section 7 concludes the paper and proposes further work.

2 Scenario

A motivating use-case to identify requirements and to illustrate the benefits of the proposed ASC architecture is an Intelligent Transportation System (ITS) scenario [18]. The scenario consists of a highway with fast moving vehicles communicating with a fixed, geographically distributed infrastructure, as illustrated in figure 1. Along the highway there are so called Road Site Units (RSU) responsible for either passing safety and traffic information to the vehicles or receiving information from the vehicles and pass it to the system. RSUs exchange information via dedicated short range communication protocols (DSRC [21]) and are installed along the road network in 2-3 km distance of each other. They are connected in a meshed wired broadband network in order to assure scalability and increase fault-tolerance, figure 2.

Information exchanged in the system mainly concerns the traffic itself and the messages are published by e.g. the Traffic Control Centre (TCC), radio stations, the police, weather stations, the road maintenance depot, and of course the vehicles themselves. Messages exchanged or events generated depend on the role and may contain information about traffic restrictions and warnings (wrong-way-driver, speed limits, redirections,...), traffic density (the number of cars and their speed within a specific range,...), traffic congestions (location, length, duration, state updates,...), accidents (location, number of cars involved, blocked lanes, state updates,...), road conditions (wet, dry, temperature, number and location of road holes, humidity, hydroplaning warnings,...), current weather conditions (fog) and forecasts, or vehicle related information (acceleration statistics, break hits, sudden use of breaks, average and current speed, passed police control points, car condition, accident alert,...). The published data is geo-located and

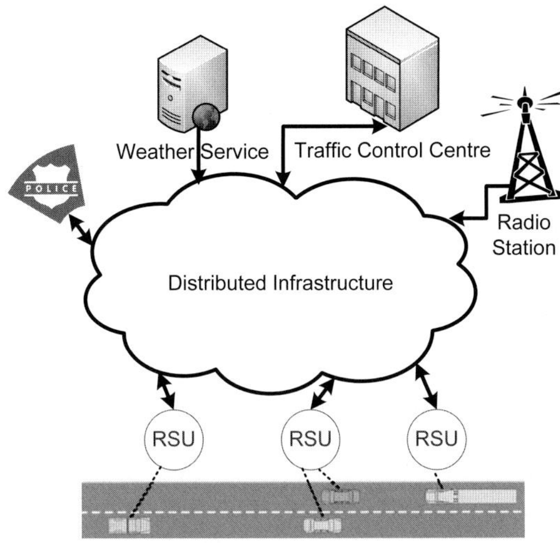

Fig. 1. Publishers and Subscribers in an Intelligent Transportation System

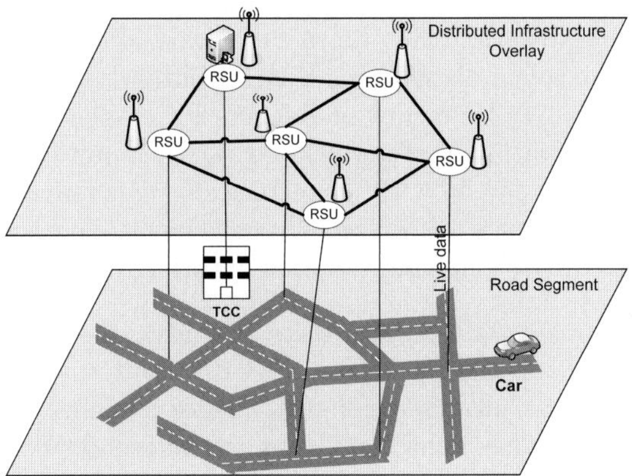

Fig. 2. Mapping of RSUs from the road network onto the meshed network

its relevance in space and time is limited to a certain region, moving direction and period of time. Data belonging to a specific region needs to be queried and updated frequently as vehicles provide new information to the RSU and need the latest data from a RSU situated in the connectivity range.

A subscriber my be a e.g. vehicle driving at high speed or a road worker in field service. Connectivity between the RSU and the passing by vehicles is characterized by a limited bandwidth, communication range, and connectivity window (ca. 300KB/sec for 2-3 sec at 100km/h in case of a single vehicle) allowing the exchange of small and a few messages only [13]. The received events can be used to generate statistics such as about the average speed/lane at coming road segments, the number of vehicles/hour per direction in upcoming road segments, the distribution of vehicles over specific road segments, the average distance between vehicles, whether groups of vehicles decrease or increase their speed. This kind of information can be used to adapt driving behaviour since drivers are informed about occurrences and actions in upcoming road segments. Road workers in field can use this information to prepare themselves and take precautions ahead in case of increasing traffic density. Road workers may be also interested in statistics like the average temperature and the actual temperature curve of a specific road segment over a specific period of time. Therefore, subscribers are interested in information which a) is represented by the very last event delivered by the pub/sub medium, b) is represented by an aggregated set of events, or c) is a prioritized set of the delivered events. In a special case events can even cancel each other and should not be delivered at all. A vehicle driver does not need to be informed about a wrong-way driver if that driver has already left the road. Additionally, in case of some extraordinary events additional information may be requested from a third person, either to enrich or to verify the event. This would require further network connection.

3 Research Questions

In this paper, we propose the ASC approach of extending the Space Container idea [14] and the concept of distributing Space Containers via Distributed Hash Tables [15], [16], [17], [18] by means of Aspects for efficient pub/sub scenarios in Intelligent Transportation Systems. Based on the limitations of traditional pub/sub communication media with respect to efficient delivery in network environments with mobile peers and on recent projects with industry partners from telematics, we derived the following research questions:

R.1 - The concept of ASC: Investigate the advantages and limitations of the idea of Aspects on top of Space Containers distributed via Distributed Hash Tables, and whether the proposed concept allows reducing the complexity of application implementations. What are the major differences between the aspect-oriented approach and the traditional pub/sub form of durable message delivery?

R.2 - Efficient communication with peers: Investigate to what extent the usage of Aspects in a Space Container helps to decrease communication time within the connectivity time window and as a result increase efficiency of information exchange between mobile peers and the pub/sub medium. Is the proposed concept, combining a DHT lookup and a Space Container access, still fast enough to deliver essential messages aggregated before within the connectivity time window?

For investigating these research issues, we gathered requirements from a set of reasonable industry case studies in the ITS. Then we designed and implemented a framework based on Pastry [19] and the Aspect-oriented Space Container[1] implementation.

[1] To be downloaded at `http://www.mozartspaces.org`

4 Related Work

This section gives an overview about the pub/sub paradigm and its most common (prototype) implementations, and about pub/sub systems in mobile networks with respect to guaranteed delivery. At last, the concept of databases is analysed in order to explain why the concept of Aspect-oriented Space Containers is more appropriate for the scenario than databases.

4.1 Publish/Subscribe Systems

The pub/sub paradigm defines two types of clients: *publishers* generating events, and *subscribers* receiving notifications of events they have previously subscribed for. This type of messaging paradigm allows a strong decoupling of publishers and subscribers in time and space. Furthermore, it enables asynchronous and anonymous communication between publishers and subscribers [1].

There are two types of pub/sub systems: *topic*-based and *content*-based [1]. In a topic-based pub/sub system subscribers can only register their interest on specific topics (mostly predefined) under which publishers are dispatching their generated events/messages. Whereas, in content-based pub/sub systems subscribers are not constrained to specific topics, they can define more precisely in what kind of events/messages they are interested in. On the one hand, this provides more flexibility to the pub/sub system but on the other hand the pub/sub system has to match events/messages to the subscriptions [22]. Hybrid pub/sub systems like Hermes [4], SIENA [23] or REBECA [24], [25] support both types of subscription.

Furthermore, the pub/sub system architecture can be further classified into *client-server* and *peer-to-peer*. In a client-server architecture publishers and subscribers are both clients which are connected to a network of servers. Generally, the servers temporarily store the events/messages generated by publisher-clients and forwarded them to the subscriber-clients. If a subscribed client is not directly connected to a server, where a publisher has dispatched an event, the server has to forward the event to other servers until the event reaches the server capable of delivering the event directly to the subscribed client. Gryphon [26], for example, is a context-based pub/sub system with a client-server architecture, which uses so called brokers as servers. The brokers are responsible to determine which subset of brokers it should send an event/message. In a P2P architecture each node can act as a publisher, subscriber or event-forwarder to another node. SIENA is a mixed form of a client-server and a peer-to-peer architecture, because publishers and subscribers are clients but servers are working together in a peer-to-peer topology.

As mentioned before pub/sub systems enable asynchronous and anonymous messaging between publishers and subscribers. Therefore, such systems have to be reliable and fault-tolerant. Reliable pub/sub systems guarantee that published events/messages are delivered to all its subscribers. Durable (fault-tolerant) pub/sub system are able to cope with unreachable subscribers and servers (due to network failures or crashed clients/servers). Some pub/sub systems like SIENA offer a best-effort delivery strategy, i.e. the system will periodically retry to deliver the message until the message was delivered successfully, a timeout expired or the maximum retry-count was reached.

The strong decoupling, the asynchronous and anonymous messaging provided by the pub/sub paradigm makes it very attractive to be used in mobile environments [10]. Client applications reside on a host that is moving and therefore accessing the network (composed of so called *event brokers*) from various locations [27]. The event brokers are responsible to guarantee the reliability and durability of the pub/sub system as described before. Furthermore, a protocol must exist which enables the update of a client's subscription as it is moving from one broker to another. During the client's movement undeliverable events have to be stored by the system and delivered as soon as the client reconnects to the system. When the client reconnects at another broker, all stored events have to be forwarded to that broker. The authors of [9] propose a two-phase handover (2PH) protocol, which reduces network traffic and the latency when a client reconnects to the system. One of the first pub/sub system that supports the reconnection of mobile clients is JEDI [28]. Later, existing pub/sub systems like SIENA and REBECA have been extended to support mobile clients [29], [30], [31].

However, our investigations of pub/sub systems have shown that currently most of the available systems are research prototypes which concentrate primarily on scalability and reliability rather on durability in P2P environments. Furthermore, current pub/sub systems for mobile scenarios have the disadvantage that the time needed to update a subscription and to forward all messages to the peer where the re-subscription is made, may take too long. Mobile clients in ITS scenarios are fast moving and have only a few seconds for data transmission (section 2).

In order to adequately address the scaling needs of distributed applications, over the past years there has been research on pub/sub systems making use of the scalability characteristics of Distributed Hash Tables (DHTs) [19]. This has led to several implementations of DHT-based pub/sub systems, like Scribe [32], Meghdoot [33], Willow [34], PastryStrings [35], or [36]. However, the papers aim at using DHT like for routing purposes, extended querying, efficient subscriptions, or the efficient distribution of events. In contrast to those approaches ASC focuses on the distribution of Space Containers as fault-tolerant intermediate subscribers, functioning as a scalable and efficient bridge between mobile peers with very short connectivity time and pub/sub systems. Thus, although ASC has been developed with respect to a mesh network (section 2), it does not prescribe the usage of P2P capable pub/sub systems (section 5.4).

4.2 Databases

The idea of ASC is to combine pub/sub systems with a DHT distributed data storage, the Space Containers, that is capable of aggregating events in order to allow an efficient delivery of events to mobile peers and releasing resources of the pub/sub medium. Therefore, the question is whether databases could have been such a data storage component as well, instead of Space Containers in the described scenario (section 2)? The problem is that established database products like Oracle or DB2 are heavy-weight components, and as such the peers in the network do not have the capacity to run them. An alternative would be embedded, light-weight databases, like Oracle Berkeley DB Java Edition[2], Apache Derby[3], hsqldb[4], Axion[5], H2[6], or db4o[7]. However, the drawback of databases is that they need a static data model of the entries they have to store, while Space Containers allow the usage of several different Coordinators at the same time, enabling dynamic data models. In case of db4o, being an object-oriented database, accessing an entry is performed via query-by-example, almost like in tuple spaces [37]. However, in [14] it has been shown that Space Containers allow an optimized realization of queries and coordination models.

Finally, databases aim to store and retrieve long living data, while the scenario focuses on short lived data. Additionally, triggers can be seen as aspects as well, but are meant for operations within the database itself, without the power of establishing connections to others peers in the network.

5 Architecture

This section pictures the architecture of an Aspect-oriented Space Container in detail. It describes the interfaces, supported operations, the way of executing operations, and the data-flow between the installed Aspects and the Space Container. Finally, an illustration of the integration between ASC and a pub/sub medium and the use of the architecture to support an Intelligent Application System is described.

5.1 Space Container

A Space Container [14] is a collection of entries accessible via a very few basic methods: read, take, write, shift, and destroy. A Space Container may be bounded to a maximum number of entries, and allows the usage of so called Coordinators. Coordinators enable establishing specific and optimized views on a set of the stored entries. The addressing scheme for a Space Container is an URI of the form `xvsm://mycomputer.mydomain.com:1234/ContainerName`. The Space Container reference is therefore dependent on the IP address of the localhost node that is hosting the Space Container. The protocol type `xvsm` makes the possible communication protocols transparent to the user. Depending on those types, within the platform `xvsm` may be translated to e.g. TCP & Java, specifying that communication takes place via a tcp-connection using java objects. A detailed explanation about Space Containers is given in [14].

2 `http://www.oracle.com/database/berkeley-db/index.html`
3 `http://db.apache.org/derby/`
4 `http://hsqldb.org/`
5 `http://axion.tigris.org/`
6 `http://www.h2database.com`
7 `http://www.db4o.com/`

5.2 Aspects

Space Containers realize some parts of Aspect-oriented Programming (AOP) by registering so called Aspects[8] at different points of a Space Container. Aspects are executed on the peer where the Space Container is located and can be triggered by the various operations on the container. They are triggered by operations either on a specific Space Container or on operations related to the entire set of Space Containers, called Space, rather on the according impact. The join points of AOP are called interception points (IPoints). Interception points, dependent on the Space Container operations are referred to as local IPoints, whereas interception points on Space operations are called global IPoints. Additionally, IPoints can be located before or after the execution of an operation, indicating two categories: pre and post. Therefore, the following local IPoints exist: pre/ post read, pre/ post take, pre/ post destroy, pre/ post write, pre/ post shift, pre/ post aspect appending, pre/ post aspect removing. In addition to the local IPoints the following global IPoints exist: pre/ post transaction creation, pre/ post transaction commit, pre/ post transaction rollback, pre/ post space container creation, pre/ post space container destruction. The main goal of IPoints for aspect appending and aspect removing is to enable exclusive aspects. This means, that an aspect is able to prohibit the addition of new aspects, respectively prohibit itself to be removed.

In addition to the parameters of a Space Container operation a so called Aspect Context can be passed along with every operation allowing the client to communicate with the installed Aspects to make logical decisions or to modify the semantics of the operation completely. Aspect may contain any computational logic, thus can be used to realize security (authorization and authentication), the implementation of a highly customizable notification mechanism, or the additional manipulation of already stored or incoming entries. In case multiple Aspects are installed on the same Space Container, they are executed in the order they were added. Adding and removing Aspects can be performed at any time during runtime.

5.3 Aspect-Oriented Space Container

Figure 3 shows a Space Container with three pre and three post Aspects installed. The accessing operation is executed via the Space Container Interface and passed immediately to the first pre-Aspect. The operation contains the parameters of the operation, like transactions, selectors and timeout, and the Aspect Context. The called Aspect may contain any computational logic that is needed by the user to be executed before the operation. It can either verify or log the current operation, or initiate external operations to other Space Containers or third-party services.

The central part of a Space Container is the implementation of the containers business logic which handles the storage of the entries and coordinates the coordinators. Before an operation can be executed on a Space Container it has to

[8] The complete API JavaDoc can be found at http://www.mozartspaces.org

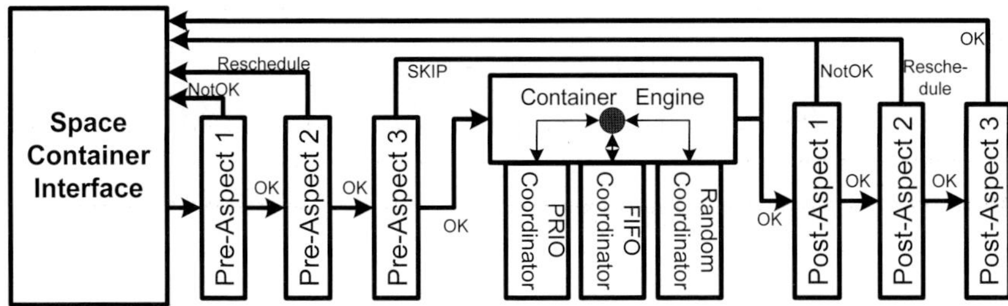

Fig. 3. Execution sequence of Aspects and data- and control-flow in a Space Container with three installed pre- and post-Aspects

pass the installed pre-Aspects. If all Aspects return OK, the container interprets the selectors of the operations and executes the operation [14]. Afterwards, all post-Aspects are executed. Depending on the result of the post-Aspects the result of the operation is either returned to the requesting peer, or the operation is rolled back.

As already mentioned, an Aspect can manipulate the execution of the operation which triggered it. This is realized by the returning values an Aspect can throw. The returned value is analysed and the execution of the operation manipulated accordingly. The following return values are supported:

- **OK:** The execution of the Aspect does not require any changes of the operation, the execution of the operation may proceed normally.
- **NotOK:** The execution of the operation is stopped and the transaction is rolled back. This kind of return value can be used by e.g. a security Aspect denying a operation if the user has not adequate access rights on the Space Container or the Space.
- **SKIP:** The operation is neither performed on the container, nor on the Spaces, nor on any following pre-Aspects. The post-Aspects are executed immediately afterwards. This return value is only supported for pre-Aspects.
- **Reschedule:** The execution of the operation is stopped and will be rescheduled for a later execution. This can be used to delay the execution of an operation until an external event occurs.

5.4 Execution of Aspect-Oriented Space Containers in Publish/Subscribe Scenarios

A reason why we recommend that a Space Container shall be used instead of a database is the fact that the number of different events in an ITS is not known beforehand and as a consequence an appropriate data model is difficult to establish. By means of Coordinators [14], a Space Container is capable of using 'dynamic' data models which can be plugged in whenever needed. A Coordinator allows different views, optimally implemented with respect to accessing requirements, on the entire data in the Space Container at the same time. As

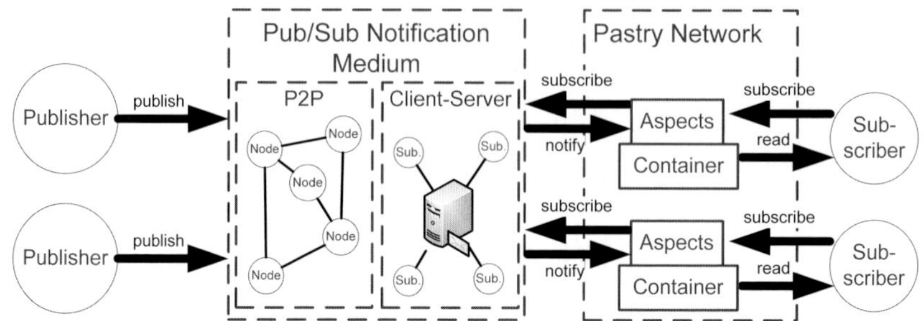

Fig. 4. The operation of Aspect-oriented Space Containers in publish/subscribe scenarios in an Intelligent Transportation System

depicted in figure 4, such a Space Container and its Aspects are deployed by means of a DHT in the RSU network. The principle is described in [17], where we described how to combine Space-Containers with an overlay network based on DHT concepts in order to a) make such Space Containers uniquely addressable in a fault-tolerant and scalable manner, and b) to replicate Space Containers in order to increase fault-tolerance and their availability.

The original subscriber (e.g. a vehicle) places its intention in receiving events from publishers by deploying a Space Container, installing Aspects and publish it in the DHT network. The Aspect registers itself as a subscriber in the pub/sub medium (P2P or client/server style) on behalf and according to the requirements of the original subscriber. From now on the Aspect will, independent of the connectivity mode of the original subscriber, receive events which are then processed by other installed pre-Aspects and the results are then stored in the Space Container. When the original subscriber re-establishes a connection to the network, it uses a read-selector to pick up the results from the Space Container. What kind of selector-type is used is completely up to the original subscriber and depends on the fact how the installed pre-Aspects work with incoming events.

If the Space Container is replicated, Aspects are replicated as well. This means that the original subscriber is subscribed as often as many replicas of that Space Container exists. This is necessary in order to avoid missing events in case one of the replicas, including the subscribed Aspect, is off-line. The way how the replicated Space Containers handle incoming events in order to stay consistent is up to the implementation of the deployed pre-Aspects. Either, the replicas are completely independent of each other and perform every operation as many times as replicas exists, or an incoming event is registered and not used for further processing until the result based on that event has been announced from a designated replica. The latter approach may be more efficient with respect to computational resources but require knowledge about group coordination.

6 Evaluation

In section 3 we defined two research questions: a) can aspects together with space containers reduce the complexity of applications being mostly off-line and

b) can aspects be used for efficient event delivery by decreasing the number of transmitted events and minimizing the communication time required by the applications. Based upon the context of the ITS scenario, introduced in section 2, we implemented a simulation in order to answer the questions.

The problem with complexity is, that it cannot be eliminated but shifted somewhere else and kept abstracted [38]. In case an application could receive all stored events it has subscribed for, it would need the processing logic embedded in the application in order to extract the information the application is really interested in. This processing logic can be outsourced into an Aspect and executed on a Space Container. This implies that application implementation has been reduced in complexity and abstracted by means of the access operations the Space Container concept offers. Additionally, the overall complexity could have been reduced as well. On the one hand each car would have retrieved all events and process them separately. On the other hand, in case of using the ASC approach, processing is done only once at the RSU.

In order to answer the second question we have implemented and evaluated an ITS scenario. Consider an application which is used to monitor the amount of vehicles passing a Road Site Unit in the last 60 seconds and calculate the average speed of these vehicles. In order to realise this functionality, every passing vehicle has to send its current speed to the RSU which stores this information and provides it to vehicles which are interested in it.

We compared the ASC approach with an implementation using durable message queues from the Java Message Service (JMS) [39]. We decided to use JMS because we have not found any downloadable P2P based pub/sub implementations supporting durable subscriptions for our simulation. JMS is an acknowledged API standard developed by Sun Microsystems and implemented by several well-known commercial and non-commercial providers, mainly according to the client-server paradigm. The JMS API provides two messaging models: *queuing* and *pub/sub*. The queueing model is a form of one-way point-to-point communication between a sender and a receiver. The receiver writes messages into a specific queue, wherefrom the receiver consumes the messages in a first-in-first-out manner. The pub/sub model works like the pub/sub paradigm introduced in section 4, but the JMS API specification prescribes that JMS-providers must be implemented reliable and durable.

In order to allow a reasonable comparison of results, the simulation consisted of a single client application, representing the vehicle, and a single peer, representing the RSU. On the one hand the peer was running a JMS server and on the other hand the ASC concept without DHT features, since no distribution is needed. As grounding for our simulation, we used the traffic statistics of the Austrian highway “A23”, which is the most frequently used highway in Austria. In 2008, 153100 vehicles used the highway daily[9] in average. Assuming that the amount of cars is consistent over the day, approximately 1.8 vehicles use the highway every second. Additionally, in case RSUs are placed every four

[9] `http://de.wikipedia.org/wiki/ Autobahn_Südosttangente_Wien` last read February, 9th 2009.

kilometres, it means that a vehicle passes the next RSU approximately after 3 minutes, based on the fact that speed is limited to 80 kilometres per hour.

Every time a vehicle drives by a RSU it reports its current speed and (if it is interested in the information) fetches the statistics from the RSU. In the JMS implementation of this scenario we used a durable message queue to store the messages. Every vehicle sends a message containing its current speed to the queue when it is passing by the RSU. Additionally, it reads all the messages which have been published by the other vehicles. Since the JMS standard does not define any way to pre-/post-process messages, the vehicles have to read all messages from the queue, count them to get the amount of cars which passed the RSU and calculate their average speed using the contents of the messages. With the figures presented in the previous paragraph the JMS queue and the physical transmission media has to deal with approximately 50 messages per seconds. The amount of data to be transmitted is limited to 300KB per second in the DSRC protocol. Since the size of the messages is between 5KB to 10KB (depending on the content) it may occur that it is physically impossible to transmit all message from the queue to the vehicle (the connection window of roadside unit is only 2-3 seconds). Since we calculated these figures, assuming the amount of cars is equally spread over the day, the amount might become larger with respect to rush hours and resulting peaks. If it is not possible for a vehicle to retrieve all messages from the message queue the size of the queue grows and any calculations done in the vehicle are not correct.

In the ASC based implementation of this scenario we used an aspect which calculates the average speed. Every vehicle passing the RSU sends its current speed to the RSU. Instead of storing all the messages in the container and providing it for interested parties, an aspect processes the messages and returns the result of the calculation. When a vehicle is interested in the traffic statistics it has to read one message only which contains the aggregated result. Therefore the amount of messages is decreased dramatically.

Figure 5 depicts the increasing amount of data within a message queue in case the messages cannot be retrieved by the vehicles within the connection time windows. The blue line shows the messages when JMS is used. The orange (horizontal) line shows the amount of messages in the ASC based implementation. In case vehicles are in transmission range, they try to fetch as many messages as possible. The only limit is the connection time window. Therefore, every time the blue line falls, messages have been retrieved. However, since - due to low transmission capabilities - not all of the messages could have been transmitted the amount of messages increases continuously. In contrast, the number of messages in the ASC based implementation is always one because only the message which already contains the processed data is stored.

In addition to the simulations described above we implemented several performance tests using ASC with a DHT implementation. Table 1 depicts the time it takes to retrieve a single entry out of 10, 100 and 1000 entries from a Space Container. As shown in the table the time to retrieve the entry is independent of the number of peers (respectively RSUs) in the DHT. This shows that the

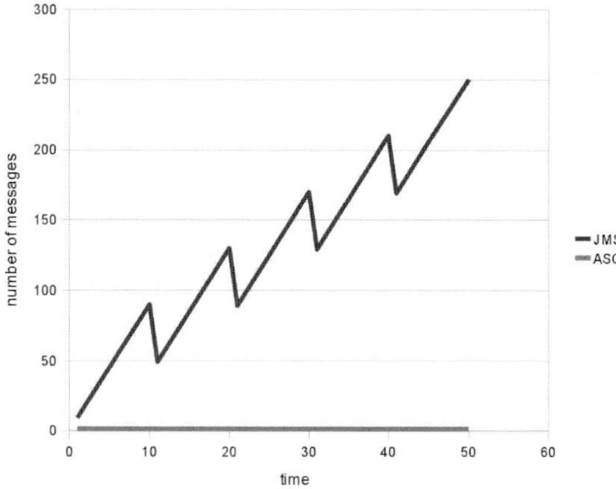

Fig. 5. The development of the size of a message queue in case data cannot be retrieved sufficiently by vehicles within the connection window

Table 1. Durations [ms] for the retrieval of 10, 100 and 1000 entries in a network with 2 to 210 peers [18]

peers	10 entr.	100 entr.	1000 entr.
2	142	208	1378
10	148	258	1002
60	169	256	1115
120	155	265	1184
180	155	231	1086
210	155	231	1086

overhead to lookup the container in the DHT and to execute a query in the Space Container does not have negative influences on the performance of the system. Additionally, the table shows that the time needed to access a Space Container is way lower than the time offered by the connectivity window.

7 Conclusion and Future Work

In this paper we described the concept of ASC, Aspect-oriented Space Containers, distributed via Distributed Hash Tables for bridging mobile peers with small connectivity window and pub/sub systems efficiently in scenarios from the Intelligent Transportation domain. We derived two research questions and answered them based on the evaluation of a scenario from the ITS:

The concept of ASC: The concept of ASC helps moving the complexity of handling events from a peer being mobile and most of the time off-line into so called Aspects of a Space Container. This approach allows the mobile peer to subscribe for events and have the Aspects handle them while the mobile peer is off-line.

Efficiency of the ASC concept: The evaluation showed that the usage of traditional notification media is not sufficient enough to deliver all events to the subscribed peers due to small bandwidth and connectivity time. The operation of Aspects and Space Containers help at preprocessing subscribed events so that the size of the transmitted data and the time needed for transmission is kept minimal. Therefore, the subscriber is capable of receiving all relevant data needed for further decision making.

Future work contains investigations regarding the way how the usage of Space Containers and Aspects change the relation between subscribers and the notification medium. Further questions that we will consider are: Is a durable notification medium still necessary if Space Containers are replicated and distributed via DHTs, since DHT use the same network as the notification medium? Does this have an effect on the semantics of durability? What is the influence on QoS coming from the number of replica set up by the subscriber? Another future work will deal with the question how to move Space Containers along the mesh of RSUs to minimize Space Container access time.

Acknowledgement

This work has been supported by the Complex Systems Design & Engineering Lab, and by the Austrian Government and the City of Vienna within the competence center program COMET.

References

1. Eugster, P.T., Felber, P.A., Guerraoui, R., Kermarrec, A.M.: The many faces of publish/subscribe. ACM Comput. Surv. 35(2), 114–131 (2003)
2. Cabrera, L., Jones, M., Theimer, M.: Herald: achieving a global event notification service. In: Proceedings of the Eighth Workshop on Hot Topics in Operating Systems, 2001, May 2001), pp. 87–92 (2001)
3. Castro, M., Druschel, P., Kermarrec, A.M., Rowstron, A.: Scalable application-level anycast for highly dynamic groups. In: Stiller, B., Carle, G., Karsten, M., Reichl, P. (eds.) NGC 2003 and ICQT 2003. LNCS, vol. 2816, pp. 47–57. Springer, Heidelberg (2003)
4. Pietzuch, P.R.: Hermes: A Scalable Event-Based Middleware. PhD thesis, Queens' College University of Cambridge (February 2004)
5. Eugster, P.: Type-based publish/subscribe: Concepts and experiences. ACM Trans. Program. Lang. Syst. 29(1), 6 (2007)
6. Chandramouli, B., Phillips, J.M., Yang, J.: Value-based notification conditions in large-scale publish/subscribe systems. In: VLDB 2007: Proceedings of the 33rd international conference on Very large data bases, VLDB Endowment, pp. 878–889 (2007)
7. Lumezanu, C., Spring, N., Bhattacharjee, B.: Decentralized message ordering for publish/subscribe systems (2006)
8. Pereira, C.M.M., Lobato, D.C., Teixeira, C.A.C., Pimentel, M.G.: Achieving causal and total ordering in publish/subscribe middleware with dsm. In: MW4SOC 2008: Proceedings of the 3rd workshop on Middleware for service oriented computing, pp. 61–66. ACM, New York (2008)

9. Wang, J., Cao, J., Li, J.: Supporting mobile clients in publish/subscribe systems. In: ICDCSW 2005: Proceedings of the First International Workshop on Mobility in Peer-to-Peer Systems (MPPS) (ICDCSW 2005), pp. 792–798. IEEE Computer Society, Washington (2005)
10. Huang, Y., Garcia-Molina, H.: Publish/subscribe in a mobile environment. Wirel. Netw. 10(6), 643–652 (2004)
11. Yoneki, E., Bacon, J.: Dynamic group communication in mobile peer-to-peer environments. In: SAC 2005: Proceedings of the 2005 ACM symposium on Applied computing, pp. 986–992. ACM, New York (2005)
12. Eichler, S.: Performance evaluation of the ieee 802.11p wave communication standard. In: VTC-2007 Fall. 2007 IEEE 66th Vehicular Technology Conference, 2007 (30 2007-October 3 2007), pp. 2199–2203 (2007)
13. Zaera, M.: Wave-based communication in vehicle to infrastructure real-time safety-related traffic telematics. Master's thesis, Telecommunication Engineering, University of Zaragoza (August 2008)
14. Kühn, E., Mordinyi, R., Keszthelyi, L., Schreiber, C.: Introducing the concept of customizable structured spaces for agent coordination in the production automation domain. In: The 8th International Conference on Autonomous Agents and Multiagent Systems (2009)
15. Bessler, S., Tomic, S., Kühn, E., Mordinyi, R., Goiss, H.D.: Sabron: A storage and application based routing overlay network for intelligent transportation systems. In: 3rd International Workshop on Self-Organizing Systems, IWSOS 2008 (2008)
16. Kühn, E., Mordinyi, R., Schreiber, C.: An extensible space-based coordination approach for modeling complex patterns in large systems. In: 3rd International Symposium on Leveraging Applications of Formal Methods, Verification and Validation, Special Track on Formal Methods for Analysing and Verifying Very Large Systems (2008)
17. Kühn, E., Mordinyi, R., Goiss, H.D., Bessler, S., Tomic, S.: Integration of shareable containers with distributed hash tables for storage of structured and dynamic data. In: 2nd International Workshop on Adaptive Systems in Heterogeneous Environments - ASHEs 2009, CISIS 2009 (2009)
18. Kühn, E., Mordinyi, R., Goiss, H.D., Bessler, S., Tomic, S.: A p2p network of space containers for efficient management of spatial-temporal data in intelligent transportation scenarios. In: International Symposium on Parallel and Distributed Computing, ISPDC 2009 (2009)
19. Rowstron, A., Druschel, P.: Pastry: Scalable, decentralized object location, and routing for large-scale peer-to-peer systems. In: Guerraoui, R. (ed.) Middleware 2001. LNCS, vol. 2218, pp. 329–350. Springer, Heidelberg (2001)
20. Kühn, E., Mordinyi, R., Keszthelyi, L., Schreiber, C., Bessler, S., Tomic, S.: Introducing aspect-oriented space containers for efficient publish/subscribe scenarios in intelligent transportation systems. In: 8th Working IEEE/IFIP Conference on Software Architecture, WICSA 2009 (2009), http://tinyurl.com/lx3lmx
21. Xu, P., Deters, R.: Using event-streams for fault-management in mas. In: IEEE/WIC/ACM International Conference on Intelligent Agent Technology, 2004 (IAT 2004). Proceedings, September 2004, pp. 433–436 (2004)
22. Liu, Y., Plale, B.: Survey of publish subscribe event systems. Technical report, Computer Science Deptartment, Indiana University (2003)
23. Carzaniga, A.: Architectures for an Event Notification Service Scalable toWide-area Networks. PhD thesis, Politecnico Di Milano (December 1998)
24. Fiege, L.: Visibility in Event-Based Systems. PhD thesis, Technischen Universitt Darmstadt (2004)

25. Zeidler, A.: A Distributed Publish/Subscribe Notification Service for Pervasive Environments. PhD thesis, Technischen Universitt Darmstadt (2004)
26. Bhola, S., Strom, R., Bagchi, S., Zhao, Y., Auerbach, J.: Exactly-once delivery in a content-based publish-subscribe system. In: DSN, pp. 7–16 (2002)
27. Caporuscio, M., Caporuscio, C.M., Carzaniga, A., Carzaniga, A., Wolf, E.L., Wolf, E.L.: Design and evaluation of a support service for mobile, wireless publish/-subscribe applications. IEEE Transactions on Software Engineering 29, 1059–1071 (2003)
28. Cugola, G., Di Nitto, E., Fuggetta, A.: The jedi event-based infrastructure and its application to the development of the opss wfms. IEEE Trans. Softw. Eng. 27(9), 827–850 (2001)
29. Nielsen, J.: Adapting the siena content-based publish-subscribe system to support user mobility. Technical report, Rutgers University - ECE department (2004)
30. Carzaniga, A., Rosenblum, D.S., Wolf, A.L.: Design and evaluation of a wide-area event notification service. ACM Transactions on Computer Systems 19, 332–383 (2001)
31. Muehl, G.: Large-Scale Content-Based Publish/Subscribe Systems. PhD thesis, TU Darmstadt (2002)
32. Castro, M., Druschel, P., Kermarrec, A.M., Rowstron, A.: Scribe: a large-scale and decentralized application-level multicast infrastructure. IEEE Journal on Selected Areas in Communications 20(8), 1489–1499 (2002)
33. Gupta, A., Sahin, O.D., Agrawal, D., Abbadi, A.E.: Meghdoot: content-based publish/subscribe over p2p networks. In: Jacobsen, H.-A. (ed.) Middleware 2004. LNCS, vol. 3231, pp. 254–273. Springer, Heidelberg (2004)
34. van Renesse, R., Bozdog, A.: Willow: Dht, aggregation, and publish/subscribe in one protocol (2005)
35. Aekaterinidis, I., Triantafillou, P.: Pastrystrings: A comprehensive content-based publish/subscribe dht network. In: ICDCS 2006: Proceedings of the 26th IEEE International Conference on Distributed Computing Systems, p. 23. IEEE Computer Society, Washington (2006)
36. Ahull, J.P., Lpez, P.G., Skarmeta, A.F.G.: Caps: Content-based publish/subscribe services for peer-to-peer systems. In: 2nd Int. Conf. on Distributed Event-Based Systems, DEBS 2008 (2008)
37. Gelernter, D.: Generative communication in linda. ACM Trans. Program. Lang. Syst. 7(1), 80–112 (1985)
38. Brooks Jr., F.P.: The mythical man-month (anniversary ed.). Addison-Wesley Longman Publishing Co., Inc., Boston (1995)
39. Hapner, M., Burridge, R., Sharma, R., Fialli, J., Stout, K.: Java message service. Technical report, Sun Microsystems, Version 1.1 (April 12, 2002)

Parallel Interconnection of Broadcast Systems with Multiple FIFO Channels*

Ruben de Juan Marín[1], Vicent Cholvi[2], Ernesto Jiménez[3], and Francesc D. Muñoz-Escoí[1]

[1] Instituto Tecnológico de Informática
Universidad Politécnica de Valencia
46022 Valencia, Spain
{rjuan,fmunyoz}@iti.upv.es

[2] Depto. de Lenguajes y Sistemas Informáticos
Universitat Jaume I, Campus de Riu Sec
12071 Castellón, Spain
vcholvi@uji.es

[3] Escuela Universitaria de Informática
Univ. Politécnica de Madrid, Ctra. Valencia, km 7
28031 Madrid, Spain
ernes@eui.upm.es

Abstract. This paper proposes new protocols for the interconnection of FIFO- and causal-ordered broadcast systems, thus increasing their scalability. They use several interconnection links between systems, which avoids bottleneck problems due to the network traffic, since messages are not forced to go throughout a single link but instead through the several links we establish. General architectures to interconnect FIFO- and causal-ordered systems are proposed. Failure management is also discussed and a performance analysis is given, detailing the benefits introduced by these interconnection approaches that are able to easily increase the resulting interconnection bandwidth.

1 Introduction

There have been multiple papers [1,2,3,4,5,6] that had devoted their attention to the interconnection of message broadcast systems. Some of them [1,2,3,4,5] were focused on causal-ordered systems, thus reducing both the size of the vector clocks [7] being used in the broadcast protocols and the amount of needed messages (since smaller groups were used). Most of them have relied on either FIFO

* This work has been partially supported by EU FEDER and the Spanish MEC under grant TIN2006-14738-C02, by EU FEDER and the Spanish MICINN under grant TIN2009-14460-C03, by IMPIVA under grant IMIDIC/2007/68, by CICYT under grant TIN2008-03687, by Bancaixa under grant P1-1B2007-44, by Comunidad de Madrid under grant S-0505/TIC-0285, and by the Spanish MEC under grant TIN2007-67353-C02-01.

R. Meersman, T. Dillon, P. Herrero (Eds.): OTM 2009, Part I, LNCS 5870, pp. 449–466, 2009.

interconnection links [1,3,5] or on causal broadcast among the interconnection servers [2].

The aim of such solutions is to enhance the scalability of the resulting broadcast mechanisms. Such scalability might be needed in different current distributed applications, like P2P applications or the data centres being used to implement *cloud computing* systems.

Other scalability efforts have been focused on other aspects of causal communication, introducing some principles that have guided the design of the interconnection solutions. One example is the usage of causal separators [8] that divide the global system into causal zones (i.e., subgroups) and reduce the size of the vector clocks needed for guaranteeing causal delivery. Another example is the solution described in [9], that also interconnects previously existing systems and ensures causal delivery, but without requiring that all messages were broadcast; i.e., point-to-point communication among different systems is also considered. To this end, such global system also relies on a set of causal servers, each one from a different local system, and using vector clocks to ensure causal delivery in such set of servers, whilst system-local communication does only rely on linear logical clocks or on physical synchronisation.

Similar efforts can be found in order to interconnect FIFO-ordered systems [3,6], although in such case the interconnection is almost trivial, since it only depends on local information from the sender node.

However none of such papers has proposed any technique for increasing the usable bandwidth of such interconnecting protocols, implementing some technique for using simultaneously several interconnecting channels able to transmit multiple messages in parallel. Note that in most cases, each broadcast system is deployed over a very fast LAN, whilst the interconnecting links are far slower. In the common case, we might assume that such solution could be provided by the network layer, using multiple paths between each pair of interconnected servers, and selecting an appropriate path per message in order to avoid congestion. But this cannot be assumed in all scenarios. For instance, the set of data centres in a cloud computing environment might use dedicated inter-centre channels; i.e., there will be a single path between each pair of centres. Thus, we do not obtain any bandwidth improvement trying to set up multiple paths in such scenario. So, in some cases, a transport or application-level parallelisation of these interconnections might enhance the overall system performance. This paper scans this alternative, providing interesting results.

The rest of the paper is organised as follows. In Section 2, we introduce our framework for the interconnection of message-passing systems. In Section 3, we show how to interconnect FIFO-ordered systems by using several interconnection links between systems. In Section 4, we introduce the architecture with which interconnect FIFO-ordered systems. Sections 5 and 6 repeat the same for causal-ordered systems, whilst Section 7 describes how process failures can be managed. Finally, Section 8 provides a performance analysis and in Section 9, we present some concluding remarks.

2 Model

In this paper, we use a model similar to the one in [6]. From a physical point of view, we consider asynchronous distributed systems made up of a set of *nodes* connected by a *communication network*. The logical system we consider consists of *processes* (executed in the nodes of the system) which interact by exchanging *messages* with one another (using the communication network). The interface between the processes and the network has two types of events [10]: by using *bc-send*$_i(m)$, process i broadcasts the message m to all processes of the system. Similarly, by using *bc-recv*$_i(m)$, process i receives the message m.

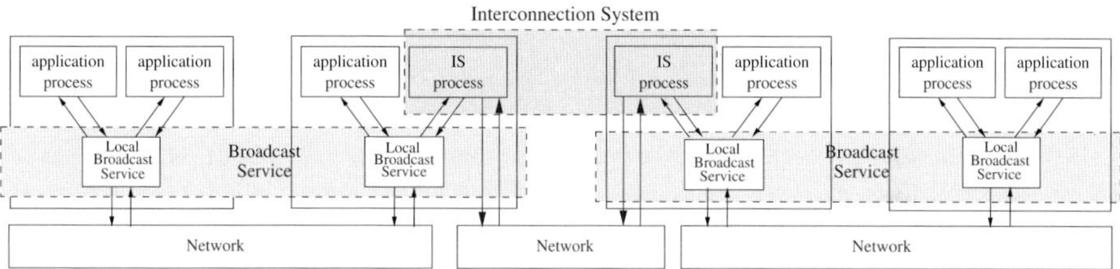

Fig. 1. Interconnection System

Failures may arise in these systems. Thus, processes may fail by crashing [11], but communication channels are assumed reliable; i.e., although temporary communication failures may arise, messages are eventually delivered to their destinations, except when such nodes fail. Additionally, all broadcast primitives described in this paper are assumed *uniform* in the sense described in [11]; i.e., the delivery orders are the *uniform FIFO order* and *uniform causal order* defined in [11, page 109]. This allows a simple failure management, as described in Section 7.

The basic broadcast service specification for n processes consists of sequences of *bc-send*$_i$ and *bc-recv*$_i$ events, $0 \leq i \leq n-1$. In these sequences, each *bc-recv*$_i(m)$ event is mapped to an earlier *bc-send*$_j(m)$ event, every message received was previously sent, and every message that is sent is received once and only once in each process. For the sake of simplicity, we also assume that any given message is sent once, at the most. This assumption does not introduce any new restriction, since it can be forced by associating a (bounded) timestamp with every send operation [12].

Following, we define *FIFO-ordered* systems, according to the ordering requirements of the broadcast services they implement.

Definition 1. *We say that a system is* FIFO-ordered *if, for all messages* m_1 *and* m_2 *and all processes* p_i *and* p_j*, if* p_i *sends* m_1 *before it sends* m_2*, then* m_2 *is not received at* p_j *before* m_1*.*

The definition of *causally ordered* systems requires us to firstly introduce the *happens-before* (denoted with $\rightarrow$) relation [13] between messages. The important property of the happens-before relation is that it completely characterises the causality relations between messages.

Given a sequence of *bc-send*$_i$ and *bc-recv*$_i$ events, $0 \leq i \leq n-1$, message m_1 is said to *happen-before* message m_2 if either:

1. The *bc-recv*$_i$ event for m_1 happens before the *bc-send*$_i$ event for m_2.
2. m_1 and m_2 are sent by the same process and m_1 is sent before m_2.

Now, we define a causally ordered system as follows.

Definition 2. *We say that a system is* causally ordered *if for all messages m_1 and m_2 and every process p_i, if m_1 happens-before m_2, then m_2 is not received at p_i before m_1 is.*

We consider systems in which each message sent must eventually be received in every process of the system. This is a very natural property (usually known as *Liveness*) which is preserved by every system that we have found in the literature. In our terminology it means that for each *bc-send*$_i(m)$ event, a *bc-recv*$_j(m)$ event will eventually occur for every process j in the system.

Now, we define what we understand by *properly interconnecting* several equally ordered systems. Roughly speaking, this consists in interconnecting these systems (without modifying any of them) by using an *interconnection system* (denoted *IS*), so that the resulting system behaves as a single one and preserves the same ordering. Such an interconnection system is made up of a set of *interconnecting system processes* (denoted *IS processes*) that execute some distributed algorithm or protocol. Each of these processes receives all the messages broadcast in its system and can itself broadcast new messages received from the interconnection link, but it cannot generate and broadcast new messages on its own. More specifically, a value broadcast by an application process in some system can only be received by an application process in another system if the interconnecting process of the latter system broadcasts it. The interconnecting processes can communicate among themselves via message passing. However, they cannot interfere with the protocol in their original local message-passing system in any way. Figure 1 presents an example of an *IS* interconnecting two systems with the above-mentioned architecture and two *IS processes.*

3 Interconnection of FIFO-Ordered Systems

By using the model introduced in the previous section, [6] provided a simple protocol to properly interconnect FIFO-ordered systems. However, the aim of such a protocol was not focused on having a very efficient protocol, but on proving that it is in fact possible to interconnect FIFO systems. Therefore, to interconnect any pair of systems, the protocol used two *IS processes*. Clearly, this could generate bottleneck problems, since all messages must pass throughout the single link formed by this pair of *IS processes*. Thus, this raised the question as to whether it is possible or not to use several *IS processes* per interconnected system. In this section, we provide an interconnecting protocol for FIFO-ordered systems that uses several *IS processes* in each system.

First, we consider the case when there are only two systems. Later, we will consider the case of several systems. Let us denote each of the FIFO ordered systems as S^k (with $k \in \{0, 1\}$). The interconnecting protocol consists of several processes, denoted isp_v^k (with $k \in \{0, 1\}$ and v denoting the *IS process* within system S^k), that are part of each of the two systems. Each such isp_v^k process is responsible for propagating to S^{1-k} all messages being broadcast by any process in $set_w(isp_v^k)$. Such transfer set $set_w(isp_v^k)$ includes the set of processes in S^k whose messages should be propagated to a different IS process in S^{1-k}, where subindex w denotes the set number within isp_v^k. For instance, in Fig. 3, isp_2^1 has two associated transfer sets. Thus, $set_1(isp_2^1)$ is connected to isp_2^0, whilst $set_2(isp_2^1)$ is connected to isp_3^0. Note that the number of *IS processes* may be different in S^0 and in S^1.

These interconnecting processes are only in charge of the interconnecting protocol. It is worthwhile remarking that each isp_v^k is part of the system S^k and, for that reason, can use the communication system implemented in S^k. Note also that the introduction of those processes does not require any modification of the original systems. We consider that the set of processes in the resulting system S^T includes all the processes in S^0 and S^1, with the exception of the *IS processes*, which are only used to interconnect S^0 and S^1.

Each isp_v^k process executes two concurrent atomic tasks, namely $Propagate\_out(isp_v^k, m)$ and $Propagate\_in(isp_v^k, m)$ (atomicity is needed in order to avoid race conditions).

- $Propagate\_out(isp_v^k, m)$ transfers the message m issued by a process in $set_w(isp_v^k)$ to $S^{\overline{k}}$ (we use $\overline{k}$ to denote $1 - k$). Each process in system S^k (except for the *IS processes*) must be included in one transfer set (associated with only one *IS process*). Furthermore, the transfer of messages from processes in $set_w(isp_v^k)$ is performed to a single *IS process* in $S^{\overline{k}}$, denoted $link_w(isp_v^k)$. However, an *IS process* may transfer messages to many *IS processes* and receive transfers from many of them, but they are not necessarily the same.
  Both $set_w(isp_v^k)$ and $link_w(isp_v^k)$ are set up prior to running the protocol.
- $Propagate\_in(isp_v^k, m)$ forwards the messages received from $S^{\overline{k}}$ to within S^k. Note that when isp_v^k receives a transfer, it performs the broadcast to the whole set of processes in system S^k, regardless of the transfer sets these processes belong to.

Fig. 2 shows the implementation of the $Propagate\_out(isp_v^k, m)$ and $Propagate\_in(isp_v^k, m)$ tasks.

It must be noted that the link between pairs of *IS processes*, one in each system, needs to be FIFO-ordered. Figure 3 shows an illustrative example of how transfer links are established between two interconnected FIFO systems. Each *IS process* isp_v^k is in charge of transferring the messages issued by processes in $set_w(isp_v^k)$ to system $S^{\overline{k}}$. There are three *IS processes* in system S^0 and two

$Propagate\_out(isp_v^k, m)$ :: task which is activated once $bc\text{-}recv_{isp_v^k}(m)$ is executed	$Propagate\_in(isp_v^k, m)$ :: task which is activated immediately after message m is received from $S^{\overline{k}}$

begin
  if m was sent by a process in $set_w(isp_v^k)$
    then transfer m to $link_w(isp_v^k)$
end
	begin
	  $bc\text{-}send_{isp_v^k}(m)$
	end

Fig. 2. The interconnecting protocol in isp_v^k

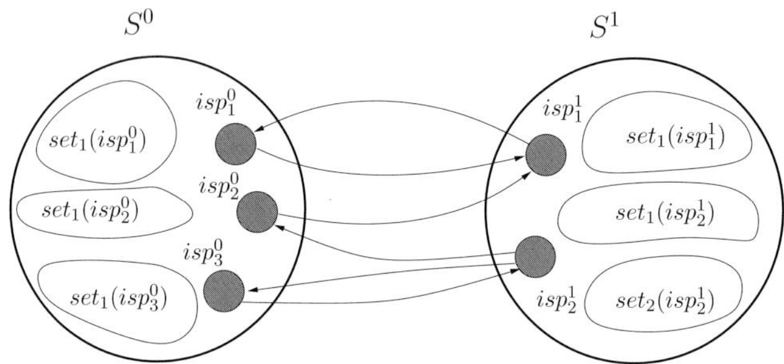

Fig. 3. Example of the interconnecting protocol for two systems

IS processes in system S^1. Both isp_1^0 and isp_2^0 transfer messages to isp_1^1, and isp_3^0 transfers messages to isp_2^1. In turn, isp_1^1 transfers messages to isp_1^0 and isp_2^1 transfers messages both to isp_2^0 and isp_3^0.

The following theorem shows that the system S^T, obtained by connecting any two FIFO-ordered systems S^0 and S^1 by using the above-mentioned interconnecting protocol, is also FIFO ordered.

Theorem 1. *Any two FIFO-ordered systems can be properly interconnected by using the protocol in Fig. 2.*

Proof. By contradiction. Assume there are two messages, m_1 and m_2, sent in that order by, say, process p_i in system S^0. Now, assume they are received by, say, process p_j in system S^1 in the reverse order.

Since S^1 is a FIFO-ordered system, m_2 must have been sent by some *IS process* in S^1 before m_1. Therefore, since the two systems are connected by a FIFO-ordered communication channel, we have that m_2 must have been transferred by some *IS process* in S^0 before m_1. This implies that, since S^0 is a FIFO-ordered system system, m_2 must have been sent (by p_i) before m_1. Thus, we reach a contradiction.

Note that the same interconnecting protocol can be used to properly interconnect any number of FIFO-ordered systems. This can be easily shown by induction on the number of systems. Let S^T denote the resulting system. For $n = 1$ the claim is clearly true, since $S^T = S^0$. For $n = 2$ it is immediate from Theorem 1.

Now, assume that we can obtain a FIFO-ordered system S' by properly interconnecting the systems $S^0, S^1, ..., S^{n-2}$. Then, from Theorem 1, we can properly interconnect S' and S^{n-1} to obtain a FIFO-ordered system S^T.

Similarly to what happened with the interconnection protocol proposed in [6], our interconnecting protocol should not affect the *response time* a process observes when issuing a broadcast operation, since its broadcast protocol is not affected by the interconnection. The latency (i.e., the time until a broadcast value is visible in any other process) is also the same.

However and contrary to the interconnection protocol proposed in [6], we can now avoid bottleneck problems due to the *network traffic*, since messages are not forced to go through a single link but through the several links we establish.

4 An Architecture to Interconnect FIFO-Ordered Systems

In this section, we describe a general architecture to interconnect FIFO-ordered systems. Such an architecture can be built following these steps:

Step 1: For each process p in system S^k, choose an *IS process* in system S^k. Call such a process $isp(p)$.

Step 2: For each $isp(p)$, set up a series of paths to some *IS processes*, denoted $paths(p)$. A *path* is formed by a series of subsequent FIFO-ordered links that connect a pair of *IS processes*. Such paths should have only one *IS process* per system they interconnect. Note that different paths (either from the same *IS process* or not) may share some of their links.

Step 3: Transfer the messages issued by process p (to other systems) by using $isp(p)$ through $paths(p)$.

Step 4: When an *IS process* receives a transfer, it broadcasts that message to every process within its own system.

The correctness proof of the above-mentioned architecture is very similar to the proof of Theorem 1 (only S^0 and S^1 must be changed by two arbitrary pairs of systems, say S^k and $S^{k'}$), and we omit it here.

Note that the protocol proposed in the previous section fits into the proposed architecture. However, other interconnection protocols that adhere to the proposed architecture could be implemented. Fig. 4 shows an illustrative example with four systems and three different ways of interconnecting them. In the example, we show the case where three *IS processes* in system S^0 (denoted isp_1^0, isp_2^0 and isp_3^0) are respectively used to transfer the messages issued by processes p, q and r in S^0 (i.e., $isp(p) = isp_1^0$, $isp(q) = isp_2^0$ and $isp(r) = isp_3^0$). As can be seen, isp_1^0 sets up three links directly to isp_1^1, isp_1^2 and isp_2^3 (in red). Moreover, isp_2^0 establishes a link to isp_1^2; then, this one establishes another link to isp_2^1, and this one to isp_2^3 (in black). Finally, isp_3^0 establishes a link to isp_1^3; then, this one establishes two links, one to isp_1^2 and another one to isp_2^1 (in blue).

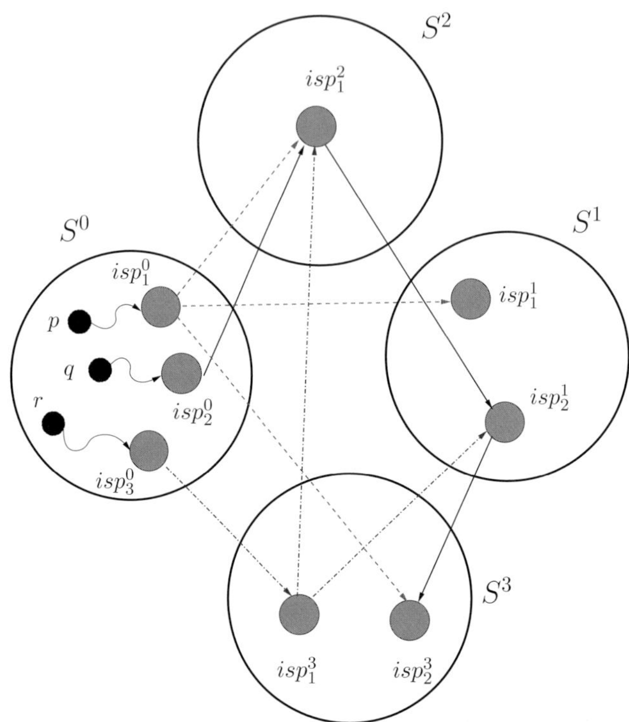

Fig. 4. Example of the architecture to interconnect FIFO-ordered systems

5 Interconnection of Causal-Ordered Systems

Contrary to what happens with totally ordered systems, that can not be interconnected in any way [6][1], and similar to what happens with FIFO ordered systems, causally ordered systems can always be properly interconnected. As in the case of FIFO ordered systems, in order to avoid bottleneck problems it would be interesting to design an interconnecting protocol that uses several *IS processes* at each system. Unfortunately, the next theorem shows than in causally ordered systems this is not possible.

Theorem 2. *Any two causally ordered systems cannot be properly interconnected if there is more than one IS process at each system, and such IS processes are not coordinated.*

Proof. By contradiction. Let us now assume the existence of a protocol that properly interconnects two causally ordered systems S^0 and S^1 such that there

[1] This result does not contradict the *FIFO forwarding* theorem of [3] that states that total order systems can be interconnected with a FIFO total order interconnecting protocol, since such a protocol needs to be intrusive; i.e., it needs to modify the regular behaviour of the local total order protocol in each system, and such degree of intrusiveness is not allowed in our system model assumptions.

are two IS processes isp_1^0 and isp_2^0 in S^0 (not coordinated in any form) and two IS process isp_1^1 and isp_2^1 in S^1 (not coordinated in any form).

Assume that a process i in S^0 issues message m. Let isp_1^0 the *IS process* that will transfer such message to isp_1^1. Now, consider a process j in S^0 that receives message m and after it, issues message m' that will be transferred, by means of isp_2^0, to isp_2^1.

It could happen that isp_2^0 transfers message m' to isp_2^1 before isp_1^0 transfers message m to isp_1^1. If isp_2^1 sends message m' before isp_1^1 sends message m, it could happen that some process in S^1 receives m' before m. This breaks causality and we reach a contradiction.

As a consequence of this theorem, if we want to interconnect causally ordered systems we are forced to use only one *IS process* per system (there are multiple samples of such protocols [3,2,4,9,6]), or to coordinate in some way such *IS processes*. Let us explore this second alternative.

As shown in the proof of Theorem 2, when multiple interconnection links are used, we need to guarantee that causally related messages are delivered in the destination system in the appropriate order. Concurrent messages do not introduce any problem, they can be delivered without any constraint. At a glance, the resulting interconnecting protocol should take care of ensuring an appropriate delivery order for causally related messages.

Most causally ordered interconnection protocols based on a single *IS process* per system simply relied on a single FIFO link in order to implement the interconnection [3,2,6] of two causal systems. If multiple *IS processes* are used, with multiple interconnection links, we might ensure that all such links deliver all forwarded messages in a global FIFO order (i.e., the messages are delivered in the receiver system in the same order they were sent from the sender system). This trivially ensures that the semantics of Theorem 2 are maintained, and also complies with the *FIFO forwarding* theorem of [3].

In order to comply with this requirement, the interconnection protocol presented in Section 3 is taken as a basis and is extended in the following way:

1. In each system S^k, one of its *IS processes* is selected as a sequencer with a deterministic criterion; e.g., that with the lowest node identifier. Let us name isp_{seq}^k such *IS process*. It maintains the number of broadcast messages, in a local variable *seq_num*, and it will assign a sequence number to each message broadcast by such system. This follows the same principle described in [14] in order to implement the Isis ABCAST protocol (with causal total order guarantees) on top of its CBCAST one (with reliable causal delivery). But there is a big difference in our approach. We do not want to extend the underlying causal broadcast protocol being used in the local system. Instead of this, we will tag with such sequence numbers the messages being forwarded by the *IS processes*; i.e., such sequence numbers are internally maintained in the interconnecting protocol, and they are completely unknown in the causal broadcast protocols being used in each interconnected system.
2. When a message m sent by any process p_i^k of the local system is delivered in its associated *IS process* (i.e., $isp(p_i^k)$), such $isp(p_i^k)$ waits until isp_{seq}^k sends

to $isp(p_i^k)$ the appropriate sequence number for m; i.e., $sn(m)$. Once $sn(m)$ is known, $isp(p_i^k)$ sends $\langle m, sn(m)\rangle$ through its interconnection link to $S^{\overline{k}}$.
3. In the system $S^{\overline{k}}$ that plays the receiver role for m, all *IS processes* also need to maintain a local variable *received* that accumulates the amount of received messages from S^k. To this end, *received* was initialised to zero, and it is increased each time a message being sent by any $isp_x^{\overline{k}}$ is delivered.
4. Once $\langle m, sn(m)\rangle$ is received by its associated $isp_j^{\overline{k}}$, such process will causally broadcast m in $S^{\overline{k}}$ as soon as $sn(m) = received{+}1$ holds in $isp_j^{\overline{k}}$. This ensures FIFO global delivery of all messages forwarded through all interconnection links between S^k and $S^{\overline{k}}$, but they can be propagated in a parallel way, so the bandwidth of such system interconnection can be greatly enhanced.

The resulting interconnecting protocol is summarised in Figure 5, as a set of four atomic concurrent tasks.

$Sequence\_out(isp_{seq}^k, m)$ :: task which is activated once $bc\text{-}recv_{isp_{seq}^k}(m)$ is executed begin $sn(m)$ =++seq_num send $\langle id(m), sn(m)\rangle$ to $isp(sender(m))$ end	$Receive(isp_v^k, m)$ :: task activated once m is received from any $isp_j^{\overline{k}}$ begin $received$++ end
$Propagate\_out(isp_v^k, m)$ :: task which is activated once $bc\text{-}recv_{isp_v^k}(m)$ is executed begin if m was sent by a process in $set_w(isp_v^k)$ then wait for receiving $\langle id(m), sn(m)\rangle$ transfer $\langle m, sn(m)\rangle$ to $link_w(isp_v^k)$ end	$Propagate\_in(isp_v^k, \langle m, sn(m)\rangle)$ :: task activated once message $\langle m, sn(m)\rangle$ is received from $S^{\overline{k}}$ begin wait until $sn(m)$ =$received$+1 $bc\text{-}send_{isp_v^k}(m)$ end

Fig. 5. The interconnecting protocol in S^k

The following theorem shows that the system S^T, obtained by connecting any two causal-ordered systems S^0 and S^1 by using this interconnecting protocol is also causal-ordered.

Theorem 3. *Any two causal-ordered systems can be properly interconnected by using the protocol in Fig. 5.*

Proof. By contradiction. Assume there are two messages m_1 and m_2 sent in system S^0 and verifying that $m_1 \rightarrow m_2$. Now, assume they are received by, say, process p_j in system S^1 in the order $m_2 < m_1$.

Due to task *Sequence_out* in S^0, it is guaranteed by the interconnecting protocol that if $m_1 \rightarrow m_2$ then $sn(m_1) < sn(m_2)$. Due to tasks *Receive* and *Propagate_in* in S^1, no process in S^1 will be able to deliver m_2 before m_1 since *Propagate_in* compels that the isp_v^1 process that receives m_2 does not broadcast such message in S^1 until it has delivered m_1 (due to their sequence numbers order, and the management of variable *received* in both tasks). If so happens, $m_1 \rightarrow m_2$ also holds in S^1 and no process can break such causal order, since it is assumed that S^1 is a causal broadcast system. As a result, all S^1 processes deliver m_1 before delivering m_2 and this raises a contradiction with the assumption given in the previous paragraph, proving thus the theorem.

6 An Architecture to Interconnect Causal-Ordered Systems

The interconnecting protocol presented in Sect. 5 is able to interconnect two causal systems. Such interconnection mechanism could be easily extended to multiple causal systems. Thus, if there is a set of causal systems $\{S^o, S^1, ..., S^{n-1}\}$ that have been interconnected in order to achieve a global causal-order system S^T, then we can interconnect another system S^n by setting one or several interconnecting links between itself and one of the systems that belong to S^T. Note that in case of setting multiple links, such links can not be set with different S_i, S_j systems of S^T since this will define cycles in the resulting global system. When a cycle exists, there will be at least two different paths for connecting two different nodes (i.e., causal-ordered subsystems) in such global system. If two paths are available in order to interconnect two different systems S^n and S^j, Theorem 2 arises again, and the implicit coordination between the *IS processes* chosen in the sending system (e.g., S^n) disappears, since each message travels along a different path and the sequence numbers of two causally related messages in S^n is not maintained in the receiver system S^j. Note that in order to preserve such dependencies in the sequence numbers all messages should be forwarded along the same path.

These constraints can be formalised in the following Lemma and Theorem.

Lemma 1. *Given a set of N causal systems $S^T = \{S^o, S^1, S^2, ..., S^{n-1}\}$ interconnected in pairs with the interconnecting protocol shown in Fig. 5, S^T is properly interconnected if it does not contain any link cycle.*

Proof. By contradiction. Let us assume that S^T is a causal-ordered global system and that there exists a cycle in S^T, and that at least two of its systems S^i and S^j belong to such cycle. If so happens, there are at least two different FIFO paths $path_1$ and $path_2$ for interconnecting S^i and S^j. At least one of such paths has a length greater than one link. Note that otherwise both paths would have been the same (a single link connecting directly S^i and S^j).

Let us assume that processes in S^i have sent two different messages m_1 and m_2 causally related in the following way $m_1 \rightarrow m_2$. So, m_1 is forwarded to S^j before m_2, but using a different path. For instance, m_1 was forwarded along $path_1$ whilst m_2 was along $path_2$. Since both paths are FIFO ordered, but they are not coordinated in order to ensure a global causal order (recall that such coordination was ensured for a single link, but not for different paths of multiple links that have traversed through different systems), it is possible that m_2 be delivered in S^j before m_1 is delivered. This breaks the causal order, and contradicts the initial assumption of S^T being a causal global system. Thus, this proves the lemma.

Theorem 4. *n causal-ordered systems $S^0, S^1, ..., S^{n-1}$, can be pair-wise interconnected with our causal IS protocol to obtain a system S^T that is also causal-ordered.*

Proof. We use induction on n to show the result. For $n = 1$ the claim is trivially true. Then, if we have a causal-ordered system S' by interconnecting systems $S^0, S^1, ..., S^{n-2}$, then we can interconnect S' and S^{n-1} in a pair-wise manner following both Theorem 3 and Lemma 1, and the resulting system S^T is causal-ordered, proving this theorem.

7 Fault Tolerance

The interconnecting protocols previously outlined can easily tolerate failures. Note that most recoverable applications (e.g., replicated databases [15]) demand *uniform* [11] delivery for broadcast messages. This means that a message is not delivered until the group communication system can ensure that it has been received by all message target processes[2]. Moreover, such message can not be "*garbage recycled*" in the sender process until such stable delivery is ensured. In [2], such uniform delivery is also taken as the key principle in order to achieve fault tolerance.

The rules to follow are these:

- An *IS process* isp^i_v does not report the uniform delivery of a message m broadcast in its system S^i until it gets a uniform confirmation from all other *IS processes* isp^j_w to which it previously forwarded m.
- A message m that has been forwarded to a system S^j is reported as uniformly delivered in such system S^j following the regular protocol being used in S^j. This means that the receiving *IS process* isp^j_w knows about such message delivery at that time, and needs to report such issue to its sender isp^i_v once this step is completed.
- If such receiving isp^j_w fails, it will be replaced by another process in S^j that will play such isp role. Two different scenarios arise:
  - The old isp^j_w was able to broadcast all messages received from isp^i_v. If so happens, the new isp^j_w will be able to report such messages as uniform, using the regular protocols of S^j. No problem arises in this case.
  - The old isp^j_w failed before being able to broadcast all messages received from isp^i_v. If so happens, system S^j does not know anything about such messages and the new isp^j_w will be unable to report any of such messages as uniformly delivered. If so happens, isp^i_v will forward again such unreported messages to (the new) isp^j_w process, once a given time-out is exhausted. Moreover, in case of interconnecting causal systems, the other *IS processes* in S^j will be blocked waiting for the delivery of those missed messages, and they will be able to tell such new isp^j_w which were

[2] In modern group communication systems [16], this constraint is relaxed: the message can be delivered as soon as it is received and complies with the intended order semantics. Later on a uniform/safe notification is delivered to the receiver process, indicating that such message delivery is already uniform/safe.

the $sn(m)$ of such messages. So, isp_w^j will be able to ask isp_v^i for the messages associated to those $sn(m)$.

- If one of the forwarding isp_v^i nodes crashes, a new isp_v^i process will be created. If there were some forwarded messages not yet reported as delivered, such new isp_v^i process knows which were such messages and it forwards them again and resets their time-outs. So, such messages are appropriately managed.

These rules avoid any message loss, and uniform delivery ensures that all broadcast messages are eventually delivered to all their destination processes. So, node failures are easily overcome.

8 Performance Analysis for Causal-Ordered Systems

The usage of FIFO interconnecting links in order to implement interconnection protocols for two causal and/or FIFO broadcast systems have been previously proposed in several papers [3,4,6]. Note also that the daisy architecture described in [2] also becomes a single FIFO link when only two causal systems need to be interconnected.

However, none of such papers explored the alternative of using more than one *IS process* per system. In most cases, the intra-system links will be far more efficient than the inter-system ones. Imagine, for instance that each system is deployed in a given laboratory or enterprise site, using a fast LAN (e.g., SCI has a bandwidth of 20 Gbps, and there are also 10Gb Ethernet LANs nowadays, with delays far below one ms in both cases), whilst inter-system links might have the regular bandwidth and delays of a WAN (less than 100 Mbps of bandwidth and more than 50 ms of delay, in most cases). As a result, the interconnecting protocols and links could be easily overloaded using a negligible workload in the systems being interconnected, since the latter are two orders of magnitude faster than the former. So, the parallelisation of such interconnections is able to multiply the resulting bandwidth without increasing the transmission delays. Such benefit is directly applicable to the protocol described in Sect. 3. It does not need any further analysis, since the interconnecting protocol does not demand any synchronisation among the *IS processes* of the systems being interconnected.

Let us concentrate in the analysis of the causal interconnection protocol described in Section 5, assuming that only two causal-ordered systems need to be interconnected. To this end, the following parameters are needed:

- *n_isps*: Number of *IS processes* in the sender system; i.e., number of interconnection links being used. Since there is a single sequencer process needed for synchronising all *IS processes* of such sender system, and such synchronisation requires a single additional message in some cases (when the *IS process* that forwards the message is not the sequencer), we need this parameter in order to set the probability of requiring such extra message.
- *ibw*: Intra-system bandwidth (in Mbps).
- *id*: Intra-system message transmission delay (in seconds).

- *sr*: The average sending rate at one broadcast system; i.e., the number of messages sent per time unit (in seconds).
- *ms*: Message size, in Mb (megabits). So, the product $sr \times ms$ is an expression that provides the required bandwidth (in Mbps). So, the following should be ensured:

$$sr < \frac{ibw}{ms} \tag{1}$$

- *ebw*: Inter-system link bandwidth (in Mbps). This parameter sets an important constraint in the global system, since the interconnection will usually be the bottleneck of such system. Such constraint is:

$$sr < \frac{ebw \times n\_isps}{ms} \tag{2}$$

- *ed*: Inter-system message transmission delay (in seconds).

Using such parameters, sections 8.1 and 8.2 evaluate the optimal number of interconnecting links and compare the performance of our parallelised interconnection with the daisy architecture proposed in [2], respectively.

8.1 Optimal Number of Links

In order to find out which is the optimal number of interconnecting links, let us explore the time needed for broadcasting a message in the whole system and how such time depends on the number of links. Thus, the implementation of a global causal-ordered broadcast only needs a system-local reliable broadcast protocol complemented with tagging all messages with vector clocks and considering such clocks in the delivery step. Such kind of protocol [14] can be implemented using a single round of messages; i.e., if there are n nodes in a system, only $n-1$ messages are needed.

Once each message is delivered in its associated *IS process*, such process needs to wait for the sequence number that should tag such message. This sequence number is sent by the sequencer process using a point-to-point message. Note, however, that the sequencer is also an *IS process*. So, this additional message is only needed with a probability of $1 - \frac{1}{n\_isps}$. Moreover, such message is smaller than all other messages considered in the next expressions, requiring thus a smaller transmission time. However, we have not considered such issue in those expressions.

Later, the message is forwarded through the interconnection link and re-broadcast in the receiving system. This implies, again, a single round of messages. So, if no additional workload is considered, the minimal time needed for receiving a message broadcast by system S_i in a node of system S_j (*trans_time*) is:

$$trans\_time = \left(3 - \frac{1}{n\_isps}\right) \times \left(id + \frac{ms}{ibw}\right) + \left(ed + \frac{ms}{ebw}\right) \tag{3}$$

Note that the constant value 3 expresses that there is a minimal number of 2 intra-system hops (the regular single broadcast round in each of the two interconnected systems) but it may reach 3 hops in the worst case (i.e., when all messages should be tagged using an explicit point-to-point message sent by the sequencer). Thus, with a negligible workload, the optimal number of *IS processes* per system is 1, since this eliminates the need of a synchronisation (intra-system) message (i.e., that carrying the sequence number), as it can be seen in expression (3).

However, when workload is considered, we could use a queueing model [17] in order to represent such system, but as it is already mentioned in [17, Chapter 5], in a queueing network we only need to identify its bottleneck centre in order to set upper bounds to the global system throughput. In our system, and with the assumptions given above, such bottleneck centre is the interconnecting channel. In the best case, the bandwidth of the interconnecting links could be the same as the internal bandwidth of each interconnected system; i.e., constraints (1) and (2) generate the same thresholds when n_isps is 1. If so happens, the optimal number of sender *IS processes* will be one, as seen above in (3). Otherwise, we need to set as many interconnecting links as given by expression (4):

$$n\_isps = \left\lceil \frac{ibw}{ebw} \right\rceil \tag{4}$$

Note that both id and ed are modeled as "*delay centres*" [17]; i.e., a link can be shared by multiple messages being forwarded along it, and we do not need a queue in order to model such delay. As a result, they do not appear in expression (4), where only queueing servers need to be considered. Indeed, such queueing centres are modelling the bandwidth of each kind of link. They have a service demand of $\frac{ms}{ibw}$ seconds in the intra-system communications and $\frac{ms}{ebw}$ seconds in the inter-system links. This explains how expression (4) is derived: its target is to balance the serving time of both kinds of servers, and this can be achieved increasing the number of interconnecting links.

8.2 Comparison with Other Solutions

To our knowledge, no other paper has proposed a parallelised interconnection of causal-ordered broadcast systems. However, the technique described in [2] can be considered as a close approach. Despite proposing a single interconnecting server for each existing system, it recommends that systems were split into multiple subsystems when they have grown excessively. So, this is an indirect way of introducing multiple interconnecting servers in each original system. Moreover, this provides the advantage of reducing the size of the vector clocks being used in each subsystem for their local broadcasts. In order to implement the global interconnection, the daisy architecture builds an upper-layer causal-ordered system composed by the interconnecting servers of all broadcast systems. Each time a message is broadcast in one of the systems, its interconnecting server re-broadcasts such message to all other interconnecting servers, who broadcast again such message to all nodes in their respective systems. So, such global broadcast consists of three different causal broadcast interactions.

Let us compare the daisy architecture with our solution described in Sect. 5. To this end, let us assume a global system where there are initially two causal-ordered broadcast systems S^0 and S^1, with $3n$ nodes each one. The intra-system bandwidth and link delays are *ibw* and *id*, respectively, whilst the inter-system bandwidth and link delays are *ebw* and *ed*, respectively. We also assume that $ibw > ebw$ and $id < ed$. Due to the size of such systems, each of them has been divided into three sets of n nodes per set using our approach, or into three separate new systems (S^{00}, S^{01}, S^{02}, and S^{10}, S^{11}, S^{12}) using the daisy architecture, again with n nodes per system.

In such scenario, the broadcast of a message m sent in S^0 with our solution implies:

- $3n - 1$ messages in order to broadcast such message into S^0.
- One additional message (in the worst case) in order to assign a sequence number to such message and notify it to the associated *IS process*.
- One inter-system message through the interconnecting link.
- $3n - 1$ messages in order to broadcast m into S^1.

Globally, this has required $6n-1$ messages (that might be $6n-2$ if the *isp* being used in S^0 is also the sequencer process) transmitted through intra-system links and one single message traversing inter-system (and slower) links. Additionally, it has required three hops in the best case, or four, in the worst one.

On the other hand, with the daisy architecture, such broadcast needs these messages:

- Let us assume that the sender of message m belongs to S^{00}. It requires $n-1$ messages in order to broadcast m into such system.
- As a result, m can be broadcast in the system composed by all interconnecting servers of the six newly created broadcast systems. Five messages are needed to this end. Two of such messages forward m to other systems that initially belonged to S^0. So, they are fast messages. On the other hand, the other three need to use the assumed slow interconnecting links.
- Into each system, $n-1$ messages are needed to locally re-broadcast m. Since there are five systems of this kind, $5n - 5$ messages are needed in this step.

At the end, this architecture needs the same global amount of point-to-point messages; i.e., $6n - 1$ messages. But in our approach, only one of such messages need to use the slow links, whilst in the daisy architecture, three messages have been forwarded through such links. This implies that with additional workload, such slow interconnecting links will be saturated sooner using the daisy approach. If the difference between the original intra-system and inter-system bandwidths is important, our solution guarantees better scalability than a daisy architecture.

On the other hand, the daisy architecture needs only three logical hops to broadcast a message between different systems, whilst our approach might need four logical hops in some cases. However, using our solution only one hop is needed into each of the initial systems S^0 and S^1 in order to locally broadcast a given message, whilst the daisy architecture introduces also three hops between processes located in different parts of such original systems.

9 Conclusions

In this paper, we have studied the interconnection of broadcast systems that are either FIFO or causally ordered. We have provided interconnection protocols that can use several interconnection links between systems, which avoid bottleneck problems due to the network traffic, since messages are not forced to go through a single link but throughout the several links we establish. Furthermore, we have proposed a general architecture with which to interconnect multiple broadcast systems. The usage of multiple interconnection links is specially convenient when scalability is a must and such interconnection links provide a limited bandwidth (compared to that of intra-system links).

References

1. Adly, N., Nagi, M.: Maintaining causal order in large scale distributed systems using a logical hierarchy. In: Proc. IASTED Int. Conf. on Applied Informatics, pp. 214–219 (1995)
2. Baldoni, R., Beraldi, R., Friedman, R., van Renesse, R.: The hierarchical daisy architecture for causal delivery. Distributed Systems Engineering 6(2), 71–81 (1999)
3. Johnson, S., Jahanian, F., Shah, J.: The inter-group router approach to scalable group composition. In: Intl. Conf. on Distr. Comp. Syst (ICDCS), Austin, TX, USA, pp. 4–14. IEEE-CS Press, Los Alamitos (1999)
4. Laumay, P., Bruneton, E., De Palma, N., Krakowiak, S.: Preserving causality in a scalable message-oriented middleware. In: Guerraoui, R. (ed.) Middleware 2001. LNCS, vol. 2218, pp. 311–328. Springer, Heidelberg (2001)
5. Fernández, A., Jiménez, E., Cholvi, V.: On the interconnection of causal memory systems. J. Parallel Distrib. Comput. 64(4), 498–506 (2004)
6. Álvarez, A., Arévalo, S., Cholvi, V., Jiménez, E., Fernández, A.: On the interconnection of message passing systems. Information Processing Letters 105(6), 249–254 (2008)
7. Mattern, F.: Virtual time and global states of distributed systems. In: Cosnard, M., et al. (eds.) Proc. Workshop on Parallel and Distributed Algorithms, pp. 215–226. North-Holland / Elsevier (1989), Reprinted in Yang, Z., Marsland, T.A. (eds.): Global States and Time in Distributed Systems, pp. 123–133. IEEE, Los Alamitos (1994)
8. Rodrigues, L., Veríssimo, P.: Causal separators for large-scale multicast communication. In: Intl. Conf. on Distr. Comp. Syst (ICDCS), Vancouver, Canada, May 1995, pp. 83–91. IEEE-CS Press, Los Alamitos (1995)
9. Kawanami, S., Enokido, T., Takizawa, M.: A group communication protocol for scalable causal ordering. In: 18th Intnl. Conf. on Adv. Inform. Netw. and Appl (AINA), Fukuoka, Japan, March 2004, pp. 296–302. IEEE-CS Press, Los Alamitos (2004)
10. Attiya, H., Welch, J.: Distributed Computing Fundamentals, Simulations and Advanced Topics. McGraw Hill, New York (1998)
11. Hadzilacos, V., Toueg, S.: Fault-tolerant broadcasts and related problems. In: Mullender, S. (ed.) Distributed Systems, 2nd edn., pp. 97–145. ACM Press, New York (1993)
12. Haldar, S., Vitányi, P.M.B.: Bounded concurrent timestamp systems using vector clocks. Journal of the ACM 49(1), 101–126 (2002)

13. Lamport, L.: Time, clocks, and the ordering of events in a distributed system. Commun. ACM 21(7), 558–565 (1978)
14. Birman, K.P., Schiper, A., Stephenson, P.: Lightweight causal and atomic group multicast. ACM Trans. Comput. Syst. 9(3), 272–314 (1991)
15. Kemme, B., Bartoli, A., Babaoglu, Ö.: Online reconfiguration in replicated databases based on group communication. In: Intl. Conf. on Dependable Systems and Networks (DSN), Göteborg, Sweden, July 2001, pp. 117–130. IEEE-CS Press, Los Alamitos (2001)
16. Chockler, G., Keidar, I., Vitenberg, R.: Group communication specifications: a comprehensive study. ACM Comput. Surv. 33(4), 427–469 (2001)
17. Lazowska, E.D., Zahorjan, J., Graham, G.S., Sevcik, K.C.: Quantitative System Performance: Computer System Analysis Using Queueing Network Models. Prentice-Hall, Inc., Englewood Cliffs (1984)

Revising 1-Copy Equivalence in Replicated Databases with Snapshot Isolation*

Francesc D. Muñoz-Escoí[1], Josep M. Bernabé-Gisbert[1], Ruben de Juan-Marín[1], Jose Enrique Armendáriz-Íñigo[2], and Jose Ramon González De Mendívil[2]

[1] Instituto Tecnológico de Informática
Univ. Politécnica de Valencia
Camino de Vera, s/n
46022 Valencia, Spain
{fmunyoz,jbgisber,rjuan}@iti.upv.es

[2] Depto. de Ing. Matemática e Informática
Univ. Pública de Navarra
Campus de Arrosadía, s/n
31006 Pamplona, Spain
{enrique.armendariz,mendivil}@unavarra.es

Abstract. Multiple database replication protocols have used replicas supporting the snapshot isolation level. They have provided some kind of one-copy equivalence, but such concept was initially conceived for serializable databases. In the snapshot isolation case, due to its reliance on multi-versioned concurrency control that never blocks read accesses, such one-copy equivalence admits two different variants. The first one consists in relying on sequential replica consistency, but it does not guarantee that the snapshot used by each transaction holds the updates of the last committed transactions in the whole replicated system, but only those of the last locally committed transaction. Thus, a single user might see inconsistent results when two of her transactions have been served by different delegate replicas: the updates of the first one might not be in the snapshot of the second. The second variant avoids such problem, but demands atomic replica consistency, blocking the start (i.e., in many cases, read accesses) of new transactions. Several protocols of each kind exist nowadays, and most of them have given different names to their intended correctness criterion. We survey such previous works and propose uniform names to these criteria, justifying some of their properties.

1 Introduction

Consistency has been thoroughly studied in parallel and distributed systems, mainly in those with shared memory, generating a set of consistency models [1,2].

* This work has been partially supported by EU FEDER and the Spanish MEC under grant TIN2006-14738-C02, by EU FEDER and the Spanish MICINN under grant TIN2009-14460-C03 and by IMPIVA under grant IMIDIC/2007/68.

R. Meersman, T. Dillon, P. Herrero (Eds.): OTM 2009, Part I, LNCS 5870, pp. 467–483, 2009.

A replicated database can be considered as an example of such kind of systems, since all replicas hold copies of the same data that should be kept consistent, building thus a specialized kind of logical shared memory, although commonly implemented in a shared-nothing set of nodes. *One-copy serializability* [3] (a.k.a.*1SR*) has been the commonly accepted correctness criterion for replicated databases, since it was enough relaxed for ensuring good performance and strict-enough for guaranteeing a comfortable replica consistency model for the application programmer. But this 1SR single concept encompasses two different issues. First, the *isolation level* being responsible for the isolation consistency among all concurrent transactions being executed in the system and, second, the *replica consistency*; i.e., the degree of admissible divergence among the states of all replicas.

In the last twenty years several things have changed that yield some opportunities for revising such issues. Regarding the first one (isolation), new levels have been defined [4,5], being *Snapshot Isolation* (SI) one of the most important, since it is quite close to the serializable level and it does not need to block read accesses due to its reliance on multi-versioned concurrency control mechanisms. Multiple *Database Management Systems* (i.e., DBMSs) –like Oracle, PostgreSQL, MS SQL Server,...– support the SI level, and even some of them label it as serializable although there are some isolation anomalies [4] that can not be avoided with SI. Despite this, with some care [6] a DBMS supporting SI is able to ensure serializable isolation. Considering the second issue (replica consistency), 1SR is considered [1] equivalent to a *sequential* [7] consistency model, but some applications might require either stricter models like the *linearizability* semantics proposed in [8](or *atomic* memory model; we will use this latter term in the sequel, since linearizability is a correctness condition for objects that enforces a non-blocking behavior and such characteristic might not be achieved in a database replication system. Some papers, e.g. [9], refer to both concepts as synonyms when they are applied to *distributed shared memory*, or DSM, systems.) –providing thus an easier programming model– or more relaxed models like the *cache* [10] one, improving thus system performance. So, it is interesting to analyze how such other replica consistency models could be combined with snapshot isolation in order to generate new *isolation+replica consistency* models for replicated databases.

Thus, the contributions of this paper consist in surveying different combinations between the snapshot isolation level and replica consistency models (adding cache and atomic semantics to the traditionally accepted sequential consistency). Note that such combinations have been already used in previous works (e.g., both the *Strong SI* level in [11] and the *Conventional SI* of [12] can be considered a combination between the SI isolation level and the atomic replica consistency model), but using different names in order to refer to the same things. As a result of this survey, we provide a new taxonomy of combined consistency models as our second contribution. As a final contribution we show that this new taxonomy is able to justify some protocol properties that are difficult to prove otherwise.

The rest of this paper is structured as follows. Section 2 describes the assumed system model, integrating transactions –and, as a result, isolation– in the traditional replica consistency models. Section 3 summarizes the isolation and replica consistency models commonly assumed in modern database replication protocols. Later, Section 4 presents a new taxonomy of models that combine both isolation and replica consistencies, although restricted in this paper to *snapshot* isolation. Section 5 shows that such taxonomy easily justifies some unproven properties of modern database replication protocols. Finally, Section 6 concludes the paper.

2 System Model

We consider a distributed system composed by a set N of nodes, interconnected by a network. Each system node has a local DBMS able to manage the *snapshot isolation* (SI) level. Replication is being managed by a middleware layer using some *Read-One Write-All-Available* (ROWAA) [3] replication protocol that should take care of guaranteeing the intended global isolation level and of ensuring some replica consistency model. Modern replication protocols [13] are based on executing transactions in a single delegate replica and propagating later the transaction updates in FIFO total order to all other replicas in the commit procedure. We assume such behavior in this paper.

Replica consistency models could be any of the traditional DSM ones. It is worth noting that these latter models have usually been specified considering as relevant events both *read* and *write* operations applied to a given set of variables shared among multiple processes. Such specifications cannot be trivially migrated to a system where the reads and updates are made in the context of transactions, since such transactions encompass multiple individual operations of that traditional kind. So, in order to discuss such replica consistency models, we need to adapt the transaction concept to the equivalent sequence of read and write operations. This is feasible when a strict-enough isolation level is being assumed.

Thus, we will transform transactions into sequences of operations in the following way: each transaction read access will be logically advanced till the transaction start time, sharing all read operations the same logical time, whilst all update accesses will be logically put at the transaction commit time, as a single multi-variable write operation. Note that only database accesses are being considered here. So, once an update on a given item has been made, the application program will be able to know which will be that item new value, e.g., using local variables to this end, without requiring another physical read access on the database item. If any of such read accesses is being made in a given transaction, it will be eliminated in the resulting mapping. As a consequence, such a reordering of read and write events can be made on all transactions. Note also that only committed transactions need to be considered, since aborted ones do not have any effect on the database state once they have terminated.

3 Isolation and Replica Consistency

Although there are multiple isolation levels [4,5] supported by modern DBMSs, only few of them have deserved attention in order to implement database replication protocols. Concretely, such isolation levels are the *serializable* [3] and the *snapshot* [4] ones. Other more relaxed levels have not been widely considered in research papers. In the best cases, they have been included [4,5] in order to carefully specify such levels or for supporting them [14] in order to provide replication transparency. In this paper, as its title suggests –and due to space constraints–, we will only focus on the *snapshot* isolation level, although the general principles described here could be also applied to other isolation levels (like the serializable one) generating thus a broader taxonomy.

A similar scenario can be found regarding replica-consistency models. Although there are many models that could be considered [1,2], only a few of them (cache, sequential and atomic) have been assumed in database replication. For instance, strict consistency levels (e.g., the atomic one) ensure consistency in some kinds of applications where a single client is able to access different copies of a given item in a short interval, like in three-tiered web applications [15]. In such context, without atomic consistency, the last client request might be executed before the effects of previous requests were applied in its serving replica. So, clients could get inconsistent replies. On the other hand, new data management trends [16,17] (e.g., in the *cloud computing* field) suggest that temporary inconsistencies should be afforded by modern applications, and that this should be considered a strong requirement when scalability is a must. So, they advocate for relaxed consistency models based on asynchronous update propagation. As a result, these two opposite trends show that multiple replica consistency models need to be considered.

Complete specifications of such isolation levels and replica-consistency models can be found in [5] and [9], respectively. We will summarize them in the following two sections.

3.1 Isolation

In order to specify an isolation level, most works [3,4,5] have used *transaction histories* composed of a partial order of transactions' events and a total order on the committed item versions generated by such transactions. Taking such histories as a base, a *dependency* [4] or *serialization* [5] *graph* is built, using transactions as its nodes and transaction dependencies as its edges. Several isolation phenomena are specified (describing which set of dependencies must exist in the graph for each kind of phenomena). Finally, an isolation level is respected when each considered transaction avoids a given subset of phenomena.

Thus, following [5]'s conventions, the possible transaction dependencies ($T_i \rightarrow T_j$) are:

- T_j *directly write-depends on* T_i ($T_i \xrightarrow{ww} T_j$) when T_i installs X_i and T_j installs X's next version.
- T_j *directly read-depends on* T_i ($T_i \xrightarrow{wr} T_j$) when T_i installs X_i, T_j reads X_i or T_j performs a predicate-based read, X_i changes the matches of T_j's read, and X_i is the same or an earlier version of X in T_j's read.

- T_j *directly anti-depends on* T_i ($T_i \xrightarrow{rw} T_j$) when T_i reads X_h and T_j installs X's next version or T_i performs a predicate-based read and T_j overwrites this read.
- T_j *start-depends on* T_i ($T_i \xrightarrow{s} T_j$) when $c_i < s_j$; i.e., when it starts after T_i commits. When start dependencies are considered, the resulting graph is named a *start serialization graph* (SSG(H)).

And the phenomena to be considered in order to specify SI are:

- *G1a: Aborted Reads.* A history H shows phenomenon G1a if it contains an aborted transaction T_1 and a committed transaction T_2 such that T_2 has read some object modified by T_1.
- *G1b: Intermediate Reads.* A history H shows phenomenon G1b if it contains a committed transaction T_2 that has read a version of object X written by transaction T_1 that was not T_1's final modification of X.
- *G1c: Circular Information Flow.* A history H exhibits phenomenon G1c if its serialization graph contains a directed cycle consisting entirely of dependency (i.e., write-dependencies or read-dependencies) edges.
- *G-SIa: Interference.* A history H exhibits phenomenon G-SIa if SSG(H) contains a read/write-dependency edge from T_i to T_j without there also being a start-dependency edge from T_i to T_j.
- *G-SIb: Missed Effects.* A history H exhibits phenomenon G-SIb if SSG(H) contains a directed cycle with exactly one anti-dependency edge.

A history respects the snapshot isolation level when G1a, G1b, G1c, G-SIa and G-SIb phenomena are proscribed. This has been ensured in stand-alone database systems using multi-versioned concurrency control, combined in some cases with write locks.

Transaction Validation Rules. In a replicated setting, a database replication protocol is needed. As already outlined above (in Section 2), most of these protocols execute initially the transactions in a single delegate node, collecting their writeset when commit is being requested (but before processing such request) and multicasting it to all replicas (usually, in FIFO total order) [18,19]. Once such updates are delivered at their target nodes, a validation stage is executed. If transactions overcome such validation, they are applied in the underlying database. To this end, the concurrency control mechanisms commented for stand-alone databases could still be useful; i.e., they may provide information about conflicts between the transaction whose writeset is being applied (and that should be committed) and other in-course transactions that might block such writeset application. If so arises, such local in-course transactions are aborted.

Using this approach, the replication protocol needs only be concerned with the validation rules followed in order to guarantee an isolation level. Thus, the validation rules for snapshot isolation [20,12,21] consist in detecting write-write conflicts between the transaction being validated and other already-committed concurrent transactions (considering concurrent those pairs of transactions that do not have a start dependency). To this end, all replication protocol classes need to propagate transaction writesets and the logical transaction start timestamps. Thus, phenomena G1a, G1b and G-SIa are avoided by the local concurrency control mechanisms used in each replica, whilst the validation rule prevents phenomena G1c and G-SIb from appearing, since no cycle of dependencies is allowed by such rule.

3.2 Consistency Models

The three replica-consistency models that have been used in common database replication protocols have been informally specified [1] as follows:

- *Atomic consistency*: Operations take effect at some point in an operation interval. Such intervals divide time into non-overlapping consecutive slots. This implies that when a process has read a given value (or version) of a concrete item, no other process could read afterward any of such item's previous versions.

  We do not demand a strict compliance to the atomic consistency semantics in this paper since they are difficult to achieve in a practical deployment, but at least that the following property is guaranteed avoiding thus the problems mentioned in [15]: *If a single client forwards a transaction t_a to a replica r_i and gets the result of t_a, any other transaction t_b sent later by this same client to any other replica r_j should be able to read the updates caused by t_a*, assuming that no other transaction is submitted to the system between t_a and t_b.
- *Sequential consistency*: This model was defined in [7] as follows: *The result of any execution is the same as if the operations of all the processors were executed in some sequential order, and the operations of each individual processor appear in this sequence in the order specified by its program.* So, it can be implemented using FIFO total order for applying all write operations in system replicas.

  Note that this does not avoid the problem outlined in [15], since sequential consistency ensures that all updates will be applied following the same sequence in all replicas. However, if replica r_j is overloaded and holds a long queue of pending updates (to be applied in the database), it might serve the first read accesses of t_b before applying the updates of t_a and, of course, before locally committing t_a.
- *Cache consistency* [10]: This model only requires that accesses are sequentially consistent on a *per-item* basis.

  There are some replication protocols (as listed in Section 4.1) that are able to comply with the requirements of this model but provide a consistency slightly higher, but that does not correspond to any already specified model. Such protocols are based on total order update propagation, but they allow that writeset application breaks such total order when writesets do not conflict (i.e., there are no write-write conflicts) with any of the previously-delivered but not-yet-committed transactions. Note that this ensures a per-item sequential consistency (as requested in the *cache* model), but also a per-transaction-writeset consistency (i.e., we can not commit half of a writeset WS_A before writeset WS_B and the other half of WS_A afterward), although not a complete sequential consistency.

Note that some authors [22] only admit the *atomic* and *sequential* consistency models when talking about 1-copy-consistency. So, in such context, a *cache* system is not allowed. However, we include it in this paper in order to state that other models more relaxed than the sequential one could make sense if performance is a must.

4 One-Copy Equivalence

One-Copy Equivalence was introduced in [3], tailored for the serializable isolation level, as a correctness criterion for replicated databases. Its aim is to ensure that the interleaved execution of multiple transactions in a replicated database system were equivalent to a unique (and serial, for that isolation level) execution in a logical single computer.

Our proposal consists in the extension of such equivalence concept to other isolation levels, considering also different replica-consistency models. There have been many proposals of this kind, but they were focused in a single isolation+replica consistency combination. For instance, we can find multiple definitions [12,21,11] of what should be understood as 1C-SI, and none of them does exactly match the others. So, it seems appropriate to carefully state how such different kinds of consistency (isolation-related and replica-related) can be merged, giving neutral names to the resulting combinations, in order to promote their acceptance. Our proposal explicitly states which is the actual combination of replica-consistency model and isolation level when a *one-copy equivalence* is being provided. If we only consider the currently existing SI database replication proposals, the resulting models are summarized in Table 1. Note that modern database replication approaches are based on FIFO total order update propagation and application into the replicas. This naturally provides a sequential replica-consistency model. So, we do not consider mandatory to specify such replica-consistency model when stating a one-copy equivalence model, being it the default one.

In Sections 4.1 through 4.3, we survey some of the existing database replication protocols that support each one of such isolation+consistency models, presenting the name they associate to such combination of isolation and consistency. The aim of these sections is to illustrate that, in some cases, a given combination has received different names in different papers. So, it seems convenient to propose and promote a standard name for such models.

Table 1. One-Copy SI Equivalence Models

Replica-Consistency Model		
Cache	Sequential	Atomic
1C-Cache-SI	1C-SI	1C-Atomic-SI

4.1 1C-Cache-SI Model

Lin et al. [21] proposed a protocol named SRCA-Opt that implements the 1C-Cache-SI model. Note however that such protocol is not proposed as the main contribution of such paper. Indeed, its aim is to propose a valid 1C-SI criterion (that follows also our assumptions and ensures 1C-Sequential-SI) and a protocol that implements such criterion (its SRCA-Rep one). In its performance evaluation section, they propose a variant of such SRCA-Rep protocol with relaxed replica consistency, and the authors are aware of that issue. One of the aims of such evaluation is to compare how these two kinds of replica consistency (sequential and cache) are able to cope with increasing workloads. Their results show that with relaxed replica consistency, and for the benchmark used in such paper with an update-intensive workload, SRCA-Opt (i.e., the protocol with cache consistency) always provides shorter transaction completion times than SRCA-Rep (i.e., the variant with sequential consistency), and such differences increase with the load.

Note, however, that in such paper no reference is given to the kind of replica consistency being guaranteed by such SRCA-Opt protocol. Indeed, a single comment is given referring to the relaxation provided:

> *However, in update intensive workloads, SRCA-Opt might be a better alternative even it does not provide full 1-copy-SI. This might be comparable with approaches in centralized systems where at high workloads lower levels of isolation are chosen (e.g., READ COMMITTED) to speed up performance.*

And, it simply compares such performance improvement with that achievable by relaxed isolation levels; i.e., it indirectly says that, among others, there are two ways for improving performance: to relax replica consistency and to relax isolation. This confirms that both kinds of consistency (isolation-related and replica-related) should be considered in order to specify a *one-copy equivalence* model.

4.2 1C-(Sequential-)SI Model

The first SI replication protocols [20,21,12] were based on sequential replica consistency, since such consistency model was already widely accepted for the serializable isolation level as part of its 1SR correctness criterion. Such protocols used the ROWAA approach and propagated transaction updates in a single total-order broadcast, eliminating the need of a traditional distributed commit protocol (either two-phase or three-phase), following the principles of the last protocol variant suggested by [18], that was proven correct for implementing 1SR in [19].

Lin et al. [21] have the merit of being the first providing a sequential specification for SI replication, and naming it as 1C-SI (as suggested in the current paper), based on the 1SR one given in [3]. Moreover, in that same paper they propose a 1C-SI protocol named SRCA-Rep and analyze its performance, as we have discussed above. However, they did not explicitly state in any part of their specification that the replica consistency being assumed was the sequential one.

On the other hand, [12] was the first paper that discussed the differences between the guarantees being provided by common SI replication protocols and those provided by stand-alone SI databases, since the notion of *latest snapshot* is different in those environments. Thus, they identify two SI variants:

- *Conventional Snapshot Isolation* (CSI) refers to stand-alone implementations, where it is trivial to guarantee that such *latest snapshot* being used by every starting transaction corresponds to the one generated by all previously committed transactions. Elnikety et al. [12] do not recommend CSI for replicated settings, since this might require blocking the start of new transactions in order to guarantee that their delegate replicas receive the updates of all previously committed transactions.
- *Generalized Snapshot Isolation* (GSI) refers to the common replicated scenario guaranteeing sequential replica consistency. In it, the delegate replica where a transaction is executed does not need to maintain all the updates of all previously committed transactions. This may happen when such updates are still in transit or buffered in such receiving replica but not yet applied. So, the *latest snapshot* for a given transaction will be that of its delegate serving replica, but such latest snapshot does not hold all the updates generated by all transactions committed in the system.

Thus, [12] proposed GSI as a new concept, distinguishing it from CSI, but such paper did not state that their differences were derived from different consistency models (sequential for GSI and atomic for CSI).

Similar differences were identified by [11]. However, such paper proposed other names in order to refer to similar concepts. Thus, its *Global Weak Snapshot Isolation* refers to our 1C-SI model, whilst *Global Strong Snapshot Isolation* refers to our 1C-Atomic-SI model. Its authors carefully identify the main problems of both models: usage of past snapshots in 1C-SI and expensive/blocking protocols in 1C-Atomic-SI, agreeing with the proposal of [12]. In order to solve such problems, [11] proposes an intermediate model: *Strong Session Snapshot Isolation*, that does only require *strong SI* semantics among the transactions of a given session, whilst transactions belonging to different sessions do only require *weak SI* semantics. A session holds the transactions being generated by a given client.

Wu and Kemme [23] provide a thorough description of the database concurrency control needed in order to implement SI in PostgreSQL. They propose a database replication protocol embedded into the DBMS core, guaranteeing thus a very good performance since transaction validation in the replication protocol can delegate many of its functions to the original DBMS concurrency control. The paper also provides a detailed justification about how replica consistency is being guaranteed by such replication protocol, although it does not mention that the resulting consistency is sequential.

A similar principle was used in [24]; i.e., to delegate conflict evaluation in the certification step to the underlying DBMS, but implementing the protocol in a middleware layer, enhancing thus the portability and maintainability of the resulting protocol, without severely compromising its performance.

In [13], a classification of modern database replication protocols based on total order was given. There are four main classes: active, passive, certification-based, and weak voting. In the *active* one, all transaction operations need to be delivered to all replicas before such transaction execution is started. Once this is done, each replica is able to directly execute such sequence of operations and to commit or abort locally such transaction without further interaction with other replicas. To this end, the transaction logic should be deterministic. In the *passive* variant, all transactions are completely executed in a single primary replica that propagates the transactions' updates at commit time to all other backup replicas. In the *certification-based* class, a transaction is executed in a single delegate replica (but different transactions may select different delegates). Once the transaction requests its commit, its writeset is collected and multicast to all replicas in total order. At delivery time, such writeset is certified against all other delivered and concurrent writesets (i.e., those belonging to transactions that committed while the transaction being certified was executed). If a write-write conflict is found, the transaction being certified is aborted. Otherwise, it is committed. Note that all replicas hold the same historic list of previously delivered writesets. So, they are able to certify each transaction without exchanging more messages with other replicas. Finally, in the *weak voting* variant, transactions are served by delegate replicas like in the *certification-based* class, and their writesets are multicast to all other replicas at commit time, but the conflict evaluation is only done in the delegate replica. In this case, no historic list of previously delivered writesets is needed. This protocol family is able to check for conflicts when each one of the remote transactions accepted by the protocol is being committed in each replica. If such commit is blocked by a local transaction that maintains a write lock on any of the updated items, such local transaction is immediately aborted. So, when the writeset of a transaction t_i is delivered in its delegate replica, if t_i has not been aborted by any previously committed remote transaction, its delegate replica will reliably broadcast a $commit(t_i)$ message to all replicas; otherwise, it broadcasts an $abort(t_i)$ one. All replicas act according to such final message in order to determine the fate of t_i.

Most of the protocols cited up to now belong to the *certification-based* class [20,21,12,23,24], since it does not demand a second broadcast in order to certify a transaction. According to [13], both *certification-based* and *weak voting* classes are able to provide the best transaction completion time. The *weak voting* class has the advantage of removing the need of a historic list of delivered writesets for certifying transactions. Note that such list could demand a lot of memory in case of dealing with long transactions, although this seldom arises. However, it introduces the problem of needing two separate broadcasts for managing each transaction. Despite this, its usage has been considered in some papers. Thus, the voting protocol of [25] is a sample of this class and it ensures 1C-SI, since the *uniform data store* assumed in such paper was implemented using PostgreSQL, whilst its non-voting protocol belongs to the *certification-based* class. Another example of weak protocol can be found in [26], where three different correctness criteria are supported: 1C-sequential-SI, 1C-atomic-SI and 1SR [3]. To this end,

1C-sequential-SI is implemented with the solution presented in the previous paper, and 1C-atomic-SI is supported extending such protocol with the pessimistic approach of [15] (detailed in Section 4.3). On the other hand, the third criterion is the traditional 1SR (i.e., 1C-sequential-serializable using our naming conventions), that can be implemented extending the 1C-sequential-SI variant with the mechanisms suggested in [6].

Finally, there have been other papers following the *passive* replication protocol class. The Ganymed middleware [27] is one of such systems. Its scheduler demands each transaction to be tagged with a *read-only* or *update* label. Depending on its label, transactions should be forwarded to the primary replica (the update ones) or can be directly served by any secondary replica (read-only transactions). This approach simplifies concurrency management, since write-write conflicts may only arise in the primary replica and can be dealt with the underlying DBMS concurrency control mechanisms without requiring any inter-replica interaction. On the other hand, since read-only transactions are served by secondary replicas, they do not interfere with the update load service and thus scalability is greatly enhanced. This same architecture was assumed in [11].

4.3 1C-Atomic-SI Model

As we have seen in Section 4.2, at least two papers have referenced this model: [12] for its CSI, and [11] for its Strong SI. The former provided an analytical evaluation based on an abstract protocol that added a blocking starting step to the concrete protocol presented for supporting GSI, but none of these papers presented a complete algorithm that implements 1C-Atomic-SI.

An implementation of this model needs to re-force a 1C-SI protocol in order to support atomic replica consistency instead of sequential replica consistency, but this is not easy. There have been some general protocols (independent of the isolation level being supported) able to provide such kind of support. The first approaches can be found in [15], where two different protocols based on DBSM [28] were described. Note that DBSM was able to support both *snapshot* and *serializable* isolations, as described in [29]. Thus, the first protocol of [15] uses an optimistic evaluation: read accesses are considered in the commit-time evaluation steps of the protocol, so both read-only and read-update transactions might abort if they have accessed past versions of their items. In its second protocol a pessimistic approach is used. It is based on multicasting in total order a transaction START message that ensures that each transaction is able to get its intended last snapshot; i.e., that all previously finished transactions have delivered their writesets in the delegate replica of the starting transaction. Let us assume that r_i is such delegate replica for a starting transaction t_i. This approach is able to approximate atomic semantics, since it ensures that when t_i is allowed to start, all transactions that were in their committing step have been able to deliver their writesets in r_i; i.e., there will not be any transaction being terminated that had their writeset "*in transit*" and that will not be known in the snapshot taken by t_i. Note that such "*in transit*" transactions might have terminated in some of the replicas and could be read by other starting transactions. So, their effects need to be present in the snapshot of t_i in order to follow the atomic consistency semantics. However, to precisely implement this solution we need to delay again the start of t_i until all such delivered writesets have been positively certified and applied in r_i, and this might imply a long interval in overloaded replicas.

A second paper that ensures 1C-Atomic consistency for replicated databases is [30]. Its *Write-Consensus Read-Quorum* (WCRQ) protocol is based on read-write quorums, minimizing thus the communication costs for guaranteeing such kind of replica consistency. To this end, writesets are broadcast to all database replicas, but in order to accept an update transaction, only a write-quorum of positive acknowledgments is needed. On the other hand, read-only transactions need to be checked in a read-quorum of replicas, and they are accepted when all such replicas return a positive acknowledgment. Let us note that such positive acknowledgments require that the versions read or written respect all the constraints imposed by the atomic consistency model; i.e., that the read values correspond to the latest existing item versions, and that the write order is the same in all replicas.

Another algorithm supporting snapshot isolation and atomic consistency was presented in [31], where the 1C-Atomic-SI model was named *Strict Snapshot Isolation* (SSI). However, instead of using a blocking step as suggested in [12], this last solution uses an optimistic approach: all transactions are allowed to start without blocking, although they need to multicast in total order a message in order to find out whether such starting phase complies with SSI; i.e., when such START message is delivered, the logical transaction starting point is set. When the transaction requests commitment, its readset is compared against those writesets delivered before its START message but not included in its snapshot. If a non-empty intersection is found, the transaction is aborted. This approach eliminates all the delays presented in our description of the pessimistic protocol of [15], although at the price of aborting all transactions that need to read any of the items updated in such hypothetical blocking interval.

Finally, [32] presents an interesting set of results discussing the overhead implied by supporting several correctness criteria in a single protocol, showing that stricter criteria do not always imply any noticeable overhead. Such work is focused on the *serializable* isolation level, so it does not explicitly cover the target of this section (1C-Atomic-SI protocols), but it includes three different correctness criteria for *serializable* isolation: 1SR (i.e., 1C-sequential-serializable, following our naming recommendations), 1C-session-serializable (based on the specifications given in [11]), and *strong serializable* [33] (that would correspond to 1C-atomic-serializable with our recommendations). The mechanisms needed for assuring the 1C-atomic-serializable model are similar to those already described above: to multicast in total order a message that allows the transaction start, even in read-only transactions.

5 Applicability

One of the results of this paper consists in finding two different and complementary issues in the consistency being ensured by database replication protocols: isolation consistency and replica consistency. Different papers have given different names to the correctness criteria assumed in their protocols. A first aim of this paper is to propose a uniform naming for those criteria. We also consider that traditional database replication protocols ensure sequential consistency and that such fact should be proved. On the other hand, this also allows to revisit DSM systems, looking for theoretical results that justify some of the properties exhibited by some database replication variants. These issues are detailed in Sections 5.1 to 5.3.

5.1 Naming

Table 2 summarizes the names given to their assumed correctness criteria in all surveyed papers where some SI-related database replication protocol is described. Note that in some papers more than one protocol is proposed or more than one isolation level is being supported by a single protocol. We only consider sequential (assumed as default) and atomic (abbreviated as "at") replica consistency in that table, but we also add the serializable (abbreviated as SER) isolation level for completeness, since some protocols or papers also discuss it.

In such table, GSI stands for *Generalized Snapshot Isolation* and CSI for *Conventional Snapshot Isolation*, whilst the well-known 1SR acronym means 1C-serializability.

Note that for the *serializable* isolation level there is no possible ambiguity since the concept of 1SR [3] (1C-sequential-serializable) has been widely accepted. On the other hand, when *snapshot* isolation is considered in a replicated context, multiple names have been used and referring only to SI is ambiguous. The aim

Table 2. Names given to the correctness criteria

Paper		1C-SI	1C-at-SI	1C-SER	1C-at-SER
Bernstein et al., 1987	[3]	–	–	1SR	–
Daudjee & Salem, 2006	[11]	weak SI	strong SI	–	–
Elnikety et al., 2005	[12]	GSI	CSI	–	–
Juárez et al., 2007	[26]	GSI	SI	SER	–
Kemme & Alonso, 2000	[20]	SI	–	SER	–
Lin et al., 2005	[21]	1C-SI	–	–	–
Muñoz-Escoí et al., 2006	[24]	GSI	–	–	–
Plattner & Alonso, 2004	[27]	SI	–	–	–
Salinas et al., 2008	[31]	GSI	strict SI	–	–
Wu & Kemme, 2005	[23]	SI	–	–	–
Zuikevičiūtė & Pedone, 2005	[29]	SI	–	1SR	–
Zuikevičiūtė & Pedone, 2008	[32]	–	–	1SR	strong SER

of our paper is to avoid such ambiguity, promoting a naming that refers to both kinds of consistency when a correctness criterion is used.

5.2 Justification of Some Protocol Properties

An example of such properties is one of the propositions presented (but not proved) in [12] regarding *Conventional Snapshot Isolation*(CSI) (i.e., 1C-Atomic-SI). Note that such paper assumed a pessimistic protocol behavior, and its *Proposition 3* says: *There is no non-blocking implementation of CSI in an asynchronous system, even if database sites never fail.* On the other hand, GSI (i.e, 1C-Sequential-SI) does not demand such blocking implementations. In such scope, "*blocking*" means that transactions are prevented from starting for some time. That proposition can be directly proven using some of the results generated in the *atomic* replica consistency model. For instance, [34] proves that in a *linearizable* (i.e., atomic) replica consistency model, the minimum worst-case time for a read operation is at least $u/4$ whilst the worst-case time for a write operation is at least $u/2$, being u the uncertainty in the message transmission delay ($u > 0$). So, both kinds of operations have a blocking interval in such model. On the other hand, the same paper proves that in a sequentially consistent system either read or write operations can be immediately completed, but not both. In practice, in the database replication field, read operations can be immediately served by the local replica, whilst write operations demand a blocking interval to deliver the updates being propagated by a total order broadcast. So, such results are able to directly justify the blocking differences between the GSI and CSI concepts presented in [12].

Note that there may be database replication protocols that overcome such blocking constraint; for instance, those based on optimistic management [31,15]. However, this is achieved by transactions that in their certification phase will be sanctioned to abort if they have violated the replica consistency model properties; i.e., they do not block at their start, but if they read data from an obsolete snapshot, they will abort. The blocking behavior of a pessimistic management prevents such kinds of abort from happening, ensuring always that the adequate snapshot is being read.

5.3 Sequential Consistency

Mosberger [1] states the following referring to the sequential consistency model and one-copy serializability:

> *In a sequentially consistent system, all processors must agree on the order of observed effects.*
> *This is equivalent to the one-copy serializability concept found in work on concurrency control for database systems [3].*

However, such paper does not prove the second sentence, although such proof is almost immediate (and have been widely assumed as such) when the 1SR definition is consulted.

Such definition says (Theorem 8.3 in [3, Page 275]):

> Let H be an RD history. If H has the same reads-from relationships as a serial 1C history H_{1C}, where the order of transactions in H_{1C} is consistent with SG(H), then H is 1SR.

Given that:

1. An RD history H is a *complete replicated data history* [3, Pages 271-272] that includes all the operations executed in every system replica and that maintains the execution order in every transaction and replica.
2. The RD history definition also compels to maintain the order between conflicting operations being executed by different transactions. A pair of operations conflict if at least one of them is a write.
3. The 1SR definition uses a logical H_{1C} history whose transaction order is consistent with that of H, with the same aim as the "*...in some sequential order...*" clause of the sequential consistency definition; i.e., to define a logical global order that matches the program order of each system process.

...such three elements set a clear correspondence between the consistency issues of 1SR and those of the sequential consistency model, justifying the statements given in [1].

6 Conclusions

Sequential consistency was assumed in the first one-copy equivalence targeted for replicated databases: one-copy serializability. Snapshot isolation is another isolation level widely used in modern applications since it does not need to block read operations and is also able to ensure serializability with some care. Thus, multiple database replication protocols have supported snapshot isolation in a replicated environment, but there is no consensus on what should be understood as one-copy equivalence when such isolation level is used. Almost each paper has given a different name to its assumed correctness criterion.

We have surveyed multiple database replication papers that provide such isolation level, and we have proposed a taxonomy in order to refer to one-copy equivalence. To this end, multiple variants have been distinguished depending on the assumed replica consistency model. This permits an easy justification of some protocol properties, since they depend on both the isolation level and the consistency model. Existent properties in such two fields are able to justify the behavior of such protocols.

References

1. Mosberger, D.: Memory consistency models. Operating Systems Review 27(1), 18–26 (1993)
2. Adve, S.V., Gharachorloo, K.: Shared memory consistency models: A tutorial. IEEE Computer 29(12), 66–76 (1996)

3. Bernstein, P.A., Hadzilacos, V., Goodman, N.: Concurrency Control and Recovery in Database Systems. Addison Wesley, Reading (1987)
4. Berenson, H., Bernstein, P., Gray, J., Melton, J., O'Neil, E., O'Neil, P.: A critique of ANSI SQL isolation levels. In: Proc. of the ACM SIGMOD International Conference on Management of Data, San José, CA, USA, pp. 1–10. ACM Press, New York (1995)
5. Adya, A.: Weak Consistency: A Generalized Theory and Optimistic Implementations of Distributed Transactions. PhD thesis, Massachusetts Institute of Technology, Cambridge, MA, USA (1999)
6. Fekete, A., Liarokapis, D., O'Neil, E.J., O'Neil, P.E., Shasha, D.: Making snapshot isolation serializable. ACM Trans. Database Syst. 30(2), 492–528 (2005)
7. Lamport, L.: How to make a multiprocessor computer that correctly executes multiprocess programs. IEEE Trans. Computers 28(9), 690–691 (1979)
8. Herlihy, M., Wing, J.M.: Linearizability: A correctness condition for concurrent objects. ACM Trans. Program. Lang. Syst. 12(3), 463–492 (1990)
9. Cholvi, V., Bernabéu, J.: Relationships between memory models. Information Processing Letters 90, 53–58 (2004)
10. Goodman, J.: Cache consistency and sequential consistency. Technical Report 61, IEEE Scalable Coherence Interface Working Group (1989)
11. Daudjee, K., Salem, K.: Lazy database replication with snapshot isolation. In: 32nd International Conference on Very Large Data Bases, Seoul, Korea, pp. 715–726. ACM, New York (2006)
12. Elnikety, S., Zwaenepoel, W., Pedone, F.: Database replication using generalized snapshot isolation. In: Symposium on Reliable Distributed Systems, Orlando, FL, USA, pp. 73–84. IEEE-CS Press, Los Alamitos (2005)
13. Wiesmann, M., Schiper, A.: Comparison of database replication techniques based on total order broadcast. IEEE Trans. on Knowledge and Data Engineering 17(4), 551–566 (2005)
14. Bernabé-Gisbert, J.M., Salinas-Monteagudo, R., Irún-Briz, L., Muñoz-Escoí, F.D.: Managing multiple isolation levels in middleware database replication protocols. In: Guo, M., Yang, L.T., Di Martino, B., Zima, H.P., Dongarra, J., Tang, F. (eds.) ISPA 2006. LNCS, vol. 4330, pp. 511–523. Springer, Heidelberg (2006)
15. Oliveira, R.C., Pereira, J., Correia Jr., A., Archibald, E.: Revisiting 1-copy equivalence in clustered databases. In: Symposium on Applied Computing, Dijon, France, pp. 728–732. ACM Press, New York (2006)
16. Finkelstein, S., Brendle, R., Jacobs, D.: Principles for inconsistency. In: 4th Biennial Conf. on Innovative Data Systems Research (CIDR), Asilomar, CA, USA (2009)
17. Helland, P., Campbell, D.: Building on quicksand. In: 4th Biennial Conf. on Innovative Data Systems Research (CIDR), Asilomar, CA, USA (2009)
18. Agrawal, D., Alonso, G., El Abbadi, A., Stanoi, I.: Exploiting atomic broadcast in replicated databases. In: 3rd International Euro-Par Conference, Passau, Germany, pp. 496–503. Springer, Heidelberg (1997)
19. Pedone, F., Guerraoui, R., Schiper, A.: Exploiting atomic broadcast in replicated databases. In: Pritchard, D., Reeve, J.S. (eds.) Euro-Par 1998. LNCS, vol. 1470, pp. 513–520. Springer, Heidelberg (1998)
20. Kemme, B., Alonso, G.: A new approach to developing and implementing eager database replication protocols. ACM Transactions on Database Systems 25(3), 333–379 (2000)
21. Lin, Y., Kemme, B., Patiño-Martínez, M., Jiménez-Peris, R.: Middleware based data replication providing snapshot isolation. In: SIGMOD Conf. on Management of Data, Baltimore, MD, USA, pp. 419–430. ACM Press, New York (2005)

22. Guerraoui, R., Garbinato, B., Mazouni, K.: The GARF library of DSM consistency models. In: ACM SIGOPS European Workshop, pp. 51–56 (1994)
23. Wu, S., Kemme, B.: Postgres-R(SI): Combining replica control with concurrency control based on snapshot isolation. In: 21st Intl. Conf. on Data Eng. (ICDE), Tokyo, Japan, pp. 422–433. IEEE-CS Press, Los Alamitos (2005)
24. Muñoz-Escoí, F.D., Pla-Civera, J., Ruiz-Fuertes, M.I., Irún-Briz, L., Decker, H., Armendáriz-Iñigo, J.E., González de Mendívil, J.R.: Managing transaction conflicts in middleware-based database replication architectures. In: Symposium on Reliable Distributed Systems, pp. 401–410 (2006)
25. Rodrigues, L., Miranda, H., Almeida, R., Martins, J., Vicente, P.: The GlobData fault-tolerant replicated distributed object database. In: 1st EurAsian Conf. on Information and Communication Technology, Shiraz, Iran, pp. 426–433. Springer, Heidelberg (2002)
26. Juárez-Rodríguez, J.R., Armendáriz-Iñigo, J.E., González de Mendívil, J.R., Muñoz-Escoí, F.D., Garitagoitia, J.R.: A weak voting database replication protocol providing different isolation levels. In: 7th Intl. Conf. on New Technologies of Distr. Syst., Marrakesh, Morocco, pp. 261–268 (2007)
27. Plattner, C., Alonso, G.: Ganymed: Scalable and flexible replication. IEEE Data Eng. Bull. 27(2), 27–34 (2004)
28. Pedone, F.: The Database State Machine and Group Communication Issues. PhD thesis, École Polytechnique Fédérale de Lausanne, Lausanne, Switzerland (1999)
29. Zuikevičiūtė, V., Pedone, F.: Revisiting the database state machine. In: VLDB Workshop on Design, Implementation and Deployment of Database Replications, Trondheim, Norway (2005)
30. Rodrigues, L., Carvalho, N., Miedes, E.: Supporting linearizable semantics in replicated databases. In: 7th IEEE International Symposium on Networking Computing and Applications, Cambridge, MA, USA, pp. 263–266. IEEE-CS Press, Los Alamitos (2008)
31. Salinas, R., Muñoz-Escoí, F.D., Armendáriz-Íñigo, J.E., González de Mendívil, J.R.: A performance analysis of g-bound, a consistency protocol supporting multiple isolation levels. In: Meersman, R., Tari, Z., Herrero, P. (eds.) OTM-WS 2008. LNCS, vol. 5333, pp. 914–923. Springer, Heidelberg (2008)
32. Zuikevičiūtė, V., Pedone, F.: Correctness criteria for database replication: Theoretical and practical aspects. In: 10th International Symposium on Distributed Objects, Middleware and Applications, Mexico, pp. 639–656. Springer, Heidelberg (2008)
33. Breitbart, Y., Garcia-Molina, H., Silberschatz, A.: Overview of multidatabase transaction management. VLDB J. 1(2), 181–239 (1992)
34. Attiya, H., Welch, J.L.: Sequential consistency versus linearizability. ACM Trans. Comput. Syst. 12(2), 91–122 (1994)

TMBean: Optimistic Concurrency in Application Servers Using Transactional Memory

Lucas Charles, Pascal Felber, and Christophe Gête

Computer Science Department, University of Neuchâtel,
Rue Emile-Argand 11, CH-2009 Neuchâtel, Switzerland

Abstract. In this experience report, we present an evaluation of different techniques to manage concurrency in the context of application servers. Traditionally, using entity beans is considered as the only way to synchronize concurrent access to data in Jave EE and using mechanism such as synchronized blocks within EJBs is strongly not recommended. In our evaluation we consider the use of *software transactional memory* to enable concurrent accesses to shared data across different session beans. We are also comparing our approach with using (1) entity beans and (2) session beans synchronized by a global lock.

Keywords: Application Server, Java EE, EJB, Software Transactional Memory.

1 Intoduction

Multicore CPUs have become common nowadays, but taking advantage of their processing power is still considered a hard task. There is a big gap between the scientific community and specialized vendors on one hand, and end developers with domain-specific skills but limited experience with multi-threaded programming on the other hand. In order to help programmers to fully take advantage of multicore architectures, we have to provide them with the right tools that are both efficient and easy to use.

The traditional approach to dealing with concurrency is based on locks. It offers a fine-grained control over the different resources to protect from data races, but has the drawback of being error prone. The programmer faces the trade-off between performance and safety: fine-grained locking is efficient but prone to complex issues like deadlocks or priority inversions, while coarse-grained locking is safe but does not scale.

Another approach to that gained significant attention from the research community is optimistic concurrency control. It relies on the assumption that the number of updates to shared data is by far lower than the number of read accesses, and conflicts are rare. Transactional memory (TM) [9,16] belongs to this category. In transactional memory, critical sections are expressed as atomic blocks performed as *transactions*. At runtime, these transactions may be executed concurrently based on the optimistic expectation that the set of values

R. Meersman, T. Dillon, P. Herrero (Eds.): OTM 2009, Part I, LNCS 5870, pp. 484–496, 2009.

read by one transaction will not overlap with the set of values written by another concurrent transaction. If no conflict occurs, concurrent transactions can commit, otherwise some of them must rollback and restart their execution. TM has been shown to scale well on multiple cores when the data access pattern behaves "well," i.e., when few conflicts are induced [1,11].

Application servers typically serve multiple clients at the same time. When running on a multicore CPU, they should be able to straightforwardly process concurrent requests from different users. However, it is not trivial to have multiple users access the same stateful application object in a consistent manner without serializing requests. In the context of Java Enterprise Edition (Java EE), while it is common to serve multiple clients accessing the same database, it is far less common to develop applications that allow some concurrent collaboration between Java beans (EJBs).

For several years now, the only way considered safe in a Java EE application to synchronize data among beans, is to use *entity beans*. This approach is using an underlying database which adds significant overhead. It is a common trap for programmers to try to use lock-based synchronization within EJBs, since it is highly error prone, in particular within an ongoing transaction that could abort and produce deadlocks. Using entity beans have the real advantage of providing a certain degree of abstraction, as of how data are shared, but the overhead of the underlying database might be prohibitive if entity beans are solely used for synchronizing concurrent threads. In that case, more lightweight and scalable techniques are desirable, such as those provided by software transactional memory (STM).

In this experience report, we evaluate the use of STM to implement concurrent Java beans in an application server. We present a benchmark in the form of a concurrent Java EE application and we compare different concurrency control strategies based on locks, entity beans and STM. The results support our claim that an STM can provide better scalability than the other strategies while being very simple to use.

The rest of this paper is structured as follows. We discuss related work in Section 2. In Section 3, we describe how concurrency is dealt in the context of a Java EE application. Section 4 explains how optimistic concurrency can be introduced in an application server to take advantage of an existing STM. We describe our benchmark in the Section 5 and discuss experimental results in Section 6. We finally conclude in Section 7.

2 Related Work

While transactional memory has gained much attention recently with multicore machines becoming ubiquitous, it has been first proposed more than a decade ago. Shavit and Touitou described the first STM implantation as a non-blocking mechanism [16], but STM really reached a sufficient level of usability when Herlihy *et al.* proposed DSTM [8], the first STM that was able to manage dynamic data structures. Java support for transactional memory was first introduced by Harris and Fraser [6]. They have modified a Java virtual machine to support the notion of *atomic blocks*, which as opposed to synchronized blocks are deadlock free and rely upon optimistic concurrency control.

In our evaluation, we use the LSA [13] STM library for Java. It uses a multi-version design (i.e., multiple versions of shared objects are available for increasing the likelihood of successful commits) with a time-based STM algorithm that was shown to be among the most efficient currently known.

Most related to our study, Cachopo *et al.* [2] have also used a multi-version STM to implement a fully distributed Web application. Their application uses STM for local synchronization but relies on a central database to synchronize the data shared among different machines. Results show that STM provides interesting benefits over a database-only approach.

Finally, the latest EJB 3.1 specification introduces a new type of *singleton* session bean that can be accessed concurrently. The approach we present here covers use cases that a singleton session bean should support, the fault tolerance left aside.

3 Application Servers and Concurrent Beans

This section briefly summarizes the architecture of Java EE application servers and describes the different options for executing concurrent beans in such architectures.

3.1 Java EE Architecture

Java EE is a specification of a multi-tier application that provides four different types of components, called "beans", to serve clients requests. On the business side it provides stateless and stateful session beans, while on the storage side it provides entity bean that are representing persistent data maintained in a database. A fourth type, message driven beans, allows asynchronous communication.

Only entity beans may safely be shared by multiple clients. As the clients might change the same data, entity beans often work within transactions. To control the usage of transactions in Java EE, several transactions attributes are defined that indicate whether operations can, must, or cannot be executed within a (new or existing) transaction. When experimenting with STM and lock-based shared session beans, we forbid transactional execution to avoid possible problems, for instance, Java EE transactions aborting while in an atomic block or a critical section.

3.2 Concurrency in Java EE

Java EE recommends that every shared access has to be done through entity beans. This approach has the advantage of lowering considerably the difficulty of concurrency management. By forcing every entity bean to be accessed only within a transaction, it ensures isolation between concurrent accesses and thus provides to the end programmer a high level abstraction of the application synchronization. The major limitation of this approach is the overhead introduced by the back-end database, especially in case persistence is not necessary and entity beans are only used for synchronization purposes.

Stateless session beans, as defined by the Java EE specification, are not bound to serve the same client. Moreover there is no guarantee that a specific client will always be served by the same bean. Such a loose constraint allows a stateless session bean container to decide how many beans are needed to serve incoming requests. Despite the name "stateless", developers are not prevented from storing values. Nor are they from sharing and accessing data concurrently, within a single Java virtual machine, by declaring static fields. This is the approach we use in our evaluation for experimenting with shared session beans.

In contrast, stateful session beans are bound to serve only one client and they remain alive as long as the client holds a reference to them. Stateful session beans keep a conversational state, which will persist for the duration of the session between the client and the server. A traditional example of such a conversational state is an online shop cart, where selected items are temporarily saved for the duration of the session. The semantics of such beans make them inappropriate for accessing shared data concurrently, and we do not consider them in our evaluation.

While it is discouraged to use synchronized blocks or explicit locks within session beans, they are still of some use when considering multi-threaded scenarios. A major problem with locks is their lack of composability, which could lead for instance to deadlock situations and hang the whole application server if not used properly. In the next section, we present a new approach to overcome this situation based on software transactional memory, which allow us to provide both scalability and composable synchronization.

4 Optimistic Concurrency Using Transactional Memory

In this section, we describe our approach based on transactional memory for executing concurrent beans in a Java EE application server. We first recall the principles underlying STM, then describe the LSA-STM library used in our implementation, and finally discuss how to enable STM within the application server to provide TM-enabled beans.

4.1 Software Transactional Memory

Software transactional memory (STM) [16] has been introduced as a means to support lightweight transactions in concurrent applications. It provides programmers with constructs to delimit transactional operations and implicitly takes care of the correctness of concurrent accesses to shared data. Typically, data accesses that are transactional are tagged by the programmer (or the compiler) and they execute speculatively. For instance, writes are buffered in thread-local storage and they are only committed to shared memory if the transaction completes successfully. In case of conflict, e.g., because two transactions write the same object, a specific component of the STM called *contention manager* decides upon the conflict resolution strategy, which usually implies aborting one of the transactions. Aborted transactions are automatically restarted until they eventually commit.

This type of optimistic concurrency control is particularly appropriate if conflicts are rare and unpredictable, i.e., the use of pessimistic approaches based on locks would limit scalability by serializing critical regions most often unnecessarily. Another important advantage of STM over lock-based synchronization is its composability [7]. STM has been an active field of research over the last few years (e.g., [8,6,10,12,15,3,14,4,11]).

4.2 LSA-STM

LSA-STM is a Java implementation of the *lazy snapshot algorithm* (LSA) [13], a time-based STM algorithm. LSA-STM allows building a consistent view of the objects accessed by the transaction using efficient *invisible* reads (i.e., a read by a transaction is not immediately visible to other transactions and read-write conflicts are detected only at commit time) while avoiding the quadratic cost of incremental validation incurred by previous algorithms (e.g., [12]) to guarantee that transactions have a consistent view of accessed data. LSA builds consistent linearizable snapshot in a very efficient manner. It provides an object-based API, i.e., the granularity of data accesses and conflict detection is a Java object. It can maintain multiple versions of shared objects and guarantees that read-only transactions never need to abort if sufficiently enough object versions are kept. Transaction demarcation is performed at the level of individual methods using Java annotations and accesses to shared objects are instrumented automatically using aspect-oriented programming. LSA-STM is freely available from `http://tmware.org/`, together with a word-based C implementation of LSA (TinySTM) [4].

4.3 TMBeans: STM-Based Concurrent Beans

To provide STM-based concurrency in EJBs, we used stateless session beans augmented with specific transactional objects that represent shared state. These transactional objects are declared as static fields, i.e., they are shared among instances of the beans. This implementation has been realized as a proof of concept, our final goal is to define a new type of beans, *TMBeans*, that would transparently manage shared state and transactional operation executions. The introduction of a new type of bean is following the logic behind the introduction of a singleton bean in the EJB 3.1 specification.

Each time a client invokes an operation on a TMBean, a new STM transaction is started to process the client request. Multiple clients accessing the same TMBean will execute concurrent transaction that may conflict and, hence, abort and restart. If there is no conflict, transactions will execute in parallel and commit. As TMBeans are not persistent and do not use a back-end database, they are expected to have lower overhead than entity beans.

It is important to notice that the transactions used by LSA-STM are not aware of the transactions provided by the application server and thus do not interfere with them. In fact, LSA-STM provides it's own transactions and transactional objects and thus does not use the JTA implementation provided by JBoss.

5 Benchmark

Our evaluation is based on a benchmark which represents a client-server version of a crossword game, where the clients will try to solve concurrently the same grid, i.e., they will pick random words and write their solutions in each of the associated cell.

5.1 The Crossword Benchmark

The game itself was developed as a web application using a client-server architecture in which clients can access the server through a web browser. While this approach allows real users to interact with the application, it is not suitable for testing purpose. Indeed, to load the server we needed a non-interactive application that could a sufficient amount of concurrent requests. To that end, we developed a standard Java application that simulates a group of users. To simplify the benchmark, we assume that words entered by the users are always correct, and we allow entering infinitely many times the same words so that the test can continue even if the grid is complete. The size grid is configurable and the average length of words is function of the grid size.

5.2 Test Application

Figure 1 illustrates the test environment, it mainly consists of two machines connected over a LAN network: a 16-core machine running the JBoss application server that executes the crossword application, and a multi-threaded client application that performs concurrent requests. The server allows clients to enter information in the grid, at the granularity of a word or a single letter. It also gathers various runtime information in order to compute statistics over the different test runs.

As the test focuses on measuring the throughput of the different locking techniques and to avoid the grid to be completed before the end of the experiment, simulated clients do not perform any computation to determine what are the

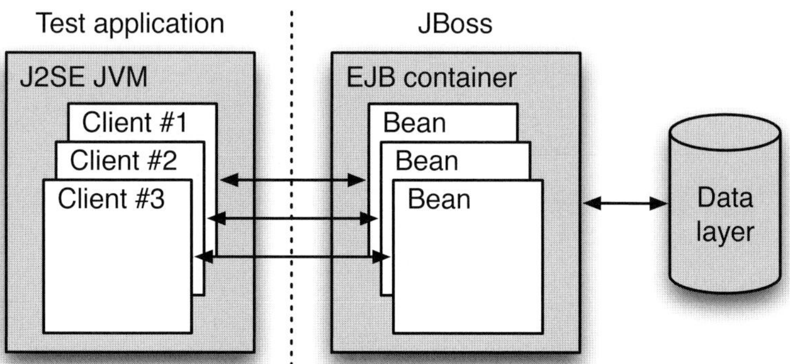

Fig. 1. Overview of the test environment

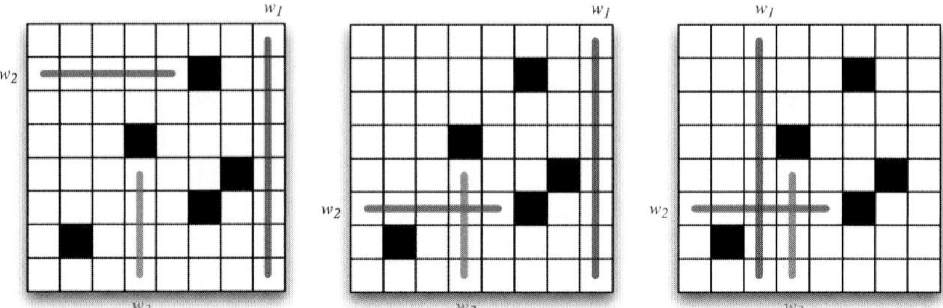

Fig. 2. Three transactions concurrently entering words w_1, w_2, and w_3 in the crossword application. **Left.** Transactions do not conflict and can all commit. **Center.** Transactions entering words w_1 and w_2 conflict and only one can commit. **Right.** All transactions conflict, but aborting the one entering w_2 allows the two others to commit.

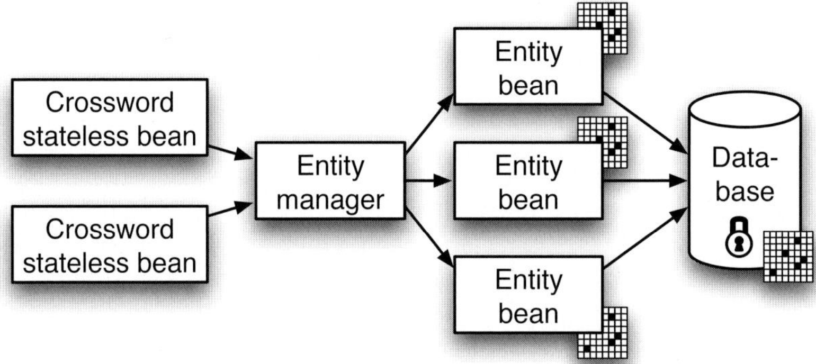

Fig. 3. Architecture overview of the test application using entity beans

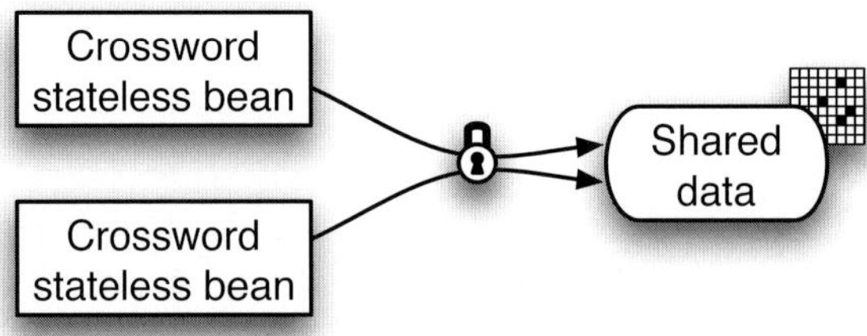

Fig. 4. Architecture overview of the test application using a global lock

suitable words or letters in the grid. Instead the task of a client consists of three steps. First, it chooses randomly a cell in the grid and a direction. Second, it reads the selected word in the grid, which implies a read access in the transaction. Finally, either the word is not yet in the grid and the client writes it, or

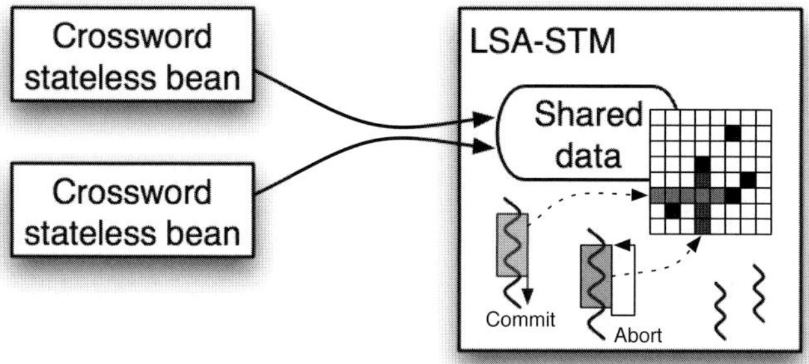

Fig. 5. Architecture overview of the test application using an STM

the word is already present and the client erases it. Therefore, in every case a client performs a read followed by a write.

As shown in Figure 2, several transactions accessing the same cell in the grid conflict and one has to abort and restart. When more than two transactions conflict, it is desirable to abort as few transactions as possible to let other complete their execution (see the right side of the figure). This choice is difficult to take in practice as contention managers only deal with pairwise conflicts. In our implementation we use the *greedy* contention management policy [5]. Note that, to simplify the procedure, we consider two write accesses to the same cell as conflicting even if they write the same letter.

5.3 Concurrency Control Strategies

We evaluate three concurrency control strategies in our tests. First, we use entity beans to orchestrate simultaneous requests from the clients (see Figure 3). Second, we use a simple approach with a global lock that protects the shared grid, stored as a static field of a stateless session bean class (see Figure 4). Finally, we evaluate the TMBeans approach, with a shared object protexted by STM transactions executed by multiple stateless session beans (see Figure 5).

5.4 Experimental Setup

Our experiment has been conducted on the server side with four quad-core AMD Opteron processors (16 cores) clocked at 2.20 GHz equipped with 8 GB of ram. The environnment used was running JBoss 4.0.5 on top of openSUSE 10.3. On the client side, we used a dual-core Intel processor clocked at 2.80 Ghz equipped with 2 GB of ram running Windows XP. The fact that a single machine hosts all the clients has no impact, as the network latency dominates the overhead of context switches on the client.

Three different criteria influence the degree of contention during a test. First obviously, the number of clients accessing the server: the more concurrent accesses we have, the higher the contention will be. Second, the size of the grid, since the probability that two transactions conflicts, i.e. pick two words with at least one letter in common, is lower with a larger grid. Third, the network latency being more important than the cost of in memory operations, every request will be processed far quicker than the round-trip-time of a message to the server; thus a transaction is likely to commit before a conflicting request from another client reaches the server. In order to circumvent this third point and highlight the benefits of optimistic concurrency, we introduced a configurable delay between each letters of a word entered in the grid. This allows us to simulate longer transactions and increase the likelihood of conflicts.

6 Results

In this section we compare the results we obtained from our benchmark. The results were reported by the clients at the end of every test run. The duration of each run is one minute and the graphs represent the average of 5 runs.

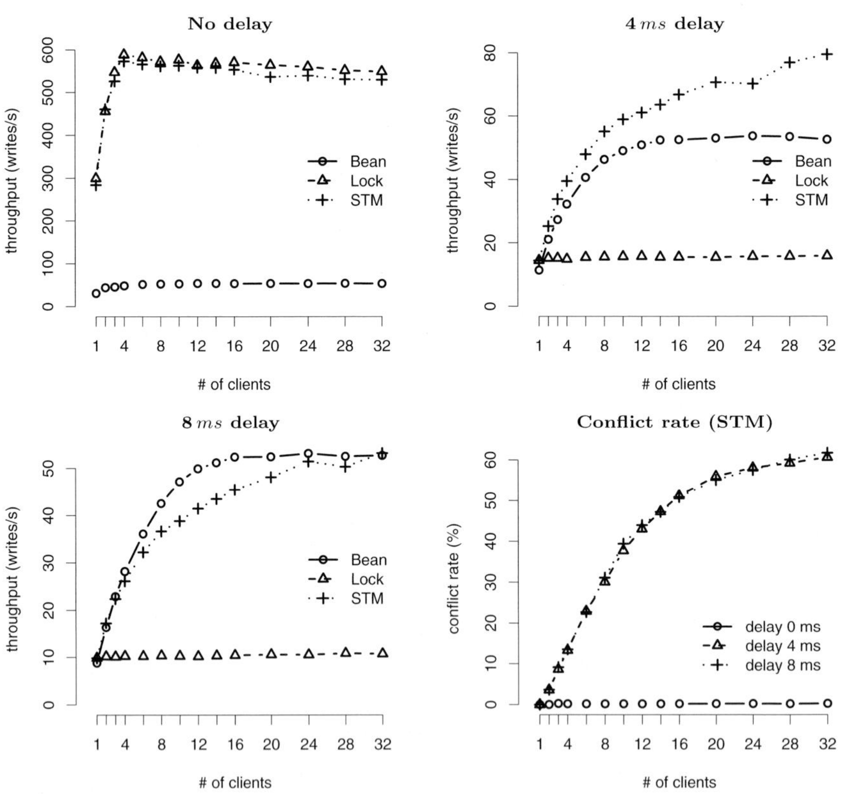

Fig. 6. Influence of the delay under high contention, with a 10x10 grid. **Top left.** No delay. **Top right.** $4\,ms$ delay between letters. **Bottom left.** $8\,ms$ delay between letters. **Bottom right.** Conflict rate for the STM strategy.

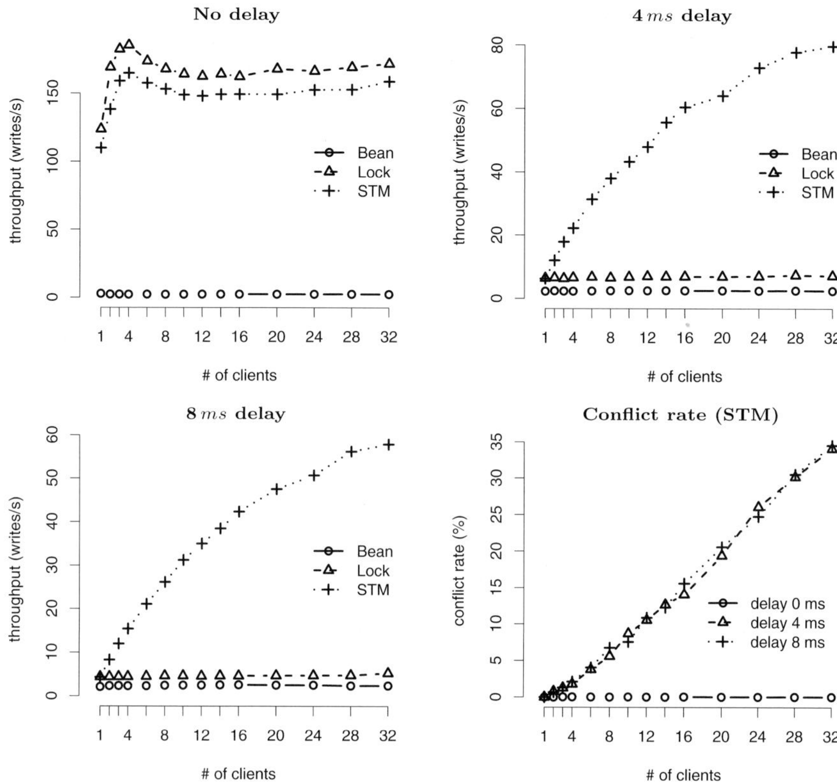

Fig. 7. Influence of the delay under low contention, with a 50x50 grid. **Top left.** No delay. **Top right.** 4 *ms* delay between letters. **Bottom.** 8 *ms* delay between letters. **Bottom right.** Conflict rate for the STM strategy.

6.1 Influence of Transaction Duration

As mentioned above, the role of the delay is to avoid the serialization effect the network has on parallel requests. Figure 6 shows the performance of the three concurrency control strategies on a small grid (10x10) with delays of 0, 4, and 8 *ms* and a variable number of clients. With no delay, and thus almost no contention as latency of the network is much higher than the processing of a transaction, we observe that the STM-based and the lock-based implementations perform similarly (although the latter seems to be slightly faster than the former because it does not incur the cost of transactional memory accesses). Both approaches are efficient as they only write to memory. In contrast, entity beans have a significantly lower throughput due to the cost of accesses to the database.

When increasing the delay to 4 *ms*, unsurprisingly the global lock strategy does not scale well. STM performs best with entity beans a close second, because the delay dominates the cost of database accesses. The trends are similar with 8 *ms* delays, but this time entity beans perform slightly better than STM. This can be explained by the fact that, under high contention, many STM transactions will abort and restart, thus reducing the overall throughput. Despite this limitation, we observe that the STM strategy still scales well up to 32 concurrent clients.

To validate our observations, we have observed the conflict rate for the STM strategy with the three delay values used in our tests (see Figure 6, bottom right). We observe that with no delay, there are almost no conflicts independently of the number of clients as transactions are much shorter than network communications. With delays of 4 and 8 *ms*, the conflict rate increases up to 60%, i.e., more than half of the transactions abort once before completing successfully.

Figure 7 shows the same data but for a 50x50 grid, i.e., with less contention. In comparison to the small grid, we observe here that entity beans perform poorly with all delay values. This can be explained by the fact that larger grids have longer words on average, and the per-letter access cost of entity beans impairs scalability. We also observe that scalability of the STM strategy is better (closer to linear), and the conflict rate remains below 35%.

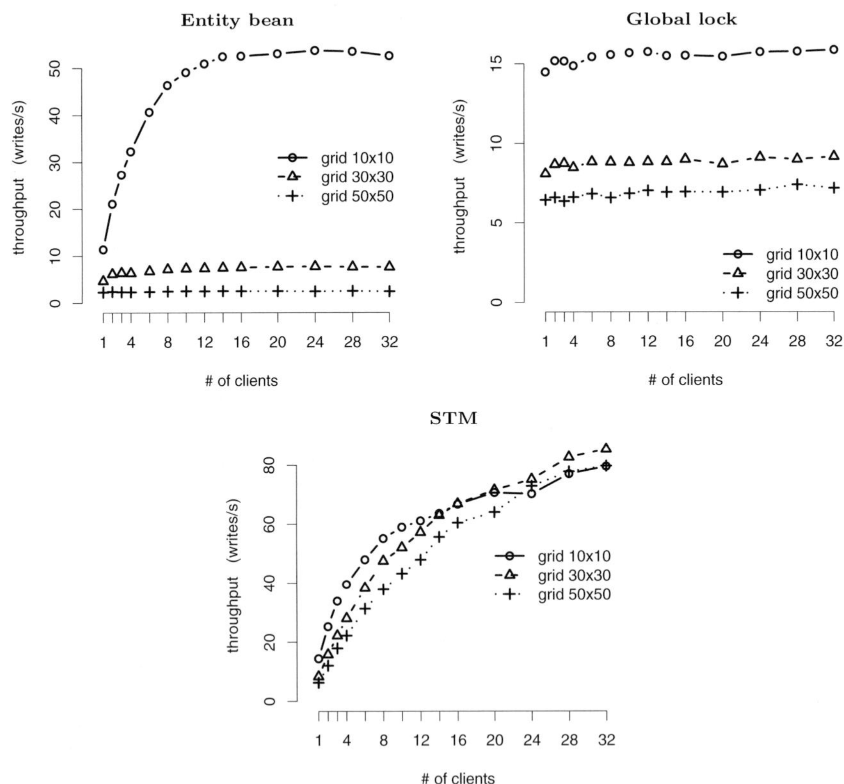

Fig. 8. Influence of the grid size with 4 *ms* delay between letters. **Top left.** Entity bean. **Top right.** Global lock. **Bottom.** STM.

6.2 Influence of Grid Sizes

The size of the grid has an important impact on the throughput as it influences both the probability for two transactions to conflict and the amount of data to be written (as words will be longer on average). In Figure 8, we compare three grid sizes (10x10, 30x30, 50x50) with a $4\,ms$ delay between entering the letters of a word. One can first observe that the STM strategy performs noticeably better than the other approaches with all configurations. Entity beans scale reasonably well with small grids, but throughput quickly saturates with large grids. The performance of the lock-based approach remains flat as client requests are serialized. Only the STM strategy scales well independently of the grid size.

7 Conclusion

In this experience report, we have presented an evaluation of two traditional techniques for managing concurrency in the context of Java EE Web applications—entity beans and coarse grained locking and compared them with a novel technique that uses software transactional memory. Our benchmark allowed us to pinpoint the benefits of using optimistic concurrency control in scenarios where conflicts are rare. In most of our experiments, STM-based beans showed better performance and scalability than the other approaches. These results demonstrate the potential benefits of providing a new form of STM-enabled beans, *TMBeans*, in application servers to enable programmers to develop lightweight concurrent stateful beans.

Acknowledgements. This work was supported in part by European Union grant FP7-ICT-2007-1 (project VELOX).

References

1. Adl-Tabatabai, A.-R., Kozyrakis, C., Saha, B.: Unlocking concurrency. Queue 4(10), 24–33 (2007)
2. Cachopo, J., Rito-Silva, A.: Versioned boxes as the basis for memory transactions. In: Proceedings of SCOOL (2005)
3. Dice, D., Shalev, O., Shavit, N.: Transactional Locking II. In: Proceedings of DISC (September 2006)
4. Felber, P., Riegel, T., Fetzer, C.: Dynamic performance tuning of word-based software transactional memory. In: Proceedings of PPoPP (February 2008)
5. Guerraoui, R., Herlihy, M., Pochon, S.: Toward a theory of transactional contention managers. In: Proceedings of PODC (July 2005)
6. Harris, T., Fraser, K.: Language support for lightweight transactions. In: Proceedings of OOPSLA (October 2003)
7. Harris, T., Herlihy, M., Marlow, S., Peyton-Jones, S.: Composable memory transactions. In: Proceedings of PPoPP (June 2005)
8. Herlihy, M., Luchangco, V., Moir, M., Scherer III, W.N.: Software transactional memory for dynamic-sized data structures. In: Proceedings of PODC (July 2003)

9. Herlihy, M., Moss, J.E.B.: Transactional memory: Architectural support for lock-free data structures. In: Proceedings of ISCA (1993)
10. Scherer III, W.N., Scott, M.L.: Advanced contention management for dynamic software transactional memory. In: Proceedings of PODC (July 2005)
11. Larus, J., Kozyrakis, C.: Transactional memory. Communication of the ACM 51(7), 80–88 (2008)
12. Marathe, V.J., Scherer III, W.N., Scott, M.L.: Adaptive software transactional memory. In: Proceedings of DISC (2005)
13. Riegel, T., Felber, P., Fetzer, C.: A lazy snapshot algorithm with eager validation. In: Proceedings of DISC (September 2006)
14. Riegel, T., Fetzer, C., Felber, P.: Time-based transactional memory with scalable time bases. In: Proceedings of SPAA (June 2007)
15. Saha, B., Adl-Tabatabai, A.-R., Hudson, R.L., Minh, C.C., Hertzberg, B.: McRT-STM: a high performance software transactional memory system for a multi-core runtime. In: Proceedings of PPoPP (2006)
16. Shavit, N., Touitou, D.: Software transactional memory. Distributed Computing 10(2), 99–116 (1997)

Optimizing Data Management in Grid Environments

Antonis Zissimos, Katerina Doka, Antony Chazapis, Dimitrios Tsoumakos, and Nectarios Koziris

National Technical University of Athens
School of Electrical and Computer Engineering
Computing Systems Laboratory
{azisi,katerina,chazapis,dtsouma,nkoziris}@cslab.ece.ntua.gr

Abstract. Grids currently serve as platforms for numerous scientific as well as business applications that generate and access vast amounts of data. In this paper, we address the need for efficient, scalable and robust data management in Grid environments. We propose a fully decentralized and adaptive mechanism comprising of two components: A Distributed Replica Location Service (*DRLS*) and a data transfer mechanism called *GridTorrent*. They both adopt Peer-to-Peer techniques in order to overcome performance bottlenecks and single points of failure. On one hand, DRLS ensures resilience by relying on a Byzantine-tolerant protocol and is able to handle massive concurrent requests even during node churn. On the other hand, GridTorrent allows for maximum bandwidth utilization through collaborative sharing among the various data providers and consumers. The proposed integrated architecture is completely backwards-compatible with already deployed Grids. To demonstrate these points, experiments have been conducted in LAN as well as WAN environments under various workloads. The evaluation shows that our scheme vastly outperforms the conventional mechanisms in both efficiency (up to 10 times faster) and robustness in case of failures and flash crowd instances.

1 Introduction

One of the most critical components in Grid systems is the data management layer. Grid computing has attracted several data-intensive applications in the scientific field, such as bioinformatics, physics or astronomy. To a great extent, these applications rely on analysis of data produced by geographically disperse scientific devices such as sensors or satellites etc. For example, the Large Hadron Collider (LHC) project at CERN [1] is expected to generate tens of terabytes of raw data per day that have to be transferred to academic institutions around the world, in seek of the Higgs boson. Apart from that, business applications manipulating vast amounts of data have lately started to invade Grid environments. *Gredia* [2] is an EU-funded project which proposes a Grid infrastructure for sharing of rich multimedia content. To motivate this approach, let us consider

R. Meersman, T. Dillon, P. Herrero (Eds.): OTM 2009, Part I, LNCS 5870, pp. 497–512, 2009.

the following scenario: News agencies have created a joint data repository in the Grid, where journalists, photographers, editors, etc can store, search and download various news content. Assume that just minutes after a breaking news-flash (e.g., the riots in Athens), a journalist on scene captures a video of the protests and uploads it on the Grid. Hundreds of journalists and editors around the world need to be able to quickly locate and efficiently download the video in order to include it in their news reports. Thus, it is imperative that, apart from optimized data transfer, such a system should be able to cope with high request rates – to the point of a flash crowd.

Faced with the problem of managing extremely large scale datasets, the Grid community has proposed the Data Grid architecture [13], defining a set of basic services. The most fundamental of them are the Data Transfer service, responsible for moving files among grid nodes (e.g., *GridFTP* [7]), the Replica Location service (*RLS*), which keeps track of the physical locations of files and the Optimization service, which selects the best data source for each transfer in terms of completion time and manages the dynamic replica creation/deletion according to file usage statistics.

However, all of the aforementioned services heavily rely on centralized mechanisms, which constitute performance bottlenecks and single points of failure: The so far centralized RLS can neither scale to large numbers of concurrent requests nor keep pace with frequent updates performed in highly dynamic environments. GridFTP fails to make optimal use of all bandwidth resources in cases where the same data must be transferred to multiple sites and does not automatically maximize bandwidth utilization. Even when using multiple parallel TCP channels, a manual configuration is required. Most importantly, GridFTP servers face the danger of collapsing under heavy workload conditions, making critical data unavailable.

In this paper, we introduce a novel data management architecture which integrates the location service with data transfer under a fully distributed and adaptive philosophy. Our scheme comprises of two parts that cooperate to efficiently handle multiple concurrent requests and data transfer: The *Distributed Replica Location Service (DRLS)* that handles the locating of files and *GridTorrent* that manages the file transfer and related optimizations. This is pictorially shown in Figure 1.

DRLS utilizes a set of nodes that, organized in a DHT, equally share the replica location information. The unique characteristic of the DRLS is that, besides the decentralization and scalability that it offers, it fully supports updates on the multiple sites of a file that exist in the system. Since in many dynamic applications data locations change rapidly with time, our Byzantine-tolerant protocol guarantees consistency and efficiently handles updates on the various data locations stored, unlike conventional DHT implementations. GridTorrent is a protocol that, inspired by BitTorrent, focuses on real-time optimization of data transfers on the Grid, fully supporting the induced security mechanisms. Based on collaborative sharing, GridTorrent allows for low latency and maximum bandwidth utilization, even under extreme load and flash crowd conditions. It allows transfers from multiple sites to multiple clients and maximizes performance by piece exchange among the participants. A very important characteristic of the proposed architecture is that it is designed to interface and exploit well-defined and deployed Data Grid components and protocols, thus being completely backwards compatible and readily deployable. This work includes an experimental section that includes a real implementation of the system and results over both LAN and WAN environments with highly dynamic and adverse workloads.

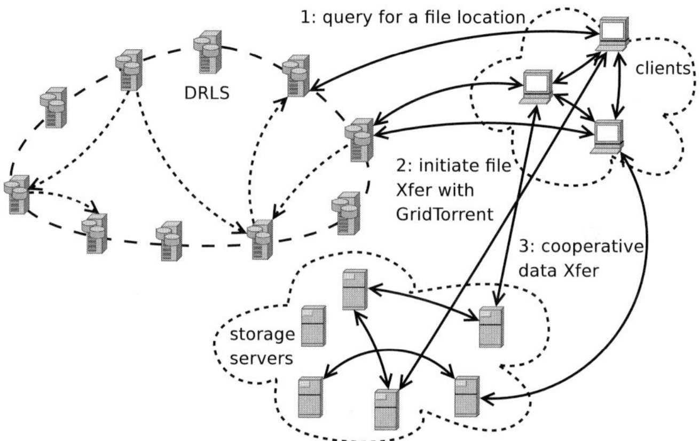

Fig. 1. Pictorial description of the proposed architecture and component interaction. Although DRLS nodes and Storage Servers appear to be a separate physical entity, it is possible to coexist in order to exploit all the available resources.

2 Current Status

In this section we overview the related work in the area of data management. We first go through existing practices for the Replica Location and Data Transfer services. Next, we present a brief description of the BitTorrent protocol, which is the basis of the proposed GridTorrent mechanism and we finally mention other relevant data transfer mechanisms.

2.1 Locating Files

Centralized Catalog and Giggle. In Grid environments, it is common to maintain local copies of remote files, called *replicas* [23] to guarantee availability and reduce latencies. To work with a file, a Grid application first asks the RLS to locate corresponding *replicas* of the requested item. This translates to a query towards the *Replica Catalogue*, which contains mappings between *Logical File Names* (LFNs) and *Physical File Names* (PFNs). If a local replica already exists the application can directly use it, otherwise it must be transferred to the local node. This initial architecture posed limitations to the scalability and resilience of the system. Efforts on distributing the catalog resulted in the most widespread

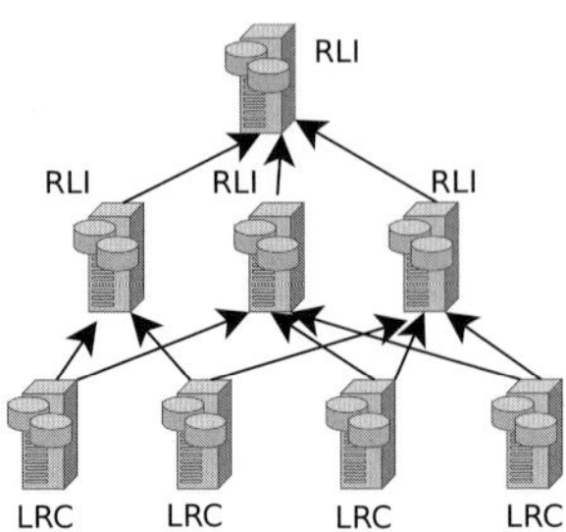

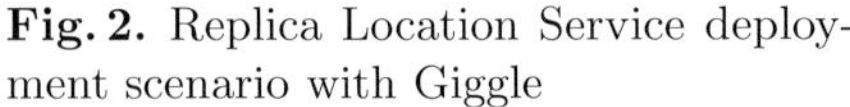

Fig. 2. Replica Location Service deployment scenario with Giggle

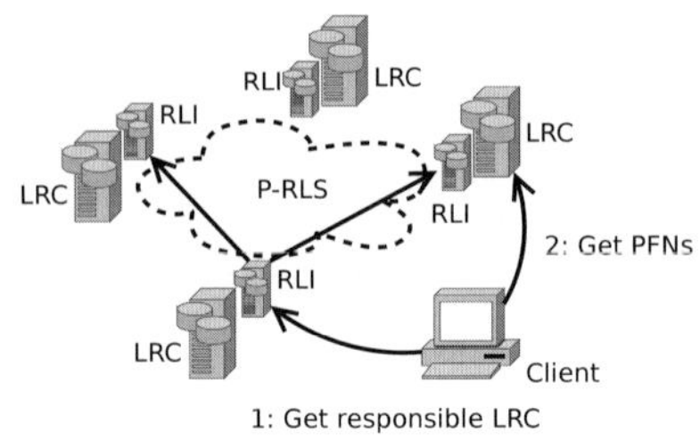

Fig. 3. Replica Location Service deployment scenario with P-RLS

solution currently deployed on the Grid, the Giggle Framework [14]. To achieve distribution, Giggle proposes a two-tier architecture, comprising of the *Local Replica Catalogs* (*LRC*s), which map LFNs to PFNs across a site and the *Replica Location Indices* (*RLI*s), which map LFNs to LRCs (Figure 2).

Distributed Replica Location Service (DRLS). Still, the centralized nature of the catalogs remains the bottleneck of the system, when the number of performed searches increases. Furthermore, the updates in the LRCs induce a complex and bandwidth-consuming communication scheme between LRCs and RLIs. To this end, in [12] we proposed a RLS based on a Distributed Hash Table (DHT). The underlying DHT is a modified Kademlia peer-to-peer network that enables mutable storage. In this work, we enhance our solution by exploiting *XOROS* [11], a DHT that provides a Byzantine-tolerant protocol for serializable data updates directly at the peer-to-peer level. In this way, we can store static information such as file properties with the traditional distributed hash table put/get mechanism, as well as dynamic information such as the actual LFN to PFN mappings with an update mechanism that ensures consistency.

Related Work. Peer-to-peer overlay networks and corresponding protocols have already been incorporated in other RLS designs. In [10], Min Cai *et al.*, have replaced the global indices of Giggle with a Chord network, producing a variant of Giggle called P-RLS. A Chord topology can tolerate random node joins and leaves, but does not provide data fault-tolerance by default. The authors choose to replicate the distributed RLI index in the *successor set* of each *root node* (the node responsible for storage of a particular mapping), effectively reproducing Kademlia's behavior of replicating data according to the replication parameter κ. In order to update a specific key-value pair, the new value is inserted as usual, by finding the *root node* and replacing the corresponding value stored there and at all nodes in its *successor set.* While there is a great resemblance to this design and the one we propose, there is no support for updating key-value pairs directly in the peer-to-peer protocol layer. It is an open question how the P-RLS design would cope with highly transient nodes. Frequent joins and departures in the Chord layer would require nodes continuously exchanging key-value pairs in order to keep the network balanced and the replicas of a particular mapping in the correct successors. Our design deals with this problem, as the routing tables inside the nodes are immune to participants that stay in the network for a very short amount of time. Moreover, our protocol additions to support mutable data storage are not dependent on node behavior; the integrity of updated data is established only by relevant data operations. Finally, the P-RLS approach retains the two-tier Giggle architecture, since the actual LFN to PFN mappings are still kept in Local Replica Catalogs imposing a bottleneck for the whole system with no support for load-balancing and failover mechanisms. In another variant of an RLS implementation using a peer-to-peer network [21], all replica location information is organized in an unstructured overlay and all nodes gradually store all mappings in a compressed form. This way each node can locally serve a query without forwarding requests. Nevertheless, the amount of data (compressed or not) that has to be updated throughout the network each time, can grow to such a large extent, that the scalability properties of the peer-to-peer overlay are lost. In contrast to other peer-to-peer RLS designs, we envision a service that does not require the use of specialized servers for locating replicas. According to our design, a lightweight DHT-enabled RLS peer can even run at every node connected to the Grid.

2.2 Transferring Files

The GridFTP Protocol. The established method for data transfer in the Grid is *GridFTP* [7], a protocol defined by the Global Grid Forum and adopted by the majority of the existing middleware. GridFTP extends the standard FTP, including features like the *Grid Security Infrastructure* (GSI) [17] and third-party control and data channels. A more distributed approach of the GridFTP service has lead to the *Globus Stripped GridFTP* protocol [8], included in the current release of the Globus Toolkit 4 [3]. Transfers of data striped or interleaved across multiple servers, partial file transfers and parallel data transfers using multiple TCP streams are some of the newly added features.

The GridTorrent Approach. Yet, the GridFTP protocol is still based on the client-server model, inducing all the undesirable characteristics of centralized techniques, such as server overload, single points of failure and the inability to cope with flash crowds. We argue that large numbers of potential downloaders together with the well-documented increase in the volume of data by orders of magnitude stress the applicability of this approach. We propose a replica-aware algorithm based on the P2P paradigm, through which data movement services can take advantage of multiple replicas to boost aggregate transfer throughput. In our previous work [27] there were made some preliminary steps towards this direction. A first GridTorrent prototype was implemented and one could use the Globus RLS and various GridFTP storage servers to download a file, as well as exploit other simultaneous downloaders, thus making a first step towards cooperation. Nevertheless, a core component of every Grid Service, the Globus Security Infrastructure (GSI) wasn't integrated with our previous prototype.

Furthermore, in torrent-like architectures like GridTorrent there is the inherent problem of not being able to upload a file unless there are downloaders interested in the specified file. To tackle this problem we introduce the GridTorrent's control channel, a separate communication path that can be used to issue commands to remote GridTorrent servers. Thus, in order to upload a file several GridTorrent servers are automatically notified and after the necessary authentication and authorization phases, the file is uploaded to multiple servers simultaneously and more efficiently. There is no need for the user to issue another set of commands for replication, because this is handled by GridTorrent. Finally, in order to scale to larger deployments our prototype is integrated with the aforementioned DRLS. In the present work, we extend GridTorrent and propose a complete architecture which can be directly deployed in a real-life Grid environment and integrate with existing Grid services.

The BitTorrent Protocol. Our work as well as other related work on the area rely on the BitTorrent protocol [15]. BitTorrent is a peer-to-peer protocol that allows clients to download files from multiple sources while uploading them to other users at the same time, rather than obtaining them from a central server. Its goal is to reduce the download time for large, popular files and the load on servers that serve these files. BitTorrent divides every file in *piece* and each piece in *blocks*. Clients find themselves through a centralized service called the *tracker* and can exploit this fragmentation by simultaneously downloading blocks from many sources. Useful file information is stored in a *metainfo* file, identified by the extension `.torrent`. Peers are categorized in *seeds* when they already have the whole file and *leechers* when they are still downloading pieces. The latest version of the BitTorrent client [4] uses a Distributed Hash Table (DHT) for dynamically locating the tracker responsible for each file transaction. Note that, in contrast to the Data Management architecture presented here, BitTorrent does not yet use a DHT for storing and distributing file information and metadata. The corresponding .torrent files still have to be downloaded from a central repository, or manually exchanged between users. The data transfer component of our architecture, GridTorrent, enhances the BitTorrent protocol with new features in order to make it compatible with existing Grid architectures. Moreover, new functionality is added, so as to be able to instruct downloads to remote peers. Finally, the tracker, which constitutes a centralized component of the BitTorrent architecture is replaced by DRLS, eliminating possible performance bottlenecks and single points of failure.

Related Work Using BitTorrent. A related work that is based in torrent-like architecture for data transfers in Grid environments can be found in GridTorrent Framework [18], which cites our previous work and therefore should not be confused our proposed architecture. The authors of GridTorrent Framework focus on a centralized tracker to provide information for the available replicas, but also use the tracker to impose security policies for data access. Their work also extend to the exploitation of parallel TCP streams between two single peers in order to surpass the limitations of the TCP window algorithm and saturate high bandwidth links. Nevertheless, the Framework's centralized design suffers of all the undesirable characteristics of centralized techniques, while the lack of integration with standardized Grid components remains a substantial disadvantage. A similar work is presented in [25], where the authors compare BitTorrent to FTP for data delivery in Computational Desktop Grids, demonstrating that the former is efficient for large file transfers and scalable when the number of nodes increases. Their work is concentrated in application environments like SETI@Home [16], distributed.net [5] and BOINC [9] where methods like cpu scavenging are used to get temporary resources from Desktop computers. In contrast to GridTorrent, their prototype uses centralized data catalog and repository, fails to communicate with standard Grid components like GridFTP and RLS, lacks the support of Globus Security Infrastructure and doesn't tackle the problem of efficient file upload in multiple repositories.

Other Data Transfer Mechanisms. The efficient movement of distributed volumes of data is a subject of constant research in the area of distributed systems. Various techniques have been proposed, apart from the ones mentioned above, centralized or in the context of the peer-to-peer paradigm. Kangaroo [24] is a data transfer system that aims at better overall performance by making opportunistic use of a chain of servers. The Composite Endpoint Protocol [26] collects high-level transfer data provided by the user and generates a schedule which optimizes the transfer performance by producing a balanced weighting of a directed graph. Nevertheless, the aforementioned models remain centralized. Slurpie [22] follows a similar approach to BitTorrent, as it targets bulk data transfer and makes analogous assumptions. Nonetheless, unlike BitTorrent, it does not encourage cooperation.

3 GridTorrent

GridTorrent, a peer-to-peer data transfer approach for Grid environments, was initially introduced in [27]. Based on BitTorrent, GridTorrent allows clients to download files from multiple sources while uploading them to other users at the same time, rather than obtaining them from a central server. Using BitTorrent terminology, GridTorrent creates a swarm where leechers are users of the Grid downloading data and seeds are storage elements or users sharing their data in the Grid. The cooperative nature of the algorithm ensures maximum bandwidth utilization and its tit-for-tat mechanism provides scalability in heavy load conditions or flash crowd situations. More specifically, GridTorrent exploits existing infrastructure since GridFTP repositories can be used as seeds with other peers downloading from them using the GridFTP partial file transfer capability. The *.torrent* file used in BitTorrent is replaced by the already existing RLS. In order to start a file download only the file's unique identifier (*UID*) is required, which is actually the content's digest. The rest of the information can be extracted from the RLS using this UID. Finally, GridTorrent makes the BitTorrent's tracker service obsolete and integrates its functionality in the RLS. Therefore, all the peers that participate in a GridTorrent swarm are also registered in the RLS, so that they are able to locate each other. In the following paragraphs we analyze the further enhancements we have developed in GridTorrent.

3.1 Security

In a Grid environment, only authenticated users are considered trustworthy of serving or downloading file fragments. Moreover, encryption is provided for the transfer of sensitive information. In order to guarantee security, our data transfer mechanism implements the Globus Grid Security Infrastructure (GSI). Currently, GridTorrent deploys the standard GSI mechanisms, in terms of authentication, integrity and encryption. A Java TCP socket is created and wrapped, along with the host credentials, as a grid-enabled socket. This is performed when the plain socket passes through the createSocket method of the GssSocketFactory of the globus GSI API. Thus, an appropriate socket is created, with respect to the input parameters that enable encryption, message integrity, peer authentication or none of the above, according to the user's preferences.

3.2 Control Channel

In GridTorrent, peers communicate with each other and exchange information regarding the current file download according to the protocol. A novel feature of GridTorrent, not found in BitTorrent protocol, is the ability of a peer to issue commands to remote peers. We call this feature *control channel*, because it is similar to the GridFTP's control channel. This feature overcomes the BitTorrent disadvantage of not being able to upload data before another peer is interested to download them, which is common practice for a peer-to-peer network, but not applicable to Grid environments. In detail, the GridTorrent control channel supports the following commands:

`Start.` [*UID*] [*RLS*] Starts downloading the file with the given UID, getting publishing information from the given RLS.

`Start.` [*filename*] [*RLS*] Starts sharing the existing file determined by the given local filename. RLS will be used for publishing information regarding the download.

`Stop.` [*UID*] Stops an active file download. Takes as a parameter the UID of the file to stop downloading.

`Delete.` [*filename*] Deletes a local file.

`List.` Lists all active file downloads of the node.

`Get.` [*UID*] Gets statistics about an active file download regarding messages exchanged and data transfer throughput. Takes the UID of the file as a parameter.

`Shutdown.` Shuts down the GridTorrent peer.

4 Replica Location Service

The RLS used in GridTorrent stores two types of metadata: static information (file properties) and dynamic information (peers that have the file or part of it).

In our design, we select a set of attributes required to initiate a torrent-like data transfer. Therefore, the file properties stored in the RLS are the following:

Logical filename (LFN): This is the name of the stored file. This name is supplied by the user to identify his file.
File size: The total size of the file in bytes.
File hash type: The type of the hashing used to identify the whole file data. Hashing is enabled in this level to ensure data consistency.
File hash: The actual file data hash. It is also used as a UID for each file.
Piece length: The size of each piece in which the file is segmented. The piece is the smallest fraction of data that is used for validating and publishing purposes. Upon a complete piece download and integrity check, other peers are informed of the acquisition.
Piece hash type: The type of the hashing used to identify each piece of the file. Hashing is enabled in this level to facilitate partial download and resume download operations.
Piece hash: The actual piece data hash. All the hashes of all the pieces are concatenated starting from the first piece.

Besides the file properties, the RLS also stores a list of all the physical locations where the file is actually stored. This is described by a physical filename (PFN). A physical filename has the following form:

`protocol://fqdn:port/path/to/file`

where `protocol` is the one that is used for the data transfer. Currently the supported protocols are `gsiftp` (GridFTP) and `gtp` (GridTorrent). The fully qualified domain name `fqdn` is the DNS registered name of the peer and it is followed by the peer's local path and the local filename.

4.1 Distributed RLS

RLS as a core Grid service must use distribution algorithms with unique scalability and fault-tolerance properties–assets already available by peer-to-peer architectures. To this end, in [12] we proposed a Replica Location Service based on a Distributed Hash Table (DHT). The underlying DHT is a modified Kademlia peer-to-peer network that enables mutable storage. We enhance this work by exploiting the XOR Object Store (XOROS) [11], a DHT that provides serializable data updates to the primary replicas of any key in the network. XOROS uses a Kademlia [19] routing scheme, along with a modified protocol for inserting and looking up values, that accounts for dynamic or Byzantine behavior of overlay participants. The put operation allows either an in-place update, or a read-modify-write via a single, unified transaction, that consists of a mutual exclusion mechanism and an accompanying value propagation step. GridTorrent has a modular architecture that enables the use of different types of Replica Location Service per swarm. More specifically, when a user initiates a file transfer, he must also supply the RLS URL, which has the following form:

`protocol://fqdn:port`

Table 1. Security overhead in the overall file transfer

configuration	mean time (sec)	overhead
authentication	43,3	0%
authentication + integrity check	44,3	2%
authentication + encryption	55,3	27%

Currently the supported protocols are `rls` (Globus RLS) and `drls` (Distributed RLS based on XOROS), so GridTorrent parses the URL to load the corresponding RLS implementation. One advantage of the above modification is the use of already implemented features to model our solution, preserving the backwards compatibility with the existing Grid Architecture. Therefore, the proposed changes in the current Grid Architecture not only enhance the performance of data transfers, but also seamlessly integrate with the current state-of-the-art in Grid Data Management.

5 Implementation and Experimental Results

Our GridTorrent prototype implementation is entirely written in Java. The GridTorrent client has bindings with Globus Toolkit 4 libraries [3] and exploits the GridFTP client API, the Replica Location Service API and the Grid Security Infrastructure API. These bindings enrich our prototype with the abilities to use existing grid infrastructure, such as data stored in GridFTP servers, metadata stored in Globus RLS and x509 certificates that are already issued to users and services for authentication, authorization, integrity protection and confidentiality. For the experiments we started GridTorrent to a number of physical nodes and issued remote requests through the control channel, to initiate and monitor the overall file transfer.

5.1 GridTorrent Security and Fault-Tolerance Performance

We first test the effect that Grid Security has in the overall data transfer process by monitoring the time needed for the transfer of a 128MB file. We distinguish three different configurations for Globus GSI:

Authentication only: This is a simple configuration where both sides need to present a valid x509 certificate signed from a Certificate Authority that is mutually trusted.

Integrity check: In this configuration besides the mutual authentication, the receiver verifies all messages to prevent man-in-the-middle attacks.

Encryption: This is the most secure configuration, where apart form mutual authentication and integrity check, every message is also encrypted.

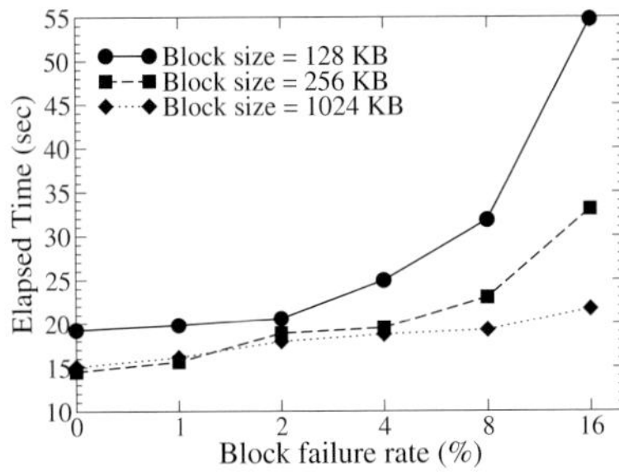

Fig. 4. Average time of completion over various number of failure rates and block sizes

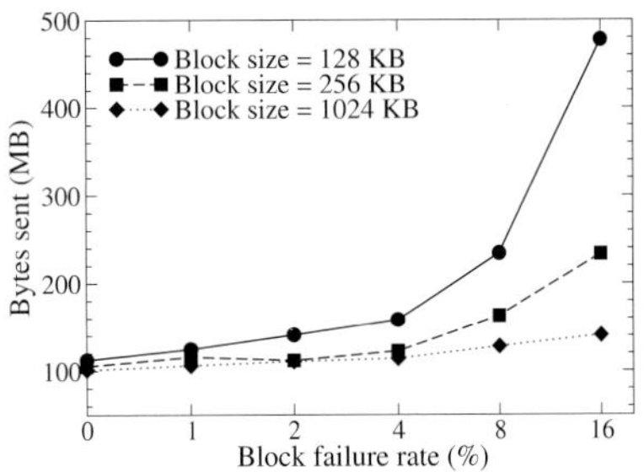

Fig. 5. Average size of uploaded data from leechers only over various number of failure rates and block sizes

The test is executed 100 times between a pair of peers (different each time) inside the same LAN. As shown in Table 1, only the Globus GSI configuration that enables encryption has considerable (about 30%) cost on the file transfer latency. This overhead is natural, because when encryption is enabled every message is duplicated in memory and parsed by a cpu-intensive cryptographic algorithm.

We continue our experiments by testing GridTorrent's tolerance in an error-prone network. For this purpose we use a single server acting as seed for a file of 128MB and 16 clients that simultaneously download the file. After extensive testing we have tuned GridTorrent to use a piece size of 1024KB. In GridTorrent, just like BitTorrent, hashes are kept in a per piece basis, and peers exchange a smaller fraction of data called block. To simulate the failure rate, every peer (leecher or seed) makes a decision to sent altered blocks based on a random uniformly distributed function, without enabling any globus security option. The results are presented in Figures 4 and 5. First of all, in all cases the download completes with an acceptable overhead, in contrast to GridFTP which has no mechanism of protection against these kinds of errors. Furthermore, we notice that as the failure rate increases transfers with smaller block sizes are more heavily affected, because one bad block causes the retransmission of all the blocks in a certain piece. So in cases where block size is $\frac{1}{8}$ of the piece size, the slowdown is 3 to 4 times in comparison with the case of a block the size of a piece and in failure rates up to 16%.

5.2 GridTorrent vs. GridFTP Performance

In this experiment we compare the performance of the GridTorrent prototype against the current GridFTP implementation in both Local and Wide Area Network environments. Specifically, we increase the number of concurrent requests over a single 128MB file from different physical nodes. Results for different file sizes (up to 512MB) are qualitatively similar. We measure the minimum, maximum and average completion time of this operation on all requesters. Our setup assumes a single server that seeds this file and up to 32 physical machines that issue simultaneous download requests. For the LAN experiments, we use our

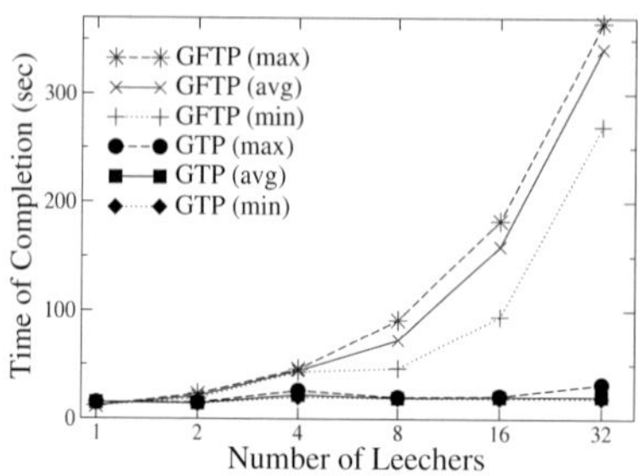

Fig. 6. Min, max and average time of completion for both GFTP and GTP over various number of downloaders in the LAN setting

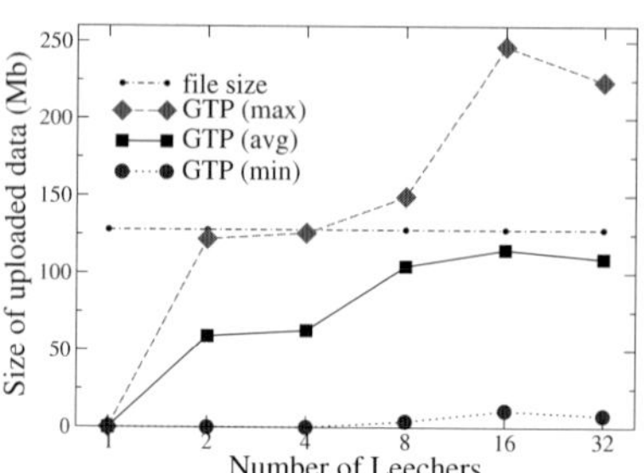

Fig. 7. Min, max and average size of uploaded data from leechers only in the LAN setting

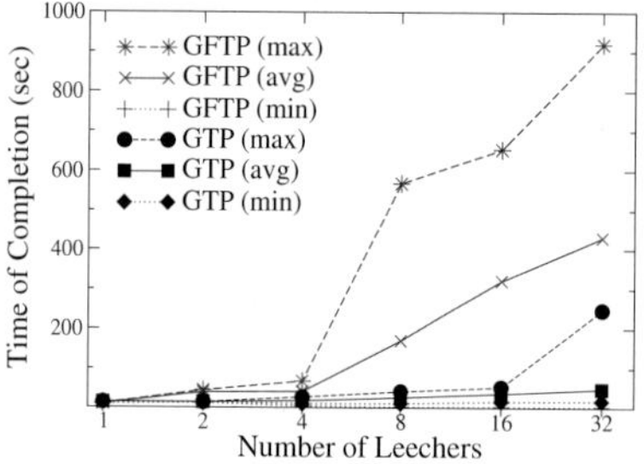

Fig. 8. Min, max and average time of completion for both GFTP and GTP over various number of downloaders in the WAN setting

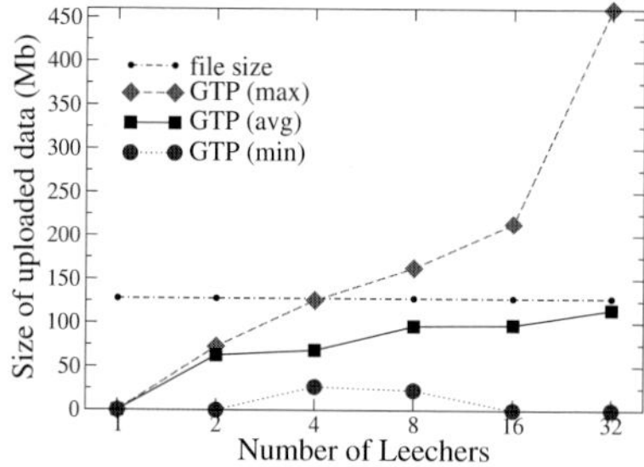

Fig. 9. Min, max and average size of uploaded data from leechers only in the WAN setting

laboratory cluster infrastructure with gigabit ethernet interconnect. For the WAN experiments, we allocate the same amount of nodes in PlanetLab [20, 6]. In this environment, there exist several heavily loaded nodes, geographically distributed with various network latencies and bandwidth constraints. Obviously, PlanetLab offers an environment more similar to a real world Grid environment, where requests may occur from different places over the globe using personal computers. Location information on the file, the list of peers that obtain or are currently downloading the file, as well as other file metadata are stored in DRLS, located in a single machine which simulates 30 nodes in a XOROS DHT.

In Figure 6, we present the completion times for the LAN setup. We notice that GridTorrent can be over 10 times faster than GridFTP in all measured times. This occurs for the largest number of leechers. GridFTP cannot enforce cooperation among nodes; thus a single server must accommodate all clients in an serialized manner. One would expect that GridFTP would not be affected by the flash crowd effect in a LAN, especially with the Gigabit ethernet connectivity, but this is not the case. GridTorrent shows remarkable performance in all

Table 2. The effect of the κ parameter in DRLS

κ	α	ϵ	Average Messages	Mean latency (sec)
20	3	2	44	1.37
15	3	2	42	1.06
10	3	2	30	0.83
5	3	2	22	0.61

three metrics, as they remain unaltered by the increase in requests. Our method can be readily employed to sustain flash crowd effects as, due to the increasing cooperation among peers, it effectively reduces the load of the single server and provides adaptive portions of the file to the rest of the nodes. In Figure 7, we present this cooperation in terms of bytes sent exclusively among the leechers. We notice that, as the number of leechers increase, this traffic increases, showing the clients' active part in this process. On average, each leecher seems to be responsible for sending almost one file's worth of data to the other leechers and no more than two times the file size in maximum.

Figure 8 summarizes our results from the WAN setup. It is evident that GridFTP cannot cope with increasing transfer loads in a real world environment. GridFTP's minimum times remain constantly low and close to GridTorrent's due to the fact that there always exists at least one leecher close to the single server that downloads the file faster. Furthermore, we register a major difference in GridFTP's maximum, minimum and average times (e.g., for 32 leechers the last one receives the file 30 times slower than the faster one and about 2 times slower on average). This large variance is due to the protocol's inability to cope with heterogeneity – small number of close nodes finish early while the rest of the clients that are not close to the server are drastically affected.

In GridTorrent, the closest nodes that finish faster are exploited and upload data to the remaining ones, decreasing the overall completion time that gracefully scales with the number of simultaneous leechers. Our method is 3 to 10 times faster both on average and in the worst case, while it exhibits very small variation between the three reported metrics. In Figure 9, we can see the level of cooperation between the leechers as they increase in numbers. We clearly notice a greater variance in the bytes sent by each peer compared to the LAN setting. This shows how adaptive GridTorrent is: Close nodes that finish early contribute to the other peers more than average, while the are few nodes that finish late and cannot share interesting data with the rest of the peers. The WAN experiment depicts in the best way why our protocol is a robust, bandwidth efficient means of file transfer that vastly outperforms current practices.

5.3 DRLS Performance

To evaluate the DRLS implementation we created a scenario where 64 peers, storing about 1000 items perform random lookups and updates at increasing rates. Measuring the mean number of messages and time required for each operation reveals that in the absence of node churn, results remain almost constant, even when constantly doubling the request rate from 1 operation every 6 seconds up to 10 $\frac{operations}{sec}$. This suggests that the underlying XOROS protocol scales to flash-crowd usage patterns as expected. When participants start to leave and new ones enter some messages get lost, so nodes have to wait for timeouts to expire before proceeding with a command. Nevertheless, as the node population settles and routing tables are updated, the performance characteristics return to the expected levels. An interesting find is that during periods of churn, a higher request rate may result in more messages, but this helps nodes react quicker to overlay changes and refresh their routing tables faster.

During this series of experiments we have also investigated the impact of the various DHT parameters: κ, which controls the number of replicas kept for each data item and sets the quorum size for the mutex protocol, α, which defines how many parallel messages can be in-flight during an operation and ϵ, which marks the number of peers that may exhibit arbitrary behavior or fail before a request is completed. As expected, the replication factor κ plays the most important role in shaping both message count and latency. Table 2 summarizes the results for multiple runs of the aforementioned scenario, with different values of κ. Lowering κ reduces the number of nodes that should be contacted in each operation, thus causing the overall latency to drop. However, mean latency is not directly proportional to the number of messages, as a lot of communication is done in parallel. Dividing the latency numbers with the average messaging cost of 80 msec results in the mean number of messages have to be sent in serial order, either due to the protocol or the α parameter. When the network is small, like the case of 64 nodes, we believe that a replication factor of 5 should be enough. On the other hand, when deploying DRLS to a massive number of participants (i.e. a "Desktop Grid"), keeping κ to the default value of 20 can help avoid data loss in case of sudden network blackouts or other unplanned and unadvertised peer problems, even if the messaging cost is higher.

6 Conclusion

In this paper, we describe a P2P-based data management architecture that comprises of GridTorrent and DRLS. GridTorrent is a cooperative data transfer mechanism that maximizes performance by adaptively choosing where each node retrieves file segments from. DRLS is a distributed replica service which is based on a modified kademlia DHT, allowing efficient processing even during node churn. Our proposed solution is compatible with the current Data Grid architecture and can be utilized without any changes by already deployed middleware. Experiments conducted both in LAN and WAN environments (the PlanetLab infrastructure), show that GridTorrent vastly outperforms GridFTP, being up to 10 times faster. Moreover, experiments on DRLS in a dynamic environment show that the benefits of a peer-to-peer network can be readily exploited to provide a scalable Grid service without significant loss in performance. DRLS is able to provide reliable location services even when the load rates multiply.

References

1. The Large Hadron Collider, http://lhc.web.cern.ch/lhc/
2. The GREDIA Project, http://www.gredia.eu/
3. The official site of Globus Toolkit, http://globus.org/toolkit
4. The official BitTorrent client, http://www.bittorrent.org
5. Distributed.net, RSA Labs 64bit RC5 Encryption Challenge, http://www.distributed.net
6. PlanetLab: An open platform for developing, deploying, and accessing planetary-scale services, http://www.planet-lab.org/
7. Allcock, B., Bester, J., Bresnahan, J., Chervenak, A.L., Foster, I., Kesselman, C., Meder, S., Nefedova, V., Quesnel, D., Tuecke, S.: Data management and transfer in high-performance computational grid environments. Parallel Computing 28(5), 749–771 (2002)
8. Allcock, W., Bresnahan, J., Kettimithu, R., Link, M., Dumitresku, C., Raicu, I., Foster, I.: The globus striped gridftp framework and server. In: Proceedings of the ACM/IEEE Conference on Supercomputing, SC 2005 (2005)
9. Anderson, D.: Boinc: A system for public-resource computing and storage. In: Proceedings of the 5th IEEE/ACM International Workshop on Grid Computing (2004)
10. Cai, M., Chervenak, A., Frank, M.: A peer-to-peer replica location service based on a distributed hash table. In: Proceedings of the 2004 ACM/IEEE conference on Supercomputing, Pittsburgh, PA (November 2004)
11. Chazapis, A., Koziris, N.: Xoros: A mutable distributed hash table. In: Proceedings of the 5th International Workshop on Databases, Information Systems and Peer-to-Peer Computing (DBISP2P 2007), Vienna, Austria (2007)
12. Chazapis, A., Zissimos, A., Koziris, N.: A peer-to-peer replica management service for high-throughput grids. In: Proceedings of the 2005 International Conference on Parallel Processing (ICPP 2005), Oslo, Norway (2005)
13. Chervenak, A., Foster, I., Kesselman, C., Salisbury, C., Tuecke, S.: The data grid: Towards an architecture for the distributed management and analysis of large scientific datasets. Journal of Network and Computer Applications (2000)
14. Chervenak, A., Palavalli, N., Bharathi, S., Kesselman, C., Schwartzkopf, R., Stockinger, H., Tierney, B.: Performance and Scalability of a replica location service. In: Proc. of the 13th IEEE International Symposioum on High Performance Distributed Computing Conference (HPDC), Honolulu (June 2004)
15. Cohen, B.: Incentives build robustness in bittorrent. In: Workshop on Economics of Peer-to-Peer Systems, Berkeley, CA, USA (June 2003)
16. Sullivan III, W.T., Werthimer, D., Bowyer, S., Cobb, J., Gedye, D., Anderson, D.: New major seti project based on project serendip data and 100,000 personal computers. In: Astronomical and Biochem ical Origins and the Search for Life in the Universe, Proc. of the Fifth Intl. Conf. on Bioastronomy (1997)
17. Foster, I., Kesselman, C., Tsudik, G., Tuecke, S.: A security architecture for computational grids. In: Proceedings of the 5th ACM conference on Computer and communications security, pp. 83–92. ACM Press, New York (1998)
18. Kaplan, A., Fox, G., von Laszewski, G.: Gridtorrent framework: A high-performance data transfer and data sharing framework for scientific computing. In: Proceedings of GCE 2007, Reno, Nevada (2007)
19. Maymounkov, P., Mazières, D.: Kademlia: A peer-to-peer information system based on the xor metric. In: Druschel, P., Kaashoek, M.F., Rowstron, A. (eds.) IPTPS 2002. LNCS, vol. 2429, p. 53. Springer, Heidelberg (2002)

20. Peterson, L., Anderson, T., Culler, D., Roscoe, T.: A blueprint for introducing disruptive technology into the internet. In: Proceedings of HotNets–I, Princeton, NJ (October 2002)
21. Ripeanu, M., Foster, I.: A decentralized, adaptive, replica location service. In: Proceedings of the 11th IEEE International Symposium on High Performance Distributed Computing (HPDC-11 2002), Edinburgh, UK (July 2002)
22. Sherwood, R., Braud, R., Bhattacharjee, B.: Slurpie: A cooperative bulk data transfer protocol. In: Proceedings of IEEE INFOCOM (March 2004)
23. Stockinger, H., Samar, A., Holtman, K., Allcock, B., Foster, I., Tierney, B.: File and object replication in data grids. Cluster Computing 5(3), 305–314 (2002)
24. Thain, D., Basney, J., Son, S.-C., Livny, M.: The kangaroo approach to data movement on the grid. In: Proceedings of the Tenth IEEE Symposium on High Performance Distributed Computing, HPDC10 (2001)
25. Wei, B., Fedak, G., Cappello, F.: Collaborative data distribution with bittorrent for computational desktop grids. In: Proceedings of the 4th International Symposium on Parallel and Distributed Computing, ISPDC 2005 (2005)
26. Weigle, E., Chien, A.A.: The composite endpoint protocol (cep): Scalable endpoints for terabit flows. In: Proceedings of the IEEE International Symposium on Cluster Computing and the Grid, CCGrid 2005 (2005)
27. Zissimos, A., Doka, K., Chazapis, A., Koziris, N.: Gridtorrent: Optimizing data transfers in the grid with collaborative sharing. In: Proceedings of the 11th Panhellenic Conference on Informatics, Patras, Greece (2007)

CA3M: A Runtime Model and a Middleware for Dynamic Context Management

Chantal Taconet[1], Zakia Kazi-Aoul[2], Mehdi Zaier[1], and Denis Conan[1]

[1] Institut Télécom; Télécom SudParis; CNRS UMR SAMOVAR
9 Rue Charles Fourier, 91011, Évry Cedex, France
[2] ISEP; 28 rue Notre Dame des Champs, 75006 Paris Cedex, France
{Chantal.Taconet,Mehdi.Zaier,Denis.Conan}@it-sudparis.eu, zkazi@isep.fr

Abstract. In ubiquitous environments, context-aware applications need to monitor their execution context. They use middleware services such as context managers for this purpose. The space of monitorable entities is huge and each context-aware application has specific monitoring requirements which can change at runtime as a result of new opportunities or constraints due to context variations. The issues dealt with in this paper are *1)* to guide context-aware application designers in the specification of the monitoring of distributed context sources, and *2)* to allow the adaptation of context management capabilities by dynamically taking into account new context data collectors not foreseen during the development process. The solution we present, CA3M, follows the model-driven engineering approach for answering the previous questions: *1)* designers specialised into context management specify context-awareness concerns into models that conform to a context-awareness meta-model, and *2)* these context-awareness models are present at runtime and may be updated to cater with new application requirements. This paper presents the whole chain from the context-awareness model definition to the dynamic instantiation of context data collectors following modifications of context-awareness models at runtime.

Keywords: ubiquity, context-awareness, meta-modelling, model at runtime.

1 Introduction

Nowadays, the use of mobile devices (*e.g.*, smart phones, PDA) is getting more and more popular. Mobile devices are not only used as simple phones as the range of distributed applications developed for mobile devices increases drastically. These trends indicate that more and more users depend on ubiquitous applications for their daily life.

Ubiquitous applications must gain in capabilities to adapt themselves and also to manage their autonomy. The term *context-aware application* appeared in 1994 [1]. Since then, many models that describe applications contexts have been proposed, many context-aware middlewares and services have been designed, and many context-aware applications have been implemented. However, about

R. Meersman, T. Dillon, P. Herrero (Eds.): OTM 2009, Part I, LNCS 5870, pp. 513–530, 2009.

15 years later, some improvements in software engineering are still necessary to enable ubiquitous context-aware applications to be easy to design, to implement and to reconfigure.

Context-aware applications need both *1)* to collect high-level observations meaningfull to them and *2)* to identify situations under which they need adaptations. High-level observations may be computed from different distributed sources such as operating systems, user profiles and environment sensors. Context managers are services in charge to compute those high-level observations [2,3,4,5]. In this paper, we want to enable applications to dynamically make use of new context managers coming from different frameworks in open environments. The design of context-aware applications for ubiquitous environments has seen the advent of a new stakeholder in the application design process that we call the "context-awareness designer" in the sequel of the paper. Context-awareness designers need the support of model engineering to manage the wide diversity of context data in ubiquitous environments. Several context modelling approaches such as context profiles and context ontologies have been surveyed in [6]. Specifying application context-awareness with Model Driven Engineering (MDE) enables designers to draw links between the context models and the application models. Context-Awareness Domain Specific Models (DSM) further the MDE approach and promote context-awareness models to express the dynamic variability due to context changes [7,8]. However, context-awareness models still lack the ability to define the contracts which link context-aware applications to context managers. Thus, the approach promoted by the paper is to use MDE for guiding context-aware application designers in the specification of the monitoring of distributed context sources.

MDE is often used statically to automate code generation. Recent research works propose to reify models at runtime [9]. This is the direction we follow in this paper in order to take into account new ubiquitous environments with new context sources requirements not foreseen during the development process. Models at runtime contribute to maintain consistency across design time decisions and runtime adaptations by enabling models to be updated at runtime.

We propose CA3M, a Context-Aware Middleware based on a context-awareness Meta-Model for the following goals. Firstly, CA3M guides context-aware application designers in the specification of the monitoring of distributed context sources. Secondly, CA3M enables applications to adapt their behaviours by dynamically taking into account new context data sources not foreseen during the development process. In our approach, in addition to classical application models such as UML models, a context-awareness model is built by the context-awareness designer. This model may be updated during the application lifetime —*i.e.*, design, deployment, runtime. At runtime, the CA3M middleware dynamically constructs bridges between context-aware applications and context managers to reify the monitoring elements of the context-awareness model of the application.

The outline of the paper is as follows. In Section 2, we motivate and give the objectives of CA3M through an illustrative scenario. Then, we present an overview of CA3M in Section 3. In Section 4, we introduce the CA3M meta-model and show the corresponding context-awareness models for the motivating scenario. We detail the implementation and provide experimental evaluation considerations in Sections 5 and 6. Then, we compare our contribution with regard to related work on context-awareness modelling and context management middleware in Section 7. Finally, we conclude and present some perspectives of our work in Section 8.

2 Motivation and Objectives

We begin this section by introducing the terminology used in the sequel of the paper. Then, we present an illustrative scenario in Section 2.2 and finally we bring out our objectives in Section 2.3.

2.1 Terminology

We present here terminology adopted for CA3M and especially in the CA3M meta-model. Some of the following concepts such as entity and situation were already present in Dey's context definition [10]. We have chosen simple definitions easy to manipulate by a context-awareness designer.

An *entity* is an element representing a physical or logical phenomenon (person, concept, etc.) which can be treated as an independent unit or a member of a particular category, and to which "observables" may be associated. A mobile device is an entity. An *observable* is an abstraction which defines something to watch over (observe). The battery level of a mobile device is an observable. Some observables may be computed from other observables, they are called *interpreted observables*. An *observation* represents the state of an observable at a given time. It is obtained from a context-manager named *collector* in the sequel of the paper. An *adaptation situation* is an observable which allows to track down a change of state in the space of the information of context. This change of state may require a reaction in the system. Such a reaction is called an *adaptation*. A mobile device battery state is an example of observable which may take a finite number of values (*e.g.*, LowBattery, AlmostLowBattery or NormalBattery) and each state may lead to a different application behaviour. The link between an observable and an application is defined through a context-awareness contract. *A context-awareness contract* may define for example the quality of context required by the application as well as the mode of communication between the application and the collector —*i.e.*, observation or notification.

2.2 Mobile-Chat Application Scenario

We illustrate the dynamicity of an application context-awareness with a mobile-chat application. This scenario brings into play distributed context sources. It shows how the application behaviour is adapted according to context changes and highlights the necessity of context-awareness modification at runtime.

Eric is a student in physics travelling from Paris to Geneva by train. He has arrived in Paris train station and must wait a couple of minutes before going on board. He decides to discuss with a friend of him named Susan. He *launches the mobile-chat application. The basic version of this application is already downloaded in his mobile. This version may be extended at runtime by several plugins if necessary. Eric uses the WiFi connection offered in the train station. The "outdoor mode" allows Eric and Susan to use the voice option. On Susan demand, her location is shown on Eric's mobile as well as her distance from Eric. For this demand, the plugin "monitoring a peer location" is added at runtime by the mobile chat application already running on Eric's mobile. New observables (*e.g.*, Eric location, Susan location, and Eric and Susan distance) are added as well for this new plugin.*

Thanks to the good network connection between Eric and Susan, they can add the video option to the mobile-chat application. After a few minutes, another friend, named Rob, joins the conversation. Rob is also in an "outdoor mode" and has a good network link quality.

Eric knows that if the battery level goes down, the video option will be disabled for everyone, so he plugs in his mobile. Then, he wants to share an mp3 file with his two friends. During the file transfer, the link quality of the connection of Susan goes down. The file is then put into the pipe queue and will be transferred later. The video option is switched off for Susan but kept "on" between Eric and Rob.

*Now, it's time for Eric to go on board. As soon as he enters the train, both his cell phone and his laptop switch to the "indoor mode" in order not to disturb the other passengers with a voice communication. Eric sits down and plugs in his laptop. The conversation with his friend was not over. However, the "indoor mode" forced the application to move from a voice-based communication to a text-based communication. In addition, the WiFi connection is automatically replaced by a 3G connection. In this situation —*i.e.*, Eric using a 3G connection—, the video option cannot be used between Eric and his peers.*

This scenario justifies the context-awareness of the mobile-chat application in order to cope with different user profiles and preferences, different terminal capabilities, and different elements of the environment in a distributed setting. The context-awareness designer may define the environment which should be taken into account before and during the execution of the mobile-chat application. Context-awareness on Eric's mobile depends not only on Eric's context, but also on Susan's and Rob's ones (*e.g.*, the kind of the connection on both sides may be used to evaluate the quality of the link between Eric and his peers). Furthermore, the context-awareness model has to be updated at runtime. Indeed, new entities and new observables have to be added as new friends are integrated into the chat. The context-awareness model may change as well when new plugins are added to the basic version of the mobile-chat. For example, the geolocation plugin needs the monitoring of other elements such as the location of a peer and the distance between two peers.

2.3 Objectives

The previous scenario illustrates three main objectives handled in the design of CA3M.

Firstly, the scenario puts the stress on distributed monitoring. For instance, the basic ubiquitous mobile-chat application instantiated on Eric'c mobile terminal needs to monitor both Eric's and Susan's contexts. The decision concerning the video option depends on distributed context data. Therefore, context-awareness designers should be able to model distributed observations.

Secondly, a context-awareness model can specify the initial kind of observables to monitor (*e.g.*, link quality). However, at design time, the models can indicate neither on which computer they should be observed nor how many instances should be taken into account (*e.g.*, the number of persons involved in the chat is unknown). New plugins may also modify the monitoring requirements and lead to model updates. More generally, the adaptation may need runtime reconfigurations because of new execution conditions (which include the availability of new context sources). Subsequently, it is necessary to enable the context-awareness model to evolve during runtime.

Thirdly, for a given observable, several context sources may be utilised. The sources come from several providers, and may have different application programming interfaces (API). For instance, the user location may be measured by the GPS (Global Positioning System) of the user's mobile device or may be obtained by the nearest GPS found. It should not be the role of the context-awareness designer to choose the concrete collector to be used. The concrete collector is unknown at design time and is chosen afterwards and even at runtime. The middleware should be able to provide a meaningfull observation to the application for several collectors with different APIs.

3 CA3M Overview

CA3M is a framework for both the design and the execution of context-aware applications. We briefly describe these two parts in Sections 3.1 and 3.2.

3.1 Design Overview

Figure 1 depicts the CA3M context-awareness design process with, from left to right, the stakeholders, the activities, and the resulting artefacts.

The figure distinguishes roughly two kinds of activities: *(1) context specification and design* and *(2) application design*. The *context specification and design* comprises the design of collectors and the specification of contexts. It produces two kinds of artefacts: implementations and models. The presentation of this modelling task is out of the scope of the paper and can be found in [11]. Of course, the APIs of the collectors vary and this task should benefit from standardisation actions. Examples of proposed standards for modelling collectors are SensorML [12] for sensors and CIM [13] for operating system resources.

In this paper, we focus on the lower part of Figure 1, that is on application design. We divide this activity into two large-grain tasks to promote a new stakeholder: the context-awareness designer. The application designer produces the application model and classes. The context-awareness designer produces context-awareness models as explained in Section 4. In summary, we apply the principle

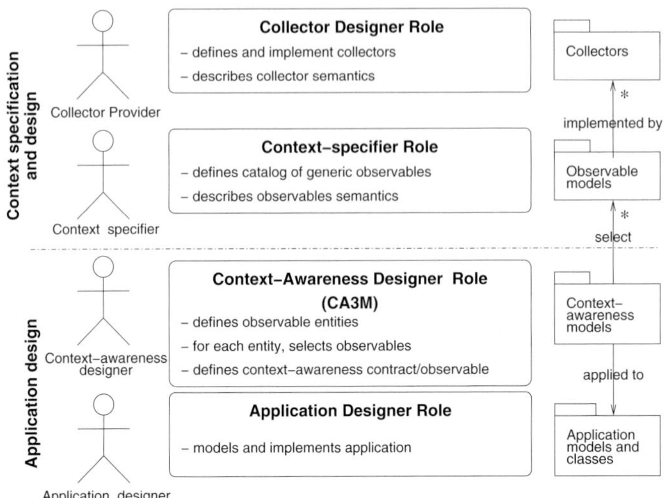

Fig. 1. Separation of design tasks for producing context-aware applications

of separation of concerns twice: *1)* separation of context data providers from context data users (context designers and context-awareness designers), and *2)* separation of application concerns from context-awareness concerns when designing and executing the application. In addition, context-awareness models are built at design time in order to be manipulated at runtime.

Context-awareness models have to conform to a specific meta-model presented in Section 4.1. This meta-model introduces meta-classes such as entities, observables, collectors, and context-awareness contracts. An instance of the context-awareness model is created at runtime which may be updated at runtime for example as new plugins are added to the mobile-chat application.

3.2 Runtime Overview

Figure 2 shows the runtime architecture of CA3M. The architecture is divided into the context-aware application, CA3M, the distributed collectors, and the distributed context sources. A context-aware application accesses context management mechanisms through CA3M. CA3M drives the monitoring of the environment according to the context-awareness model. The distributed collectors provide context management and context interpretation. They collect data from distributed context sources, and they interpret and compute new high-level observations. The observations may be obtained from entities at different levels of the architecture: the system level, the network level, the environment level, but also the software level including observations of the context-aware application itself. CA3M corresponds to knowledge manager (the "K") of the K-MAPE autonomic computing loop presented in [14]. CA3M model manager is in the knowledge part of the loop, the collector is in the monitoring and analysing part

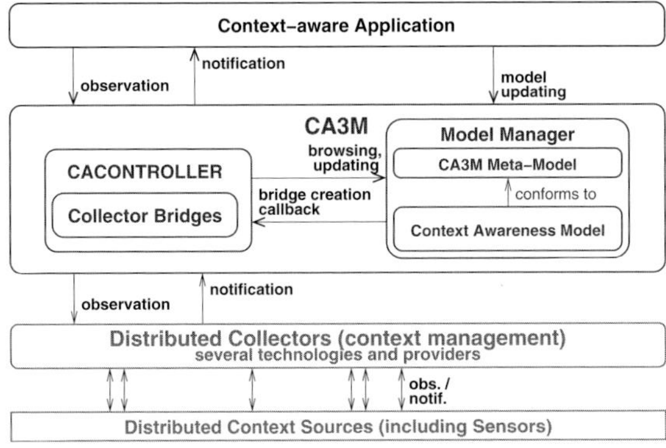

Fig. 2. Runtime architecture of CA3M

of the loop, the CAController is in the planning part, and the application is of course the execution part.

In this architecture, at any level, the interactions may be either top-down, or bottom-up. The top-down interactions correspond to the observation mode: the upper level synchronously requesting an observation. In the observation mode, since the upper level initiates the exchange, it controls the interruption of the application service. The drawback of this mode is that the upper level must know when an interaction is relevant *w.r.t.* its current context execution, that is when there is a significant probability of a meaningful context change happening.

The bottom-up interaction is called the notification mode. In this latter case, the contract defines when the lower level notifies the upper level: periodically or when the observation goes past a given threshold from the last notified value, or even at any change of the observation value. For instance, a notification may be sent to the application if the battery state changes (*e.g.*, from "Normal-Battery" to "AlmostLowBattery"). In the notification mode, the application is less impacted by the monitoring of its environment than in the observation mode, provided that the application is able to express its contract. The context-awareness contract includes application operations to be called on notification. In conclusion, following [2], we decide to provide the two modes of interaction.

The collector bridge is defined following the bridge pattern [15]. The objective of a bridge is to decouple an abstraction from its implementation so that the two can vary independently. We use the bridge pattern for several reasons. Firstly, there may be plenty of collector implementations for a given observable. Secondly, collector may have slightly different APIs and we do need to decouple the application code from the collector implementation interfaces. Finally, we want runtime binding to the collector implementations and we wish to hide some tricky parts of the collector interfaces to the application programmer. The interfaces of the collector bridge abstraction is presented in Section 5.1.

CA3M comprises the CAController and the Model Manager. The CAController is in charge of binding to concrete collectors, creating collector bridges, and notifying the application when necessary as defined by the contracts. The model manager loads the application context-awareness model. The model manager handles query requests about the structure of the model as well as update requests to deal with runtime modifications of the model.

There are two kinds of interactions between the CAController and the Model Manager. Firstly, the CAController may use query operations for browsing the model. This kind of interaction is necessary after the initial loading of the model to create a bridge for each observable foreseen at the design phase. Secondly, in the case of model change, such as the addition or removal of an observable, the model manager triggers callbacks to the CAController in order to create or remove a collector bridge.

4 Context-Awareness Modelling

We present in this section a generic and extensible way to model the context-awareness of any application using the MDE approach. We describe the context awareness meta-model in Section 4.1. We illustrate our solution by modelling the context-awareness of the mobile-chat application in Section 4.2.

As shown in Figure 1, the context-awareness meta-model depends on the observable meta-model. This enables us (*i*) to share observable models between several context-aware applications and (*ii*) to exploit several observable models coming from different providers. The *observable meta-model* defines the observable and the interpreted observable concepts allowing them to be independent from applications. Thus, each observable model is then a catalog of pre-defined observables at the disposal of context-awareness designers. A context-awareness designer selects observables from one or several observable models which are relevant for an application and links the observables to entities. The concepts manipulated by the context-awareness designers are defined in the *context-awareness meta-model* which is detailed in the rest of this section.

4.1 CA3M Context-Awareness Meta-model

Figure 3 describes the CA3M context-awareness meta-model. The ContextAware System meta-class is the entry point of this meta-model. The left part of the meta-model defines the entities, their observables, the links between entities, the interpreted observables and the adaptation situations. The right part of the meta-model defines the context-awareness contracts. A context-awareness contract may be specialised for different context-awareness control mechanisms. A context-awareness contract is associated to an observable. For a notification contract, a context-awareness contract defines the events which trigger notifications and the class in the application model to be called in case of notification.

Entity represents a logical or physical element to be observed, *e.g.* a device. It allows a context-aware system to differentiate several distributed observables from different physical or virtual entities, *e.g.* the bit rate of two devices. An entity may be linked to another entity through the EntityRelation meta-class. An entity is linked to several Observable. An InterpretedObservable is linked to several source observables through the derivedFrom association, *e.g.* the battery

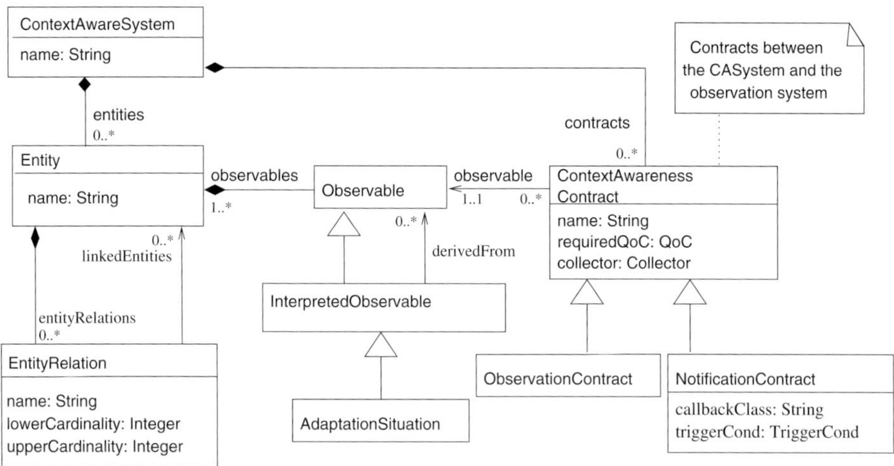

Fig. 3. The context-awareness meta-model

state is derived from the battery level and the battery plugged observables. The type of the observables necessary to compute an interpreted observable is provided by the observable meta-model, not described in this paper (see [11]). When an interpreted observable is added to the model, the type of the source observables (defined by the derivedFrom association) are verified with the source types defined in the observable model. As type of the sources are defined in the observable model, source observables may be omitted from the context-awareness model. For example, the battery state, the battery level and the battery plugged observables are at evidence linked to the same entity (the user device). The battery level and the battery plugged observables are shown in dotted line in Figure 4 to show that they may be omitted from the context-awareness model. An AdaptationSituation is a kind of observable which has the characteristic to take a finite number of domain values and which is used to identify adaptation situations.

4.2 Mobile-Chat Context-Awareness Model

Figure 4 models the context-awareness of the basic mobile-chat application executing on Eric's mobile. This model conforms to the context-awareness meta-model presented in Figure 3. UserDevice, Environment and PeerDevice are the entities.

On each device, the observables BitRate and NetworkType are inputs to compute the interpreted observable LinkQuality. The context-awareness designer models the adaptation situation PeerLinkQuality computed from the interpreted observables (*i*) link quality of UserDevice and (*ii*) link quality of PeerDevice. For example, the adaptation situation value `VideoQuality` means that the video option is authorised in accordance with the link quality between the two

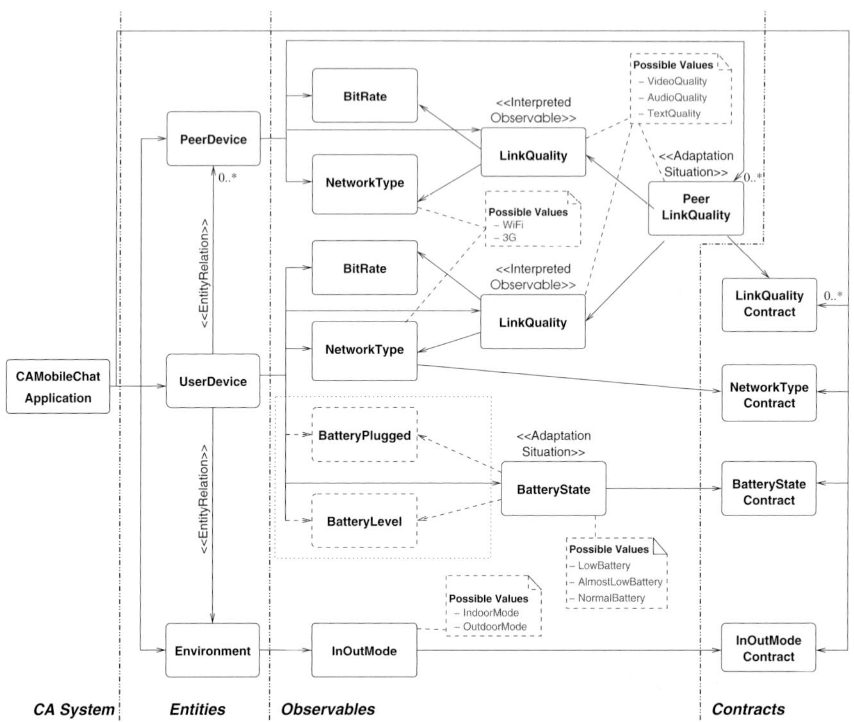

Fig. 4. The context-awareness model for the mobile-chat application

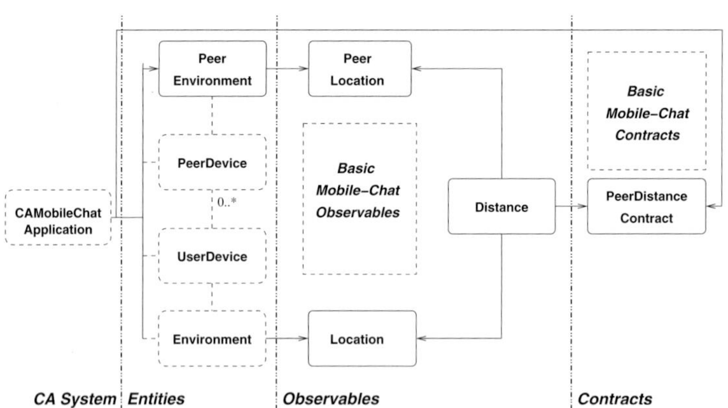

Fig. 5. Model extensions for the plugin location

parties. Several adaptation situations PeerLinkquality are instantiated during the execution (*e.g.*, one for Eric-Susan's link).

The possible values of the interpreted observable BatteryState (computed from its source values) can be `NormalBattery`, `AlmostLowBattery` and `LowBattery` (where: $NormalBattery = BatteryPlugged \vee BatteryLevel > 20\%$; $AlmostLowBattery = \neg BatteryPlugged \wedge BatteryLevel < 20\% \wedge BatteryLevel > 10\%$; $LowBattery = \neg BatteryPlugged \wedge BatteryLevel < 10\%$). The observable InOutMode is associated to the entity Environment. According to Eric's scenario, it takes two possible values: `IndoorMode` and `OutdoorMode`.

Several context-awareness contracts are represented in this figure. LinkQualityContract, BatteryStateAdaptContract, NetworkTypeContract and InOutModeContract are notification contracts. The class of the application model refered by each contract has to implement a notification interface for the callbacks.

Figure 5 shows the elements added to the model when the *monitoring a peer location* plugin is added to the application. Note that the initial model instance has already changed during the execution when Rob is added to the discussion. Here, another PeerDevice entity is instantiated and the link Eric-Rob is created on the fly.

5 CA3M Prototype Implementation

In this section, we present the CA3M implementation. We describe the CA3M class diagram in Section 5.1. Then, we explain modelling implementation choices in Section 5.2. As COSMOS is the context manager chosen for our evaluation, we describe COSMOS collector bridge in Section 5.3.

5.1 CA3M Class Diagram

Figure 6 depicts the CA3M UML class diagram representing the interfaces between a context-aware application and CA3M. We explain how to use CA3M in notification and in observation modes, and how to modify the application model to add or remove entities and observables.

The CAController may use either a static model (class CAControllerStaticModel) or a dynamic model (class CAControllerDynamicModel). In this article, so far, we focused on the dynamic model which allows CA3M to modify the model at runtime through the interface ICAModelUpdate. But if the mobile device, because of memory or library constraints, cannot afford the dynamic one, the static one may be used instead. In this latter case, CAControllerStaticModel is produced statically by transformation for a given model and the model cannot be updated at runtime.

In the mobile-chat scenario, we present the dynamic case. Each time a new peer user connects to the chat, a new PeerDevice entity and its associated observables are added to the model. Next, CA3M automatically creates a new bridge for each new observable, each bridge providing the interface ICollector for the observation mode. For the notification mode, the class defined in the notification contract should provide the interface ICollectorNotification.

5.2 Modelling Implementation Choices

The most popular meta-modelling languages for defining DSMs are MOF (*Meta-Object-Facility*) [16], ECORE from *Eclipse Modelling Framework* (EMF) [17]

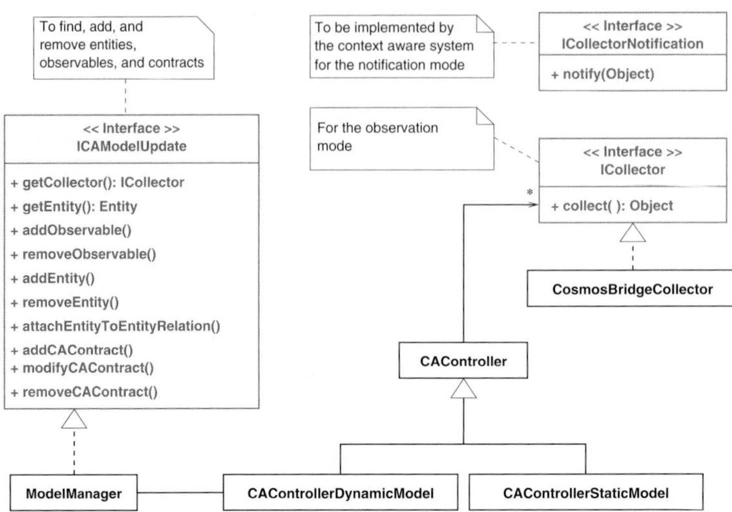

Fig. 6. CA3M related application interfaces

and UML Profile [18]. Designing the context-awareness model as a UML profile was not possible because with UML profiles we could not define associations between profile meta-classes. Between MOF and EMF, we have chosen EMF because of the availability of many EMF tools.

For the static CAController, we use ECORE models for transformation purpose to generate CAControllerStaticModel classes. When possible, the dynamic CAController is used instead. At runtime, the model manager loads an application context-awareness model, accessed and updated through the EMF generated API. Through this API, new entities, observables, and contracts may be added to the model at runtime. Thanks to an EMF adapter, insertions of observables trigger the creation of bridge collectors.

5.3 CA3M Bridge Illustrated with COSMOS Collector Bridge

CA3M architecture allows the CAController to be interfaced with several context management frameworks. The constraints on the collector framework are the following ones. The collector framework should provide notification and observation modes and be able to compute high-level observations from distributed observations. At least one collector bridge has to be implemented per collector framework to wrap collector API with CA3M API. The bridge class has to be designed to enable the bridge to work with any observation class. Several bridges may be implemented according to the kind of binding from the bridge to the collector. We design two bindings: one for the connection to an external collector and the other one for an instantiation of the collector into CA3M.

For our evaluation, we have interfaced CA3M with the COSMOS framework [19]. COSMOS offers tools to collect, interpret and process context data. We have chosen COSMOS because it provides developers with the ability to define new finely tuned interpreted observables. The basic structuring concept of COSMOS is the *context node*. The architecture of a context node is component-based. For instance, the component (or context node) BatteryState is the composition of the context nodes BatteryPlugged and BatteryLevel with a context operator realising the logical and comparison operations mentioned previously. Observation and notification modes are provided through Pull and Push interfaces, respectively.

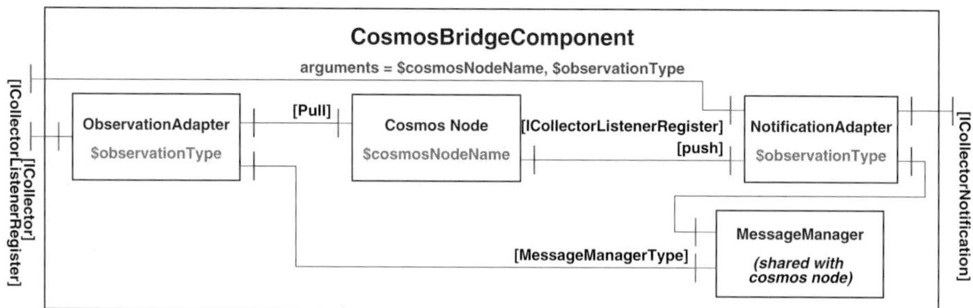

Fig. 7. The COSMOS Bridge component

Figure 7 presents the COSMOS bridge component. The bridge hides COSMOS specific interfaces and COSMOS internals using the FRACTAL component model [20] and Dream [21] message management for context reporting. It includes adapters to transform COSMOS specific message chunks to application observation types. The bridge provides CA3M interfaces for collecting observations and registering notification handlers. It forwards notifications to all the registered ICollectorNotification interfaces.

6 Experimental Evaluation

6.1 Performance Evaluation of the Prototype

We present the overhead incurred by using CA3M compared to a direct usage of COSMOS collectors by the application. We have conducted performance measurements on a laptop PC with the following software and hardware configuration: 2.8GHz processor, 512MB of RAM, GNU/Linux Fedora 9, Java Virtual Machine Sun JDK 1.6, COSMOS 0.1.5, FRACTAL implementation Julia 2.5.2. Each test was run 1000 times. A garbage collection and a warm-up phase occurred before each measure.

The Table 1 presents the average measures with a 95% confidence interval. The table includes the following overhead measurements. The two first lines presents the time necessary for the initial model loading. Model loading concerns reading the EMF model file and reifying the concepts in EMF objects. The third line gives the overhead time for the collector instantiation, the overhead comes from the bridge instantiation. The fourth line presents the overhead time for an observation. The overhead comes from the transformation of an observation from a Cosmos Chunk to an application Java object. Lines five and six give the overhead concerning the memory usage after all bridge instantiations (the memory usage is given by the difference between `totalMemory` et `freeMemory` in the JVM). The CA3M overhead follows from model management (`emf` library), and bridge management (especially the reflexivity necessary to interconnect a COSMOS bridge to any kind of context node).

Table 1. CA3M overhead measurements

1-—model load (1 entity, 1 observable, 1 contract)	$723ms \pm 4$
2-—model load (100 entities, 100 observables, 100 contracts)	$785ms \pm 5$
3-—instantiation overhead (1 node)	$141ms \pm 8$
4-—observation overhead	$0.439ms \pm 0,003$
5-—memory overhead (1 entity, 1 observable, 1 contract)	$0,56MB \pm 0,05$
6-—memory overhead (100 entities, 100 observables, 100 contracts)	$1,26MB \pm 0,07$

6.2 CA3M Experimentation

We evaluate the context-awareness meta-model with the mobile-chat application and also with a mobile commerce application. For these applications, we are able to share common ubiquitous observables (*e.g.*, BatteryState) defined in an ubiquitous observable model. In a second step, we define each application context-awareness models. In the two applications, we are able to reconfigure application context-awareness both statically and at runtime. The complete chain from context-awareness modelling to context collector bridge creation, observation and notification is tested. We validate the whole process with the COSMOS context manager. Through these experimentations, the following criteria are validated. The observable model is reusable through different applications. CA3M enable designers to define interpreted observables computed from distributed source observables. A context-awareness model may be modified at runtime. And a collector bridge may be instantiated at runtime enabling runtime collector binding.

7 Related Work

Since CA3M links context-management frameworks, context-awareness models and context-aware middleware, the related work section deals with these three kinds of research works. Many context management framework [3,4] have been designed without context modelling. Due to the variety of contexts to be collected and analysed, we argue that context management needs the support of abstract context modelling.

Several research works address context-awareness with the MDE approach. ContextUML [22] was one of the first DSM for context-awareness. It presents a context-awareness meta-model for Web services. ContextUML models associate context constraints with filtering operations which should be applied to input or output messages of a Web service. CAPPUCINE [8] describes an MDE approach for dynamically producing product lines according to context information. Those models put the stress on adaptation mechanisms rather than context modelling. Those models are used for transformation purpose. CA3M, presented in this paper, also defines a DSM for adding a context-awareness concern in the application modelling stage. CA3M presents two main differences with the above research works. Firstly, CA3M modelling concepts enable application designers to express complex situations computed from distributed context observations. Secondly, the context-awareness model is not only used for transformation purpose, but is also available at runtime. Therefore, the application or the middleware are able to add new observables at runtime. In addition, Reichle & *al.* [23] proposes ontology-based context model and model framework, and the ontology model is present at runtime. Context model interactions are performed using CQL (Context Query Language). Some concepts present in this model such as entity and observables are similar to the CA3M concepts. Other ones such as context-awareness contracts do not exist and the framework does not propose the notification mode. Finally, [24,25] consider context-awareness with model at runtime for application adaptation purpose but not for monitoring adaptation.

Several middleware solutions have been proposed for managing context-awareness. CARISMA [26] proposes to define context-aware profiles for tuning the behaviour of middleware services according to context information. These profiles are available at runtime. Applications may modify these profiles through a reflexive API. Profiles for context management are not addressed by CARISMA. CA3M offers also a reflexive API, not only to modify application behaviours, but also to add new observables which will be collected by new context collectors. Context-Toolkit [4] allows application developers to attach application handlers to context widgets. Context Toolkit triggers call-back to these handlers when context values change. The links between the handlers and the collectors are programmed rather than modelled in notification contracts as this is the case in CA3M. RCSM [27] offers CA-IDL, a language to define context situations and to specify the operations to trigger in these situations. Adaptation situations are evaluated by RCSM which triggers reactive operations at runtime. RCSM limitations come from the limited number of context information available in the CA-IDL grammar. The triggering of operations and the context monitoring are defined statically at compilation time.

Another issue of this article is to be able to be connected to several context management frameworks with different APIs. This may be handled statically by model transformation such as proposed in several context-awareness MDE works such as [28]. But this approach does not enable the middleware to choose the collectors at runtime. In MUSIC [5], the context-management framework includes an observation and a notification mode and the framework may be compared to the CA3M CAController. As in CA3M, they consider the dynamically insertion and removal of so called context plugins in the middleware. When a new collector is connected by the framework, the collector framework provides an archive which includes a wrapper between the collector and the framework. As a consequence, one wrapper has to be provided for each collector. In CA3M, we provide instead a bridge for all the collectors of the same family (*e.g.*, one bridge for all the COSMOS collectors). Thanks to bridges, CA3M may be connected to various kinds of context management frameworks more easily.

8 Conclusion and Perspectives

In this article, we have presented the CA3M middleware. Our contributions are the following ones. CA3M provides tools to easily reconfigure the context-awareness of ubiquitous applications. Firstly, we have defined a context-awareness meta-model which defines concepts chosen for a new stakeholder that we call the context-awareness designer. This meta-model defines entities, observables, and context-awareness contracts. It allows designers to model distributed observables. We validate this meta-model through the definition of models for several ubiquitous applications. Secondly, we provide a middleware which, based upon an ubiquitous application context-awareness model, is able to connect to different context-manager collectors. CA3M offers two kinds of interactions between context managers and applications: the observation mode and the notification mode. We consider the design of bridges for binding to various context managers and we have validated and evaluated the approach building the bridge component for the COSMOS context manager. In summary, the main contribution of our proposition is to enable to update application context-awareness models at runtime. The advantage is to allow autonomous context-awareness.

This work may be extended in several directions. The MDE approach may be used not only to connect applications to collectors but also to produce high-level collectors from existing lower-level collectors. In addition, we have developped a bridge for the COSMOS context manager and we plan to validate our approach with additional context managers. We also believe that the model at runtime approach can benefit to a better choosing of collectors at runtime using a collector discovery service. Last but not least, we plan to extend CA3M to deal with adaptation mechanisms for changing the behaviour or the structure of applications. We intend to add various context-awareness adaptation contracts in the model and adaptation mechanisms in the middleware.

References

1. Schilit, B., Theimer, M.: Disseminating Active Map Information to Mobile Hosts. IEEE Network 8(5), 22–32 (1994)
2. Baldauf, M., Dustdar, S., Rosenberg, F.: A Survey on Context Aware Systems. International Journal of Ad Hoc and Ubiquitous Computing 2(4), 263–277 (2007)
3. Coutaz, J., Crowley, J., Dobson, S., Garlan, D.: Context is Key. CACM 48(3), 49–53 (2005)

4. Dey, A., Salber, D., Abowd, G.: A Conceptual Framework and a Toolkit for Supporting the Rapid Prototyping of Context-Aware Applications. Special issue on context-aware computing in the Human-Computer Interaction Journal 16(2-4), 97–166 (2001)
5. Paspallis, N., Rouvoy, R., Barone, P., Papadopoulos, G., Eliassen, F., Mamelli, A.: A Pluggable and Reconfigurable Architecture for a Context-aware Enabling Middleware System. In: Meersman, R., Tari, Z. (eds.) OTM 2008, Part I. LNCS, vol. 5331, pp. 553–570. Springer, Heidelberg (2008)
6. Strang, T., Linnhoff-Popien, C.: A context modeling survey. In: UbiComp Workshop on Advanced Context Modelling, Reasoning and Management, Nottingham/England, September 2004, pp. 34–41 (2004)
7. Bencomo, N., Sawyer, P., Blair, G., Grace, P.: Dynamically Adaptive Systems are Product Lines too: Using Model-Driven Techniques to Capture Dynamic Variability of Adaptive Systems. In: 2nd International Workshop on Dynamic Software Product Lines (DSPL 2008), Limerick, Ireland (September 2008)
8. Parra, C., Blanc, X., Duchien, L.: Context Awareness for Dynamic Service-Oriented Product Lines. In: 13th International Software Product Line Conference (SPLC), San Francisco, CA, USA (August 2009)
9. Bencomo, N., Blair, G., France, R., Munoz, F., Jeanneret, C. (eds.): 3rd Workshop on Models @ runtime, Toulouse, France (September 2008)
10. Dey, A.: Providing Architectural Support for Building Context-Aware Applications. PhD thesis, College of Computing, Georgia Institute of Technology (December 2000)
11. Taconet, C., Kazi-Azoul, Z.: Context-Awareness and Model Driven Engineering: Illustration by an e-commerce application scenario. In: CMMSE Workshop on Context Modeling and Management for Smart Environments, London, UK, November 2008, pp. 864–869 (2008)
12. Open Geospatial Consortium: Opengis sensor model language (sensorml): Implementation specification, version 1.0.0. OpenGIS Implementation Specification (July 2007)
13. Distributed Management Task Force: Common information model (cim): Infrastructure specification, version 2.3 final. OpenGIS Implementation Specification (October 2005)
14. Kephart, J., Chess, D.: The Vision of Autonomic Computing. IEEE Computer 36(1) (January 2003)
15. Gamma, E., Helm, R., Johnson, R., Vlissides, J.: Design Patterns: Abstraction and Reuse of Object-Oriented Design. Addison-Wesley Professional Computing Series. Addison Wesley Professional, Reading (October 1993)
16. Object Management Group: Meta Object Facility (MOF) Core Specification Version 2.0. OMG document formal/06-01-01 (January 2006)
17. Budinsky, F., Merks, E., Steinberg, D.: Eclipse Modeling Framework 2.0, March 2008. Addison Wesley, Reading (2008)
18. Object Management Group: UML 2.0 Superstructure Specification v2.1.1. OMG documents formal/2007-02-05 (February 2007)
19. Conan, D., Rouvoy, R., Seinturier, L.: Scalable Processing of Context Information with COSMOS. In: Indulska, J., Raymond, K. (eds.) DAIS 2007. LNCS, vol. 4531, pp. 210–224. Springer, Heidelberg (2007)
20. Bruneton, E., Coupaye, T., Leclercq, M., Quema, V., Stefani, J.B.: The Fractal Component Model and Its Support in Java. Software—Practice and Experience 36(11), 1257–1284 (2006)

21. Leclercq, M., Quema, V., Stefani, J.B.: DREAM: a Component Framework for the Construction of Resource-Aware, Configurable MOMs. IEEE Distributed Systems Online 6(9) (September 2005)
22. Sheng, Q., Benatallah, B.: ContextUML: A UML-Based Modeling Language for Model-Driven Development of Context-Aware Web Services. In: ICMB, Sydney, Australia, July 11-13, pp. 206–212 (2005)
23. Reichle, R., Wagner, M., Khan, M., Geihs, K., Lorenzo, J., Valla, M., Fra, C., Paspallis, N., Papadopoulos, G.: A Comprehensive Context Modeling Framework for Pervasive Computing Systems. In: Meier, R., Terzis, S. (eds.) DAIS 2008. LNCS, vol. 5053, pp. 281–295. Springer, Heidelberg (2008)
24. Cetina, C., Giner, P., Fons, J., Pelechano, V.: A Model-Driven Approach for Developing Self-Adaptive Pervasive Systems. In: Models@runtime 2008, Toulouse, France, September 2008, pp. 97–106 (2008)
25. Occello, A., Dery-Pinna, A., Riveill, M.: A Runtime Model for Monitoring Software Adaptation Safety and its Concretisation as a Service. In: Models@runtime 2008, Toulouse, France, September 2008, pp. 67–76 (2008)
26. Capra, L., Blair, G., Mascolo, C., Emmerich, W., Grace, P.: Exploiting Reflection in Mobile Computing Middleware. Mobile Computing and Communications Review 1(2) (2003)
27. Yau, S.S., Karim, F.: An Adaptive Middleware for Context-Sensitive Communications for Real-Time Applications in Ubiquitous Computing Environments. Real-Time Systems, 29–61 (2004)
28. Ayed, D., Delanote, D., Berbers, Y.: MDD Approach for the Development of Context-Aware Applications. In: Kokinov, B., Richardson, D.C., Roth-Berghofer, T.R., Vieu, L. (eds.) CONTEXT 2007. LNCS, vol. 4635, pp. 15–28. Springer, Heidelberg (2007)

Engineering Distributed Shared Memory Middleware for Java

Michele Mazzucco[1,3,⋆], Graham Morgan[2], Fabio Panzieri[3], and Craig Sharp[2]

[1] University of Cyprus, Nicosia, CY 1678, Cyprus
[2] Newcastle University, Newcastle upon Tyne, NE17RU, UK
[3] University of Bologna, Bologna, 40127, Italy

Abstract. This paper describes the design, implementation and initial evaluation of an object-based Distributed Shared Memory (DSM) middleware system for Java. The resulting implementation allows the construction of event-based distributed systems using a simple programming model, allowing applications to be deployed without hardware or communication channel assumptions. Our implementation utilises standard, freely available, Message Oriented Middleware (MOM). This approach eases DSM development as many reliability and scalability issues associated to DSM may be handled by MOM. In addition to an implementation description, we provide performance results of a prototype system on a Local Area Network.

1 Introduction

The evolution of middleware architectures has provided developers with enabling technologies, easing the implementation of large-scale distributed applications deployed in heterogeneous environments. Such middleware identify the remote procedure call (RPC) as the mechanism within which transparency of distribution is achieved. This has the result of making the interface and associated implementation the unit of distribution across a middleware platform. For clarity, we consider objects as the unit of distribution as this is by far the most popular approach supported in middleware.

Distributed Shared Memory. (DSM) systems attempt to provide a higher level of abstraction to the developer than that found in middleware where developers knowingly incorporate RPCs into their applications. In such systems transparency of distribution is afforded via the access of shared memory. Irrelevant of where a client access occurs, or where the shared resource is located, the developer views such an access as simply a local access of a local resource within the regular programming style of the implementation language being used. As such, the appropriate utilization of required services (*e.g.*, location and discovery) is handled by the DSM run-time that transparently intercepts user access attempts to remote memory addresses and translates them into the appropriate messages.

Developing a DSM system for use with object-oriented middleware and providing distributed application deployment in heterogeneous environments would be beneficial.

⋆ Michele was partly funded by the European Commission under the Seventh Framework Programme through the SEARCHiN project (Marie Curie Action, contract number FP6-042467).

R. Meersman, T. Dillon, P. Herrero (Eds.): OTM 2009, Part I, LNCS 5870, pp. 531–548, 2009.

Developers would program their applications without the hindrance incurred from using the services required for distributed object implementation. This would abstract the bulk of the required distributed service architecture currently used directly by developers in RPC based middleware into the DSM system itself. We now term such a DSM system *'DSM Middleware'*.

If DSM middleware is to be successfully deployed and used by distributed application developers (who would otherwise use object-oriented middleware) the benefits associated with object oriented middleware must be maintained. These benefits include:

- Platform independence: reliance should not be directly placed on hardware or operating system services, allowing development in heterogeneous environments;
- Ease of programming: like RPC in object-oriented middleware, DSM middleware should not require significant changes in programming style to accommodate distribution;
- Run-time deployment: to permit evolving software solutions, the addition of software artifacts should be allowed at run-time, and not be restricted by compile-time decisions.

The principal contribution of this paper consists of providing a practical implementation of a DSM protocol in order to support the construction of event-based application systems. In addition, our architecture can use non-reliable communication support (*e.g.*, the Internet), where packets may be lost, or experience unpredictable delays.

2 Design Issues

Before we can clearly identify a suitable approach for the provision of DSM middleware, we must first explore general approaches to DSM implementation. Within such approaches we identify themes of development that may be suitably tailored, or used "as is" within our own DSM middleware. We consider suitability based on the benefits of object-oriented middleware listed in the previous section. We divide this section into three further subsections based on the basic design choices of a DSM:

1. Implementation level, *i.e.*, where within existing middleware is it most appropriate to implement DSM;
2. Consistency model, *i.e.*, how best to afford sufficient consistency of a shared resource without hindering performance;
3. Communications, *i.e.*, how to enact communications appropriately to provide the propagation of state changes associated with DSM updates.

2.1 Implementation Level

A number of alternatives exist for determining the level of abstraction where a DSM implementation is to be deployed: from systems that maintain consistency entirely in hardware to those that exist entirely in software. Considering our requirement of a middleware solution, we cannot guarantee homogeneous hardware support; we focus exclusively on software supported DSM systems. Software DSM systems can be split into three classes: page-based, variable-based, and object-based. In each of these approaches our concern is where and how transparency of remote access is introduced:

1. Page-based implementations use the *memory management unit* (MMU) to trap remote access attempts;
2. Variable-based run-times require custom compilers to add special instructions to program code in order to detect remote access requests;
3. Object-based systems use special programming language features to determine when the memory of a remote machine is to be accessed.

Due to platform dependencies (*e.g.*, operating system), we cannot consider page-based solutions as an adequate approach for a heterogeneous solution to DSM middleware. Although variable-based solutions may be possible (given that a certain degree of platform independence is provided) the compile time requirements restrict the ability to introduce new types during run-time. This leaves the possibility of object based solutions. Although this approach seems tightly coupled to a particular programming language, a degree of platform independence is afforded beyond that offered by page-based systems. In addition, if the language in question supports the introduction of new types during run-time, then this desirable feature may be incorporated into the DSM middleware.

2.2 Memory Consistency Model

Choosing an appropriate memory consistency model presents a trade-off between minimizing access order constraints and the complexity of the programming model: strict memory models (*e.g.*, sequential [14]) reduce complexity from the programmer's perspective but are achieved at the expense of performance; increased message passing coupled with the locking of resources is required. On the other hand, weak memory models (*e.g.*, release consistency [12]) grant improved performance, but allow the memory to return unexpected values. It is the duty of the programmer, using explicit synchronization techniques, to provide algorithmic semantics equivalent to the sequential model.

Given the type of DSM middleware under consideration, the most important design choice, in terms of scalability, is where to physically store memory. If a *single reader/single writer* algorithm is in operation, *i.e.*, if such a storage space is consigned to a single location, there is a greater potential for memory contention issues to arise, commonly resulting in bottlenecks. Furthermore, data may be geographically separated from an accessing process to such an extent that latency of message exchange may be sufficiently high as to hinder performance.

In order to realise a scalable solution for DSM middleware, a compromise must be reached regarding the consistency of memory against the performance incurred from using such memory. One design option would be to replicate shared memory across the DSM middleware, affording local access when appropriate, while seeking to maintain a degree of consistency across replicas to ensure successful application operation.

2.3 Communication Channel

To ensure availability for the widest audience, a developer must rely on standard protocols such as those governing public access network traffic (*e.g.*, TCP/IP for the Internet). As existing middleware provides a convenient and practical communication abstraction for developers over such protocols, it would be imprudent not to exploit such middleware.

As RPC is the primary mechanism for enacting communication within existing middleware, one must consider RPC as a suitable communication mechanism on which to construct DSM middleware. Using RPC requires an initialised and maintained communication stream between sender and receiver, either throughout a call or for as long as RPC participants hold references to each other (usually sender holding reference to receiver). This tightly coupled approach to communication is satisfactory for small numbers of participants but does not scale to support hundreds or thousands of participants. RPC used in such a manner is not scalable, as the management of connections at both client and server would be a substantial drain on available processing resources.

Middleware developers have tackled this scalability issue by abstracting away the one-to-one communication model of RPC in favour of a many-to-many solution. This is achieved by providing messaging services that decouple sender from receiver (*e.g.*, the sender does not know who is receiving its messages [8]). In distributed systems such an approach to message exchange is encapsulated in Message Oriented Middleware (MOM) [17] with the Java Message Service (JMS) [28] providing an example implementation in Java.

In MOM, senders publish their messages onto well-known message channels (topics or queues, depending on the model in operation), while receivers express interest in receiving messages from such channels. The use of MOM allows additional services (such as message ordering) to be abstracted away from the concern of the programmer to the systems level.

2.4 Memory Design

To provide a DSM middleware for use by developers, a suitable approach to implementation would be object-based, while affording run-time introduction of software objects. Although not ideal (given that language dependency persists) such an approach would provide a significant degree of platform independence. The Java Programming Language offers a semi-platform independent solution as the Java Virtual Machine (JVM) is available on most platforms and widely used by existing RPC middleware solutions. In addition, the reflective qualities of Java coupled with the serializability of object instances, allow the introduction of new object types at run-time.

Once the decision is taken to model the shared memory abstraction as a collection of shared objects, two further choices present themselves: (i) the algorithm to use and (ii) the memory consistency model.

As we are deploying our DSM on multiple machines using copies of objects (with a view to eventual distribution over a wide geographical area), we need to employ replication techniques that minimise the possibility of bottleneck and excessive access delays due to network latencies. This raises a significant challenge in ensuring consistency of data across replicated memory locations; hence an agreement protocol will be required to maintain a degree of consistency. This subject however, presents complexity in distributed systems owing to the unavailability of an absolute global time. The lack of such temporal-synchronisation means it is not always possible to determine the order in which events occurred due to the asynchronous nature of the network (message delays are bounded but unknown) coupled with the non-determinism of multi-threaded process execution on preemptive operating systems. Our approach is quite similar to that proposed by [9] such that we exploit the ordering features provided by the communication channel – in particular, given two messages m_1 and m_2, if $m_1 \rightarrow m_2$[1] then all processes will receive m_1 before receiving m_2 – while the run-time system uses an algorithm which allows the update of remote replicas.

Extensive research has been carried out in the area of memory consistency models; for the purposes of our work, we adopt and experiment with (i) the sequential and (ii) the Pipelined RAM (PRAM) consistency models [16]. These models offer a compromise between programming constraints and implementation complexity, allowing us to evaluate the feasibility of our approach at an early stage.

An integral component of any agreement-protocol is support for ordering and reliable message delivery. By choosing a MOM solution for our communication channel we not only provide scalability for message exchange, but the opportunity to use as required, associated services potentially providing ordering and reliability guarantees; this will ease the overall development of the DSM agreement protocol. Having identified Java as a suitable implementation language for our DSM, so we choose JMS as our MOM technology. Consequentially, our system provides the following main features:

1. The combination Java/object-based run-time allows us to deploy our DSM in heterogeneous environments (possessing a Java Virtual Machine);
2. By utilizing existing middleware services in our DSM middleware we aim to provide QoS guarantees (*e.g.*, atomic delivery order of messages, exactly-once semantics);
3. The use of Java and MOM, coupled with our DSM middleware (constructed using existing middleware techniques) means we can grant the developer run-time introduction of new types;
4. Our *multiple reader/multiple writer* algorithm uses both partitioning and replication. Read operations are local (see Figure 1), while nodes interact only with a (dynamic) subset of shared objects, as shown in Figure 2. Furthermore, different processes can operate on the same object concurrently, as processes read/write their own local copies, thus increasing the degree of parallelism;
5. Distributed applications do not interact with the memory via an object's methods (this is by far the most common approach used by object-based DSM implementations). Instead, our system provides the same primitives of page-based systems (*i.e.*, `read` and `write`).

3 Implementation

In our system, interactions between actual shared object instances are not the concern of the application developer; rather, he or she interacts with "wrapper" objects, which provide the abstraction of DSM as depicted in Figure 2. This approach originates from the technique used by page–based implementations. In order to recreate similar behaviour, the shared memory abstraction is based on two elements; namely, wrapper objects and memory addresses. The wrapper object is the DSM coherence unit while a memory address unequivocally locates a wrapper (given a replicated wrapper object o, all replicas

[1] The $\rightarrow$ symbol indicates the "happened before" relation as defined in [13].

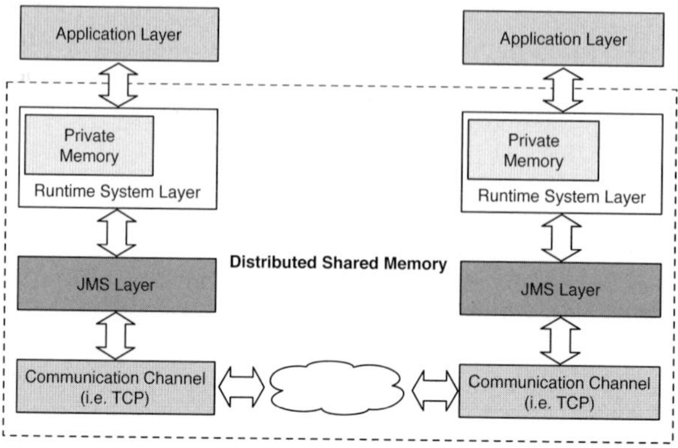

Fig. 1. DSM architecture. The cache allows for local reads

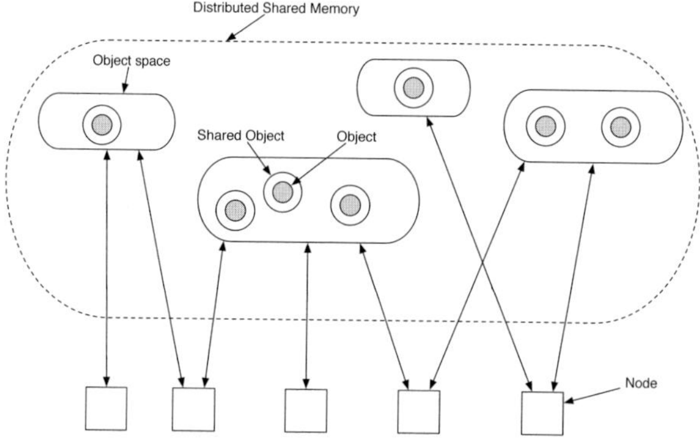

Fig. 2. The DSM uses both replication and partitioning to reduce the number of exchanged messages

share the same address $Addr(o)$). The main advantage of this scheme is that it supplies a single system image; all processes reading (or writing) the memory address x read (or write) the same item.

3.1 Local Memory

Even though the proposed scheme is object-based, it uses some techniques adopted by page-based protocols. The main difference is that the Java programming language does not allow the programmer to directly manipulate the physical memory. This is a shortcoming in the sense that it adds overhead, but also an advantage in terms of providing

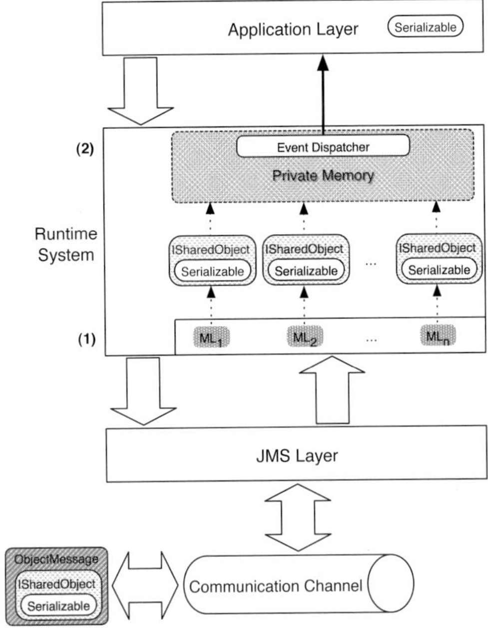

Fig. 3. Updates management

platform independence. In order to solve this limitation we decided to implement the local memory abstraction through two hash tables, $\langle k, v \rangle$:

1. `memory`, acting as local cache where the keys are memory addresses and the values are `ISharedObject` instances, specifically wrapper objects. As the objects stored in the DSM are transmitted over the network we require the content to be `Serializable`.
2. `subscriptions`, storing subscribed topics. The key is the topic name while the value is an object containing all JMS objects needed during communication phases. Hence a "topic" in this discussion, represents a list of update-messages having relevance to specific shared-object replicas.

Since the system is asynchronous in nature, shared memory management challenges comprised of (i) how to update the local cache, and (ii) how to notify the application when updates happen. The run-time system we propose, depicted in Figure 3, consists of a `MessageListener` object for every subscribed topic and the *event dispatcher*. The event dispatcher is based on the *Observer* design pattern [11], which guarantees that every time the subject is updated, all observers are notified automatically and transparently. Consequently, the shared memory API is composed of three methods:

1. `void write(ISharedObject)`: writes a wrapper object to the shared memory. The address to write is contained in the argument;
2. `void read(Address)`: reads the specified shared memory address. In using the event dispatcher, this method need not return a value, as the application will be

notified as soon as the data becomes available (averting the possibility of reading an address that has not yet been written);

3. `void deleteLocal(Address)`: locally deletes the memory zone bound with the specified memory address, resulting in the topic specified by the function argument to become un-subscribed.

3.2 Remote Data

Our approach allows the run-time system to distinguish between local and remote access attempts. However, due to the introduction of the Observer design pattern, access attempts use this notification system whether local or remote reads are taking place. Remote access requests are satisfied using a three-step algorithm: Transmission request to the topic T is bound with the memory address; Local memory synchronization, *i.e.*, creation of a new replica; Subscription of the topic T.

Since memory is physically distributed, interaction between system components occurs only through messages. During the step number 3.2 the synchronization protocol must guarantee that the requesting node will receive only one (correct) reply in order to maintain the consistency among replicated data. The solution to this issue is the solution to the consensus problem.

3.3 Agreement Protocol

Several definitions of the consensus problem can be found in literature. For the purpose of our discussion we state the consensus problem in the following terms: given a collection of processes $p_1,\ldots,p_n$ $(n > 1)$ communicating via message passing, every process begins in the undecided state and proposes a single value. Following a deterministic protocol, at some point during its computation a process must irreversibly decide on a single value, v_i, drawn from a set $V = \{v_1,\ldots,v_n\}$. If every correct process proposes a value then an algorithm is a consensus protocol only if it satisfies the following three properties:

1. *Termination*: every correct process eventually decides a value;
2. *Agreement*: all correct processes decide the same value;
3. *Integrity*: if the correct process p_j decides v_i, then some correct process has proposed that value.

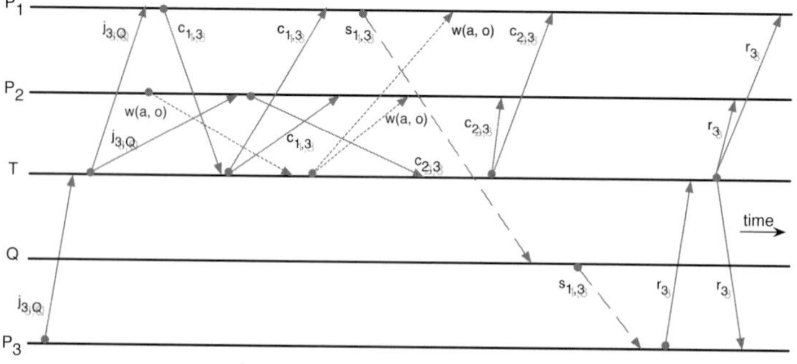

Fig. 4. Naive agreement protocol

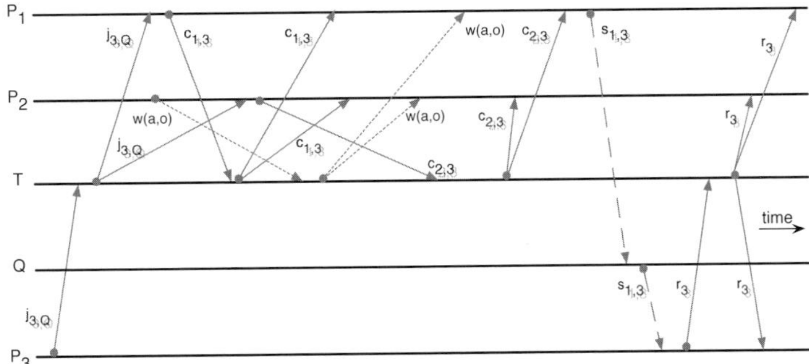

Fig. 5. The agreement protocol. The synchronization message $s_{1,3}$ is sent only after all proposals are received.

A protocol run starts when a process p_i would like to join a topic T (*e.g.* when a remote memory access attempt is made): p_3 creates a temporary queue Q needed to receive the synchronization message and publishes to T a message containing its identifier and Q's identifier (message $j_{3,Q}$ in Figure 4).

When a process p_j ($\forall j \in T$) falling into T receives a request, it immediately ceases outgoing memory communications to T's channel (further requests will be buffered) and publishes to the topic its own proposal (messages $c_{1,3}$ and $c_{2,3}$). To find an agreement, our protocol exploits the delivery order warranty provided by JMS. Since all nodes belonging to the same subsystem receive messages in the same order, the leader is the sender of the first received message. The elected process continues by sending a synchronization message to the queue Q containing all `ISharedObjects` bound to T.

At first glance, the protocol described above looks correct. However, if used, it would cause coherence problems under certain circumstances. As illustrated in Figure 4 after the receipt of the message $s_{1,3}$, P_3's cache differs from those of P_1 and P_2, because the leader P_1 sent the synchronization message before receiving P_2's update.

The solution requires that the coordinator send the synchronization message only when all proposals are received (message $s_{1,3}$ in Figure 5). Since processes stop outgoing memory communications as soon as they receive a join request, the causal delivery order guarantees that *all* updates (marked as $w_{address,object}$) are propagated *before* the leader receives the last proposal. When the initiator node receives the synchronization message, it updates its own memory, deletes the queue Q, subscribes to the topic T, and finally publishes a message ending the protocol, marked as r_3. When processes receive an r message from the channel T, they recommence outgoing memory communications to T.

Finally, if a process is no longer interested in a topic T, a protocol allowing that node to leave T is used. As this algorithm is very similar to the previous one, its details have been omitted for the sake of brevity. The main difference with the leaving protocol however is that the number of available nodes is decremented.

This proposed scheme does have one serious shortcoming however; since the coordinator process must wait until all proposals are received, it has to know how many processes belong to T. Hence, due to the findings of the FLP theorem [10], this protocol cannot cope with crashes (this may be solved however with group membership protocols like JGroups [5] or Project Shoal from Glassfish [1]).

3.4 Memory Updates

To distinguish between new objects and updates the `ISharedObject` data type is extended by two interfaces, namely `INewData` and `IUpdate`.

Within synchronization messages (see figure 5), the requesting process receives a hash table containing only `INewData` objects. Thus, the message content is stored into the memory hash table as it is; every time the local cache is updated, registered observers are notified.

During updates matters become more complex; the proposed scheme exploits the observation that when operations are commutative they can be reordered without affecting the final state [20, 30]. If the received message contains an "entire" object then the content is stored in the local cache (this could mean for example, that operations are not commutative) while if the message content is an update then the original object is modified using *reflection* techniques.

In the case of updates, the DSM low-level behavior differs according to the memory consistency model in operation. If the sequential consistency model is used, then the update is sent over the communication channel and the local memory is updated only when the message is received (a topic subscription could be required). The PRAM model requires instead that the new value be written to the local memory before updating remote copies; in order to avoid coherence problems, the following actions are required: The old value is saved so that in the case of communication problems it can be restored; The node which updates the shared memory cannot receive its own message (as is the case in the sequential model), otherwise the memory would be updated twice for each `write` invocation.

4 Experimental Results

A number of performance tests were conducted on our prototype architecture. This section presents the preliminary results of those tests.

Our testbed environment consisted of a cluster composing Pentium 4 PCs running Linux 2.6.10 and JVM Sun 1.5.0_02 and connected by a 100 Mbit Fast Ethernet LAN. The JMS provider we used was Joram 4.3.1 [21] with each node deployed on a Pentium 4 3.0 GHz with 1 GB of RAM, while clients were deployed on Pentium 4 2.4 GHz with 512 MB of RAM. Before discussing the results of the tests, it must be stressed that they were carried out on a shared network with shared servers. Therefore the occurrence of unpredictable delays due to unrelated network traffic was a possibility, however this did present the opportunity to test the DSM system performance in a realistic environment.

A "benchmark" application suite was developed to determine the overhead of components, the agreement protocol cost and the cost of updates. A detailed description of the benchmarks is given during the tests discussion.

4.1 Components Overhead

The aim of this test was to measure the overhead incurred by each component. We measured the performance differences between (*i*) TCP and (transient) JMS, (*ii*) transient and persistent JMS, (*iii*) 'local' and 'remote' connections, and (*iv*) JMS and our DSM.

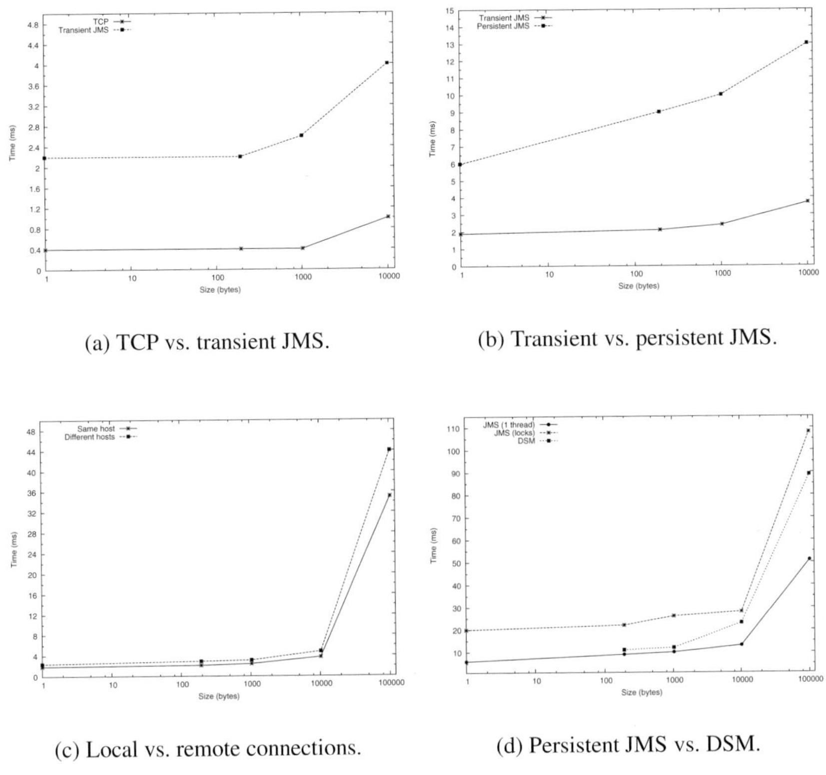

(a) TCP vs. transient JMS.

(b) Transient vs. persistent JMS.

(c) Local vs. remote connections.

(d) Persistent JMS vs. DSM.

Fig. 6. Components cost

The overhead was calculated as round-trip time depending on the message size. The results show the mean values of 1000 measurements.

TCP vs. JMS. Figure 6(a) shows the overhead introduced by JMS. While TCP provides *at-most-once* semantics with FIFO order, this JMS configuration (single server, transient) guarantees the same semantic but a causal atomic delivery order. Measurements show that the overhead is approximately the same in absolute terms and thus its impact decreases as the message size increases. JMS is 11 times more expensive than TCP when a single byte is sent and four times more expensive when the message is 10 KB large.

An additional point of interest regarding the difference between 1 KB and 10 KB messages is that since the network MTU is 1500 bytes, the first message is contained within one IP packet while the second is split into seven fragments. In these conditions the cost of a TCP send operation grows by a factor of 2.5 while the JMS grows by only 1.5.

Transient vs. Persistent mode. This test measured the overhead needed to guarantee an *exactly-once* delivery semantic. When persistent mode is used in transit, messages are not lost due to a JMS provider failure. In this scenario the JMS provider is distributed among three nodes: Figure 6(b) shows that the overhead is indeed quite noticeable; the persistent mode is approximately four times more expensive than the transient, with an increasing trend.

Transparency cost. The aim of this third experiment was to determine the outcome of JMS clients connecting to a server, then sending messages to a topic deployed on another server (usually JMS clients create only one connection). This test was carried out by running a distributed JMS configuration composed of three nodes configured in transient mode. Figure 6(c) shows that the difference is approximately 26%, with an increasing trend; as shown in Figure 7. This difference explains the high overhead introduced by the run-time system when large messages are handled. Unfortunately there is no immediate solution to this problem. The use of a distributed architecture requires some form of synchronization, and this happens by message exchange. This is price for providing location transparency, and is related to the naming service (JNDI).

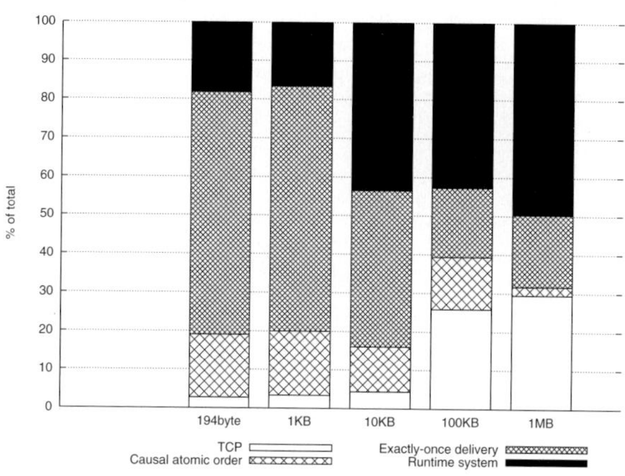

Fig. 7. Components overhead for different message sizes

JMS vs. DSM The last experiment in this set of tests measures the overhead introduced by the run-time system. The communication channel is provided by three Joram nodes, configured in persistent mode. The results shown in Fig. 6(d) include the following scenarios: (*i*) a JMS client using a single thread to receive and publish messages, (*ii*) a multi-threaded client using one thread as publisher and another thread as subscriber, and (*iii*) a client using the facilities provided by the run-time system. Messages are sent by the `write()` primitive while incoming messages are handled by the event dispatcher. In this experiment, once the message has been published, the client waits for its receipt and thus the single-thread version performs better; unfortunately in a real scenario the two operations are handled individually. The dual-threaded application can be modelled in the form of a "producer-consumer" with a bounded buffer of one-item capacity; the overhead needed to synchronize the two threads is very high (111% for 100 KB messages). Finally the DSM guarantees the coherence of replicated data as well as a synchronization mechanism. Results show that the run-time system performs well only when messages are sufficiently small (1 KB).

4.2 Agreement Protocol

The second test evaluates the cost of shared memory synchronization. Assuming a system composed of n nodes in which k form part of the sub-set ($k <= n$), all nodes receive $k+2$ messages and publish only one message (except for the coordinator, which sends two messages). These experiments were repeated five times. Average results, as depicted in Figure 8, show that external factors are much more important than the number of nodes involved in the election. As already mentioned, both network and computation resources were shared during tests, explaining why 1371 ms are needed to allow the admission of the node number 28 while the protocol requires only 350 ms when nodes are 31.

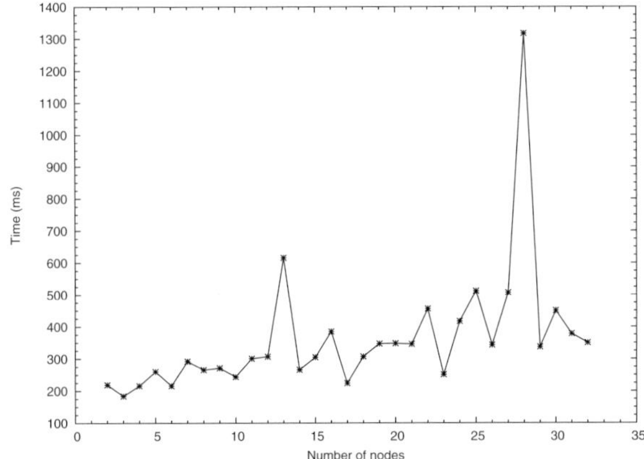

Fig. 8. Cost of memory synchronizations for different network sizes

4.3 Updates

This section shows the update test results related to the sequential consistency model for objects of 194 bytes. The performance of the PRAM is not shown because local memory is updated with no communication overhead.

The JMS provider is composed of three persistent nodes while DSM processes interact through a clustered topic. The clustered topic abstraction supplies a fault tolerant mechanism but does not provide any form of load balancing. This set of experiments aims to answer to the following questions:

1. What is the effect of varying the number of nodes (1, 4, 8, 16 and 32)?
2. How does the system react to increasing the amount of requests per node from 125 to 500, using increments of 125?
3. How does the system behave with respect to the system configuration (25%, 50%, 75% and 100% of nodes falling into the same sub-unit)?

Average cost depending on the number of nodes. The results of this test were not particularly significant given the rapidness of nodes' attempts to update shared memory. Executing this test presents problems as it interferes with the provider's ability to

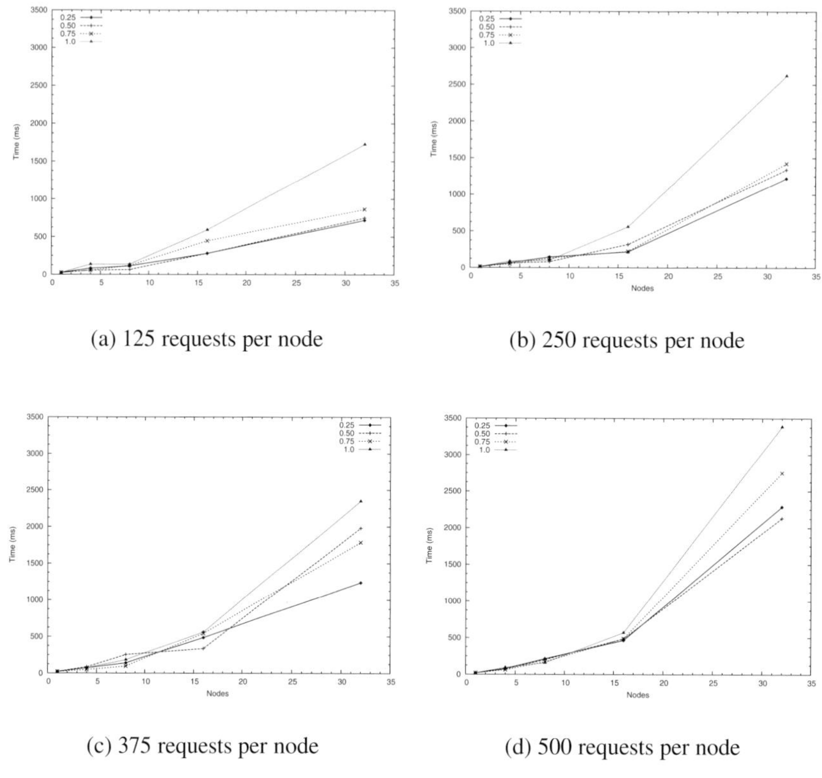

(a) 125 requests per node

(b) 250 requests per node

(c) 375 requests per node

(d) 500 requests per node

Fig. 9. Updates

handle requests; the significant difference between the values shown in Figure 9 is due to the way the topic abstraction is implemented, and thus adding more machines to the provider only partially mitigates the issue.

Average cost depending on the number of requests. The results demonstrated that when the number of clients is four, there is no substantial difference. If the number of nodes continues growing however, then the provider becomes inundated with requests and consequentially is no longer able to satisfy those requests expeditiously. With regard to observed spurious values: when the number of nodes is high they can be attributed to external factors, but if the number is low (and thus the throughput is high) it becomes fundamental where the connection is created.

Average cost depending on the system configuration. As expected, the configuration which features all nodes belonging to the same subunit, consistently performs the worst due to its semantic being equivalent to broadcast communications. The configuration which features 50% of all nodes belonging to the same subsystem performs relatively well. The other two tests were carried out concurrently and thus, since the JMS provider knows the number of both publishers and subscribers it can allocate more resources where they are needed.

5 Related Work

In this section we compare our DSM middleware system with other DSM systems featuring a design and approach comparative to ours.

Previous systems typically do not address problems related to the communication channel. In particular, several implementations assume that packets are delivered in the correct order while messages cannot be lost; TreadMarks [2] and IVY [15], for instance, use a combination of UDP and timeouts to maintain coherence among replicas. While this shortcoming is not a significant hindrance in LAN environments (where message latency is much lower compared to the Internet), as a design choice this is particularly critical since we don't make any assumption about the communication channel.

The most popular DSM systems have been built with a predefined operating system/hardware architecture combination (*e.g.*, TreadMarks, IVY, Brazos [26]) or relying on modified compilers (*e.g.*, Munin [7], Shasta [22], Jackal [29]), including approaches based on the Java language with modifications required within the JVM (*e.g.*, Java/DSM [31]).

Maintaining memory coherence in our DSM middleware bears similarities to the solution adopted by TreadMarks, there are several differences however. For example, at the implementation level, our DSM system is object-based while TreadMarks is page-based. Consequentially, TreadMarks may be affected by false sharing and fragmentation problems which require additional protocols [3] in order to be minimized. In addition, at the communication level, TreadMarks is based on the exchange of UDP messages while our solution relies on JMS.

Scalability is rarely confronted in existing systems similar to ours. For example, TreadMarks propagates updates to all nodes while our solution "hits" interested nodes only. JDSM [25] uses a centralized node (the Cluster Manager) to intercept requests, creating a bottleneck as the number of nodes increases.

The use of group communication has precedents. Orca [4] uses group communications to implement a sequentially consistent object-based DSM. Orca, like the system described in this paper, relies on features of the communication channel (reliable, total ordered broadcast) to implement transparent replication as well as to maintain data-object consistency. Brazos uses two main system threads to reduce communication overhead and uses a selective multicast to reduce communication traffic. Finally, the work described in [24] presents an object-based DSM, designed as an extension to the .NET Framework. It provides a causally consistent memory model where causal relationship is achieved through vector logical clocks sent with every message.

Java has been used previously to implement a DSM system. Java/DSM [31] addresses problems that arise when a heterogeneous set of nodes are used, however it modifies the memory management of the JVM and thus it is not portable. As in our system, JDSM does not modify the JVM, however it is quite different from the solution we propose. First, it requires shared objects to be declared during the initialization step. Second, its architecture (composed of a Cluster Manager, Server and Client) requires that requests be sent from the Client to the Cluster Manager and then forwarded to one of the available Servers. However, over-involvement of the Cluster Manager in this scenario produces a potential bottleneck, hindering scalability. Finally, node interactivity can use three different communication protocols, TCP/IP socket, PM and VIA. Again, the lack of MOM in a middleware environment leaves DSM middleware without services that can be incorporate at little cost (*e.g.*, transactions).

Java Past Set (JPS) [19] differs from our system primarily because shared objects are not replicated, hence for reasons explained in section 2.3, scalability problems arise when the number of accessing clients increase. Data location is achieved with *element object descriptors* and can be distributed according to several policies (*i.e.* place the object with the first node that requested it or try to distribute the same number of objects to every node). Thus in JPS, data distribution can happen in several ways, while we replicate data on demand and remove replicas when they are no longer needed. A commonality between JPS and our DSM is the data update phase, defined as *user redefinable memory semantics*, allowing the application programmer to define the memory operation semantic.

Finally, the only known system designed to host multi-player games is Plurix [23], providing the illusion of a shared memory by utilising a custom operating system and Java compiler; the resulting architecture works exclusively on Intel platforms while the memory coherence unit is a page of virtual memory.

6 Conclusions and Future Work

In this paper we have presented distributed shared memory middleware. In comparison to existing approaches, this platform provides greater transparency to the application programmer. Firstly, it has been identified that the proposed protocol does not prerequisite any hardware or software architectural assumptions (the features of Java handle the differences). Secondly, the run-time system transparently handles the message passing details. Thirdly, this system is able to use non-reliable channels, where packets can be lost and delays are of an arbitrary length.

To summarize, we have shown the applicability of group communications, replication, caching and interest management techniques to support the construction of event-oriented distributed systems. The next step is to conduct a more extensive plan of tests. In particular, we plan to deploy our middleware solution in a WAN environment such as PlanetLab or the Grid, where packets can be lost or delayed.

There remain problems to address: client connection to the JMS provider for example, which is responsible for the poor performance observed in certain circumstances, or issues of fault tolerance during memory synchronization. Differing approaches to the latter problem have been proposed, however some techniques (*e.g*, [6]) are not scalable while others (*e.g*, [27]) allow only one crash occurrence. A possible solution requires more realistic assumptions about the communication channel to solve the consensus problem (or alternatively to use a randomized algorithm). If it is possible to solve consensus, it becomes feasible to adopt an approach similar to that presented in [18] to periodically save the memory content in stable storage, while the use of failure detectors would inform the run-time when the memory has to be restored (our system uses a write-update algorithm).

References

[1] Abdelaziz, M., et al.: Project Shoal, a dynamic Java clustering framework, https://shoal.dev.java.net
[2] Amza, C., Cox, A.L., Dwarkadas, S., Keleher, P., Lu, H., Rajamony, R., Yu, W., Zwaenepoel, W.: TreadMarks: Shared Memory Computing on Networks of Workstations. IEEE Computer 29(2), 18–28 (1996)
[3] Amza, C., Cox, A.L., Dwarkadas, S., Jin, L.-J., Rajamani, K., Zwaenepoel, W.: Adaptive Protocols for Software Distributed Shared Memory. Proceedings of the IEEE, Special Issue on Distributed Shared Memory Systems 87(3), 467–475 (1999)
[4] Bal, H.E.: Orca: A Language for Distributed Programming. ACM SIGPLAN Notices 25(5), 17–24 (1990)
[5] Ban, B., et al.: JGroups, a toolkit for reliable multicast communication, http://www.jgroups.org
[6] Cabillic, G., Muller, G., Puaut, I.: The Performance of Consistent Checkpointing in Distributed Shared Memory Systems. In: Proceedings of the 14th IEEE International Symposium on Reliable Distributed Systems (SRDS 1995), September 1995, pp. 96–105 (1995)
[7] Carter, J.B., Bennett, J.K., Zwaenepoel, W.: Implementation and Performance of Munin. In: Proceedings of the 13th ACM Symposium on Operating Systems Principles (SOSP 1991), pp. 152–164. ACM Press, New York (1991)
[8] Eugster, P.T., Felber, P.A., Guerraoui, R., Kermarrec, A.-M.: The Many Faces of Publish/Subscribe. ACM Computing Surveys (CSUR) 35(2), 114–131 (2003)
[9] Fekete, A., Kaashoek, M.F., Lynch, N.: Implementing Sequentially Consistent Shared Objects using Broadcast and Point-To-Point Communication. Journal of the ACM (JACM) 45(1), 35–69 (1998)
[10] Fischer, M.J., Lynch, N.A., Paterson, M.S.: Impossibility of Distributed Consensus with One Faulty Process. Journal of the ACM (JACM) 32(2), 374–382 (1985)
[11] Gamma, E., Helm, R., Johnson, R., Vlissides, J.: Design Patterns: Elements of Reusable Object-Oriented Software. Addison Wesley, Reading (1995)
[12] Gharachorloo, K., Lenosk, D., Laudon, J., Gibbons, P., Gupta, A., Hennessy, J.: Memory Consistency and Event Ordering in Scalable Shared-Memory Multiprocessors. In: Proceedings 17th Annual International Symposium on Computer Architecture, pp. 15–26. IEEE Computer Society, Los Alamitos (1990)
[13] Lamport, L.: Time, Clocks, and the Ordering of Events in a Distributed System. Communications of the ACM 21(7), 558–565 (1978)
[14] Lamport, L.: How to Make a Multiprocessor Computer that Correctly Executes Multiprocess Programs. IEEE Transactions on Computers C-28(9), 690–691 (1979)
[15] Li, K., Hudak, P.: Memory Coherence in Shared Virtual Memory Systems. ACM Transactions on Computer Systems 7(4), 321–359 (1989)
[16] Lipton, R.J., Sandberg, J.S.: PRAM: A Scalable Shared Memory. Technical Report CS-TR-180-88, Dept. of Computer Science, Princeton University (September 1988)
[17] Menasce, D.A.: MOM vs. RPC: Communication Models for Distributed Applications. IEEE Internet Computing 9(2), 90–93 (2005)
[18] Morin, C., Kermarrec, A.-M., Banatre, M., Gefflaut, A.: An Efficient and Scalable Approach for Implementing Fault-Tolerant DSM Architectures. IEEE Transactions on Computers 49(5), 414–430 (2000), citeseer.ist.psu.edu/morin97efficient.html
[19] Pedersen, K.S., Vinter, B.: Java PastSet: A Structured Distributed Shared Memory System. IEEE Proceedings – Software 150(2), 147–153 (2003)
[20] Pu, C., Leff, A.: Replica Control in Distributed Systems: An Asynchronous Approach. In: Proceedings of the 1991 ACM SIGMOD International Conference on Management of Data (SIGMOD 1991), pp. 377–386. ACM Press, New York (1991)

[21] ScalAgent Distributed Technologies. JORAM: Java Open Reliable Asynchronous Messaging (2005), http://joram.objectweb.org
[22] Scales, D.J., Gharachorloo, K., Thekkath, C.A.: Shasta: a Low Overhead, Software-Only Approach for Supporting Fine-Grain Shared Memory. In: Proceedings of the 7th International Conference on Architectural Support for Programming Languages and Operating Systems (ASPLOS-VII), pp. 174–185. ACM Press, New York (1996)
[23] Schöttner, M., Wende, M., Göckelmann, R., Bindhammer, T., Schmid, U., Schulthess, P.: A Gaming Framework for a Transactional DSM System. In: Proceedings of the 3rd IEEE/ACM International Symposium on Cluster Computing and the Grid (CCGrid 2003), Tokyo, Japan, May 2003, pp. 502–509. IEEE Computer Society, Los Alamitos (2003), http://www.plurix.de/
[24] Seidmann, T.: Distributed Shared Memory Using The NET Framework. In: Proceedings of the 3rd IEEE/ACM International Symposium on Cluster Computing and the Grid (CCGrid 2003), Tokyo, Japan, pp. 457–462. IEEE Computer Society, Los Alamitos (2003)
[25] Sohda, Y., Nakada, H., Matsuoka, S.: Implementation of a Portable Software DSM in Java. In: Proceedings of the 2001 Joint ACM-ISCOPE Conference on Java Grande (JGI 2001), pp. 163–172. ACM Press, New York (2001)
[26] Speight, E., Bennett, J.K.: Brazos: A Third Generation DSM System. In: Proceedings of the First Usenix Windows NT Symposium, August 1997, pp. 95–106 (1997), http://www-brazos.rice.edu/brazos
[27] Sultan, F., Iftode, L., Nguyen, T.: Scalable Fault-Tolerant Distributed Shared Memory. In: Proceedings of the 2000 ACM/IEEE conference on Supercomputing (Supercomputing 2000), pp. 20–32. IEEE Computer Society, Los Alamitos (2000)
[28] Sun. Java Message Service. Sun Microystems, Version 1.1 (April 2002)
[29] Veldema, R., Hofman, R.F.H., Bhoedjang, R.A.F., Bal, H.E.: Runtime Optimizations for a Java DSM Implementation. In: JGI 2001: Proceedings of the 2001 joint ACM-ISCOPE conference on Java Grande, pp. 153–162. ACM Press, New York (2001)
[30] Wuu, G.T., Bernstein, A.J.: Efficient Solutions to the Replicated Log and Dictionary Problems. In: Proceedings of the 3rd Annual ACM Symposium on Principles of Distributed Computing (PODC 1984), pp. 233–242. ACM Press, New York (1984)
[31] Yu, W., Cox, A.L.: Java/DSM: A Platform for Heterogeneous Computing. Concurrency - Practice and Experience 9(11), 1213–1224 (1997)

CLON: Overlay Networks and Gossip Protocols for Cloud Environments*

Miguel Matos[1], António Sousa[1], José Pereira[1], Rui Oliveira[1], Eric Deliot[2], and Paul Murray[2]

[1] Universidade do Minho, Braga, Portugal
{miguelmatos,als,jop,rco}@di.uminho.pt
[2] HP Labs, Bristol, United Kingdom
{eric.deliot,pmurray}@hp.com

Abstract. Although epidemic or gossip-based multicast is a robust and scalable approach to reliable data dissemination, its inherent redundancy results in high resource consumption on both links and nodes. This problem is aggravated in settings that have costlier or resource constrained links as happens in Cloud Computing infrastructures composed by several interconnected data centers across the globe.

The goal of this work is therefore to improve the efficiency of gossip-based reliable multicast by reducing the load imposed on those constrained links. In detail, the proposed CLON protocol combines an overlay that gives preference to local links and a dissemination strategy that takes into account locality. Extensive experimental evaluation using a very large number of simulated nodes shows that this results in a reduction of traffic in constrained links by an order of magnitude, while at the same time preserving the resilience properties that make gossip-based protocols so attractive.

1 Introduction

Cloud Computing is an emerging paradigm to deliver IT services over the Internet, ranging from low level infrastructures to application platforms or high level applications. It promises elasticity, the ability to scale up and down according to demand, and the notion of virtually infinite resources in a pay-per-use business model.

However, there are several pending issues to solve in order to consolidate this paradigm, such as availability of service, data transfer bottlenecks and performance unpredictability [3]. Another crucial issue is the management of the underlying infrastructure as the Cloud provider needs to be able to properly meter, bill, and abide by the Service Level Agreements of its customers among other essential management operations.

The ongoing *Dependable Cloud Computing Management Services* project [1] aims to offer strong low level primitives in order to leverage the management of

* This work is supported by HP Labs Innovation Research Award, project DC2MS (IRA/CW118736).

R. Meersman, T. Dillon, P. Herrero (Eds.): OTM 2009, Part I, LNCS 5870, pp. 549–566, 2009.

Cloud infrastructures. We identified Reliable Multicast as an important building block to the management of such infrastructures as it offers strong abstractions on top of which other essential services could leverage, such as data aggregation, consensus and the dissemination of customer-related information. Unfortunately, due to the characteristics of a typical Cloud scenario, existing proposals are not able to properly address the problem of reliable dissemination in a highly scalable and resilient fashion. This is due to the underlying network infrastructure and to the assumptions and requirements about the dynamics of the environment.

The Cloud infrastructure is composed by several data centers spread worldwide and organized in a federation. The members of the federation are interconnected by long-distance expensive WAN links with high aggregate bandwidth demands, while the links that internally connect its components typically have less stringent requirements. The communication demands intra-data center and inter-data center are very different, both in terms of latency and bandwidth required to provide a reliable service, and in the need of timeliness of information available across the federated infrastructure. In a smaller scope, this can be also observed in the architecture of a single data center, as collections of nodes are also grouped in a federated manner. The increasing aggregate bandwidth demand could be alleviated, but not solved, by using a fat tree network layout [2], where leaf nodes are grouped in a way to mitigate the load imposed on the individual network devices, while at the same time providing transparent load balancing and failover capabilities among those devices.

On the other hand these scenarios are highly dynamic with nodes constantly joining and leaving the system due to failures or administrative reasons, and as such the assumption of a stable system does not hold. In fact in systems of this scale, failures are commonplace as has been presented in [19], which studies the pattern of hard drive disk failures in very large scale deployments.

The goal of this paper is therefore to build a reliable multicast service that is able to cope with the requirements of a cloud environment, namely its massive scale, the dynamics of the infrastructure where nodes constantly join and leave the system, the inherently federated infrastructure where the aggregate bandwidth requirements vary considerably, while offering strong reliability even in the presence of massive amounts of failures as demanded by an infrastructure that needs to run 24/7. This is addressed at two distinct levels: the Peer Sampling Service which follows a flat approach that does not rely on special nodes or global knowledge but instead takes into account locality at construction time; and the dissemination protocol, which is also locality aware and can be configured to clearly distinguish between transmission to remote or local neighbors. By disseminating on top of the right overlay, and carefully choosing which strategy to use on a per node basis, we are able to reduce bandwidth consumption on undesirable links without impairing the resilience and reliability of both the overlay and the dissemination.

The rest of this paper is organized as follows: Section 2 introduces the concepts used throughout the paper; Section 3 describes existing protocols, how they relate to our work and why they fail do meet the requirements pointed above; Section 4 presents our proposal to address the aforementioned problems; Section 5 describes the experimental evaluation conducted and finally Section 6 concludes the paper.

2 Background

Reliable Multicast is an important building block in distributed systems as it offers a strong abstraction to a set of processes that need to communicate reliably. The overlay is a fundamental concept to Reliable Multicast as it abstracts the details of the underlying network by building a virtual network on top of it, which can be seen as a graph that represents the 'who knows who' relationship among nodes.

To construct those overlays two main approaches exist: structured and unstructured protocols. The former is frugal in resource consumption of both nodes and links, but is highly sensitive to churn. This is because the overlay is built as a spanning tree that takes into account optimization metrics such as latency or bandwidth that is built before-hand and thus can take advantage of nodes and links with higher capacity. However, due to this pre-building, upon failures the tree must be rebuilt, precluding the dissemination while this process takes place. As such, in highly dynamic environments where the churn rate is considerable, the cost of constantly rebuilding the tree may become unbearable. Furthermore, nodes closer to the root of the tree handle most of the load of the dissemination thus impairing scalability. On the other hand, in the unstructured approach links are established more or less randomly among the nodes, without any efficiency criteria. Thus, to ensure that all nodes are reachable, links need to be established with enough redundancy, which has a significant impact on the overlay. First, as the overlay is redundant each node receives multiple copies of a given message through its different neighbors due to the existence of multiple implicit dissemination trees. While this is undesirable from an efficient resource usage point of view, it yields strong properties: reliability, resilience and scalability. The first two come naturally from the inherent redundancy in the establishment of links: as there are multiple dissemination paths available, failure does not impair the successful delivery of a given message as it will be routed by some other path. Furthermore, as there is no implicit structure on the overlay the churn effect is mitigated as there is no need to global coordination or rebuilding of the overlay. By requiring each node to know only a small subset of neighbors the load imposed on each one in the maintenance of the overlay and in the dissemination is minimized, which allows those protocols to scale considerably.

The key overlay properties, according to [11] are: Connectivity, that indicates node reachability and attests the robustness of the overlay; Average Path Length, that is the average number of hops separating any two nodes and is related to the overlay diameter; Degree Distribution, which is the number of neighbors of each node, and measures a node reachability and its contribution to the connectivity; and Clustering Coefficient, which measures the closeness of neighbor relations, and is related to robustness and redundancy.

The mechanism used to construct the overlay in the unstructured approach is known as the Peer Sampling Service [11], and several works before have focused on building such a service in a fully decentralized fashion [8,16,20,14,15,9]. With the abstraction provided by the Peer Sampling Service, peers wishing to disseminate messages, simply consult the service to obtain a subset of known neighbors, and forward those messages to them.

Dissemination on top of unstructured overlays typically uses the epidemic or gossip-based approach. This approach relies on the mathematical models of epidemics [4]: if each infected element spreads its infection to a number of random elements in the universe, then all the population will be infected with high probability. The amount of elements that need to be infected by a given element - the fanout - is a fundamental parameter of the model, below that value the dissemination will reach almost none of the population, and above it it will reach almost all members. The gossip process, i.e. the decision of when and how to send the message payload to the neighbors may follow several approaches [13], which we describe next. The most common gossiping strategy is eager push, in which peers relay a message as soon as received to a number of targets for a given number of rounds, and is used by several well known protocols [7,12,18]. The major drawback of this strategy is the amount of bandwidth required, as multiple message copies are received by nodes. Oppositely, in lazy push, peers forward an advertisement of the message instead of the full payload. Peers receiving the advertisement could then ask the source for the payload, and achieve exactly once message payload delivery. Assuming that the message payload tends to be much larger than an advertisement with its id, this strategy drastically reduces the bandwidth requirement of the previous strategy but increases the latency of the dissemination process, as three communication steps are needed to obtain the payload. Furthermore, this also has an impact on reliability as the additional communication steps increases the time window to network and node faults. A different strategy relies on pulling, where nodes periodically ask neighbors for new messages. In the eager variant, when a node asks for news to a neighbor, the latter will send all new known messages to the petitioner. In contrast, in the lazy approach the node that receives the news request only sends new message ids. The petitioner would then be able to selectively pull messages of interest.

The eager versus lazy strategy is clearly a trade-off between bandwidth and latency, while the difference between a push and pull scheme is more subtle. In push, nodes behave reactively to message exchanges, while on pull nodes behave in a proactive fashion by periodically asking for news. Thus, in an environment where messages are generated at low rates, a push strategy has no communication overhead, while the pull approach presents a constant noise due to the periodic check.

3 Related Work

In this section we will briefly describe several unstructured overlay construction algorithms, and analyse how they relate to our work.

Scamp [8] is a peer-to-peer membership service with the interesting property that the average view size converges naturally to the adequate value by using local knowledge only. This is achieved by integrating nodes in the local view with a probability inversely proportional to the view size and by sending several subscription requests for each node that joins the overlay. With this mechanism the protocol ensures that the view size converges to $(c + 1)log(N)$, where c is a protocol parameter related to fault tolerance and N is the system size. As pointed by its authors, Scamp is oblivious to locality and is a reactive protocol as it does not do any effort on the evolution of the overlay.

Cyclon [20] is a scalable overlay manager that relies on a shuffling mechanism to promote link renewal among neighbors. In opposition to Scamp, Cyclon is a proactive protocol that continuously tries to enhance the overlay by means of periodic executions of the shuffle mechanism. The shuffle operation is very simple: each peer selects a random subset of peers in its local view and chooses an additional peer to which it will send this set. The receiving node also selects a subset of known neighbors and sends it to the initial node. After the exchange, each node discards set entries pointing to themselves and includes the remaining peers on their views, discarding sent entries if necessary. By including links in the exchange set accordingly to their age, the protocol provides an upper bound on the time taken to eliminate links to dead nodes.

HyParView [14] also uses shuffling to build the overlay. However, each node maintains two views: a small active view with stable size used for message exchange; and a larger passive view maintained by shuffling and used to restore the active view on the presence of failures. By relying on a large passive view, the protocol is able to cope with massive failures, and by using a small sized active view the redundancy of message transmissions is reduced.

The Directional Gossip Protocol [15], aims at providing dissemination guarantees in a WAN scenario. To accomplish this, the authors adopt a two-level gossip hierarchy: the lower level runs a traditional gossip protocol in the LAN, and the other level is responsible for gossiping among LANs, through the WAN links. The latter is achieved by using gossip servers, for each LAN there is a selected gossip server that is internally seen as yet another process. When a server receives a message from its LAN, it sends the message to the known gossip servers of the other LANs. When receiving an external message, it disseminates the message internally using traditional gossip protocols. While this protocol achieves good results in the amount of messages that cross WAN links, it relies on the undesirable selection of nodes with special roles, the gossip servers.

The Localizer algorithm [16] builds on the work done in Scamp by constantly trying to optimize the resulting overlay according to some proximity criteria. Periodically, each node chooses two nodes randomly, computes the respective link cost and sends those values to both. The receivers reply with their respective degrees and additionally, one of the nodes sends the estimate cost of establishing a link with the other. The initiator locally computes the gain of exchanging one of its links with one between the other nodes and, if desirable, the exchange is performed with a probability p, given by a function which weights the trade-off between the closeness to an optimal configuration and the speed of convergence. Localizer has not been deeply studied in presence of high churn rates and requires the interaction among three nodes to work properly.

HiScamp [9], is a hierarchical protocol that leverages on the work done in Scamp, by aggregating nodes into clusters according to a distance function. Joining nodes contact a nearby node and, based on the distance function, either join an existing cluster or start a new one. The protocol uses two views, an inView which is used to handle subscriptions, and a hView used to disseminate messages. The hView has as many levels as the hierarchy, where the lowest level contains gossip targets in the same cluster, and the other levels contain targets on each hierarchy level. The inView has one lesser level than the hierarchy, which is common to all nodes in the same level, and contains all nodes belonging to that level. Each cluster is seen as an individual abstract node on the next level, and each level runs a Scamp instance that manages its overlay. To avoid the single point of failure of having a single node representing a cluster, an algorithm is run periodically to ensure that a given cluster is represented by more than one node. HiScamp effectively reduces the stress imposed on long distance links, but at the cost of decreased reliability.

Araenola [17] relies on the properties of k-regular graphs to build overlays with a constant degree K. The protocol thus imposes a constant overhead on each node, with low latency and high connectivity. A recent extension proposes a mechanism for biasing the overlay to mimic a given network topology that works by finding nearby peers and establishing additional links to them up to a certain protocol parameter.

Scamp, Cyclon and HyParView are flat protocols that do not take into account locality and therefore fail to cope with the requisite of distinguishing links characteristics. This is important as we want to reduce the load imposed on the long-distance links that connect the members of the federation to increase the aggregate bandwidth available to them. Nonetheless they are highly resilient to churn and failures of links and nodes, and address our reliability concerns. On the other hand, Localizer, Directional Gossip and HiScamp are protocols that take into account locality and therefore are able to reduce the stress imposed on the long-distance links but unfortunately they are sensible to churn and failures as the experimental evaluation of the respective papers attest. This weakness precludes their use in scenario with requirements such as ours, and is due to the reliance on nodes with a special role to handle locality. Upon failure of those nodes, new nodes need to be selected for the special role to guarantee the connectivity of the members of the federation. However, this is hard to achieve in a fully distributed and dynamic environment as it requires some sort of distributed agreement to elect which nodes are special. Even if the special nodes are chosen in a probabilistic fashion, it is not clear how to get them to know each other and how to properly handle their failures, as it will imply some a-priori knowledge of which nodes are on the other locations of the federation to establish the long-distance links with them. Areanola relies on post optimizations to the original overlay, and relies on the establishment of a constant number of additional links to handle locality. While due to the properties of k-random graphs the constant number of links of the base algorithm does not impair reliability, it is not clear if this is the case in the network-aware extension.

There are other protocols [10,6] that are able to manipulate the probabilities of infecting neighbors based on metrics such as interests. However, due to this they do not consider delivering messages to all participants of the universe, and as such we do not cover them in detail.

4 Clon Protocol

In this section, we describe our Reliable Multicast service, whose goal is to address the reliability and resilience requirements of a Cloud scenario, while coping with its aggregate bandwidth demands. Instead of starting with an hiearchical approach as the ones presented above and improving its resilience to churn and faults, we rely on the resilience of the unstructured flat approach and improve it to approximate the desirable perfomance metrics, namely with respect to the bandwidth requirements on the costlier long-distance links.

This is achieved at two distinct levels, the Peer Sampling Service and the dissemination process. First, the peer sampling service builds an overlay that mimics the structure of the underlying network but without relying on special nodes as in the approaches presented previously or in any type of global knowledge. With this approach our protocol is able to tolerate considerable amounts of failures and be resilient to churn as the traditional flat protocols. Finally, the dissemination protocol builds atop the Peer Sampling Service and is responsible for the actual exchange of messages between peers. This protocol supports different dissemination strategies that could be used to achieve different latency versus bandwidth trade-offs without endangering correctness. Additionaly, the dissemination also takes into account locality further reducing the load imposed on the costlier links.

4.1 Peer Sampling Service

The Peer Sampling Service uses the same philosophy of the Scamp [8] protocol, namely the probabilistic integration of nodes and the injection of several subscription requests, which allows the average node degree to adjust automatically with the system size. However, instead of relying only on the view size of the node integrating the joiner, the protocol relies on an oracle to manipulate the view size perceived by the integration routine. Thus, nodes could be integrated with different probabilities based on the locality of the joiner, but nonetheless maintain the convergence and adaptability to varying system sizes. In detail, the oracle should provide higher virtual view sizes to remote nodes and therefore reduce their probability of integration, or lower virtual view sizes to achieve the opposite result. By properly configuring the oracle it is possible to manipulate the views of the nodes in order to achieve desirable configurations, namely have them know mostly local nodes and some remote nodes and thus bias the overlay

```
1  upon init
2    contact = getContactNode()
3    send(contact, Subscription(myself))
4
5  proc handleSubscription(nodeId)
6    for n ∈ view
7        send(n, Join(nodeId))
8    for (i=0; i < c; i++)
9        n = randomNode(view)
10       send(n, Join(nodeId))
11
12 proc handleJoin(nodeId)
13   keep = randomFloat(0,1)
14   keep = Math.Floor(localityOracle(viewSize, nodeId)) * keep)
15   if (keep == 0) and nodeId ∉ view
16     view.Add(nodeId)
17   else
18       n = randomNode(view)
19       send(n, Join(nodeId))
```

Listing 1.1. CLON protocol: Peer Sampling Service

to the underlying network topology. As all the nodes contribute to the network awareness, the protocol retains its reliability in face of faults as it does not depend on special nodes to handle it, while at the same time reducing the load imposed on the long-distance links, as fewer links to remote nodes are established when comparing to the traditional flat approaches.

The rest of this section describes the Peer Sampling Service developed that can be observed in Listing 1.1.

Upon boot, a node obtains a contact node from an external mechanism and sends a *Subscription* request to it (lines 1 to 3).[1] The receiver of the subscription creates a *Join* request and forwards it to all nodes in its view, and to c additional random nodes (lines 5 to 10). A node receiving a join request (lines 12 to 19) generates a random seed and weights it with the value returned by the *localityOracle*. The oracle receives the view size and the id of the node joining the overlay and should return a value indicating the preference that should be given to the integration of the joiner, as pointed previously. If the calculation in line 14 yields zero, the joiner is integrated into the view of the node, otherwise the subscription is forwarded to a random node.

This service offers two calls to the dissemination protocol *PeerSampleLocal* and *PeerSampleRemote* that return a set of local and remote nodes, respectively.

4.2 Dissemination Protocol

This section presents the dissemination protocol which improves the work in [5]. The original protocol combines eager and lazy push strategies to achieve desirable bandwidth/latency trade-offs in a single protocol without endangering correctness. It is divided in two main components: one responsible for the selection of the communication targets, and the other for the actual point-to-point communication and selection of the transmission strategy. In this work we bring locality awareness to the dissemination process by introducing distinct round types to remote and local nodes, and by reordering the queue of pending lazy requests.

[1] In fact the problem of how to know the initial contact node is still a pending issue.

```
1  initially
2   K = ∅ /*known messages*/
3
4  proc Multicast(d)
5   Forward(mkdId(),d,0,0)
6
7  proc Forward(i,d,rl,rr)
8   Deliver(d)
9   K = K ∪ {i}
10  P = ∅
11  if rr < maxRRemote
12    P = P ∪ PeerSampleRemote(remoteFanout)
13  if rl < maxRLocal
14    P = P ∪ PeerSampleLocal(localFanout)
15  for each p ∈ P
16    L-Send(i,d,rl+1,rr+1,p)
17
18 upon L-Receive(i,d,rl,rr,s)
19   if i ∉ K
20     if isRemote(s)
21       ri = 0
22     Forward(i,d,rl,rr)
```

Listing 1.2. Dissemination Protocol: Peer Selection

The rationale behind the introduction of different round types is that the number of nodes in a given local area and the number of local areas in the system will likely differ by some orders of magnitude (for example a scenario with 5 local areas and 200 nodes on each area), and therefore the number of rounds necessary to infect each one of those entities is quite different. Instead of coping with the necessity of reaching all the nodes with a higher global round number, we split that in a local round and a remote round. As such, it is possible to infect some nodes on the remote areas and stop disseminating to them, and let the local dissemination infect the remaining local nodes, thus reducing the number of messages that traverse the long-distance links. This flexibility allows the dissemination of the message only to a given portion of the population without wasting resources to send it to the other portion.

The peer selection algorithm is shown in Listing 1.2. Initially, the algorithm is started with an empty set of known messages, that is used to avoid delivering duplicates to the application via the *Deliver* upcall. An application wishing to send a message calls the *Multicast* primitive on line 4 that generates a unique message id, initializes the rounds to zero and *Forward* the message. In *Forward*, the message id is added to the known set of messages, and the protocol enters the peer selection phase from line 11 to 14, where the distinction between remote and local peers is made. The amount of peers specified by the respective fanouts is collected independently, if the respective round number has not expired, and then the $L - Send$ primitive of the point-to-point communication strategy selection is invoked for all the collected peers. When the point-to-point communication layer delivers a message to this level, via the $L - Receive$ upcall, the message id is checked against the known set of ids and, if the message is new, it is forwarded. The last important remark is the reset to the local message round if the message comes from a remote origin (lines 20 and 21). This is because messages being received remotely have a local round count that is meaningless to this local area and, therefore, must be reset to zero for dissemination to be successful locally. The *isRemote* oracle must then indicate whether the origin of the message is considered to be remote or not.

```
1  initially
2    ∀i: C[i] = ⊥
3    R = ∅
4
5  proc L−Send(i,d,rl,rr,p)
6    if isEager(i,d,rl,rr,p)
7        send(p,MSG(i,d,rl,rr))
8    else
9        C[i] = (d,rl,rr)
10       send(p,IHAVE(i))
11   R = R ∪ {i}
12
13 proc handleIHAVE(i,s)
14   if i ∉ R
15     QueueMsg(i,s)
16
17 proc handleMSG(i,d,rl,rr,s)
18    if i ∉ R
19      R = R ∪ {i}
20      Clear(i)
21    L−Receive(i,d,rl,rr,s)
22
23 proc handleIWANT(i,s)
24    (d,rl,rr) = C[i]
25    send(s,MSG(i,d,rl,rr))
26
27 forever
28   (i,s) = ScheduleNext()
29    send(s,IWANT(i,myself))
30
31 proc QueueMsg(i,newSource)
32   if i ∉ Queue
33     Queue.add(i,newSource)
34   else
35     (i,oldSource) = Queue.get(i)
36     Queue.add(i,newSource)
37     if isCloser(newSource,oldSource)
38       Queue.swap(newSource,oldSource)
```

Listing 1.3. Dissemination Protocol:Point-to-Point Communication

In Listing 1.3 it is possible to observe the point-to-point communication part of the protocol. The $L - Send$ function called by the previous layer queries the strategy oracle *isEager* to infer whether the message should be sent in a eager or lazy approach. If the message should be sent eagerly, then a MSG is sent with the actual payload, otherwise an advertisement with the id is sent via $IHAVE$. Messages sent lazily may then be retrieved with the $IWANT$ call that will send the actual payload to the requester (lines 23 to 25). If an advertisement is received (lines 13 to 15), the message is queued to be retrieved in a point in the future via the *ScheduleNext* function. The requests are added to the retrieval queue in an order that puts request to local nodes first. Thus, if a request is already scheduled to retrieval from a remote node and the incoming advertisement is from a local node, the requests order is swapped (lines 31 to 38). This simple procedure further reduces the transmissions on remote links as the payload retrieval is first attempted on local nodes. The *isCloser* oracle simply compares the distance between the two potential sources and can be built upon the *isRemote* oracle defined previously. Upon reception of a new message (lines 17 to 21), the message id is added to the set of known messages, any pending requests on the message are cleared, and the payload is delivered to the peer selection layer via the $L - Receive$ upcall.

Finally, this layer of the CLON service will offer a *Multicast* primitive which an application could use to disseminate messages, and a *Deliver* upcall which will be used to deliver disseminated messages to the application.

5 Evaluation

This section describes the experimental evaluation conducted in order to verify that the devised protocols address the requirements presented in Section 1.

All experiments have been run on a simple round-based simulator and assuming 1000 nodes distributed evenly among 5 local areas that are connected to each other by long-distance links, i.e. links were we want to reduce the bandwidth consumption whithout impairing reliability. As CLON is based on a flat approach, we compare it with the Scamp protocol, to assess the performance improvement that CLON brings while offering the same resilience to faults.

5.1 Peer Sampling Service

In the first experiment, depicted in Figure 1, we analyse the properties of the overlays generated by Scamp and CLON , and the impact of each one in the reduction of the load imposed on the long-distance links. To access the reliability of both protocols in the presence of failures we devised three drop strategies that randomly remove nodes from the overlay without healing, from 0% to 100%, in steps of 10%. The strategy $UniformDrop$ drops nodes from the universe of nodes in a random fashion. Additionally, $OneAreaDrops/TwoAreaDrops$ remove nodes from one/two pre-selected local areas to access the contribution of a particular local area to the overall connectivity. Both protocols are configured with the $c = 6$ which makes the average view size 9. After observing the connectivity level of Scamp with this configuration, we configured the *localityOracle* of CLON such that the protocol provides the same reliability level as Scamp. It is important to notice that the oracle configuration should take into account the way the contact node is chosen. In our experiments the contact is chosen randomly across the set of existing nodes, and thus nodes will receive four more times subscriptions from remote nodes than local ones, as in this scenario we have four times more nodes than local ones. Therefore, if we want to have the same amount of local and remote nodes in a view, the oracle should increase by four times the virtual view sizes when receiving subscriptions for remote nodes to compensate for the greater amount of remote subscriptions received, and, therefore, match the desired ratio of remote and local nodes in a view.

Figure 1a depicts the evolution of both protocols when applying the dropping strategies presented above. As Scamp is not locality-aware, its views are composed, on average, by 6.8 and 2.2 remote and local nodes respectively, which reflects the fact that we have much more remote nodes than local ones. On the other hand, we observed that to maintain the same reliability level in CLON , it is only necessary to have views with 2.7 and 6.3 remote and local nodes respectively. It is important to note that the composition of the view is an essential metric to obtain the desired reliability level while at the same time reduce the load imposed on the long distance links. In fact, simply by changing the ratio of remote/local nodes in the view of the nodes, it is possible to directly affect the number of messages that traverse the long distance links, and consequently the load imposed on them. As it is possible to observe, with this configuration

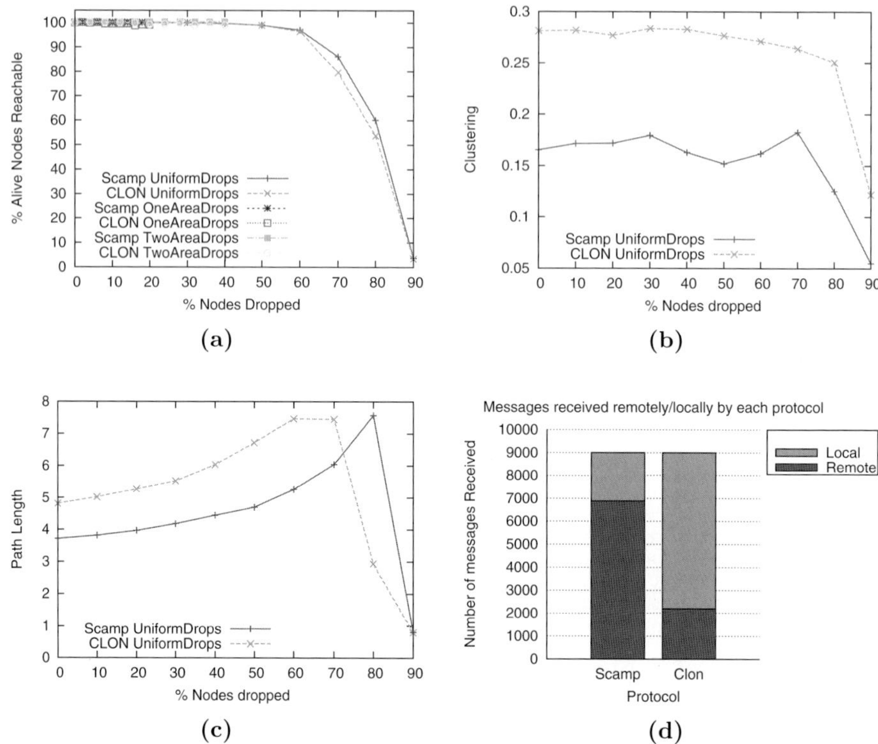

Fig. 1. Overlay properties and impact on the dissemination with a flooding protocol

both protocols are able to tolerate up to 60% node drops, in the $UniformDrop$ strategy, without compromising the reachability of the alive nodes. Nonetheless, after 70% failures both protocols fail to achieve the desired reachability level of $> 90\%$. The impact of localized failures in a given local area has no practical impact on the reachability, which means that local areas are inter-connected with enough redundancy to tolerate those localized failures as can be observed in the $OneAreaDrop/TwoAreasDrop$ dropping strategies. The lines depicting the connectivity of $OneAreaDrop$ and $TwoAreasDrop$ only go to 20% and 40% respectively, because that is the amount of global failures that a complete failure of one/two local areas represent.

On Figure 1b depicts the evolution of the clustering coefficient in presence of increasing drop rates. As expected, the clustering coefficient of CLON is a little higher than that of Scamp, because CLON tends to mimic the underlying network structure, which is inherently clustered among the local areas.

The average path length of the overlay could be observed in Figure 1c, and shows that CLON has a slightly larger average path length than Scamp. This comes directly from the fact that due to the lower ratio of known remote nodes not all the local areas are reachable directly by all the nodes and, therefore some nodes need some extra hops to reach the entire overlay.

Finally, Figure 1d depicts the number of messages exchanged when disseminating on top of both overlays, before applying any drop strategy. To this end, we used a naive eager push dissemination protocol that just floods all its known neighbors in an infect and die fashion. Every alive node multicast exactly one new message and after all messages have been delivered we count the average number of messages by each node received through remote and local neighbors. The amount of total messages received in both protocols is around 9000, which reflects the injection of 1000 new messages on the system, one by each node, and the fact that they are sent on average 9 times by each node, the average view size. As it is possible to observe, Scamp receives around 7000 messages through remote nodes, which reflects the average number of remote neighbors each node has. On the contrary, due to the biasing to the underlying network that gives preference to local links over remote ones, CLON receives around 2000 messages via remote neighbors, thus, being able to achieve a reduction of more than 70% on the amount of messages that traverse the long-distance links, while tolerating the same amount of failures.

In the next experiment, depicted in Figure 2, we made the biasing to the underlying network more aggressive by further reducing the number of remote nodes known on average. While previously the goal has to achieve the reliability level of Scamp, which in this configuration tolerates around 60% global node failures, here we intended to tolerate up to 20% and 30% failure rates, while guarantying that more than 99% of the alive nodes are reachable. On the left we have the reachability, as in Figure 1a, and on the right the respective amount of messages received locally and remotely. The labels Scamp and CLONI refer to the previous configuration for reference purposes, while CLONII and CLONIII are configured to tolerate up to 30% and 20% global failures respectively. As it is possible to observe, CLONII (blue line on the left plot) tolerates up to 30% global failures with more than 99% confidence with a slight reduction on the number of remotely received messages (right plot) when compared to the original configuration (CLONI). On the other hand, the configuration CLONIII tolerates up to 20% total failures with the same confidence level but further

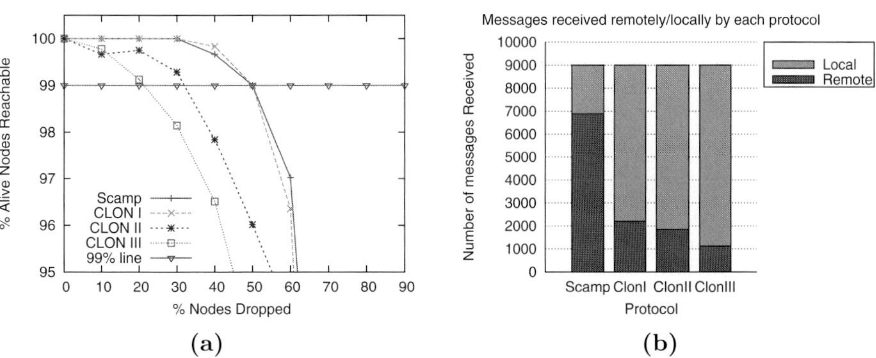

Fig. 2. Different oracle configurations

reduces the number of messages received remotely to slightly more than 1000. This experiment shows that with the adequate oracle configuration CLON is able to offer different reliability guarantees and consequently directly impact the load imposed on the long-distance links. This trade-off is related to the amount of remote neighbors a given node has, with smaller values the reliability decreases as a local areas are more prone to become disconnected from each other but the number of messages that traverse the long-distance links is reduced. Nonetheless, with a reliability level equivalent to standard flat overlay management protocols such as Scamp, CLON is able to significantly reduce the load imposed on the long-distance links, and contribute to increase the aggregate bandwidth available on them, as per the requirements presented in Section 1.

5.2 Dissemination Protocol

In this section, we evaluate the proposed dissemination protocol. The protocol is run atop the Peer Sampling Service with the same configuration of the first experiment. To fully understand the advantages of the strategy combination, it is important to assume that message advertisements are considerable smaller than the actual payload, which holds true on many real scenarios. If this assumption does not hold, the oracle could just be configured to a pure eager strategy as in this case the lazy strategy has no advantage over the eager one.

The results obtained are depicted in Figure 3. On the left the dissemination protocol is configured with a simple strategy: it uses eager push to local nodes and lazy push to remote ones. The rationale behind this strategy is to disseminate very fast on local areas by means of the eager strategy, and send advertisements to remote areas. When a given local area lazily receives the payload of a new message, it will spread it quickly inside it. As more local nodes receive the payload eagerly transmitted, the queued requests for those payloads via remote nodes are dropped and, therefore, a considerable amount of remote transmissions is avoided. Although the latency is considerable in this configuration due to the lazy pushing to remote nodes, the objective is to observe the lower bound on the number of messages that traverse the long-distance links. The improvement brought by this configuration of the dissemination protocol is clear because in both Scamp and CLON the number of messages received through remote neighbors (red bar) significantly decreases. In Scamp this value goes down from around 7000 with the flooding protocol to slightly more than 2000, whereas in CLON the value goes down to around 600 messages received remotely, an improvement of an order of magnitude if we consider the combination of Scamp with a flooding gossip protocol. As expected, the number of messages received locally (green bar) is much higher than Scamp due to the biasing to the underlying network, which gives preference to the local links over the remote ones. Finally, the number of advertisements received is considerably higher in Scamp than in CLON , again due to Scamp having much more remote neighbors on average. The reduced number of announcements on CLON is due to two facts: the average number of known remote nodes is smaller than Scamp and, therefore fewer announcements are generated; and only a fraction of payloads are lazily pushed (and therefore the $IWANT$ announcements sent) due to the prior infection by local nodes.

Finally, in Figure 3b, we analyse the bandwidth/latency trade-offs offered by the dissemination protocol. The goal is to observe the impact of the chosen payload transmission strategy (by means of the $isEager$ oracle, see Listing 1.3) on the latency and bandwidth consumption of the dissemination process. To this end we run a set of experiments where the $isEager$ oracle returns False if and only if the target node is external and the external round is below a given threshold. The rationale is to transmit the message payloads eagerly for a certain number of rounds and then fall back to lazy strategy. In the experiment we varied the TTL value from 0 to 9, and for each value we run the dissemination protocol on top of the same overlay of the first experiment of the previous Section. On the X axis it is possible to observe the different TTL used for each run. As such, on the leftmost part of the axis we have a completely lazy strategy that becomes gradually eager as we move to the right. On the left Y axis we measure the bandwidth consumption, blue line, with respect to the number of message payloads transmitted over the long distance links. On the right Y axis we measure the latency of the dissemination, green line, in the number of hops necessary to infect all nodes in the overlay. For instance, in the completely lazy strategy, i.e. when lazy after the round zero, nodes receive on average slightly more than 600 messages through remote links, which in fact is the experiment of the left. With this configuration the latency to infect all nodes is 11 hops. As expected, the bandwidth increases with the eagerness to transmit the payloads as more redundant messages are sent, while the latency decreases, as messages reach all nodes quicker without the additional roundtrips of a lazy strategy. It is interesting to notice that in this scenario the latency reaches its minimum after the 4th round when it becomes closer to the overlay diameter. On the other hand, the bandwidth tends to stabilize only around the 7th round. Therefore, in this scenario using a eager strategy for more than four rounds will only waste bandwidth without bringing any improvement on the latency of the dissemination process.

The point where the two lines intersect presents an interesting trade-off because it is when the bandwidth overhead required for the dissemination is small, with a moderate latency penalty. This accounts for around 1000 message payloads received remotely which is half of the value obtainable with a flooding protocol as the one used in Figure 1d. In fact, even if the latency should be reduced to a minimum, for instance by switching to the lazy push strategy after the 3rd round, the impact on the bandwidth consumption on the long-distance links is still attractive as only around 1400 remote messages are received.

To finalize, we showed that by combining eager and lazy push strategies, it is possible to considerably reduce the stress put on resource constrained links. For instance, from the initial setting of having a flooding protocol on top of Scamp to a combined dissemination protocol on top of CLON , it is possible to achieve a reduction on the number of payloads transmitted from more than 7000 messages to around 1400, while offering an attractive latency value. Furthermore,

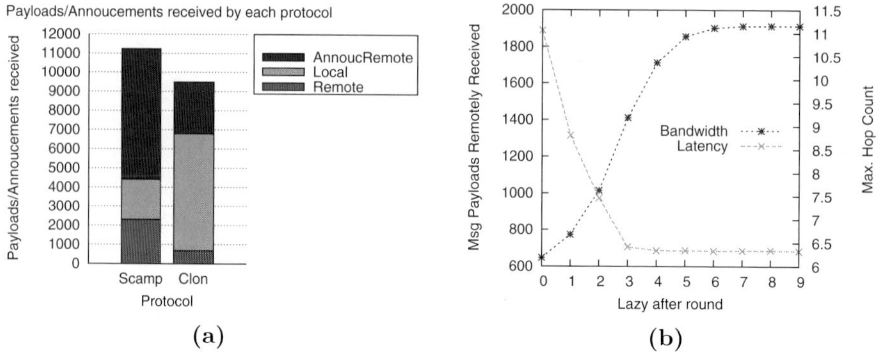

Fig. 3. Latency/Bandwidth trade-off

the excess of locally received messages on CLON , could have been mitigated by also using the eager/lazy strategy combination.

6 Conclusion

On this work we presented CLON , a Reliable Multicast Service that aims to cope with the requirements of a Cloud environment, namely the reduction of the bandwidth consumption on undesirable costlier links, such as inter data center links, while coping with the reliability and resilience required in such environments. We addressed the problem at two different levels: the Peer Sampling Service which is related to the construction and maintenance of the overlay network, and the dissemination protocol which disseminates messages on top of the built overlay.

By taking into account locality at construction time, and by refusing to rely on nodes with special roles to handle locality, we obtained an overlay that is biased to the network topology without compromising the key properties that ensure the reliability and robustness of unstructured protocols. On the dissemination protocol, we also take into account locality by having different rounds for remote and local nodes, and by offering to the programmer different latency/bandwidth trade-offs on a single protocol by configuring an oracle to behave accordingly to the desired policy.

By relying on oracles to configure the protocol's different trade-offs, and abstracting those particular configurations out of the model, we obtained a highly configurable service that can be used on a wide range of scenarios from the low level infrastructure management in a Cloud environment to an added-value service offered to client applications.

The experimental results obtained are promising as we achieved a reduction of an order of magnitude on the number of message payloads that traverse the long distance links.

References

1. DC2MS: Dependable Cloud Computing Management Services (2008), http://gsd.di.uminho.pt/projects/projects/DC2MS
2. Al-Fares, M., Loukissas, A., Vahdat, A.: A scalable, commodity data center network architecture. SIGCOMM Computer Communication Review 38(4), 63–74 (2008)
3. Armbrust, M., Fox, A., Griffith, R., Joseph, A.D., Katz, R.H., Konwinski, A., Lee, G., Patterson, D.A., Rabkin, A., Stoica, I., Zaharia, M.: Above the clouds: A berkeley view of cloud computing. Technical Report UCB/EECS-2009-28, EECS Department, University of California, Berkeley (February 2009)
4. Bailey, N.: The Mathematical Theory of Infectious Diseases and its Applications, 2nd edn. Hafner Press (1975)
5. Carvalho, N., Pereira, J., Oliveira, R., Rodrigues, L.: Emergent structure in unstructured epidemic multicast. In: Proceedings of the 37th Annual IEEE/IFIP International Conference on Dependable Systems and Networks, pp. 481–490. IEEE Computer Society, Washington (2007)
6. Eugster, P., Guerraoui, R.: Hierarchical probabilistic multicast. Technical Report LPD-REPORT-2001-005, Ecole Polytechnique Fédérale de Lausanne (2001)
7. Eugster, P., Guerraoui, R., Handurukande, S., Kouznetsov, P., Kermarrec, A.-M.: Lightweight probabilistic broadcast. ACM Transactions on Computer Systems 21(4), 341–374 (2003)
8. Ganesh, A., Kermarrec, A.-M., Massoulié, L.: Scamp: Peer-to-peer lightweight membership service for large-scale group communication. Networked Group Communication, 44–55 (2001)
9. Ganesh, A., Kermarrec, A.-M., Massoulié, L.: Hiscamp: self-organizing hierarchical membership protocol. In: Proceedings of the 10th workshop on ACM SIGOPS European workshop, pp. 133–139. ACM, New York (2002)
10. Hopkinson, K., Jenkins, K., Birman, K., Thorp, J., Toussaint, G., Parashar, M.: Adaptive gravitational gossip: A gossip-based communication protocol with user-selectable rates. IEEE Transactions on Parallel and Distributed Systems 99(1) (2009)
11. Jelasity, M., Guerraoui, R., Kermarrec, A.-M., van Steen, M.: The peer sampling service: experimental evaluation of unstructured gossip-based implementations. In: Proceedings of the 5th ACM/IFIP/USENIX International Conference on Middleware, pp. 79–98. Springer-Verlag New York, Inc., New York (2004)
12. Kaldehofe, B.: Buffer management in probabilistic peer-to-peer communication protocols. In: Proceedings of the 22nd International Symposium on Reliable Distributed Systems, October 2003, pp. 76–85 (2003)
13. Karp, R., Schindelhauer, C., Shenker, S., Vocking, B.: Randomized rumor spreading. In: Proceedings of the 41st Annual Symposium on Foundations of Computer Science, p. 565. IEEE Computer Society, Washington (2000)
14. Leitão, J., Pereira, J., Rodrigues, L.: Hyparview: A membership protocol for reliable gossip-based broadcast. In: Proceedings of the 37th Annual IEEE/IFIP International Conference on Dependable Systems and Networks, pp. 419–428. IEEE Computer Society, Los Alamitos (2007)
15. Lin, M., Marzullo, K.: Directional gossip: Gossip in a wide area network. In: Hlavicka, J., Maehle, E., Pataricza, A. (eds.) EDDC 1999. LNCS, vol. 1667, pp. 364–379. Springer, Heidelberg (1999)
16. Massoulié, L., Kermarrec, A.-M., Ganesh, A.: Network awareness and failure resilience in self-organising overlay networks. In: Proceedings of the 22nd Symposium on Reliable Distributed Systems, pp. 47–55 (2003)

17. Melamed, R., Keidar, I.: Araneola: A scalable reliable multicast system for dynamic environments. In: IEEE International Symposium on Network Computing and Applications, pp. 5–14 (2004)
18. Pereira, J., Rodrigues, L., Oliveira, R., Kermarrec, A.-M.: Neem: Network-friendly epidemic multicast. In: Proceedings of the 22nd Symposium on Reliable Distributed Systems, pp. 15–24. IEEE, Los Alamitos (2003)
19. Schroeder, B., Gibson, G.A.: Disk failures in the real world: what does an mttf of 1,000,000 hours mean to you? In: Proceedings of the 5th USENIX conference on File and Storage Technologies, pp. 1–16. USENIX Association, Berkeley (2007)
20. Voulgaris, S., Gavidia, D., Steen, M.: Cyclon: Inexpensive membership management for unstructured p2p overlays. Journal of Network and Systems Management 13(2), 197–217 (2005)

A Solution to Resource Underutilization for Web Services Hosted in the Cloud

Dmytro Dyachuk and Ralph Deters

MADMUC lab
Department of Computer Science
University of Saskatchewan
Saskatoon, Saskatchewan
S7N 5C9 Canada
dmytro.dyachuk@usask.ca, deters@cs.usask.ca

Abstract. At the moment the service market is experiencing a continuous growth, as services allow easy and quick enhancing of new and existing applications. However, hosting services according to a common on-premise model is not sufficient for dealing with erratic, spike-prone service loads. A new more promising approach is hosting services in the cloud (utility computing), which enables dynamic resource allocation. The last provides an opportunity to meet average response time requirements even in case of long-term fluctuating loads. Unfortunately, in the presence of short term fluctuations the resources utilization has to stay under 50% in order to achieve response time of the same order as job sizes.

In this work we suggest to compensate the problem of underutilization caused by hosting low-latency services by means of allocating the remaining resources to time insensitive service requests. This solution uses load balancing combined with admission control and scheduling application server threads. The proposed approach is evaluated by means of experiments with the prototype conducted with Amazon's EC2. The experimental results show that the servers utilization can be increased without penalizing low-latency requests.

1 Introduction

After being introduced in 2000, Web Service technology [12] started penetrating the world of the distributed computing. Web Services excelled in compare to the other competitors mostly due to using platform independent XML based technologies. Simple Object Access Protocol (SOAP) [22] is employed as a message encoding protocol and Web Service Description Language WSDL [13] is used as a language for specifying the protocol of interacting with services. XML based protocols enable implementing large platform and language independent distributed systems. Moreover, Web Services as opposed to Common Object Request Broker Architecture (CORBA) middleware offer loser coupling and use simpler message encoding protocol, which is supported a larger number of programming languages. Even though Web Services may seem a superior approach to CORBA,

R. Meersman, T. Dillon, P. Herrero (Eds.): OTM 2009, Part I, LNCS 5870, pp. 567–584, 2009.

the heterogeneity of Web Service brought by XML adversely impacts their performance. And thus CORBA is still often favored for implementing low-latency and/or real-time applications [14].

At the moment the service market is experiencing a continuous growth, as services allow easy and quick enhancing new and existing applications. Among the offered services, a substantial percentage is constituted by low-latency data services[1]. Some of these services have real-time properties implying that a response from a service will contain information which is valid for a predefined amount of time (i.e. a stock quote service). Thus, it is crucial for real-time data services to ensure that a request is being processed within a given time constraint and the response from a service will be delivered in a timely manner.

In this paper we address the issue of resource underutilization for low-latency data services hosted in the utility computing cloud. Low-latency service can be defined a service whose response time closely approaches request execution time, where request execution time is the amount of time for processing a single request.

Request arrival rate has a strong impact on services response time, the higher is the request rate the higher is the response time. In order to keep the average response time under a certain threshold the arrival rate has to be limited as well. Unfortunately, in order to achieve low-latency response time the utilization of a server hosting a service should stay markedly lesser than 50% [fig. 2, 3].

In this work we suggest eliminating the problem of underutilization caused by hosting low-latency services by means of allocating remaining resources to time insensitive service requests. Time insensitive requests can represent a service with no response time guarantees, for instance low-priced services, services used for free public consumption, offered without Service Level Agreements(SLA), etc. The solution consists of three layers: load balancing, admission control/request scheduling and thread scheduling. A load-balancing component distributes incoming requests among servers in a server pool. After that the admission control component decides if a service request can be accepted into the service, otherwise it will be put into a waiting queue, where time critical requests (low-latency) are preceding the time insensitive (regular) requests. As for the last step the priority of a thread assigned to handling an accepted request is being altered. In this work we present experimental results conducted with Amazon's EC2[1] and compare them with reference results obtained from the queuing theory.

The paper is structured as follows. In the second section we discuss three mainstream approaches for hosting services which are exposed for massive public consumption. The underutilization problem emerging while attempting to guarantee low-latency response times is presented in section three. An approach for tackling the problem of underutilization as well as its implementation details can be found in section four. Results of the experiments are located in section five. The last three sections contain a conclusion as well as an outlook on the related and future work.

[1] Companies like Strike Iron (http://www.strikeiron.com/) provide a number of services for accessing geographical, financial data, etc.

2 Hosting Services

At current, three main approaches can be distinguished for hosting services.

- On-premise (or in-house) hosting
- Utility computing
- Platform-as-a-Service hosting

On-premise hosting implies that a service provider is responsible for purchasing physical equipment, like servers, network bandwidth in order to host a service. This model has a set of advantages as a service provider has the flexibility of choosing the most suitable network topology and appropriate hardware configurations as well as an operating system. However, one of the key providers responsibilities is to perform an analysis of possible service loads and allocate an appropriate number of servers. The drawbacks of this methodology consists in the fact that often capacity planning is performed on the base of the worst case analysis. The last suggest allocating resources with regard to the maximum load. And as a result most of the time servers can be highly underutilized, what in its turn negatively reflects on the running costs. Moreover, on-premise infrastructure is vulnerable to erratic workloads which can be characterized by the presence of spikes in request arrival rates. The spike arrivals often yield to a significant response time increase or even denying services for some requests. Besides that, in case clients are distributed over a large space area on-promise computing requires significant investments for implementing low-latency services. The communication with services happens over the Internet and the time for delivering a request to services itself has a strong correlation with the percentage of lost packets, packet latency and network throughput itself. All the aforementioned factors depend on the physical distance from a service to a client. Thus, the larger the distance is the more packets are being lost, more hops have to be made and lesser is the throughput. Consequently, a service provider is behooved to place servers in different parts of a continent or even on different continents.

Utility computing in contrast to on-promise hosting provides the ability of "renting' servers in the cloud instead of purchasing physical machines. Thus, utility computing is often seen as an affordable alternative to the expensive on-premise approach. With utility computing it became possible to increase/decrease server pools dynamically with regard to service load (request arrival rate). Utility computing also looks more attractive for implementing low-latency services, as service provider can decide on the geographical location of server pools and host a service closer to clients. For instance, with Amazon's EC2 [1] the resources can be allocated in any of the four availability zones: three regions in North America and one in Europe. However, it is important to realize that service provider is fully responsible for implementing load balancing as well as fault-tolerance mechanisms.

Utility computing services may differ in terms of supported platforms as well as in term of the resources offered. But usually the choice of servers configuration is limited to a number of the predefined ones. The situation is similar in terms of the supported operating systems. Utility computing is viewed as a commodity and thus the price for its uasage is based on the server hours, amount of incoming/outgoing traffic, volume of persistent storage, etc.

Platform-as-a-Service (PaaS) is the latest technology in comparison to the other two. PaaS providers usually take responsibility for maintaining underlying operating system, allocating sufficient amount of resources as well as implementing load balancing. However, there is a number of constraints imposed on hosted services. Those constraints emerge mostly due to security precautions, ensuring proper isolation when several services are sharing a host, as well as correct operating of services and enabling their management. For example, Google-App [4] engine allows hosting applications written in Python with no access to file I/O, the running time of each request is limited to one second and the proprietary storage (BigTable[4]) has to be used as a means for storing persistent data. The costs for hosting a service in such environment can be based on the number of requests served, consumed CPU time, amount of storage used, etc.

The last two methods for hosting services are often referred as hosting service in the Cloud. We believe that due to numerous limitation imposed by PaaS providers service providers will favor utility computing.

3 Resource Utilization in the Cloud

Works [25,26] have shown that loads on the Internet systems are highly irregular, low request arrival rates frequently intermittented by high request rates. The irregularity introduced by load adversely reflects on the response time, causing abnormally high service response time during the peak load and resource underutilization during the low loads. The abnormally high response time can negatively affect the reputation of a service or even cause waiting timeouts on a client side, what in its turn will decrease perceived reliability of the service.

Utility computing made it possible to adjust systems capacities dynamically to load fluctuations and thus to avoid the aforementioned problem. However, there is a latency associated with allocating new servers, and thus this approach allows compensating loads fluctuations only within 10-20 minutes range. Therefore, it would be a natural solution to allocate more resources, just enough to sustain the load, while the new servers are being added.

The number of request arrivals to an Internet system can be represented by means of a Poisson process[20]. A Poisson process sufficiently describes short-term arrival rate fluctuations and therefore we employ it for describing the number of request arrivals per time unit.

Let's define job size S or service time the amount of time required for processing a specific request(job) on an idle resource. For a CPU bound service this can represent the amount of the CPU time required for processing a request. Service response time T will be used as the time required for processing a request in the presence of other requests, response time T encompasses the sum of waiting time W and service time S: $T = W + S$.

Figure 1 contains a chart showing the analytical results for modeling the response time of a service running on a single core machine with an average

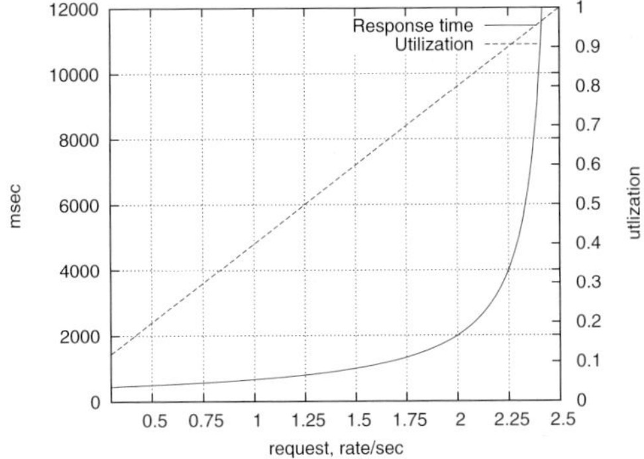

Fig. 1. Response Time (M/M/1), s=400msec

service time of 400 msec. The results are obtained according to the following formula $\overline{T} = S/(1 - \lambda S)$ [11], where λ is the average request arrival rate and $\overline{T}$ is the mean response time[2].

A chart on figure 2 shows the dependence of a service response time from its request arrival rate. As one can see on both charts [fig. 1 and 2] response time rapidly increases as the load approaches the capacity limit (2.5 req/sec). The utilization of the CPU by the service and analytical utilization can be seen on charts [fig. 1 and 3]. Therefore, for service providers it becomes of a great importance to keep the arrival rate so that utilization will not exceed 70%. Thus, on average 30% of the resource will have to idle. The situation gets worse if the service provider intends to provide low-latency services. In this case the arrival rate most likely will have to be limited to 20%-30% of the service's capacity. In this paper we propose allocating the remaining percent of the resource to the time insensitive requests and thus increase the utilization of service hosts without affecting the critical requests.

One can distinguish two different approaches for prioritizing critical requests, preemptive and non-preemptive [11]. In the preemptive case service suspends processing non-critical requests as soon as a critical request arrives. In the non-preemptive case the requests are placed in a waiting queue and the critical requests are being served first. The obvious advantage of the first approach is that the critical job does not have to wait for a service to become free.

An impact of the waiting time can be seen on figure 4. The chart shows how preemption affects a service with a mean service time of 400 msec. Critical requests are arriving at the average rate of 0.5 req/sec, the non-critical requests arrive at the rate increasing from 0 to 2.0 req/sec, the x axis shows the total arrival rate. As it can be seen in non-preemptive case critical requests response time increases with the growth of the regular requests arrival rate, while in the preemptive case the arrival rate of the regular requests does not affect the

[2] In spite the fact that this formula is derived for M/M/1/FCFS queuing system it is also valid for M/M/1/PS, which more accurately models processor sharing systems [11].

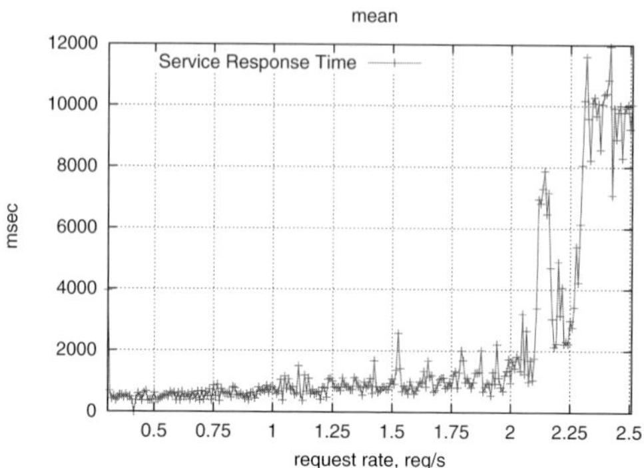

Fig. 2. Service Response Time

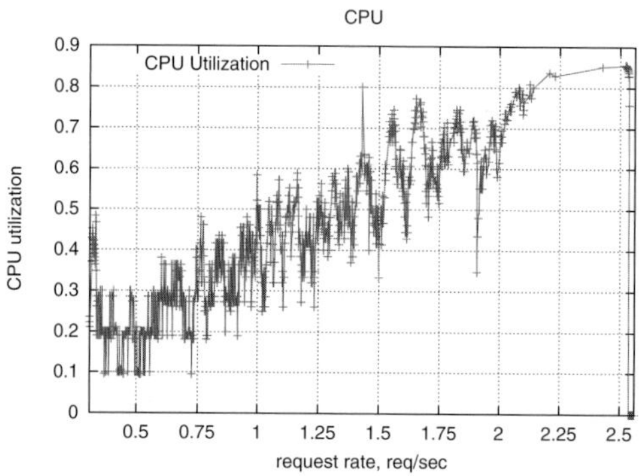

Fig. 3. CPU Utilization

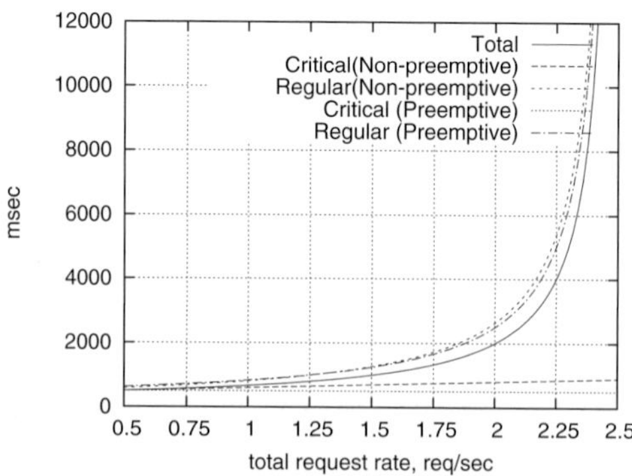

Fig. 4. Response time of preemptive and non-preemptive service

response time of the critical ones. Therefore, this work is based on preemptive priority scheduling.

4 Solution Architecture

In order to implement a scalable service hosted in the Cloud usually a two-tier architecture is employed. The first tier consists of a load-balancing reverse proxy. The second tier is constituted of a set of servers forming a server pool [fig. 5]. The key proxy roles is to hide servers from a direct client access and to distribute load among servers. The proxy is governed by a load balancing policy. The most common load balancing policies include but are not limited to Weighted Round Robin [17], Least Connections, Sticky Sessions, etc.

In this work we employ the Round Robin policy due its wide popularity and low overhead. Round Robin operates by equally distributing the load by forward incoming requests to servers in turns. Thus after a request has arrived at the load balancer it will be forwarded to a server.

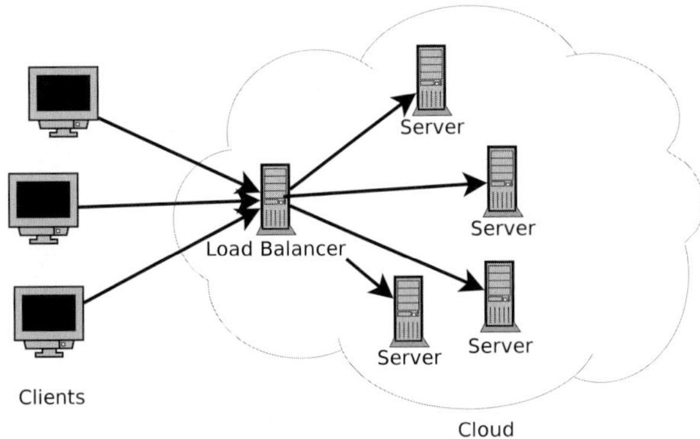

Fig. 5. Architecture for services running in the cloud

In case a request can not be accepted into service immediately it will be placed into a waiting queue. The acceptance decision is made by an admission control mechanism which is used as a means for preventing thrashing effect [15]. Thrashing can emerge in systems where requests are handled concurrently in threads [16]. Requests residing in the queue are ordered according to their priorities. After the request has been admitted in service, the priority of a handling thread is altered.

Each request has priority p and weight w associated with it. The last two parameters are extracted from the SOAP header [fig. 7] which is mandatory for every service request. The maximum priority is denoted as $\hat{p}$ and is used to mark the most critical requests, which do not have to wait for an admission.

Weight is a parameter which can be used for managing admission control on a fly, for instance for implementing adaptive admission control [16]. In this work

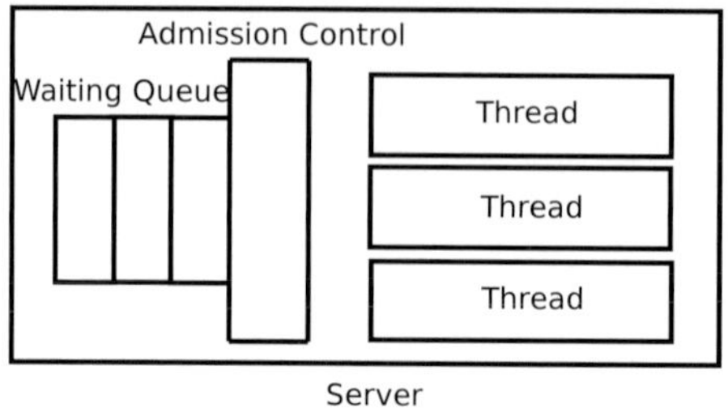

Fig. 6. Architecture of a hosting server

all the requests have equal weight w_c thus the maximum number of requests which can be processed in parallel is R_0/w_c, where is R_0 is a constant describing server's capacity. If to note R as the amount remaining resources then the algorithm for implementing scheduling and admission control is the following:

if *there is a request* i *in the waiting queue, such that* $p_i = \hat{p}$ ***then***
  $R = R - w_i$
  start processing request i *and set the priority of a processing thread to* p_i
else
  i = *request with the highest priority in the waiting queue*
  if $w_i \leq R$ ***then***
    $R = R - w_i$
    start processing request i *and change the priority of a processing thread to* p_i
  end if
end if

After request x has been processed the amount of the remaining resources is updated: $R = R + w_x$. The admission control and scheduling is invoked every time a new request arrives or a request leaves the service.

```
<soapenv:Header>
<tns:weight xmlns:tns="http://jobid.usask.ca"> 10 </tns:weight>
<tns:priority xmlns:tns="http://jobid.usask.ca"> 90 </tns:priority>
...
</soapenv:Header>
```

Fig. 7. SOAP header

5 Experiments

We have built a prototype in order to test if the proposed approach allows better utilization of service servers without impacting response time for low-latency requests.

5.1 Prototype Implementation

The prototype was implemented in Java and C. We used Apache Tomcat 5.5 [3] as a servlet container, Apache Axis2 as a Web Services middleware, HAProxy 1.3.12 [5] as a load balancer, and Linux (kernel 2.6.21.7-2.fc8xen) as an operating system. Sun's JRE 1.6 [6] was chosen as a Java runtime.

HAProxy is a high performance solution for implementing load balancing while supporting high-availability and proxying. HAproxy can operate on TCP as well as on HTTP levels. In this work we assume that services are stateless and therefore incoming service requests can be forwarded to any of the servers hosting a service and thus use regular HTTP load balancing.

Admission control and waiting queue scheduling mechanisms were implemented in Java and deployed as an Axis2 module, which can be used with an arbitrary Axis2 Web Service.

It is important to note that in modern Java runtimes user threads are mapped one-to-one to kernel threads. Consequently, user thread priorities are directly translated into the kernel thread priorities. Unfortunately, Sun's JRE Linux provides very limited abilities for modifying priorities of user threads within Java environment itself. It is not possible to set threads priorities to real-time nor alter the thread scheduling policy. Therefore, we had to develop our own Linux native library in C for setting arbitrary kernel thread priorities. The library interacts with scheduling and admission control module through Java Native Interface.

As it was mentioned before, after a job has been accepted it will be processed in a thread. During preliminary experiments we discovered that changing threads priorities at runtime results in additional delay and creates extra overhead. Therefore, the default configuration of Tomcat server was altered to contain two connectors and consequently two thread pools. One connector was responsible for serving regular requests and the other was accepting low-latency ones. The priorities in the thread pools were set immediately after the server has booted up and were not altered during the runtime.

One of the key goals in implementing the proposed idea was to decouple the described tools from a specific application server. At current the module was tested with Tomcat 5.5 which employs leader-follower model for handling incoming connections. But the developed scheduling module can be adapted for integrating with master-slave connectors without extensive modifications.

5.2 Prototype Evaluation

The proposed solution was tested with a Web Service running on Axis2+Tomcat platform. The services business logic consisted of recursive Fibonacci numbers calculations. Calculating Fibonacci numbers is a stack intensive operation which mimics method invocations typical for most services. The service contained a single method which required 400msec of CPU time. In order to eliminate the noise created by various jobs size and focus on the impact of arrivals and scheduling the jobs sized were set to constant size.

Table 1. Loads created by critical and regular requests

	One Server	Two Servers	Four Servers
$\lambda_{critical}$	0.5	1.0	2.0
$\lambda_{regular}$	0.5..2.0	0.5..4.0	0.5..8.0
λ_{total}	1.0..2.5	1.5..5	1.5..8.0

Table 2. Response time statistics (low-latency requests)

	One Server	Two Servers	Four Servers
Mean	443.4	405.2	402.5
95%	648.15	484.50	434.04
Max.	1105.1	627.0	594.6

Load balancer as well servers were running on Amazon's EC2 [1] single-core machines with 1.7 GB of RAM (m1.small). The service infrastructure encompassed a load balancer and a server pool. The load balancer was governed by the Round Robin scheduling policy. Three different cases analyzed when the service server pool consisted of one, two and four servers.

Requests were arriving according to a Poisson process. In each experiment the service was exposed to an increasing load created by regular requests [tbl. 1], while the low-latency requests were arriving the same rate. Moreover, low-latency requests were allocated 20% of the resources. while the rest was available for handling time insensitive requests. Such a load pattern was deliberately chosen to reflect the situation when the servers pool size is determined by the low-latency request rate. Thus, the load imposed by the regular request plays the second role and may fluctuate.

Each experiment consisted of nine phases corresponding to different arrival rates and lasted for one hour. The x axis on all charts depicts the arrival rates of regular requests. In each experiment the total load approached the services capacity and thus at the last phase the service was operating under 90% CPU utilization.

Tsung 1.3 [8] was used as a load generator. Each data sample on charts corresponds to an average value collected over a ten second time interval. Clients behavior followed an open model, which means that response time from a server did not affect arrivals of the next requests. CPU utilization was measured using standard *ps* tool available in any Linux distribution.

Single Server Case: As expected, the response time for regular requests started rapidly increasing as the total arrival rate began reaching the maximum request rate (2.0 req/sec for regular or 2.5 req/sec in total) [fig. 9]. However, the processing of the low-latency requests has not been disrupted and the service processed those requests on average in 443.4 msec [fig. 8, tbl. 2]. It is important to note, that according to the queuing theory in the preemptive case the higher priority jobs are not affected by lower priority ones. In reality, hoisting Linux threads priorities to real-time does not guarantee full preemption, unless a real-time Linux

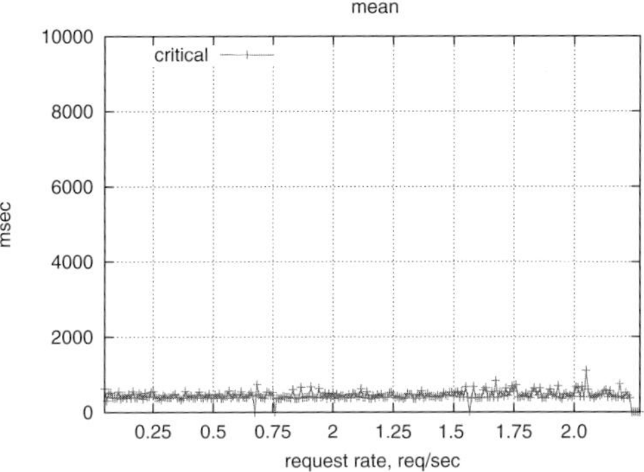

Fig. 8. Response time of critical (low-latency) service invocation. Single server case.

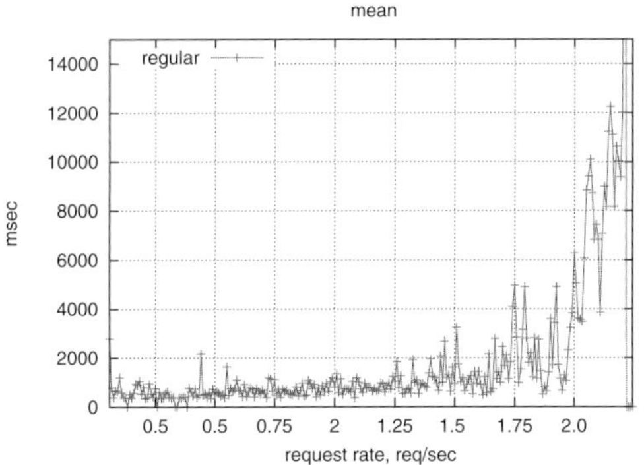

Fig. 9. Response time of regular service invocation. Single server case.

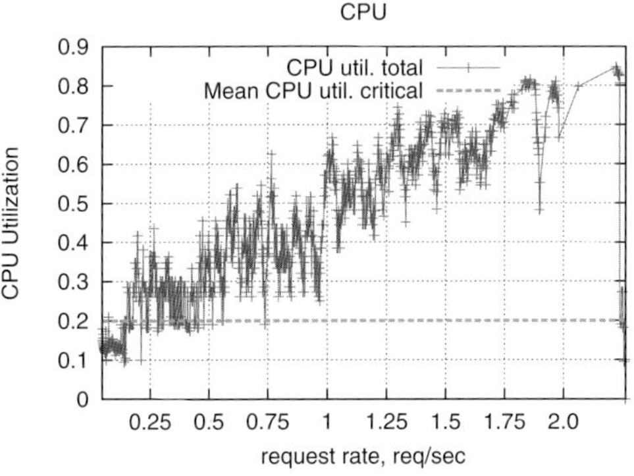

Fig. 10. CPU utlization. Single server case.

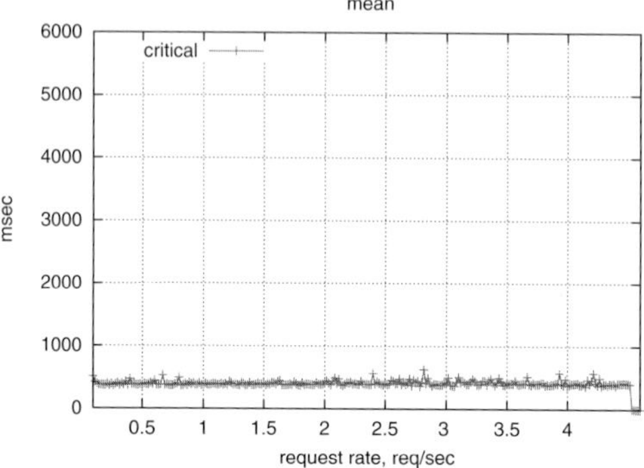

Fig. 11. Response time of critical (low-latency) service invocation. Two servers case.

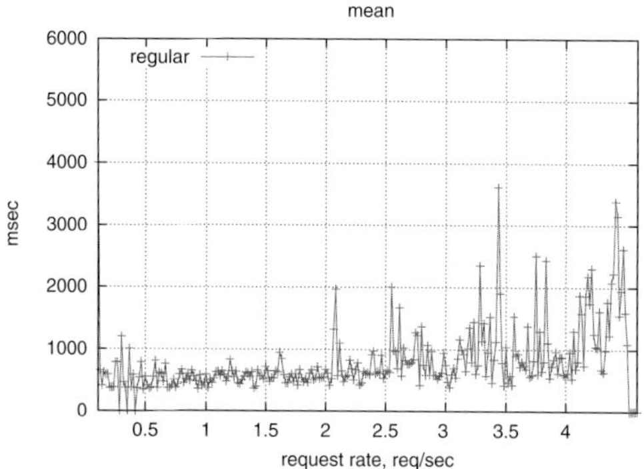

Fig. 12. Response time of regular service invocation. Two servers case.

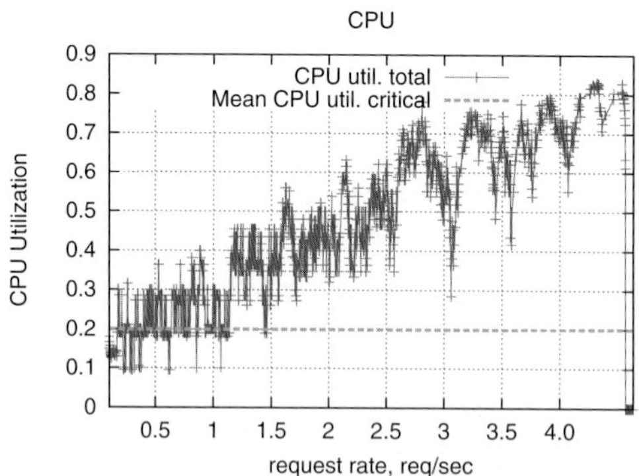

Fig. 13. CPU utilization. Two servers case.

kernel is used. Unfortunately, running Linux with a real-time kernel in the Amazon's EC2[1] at the moment is technically impossible, as the only allowed kernels are the ones provided by Amazon[1]. However, the results show that the mean

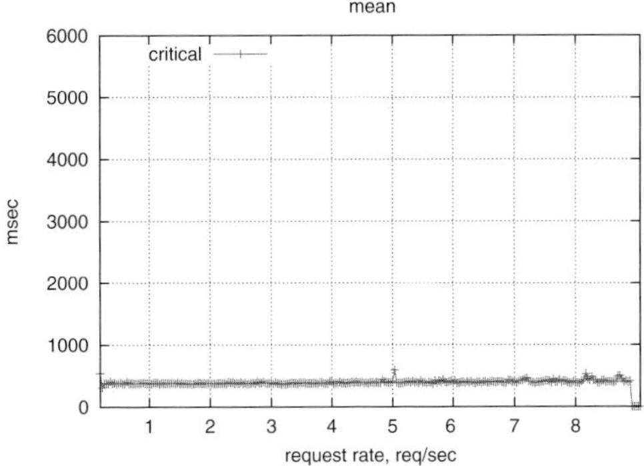

Fig. 14. Response time of critical (low-latency) service invocation. Four servers case.

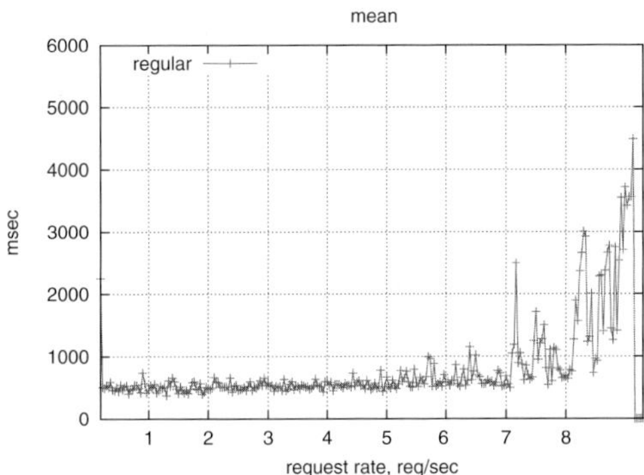

Fig. 15. Response time of regular service invocation. Four servers case.

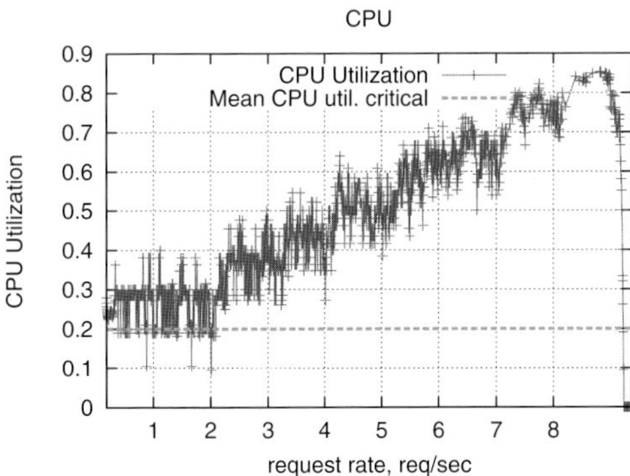

Fig. 16. CPU utlization. Four servers case.

response time for the the low-latency requests was 443 msec what is only 10.1% larger than its average size obtained using a queuing model [fig. 4] of 400.4 msec. Besides that 95% of critical requests were served in lesser than 648 msec [tbl. 2].

At same time response time of the regular requests increased from 400 msec to 14000 msec [fig. 9] or approximately 35 times.

As expected CPU utilization has increased from the initial 20% to 90% [fig. 10]. The X axis reflects the arrival rate of the non-critical requests. As it can be seen the fluctuations in the request arrivals resulted in the irregularities in the CPU utilization. Even when the total CPU utilization has reached 90%, when the time insensitive service invocations were occurring at the highest rate the response time for time sensitive service invocations remained unchanged [fig. 8]. As a result we can conclude that the suggested mechanism can be used for isolating the CPU bound critical load from the non-critical one.

Two Servers: In this case the server pool contained two servers. As it can be seen from the charts on figures 11 and 12 the service was processing all low-latency requests by 627 msec, moreover 95% of the request were served in lesser than 484.50 msec. It is important to note that with the increase of the server pool the difference between the mean response time in the reference result (400 msec) and actual result has shrunk. Since there were two servers load balancing helped to mitigate the fluctuations emerging in the Poisson arrivals.

The similar behavior is exhibited by regular requests. Their response time is still rapidly increasing as the load approaches 90% [fig. 12], nevertheless the growth rate is lesser than in the first case. The graph on figure 13 shows CPU utilization on one of the servers from the pool. The other servers experienced the same load and therefore the behavior of their CPU utilization is practically the same.

Four Servers: After scaling the load twice and doubling the resource pool the behavior exposed by critical and regular requests remains the same [fig. 12,13]. Besides that the 95% percentile of the response time for low-latency requests has got smaller as opposed to the single and two server cases. The CPU utilization has showed the same behavior as in the previous two cases [fig. 16].

6 Related Work

In work [27] Urgaonkar et al. presented an analytical model based on results from the queuing network. The model allows precise approximation of the mean response time from multi-tier systems and can be employed for performing dynamic capacity provisioning. The general principles which can be employed for performing service environment provisioning in order to meet response time guarantees were outlined by Ludwig in [21].

Another branch of works [24] focused on providing differentiated response time of a service running on a single server. In this work Siddaharta et al. implemented a proxy (Smartware) which intercepts the incoming requests, parses and classifies them according to user and client (device) type. In contrast to our work Smartware operates is a non-preemptive scheduler which is governed by a a randomized probabilistic scheduling policy derived from the lottery scheduling [28]. The experiment showed that the proposed approach outperforms the classical non-preemptive priority scheduling [11].

Real-time Java [7] provides tools for building real-time applications in Java. The environment provides a full spectrum of necessary tools, i.e. creating preemptive threads, priority inheritance protocol [23]. Unfortunately, besides the fact that this is a not a free product, it can be used only with Solaris or real-time Linux operating systems. The latter requires a special kernel, which is not supported by Amazon's EC2.

Levy et al. in [18] presented an approach for performance management of cluster-based Web Services. The authors addressed an problem of meeting response time guarantees for critical requests in cluster environments. The proposed solution employs non-preemptive Weighted Round Robin [17] load balancing schema in which weights are adjusted dynamically with respect to changes in the request arrivals rates.

Abdelhazer et al. in [10,9] proposed using control theory in order to meet predefined QoS requirements. The authors considered the problem of ensuring QoS in the environment with the fixed amount of resources. Content adaptation combined with admission control were used as a means for protecting servers from overloads. The conducted experiments showed that admission control which implements non-preemptive priority scheduling can be used as sufficient means for providing differentiated throughput.Moreover, this work is closely related to ours as it allows allocating underutilized resources to requests without performance guarantees. However, the authors focused on the throughput. It was demonstrated that their approach allows preserving the throughput of time-sensitive (premium) requests while allocating unused resources to processing time insensitive requests. The authors have enforced prioritization only at the admission control phase and thus implemented non-preemptive priority scheduling policy. Our approach differs by extending admission control with altering priorities of the threads handling incoming requests. Therefore, we implemented preemptive scheduling. In the preemptive case high priority requests have smaller waiting time as in case the resource is busy they do not have to wait. As a result, preemptive approach allows even better isolation of the high priority requests [sec. 2].

In [19] Lin and Dinda addressed the problem of underutilization by co-allocating virtual machines on the same physical host. The proposed idea is implemented as a user level scheduler VSched. VSched allows hosting virtual machines which are running applications with stringent time constraints together with virtual machines time insensitive batch processes. This implements preemptive scheduling with the isolation of time sensitive applications on the operating system level.

7 Future Work

In the future experiments we would like to replace the existing HAProxy load balancer with our own implemented in Erlang. HAProxy is a highly efficient and scalable load balancer. However, it was primarily designed for handling HTTP traffic. We plan to reimplement the basic functionality of HAProxy and to extend it in order to address specific needs of Web Services and cloud computing. Using Erlang would allow minimizing the overhead caused by load balancing and scheduling, while having our own component would give freedom for realizing more complex load balancing/scheduling policies.

In this paper we have addressed the issue of underutilization emerging in CPU bound services. In the future we would like to overcome this limitation and to consider services which may have other bottlenecks affecting service's capacity, such as memory, disk I/O, etc.

In this work we have considered the case where critical requests were guaranteed to be allocated sufficient amounts of resources in order to maintain their mean response time at the desired level. We would like to augment this approach by considering cases when there could be multiple classes of service with different Quality of Service objectives.

8 Conclusion

Hosting services in the utility computing cloud allows dynamically scaling a server pool running a service. As a result, it is possible to allocate a sufficient amount of resources to accommodate oscillating load [20]. Unfortunately, adjusting the pool size allows compensating only long-term fluctuations while the short term ones can be compensated only by allocating additional resources. Consequently, service providers must keep service utilization under 50% to provide services whose response time approaches CPU time required for processing the request. Which means that 50% of the time resources are idling. In order to compensate this loss we suggest to assign the remaining resources to processing time insensitive requests. i.e. service offered without Service Level Agreements.

The suggested idea was implemented in the form of a two-tier scheduling system integrated with load-balancing. The proposed system transparently schedules processing order of the incoming requests, while avoiding thrashing using admission control mechanisms. Moreover, the arrived requests are handled in threads running at different priorities. By altering thread priorities and performing special admission control it became possible to achieve higher server utilization without affecting the response time for low-latency requests.

A prototype of the proposed approach was implemented as an Axis2[2] module and thus can be easily integrated with an arbitrary Axis2 Web Service. The prototype was evaluated by means of experiments conducted over a scalable CPU bound service hosted in Amazon's EC2[1]. The service experienced short term load fluctuations which were modeled using a Poisson process. The load created by low-latency was fixed at 20%, while the time insensitive requests were using from 10% to 70%. The experimental results showed the response time of low-latency requests was unaffected even when the total load exceeded 90%.

Acknowledgements

The authors would like to thank the Discuss lab at the University of Saskatchewan, especially George Biswas for using some of their software libraries.

References

1. Amazon Elastic Compute Cloud (Amazon EC2), http://www.amazon.com/gp/browse.html?node=201590011
2. Apache Axis, http://ws.apache.org/axis/
3. Apache Tomcat, http://tomcat.apache.org/
4. Google App Engine, http://code.google.com/appengine
5. HAProxy. The Reliable, High Performance TCP/HTTP Load Balancer, http://haproxy.1wt.eu/
6. Java SE Runtime Environment 6.0, http://www.java.com/
7. Sun Java Real-Time System, http://java.sun.com/javase/technologies/realtime/rts/
8. Tsung, http://tsung.erlang-projects.org/
9. Abdelzaher, T., Shin, K., Bhatti, N.: User-level qos-adaptive resource management in server end-systems (2003)
10. Abdelzaher, T.F., Shin, K.G., Bhatti, N.: Performance guarantees for web server end-systems: A control-theoretical approach. IEEE Trans. Parallel Distrib. Syst. 13(1), 80–96 (2002)
11. Bolch, G., Greiner, S., de Meer, H., Trivedi, K.S.: Queueing Networks and Markov Chains: Modeling and Performance Evaluation with Computer Science Applications, 2nd edn., May 2006. Wiley/Blackwell (2006)
12. Booth, D., Haas, H., McCabe, F., Newcomer, E., Champion, M., Ferris, C., Orchard, D.: Web services architecture. Technical report, W3C (2004), http://www.w3.org/TR/ws-arch/
13. Christensen, E., Curbera, F., Meredith, G., Weerawarana, S.: Web services description language (WSDL) 1.1 (March 2001)
14. Cooper, G., DiPippo, L., Esibov, L., Ginis, R., Johnston, R., Kortman, P., Krupp, P., Mauer, J., Squadrito, M., Thurasignham, B., Wohlever, S., Wolfe, V.: Real-time corba development at mitre, nrad, tripacific and uri. In: Proceedings of the Workshop on Middleware for Real-Time Systems and Services, San Francisco, CA (1997)
15. Denning, P.J.: Thrashing: Its causes and prevention. In: Proceedings AFIPS, Fall Joint Computer Conference (1968)
16. Heiss, H.-U., Wagner, R.: Adaptive load control in transaction processing systems. In: VLDB 1991: Proceedings of the 17th International Conference on Very Large Data Bases, pp. 47–54. Morgan Kaufmann Publishers Inc., San Francisco (1991)
17. Leung, J.Y.-T. (ed.): Handbook of Scheduling: Alogorithms, Models, and Performance Analysis. Chapman & Hall/CRC, Boca Raton (2004)
18. Levy, R., Nagarajarao, J., Pacifici, G., Spreitzer, A., Tantawi, A., Youssef, A.: Performance management for cluster based web services. IEEE Journal on Selected Areas in Communications, 247–261 (2005)
19. Lin, B., Dinda, P.A.: Vsched: Mixing batch and interactive virtual machines using periodic real-time scheduling. In: SC 2005: Proceedings of the 2005 ACM/IEEE conference on Supercomputing. IEEE Computer Society, Washington (2005)
20. Liu, Z., Niclausse, N., Jalpa-Villanueva, C.: Traffic model and performance evaluation of web servers. Perform. Eval. 46(2-3), 77–100 (2001)
21. Ludwig, H.: Web services qos: External slas and internal policies - or: How do we deliver what we promise? In: Proc. 4th IEEE Int'l Conf Web Information System Eng. Workshops, pp. 115–120. IEEE CS Press, Los Alamitos (2003)
22. Mitra, N.: Soap version 1.2 part 0.

23. Sha, L., Rajkumar, R., Lehoczky, J.P.: Priority inheritance protocols: An approach to real-time synchronization. IEEE Transactions on Computers 39(9), 1175–1185 (1990)
24. Siddhartha, P., Ganesan, R., Sengupta, S.: Smartware - a management infrastructure for web services. In: Proc. of the 1st Workshop on Web Services: Modeling, Architecture and Infrastructure (WSMAI 2003), Angers, France, April 2003, pp. 42–49. ICEIS Press (2003), In conjunction with ICEIS 2003
25. Song, L., Marin, G.A.: Generating realistic network traffic for security experiments. In: SoutheastCon, 2004. Proceedings, pp. 200–207. IEEE, Los Alamitos (2004)
26. Uhlig, S., Bonaventure, O.: Understanding the long-term self-similarity of internet traffic. In: Smirnov, M., Crowcroft, J., Roberts, J., Boavida, F. (eds.) QofIS 2001. LNCS, vol. 2156, p. 286. Springer, Heidelberg (2001)
27. Urgaonkar, B., Pacifici, G., Shenoy, P., Spreitzer, M., Tantawi, A.: An analytical model for multi-tier internet services and its applications. In: SIGMETRICS 2005: Proceedings of the 2005 ACM SIGMETRICS international conference on Measurement and modeling of computer systems, vol. 33, pp. 291–302. ACM Press, New York (2005)
28. Waldspurger, C.A., Weihl, W.E.: Lottery scheduling: Flexible proportional-share resource management. In: Operating Systems Design and Implementation, pp. 1–11 (1994)

On the Cost of Prioritized Atomic Multicast Protocols*

Emili Miedes and Francesc D. Muñoz-Escoí

Instituto Tecnológico de Informática
Universidad Politécnica de Valencia
Campus de Vera s/n, 46022 Valencia, Spain
{emiedes,fmunyoz}@iti.upv.es

Abstract. A prioritized atomic multicast protocol allows an application to tag messages with a priority that expresses their urgency and tries to deliver first those with a higher priority. For instance, such a service can be used in a database replication context, to reduce the transaction abort rate when integrity constraints are used. We present a study of the three most important and well-known classes of atomic multicast protocols in which we evaluate the cost imposed by the prioritization mechanisms, in terms of additional latency overhead, computational cost and memory use. This study reveals that the behavior of the protocols depends on the particular properties of the setting (number of nodes, message sending rates, etc.) and that the extra work done by a prioritized protocol does not introduce any additional latency overhead in most of the evaluated settings. This study is also a performance comparison of these classes of total order protocols and can be used by system designers to choose the proper prioritized protocol for a given deployment.

1 Introduction

A group communication service (GCS) is a middleware component that provides a set of services that can be used as building blocks to design and build distributed systems. A GCS usually offers an atomic (i.e., total order) multicast message delivery service which enables an application to send messages to a set of destinations such that they are delivered in the same order to each destination. Group communication and total order topics have been studied for more than two decades from both a theoretical [1,2] and a practical [3,4,5,6] point of view. A useful additional guarantee a GCS may offer is priority-based delivery [7,8,9], which allows a user application to prioritize the sending and delivery of certain messages.

Such a service can be used in a scenario like the following. Consider an application that runs on top of a database replication system and is physically

* This work has been partially supported by EU FEDER and Spanish MEC under grant TIN2006-14738-C02-01, by EU FEDER and Spanish MICINN under grant TIN2009-14460-C03 and by IMPIVA under grant IMIDIC/2007/68.

R. Meersman, T. Dillon, P. Herrero (Eds.): OTM 2009, Part I, LNCS 5870, pp. 585–599, 2009.

distributed among several sites. Such systems usually follow a *constant interaction* model [10], according to which, updates made by a transaction are broadcast in total order to all the database replicas at the end of the transaction, using a single message. The order in which a set of messages corresponding to different transactions are delivered by the replicas determines the final order in which a set of transactions are applied to the database. This order has a deep impact on the evaluation of the integrity constraints defined in the database. The idea is to alter the order in which transactions are committed for achieving a favorable constraint evaluation, thus reducing the transaction abort rate. Note that the database replication protocol is able to know which database tables and fields have been accessed by a given transaction, and it is able to use such information for assigning priorities. To do so, the replication protocol should be also aware of the semantic integrity constraints defined in the database schema. MADIS [11] is an example of database replication middleware where all these issues can be managed. A transaction implementation based on stored procedures is another alternative for providing all the information needed by the replication protocol in order to assign priorities (accessed tables and fields, values being used in the updates, etc.).

Non-prioritizing total order broadcast policies have been widely studied, while, as far as we know, only a few studies exist for priority-based protocol variants. In [9] and [8], two priority-based total order protocols are presented. Low priority messages may suffer starvation if too many high priority messages are sent. The problem of message starvation is dealt with specifically in [12]. In [13,14] another common problem of this kind of protocols, known as *priority inversion*, is addressed.

In [15] we proved that total order prioritization is able to reduce transaction abort rates in replicated databases, thus showing the utility of atomic multicast prioritization. In this paper we show that atomic multicast prioritization techniques do not impose a significant overhead on the latency of the multicast messages. As a result, this reinforces the usefulness of this approach, since its advantages proved in [15] do not introduce any performance degradation.

The paper is organized as follows. In Section 2 we describe the system we use in this experimental work. In Section 3 we present three different kinds of broadcast protocols and their appropriate prioritizations mechanisms. In Section 4 we present some experimental work we have done to show that the overhead added by the prioritization techniques is not significant. Finally, we conclude the paper in Section 5.

2 System Description

The system is composed of a set Π of processes that communicate through message passing. Each process has a multilayer structure, whose topmost level is a user application that accesses a replicated DBMS, which in turn uses the services offered by a group communication system. The latter is composed of one or more group communication protocols, which use the underlying network services to send and deliver messages. In Section 4.1 we provide additional information related to the physical environment we used.

Processes run on different physical nodes and the drift between two different processors is not known. The time needed to transmit a message from one node to another is bounded but the bound is unknown. In practice, the system does not need more synchrony than that offered by a conventional network which offers a reasonable message delivery time. Process failures and network partitions may occur. However, since we are focusing on the comparison of prioritization techniques, we do not address failure handling (which can be realized by mechanisms such as group membership services and fault-tolerance protocols).

3 Algorithms

In this section we show three sketches of algorithms that implement the priority-based total order broadcast service.

In Fig. 1, we present a modification of the original fixed sequencer-based UB (i.e., unicast-broadcast) algorithm presented in [2] (underlined text shows the extensions). The main difference is that incoming messages are not immediately sequenced and sent to all the destinations but queued according to their priority and later sent.

```
SENDER:
Procedure TO-broadcast(m, prio):
    prio(m) := prio
    send m to sequencer
DESTINATIONS:
Initialization:
    nextdeliver := 1
    pending := ∅
When receive (m, seqnum):
    pending := pending ∪ {(m, seqnum)}
    while ∃ (m, seqnum) ∈ pending :
    seqnum = nextdeliver do
        deliver m
        nextdeliver++

SEQUENCER:
Initialization:
    seqnum := 1
    incoming := ∅
Parallel: when receive (m):
    insert m in incoming,
    according to prio(m)
Parallel: after initialization:
    while incoming is not empty do
        m := first message in incoming
        incoming := incoming \ {m}
        sn(m) := seqnum
        send (m, sn(m)) to all
        seqnum++
```

Fig. 1. Modification of fixed UB

A modification of the privilege-based algorithm of [2] is shown in Fig. 2. In a privilege-based algorithm, each node can not broadcast any message until it gets the token with the sending privilege. When this happens, it is able to broadcast a single message (there are variants that broadcast all buffered messages, but [16] proved that their performance is worst), transferring immediately the token to its next neighbor in a circular order. In its prioritized variant, this protocol holds in each node all pending messages (i.e., those buffered messages to be sent) ordered according to their priority, instead of in a FIFO order.

In Fig. 3, we show the modification of the communication history algorithm shown in [2]. As the original algorithm, it assumes that FIFO channels are available. These protocols are usually employed for combining both total and causal

```
SENDER (code of process p):
Initialization:
    tosend := ∅
    if p = s_1
        token.seqnum := 1
        send token to s_1
Procedure TO-broadcast (m, prio):
    insert m in tosend according to prio
When receive token:
    if tosend ≠ ∅ then
        m := first message in tosend
        send (m, token.seqnum) to destinations
        token.seqnum++
        tosend := tosend \ {m}
    send token to s_{i+1 mod n}
DESTINATIONS:
Initialization:
    nextdeliver := 1
    pending := ∅
When receive (m, seqnum):
    pending := pending ∪ {(m, seqnum)}
    while ∃ (m, seqnum) ∈ pending :
    seqnum = nextdeliver do
        deliver m
        nextdeliver++
```

Fig. 2. Modified privilege-based algorithm

```
SENDER/DESTINATION (process p):
Initialization:
    received := ∅
    delivered := ∅
    LC := {0, ..., 0}
Procedure TO-broadcast(m, prio)
    ts(m) := ++LC[p]
    prio(m) = prio
    send FIFO (m, ts(m)) to all
When receive (m, ts(m)):
    LC[p] := max(LC[p], ts(m)) + 1
    if p ≠ sender(m)
        LC[sender(m)] := ts(m)
    received := received ∪ {m}
    deliverable := ∅
    for each message m in
    received \ delivered do
        if ts(m) ≤ min_{q∈Π} { LC[q]} then
            deliverable := deliverable ∪ {m}
    deliver all messages in deliverable,
        in increasing order of
        (ts(m), prio(m), sender(m))
    delivered := delivered ∪ deliverable
```

Fig. 3. Modified communication-history algorithm

orders. To this end, they tag each sent message with a logical timestamp (in most cases, a vector timestamp) able to implement causal delivery. Our prioritized variant is able to impose a different total order where concurrent messages are sequenced according to their priority.

4 Experimental Work

In this section, we present the experimental work we have done to observe the performance of the total order protocols and evaluate the cost overhead of their prioritized versions. First of all, we describe the testbed, including the physical setting. Then, we describe the parameters and the methodology used to run the tests and finally we present and discuss the results.

4.1 Testbed

To evaluate the prioritization techniques, we implemented three total order protocols: a sequencer-based, a privilege-based and a communication history one. We also implemented their corresponding prioritized versions, according to the guidelines given in Sect. 3.

The experiments have been conducted in a system of eight nodes with an Intel Pentium D 925 processor at 3.0 GHz and 2 GB of RAM, running Debian

GNU/Linux 4.0 and Sun JDK 1.5.0. The nodes are connected by means of a 24-port 100/1000 Mbps DLINK DGS-1224T switch that keeps the nodes isolated from any other node, so no other network traffic can influence the results.

4.2 Methodology

To evaluate the performance of the prioritization techniques, in each node, the application broadcasts a series of messages to all the nodes in the system, by means of a total order protocol. The messages are broadcast at a uniform sending rate which is constant during the whole test. We have performed tests with different sending rates. Besides this, we have no other flow control mechanism neither in the application nor in the total order protocols.

Each message is tagged with a uniformly-distributed random priority which is an integer number.

The length of the messages is not fixed, but depends on the headers saved in them by the total order protocols. Nevertheless, in all the cases it is less than the MTU of the network we are using (1500 bytes), so all the application messages fit into one wire-level packet.

Each message is totally ordered and delivered by all the nodes in the system. To evaluate the performance of a given protocol, we measure the *delivery time* of each message, i.e., the time observed by the application in a given node, from the moment in which it broadcasts the message to the moment in which it receives back the message, once totally ordered.

For each message we have a delivery time and for each node we have a series of delivery times, corresponding to all the messages sent by that node. If we merge all the delivery times from all the nodes, we can compute a global mean and median delivery time. Such a mean (median) time expresses the mean (median) time needed by messages to get totally ordered.

This test is run with different total order protocols and also with their corresponding prioritized versions. With these values we analyze the dispersion of the series of delivery times. A significant difference between the mean and the median values, especially when the median is lower than the mean, implies that there is a number of (low priority) messages that have a high delivery time, which means that the prioritization mechanism is working as expected and has been able to prioritize a number of messages. Nevertheless, the mean value of the test should not exceed some bound. An excessively high value for the mean delivery time implies that too many messages are being delayed and this delay is extending their delivery times. In this case we say that the protocol became *saturated.*

In order to get more trustworthy results, we discard the first 3200 messages[1] recorded in each node. These values correspond to delivery times of messages delivered during a period of time in which the total order protocol is being initialized so the system is not yet in a steady-state regime.

[1] This number has been chosen empirically, after analyzing the behavior of the data structures managed by the total order protocol implementations.

During the execution of these tests we also analyzed two additional indicators: a) the processing time employed by the prioritization mechanisms and b) the memory use. In Section 4.4 we provide additional details.

4.3 Parameters

The considered parameters are the class of total order protocol, the number of nodes and the sending rate at which the test application broadcasts messages.

Protocol type. We have implemented three non-prioritized total order protocols and a prioritized version for each. The UB protocol is an implementation of the UB sequencer-based total order algorithm proposed by [17][2]. The TR protocol implements a token ring-based algorithm. It is similar to the ones of [5] and [6] but there is a significant difference. In the TR protocol, when a node receives the token, it broadcasts just a message, as in [16], instead of broadcasting multiple messages, as in [5] and [6]. Finally, the CH protocol is an implementation of the causal history algorithm in [2].

The corresponding prioritized versions are UB_{prio}, TR_{prio} and CH_{prio}, respectively. They have been implemented according to the techniques proposed in Sect. 3.

Sending rate. In each test, a node broadcasts messages using a uniform sending rate. We have run tests with 4 and 8 nodes and sending rates of 10, 40, 60, 80 and 100 messages sent per second and node. Note that this generates maximum global sending rates of 400 msg/s and 800 msg/s, in systems with 4 and 8 nodes, respectively.

Number of messages delivered by each node. To ease the comparison, in each test, each node receives the same sequence of messages. This sequence has 32000 messages. A test ends when all the nodes deliver those messages.

To ensure a stable operation of the protocols during a test, each node sends more messages than those strictly necessary. For instance, in a test with 4 nodes, each node would only need to send 8000 messages. In practice, as the nodes deliver messages at a rate lower than the sending rate, there is a final period of time in a test in which the system is no longer *stable*, because the queues of the protocols are getting empty and this may affect the measuring of the delivery times. Moreover, the difference between the sending rate and the delivery rate is different in each test, and depends basically on all the parameters (the protocol used in the test, the number of nodes and the sending rate itself) and this poses additional difficulties to the protocol comparison.

To solve this issue, each node sends as many messages as needed, to ensure a continuous flow of messages during the whole test. This approach also solves the lack of liveness shown by the CH and CH_{prio} protocols, as described in [2].

[2] UB stands for *Unicast-Broadcast*, as in [2].

4.4 Cost Evaluation

To evaluate the cost employed by the prioritization mechanisms, for each original protocol and its corresponding prioritized version we measure the time employed to run certain parts of both protocols. We call this time the *prioritization time*. The sections measured are semantically equivalent, so we can get comparable measures.

For instance, to evaluate the sequencer-based protocols, we measure the time lapse between the time when the sequencer starts to handle a message and the time when it broadcasts the message, once sequenced. The corresponding prioritized protocol has an equivalent section, in which prioritization takes place. Measuring the time needed to run both sections and comparing both times, we can get a very tight approximation of the time needed by the prioritization mechanism applied by the prioritized protocol.

These measures are only comparable between a given protocol and its corresponding prioritized version. For other protocol families, the parts of the protocols considered are different.

For each test, we measure the prioritization time in each node[3]. Then we compute the mean prioritization time as the mean for all the nodes. These numbers are presented in great detail in [18] and discussed in Section 4.5.

To evaluate the memory use, we analyzed how much of the total amount of memory available by the Java Virtual Machine is being used during each test by each node. In [18] we graphically represent this evolution in several settings (in systems of different sizes, with different protocols and sending rates, as explained in 4.3). Moreover, for each test, we count the number of times the Java garbage collector has been run in each node and with all of them, we compute the mean number of garbage collection runs. These numbers are explained in Section 4.5.

4.5 Results

In Table 1 we show the mean and median global delivery times (in ms), as well as the first and third quartiles in systems with 4 and 8 nodes, respectively, at different sending rates.

Delivery times in a 4-node system. In this configuration (see also Fig. 4), the UB and UB_{prio} protocols perform well at sending rates up to 80 msg/s. At 100 msg/s UB still shows low median delivery times but their dispersion is high, because the protocol is getting saturated. This can be clearly seen in Figure 4.a (median times), where both protocols are able to deliver their messages in less than 1.6 ms, although UB_{prio} needs more time than UB. Saturation is obvious when Fig. 4.b is considered (mean times), since once the 80 msg/s threshold is surpassed, both protocols increase their delivery times with an exponential trend. Note also that there are no significant differences between both variants when their mean delivery times are considered.

[3] We also discard the first 3200 messages, as explained in Section 4.2.

Table 1. Delivery times (ms) with 4 and 8 nodes

		UB	UB_{prio}	TR	TR_{prio}	CH	CH_{prio}
		4 nodes					
10 msg/s	mean	1.45	1.25	6.69	6.33	76.77	77.00
	1st q.	1.20	1.08	0.89	0.89	65.13	65.16
	med.	1.28	1.18	1.26	1.28	81.10	81.35
	3rd q.	1.36	1.26	9.30	7.52	93.38	93.44
40 msg/s	mean	1.50	1.46	1.29	1.27	17.77	17.86
	1st q.	1.11	1.09	0.72	0.72	13.13	13.10
	med.	1.24	1.31	1.02	1.02	17.12	17.01
	3rd q.	1.34	1.54	1.27	1.27	20.84	20.84
60 msg/s	mean	1.30	1.51	1.70	1.70	12.22	11.95
	1st q.	0.97	1.09	0.75	0.76	8.83	8.77
	med.	1.09	1.32	1.07	1.08	12.60	12.61
	3rd q.	1.24	1.53	1.32	1.35	12.88	12.91
80 msg/s	mean	3.43	2.20	2.36	2.75	9.13	9.29
	1st q.	1.17	1.27	0.87	0.77	4.97	4.96
	med.	1.27	1.42	1.20	1.09	8.66	8.62
	3rd q.	1.53	1.70	1.51	1.37	8.98	8.96
100 msg/s	mean	134.25	487.36	4.85	26.14	7.10	6.71
	1st q.	1.12	1.35	0.83	0.83	4.62	4.59
	med.	1.28	1.6	1.17	1.18	4.84	4.83
	3rd q.	1.79	2.75	1.51	1.52	5.14	5.20
		8 nodes					
10 msg/s	mean	1.89	11.35	2.05	2.08	90.33	90.96
	1st q.	1.34	1.53	1.37	1.39	85.21	85.40
	med.	1.53	1.73	1.86	1.87	97.62	94.21
	3rd q.	1.70	1.92	2.39	2.4	101.79	101.74
40 msg/s	mean	3.84	221.37	7.85	7.60	23.62	23.20
	1st q.	1.40	1.65	1.53	1.50	20.76	20.74
	med.	1.62	1.93	2.19	2.17	21.18	21.17
	3rd q.	2.04	2.86	2.91	2.86	21.89	21.68
60 msg/s	mean	190.82	670.48	75.53	151.49	17.09	17.22
	1st q.	1.42	1.72	1.68	1.69	12.96	12.98
	med.	1.86	2.51	2.54	2.53	13.28	13.27
	3rd q.	3.65	9.94	3.69	3.60	17.07	17.05
80 msg/s	mean	6718.52	13608.62	460.35	750.16	86.96	136.80
	1st q.	6373.32	13.32	2.24	2.21	9.02	9.18
	med.	6660.33	604.56	3.8	3.70	9.88	13.80
	3rd q.	6882.63	24776.30	340.26	34.26	65.51	237.24
100 msg/s	mean	20102.49	25264.85	5477.03	5148.22	100.05	125.82
	1st q.	14290.93	104.60	5119.25	5.14	5.78	5.70
	med.	18435.50	17349.36	5517.79	65.87	9.27	9.64
	3rd q.	23159.88	47670.92	5891.15	6908.98	58.16	145.75

The TR and TR_{prio} yield better performance numbers than sequencer-based protocols, even at 100 msg/s. At 10 msg/s the mean is slightly higher than expected although in these cases, the protocols are not saturated. When the sending rate is low, it may happen that the node which receives a token does not have any message to broadcast. In this case, it simply forwards the token to the next node in the ring. If a message is then broadcast by the application in the first node, then it will have to wait until the token arrives again to that node, thus increasing the delivery time of that particular message and also the mean delivery time. As this happens only to some messages, the delivery time of the rest of messages is low (due to the low sending rate and the low contention accessing the network). At higher sending rates this problem no longer arises. At 100 msg/s the dispersion in TR_{prio} is slightly higher as a side effect of the prioritization mechanism, as in UB_{prio}. Despite this, it is able to scale quite better than sequencer-based protocols, since both its median and mean values are much lower than the latter ones.

Regarding the CH and CH_{prio}, we can see that at low sending rates, the delivery time is very high but it decreases noticeably as the sending rate is increased. Indeed, no value has been shown for these protocols in Fig. 4.a. On the other hand, they are the single family of protocols able to decrease its delivery time when the sending rate is increased. So, whilst sequencer-based and

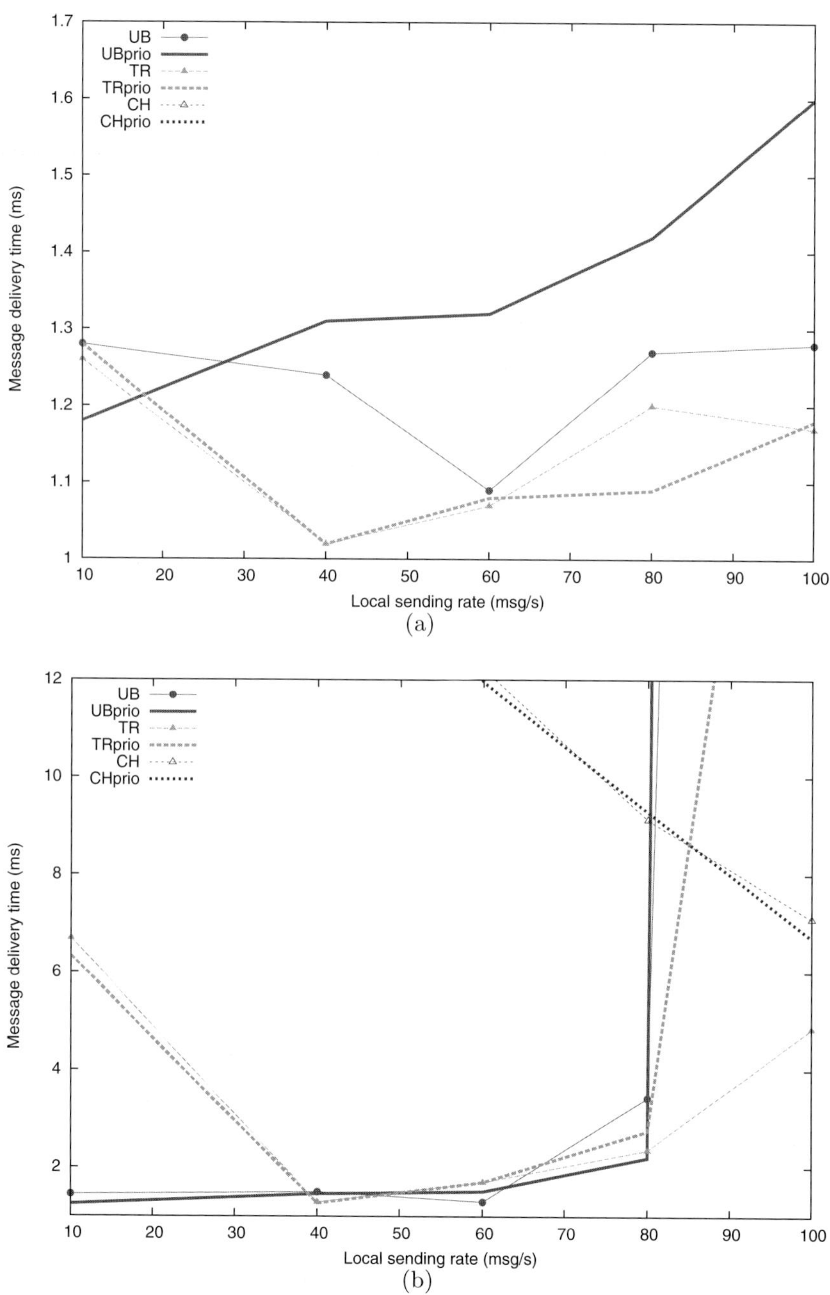

Fig. 4. Median (a) and mean (b) delivery times in a 4-node system

privilege-based protocols start to be overloaded in Fig. 4.b, the communication-history ones start to be shown in such figure, presenting a descending trend. The design of the CH protocol forces an unordered message received by a node to wait until messages are received from the other nodes. Then, the order is locally (and deterministically) decided without any other message exchange. As the sending rate is increased, messages are forced to wait less time thus reducing the global mean and median delivery time. On the other hand, we can see that the dispersion is kept low in all the cases, since their mean and median always have close values. The reason of the delay experienced by the messages is mainly because ensuring the causal property imposes a delay on each message significantly greater than the delay imposed by the prioritization mechanism. As the delay imposed by the causal ordering is similar for all the messages sent at a given sending rate, the dispersion of the delivery times is kept low.

Summarizing, in Table 1 and Fig. 4.b we can see that, in a system with 4 nodes, at sending rates up to 60 msg/s, the mean delivery time of any original (non prioritized) protocol is practically equal to the mean delivery time for the corresponding prioritized protocol, which means that the prioritization mechanisms are not imposing a noticeable overhead. Something similar happens to the median delivery times. Above 60 msg/s the numbers diverge because the load starts to be too high and then the response depends on each particular protocol, as already explained above.

Delivery times in an 8-node system. In such configuration (Fig. 5.a), we can see that UB_{prio}, and TR_{prio} offer good median delivery times at sending rates up to 60 msg/s. Moreover, these numbers are comparable to the ones for their corresponding original (non prioritized versions). At sending rates above 60 msg/s, these protocols get saturated, in varying degrees, and the delivery times start to get unpractical.

Regarding CH and CH_{prio} protocols, they show a similar trend to that found in a 4-node system. They are able to show a decreasing trend in their median delivery times and provide acceptable values at 80 and 100 messages sent per second. No other protocol is able to guarantee such delivery times at 100 msg/sec.

Considering mean delivery times (Fig. 5.b), the UB_{prio} protocol provides the highest increasing trend, starting with values at low sending rates that are only surpassed by communication-history protocols. However, since its median values are good up to 60 msg/sec, this means that this family provides the best

Table 2. Mean prioritization times (ms)

# nodes	msg/s	UB	UB_{prio}	TR	TR_{prio}
4	10	0.004115	0.013898	0.004477	0.005297
	40	0.004034	0.009834	0.004255	0.003504
	60	0.003263	0.009257	0.003515	0.003170
	80	0.003520	0.009834	0.003717	0.002175
	100	0.003376	0.011431	0.003315	0.002030
8	10	0.003741	0.014105	0.004527	0.005180
	40	0.003547	0.012120	0.005129	0.004833
	60	0.003685	0.016840	0.004877	0.003589
	80	0.003644	0.015255	0.004794	0.005024
	100	0.004063	0.015217	0.005750	0.009411

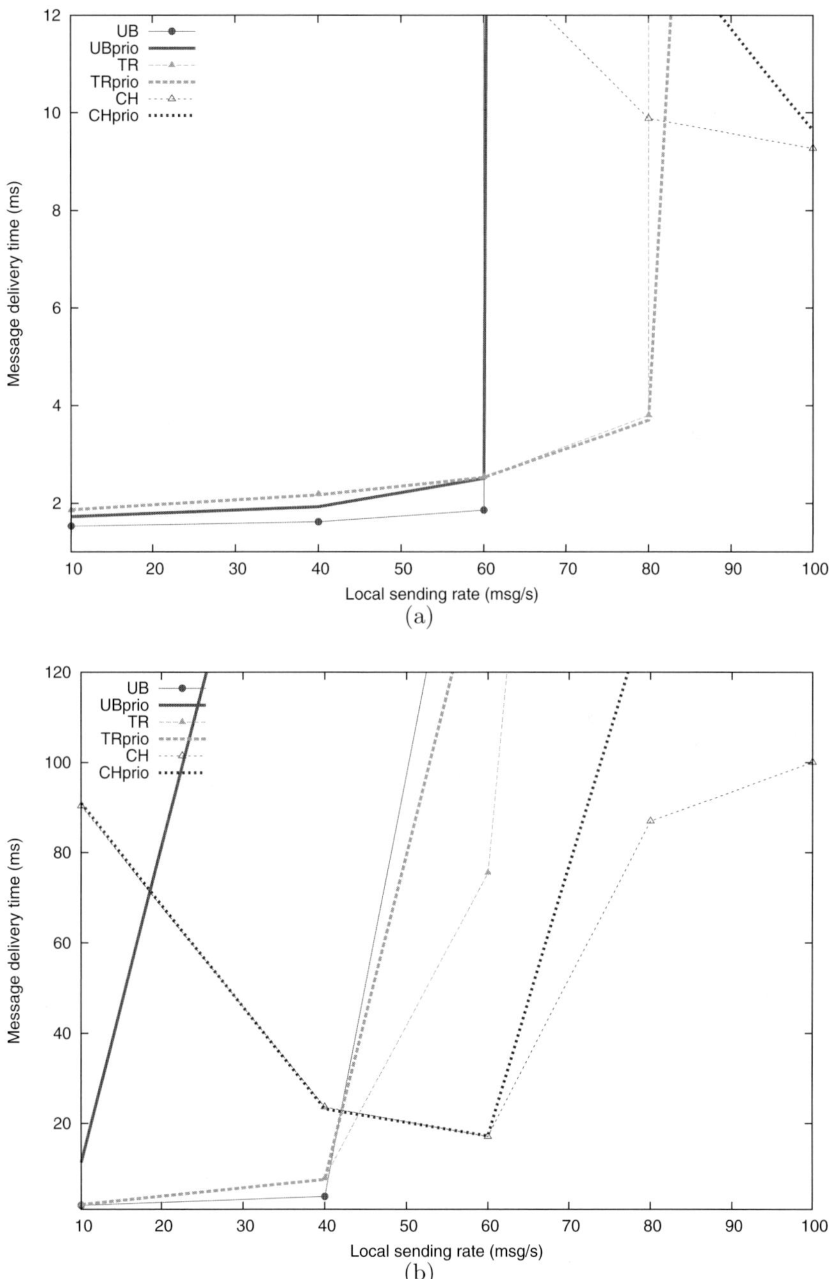

Fig. 5. Median (a) and mean (b) delivery times in an 8-node system

prioritization results in the range 10 to 60 msg/sec, and that it gets completely saturated once such range is exceeded. Both privilege-based and communication-history protocol families stand much better for high sending rates, being CH and CH_{prio} protocols the clear winners.

Prioritization time overhead. Regarding the *prioritization times* presented in Table 2, we can analyze the differences among the values of the conventional (non-prioritized) protocols and the prioritized ones. The bigger differences can be found when comparing the UB and the UB_{prio} protocols at any sending rate and with 4 or 8 nodes in the system. At a first glance it seems that the prioritization mechanisms in UB_{prio} is introducing a significant load to the original protocol. Nevertheless, we can see that in all cases, the overhead is around a few microseconds, which compared to the full delivery time (in the order of milliseconds) is negligible.

In the case of the TR and TR_{prio} protocols, the differences are smaller, and again, compared to the full delivery times, are negligible. Moreover, in some cases the prioritized protocol is able to deal with these message management steps faster than the non-prioritized protocol. So, no overhead is introduced in this protocol family.

Finally, the CH and CH_{prio} protocols cannot be studied in this regard. The prioritization overhead is negligible in this protocol, since the message delivery time (once the message has been received in their target nodes) is highly dominated by the delays introduced in order to ensure causal delivery. Thus, the data to be shown in this table, according to the criteria used for other protocols, would have been approximately a 93% of the time shown in Table 1, but it would not provide the information shown in the other protocols (prioritization overhead), but mainly the causal-related delays.

Table 3. Mean numbers of garbage collection runs

# nodes	msg/s	UB	UB_{prio}	TR	TR_{prio}	CH	CH_{prio}
4	10	43.50	54.25	975.75	985.00	51.00	51.00
	40	27.00	30.75	62.00	62.50	41.25	42.00
	60	18.50	22.25	39.25	39.50	20.50	19.75
	80	17.25	26.25	31.50	30.75	17.00	15.75
	100	17.50	22.75	23.75	23.50	17.00	15.75
8	10	35.50	41.62	228.00	230.25	52.75	53.00
	40	18.62	25.12	25.00	24.75	18.62	18.87
	60	19.87	24.87	21.00	21.75	17.37	18.00
	80	22.12	25.50	19.87	20.12	18.00	18.37
	100	23.00	27.00	19.00	19.87	18.00	18.12

Prioritization memory overhead. Memory usage is summarized in Table 3. To this end, such table presents the amount of garbage collections executed by the JVM in order to recover *free* memory that was previously assigned to dynamic data being used in the protocol. At a glance, in all cases a surprising trend is shown: the amount of garbage collections decreases when the sending rate is increased. This trend can be easily explained. Since the amount of messages being broadcast in each test is constant, the greater the sending rate is, the shortest the test length will be. So, there is nothing annoying in this behavior.

We can observe that in general, there are no big differences between the figures for the *TR* and *CH* protocols and the ones for their corresponding prioritized versions. This means that no memory-related overhead is being introduced by the prioritization mechanism in such protocols.

Significant differences exist however, among the numbers of garbage collection runs for the *UB* and those for UB_{prio}. The reason of these differences is basically the memory overhead suffered by the sequencer node which typically uses more memory than the rest of system nodes[4]. Note also that UB_{prio} has been the protocol providing the best prioritization results; i.e., for a particular workload, it was able to keep the median delivery time and both quartiles like the non-prioritized protocol, whilst its mean delivery time was quite longer. This means that there were multiple low-priority messages that needed a lot of time to be delivered. Such messages were kept in the sequencer queue, increasing the memory demands of such sequencer process, as Table 3 confirms.

On the other hand, it is also remarkable the high number of garbage collections required by both privilege-based protocols at the lowest sending rate (around 980 collections with 4 nodes, and around 230 with 8 nodes). This partially explains the long mean delivery time of such protocols in a 4-node system with a sending rate of 10 msg/sec.

In [18], we depict the evolution of the amount of free memory available for the Java Virtual Machine during each test under different settings. The figures presented in Table 3 can be contrasted against those graphical representations.

Scalability. The best protocols in this issue seem to be the communication history ones, since they demand a high sending rate in order to provide acceptable delivery times. Thus, in a 4-node system their median delivery time starts with 81 ms at 10 msg/sec and finishes with 5 ms at 100 msg/sec. The same trend is shown in an 8-node system, but in the latter no improvement is detected form 80 to 100 msg/sec. So, this protocol seems to start its overloading with a global sending rate of near 640 msg/sec. Unfortunately, its prioritization quality is the worst one, since the mean delivery time in the prioritized variant is not longer than that of the non-prioritized one. This has been already explained: these protocols also guarantee causal delivery, and they can only prioritize concurrent (i.e., non-causally-related) messages.

The second best protocols, regarding scalability, are the privilege-based ones. They are able to serve individual sending rates of 100 msg/sec in a 4-node system without any noticeable overload in both median and mean delivery times. This means that they have been able to comfortably deal with a global load of 400 msg/sec in such system. In an 8-node system, they show the first signs of overloading at 60 msg/sec (i.e., with a global load of 480 msg/sec), with a mean delivery time of 151 ms in the prioritized version and 75 ms in the non-prioritized one, whilst the median delivery time was still below 3 ms. Despite this, the prioritized variant is able to maintain a median delivery time of 66 ms with a global sending rate of 800 msg/sec.

---

[4] As stated in Section 4.2, these mean numbers are got from the numbers for all the nodes in the system, including its sequencer in case of the *UB* and UB_{prio} protocols.

The worst protocol family seems to be the sequencer-based one. Despite providing very good median delivery times in a 4-node system with all studied sending rates, it finds some problems to deal with the highest sending rate of such configuration (100 msg/sec). In the latter case, its mean delivery times (134 ms in the non-prioritized version and 487 ms in the prioritized one) reveal that there were many messages with unacceptable high delivery times. The same starts to happen with similar global sending rates in an 8-node system (at 480 msg/sec, mean delivery times exceed 190 ms in the best case), but delivery times get unaffordable values at 640 msg/sec, quite longer than those of all other protocol families at 800 msg/sec.

5 Conclusions

We have presented an experimental study in which we show that the prioritization techniques do not impose an important overhead (in terms of message delivery latency, processing time and memory use) on the original total order protocols, thus proving that, besides being easy to understand and implement, and being useful for replicated database management (as shown in [15]), the techniques are affordable in terms of performance. The main conclusion is that prioritized total order broadcast protocols are a valuable building block that can be used to improve the design and implementation of distributed applications and their performance, as well.

As a second contribution this experimental study can be seen also as a performance comparison among conventional non-prioritized total order protocols. The results of this comparison show that sequencer-based and privilege-based protocols offer a comparable performance when the number of nodes is small (4 or 8) and the individual sending rate is not too high (around 60 msg/s). As the number of nodes or the sending rate is increased the sequencer-based protocols start to get saturated and the communication history ones improve their performance. At higher sending rates, communication history protocols are the unique ones that can stand such load.

References

1. Chockler, G., Keidar, I., Vitenberg, R.: Group communication specifications: a comprehensive study. ACM Computing Surveys 33(4), 427–469 (2001)
2. Défago, X., Schiper, A., Urbán, P.: Total order broadcast and multicast algorithms: Taxonomy and survey. ACM Computing Surveys 36(4), 372–421 (2004)
3. Birman, K.P., Joseph, T.A.: Reliable communication in the presence of failures. ACM Transactions on Computer Systems 5(1), 47–76 (1987)
4. Dolev, D., Malki, D.: The Transis approach to high availability cluster communication. Communications of the ACM 39(4), 64–70 (1996)
5. Moser, L.E., Melliar-Smith, P.M., Agarwal, D.A., Budhia, R., Apadopoulos, C.L.P.: Totem: a fault-tolerant multicast group communication system. Comm. of the ACM 39(4), 54–63 (1996)

6. Amir, Y., Danilov, C., Stanton, J.R.: A low latency, loss tolerant architecture and protocol for wide area group communication. In: DSN, pp. 327–336 (2000)
7. Tully, A., Shrivastava, S.K.: Preventing state divergence in replicated distributed programs. In: 9th Symposium on Reliable Distributed Systems, pp. 104–113 (1990)
8. Nakamura, A., Takizawa, M.: Priority-based total and semi-total ordering broadcast protocols. In: 12th Intl. Conf. on Dist. Comp. Sys (ICDCS 1992), pp. 178–185 (1992)
9. Rodrigues, L., Veríssimo, P., Casimiro, A.: Priority-based totally ordered multicast. In: 3rd IFAC/IFIP workshop on Algorithms and Architectures for Real-Time Control (1995)
10. Wiesmann, M., Schiper, A., Pedone, F., Kemme, B., Alonso, G.: Database replication techniques: A three parameter classification. In: SRDS, pp. 206–215 (2000)
11. Irún-Briz, L., de Juan-Marín, R., Castro-Company, F., Armendáriz-Iñigo, E., Muñoz-Escoí, F.D.: MADIS: A slim middleware for database replication. In: Cunha, J.C., Medeiros, P.D. (eds.) Euro-Par 2005. LNCS, vol. 3648, pp. 349–359. Springer, Heidelberg (2005)
12. Nakamura, A., Takizawa, M.: Starvation-prevented priority based total ordering broadcast protocol on high-speed single channel network. In: 2nd Intl. Symp. on High Performance Dist. Comp., pp. 281–288 (1993)
13. Baker, T.: Stack-based scheduling of real-time processes. Journal of Real-Time Systems 3(1), 67–99 (1991)
14. Wang, Y., Brasileiro, F., Anceaume, E., Greve, F., Hurfin, M.: Avoiding priority inversion on the processing of requests by active replicated servers. In: Dependable Systems and Networks, pp. 97–106. IEEE Computer Society, Los Alamitos (2001)
15. Miedes, E., Muñoz, F.D., Decker, H.: Reducing transaction abort rates with prioritized atomic multicast protocols. In: Luque, E., Margalef, T., Benítez, D. (eds.) Euro-Par 2008. LNCS, vol. 5168, pp. 394–403. Springer, Heidelberg (2008)
16. Défago, X., Schiper, A., Urbán, P.: Comparative performance analysis of ordering strategies in atomic broadcast algorithms. IEICE Trans. on Information and Systems E86-D(12) , 2698–2709 (2003)
17. Kaashoek, M.F., Tanenbaum, A.S.: An evaluation of the Amoeba group communication system. In: 16th IEEE International Conference on Distributed Computing Systems (ICDCS 1996), pp. 436–448. IEEE Computer Society, Los Alamitos (1996)
18. Miedes, E., Muñoz-Escoí, F.D.: On the cost of prioritized atomic multicast protocols. Technical Report ITI-SIDI-2009/002, Instituto Tecnológico de Informática, Universidad Politécnica de Valencia (February 2009)

Evaluating Throughput Stability of Protocols for Distributed Middleware*

Nuno A. Carvalho, José P. Oliveira, and José Pereira

Universidade do Minho

Abstract. Communication of large data volumes is a core functionality of distributed systems middleware, namely, for interconnecting components, for distributed computation and for fault tolerance. This common functionality is however achieved in different middleware platforms with various combinations of operating system and application level protocols, both standardized and ad hoc, and including implementations on managed runtime environments such as Java. In this paper, in contrast with most previous work that focus on performance, we point out that architectural and implementation decisions have an impact in throughput stability when the system is heavily loaded, precisely when such stability is most important. In detail, we present an experimental evaluation of several communication protocol components under stress conditions and conclude on the relative merits of several architectural options.

1 Introduction

Communication protocols used as components in distributed systems middleware range from the ubiquitous UDP/IP and TCP/IP Internet standards to custom protocols designed to address different reliability, ordering, performance, resource usage, and resilience requirements. In particular, multiparty or group communication protocols have been traditionally implemented at the application level and been highly relevant to middleware, for instance, to keep track of operational servers in a cluster and support load balancing of processing tasks across server clusters.

There has in fact been an increasing interest in group communication protocols such as JGroups [1], Spread [2] or Appia [3] in middleware supporting current multi-tier applications, towards both higher throughput and stricter consistency requirements. An example of this trend is the distributed software transactional memory proposed for FénixEDU [4]. Instead of relying solely on the underlying shared database management system to enforce consistency across different servers, updates are propagated and implicitly ordered using group communication. Another example is consistent database replication [5]. This allows concurrent conflicting updates to be processed by different replicas without fine synchronization thus enabling high performance. However, it ensures that all transactions are serialized and thus no conflicting updates are committed, avoiding the need for reconciliation or explicit sharing easing application development.

* Partially funded by FCT through projects SFRH/ BDE/ 33304/ 2008 and PTDC/ EIA/ 72405/ 2006 (Pastramy), by PT Inovação S.A., and by ParadigmaXis S.A.

R. Meersman, T. Dillon, P. Herrero (Eds.): OTM 2009, Part I, LNCS 5870, pp. 600–613, 2009.

In fact, the motivation for this work was sparked by experimental observations when building and testing the ESCADA Replication Server,[1] a modular database replication protocol. Briefly, the testing setup used a cluster of servers, each running a PostgreSQL replica and ESCADA, with the workload of the TPC-C benchmark [6]. Under this scenario, one would observe that update dissemination would eventually slow down and the number of updates stored in memory grow. This was surprising, since the bandwidth being generated was easily achieved by a standalone benchmark of the group communication protocol in the same hardware setup.

This paper aims at explaining why the group communication within the larger application scenario would perform worse than in the standalone benchmark by testing the following two hypotheses:

- By running within a Java application with a large memory heap (i.e. the ESCADA Replication Server), the group communication protocol has to compete for memory with other threads, as the garbage collector represents an increasing share of the computation taking place.
- By running along a large number of interactive processes (i.e. instances of the PostgreSQL server) which together consume a substantial share of CPU bandwidth, the group communication has to compete for time slices.

Either way, the communication protocol would be unable to schedule events timely, for instance, to deal with window-based retransmission [7] implemented at the user level. This would prevent the protocol from fully exploiting available network bandwidth.

If true, this has an impact on architectural decisions when designing or selecting group communication protocols. Namely, a protocol made available as a library in Java should be particularly susceptible to the first. Any protocol that implements window-based mechanisms at the user level, regardless of using Java, is susceptible to the second. If true, this poses a challenge to using group communication in large servers running Java virtual machines with large heaps (e.g. application servers) or pools of interactive daemons (e.g. web or database servers).

Moreover, an in-depth knowledge of the dynamics of communication protocols in various workload conditions is also key to enabling self-managing distributed systems. In detail, being able to operate large and complex multi-tier applications depends on being able to ensure that individual system components are kept within their capacities to prevent congestion and trashing phenomena. If models underlying the creation of rule sets are unaware that communication capacity is degraded by server workloads, the resulting policies will be unable to keep the system within safe boundaries.

The rest of the paper is structured as follows. Section 2 we describe the communication protocols that we are evaluating and Section 3 we introduce our experimental setting. In Section 4 we present results that test each of the hypotheses. Finally, Section 5 discusses related work and Section 6 concludes the paper.

2 Protocols

To assess the stability of communication protocols for distributed middleware we select three kinds of protocols: point to point with network stack at kernel mode, such as

[1] http://escada.sf.net

Transmission Control Protocol (TCP) [7] or User Datagram Protocol (UDP) [8]; point to point with network stack at user mode, like LimeWire RUDP [9] or ENet [10]; and group communication protocols, such as Appia [3] or JGroups [1].

2.1 Point-to-Point in Kernel Mode

Simple point-to-point protocols implemented within the operating system kernel provide a baseline for comparison. First, the dynamics of TCP/IP in a number of environment conditions is well known and its implementation in mainstream operating systems is thoroughly tested and optimized. Second, because application level protocols are built on them, frequently on UDP/IP, and incur at least in the same overhead. Thus whenever possible, we test multiple APIs in C and Java, to discover also the impact of the Java Virtual Machine (JVM).

In detail, TCP protocol was assessed with three interfaces: the native BSD sockets interface in C, Java using `java.net` package and Java using `java.nio` package. Whenever possible, the same buffers are used for multiple I/O operations to reduce memory management overhead. In `java.nio`, direct byte buffers are used as the documentation describes them as improving performance.

The UDP protocol was evaluated in two implementations, C and Java using the `java.net` interface. Note however that UDP is not reliable and thus the amount of data sent differs from the amount of data received. This makes the tests useful only to determine baseline overhead.

Finally, the Stream Control Transmission Protocol (SCTP) [11] is aimed at combining the best features of TCP and UDP, ensuring the delivery of messages with or without order, has congestion control, allows the use of multiple streams and multihoming. These features can be switched on and off in contrast to the existing on TCP and UDP, and in this paper, a configuration similar to TCP has been selected.

2.2 Point-to-Point in User Mode

These protocols should provide an interesting indication of the cost of implementing reliability in user mode and in Java, when compared with point-to-point protocols in kernel. They should also provide an indication of the cost of group communication, when compared to such protocols.

The LimeWire application, implemented in Java, client of the Gnutella network, was the selected implementation for the evaluation of the Reliable User Datagram Protocol (RUDP), a lightweight version of TCP, whose features are: guaranteed delivery of messages, congestion control and retransmission of lost packets.

The ENet [10] protocol, reliable and in-order communication on top of UDP, was evaluated through their implementations in C and Java [12]. The evaluated versions were 1.1 and beta1, respectively.

2.3 Group Communication Protocols

Appia [3] is an open source layered communication toolkit implemented in Java providing extended configuration and programming possibilities. The Appia toolkit is composed by a core that is used to compose protocols and a set of protocols that provide group communication, ordering guaranties, atomic broadcast, among other properties. Appia is a protocol kernel that offers a clean and elegant way for the application to express inter-channel constraints. In assessing this toolkit only one process writes the data and another reads it, creating a point-to-point channel in the group membership. The evaluated version was 4.1.0.

JGroups [1] is a group communication toolkit implemented in Java, which offers reliability and group membership on top of TCP or UDP. Its most powerful feature is its flexible protocol stack, which allows developers to adapt it to exactly match their application requirements and network characteristics. Like Appia, was selected for this evaluation to measure the impact of increased network stack, particularly being it in user space. Once again, in assessing this toolkit only one process writes the data and another reads it, creating a point-to-point channel in the group membership. The evaluated version was 2.6.3 GA.

Spread [2], another group communication toolkit, consists of a library that user applications are linked with, in this evaluation our application was implemented in Java, and a binary daemon which runs on each computer that is part of the processor group. In this combined implementation, which delegates the communication work to other process, we are particularly interested in observing the impact of the Garbage Collector in throughput stability. The evaluated version was 4.0.0.

3 Experimental Setting

3.1 Hardware and Software

The experimental evaluation described in this paper was performed using two HP Proliant dual Opteron processor machines with the configuration outlined in Table 1. The operating system used is Linux, kernel version 2.6.22-16, from Ubuntu. The C programs are compiled with GCC 4.1.3 without any special flags. The Java based evaluations are compiled and run with Sun's Java 1.6.0_03. The availability of multiple CPU cores allows us to assess also the ability of protocols to take advantage of this features, which is increasingly important in current hardware configurations.

3.2 Measurements

A run consists in having one process sending messages as fast as allowed by the communication protocol (i.e. the *Writer process*), while in a different machine another process reads them also as fast as possible (i.e. the *Reader process*). No artificial delays are inserted in any of them. Although experiments have been reproduced with

Table 1. Configuration used for benchmarking

Resource	**Properties**
Processor	2 × 2.4 GHz AMD Opteron (64 bits)
RAM	4 GBytes
Operating System	Linux 2.6.12-16 (Ubuntu Kernel)
Network	Gigabit Ethernet

different sizes, this paper includes only results obtained by writing and reading data in 2000 bytes chunks. Test machines are otherwise idle, to avoid disturbing measurements.

Measurements are done concurrently by running Dstat [13] every second. Dstat is a standard resource statistics tool, which collects memory, disk I/O, network, and CPU usage information available from the operating system kernel. All measurements therefore include all load on each of the machines. These are saved to a log file and later processed off-line to extract the results presented in the paper.

Each run lasts 10 minutes. Communication starts after 4 minutes. Measurements during the first 4.5 minutes and the last 30 seconds of each run are discarded. This allows background workload generators to warm up and wind-down without impacting results. When fully automated, test runs described in this paper take approximately 10 hours to run and produce 15GBytes of log files.

3.3 Background Workload Generators

The first competing background workload generated aims at reproducing the conditions in a loaded server, in which a large number of processes or threads alternate between

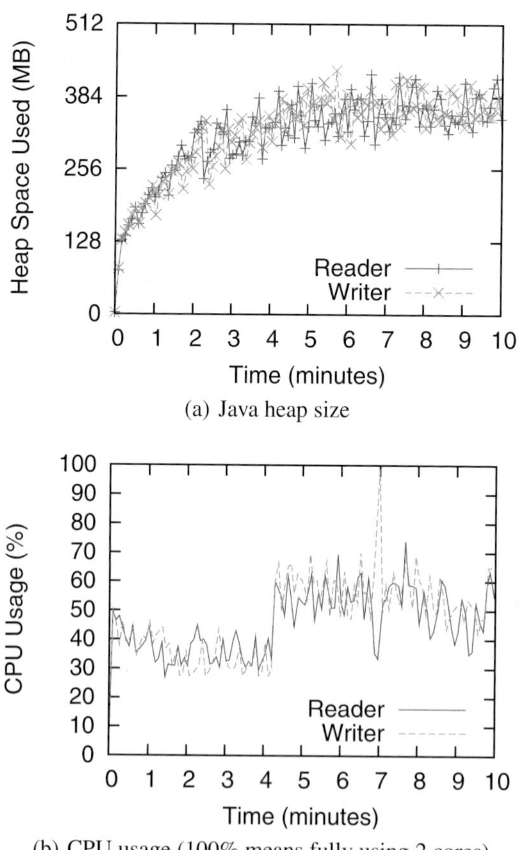

(a) Java heap size

(b) CPU usage (100% means fully using 2 cores)

Fig. 1. Resource usage with the Garbage Collector workload idle and busy periods and compete for CPU. Due to the common operating system scheduler policy of favoring interactive processes, this workload cannot be duplicated simply by having a single background process in an infinite loop. Instead, we use the operating system clock to determine at each time what share of the CPU has been used and have a pool of processes alternate between idle and busy periods to meet the desired CPU occupancy. This strategy affects all processes, because the load is imposed on the scheduler, which is also affecting processes in kernel mode.

The second competing background workload generator aims at reproducing the conditions in a loaded Java based single virtual machine server, in which a large heap is being managed. Due to common garbage collector optimizations, it is not enough to simply allocate memory in tight loop, since every allocation is short lived and favors generational garbage collectors. Instead, we build random linked structures such that probabilistically some elements become unreferenced and others are added at the same rate.

In this paper we tune the parameters of this workload generator such that it uses approximately 384MBytes out of a maximum of 512MBytes allocated to the virtual machine, as can be observed in Fig. 1(a). The resulting usage of CPU is shown in Fig. 1(b), which shows the garbage collector workload alone up to minute four and then the cumulative effect of a test run with a TCP/IP socket. The target CPU occupancy of the CPU workload generator was then set at 60%, to allow direct comparison with the garbage collector workload generator. Note that this corresponds to slightly more that the load that one to the two cores can handle.

4 Results

4.1 Unstable Protocols

Unfortunately we were unable to make all target group communication protocols run the proposed test successfully. Namely, Spread daemons would disconnect either the sender or the receiver and we were unable to finish any test run. This behavior is well known and expected, having been thoroughly discussed in the supporting mailing list, since Spread does not do end-to-end flow control and expected the application to do it. We were also unable to reliably complete test runs with the Appia protocol, although it has end-to-end flow control implemented by "memory managers". It would either block or crash with out-of-memory errors.

The same problem happened with some of the point-to-point protocols being used for comparison. Namely, both implementations of eNet would consume an ever increasing amount of memory at the sender, leading to trashing or an out-of-memory crash. The LimeWire RUDP protocol, although stable, would not be able to use a significant portion of the available bandwidth and thus does not provide an interesting comparison. We do however understand that this is most likely a design decision and not a bug, since the typical usage of Limewire RUDP will have multiple concurrent connections over a single residential network link (e.g. ADSL).

4.2 Scenario 1: No Competing Workload

Running all protocols without any competing background workload provides a baseline for later comparison as well as a first measurement of the resources required to saturate the 1GBits network. The results are shown in Fig. 2 and Fig. 3.

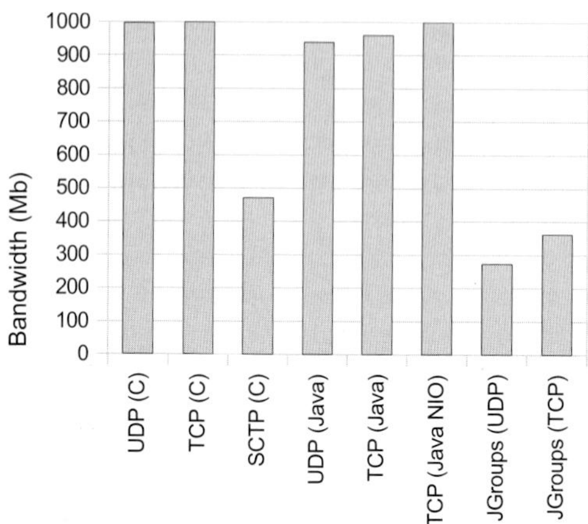

Fig. 2. Bandwidth usage without background workloads

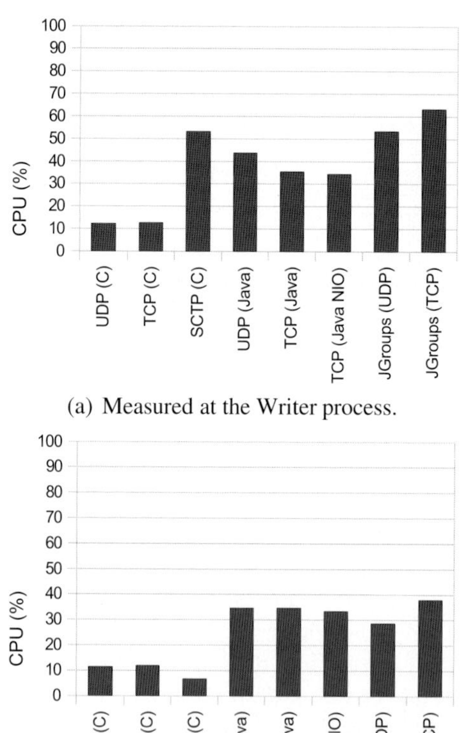

(a) Measured at the Writer process.

(b) Measured at the Reader process.

Fig. 3. CPU usage without background workload (100% represents the two CPUs of the machine)

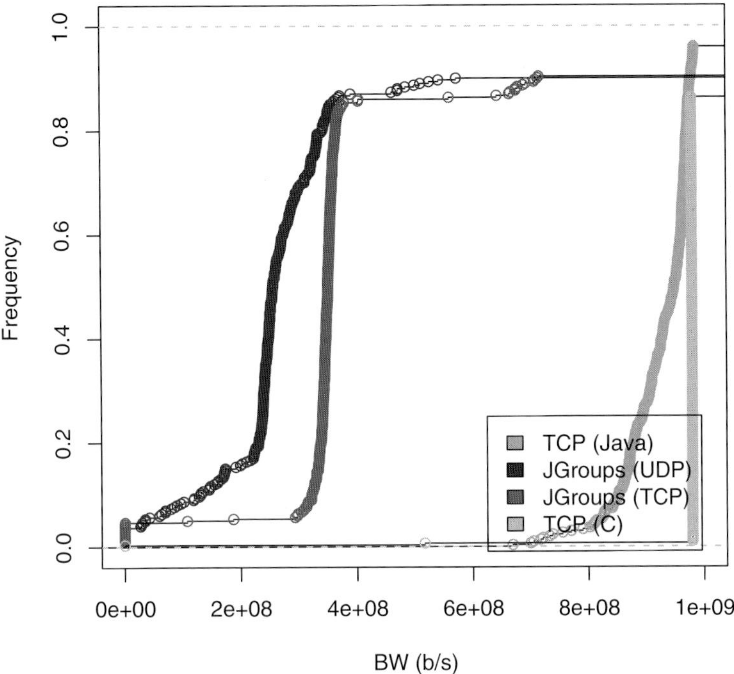

Fig. 4. Distribution of bandwidth in time periods

As shown in Fig. 2, bare TCP/IP and UDP/IP are always able to saturate the network, regardless of the API being used in C or Java. However, as seen in Fig. 3, there is a large CPU overhead when using Java even if we took care not to allocate memory for each operation, i.e. we always write from and read to the same buffers. Using the novel NIO interface with direct buffers does not make a noticeable impact.

Regarding SCTP, also implemented in the operating system kernel, it is interesting to note that it does not fully saturate the network. The reason for this seems to be that the sender is fully using one of the available CPU cores. In contrast, CPU usage seems to be asymmetric, as the receiver is much less intensive.

Both configurations of JGroups tested are unable as well to saturate the network. Again, the reason seems to be that the sender fully uses one of the two available CPU cores and is unable to exploit the second. Interestingly, the TCP/IP configuration is able to achieve slightly higher throughput and use the second CPU core to some extent. This is probably true as TCP/IP processing is done within the kernel and thus scheduled to multiple cores.

Finally, besides average throughput, it is interesting to note how each option is able to sustain such throughput stably, without variation. Fig. 4 plots the empirical cumulative distribution function of bandwidth observed in each period of time. A straight vertical line or steep slope denote low variance while a moderate slope or staircase denote high variance. It can be observed that TCP in Java is more unstable than in C, and that the UDP configuration of JGroups more unstable than the TCP one.

Lessons Learned: These results point out that one should make as much use of kernel based TCP/IP as possible and avoid Java in the implementation of group communication. Otherwise, one should make the protocol multi-threaded and account for additional CPU usage.

4.3 Scenario 2: Competing CPU Workload

As described in Section 3, this scenario adds competing background CPU workload, thus introducing scheduling latency in user level processes. The impact on average bandwidth is shown in Fig. 5, showing that all protocols have their throughput reduced.

Fig. 6 shows the same results as a fraction of the original maximum achieved with each protocol, making it easier to evaluate which protocols suffer the most. The least affected is UDP, although this is misleading since UDP is not reliable and is discarding

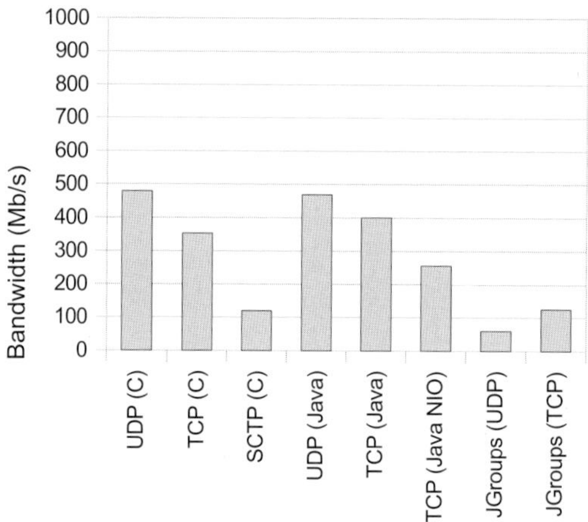

Fig. 5. Bandwidth usage with competing CPU workload

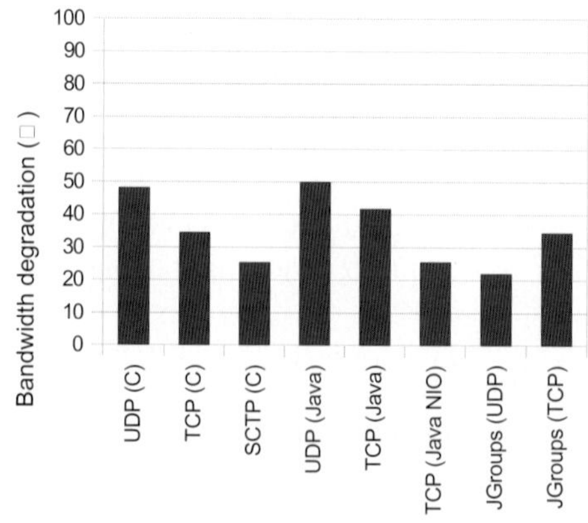

Fig. 6. Bandwidth degradation with competing CPU workload

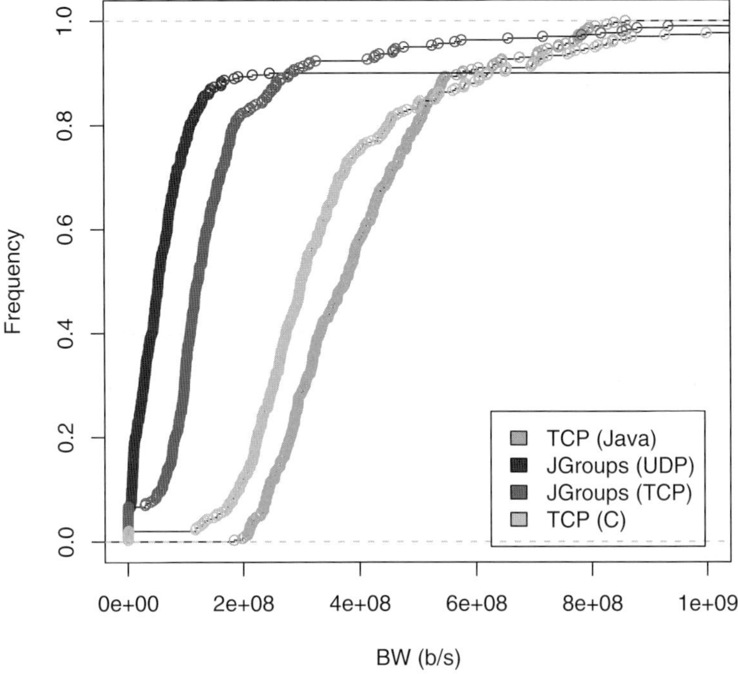

Fig. 7. Distribution of bandwidth in time periods with CPU workload

some traffic. Most interestingly, SCTP is one of the protocols that is most affected, even if it is implemented in the kernel. Moreover, the Java NIO interface performs worse that the original Java sockets interface.

Regarding group communication protocols, the UDP configuration is much more affected than its TCP counterpart. This confirms our hypothesis that it is hard to have a retransmission algorithm in user mode when there is a competing workload and consequence scheduling latency.

Finally, Fig. 7 shows that throughput stability suffers with any of the protocols, which exhibit similar variability, confirming that there is no significant disadvantage of Java in this scenario.

It would also be interesting to perform the experiments using a real time scheduling class for protocol threads. This is however not straightforward for two reasons. The protocols that need it the most, such as JGroups, are implemented as libraries and this might require elevating the privileges of the Java virtual machine as a whole, which is undesirable. Second, since the protocol is itself responsible for a substantial share of CPU usage, it could seriously degrade the performance of the entire service.

Lessons Learned: These results reinforce that one should make as much use of kernel based TCP/IP as possible in the implementation of group communication. Otherwise, one should make the protocol multi-threaded and account for additional CPU usage.

4.4 Scenario 3: Competing Garbage Collector Workload

As described in Section 3, this scenario adds competing background garbage collector workload, thus being applicable only to Java protocols. As shown in Fig. 8 the degradation of all protocols except the UDP configuration of JGroups is similar to that with the CPU workload. Recall that both workloads were tuned to consume approximately the same amount of CPU, although performing different tasks.

Fig. 9 shows the same results as a fraction of the original maximum achievable with each protocol. This shows that this workload is however highly problematic for the UDP configuration of JGroups, as it is reduced to as little as 6% of its initial capacity. This is more than enough to explain our trouble with the ESCADA Replication Server and should be worrying to anyone using group communication. Recall that this happens with a workload that consumes approximately only 384MB out of 2GB RAM and only

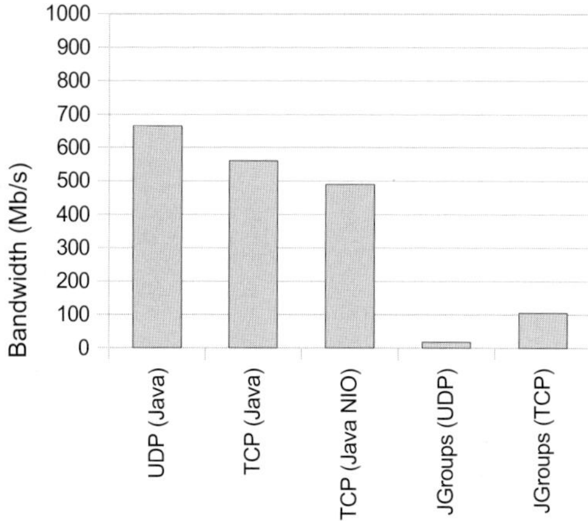

Fig. 8. Bandwidth usage with competing garbage collector workload

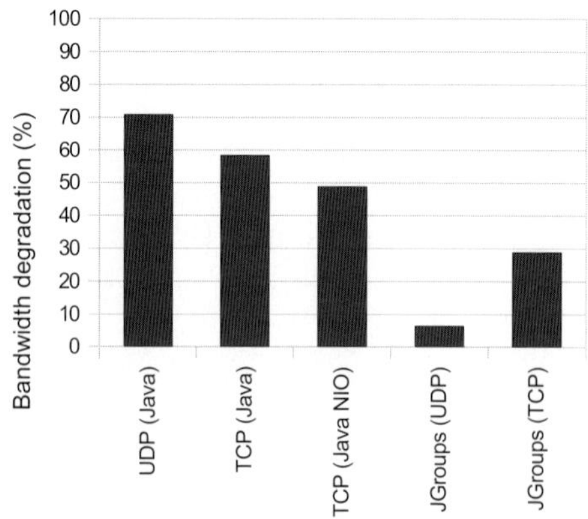

Fig. 9. Bandwidth degradation with competing garbage collector workload

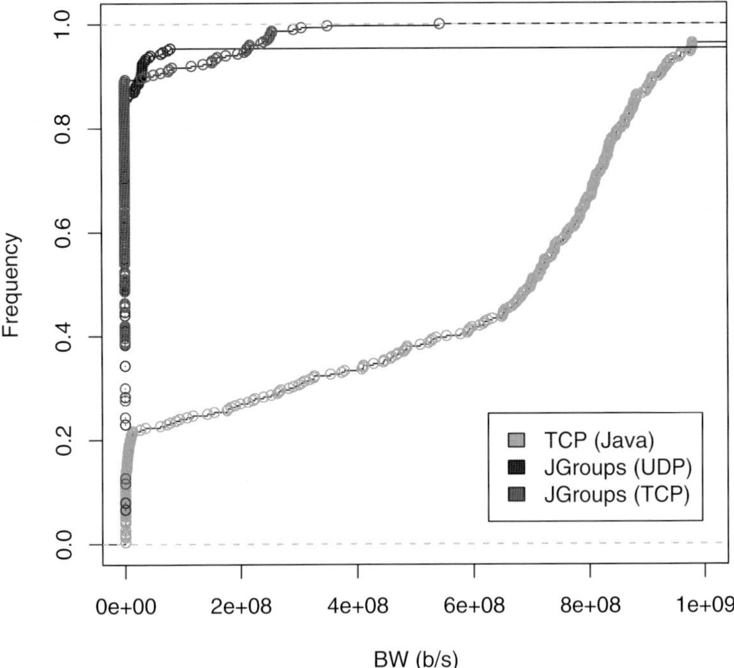

Fig. 10. Distribution of bandwidth in time periods with garbage collector workload

one of the two CPU cores available, as shown in Fig. 1(a) and Fig. 1(b), which should be the nominal load expected in many servers.

Finally, Fig. 10 when compared to Fig. 7 shows that although the impact on kernel based TCP/IP of the competing garbage collector workload seems similar to the CPU workload in terms of average bandwidth, it introduces much more variability which may cause additional trouble for timing sensitive applications.

Lessons Learned: These results show that implementing fine grained retransmission protocols in Java and deploying them as a library in large applications leads to disastrous results in terms of throughput stability.

5 Discussion

The experimental evaluation of communication protocols has been addressed by multiple previous projects focusing both the performance and dependability, including a spectrum of real and simulated stressful environments.

Some work focused on performance evaluation [14], measuring latency and throughput of protocols in different situations, targeting the overhead of architectural decisions and of the Java platform. In this paper we show that such results may be somewhat optimistic, since Java-based protocols provided as a library are more vulnerable to competing workloads and thus synthetic benchmarks will not show their actual limits in a real-world environment.

Work focused on JGroups [15] has shown that the performance the TCP-based configuration is superior to the UDP-based configuration. This is attributed to poor performance of the network switch with multicast. In this paper we take this further and show that this is the case even in a point-to-point connection when no multicast is required. Namely, we show that this might be explained also by CPU used by the protocol itself and also by competing workloads on garbage collector and CPU.

An alternative approach has focused on the effect of a slow receiver in the throughput of multicast protocols [16], showing that performance degradation of the group as a whole is unavoidable, even when using state of the art protocol mechanisms. This works was motivated by having observed the degradation in a real setting [17]. This result is highly relevant together with our contribution, showing that our results, measured with just two elements, will have a serious impact in larger groups.

Finally, previous work has targeting group communication protocols with a variety of fault injection techniques, such as memory leaks at the client application level, process hangs, abrupt crashes, and packet loss at the network level [18]. Interestingly, it concludes also that a library-based approach is more susceptible to perturbation. However, we strengthen the result showing that perturbation occurs in normal operational conditions without bugs (e.g. memory leaks) and even in very small groups.

6 Conclusions and Future Work

In this paper we set out to explain the poor performance of group communication protocols observed when there is a competing workload in machines participating in the group. Based on the hypotheses that this effect might be caused by garbage collection and scheduling latency, the first challenge overcome was to reproduce the right workload without having to setup large complex servers.

The benchmark results that were then achieved, comparing multiple protocols with varying competing background workloads lead us to conclude that a protocol with (i) a library-based design, (ii) implemented in Java, and (iii) using an user-level window mechanism on top of UDP, result in a fragile combination that cannot sustain stable high throughput in the presence of a moderate competing background load. Namely, we show a configuration in which throughput is reduced to 6% of that achievable when the system is idle. As a secondary conclusion, we have shown that there is a large performance and resource usage gap between group communication and TCP sockets, that exists even when doing a similar point-to-point communication task.

These conclusions pave the ground for future work in several directions. First, it is interesting to reproduce the results with a wider variety of experimental settings. Namely, using different operating systems and Java virtual machines. For instance, the novel Garbage-First Java garbage collector [19] might have an impact in the results. Second, it is interesting to determine exactly how competing workloads impact the throughput of protocols. Finally, since the gap between application-level protocols and bare TCP is so large, there is definitely room for improvement in the design and implementation of group communication protocols that should be explored.

References

1. Ban, B.: Design and implementation of a reliable group communication toolkit for java. Technical report, Dept. of Computer Science, Cornell University (1998)
2. Spread Concepts LLC: The Spread Toolkit, (2009) http://www.spread.org/
3. Miranda, H., Pinto, A., Rodrigues, L.: Appia, a flexible protocol kernel supporting multiple coordinated channels. In: Proceedings of the 21st International Conference on Distributed Computing Systems, pp. 707–710 (April 2001)
4. Carvalho, N., Cachopo, J., Rodrigues, L., Rito Silva, A.: Versioned transactional shared memory for the fenixedu web application. In: Proceedings of the Second Workshop on Dependable Distributed Data Management (in conjunction with Eurosys 2008). ACM, New York (2008)
5. Correia Jr., A., Pereira, J., Oliveira, R.: AKARA: A flexible clustering protocol for demanding transactional workloads. In: Proceedings of the International Conference on Distributed Objects, Middleware and Appocations, DOA (2008)
6. Transaction Processing Performance Council: TPC-C. In: TPC (2009)
7. Cerf, V., Dalal, Y., Sunshine, C.: Specification of Internet Transmission Control Program. RFC 675 (December 1974)
8. Postel, J.: User Datagram Protocol. RFC 768 (Standard) (August 1980)
9. Lime Wire LLC: LimeWire (2009), http://wiki.limewire.org/
10. Salzman, L.: ENet (2009), http://enet.bespin.org/
11. Ong, L., Yoakum, J.: An Introduction to the Stream Control Transmission Protocol (SCTP). RFC 3286 (Informational) (May 2002)
12. Vasquez, D.: JeNet (2009), https://jenet.dev.java.net/
13. Wieërs, D.: Dstat (2009), http://dag.wieers.com/home-made/dstat/
14. Baldoni, R., Cimmino, S., Marchetti, C., Termini, A.: Performance analysis of java group toolkits: A case study. In: Guelfi, N., Astesiano, E., Reggio, G. (eds.) FIDJI 2002. LNCS, vol. 2604, pp. 49–60. Springer, Heidelberg (2003)
15. Abdellatif, T., Cecchet, E., Lachaize, R.: Evaluation of a group communication middleware for clustered J2EE application servers. In: Proceedings of the 2004 International Symposium on Distributed Objects and Applications (DOA), pp. 1571–1589 (2004)
16. Birman, K.: A review of experiences with reliable multicast. Software Practice and Experience 29(9) (July 1999)
17. Piantoni, R., Stancescu, C.: Implementing the Swiss Exchange Trading System. In: IEEE International Symposium on Fault-Tolerant Computing (June 1997)
18. Pertet, S., Ghandi, R., Narasimhan, P.: Group communication: Helping or obscuring failure diagnosis? Technical report, Carnegie Mellong University (2006)
19. Detlefs, D., Flood, C., Heller, S., Printezis, T.: Garbage-first garbage collection. In: Proceedings of the 4th international symposium on Memory management, pp. 37–48. ACM, New York (2004)

Evaluating Transport Protocols for Real-Time Event Stream Processing Middleware and Applications*

Joe Hoffert, Douglas C. Schmidt, and Aniruddha Gokhale

Institute for Software Integrated Systems, Dept. of EECS,
Vanderbilt University, Nashville, TN, USA 37203
{jhoffert,schmidt,gokhale}@dre.vanderbilt.edu
www.dre.vanderbilt.edu

Abstract. Real-time event stream processing (RT-ESP) applications must synchronize continuous data streams despite fluctuations in resource availability. Satisfying these needs of RT-ESP applications requires predictable QoS from the underlying publish/subscribe (pub/sub) middleware. If a transport protocol is not capable of meeting the QoS requirements within a dynamic environment, the middleware must be flexible enough to tune the existing transport protocol or switch to a transport protocol better suited to the changing operating conditions.

Realizing such adaptive RT-ESP pub/sub middleware requires a thorough understanding of how different transport protocols behave under different operating conditions. This paper makes three contributions to work on achieving that understanding. First, we define ReLate2, which is an evaluation metric that combines packet latency and reliability to evaluate transport protocol performance. Second, we use the ReLate2 metric to quantify the performance of various transport protocols integrated with the OMG's Data Distribution Service (DDS) QoS-enabled pub/sub middleware standard using our FLEXible Middleware And Transports (FLEXMAT) prototype for experiments that capture performance data. Third, we use ReLate2 to pinpoint configurations involving sending rate, network loss, and number of receivers that show the pros and cons of the protocols.

Keywords: Pub/Sub Middleware, Data Distribution Service, Transport Protocols, Metrics.

1 Introduction

Emerging trends and challenges. *Real-time Event Stream Processing* (RT-ESP) applications support mission-critical systems (such as collaboration of weather monitoring radars to predict life-threatening weather [4]) by managing and coordinating multiple streams of event data that have (possibly distinct) timeliness requirements. Streams of event data may originate from sensors (*e.g.*, surveillance cameras, temperature probes), as well as other types of monitors

* This work is supported in part by the AFRL/IF Pollux project and NSF TRUST.

R. Meersman, T. Dillon, P. Herrero (Eds.): OTM 2009, Part I, LNCS 5870, pp. 614–633, 2009.

(*e.g.*, online stock trade feeds). These continuously generated data streams differ from streaming the contents of a data file (such as a fixed-size movie) since the end of RT-ESP data is not known *a priori*. In general, streamed file data demand less stringent delivery and deadline requirements, instead emphasizing a continuous flow of data to an application.

RT-ESP applications require (1) *timeliness* of the event stream data and (2) *reliability* so that sufficient data are received to make the result usable. Moreover, RT-ESP applications encompass multiple senders and receivers, *e.g.*, multiple continuous data streams can be produced and multiple receivers can consume the data streams. With the growing complexity of RT-ESP application requirements (*e.g.*, large number of senders/receivers, variety of event types, event filtering, QoS, and platform heterogeneity), developers are increasingly leveraging pub/sub middleware to help manage the complexity and increase productivity [18,8].

To address the complex requirements of RT-ESP applications, the underlying pub/sub middleware must support a flexible communication infrastructure. This flexibility requirement is manifest in several ways, including the following:

- Large-scale RT-ESP applications require flexible communication infrastructure due to the complexity inherent in the scale involved. As the number and type of event data streams continue to increase, the communication infrastructure must be able to coordinate these streams so that publishers and subscribers are connected appropriately. Flexible communication infrastructure must adapt to fluctuating demands for various event streams and environment changes to maintain acceptable levels of service.
- Certain types of large-scale RT-ESP applications require a flexible communication infrastructure due to their dynamic and *ad hoc* nature. These application environments incur fluctuations in resource availability as they include mobile assets with intermittent connectivity and underprovisioned or temporary assets from emergency responders. Examples of *ad hoc* large-scale RT-ESP applications include tactical information grids, *in situ* weather monitoring for impending hurricanes, and emergency response networks in the aftermath of regional disasters.

Several pub/sub middleware platforms have been developed to support large-scale data-centric distributed systems, such as the Java Message Service, Web Services Brokered Notification, and the CORBA Event Service. These platforms, however, do not support fine-grained and robust QoS. Some large-scale distributed system platforms, such as the Global Information Grid and Network-centric Enterprise Services, require rapid response, reliability, bandwidth guarantees, scalability, and fault-tolerance. Moreover, these systems are required to perform under stressful conditions and over connections with less than ideal behavior, such as latency and bandwidth variability, bursty loss, and routers quickly alternating destinations (*i.e.*, route flaps).

Solution approach → A FLEXible Middleware and Transports (FLEXMAT) Evaluation Framework. Developing such a flexible communication infrastructure is hard because it must have a detailed understanding of the capabilities that the underlying transport protocols provide. The infrastructure must also understand how these protocols behave under different operating conditions stemming from both the application-imposed workload changes, as well as system dynamics, such as failures and network congestion. Building on this understanding, QoS-enabled pub/sub middleware can help alleviate the complexity of managing multiple event streams and maintaining real-time QoS for multiple event streams in highly dynamic environments.

This paper describes the design and capabilities of the *FLEXible and Integrated Middleware and Transport Evaluation Framework* (FLEXMAT) to address these requirements. To evaluate the impact of various transport protocols that can lead to the realization of a QoS-enabled pub/sub middleware we developed *ReLate2*, which is a composite metric for FLEXMAT that considers both reliability and latency. This paper uses ReLate2 to evaluate the reliability and latency of transmitted data for various experimental configurations involving parameters such as sending rate, network loss, and number of receivers.

To facilitate the empirical benchmarking environment, and collection of the Relate2 metrics, FLEXMAT integrates and enhances the following capabilities:

- The *Adaptive Network Transports* (ANT) framework, which provides infrastructure for composing transport protocols that builds upon properties provided by the scalable reliable multicast-based Ricochet transport protocol [12]. Ricochet enables trade-offs between latency and reliability, which are needed qualities for pub/sub middleware supporting RT-ESP applications. Ricochet also supports modification of parameters to affect latency, reliability, and bandwidth usage.
- OpenDDS (`www.opendds.org`), which is an open-source implementation of the OMG Data Distribution Service (DDS) standard that enables applications to communicate by publishing information they have and subscribing to information they need in a timely manner. OpenDDS provides support for various transport protocols, including TCP, UDP, IP multicast, and a reliable multicast protocol. OpenDDS also provides a pluggable transport framework that allows integration of custom transport protocols within OpenDDS.

We apply the ReLate2 metric across various commonly used and custom FLEXMAT transport protocols. We then empirically quantify the results and analyze the pros/cons of various transport protocol configurations in the context of FLEXMAT. By capturing the insights gained from this effort, our goal is to enhance the development and validation of QoS-enabled pub/sub middleware.

Paper organization. The remainder of this paper is organized as follows: Section 2 describes a representative RT-ESP application to motivate the challenges that FLEXMAT is designed to address; Section 3 examines the structure and functionality of FLEXMAT and the ReLate2 metric we created to evaluate FLEXMAT and its adaptive transport protocol framework; Section 4 analyzes the results of experiments conducted by applying the ReLate2 metric to FLEXMAT; Section 5 compares FLEXMAT with related work; and Section 6 presents concluding remarks.

2 Motivating the Need for FLEXMAT

This section describes a representative RT-ESP application to motivate the challenges that FLEXMAT addresses.

2.1 Search and Rescue (SAR) Operations for Disaster Recovery

To highlight the challenges of providing timely and reliable event stream processing for RT-ESP applications, our FLEXMAT work is motivated in the context of supporting search and rescue (SAR) operations. These operations help locate and extract survivors in a large metropolitan area after a regional catastrophe, such as a hurricane, earthquake, or tornado. SAR operations can use unmanned aerial vehicles (UAVs), existing operational monitoring infrastructure (*e.g.*, building or traffic light mounted cameras intended for security or traffic monitoring), and (temporary) datacenters to receive, process, and transmit event stream data from various sensors and monitors to emergency vehicles that can be dispatched to areas where survivors are identified.

Figure 1 shows an example SAR scenario where infrared scans along with GPS coordinates are provided by UAVs and video feeds are provided by existing infrastructure cameras. These infrared scans and video feeds are then sent to a datacenter, where they are processed by fusion applications to detect survivors. Once survivors are detected the application will develop a three dimensional view and highly accurate position information so that rescue operations can commence.

A key requirement of the data fusion applications within the datacenter is tight timing bounds on correlated event streams such as the infrared scans coming from UAVs and video coming from cameras mounted atop traffic lights. The event streams need to match up closely so the survivor detection application can produce accurate results. If an infrared data stream is out of sync with a video data stream the survivor detection application can generate a false negative and fail to initiate needed rescue operations. Likewise, without timely data coordination the survivor detection software can generate a false positive expending scarce resources such as rescue workers, rescue vehicles, and data center coordinators unnecessarily.

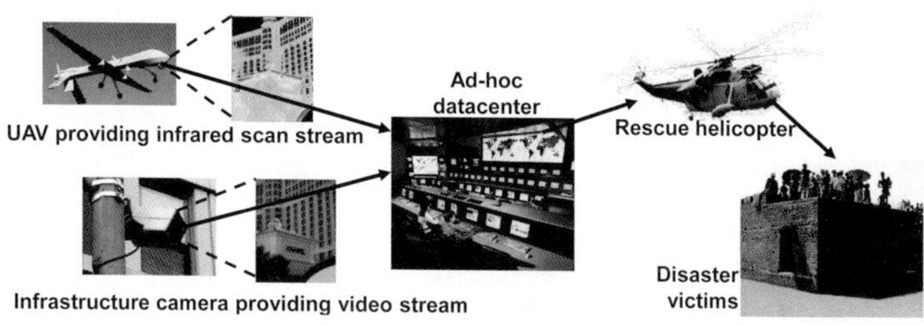

Fig. 1. Search and Rescue Motivating Example

2.2 Key Challenges in Supporting Search and Rescue Operations

Meeting the requirements of SAR operations outlined in Section 2.1 is hard due to the inherent complexity of synchronizing multiple event data streams. These requirements are exacerbated since SAR operations often run in tumultuous environments where resource availability can change abruptly. These changes can restrict the availability of resources (*e.g.*, data stream dropouts and subnetwork failure due to ongoing environment upheaval) as well as increase them (*e.g.*, network resources being added due to the stabilization of the regional situation). The remainder of this section describes four challenges that FLEXMAT addresses to support the communication requirements of the SAR operations presented above.

Challenge 1: Maintaining Data Timeliness and Reliability. SAR operations must receive sufficient data reliability and timeliness so that multiple data streams can be fused appropriately. For example, the SAR operation example described above highlights the exploitation of data streams (such as infrared scan and video streams) by several applications simultaneously in a datacenter. Figure 2 shows how fire detection applications and power grid assessment applications can use infrared scans to detect fires and working HVAC systems respectively. Likewise, Figure 3 shows how security monitoring and structural damage applications can use video stream data to detect looting and unsafe buildings respectively. Section 3.1 describes how FLEXMAT addresses this challenge by incorporating transport protocols that balance reliability and low latency.

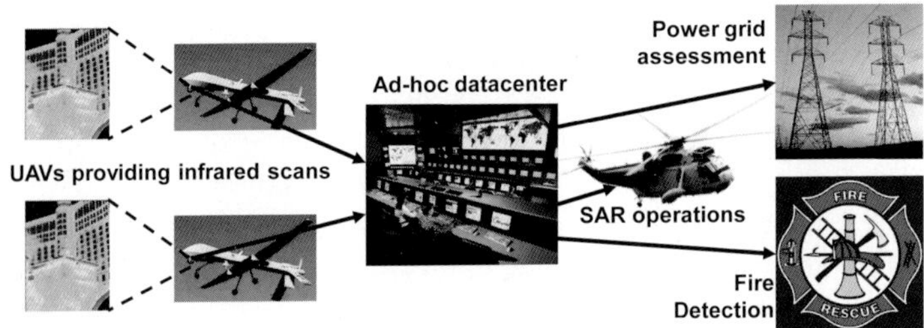

Fig. 2. Uses of Infrared Scans during Disaster Recovery

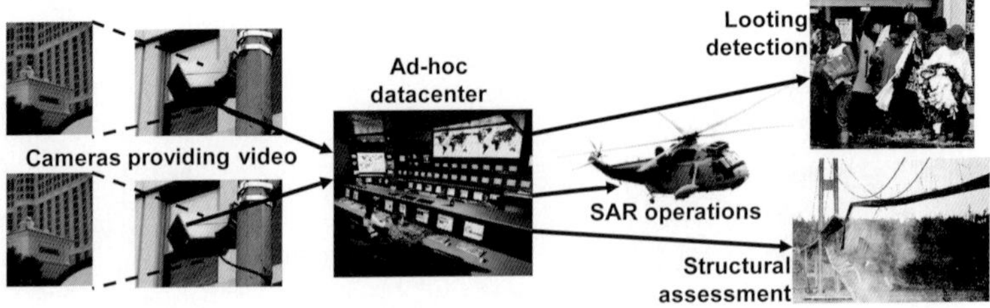

Fig. 3. Uses of Video Stream during Disaster Recovery

Challenge 2: Managing Subscription of Event Data Streams Dynamically. SAR operations must seamlessly incorporate and remove particular event data streams dynamically as needed. Ideally, an application for SAR operations should be shielded from the details of when other applications begin to use common event data streams. Moreover, applications should be able to switch to higher fidelity streams as they become available. Section 3.1 describes how we address this challenge by using anonymous QoS-enabled pub/sub middleware that seamlessly manages subscription and publication of data streams as needed.

Challenge 3: Providing Predictable Performance in Dynamic Environment Configurations. In scenarios where there is much variability and instability in the environment, such as with regional disasters, the performance of SAR operations must be known *a priori*. SAR operations tested only under a single environment configuration may not perform as needed when introduced to a new environment. The operations could unexpectedly shut down at a time when they are needed most due to changes in the environment. Section 4.2 describes how we determine application performance behavior for dynamic environments.

Challenge 4: Adapting to Dynamic Environments. SAR operations not only must understand their behavior in various environment configurations, they must also adjust as the environment changes. If SAR operations cannot adjust then they will fail to perform adequately given a shift in resources. If resources are lost or withdrawn, the SAR operations must be configured to accommodate fewer resources while maintaining a minimum level of service. If resources are added, the operations should use them to provide higher fidelity or more expansive coverage. Section 3.1 describes how we are incorporating adaptable transport protocols that can be adjusted for reliability, latency, and/or network bandwidth usage.

3 The Structure and Functionality of FLEXMAT and ReLate2

This section presents an overview of FLEXMAT, including the OpenDDS and ANT transport protocols it uses. We then describe the ReLate2 metric created to evaluate the performance of FLEXMAT in various environment configurations to support RT-ESP application requirements for data reliability and timeliness.

3.1 Design of FLEXMAT and Its Transport Protocols

FLEXMAT integrates and enhances QoS-enabled pub/sub middleware with adaptive transport protocols to provide the flexibility needed by RT-ESP applications. FLEXMAT helps resolve Challenge 2 in Section 2.2 by providing anonymous publication and subscription via the OMG Data Distribution Service (see Sidebar 1 for a brief summary of DDS). FLEXMAT is based on the OpenDDS implementation of DDS and incorporates several standard and custom transport protocols.

We chose OpenDDS as FLEXMAT's DDS implementation due to its (1) open source availability, which facilities modification and experimentation, and (2) support for a *pluggable transport framework* that allows RT-ESP application developers to create custom transport protocols for sending/receiving data. OpenDDS's pluggable transport framework uses patterns (*e.g.*, Strategy [7] and Component Configurator [6]) to provide flexibility and delegate responsibility to the protocol only when applicable.

Sidebar 1: Overview of DDS

The OMG Data Distribution Service (DDS) specifies standards-based anonymous QoS-enabled pub/sub middleware for exchanging data in event-based distributed systems. It provides a global data store in which publishers and subscribers write and read data, respectively. DDS provides flexibility and modular structure by decoupling: (1) *location*, via anonymous publish/subscribe, (2) *redundancy*, by allowing any numbers of readers and writers, (3) *time*, by providing asynchronous, time-independent data distribution, and (4) *platform*, by supporting a platform-independent model that can be mapped to different platform-specific models.

The DDS architecture consists of two layers: (1) the *data-centric pub/sub* (DCPS) layer that provides APIs to exchange topic data based on specified QoS policies and (2) the *data local reconstruction layer* (DLRL) that makes topic data appear local. This paper focuses on DCPS since it is more broadly supported than the DLRL.

The DCPS entities in DDS include *Topics*, which describe the type of data to be written or read; *Data Readers*, which subscribe to the values or instances of particular topics; and *Data Writers*, which publish values or instances for particular topics. Various properties of these entities can be configured using combinations of the 22 QoS policies. Moreover, *Publishers* manage groups of data writers and *Subscribers* manage groups of data readers.

OpenDDS currently provides several transport protocols. Other protocols for the FLEXMAT prototype are custom protocols (described below) that we integrated with OpenDDS using its pluggable transport framework.

OpenDDS Transport Protocols. By default, OpenDDS provides four transport protocols in its transport protocol framework: TCP, UDP, IP multicast (IP Mcast), and a NAK-based reliable multicast (RMcast) protocol, as shown in Figure 4. OpenDDS TCP is a reliable unicast protocol, whereas UDP is an unreliable unicast protocol. IP Mcast can send data to multiple receivers.

While TCP, UDP, and IP Mcast are standard protocols, RMcast warrants more description. It is a negative acknowledgment (NAK) protocol that provides reliability. For example, the sender sends four data packets, but the third data packet is not received by the receiver. The receiver realizes this packet has not

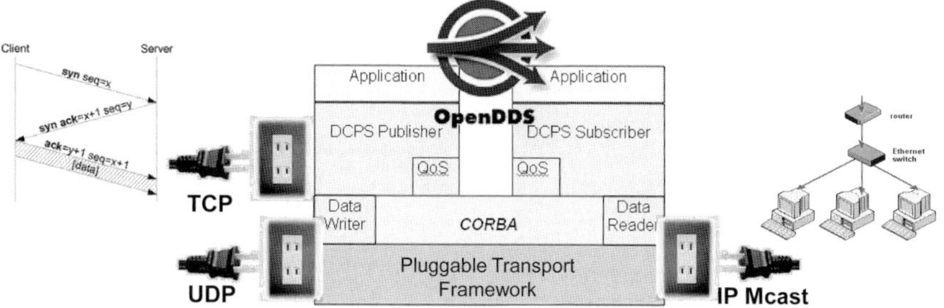

Fig. 4. OpenDDS and its Transport Protocol Framework

been received when the fourth data packet is received. At this point the receiver sends a NAK to the sender and the sender retransmits the missing data packet. The receiver sends a unicast message to the sender for loss notification and the sender retransmits the missing data packet to the receiver.

In addition to providing reliability, the RMcast protocol orders data packets. When the protocol for a receiver detects a packet out of order it waits for the missing packet before passing the data up to the middleware. The receiver must buffer any packets that have been received but have not yet been sent to the middleware. RMcast helps resolve Challenge 1 in Section 2.2 by providing reliability and timeliness for certain environment configurations.

Adaptive Network Transport Protocols. The ANT transport protocol framework supports various transport protocol properties, including multicast, packet tracking, NAK-based reliability, ACK-based reliability, flow control, group membership, and membership fault detection. These properties can be composed dynamically at run-time to achieve greater flexibility and support adaptation.

The ANT framework originally was developed from the Ricochet [12] transport protocol. Ricochet uses a bi-modal multicast protocol and a novel type of forward error correction (FEC) called lateral error correction (LEC) to provide QoS and scalability guarantees. Ricochet supports (1) time-critical multicast for high data rates with strong probabilistic delivery guarantees and (2) low-latency error detection along with low-latency error recovery.

We included ANT's Ricochet transport protocol, ANT's NAKcast protocol, which is a NAK-based multicast protocol, and ANT's baseline transport protocol in FLEXMAT. The ANT Baseline protocol mirrors the functionality of IP Mcast as described in Section 3.1. Using ANT's baseline protocol helps quantify the overhead imposed by the ANT framework since similar functionality can be achieved using the OpenDDS IP Mcast pluggable transport protocol.

Forward Error Correction (FEC). Ricochet is based on the concepts of FEC protocols. FEC protocols are designed with reliability in mind. They anticipate data loss and proactively send redundant information to recover from this loss. Sender-based FEC protocols have the sender send redundant information, as shown in Figure 5. In contrast, receiver-based FEC (a.k.a. Lateral Error

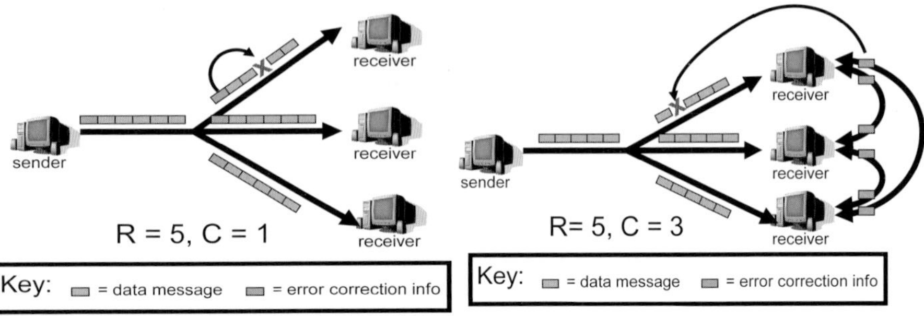

Fig. 5. FEC Reliable Multicast Protocol - Sender-based

Fig. 6. FEC Reliable Multicast Protocol - Receiver-based (LEC)

Correction (LEC)) have receivers send each other redundant information as shown in Figure 6. The Ricochet protocol we employ in FLEXMAT is an example of an LEC protocol.

Lateral Error Correction (LEC). LEC protocols have the same tunable R and C rate of fire parameters as sender-based FEC protocols. Unlike sender-based FEC protocols, however, the recovery latency depends on the transmission rate of receivers. As with gossip-based protocols, LEC protocols have receivers send out to a subset of the total number of receivers to manage scalability and network bandwidth. Moreover, the R and C parameters have slightly different semantics for LEC protocols than for sender-based FEC protocols.

The R parameter determines the number of packets a *receiver*, rather than the sender, should receive before it sends out a repair packet to other receivers. The C parameter determines the number of receivers that will be sent a repair packet from any single receiver. As described in Section 4.2, we hold the value of C constant (*i.e.*, the default value of 3) while modifying the R parameter.

The Ricochet protocol helps resolve Challenge 1 in Section 2.2 by providing high probabilistic reliability and low latency error detection and recovery. Ricochet also helps resolve Challenge 4 in Section 2.2 by supporting tunable parameters that effect reliability, latency, and bandwidth usage. We designed the ANT framework so that different transport protocols can be switched dynamically.

3.2 Evaluation Metric for Reliability and Latency

We now describe considerations for evaluating FLEXMAT's latency and reliability. We present guidelines for unacceptable percentages of packet loss for multimedia applications. We also introduce the *ReLate2* metric used to evaluate FLEXMAT empirically in Section 4.

One way to evaluate the effect of transport protocols with respect to both overall latency and reliability would be simply to compare the latency times of protocols that provide reliability. Since some reliability would be provided these protocols would presumably be preferred over protocols that provide no reliability. The reliability provided by the reliable protocols in our experiments, however, deliver different percentages of reliability. Moreover, depending upon the environment configuration the average data latency between protocols differs as well. To compare results, the level of reliability must also be quantified.

For RT-ESP applications involving multimedia, such as our motivating example of SAR operations in Section 2, over 10% loss is generally considered unacceptable. Bai and Ito [1] limit acceptable MPEG video loss at 6% while stating that a packet loss rate of more than 5% is unacceptable for Voice over IP (VoIP) users [2]. Ngatman et al. [14] define consistent packet loss above 2% as unacceptable for videoconferencing. We use these values as guidelines to develop the *ReLate2* metric that balances reliability and latency.

The 10% loss unacceptability for multimedia is due to the interdependence of packets. For example, MPEG frames are interdependent such that P frames are dependent on previous I or P frames while B frames are dependent on both preceding and succeeding I or P frames. The loss of an I or P frame therefore results in unusable dependent P and B frames, even if these frames are delivered reliably and in a timely manner.

We conservatively state that a 10% packet loss should result in an order of magnitude increase in any metric value generated. We therefore developed our ReLate2 metric to multiply the average latency by the percent packet loss as follows:

$$ReLate2_p = \frac{\sum_{i=1}^{r} l_i}{r} \times (\frac{t-r}{t} \times 100 + 1)$$

where p is the protocol being evaluated,
r = number of packets received,
l_i = latency of packet i,
and t = total number of packets sent.

We add 1 to the percent packet loss to normalize for any loss less than 1% where the metric would otherwise yield a value lower than the average latency, specifically the value 0 where all packets are delivered. This adjustment produces a ReLate2 value equal to the average latency when there is no packet loss which still accommodates meaningful comparisons for protocols that deliver all packets. Section 4.2 uses the ReLate2 metric to determine the transport protocols that best balance reliability and latency.

4 Experimental Setup, Results, and Analysis

The section presents the results of experiments we conducted to determine the performance of FLEXMAT in a representative RT-ESP environment. The experiments include FLEXMAT using multiple transport protocols with varying numbers of receivers, percentage data loss, and sending rates as would be expected with SAR operations in a dynamic environment as described in Section 2.1.

4.1 Experimental Setup

We conducted our experiments using two network testbeds: (1) the Emulab network emulation testbed and (2) the ISISlab network emulation testbed. Emulab provides computing platforms and network resources that can be easily configured with the desired computing platform, OS, network topology, and network traffic shaping. ISISlab uses Emulab software and provides much of the same functionality, but does not (yet) support traffic shaping. We used Emulab due to its ability to shape network traffic and ISISlab due to the availability of computing platforms.

As outlined in Section 2, we are concerned with the distribution of data for SAR datacenters, where network packets are dropped at end hosts [11]. The Emulab network links for the receiving data readers were configured appropriately for the specified percentage loss. The experiments in ISISlab were conducted with modified source code to drop packets when received by data readers since ISISlab does not yet support network traffic shaping.

The Emulab network traffic shaping was mainly needed when using TCP. OpenDDS does not support programmatically dropping a percentage of packets in end hosts for TCP. We therefore used network traffic shaping for TCP which only Emulab provides.

Using the Emulab environment and the *ReLate2* metric defined in Section 3.2, we next determined the protocols that balanced latency and reliability well, namely RMcast, ANT NAKcast, and ANT Ricochet. Since we could programmatically control the loss of network packets at the receiving end hosts with these protocols, we then used ISISlab due to its availability of nodes to conduct more detailed experiments involving these protocols. We obtained up to 27 nodes fairly easily using ISISlab, whereas this number of nodes was hard to get with Emulab since it is often oversubscribed.

Our experiments using Emulab and ISISlab used the following traffic generation configuration utilizing OpenDDS version 1.2.1: (1) one DDS data writer wrote data, variable number of DDS data readers read data, (2) the data writer and each data reader ran on its own computing platform, and (3) the data writer sent 12 bytes of data 20,000 times at a specified sending rate. To account for experiment variations we ran 5 experiments for each configuration, *e.g.*, 5 receiving data writers, 50 Hz sending rate, 2% end host packet loss. We used Ricochet's default C value of 3 for both Emulab and ISISlab experiments.

Emulab configuration. For Emulab, the data update rates were 25 Hz and 50Hz for general comparison of all the protocols. We varied the number of receivers from 3 up to 10. We used Ricochet's default R value of 8. As defined in Section 3.1, the R value is the number of packets received before sending out recovery data.

We used the Emulab pc850 hardware platform, which includes an 850 MHz processor and 256 MB of RAM. We ran the Fedora Core 6 operating system with real-time extensions on this hardware platform, using experiments consisting of between 5 and 12 pc850 nodes. The nodes were all configured in a LAN configuration. We utilized the traffic shaping feature of Emulab to run experiments with

Table 1. Emulab Variables

Point of Variability	Values
Number of receiving data writers	3 - 10
Frequency of sending data	25 Hz, 50 Hz
Percent end-host network loss	0 to 3 %

Table 2. ISISlab Variables

Point of Variability	Values
Number of receiving data writers	3 - 25
Frequency of sending data	10 Hz, 25 Hz, 50 Hz, 100 Hz
Percent network loss	0 to 5 %

network loss percentages between 0 and 3 percent. Table 1 outlines the points of variability for the Emulab experiments.

ISISlab configuration. We used ISISlab for experiments involving transport protocols where we could programmatically affect the loss of packets in the end hosts. By modifying the source code, we could discard packets based on the desired percentage. In particular, we focused the ISISlab experiments on the ANT NAKcast and Ricochet protocols since from the initial experiments these protocols showed the ability to balance latency and reliability. At times, OpenDDS RMcast showed the ability to balance reliability and low latency. Since its behavior was erratic for a NAK-based protocol, however, we excluded it from the detailed experiments. Table 2 outlines the points of variability for the ISISlab experiments.

ISISlab provides a single type of hardware platform: the pc8832 hardware platform with a dual 2.8 GHz processor and 2 GB of RAM. We used the same Fedora Core 6 OS with real-time extensions as for Emulab. We ran experiments using between 5 and 27 computing nodes which map to between 3 and 25 data readers respectively. All nodes were configured in a LAN as was done for Emulab. We ran experiments using Ricochet's R value of 8 and 4, as explained in Section 4.2.

4.2 Results and Analysis of Experiments

This section presents and analyzes the results from our experiments, which resolves Challenge 3 in Section 2.2 by characterizing the performance of the transport protocols for various environment configurations.

The Baseline Emulab Experiments. The initial set of experiments for the FLEXMAT prototype included all the OpenDDS protocols as enumerated in Section 3.1. These experiments used Emulab as described in Section 4.1. Our baseline experiments used 3 data readers, 0% loss, and 25 and 50 Hz update rates. As expected, all protocols delivered all data to all data readers, *i.e.*, 3 receivers * 20,000 updates = 60,000 updates.

As shown in Figures 7 and 8, the latency at times was lowest with protocols that do not provide any reliability, *i.e.*, OpenDDS UDP, OpenDDS IP Mcast, and ANT Baseline). The OpenDDS RMcast and ANT Ricochet protocols were the only ones that never produced the lowest overall average latency. As expected, average latency times decreased as the sending rate increased from 25 Hz to 50 Hz.

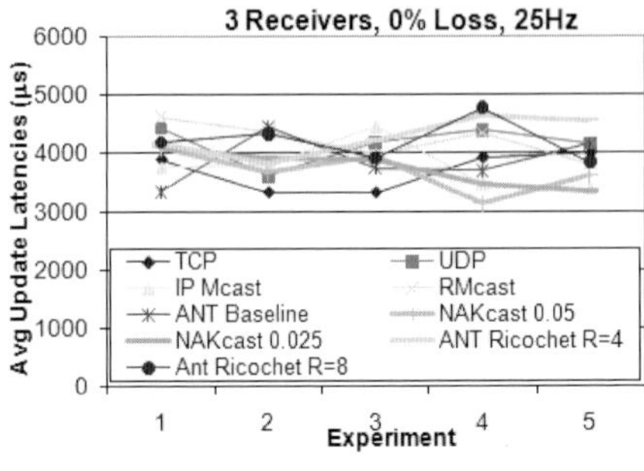

Fig. 7. Emulab: 3 readers, 0% loss, 25Hz

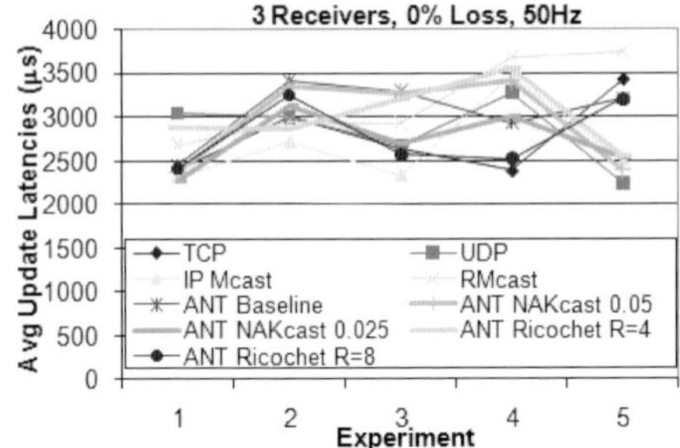

Fig. 8. Emulab: 3 readers, 0% loss, 50Hz

The next set of experiments added 1% network packet loss for the receiving end hosts. We do not include figures for the 50 Hz update rate as the data are comparable to that seen with a sending rate of 25 Hz. As shown in Figure 9, there is a clear delineation between the protocols that provide reliability and those that do not.

TCP received all updates sent, whereas ANT NAKcast and ANT Ricochet received a high percentage of updates with ANT NAKcast receiving all updates except for one experiment run where it received 59,999 out of the 60,000 updates. Both configurations of ANT Ricochet delivered a consistently high percentage of updates between 99.95% and 99.99%. UDP, IP Mcast, and ANT Baseline group together in the figure with low reliability.

We were unable to configure OpenDDS IP Mcast to use Emulab's network traffic shaping. Instead we calculated the amount of packet loss that is comparable to the other unreliable transports, *i.e.*, 1% loss. We are confident this calculation does not invalidate the values seen and used for OpenDDS IP Mcast as the values for ANT's version of IP Mcast, *i.e.*, ANT Baseline, produces similar results.

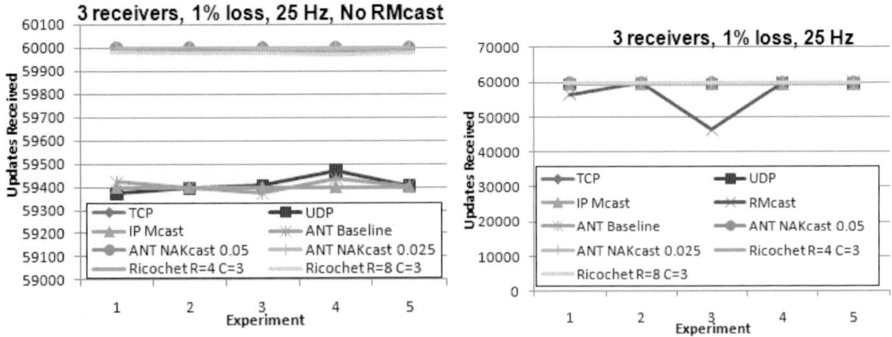

Fig. 9. Emulab: 3 readers, 1% loss, 25Hz, no RMcast

Fig. 10. Emulab: 3 readers, 1% loss, 25Hz

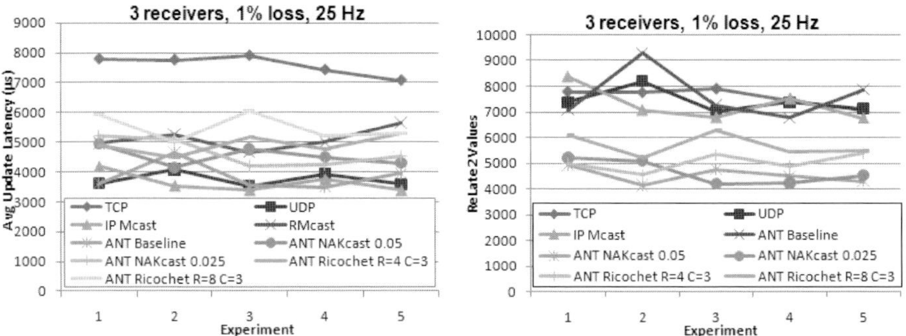

Fig. 11. Emulab: 3 readers, 1% loss, 25Hz **Fig. 12.** Emulab: 3 readers, 1% loss, 25Hz

Figure 10 shows the erratic behavior of RMcast. At times RMcast received all updates and other times it received all updates only up to a certain number and then received no additional updates. The cause of this problem was not explained by the RMcast developers. We therefore removed RMcast from further consideration.

Figure 11 highlights the latency overhead incurred by TCP. This latency is due to TCP's use of positive acknowledgments. Moreover, TCP's latency overhead increases as the amount of loss increases. All other protocols are fairly comparable with respect to latency for this environment configuration.

Figure 12 shows the ReLate2 values for all the protocols considered. We see that using ReLate2 splits the protocols that support both reliability and low latency from those that do not. The separation of the protocols using ReLate2 is more pronounced with higher levels of network loss and number of receivers.

We now analyze the results of the Emulab experiments, which involved all the transport protocols presented in Section 3.1. We utilize the ReLate2 metrics defined in Section 3.2 to evaluate the results from the initial Emulab experiments. The results show that ANT NAKcast and ANT Ricochet always produced the lowest ReLate2 values even for multiple configurations of the protocols, *i.e.*, NAKcast timeout values of 0.05 and 0.025 and Ricochet R values of 4 and 8. The protocols that support reliability but unbounded latency and the protocols that support low latency but no reliability are clearly separated from the protocols that support both low latency and reliability.

Moreover, the ReLate2 value is equal to the average latency when there is no loss, as is the case for TCP and the majority of cases for NAKcast. When NAKcast does not receive all updates, it is only missing some of the very last updates which could not be detected since no packets were received after them. The data and figures show that the ReLate2 metric is useful for evaluating protocols that balance reliability and latency.

The NAKcast and Ricochet Experiments. Our next set of experiments focused on the protocols that are best suited for balancing reliability and latency based on the ReLate2 metric (*i.e.*, ANT NAKcast and ANT Ricochet). We focus on these protocols for comparison to gain a better understanding of trade-offs between them. We provide experimental results and analyze the results. We note that if RMcast's behavior would stabilize it would also be a protocol worth evaluating for reliability and low latency.

In particular, for comparison we focused on specific configurations of NAKcast and Ricochet, *i.e.*, NAKcast with a timeout period of 0.05 seconds and Ricochet with an R value of 4. We constrained the protocols in this way because configured correctly either protocol can generally provide lower ReLate2 values than the other. However, we are interested in a relative comparison of the protocols themselves rather than reconfigurations that can make the one protocol outperform the other for a particular environment.

As noted in Section 4.1, we used the ISISlab testbed for experiments involving only ANT NAKcast and ANT Ricochet due to the availability of a larger number of hardware nodes. We were able programmatically to induce packet loss at the end hosts for these two protocols since the ANT source code is available and thus we did not require Emulab's network traffic shaping capability.

As with the Emulab experiments in Section 4.2, we began with experiments where the number of receivers and packet loss were low. We also expanded the sending rates to include 10Hz and 100Hz along with the original rates of 25Hz and 50Hz. Adding sending rates made sense as the packet loss recovery times for both of these protocols are sensitive to the update rate.

The packet loss recovery time for NAKcast is sensitive to the update rate since loss is only discovered when packets are received. If packets are received faster then packet loss is discovered sooner and recovery packets can be requested, received, and processed sooner. Likewise, the packet loss recovery time for Ricochet is sensitive to the update rate since recovery data is only sent out after R packets have been received. When packets are received sooner, recovery data is sent, received, and processed sooner.

Moreover, our results and analysis are focused on environment configurations with relatively low (*i.e.*, 1%) and high (*i.e.*, 5%) network loss combined with relatively few (*i.e.*, 3) and many (*i.e.*, 20) receivers. While we ran experiments that ran the spectrum of configurations between these bounds, the particular experiments at these limits are useful for understanding the behavior of the protocols. We show data collected while using 10Hz and 100Hz sending rates to highlight the behavorial distinctions of the protocols.

Figures 13 and 14 show that for a low number of receivers (*i.e.*, 3), a low loss percentage (*i.e.*, 1%), and low sending rate (*i.e.*, 10Hz), NAKcast, in general, has lower ReLate2 values. In fact, NAKcast 0.05 provided the lowest ReLate2 values for all of the ISISlab protocol configurations tried, i.e., NAKcast with timeout values of 0.05 and 0.025 seconds and Ricochet with R values of 4 and 8. Ricochet provided lower average update latency as the sending rate increases. We discuss this observation in more detail at the end of this section. The number of updates received remains constant across various sending rates for both protocols and we do not include those figures here.

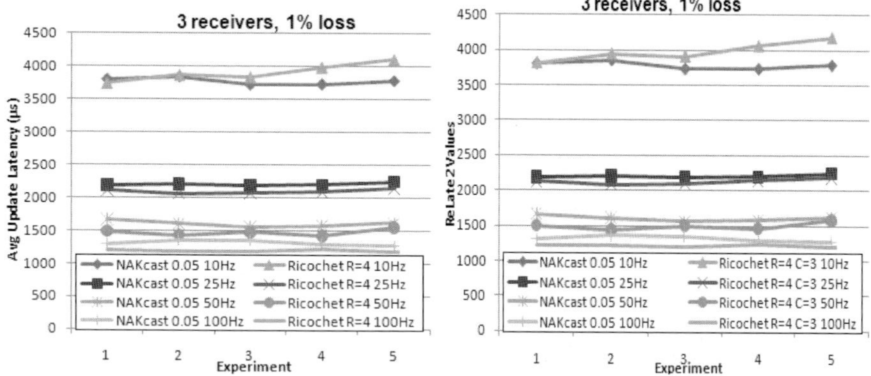

Fig. 13. ISISlab: 3 readers, 1% loss

Fig. 14. ISISlab: 3 readers, 1% loss

Figures 13 and 14 also show the reliability of Ricochet at low loss rates. This reliability can be seen by comparing the figures and noticing that the graphs appear very similar. This similarity points out that Ricochet is almost as reliable as NAKcast with reliability rates ranging from 99.97% to 99.99%. This reliability is fairly constant across the different sending rates.

Figures 15 and 16 show the effect on the protocols of increasing packet loss. In this environment configuration we have changed the network loss from 1% to 5%. We see that NAKcast performed best not only for a sending rate of 10 Hz as was the case for 1% loss but also for 25 Hz. Ricochet still provided the best ReLate2 values for sending rates of 50 Hz and 100 Hz. Moreover, while Ricochet average update latency improved over NAKcast the ReLate2 values don't reflect this as Ricochet only had better ReLate2 values for sending rates of 50 and 100 Hz. This is due to Ricochet's reliability ranging from 99.42% to 99.56% which has decreased from the experiments with 1% loss.

Figures 17 and 18 show the effect on the protocols of increasing the number of receivers. In this environment configuration we increased the number of receivers from 3 to 20. We see that now Ricochet and NAKcast performed equally well at 10 Hz where NAKcast always performed best at that rate with only 3 receivers.

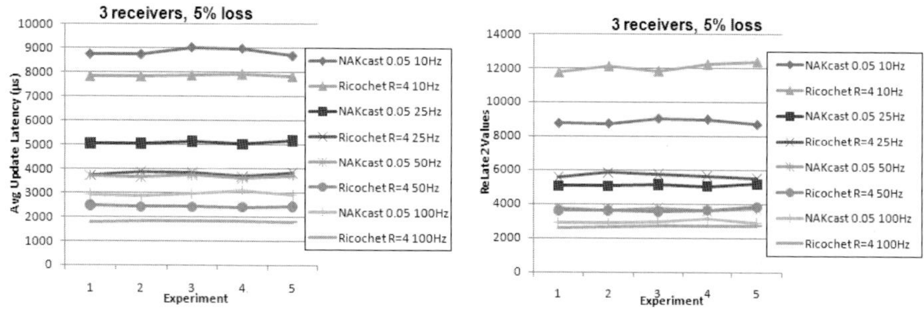

Fig. 15. ISISlab: 3 readers, 5% loss

Fig. 16. ISISlab: 3 readers, 5% loss

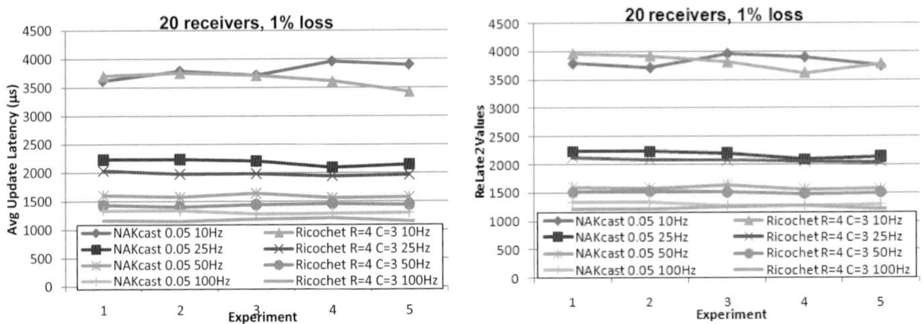

Fig. 17. ISISlab: 20 readers, 1% loss

Fig. 18. ISISlab: 20 readers, 1% loss

Ricochet provided the best ReLate2 values for the other sending rates. Moreover, Ricochet's reliability is almost as high as with only 3 receivers ranging from 99.94% to 99.96% of updates received.

Finally, Figures 19 and 20 show the effect on the protocols of increasing the number of receivers and loss rate. In this environment configuration we had 20 receivers and 5% network loss.

We see that while Ricochet had a noticeable improvement in average update latency compared to NAKcast, NAKcast offset this discrepancy with its higher reliability. For higher rates, *i.e.*, 25, 50, and 100 Hz, the ReLate2 values for Ricochet and NAKcast are comparable. NAKcast always provided the lowest ReLate2 values for 10 and 25 Hz while Ricochet always provided the lowest ReLate2 values for 50 and 100 Hz. Moreover, Ricochet's reliability is in the same range as for 3 receivers with 5% loss ranging from 99.46% to 99.55% of updates received.

The results above show that for a set protocol configuration there are performance trade-offs between NAK-based and LEC protocols. In general, NAK-based protocols performed better with a lower network loss percentage, lower sending rates, and few receivers. In this environment configuration there is no concern for NAK storms where receivers flood the sender with requests for

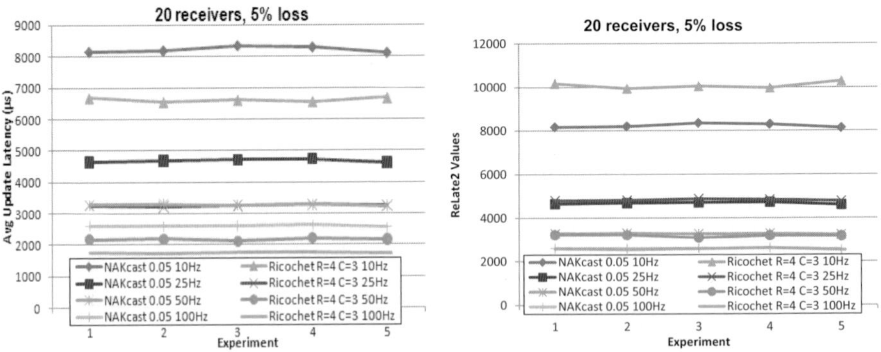

Fig. 19. ISISlab: 20 readers, 5% loss

Fig. 20. ISISlab: 20 readers, 5% loss

retransmissions. Moreover, NAK-based protocols only needed to receive one update that is out of sequence to determine loss whereas LEC protocols need to receive R updates before error detection and correction information is sent among the receivers. NAK-based protocols also delivered consistently high reliability, at the cost of higher latency for higher sending rates.

LEC protocols, however, provided better performance when network loss was higher and sending rates increased. LEC protocols did not incur increasingly more network usage as network loss and number of receivers increased. LEC protocols scaled well in the number of receivers and in network loss. LEC protocols also generally provided lower latency at the cost of small decreases in reliability.

NAKcast 0.05 provided the lowest ReLate2 values and lowest average latency for 3 receivers, 1% loss, and 10 Hz sending rate. The data make sense since the sending rate was less than the timeout period and the loss rate and number of receivers were low. If the network drops a packet the packet is as likely to be discovered in the same amount of time by NAKcast with a timeout of 0.05 as it is with a higher timeout. The sending rate is so low that increasing the NAKcast timeout to 0.025 seconds provided no benefit and indeed added overhead as timeouts are generated and checked more frequently.

5 Related Work

This section compares our work on FLEXMAT with related R&D efforts.

Performance evaluation of network transport protocols. Much prior work has evaluated network transport protocols, *e.g.*, Balakrishnan *et al.* [12] evaluate the performance of the Ricochet transport protocol with the Scalable Reliable Multicast (SRM) protocol [17]. Bateman *et al.* [13] compare the performance of TCP variations both using simulations and in a testbed. Cheng *et al.* [19] provide performance comparisons of UDP and TCP for video streaming in multihop wireless mesh networks. Kirschberg *et al.* [9] propose the Reliable Congestion Controlled Multicast Protocol (RCCMP) and provide simulation results for its performance. In contrast to our work on FLEXMAT, these evaluations specifically target the protocol level independent of any integration of QoS-enabled pub/sub middleware.

Performance evaluation of enterprise middleware. Xiong *et al.* [15] conducted performance evaluations for three DDS implementations, including OpenDDS. That work highlighted the different architectural approaches taken and trade-offs of these approaches. In contrast, to our work on FLEXMAT, however, that prior work did not include performance evaluations of DDS with various transport protocols.

Sachs *et al.* [10] present a performance evaluation of message-oriented middleware (MOM) in the context of the SPECjms2007 standard benchmark for MOM servers. The benchmark is based on the Java Message Service (JMS). In particular, the work details performance evaluations of the BEA WebLogic server under various loads and configurations. In contrast to our work on FLEXMAT, however, that work did not integrate various transport protocols with the middleware to evaluate its performance.

Tanaka *et al.* [20] developed middleware for grid computing called Ninf-G2. In addition, they evaluate Ninf-G2's performance using a weather forecasting system. The evaluation of the middleware does not integrate various protocols and evaluate performance in this context, as our work on FLEXMAT does.

Tselikis *et al.* [5] conduct performance analysis of a client-server e-banking application. They include three different enterprise middleware platforms each based on Java, HTTP, and Web Services technologies. The analysis of performance data led to the benefits and disadvantages of each middleware technology. In contrast, our work on FLEXMAT measures the impact of various network protocols integrated with QoS-enabled pub/sub middleware.

Performance evaluation of embedded middleware. Bellavista *et al.* [16] describe their work called Mobile agent-based Ubiquitous multimedia Middleware (MUM). MUM has been developed to handle the complexities of wireless hand-off management for wireless devices moving among different points of attachment to the Internet. In contrast, our work on FLEXMAT focuses on the performance and flexibility of QoS-enabled anonymous pub/sub middleware.

TinyDDS [3] is an implementation of DDS specialized for the demands of wireless sensor networks (WSNs). TinyDDS defines a subset of DDS interfaces for simplicity and efficiency within the domain of WSNs. TinyDDS includes a pluggable framework for non-functional properties, *e.g.*, event correlation and filtering mechanisms, data aggregation functionality, power-efficient routing capability. In contrast, our work on FLEXMAT focuses on how properties of various transport protocols can be used to maintain specified QoS.

6 Concluding Remarks

Developers of RT-ESP systems face a number of challenges when developing their applications for dynamic environments. To address these challenges, we have developed FLEXMAT to integrate and enhance QoS-enabled pub/sub middleware with flexible transport protocols to support RT-ESP applications. This paper defines the ReLate2 metric to empirically measure the reliability and latency of FLEXMAT as a first step to having QoS-enabled pub/sub middleware autonomically adapt transport protocols as the changing environment dictates.

The latest information and source-code for FLEXMAT and related research can be obtained at `www.dre.vanderbilt.edu/~jhoffert/ADAMANT`.

References

1. Bai, Y., Ito, M.: A new technique for minimizing network loss from users' perspective. Journal of Network Computing Appllications 30(2), 637–649 (2007)
2. Bai, Y., Ito, M.R.: A Study for Providing Better Quality of Service to VoIP Users. In: 20th International Conference on Advanced Information Networking and Applications (AINA 2006), April 2006. LNCS, vol. 3410, pp. 799–804. Springer, Heidelberg (2006)

3. Boonma, P., Suzuki, J.: Middleware support for pluggable non-functional properties in wireless sensor networks. In: IEEE Congress on Services - Part I, pp. 360–367 (July 2008)
4. Plale, B., et al.: CASA and LEAD: Adaptive Cyberinfrastructure for Real-Time Multiscale Weather Forecasting. Computer 39(11), 56–64 (2006)
5. Tselikis, C., et al.: An evaluation of the middleware's impact on the performance of object oriented distributed systems. Journal of Systems and Software 80(7), 1169–1181 (2007); Dynamic Resource Management in Distributed Real-Time Systems
6. Schmidt, D., et al.: Pattern-Oriented Software Architecture: Patterns for Concurrent and Networked Objects, vol. 2. Wiley & Sons, New York (2000)
7. Gamma, E., et al.: Design Patterns: Elements of Reusable Object-Oriented Software. Addison-Wesley, Reading (1995)
8. Eisenhauer, G., et al.: Publish-subscribe for high-performance computing. Internet Computing, IEEE 10(1), 40–47 (2006)
9. Kirschberg, J., et al.: Rccmp: reliable congestion controlled multicast protocol. In: 1st EuroNGI COnference on Next Generation Internet Networks Traffic Engineering (April 2005)
10. Sachs, K., et al.: Performance Evaluation of Message-oriented Middleware using the SPECjms2007 Benchmark. Performance Evaluation (to appear, 2009)
11. Balakrishnan, M., et al.: Slingshot: Time-critical multicast for clustered applications. In: Proceedings of the IEEE Conference on Network Computing and Applications (2005)
12. Balakrishnan, M., et al.: Ricochet: Lateral error correction for time-critical multicast. In: NSDI 2007: Fourth Usenix Symposium on Networked Systems Design and Implementation, Boston, MA (2007)
13. Bateman, M., et al.: A comparison of tcp behaviour at high speeds using ns-2 and linux. In: CNS 2008: Proceedings of the 11th communications and networking simulation symposium, pp. 30–37. ACM, New York (2008)
14. Ngatman, M., et al.: Comprehensive study of transmission techniques for reducing packet loss and delay in multimedia over ip. International Journal of Computer Science and Network Security 8(3), 292–299 (2008)
15. Xiong, M., et al.: Evaluating Technologies for Tactical Information Management in Net-Centric Systems. In: Proceedings of the Defense Transformation and Net-Centric Systems conference, Orlando, Florida (April 2007)
16. Bellavista, P., et al.: Context-aware handoff middleware for transparent service continuity in wireless networks. Pervasive and Mobile Computing 3(4), 439–466 (2007); Middleware for Pervasive Computing
17. Floyd, S., et al.: A reliable multicast framework for light-weight sessions and application level framing. IEEE/ACM Trans. Netw. 5(6), 784–803 (1997)
18. Kumar, V., et al.: Distributed stream management using utility-driven self-adaptive middleware. In: Proceedings of Second International Conference on Autonomic Computing, ICAC 2005, pp. 3–14 (June 2005)
19. Cheng, X., et al.: Performance evaluation of video streaming in multihop wireless mesh networks. In: NOSSDAV 2008: Proceedings of the 18th International Workshop on Network and Operating Systems Support for Digital Audio and Video, pp. 57–62. ACM, New York (2008)
20. Tanaka, Y., et al.: Design, Implementation and Performance Evaluation of GridRPC Programming Middleware for a Large-Scale Computational Grid. In: GRID 2004: Proceedings of the 5th IEEE/ACM International Workshop on Grid Computing, Washington, DC, USA, pp. 298–305. IEEE Computer Society, Los Alamitos (2004)

Reliable Communication Infrastructure for Adaptive Data Replication

Mouna Allani[1], Benoît Garbinato[1], Amirhossein Malekpour[2], and Fernando Pedone[2]

[1] University of Lausanne
[2] University of Lugano

Abstract. In this paper, we propose a data replication algorithm adaptive to unreliable environments. The data replication algorithm, named Adaptive Data Replication (ADR), has already an adaptiveness mechanism encapsulated in its dynamic *replica placement* strategy. Our extension of ADR to unreliable environments provides a data replication solution that is adaptive both in terms of replica placement and in terms of request routing. At the routing level, this solution takes the unreliability of the environment into account, in order to maximize reliable delivery of requests. At the replica placement level, the dynamically changing origin and frequency of read/write requests are analyzed, in order to define a set of replica that minimizes communication cost. Performance evaluation shows that this original combination of two adaptive strategies makes it possible to ensure high request delivery, while minimizing communication overhead in the system.

1 Adaptive Data Replication

Data replication is a well-known technique to increase data availability and load balancing. A data replication system can be characterized by two key policies: a *replica placement policy*, which determines how many replicas the scheme creates and where it places them, and a *replica consistency policy*, which determines the level of consistency the scheme ensures among replicas, e.g., eager consistency or lazy consistency. These policies are typically implemented on top of a communication substrate ensuring a set of properties necessary for the correctness of the data replication system. An example of communication substrate is the *group communication* abstraction [1,5,8]. In this case, the *group communication* offers a set of guaranties including adaptiveness to membership changes, message ordering, and multicast reliability. In this paper, we define as a communication substrate a routing mechanism adaptive to unreliable environment in order to use it as the basis for a replica placement solution. We are primarily concerned with replica placement; replica consistency is out of the scope of the paper.

Regarding the replica placement policy, various replica management schemes have been proposed, based on a fixed number of replicas placed in fixed locations [2,15,13]. This approach works well when the source and the frequency of read and write requests are known in advance and remain static during the execution, which then implies that clients accessing the replicas are themselves static and generate

R. Meersman, T. Dillon, P. Herrero (Eds.): OTM 2009, Part I, LNCS 5870, pp. 634–652, 2009.

a steady stream of requests. When the frequency and the source of requests are variable, however, the ability to dynamically create, move, and delete replicas is essential when it comes to devising efficient replication schemes.

In a dynamic distributed environment, replica placement significantly affects the overall performance of the replication scheme. For example, since reading a replica locally is faster and less costly than reading it remotely, a widely distributed replication scheme is particularly well suited in read-intensive environments. On the other hand, writing to a large number of replicas may be slow and increase communication costs. For this reason, a narrowly distributed replication scheme is more adequate in write-intensive environments. In addition, the occurrence of node and link failures further challenges the effectiveness and performance of the replication scheme, as it can radically compromise replica placement decisions made before the failures occurred. The problem of placing replicas in dynamic and unreliable distributed environments advocates integrating *adaptiveness* into the replication schemes.

To adapt to the dynamic behavior of the environment and the application access patterns, various solutions have been proposed in the literature [16,11,12,19]. Among these, the *Adaptive Data Replication* algorithm (ADR) described in [19] is particularly interesting, as it was shown to be *convergent-optimal* with respect to communication costs. That is, as soon as the read-write access pattern changes, ADR adapts its replication scheme to minimize the communication cost caused by the routing of access requests. Intuitively, ADR organizes replicas as a connected graph, known as the *replication scheme*, which expands or contracts as the read-write access pattern changes.

Unfortunately, this convergence towards optimality only holds under two strict conditions: (1) the network is organized as a tree—finding an optimal replication scheme was shown to be NP-complete for general topologies [20]—and (2) no process or link failures occur. Condition 1 implies that one must first build an overlay tree covering the network. Condition 2 implies that ADR ceases to work correctly as soon as a failure happens.[1] Indeed, unreliable links may cause requests to be lost, thus misleading the replica placement strategy of ADR, while node failures may break the connectivity of the replication scheme, an essential assumption for ADR to work. Conditions 1 and 2 make ADR unsuitable to unreliable large-scale distributed environments.

Contributions. In this paper, we propose an architecture that extends ADR to make it capable of dynamically reorganizing itself based on changes in the application access patterns, and on link and node failures. The new replica placement strategy relies on a specialized routing layer, which encapsulates our *adaptive request routing strategy*. The latter is based on a tree overlay that aims at maximizing the reliability of request routing, in spite of link and node failures. This tree, named the *Maximum Reliability Tree* (MRT), is a spanning tree containing the most reliable paths in the system [4].

[1] In [19], processes switch to a special failure mode until recovery occurs. As detailed in Section 6, this approach is quite different from ours.

Roadmap. The remainder of this paper is organized as follows. Section 2 formally defines our model, describes and motivates the problem solved in the paper, and sketches the architecture of our solution. Section 3 presents our *adaptive request routing* algorithm based on a spanning tree maximizing the reliability of communication paths, while Section 4 describes an extension of the *adaptive replica* algorithm defined in [19], which aims at minimizing communication costs given a read-write pattern. In Section 5, we evaluate the benefit of using our *adaptive request routing* solution in terms of performance and adaptiveness, when both the access pattern changes and failures occur. Finally, Section 6 puts the proposed approach into perspective by comparing it with the state of the art; Section 7 concludes the paper and discusses future work.

2 A Modular Approach to Adaptiveness

In this paper, we consider an asynchronous distributed system composed of processes (nodes) that communicate by message passing. Our model is probabilistic in the sense that processes may crash and links may lose messages with a certain probability. More formally, the tuple $S = (\Pi, \Lambda, C)$ completely defines the (unreliable) environment considered in this paper. With Π the set of processes and Λ is a set of bidirectional communication links. We only consider systems with a connected graph topology. Process crash probabilities and message loss probabilities are modeled as *failure configuration* C.

We then define object o as the data to replicate, while $R \subseteq \Pi$ denotes the replication scheme of o, i.e., the set of nodes holding a copy of o. Any request sent to R is either a read or a write operation. Given these definitions, our approach consists in addressing the two following questions.

Adaptive Replica Placement. Given a pattern of reads and writes to o, what nodes should be part of R in order to minimize the communication cost?

Adaptive Request Routing. Given some failure configuration C, how should read/write requests be routed to maximize reliable delivery, and thus provide the replica placement layer with accurate information?

2.1 Adaptiveness to Access Patterns

The main idea of the adaptive replica placement strategy, borrowed from [19], consists in having the replication scheme R evolve like an amoeba along the branches of some tree-based communication overlay. The replica placement is managed in a fully decentralized manner. Each process p_k in R analyzes its access pattern and independently decides to either:

1. **expand** R, by sending a copy of o to one of its neighbors in the tree that does not yet hold a replica;
2. **contract** R, by quitting the replication scheme and discarding its copy of o;
3. **switch** R, by moving o to one of its neighbors in the tree, in case p_k is the only node holding a copy of o.

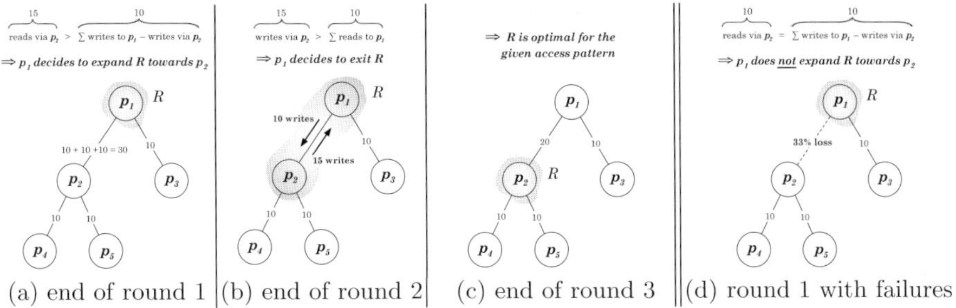

(a) end of round 1 (b) end of round 2 (c) end of round 3 (d) round 1 with failures

Fig. 1. Adaptive Replication Scheme – Example

Figures 1 (a) to (c) illustrate the behavior of the adaptive replica placement strategy on a concrete example, based on five processes. In this example, we initially have replication scheme $R = \{p_1\}$, i.e., p_1 is the only node with a copy of object o. In addition, we assume that each process periodically sends 5 reads and 5 writes to R and that they do so in synchronized rounds.[2] In the following, we define the communication cost of a round as the total number of requests transiting through any link in the system.

The situation after round 1 is shown in Figure 1 (a): p_1 received 15 reads and 15 writes from p_2 (5 reads and 5 writes from p_2 + 5 reads and 5 writes from p_4 + 5 reads and 5 writes from p_5), 5 reads and 5 writes via p_3 (5 reads and 5 writes from p_3 itself) and 5 reads and 5 writes from p_1 itself. p_1 notices that it received more reads via node p_2 (15 = 5 reads from p_2, 5 from p_4 and 5 from p_5) than the total number of writes originated from elsewhere, i.e., 10 writes (5 writes from p_1 + 5 from p_3). Based on this analysis, p_1 concludes that having a replica on node p_2 may improve the overall communication cost, which currently equals 60 (10 requests through link $l_{2,4}$ plus 10 through link $l_{2,5}$ plus 30 through link $l_{1,2}$ plus 10 through link $l_{1,3}$). This decision to expand R towards p_2 leads to the situation pictured in Figure 1 (b), with a communication cost of 55. To update all existing replicas, this change imposes however that p_1 and p_2 inform each other about the respective writes they received. At the end of round 2, p_1 finally decides to contract R by exiting the replication scheme. This situation is pictured in Figure 1 (c) and leads to a communication cost of 50. At this point, $R = \{p_2\}$ is the optimal replication scheme for the given access pattern.

2.2 Adaptiveness to Failures

To illustrate the need for adaptiveness to failures, let us revisit our example when injecting some unreliability into the system. As shown in Figure 1 (d), we inject a 33% message-loss into the link connecting p_1 and p_2, i.e., $l_{1,2}$ roughly loses one message out of three. So, at the end of the first round, p_1 compares the effective number of reads received via node p_2, which is equal to 10, with the number of all writes originated from elsewhere, which is also equal to 10. Based

[2] Synchronized rounds are only assumed to simplify our example and are by no means imposed by the replica placement strategy.

on this analysis, p_1 concludes that there is no need to expand R towards p_2 nor to switch with p_2, contrary to what happened in the setting shown in Figure 1 (a). That is, to process p_1, the replication scheme R appears to be optimal for the given access pattern, but this analysis is biased by the system unreliability, as the cost seen at p_1 is not the real cost imposed by the routing of requests. This observation clearly shows that in order to take full advantage of the adaptive replication scheme described earlier, we need to also adapt to the presence of failures when routing requests.

2.3 Solution Overview

Our solution follows the three-layer architecture pictured in Figure 2. The top layer executes the *adaptive replica placement* algorithm sketched earlier, which manages a replication scheme R changing according to the read-write pattern produced by some distributed application on top of it. The complete algorithm is described in Section 4.

The Replica Placement (RP) strategy relies on the *adaptive request routing* layer, which offers a set of communication primitives, and relies on a low-level *system layer* providing basic *best-effort* `send` and `receive` primitives. For the correctness of our solution, the best-effort aspect of the `send` primitive is hidden using a simple message resend/ack mechanism. For simplification, this retransmission mechanism was not included in the algorithm description. As detailed in Section 3, our routing solution permits to minimize the message overhead induced by this resend/ack mechanism by routing messages through the overlay including the most reliable paths covering the system: MRT. Thus, as shown in Section 5, our routing solution based on MRT induces an message overhead lower than when using any other tree overlay.

For each object o, the RP layer basically maps the corresponding replication scheme R to a dedicated group G managed by the Adaptive Routing (AR) layer. Coming to the AR layer, the latter offers the following adaptive and reliable services: (1) creation of a new group G and its announcement to all nodes in the system, (2) request routing from any node outside G to some node in G, and (3) multicasting among nodes in G. The algorithm executed by AR is detailed in Section 3.

Finally, the routing layer also relies on the system layer, not only for its send and receive primitives, but also for its ability to provide key information about nodes and links in the system. In particular, the system layer is responsible for providing an approximation of the failure rates of links and nodes, in terms of message-loss probabilities and crash probabilities respectively. That is, the system layer is capable of providing an approximation of the tuple $S = (\Pi, \Lambda, C)$ modeling the system.[3]

On Reliability and Consistency. Since our model is probabilistic, reliability should also be understood in probabilistic terms. Indeed, in the remainder of this paper, when we say for instance that our routing algorithm will *reliably* route a request, we actually mean that it will *maximize the probability* of the message

[3] In [4], we show how to use Bayesian statistical inference for this.

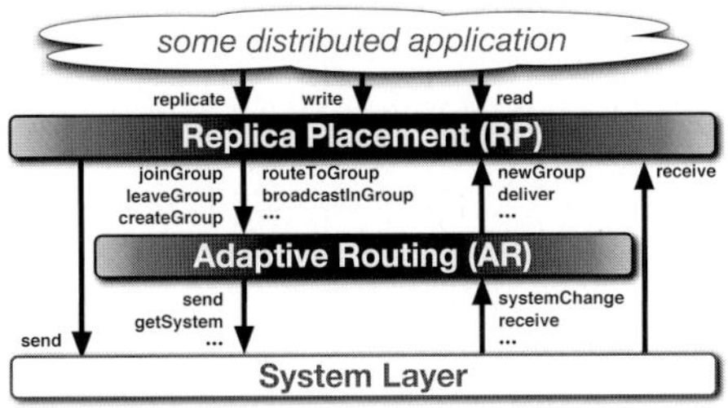

Fig. 2. Solution Architecture

reaching its destination, given the failure probabilities of system. Furthermore, as stated in Section 1, this paper is primarily focused on replica placement; replica consistency is out of its scope. As a consequence, the routing layer ensures no ordering guarantees on requests. Such a property if desired can be built on top of our system.

3 Adaptive Request Routing

For the sake of simplicity, we start by describing the lower layer of our solution: Adaptive Request Routing (AR). AR offers a reliable routing solution based on an underlying reliable overlay tree covering the whole system S named MRT. A group represents a connected subtree of this MRT. This group changes over time and our AR solution adapts accordingly. All group changes[4] are assumed to happen only at the group subtree leaves. That is, only a leaf of a group subtree can leave the group and a new node becomes a leaf of this subtree.

3.1 Interface

To create or change a group, AR provides the following primitives:

- *createGroup(gid)* enables a node to create a group of unique identifier *gid* and to be the first member of this group.
- *joinGroup(gid)* enables a node not in the group *gid* and with only **one neighbor in the group** to join this group and to be able to receive messages sent to the group members and to broadcast messages to the group members.
- *leaveGroup(gid)* called by a node in the group with only **one neighbor in the group**, it enables such a node to leave the group of id *gid* and thus stop receiving messages sent to the group.

For each group, once created, AR provides two services: (1) adaptive and reliable message routing from any node outside the group to some node in the group, (2) adaptive and reliable broadcasting among nodes in the group. These services are respectively encapsulated in the *routeToGroup(gid,m)* and *broadcastInGroup (gid,m)* primitives:

[4] By group change, we refer to an explicit leave/join to the group and not to node failure/recovery.

- *routeToGroup(gid,m)* enables a node to reliably send a message m to any member of the group *gid*.
- *broadcastInGroup(gid,m)* enables a node in group *gid* to reliably broadcast a message m to all members in the group.
- *deliver(gid, m)* works as a callback and enables a node to receive application message m.
- *newGroup(gid)* works as a callback and enables a node to receive an announcement of the new group *gid*.
- *resetGroup(gid)* works as a callback and enables a node to receive an announcement of a reset of the group *gid*, such a reset is due to an environment change.

The goal of AR is to take into account the environment unreliability by routing messages through a reliable tree named the Maximum Reliability Tree (MRT). It also adapts its communication services to group and environment changes.

3.2 Routing Algorithm

Algorithm 1 describes the main primitives provided by our communication layer. To create a group with a unique identifier *gid*, a process p_k calls the *createGroup (gid)* primitive. This primitive starts by announcing the new group to all members in the system. To do so, it broadcasts an initial message NEWGROUP through an underlying reliable tree overlay dedicated to this group. To that end, p_k first builds a tree *mrt* covering the whole system S using the $mrt()$ primitive (line 9). This primitive is responsible for building the MRT with the root passed as argument (here p_k). If the *mrt* at p_k was already computed to serve other groups, p_k simply assigns it as the tree to serve group *gid* (line 10). The MRT of a process p_k contains the most reliable paths in S connecting p_k to all other processes in Π. Defined in [4] to ensure a reliable broadcast, MRT materializes the reliable aspect of our communication model. This paper does not detail the construction technique of MRT (see [4] for details). Then, p_k calls the *propagate()* primitive to launch the broadcast of the NEWGROUP (line 11). When a process p_k receives NEWGROUP message (line 15) from a neighbor p_j, it becomes aware of the group and knows how to route messages to it. To that end, p_k saves the tree overlay built to serve the new group *gid*: $T[gid]$ and a routing direction $direction[\,]$ towards the group in $T[gid]$, which is the neighbor that forwards the NEWGROUP message to it (here p_j) (lines 16 & 17). Note that, at the source of the NEWGROUP message, this direction is set as the process itself as it represents the first member of the group (line 12). At the source and the receivers of the NEWGROUP message, a callback to the *newGroup()* primitive is performed to announce the new group to the upper layer (lines 14 & 18). Note that at any time, each member of a group *gid* knows all other members of this group. At the initialization process, the group consists only of the process that launched the NEWGROUP broadcast (line 13). Such a knowledge will be updated as soon as members start to join or leave the group.

A process p_k not in the group *gid* aiming at routing a message to any member of *gid* calls the $routeToGroup()$. This primitive simply sends the indicated message to the p_k's direction to the aimed group: $direction[gid]$ (line 24).

```
initialization:
   S ← getSystem()
   T ← ∅                                              {set of overlay trees}
   mrt ← ⊥
   direction ← ∅
   group ← ∅

procedure createGroup(gid)
   if mrt = ⊥ then
      mrt ← mrt(S, {p_k})
   T[gid] ← mrt
   propagate(T[gid], p_k, gid, NEWGROUP)
   direction[gid] ← p_k                               {root of gid is p_k}
   group[gid] ← {p_k}
   newGroup(gid)

upon receive(T_i, NEWGROUP, gid) via p_j do
   T[gid] ← T_i
   direction[gid] ← p_j
   newGroup(gid)
   propagate(T_i, p_j, gid, NEWGROUP)

procedure broadcastInGroup(gid,m)
   broadcast(gid, m)
   deliver(gid,m)

procedure routeToGroup(gid, m)
   send(gid, m) to direction[gid]

procedure joinGroup(gid)
   p_j ← direction[gid]
   send(FETCH-GROUP, gid) to p_j
   wait until receive(GROUP, gid, group_j) from p_j
   group[gid] ← group_j ∪ {p_k}
   broadcast(gid,JOIN)

upon receive(FETCH-GROUP, gid) from p_j do
   send(GROUP, gid, group[gid]) to p_j

upon receive(T_i, JOIN,gid) from p_i via p_j do
   group[gid] ← group[gid] ∪ {p_i}
   propagate(T_i, p_j, gid, JOIN)

procedure leaveGroup(gid)
   if direction[gid] = p_k then
      let p_j ∈ neighbors(gid) ∩ group[gid]
      send(NEWROOT, gid) to p_j
      direction[gid] ← p_j
   broadcast(gid,LEAVE)
   group[gid] ← ⊥

upon receive(NEWROOT, gid) do
   direction[gid] ← p_k

upon receive(T_i, LEAVE,gid) from p_i via p_j do
   group[oid] ← group[oid] \ {p_i}
   propagate(T_i, p_j, gid, LEAVE)

function group(gid)
   return group[gid]

function neighbors(gid)
   return {p_j : p_j ∈ V(T[gid]) ∧ l_{j,k} ∈ E(T[gid])}
```

Algorithm 1. Routing algorithm at p_k – Basic primitives

The second service provided by our communication layer is encapsulated in the *broadcastInGroup*() primitive. When called, this primitive calls the *broadcast()* primitive (line 21). The *broadcast()* primitive is called by a node in a group *gid* to broadcast a message among other members of *gid*.

To ensure an up-to-date knowledge about the group at each of its members, any join or leave event in this group is propagated to all members of the group. To join group *gid*, a node p_k calls the *joinGroup()* primitive. This primitive permits to p_k to obtain a view from the group it intends to join and to announce its arrival to other group members by calling *broadcast()* primitive to send a JOIN message to other members of the group.

To leave the group *gid*, a node p_k calls the *leaveGroup()* primitive. To ensure the correctness of the routing solution, each node p_k leaving a group *gid* has to check if it is the root of $T[gid]$ (line 37). If yes, p_k has to yield its root status to a neighbor in the group (line 38). Finally, p_k informs other members of the group about its leaving by sending a LEAVE message using the *broadcast()* primitive.

Algorithm 2 includes the *broadcast()* and *propagate()* primitives and callbacks delivering upper layer messages. The *broadcast()* primitive can be called by any member of a group *gid* to diffuse a message m to other members of the group. This primitive is used to diffuse both local messages to our communication layer (e.g., JOIN & LEAVE) and messages of the upper layer given by a call to the *broadcastInGroup()* primitive. This primitive first extracts T' as the subtree of $T[gid]$ covering the group (line 2). It then calls the *propagate()* primitive to send m through T' (line 3).

```
1:  procedure broadcast(gid, m)
2:     let T' ⊂ T[gid] : p_i ∈ T' ⇒ p_i ∈ group[gid]
3:     propagate(T', p_k, gid, m)

4:  procedure propagate(T', p_j, gid, m)
5:     for all p_i : link l_{k,i} ∈ E(T') ∧ j ≠ i do
6:        send(T', gid, m) to p_i

7:  upon receive(T', gid, m) from p_j do
8:     propagate(T', p_j, gid, m)
9:     deliver(gid, m)

10: upon receive(gid, m) from p_j do
11:    if (p_k ∈ group[gid]) then
12:       deliver(gid, m)
13:    else
14:       send(gid, m) to direction[gid]
```

Algorithm 2. Routing algorithm at p_k – Dissemination mechanism

3.3 Handling Failures and Configuration Changes

While being probabilistically reliable, the above communication algorithm does not ensure adaptiveness to the environment changes. Its adaptiveness is focused on group changes and on taking into account the environment unreliability. To enhance the adaptiveness of our communication solution, we extend AR as shown in Algorithm 3. This extension allows AR layer to adapt to failures that may change the system topology and to take into account new configurations, i.e., new links and processes reliability. To adapt to environment changes, either regarding the topology or the configuration, we assume, as shown in Figure 2, that AR is notified by an underlying *System Layer*. The System Layer ensures at each process, the availability of an up-to-date view about the system. The details about how this layer obtains this view can be found in [4].

When a node p_k is notified by a new system view (line 1), if p_k is the root of at least one tree of one group (line 5), then p_k has the responsibility to define a new tree for that group in order to cover the new system configuration.

By building the new tree, p_k may break the connectivity property of the group members. To reconnect the group members in a subtree of $T[gid]$, p_k defines a new set of members of the group *gid*. This group is the set of nodes in the smallest subtree of $T[gid]$ including all old members of the group. Thus, some of the new group members were previously in the group, others are added by this mechanism only to heal the subtree connecting the group members. Then, p_k calls the *propagate()* primitive to disseminate the NEWTREE message annoucing the new tree $T[gid]$ and the new reconnected group to all nodes in S (line 9). When receiving a NEWTREE message from a neighbor p_j for a group *gid*, a process p_k assigns to its $T[gid]$ the given new tree T_{gid} (line 12) and changes its routing direction to p_j (line 13). Then, p_k checks if it was added to the new group while not being previously in the old one (lines 14 & 15). In this case, p_k calls the *forceJoin* primitive to inform the upper layer about this forced join. For this, p_k indicates p_j as the neighbor in the group. Note that p_j is in the group because it is the sender of the NEWTREE message indicating to p_k that it has to join the group. Thus p_j is either a previous member of the group or a new member forced in its turn to join the group in order to reconnect it. Section 4.1, details this further.

As a member of the group, p_k integrates the new group in its group view (line 17). Finally the source and the receivers of the NEWTREE message, call the *resetGroup()* primitive to announce the group change to the upper layer (lines 10 & 19).

1: **upon** systemChange(S') **do**
2:   $S \leftarrow S'$
3:   **if** $\exists\ gid : direction[gid] = p_k$ **then**
4:     $mrt \leftarrow mrt(S, \{p_k\})$
5:     **for all** $T[gid] \in T\ :\ direction[gid] = p_k$ **do**
6:       $T[gid] \leftarrow mrt$
7:       **let** T' **be the smallest subree of** $T[gid]$ **such that** $group[gid] \subset V(T')$
8:       $group[gip] \leftarrow V(T')$
9:       propagate($T[gid]$, p_k, gid, NEWTREE, $group[gid]$)
10:       resetGroup(gid)

11: **upon** receive(T_{gid}, NEWTREE, gid, $group_{gid}$) **via** p_j **do**
12:   $T[gid] \leftarrow T_{gid}$
13:   $direction[gid] \leftarrow p_j$
14:   **if** $p_k \in group_{gid}$ **then**
15:     **if** $p_k \notin group[gid]$ **then**
16:       forceJoin(gid, p_j)
17:     $group[gid] \leftarrow group_{gid}$
18:   propagate(T_{gid}, p_j, gid, NEWTREE, $group_{gid}$)
19:   resetGroup(gid)

Algorithm 3. Routing algorithm at p_k – Adaptiveness to failures

Root failure. Note that in our AR solution, the root failure is problematic, as it represents the responsible for creating new covering tree if any changes happen. In addition, if such a root is a singleton, its failure results in the object being inaccessible. A solution for the singleton failure was proposed in [19], which we also retain in this paper. The idea is to impose a rule to the replica placement algorithm so that at any time at least two replicas of an object must be available.

When it comes to the root failure, several solutions were proposed in this context [6,9]. Similarly, we can replicate the root to improve its reliability. Details of this strategy could be found in [6].

4 Adaptive Replica Placement

The Replica Placement (RP) layer defines, for each object o to replicate, a replication scheme R changing according to the read-write pattern to o in order to move R towards the center of the read-write activity. When the read-write pattern is stable, R eventually converges towards the optimal replication scheme ensuring the minimum communication cost.

4.1 Initialization and Replica Access

The RP layer relies on the AR layer for each communication step. For each object o, AR layer manages a group that corresponds to the set of processes holding a replica of o, i.e., R. In other words, our replication scheme R at RP is seen at AR as a group of processes to which AR provides a set of communication services. RP refers to AR to get information about the neighborhood in the overlay defined by AR to serve the group of one object o by calling the *neighbors()* primitive. It also refers to AR to get a view of R using the *group()* primitive. To adapt the underlying communication solution to the R changes (i.e., to the group changes), RP informs AR about all changes in R by calling the *joinGroup*() and *leaveGroup*() primitives.

Algorithm 4 details the primitives provided by RP to any upper application. The management of replicas of an object o starts by a call to the *replicate*() primitive by the initial process p_k holding o. In this initialization step, p_k calls the *createGroup*() primitive (line 7) of the AR by indicating *oid* as the identifier of the object o for which p_k wants to create a replication scheme (or a group). As detailed in Section 3.2, the *createGroup*() primitive reliably announces the object o to all processes in the system and creates a group dedicated to this object.

The following replica placement steps (i.e., replica creation or replica discarding) are then performed cooperatively based on statistics collected locally at each process concerning the received requests. At a process p_k, for each object o of identifier *oid*, *reads[oid]* and *writes[oid]* respectively refer to the total number of reads and the total number of writes p_k received for the object o. These counters also include for each neighbor p_j of p_k (according to the overlay defined by AR to serve the group dedicated to o) the number of reads, *reads[oid,* p_j*]* and the number of writes, *writes[oid,* p_j*]* received from p_j for the object o.

```
1: initialization:
2:    reads ← ∅                                        {set of read counters}
3:    writes ← ∅                                       {set of write counters}
4:    Ω ← ∅                                            {set of local objects}
5:    leavePending ← ∅                                 {set of boolean}

6: procedure replicate(o)
7:    createGroup(o.id)

8: function read(oid) : state
9:    if p_k ∈ group(oid) then
10:      return Ω[oid].state
11:   else
12:      routeToGroup(oid,READ)
13:      wait until receive(RESPONSE,o) with o.id = oid
14:      return o.state

15: procedure write(oid, state)
16:   if p_k ∈ group(oid) then
17:      broadcastInGroup(oid,WRITE,state)
18:   else
19:      routeToGroup(oid,WRITE,state)

20: upon newGroup(oid) ∨ resetGroup(oid) do
21:   for all p_j ∈ neighbors(oid) do
22:      reads[oid, p_j] ← 0
23:      writes[oid, p_j] ← 0
24:   leavePending[oid] ← false

25: upon forceJoin(oid, p_j) do
26:   send(FETCH-OBJECT, oid) to p_j
27:   wait until receive(OBJECT, o) from p_j with o.id = oid
28:   Ω[oid] ← o

29: upon receive (FETCH-OBJECT, oid) from p_j do
30:   send(OBJECT, Ω[oid]) to p_j

31: upon deliver(oid,READ) from p_i via p_j ∈ neighbors(oid) do
32:   send (RESPONSE,Ω[oid]) to p_i
33:   reads[oid] ← reads[oid] + 1
34:   reads[oid, p_j] ← reads[oid, p_j] + 1

35: upon deliver(oid,WRITE,state) via p_j ∈ neighbors(oid) ∪ {p_k} do
36:   if p_j ∉ group(oid) then
37:      broadcastInGroup(oid,WRITE,state)
38:   else
39:      Ω[oid].state = state
40:      writes[oid] ← writes[oid] + 1
41:      writes[oid, p_j] ← writes[oid, p_j] + 1
```

Algorithm 4. Adaptive replication at p_k – Reading & Writing

When notified of a new group (line 20), process p_k becomes aware of the replication scheme R of the corresponding object o and initializes its counters *reads[ ]* and *writes[ ]* according to its set of neighbors defined by AR (lines 22 & 23).

To read or write a state at o, p_k calls, respectively, the *read()* and *write()* primitive. If p_k has a replica of o, i.e., it is a member of the group dedicated to o (lines 9 & 16), the *read()* function simply returns the state of the replica extracted from the local structure Ω. The *write()* function, in this case, calls the $broadcastInGroup()$ primitive (line 37) of the AR algorithm to broadcast the update within processes of the replication scheme of o. Otherwise, these primitives call the $routeToGroup()$ primitive to route the request to a process holding a replica of o.

When a process p_k receives a write or a read request for *oid* (lines 35 & 31), it respectively updates its local replica (line 39) or sends back the response extracted from its replica (line 32), then updates its counters accordingly (lines 40 & 41 - 33 & 34).

To suport the adaptiveness to environment changes, our RP algorithm provides the *forceJoin()* primitive (line 25). This primitive permits to an underlying

```
1:  periodically do :
2:    for all oid ∈ Ω do
3:      if p_k ∈ group(oid) then
4:        if ¬ tryExpanding(oid) then
5:          if group(oid) = {p_k} then
6:            trySwitching(oid)
7:          else if | neighbors(oid) ∩ group(oid) | = 1 then
8:            tryContracting(oid)
9:        reads[oid] ← 0
10:       writes[oid] ← 0

11: function tryExpanding(oid) : boolean
12:   success ← false
13:   candidates ← neighbors(oid) \ group(oid)
14:   for all p_j ∈ candidates do
15:     if reads[oid, p_j] > writes[oid] − writes[oid, p_j] then
16:       send(JOIN,Ω[oid]) to p_j
17:       success ← true
18:   return success

19: upon receive (JOIN,o) do
20:   Ω[o.id] ← o
21:   joinGroup(o.id)                                {inform AR about the group change}

22: procedure trySwitching(oid)
23:   if ∃ p_j ∈ neighbors(oid) : 2 × (reads[oid, p_j] + writes[oid, p_j]) > reads[oid] + writes[oid]
      then
24:     send(BE-SINGLETON,Ω[oid]) to p_j
25:     wait until receive(ACK-SINGLETON,oid) from p_j
26:     leaveGroup(oid)                              {inform AR about the group change}
27:     Ω[oid] ←⊥

28: upon receive (BE-SINGLETON,o) do
29:   Ω[o.id] ← o
30:   joinGroup(o.id)                                {inform AR about the group change}
31:   send(ACK-SINGLETON,o.id)

32: procedure tryContracting(oid)
33:   let p_j ∈ neighbors(oid) ∩ group(oid)
34:   if writes[oid, p_j] > reads[oid] then
35:     leavePending[oid] ← true                     {Trying to contract from oid}
36:     send(REQUEST-LEAVE,oid) to p_j
37:     wait until receive(REPLY-LEAVE,reply,oid) from p_j
38:     if reply then
39:       leaveGroup(oid)                            {inform AR about the group change}
40:       Ω[oid] ←⊥
41:     leavePending[oid] ← false

42: upon receive(REQUEST-LEAVE,oid) from p_j do
43:   if ¬leavePending[oid] then
44:     send(REPLY-LEAVE,true,oid)                   {Not in contract test from oid}
45:   else
46:     reply ← p_k > p_j
47:     send(REPLY-LEAVE,reply,oid)
```

Algorithm 5. Adaptive replication at process p_k – Replicas Placement communication layer to add a member to one replication scheme R. When called at a node p_k, this primitive fetches a copy of the object with the indicated unique identifier from the indicating node p_j as a neighbor in R. It then includes o to become a member of its replication scheme (line 28).

4.2 Adaptive Placement

As soon as the requests for object o start to be submitted, the RP solution adapts the replication scheme R to the read-write pattern in a decentralized manner. Starting as a singleton, R may expand (by placing new replicas in appropriate processes), switch (by changing the replica holder if R is a singleton) or contract (by retrieving replicas from a specific process, if R is not a singleton) while remaining connected. These actions are tested periodically by some processes in R based on a set of statistics concerning the received requests. Algorithm 5 gives a formalization of the adaptive replica placement detailed in [19]. Hereafter we describe this algorithm and its interaction with AR layer.

The expansion test is executed by each process p_k in R (line 3) with at least one neighbor not in R (line 13). To do so, for each neighbor p_j not in R (line 14), p_k sends a 'Join' request to p_j (line 16) if the number of reads that p_k received from p_j is greater than the total number of writes received from elsewhere (line 15). When a node p_k joins R, it informs AR about the new member of the group by calling the *joinGroup*() primitive (line 21).

If the expansion test fails (line 4) and p_k is the singleton (line 5), it executes the switch test (line 6). To switch, p_k should first find a neighbor p_j such that the number of requests received by p_k from p_j is greater than the number of all other requests received by p_k (line 23). Then p_k sends a copy of the object o to p_j with an indication that p_j becomes the new singleton of R (line 24). Before discarding its local replica (line 27) p_k must receive a confirmation from p_j to ensure a non empty replication scheme (line 25). To inform AR, p_k and p_j respectively call the *leaveGroup*() and *joinGroup*() primitives (lines 26 and 30).

The contraction test is also executed after a failed expansion test but when p_k has only one neighbor in R: p_j (line 7). To contract from R, p_k should have more writes from p_j than all received reads (line 34), In this case, p_k requests permission from p_j to leave R (line 36). If permitted (line 38), p_k discards its local replica (line 40) and informs AR (line 39). The permission request sent to p_j permits to manage a possible risk of mutual contractions. If not managed, such a risk may induce an empty replication scheme when p_k and p_j constitute all elements of R. When detected (line 45) such a conflict is resolved by a simple mechanism permitting to the process with the lower id to leave R (line 46).

5 Evaluation

The efficiency of the adaptive replica placement algorithm has been previously proved in [19]. In this section we are more concerned by the immersion of this algorithm in unreliable environment. That is, we aim at evaluating the advantage of using AR (based on MRT as the network overlay) as the underlying communication model of the RP solution.

Evaluation method. To evaluate the advantage of our reliable communication model, we define a comparison tree to MRT as a covering tree built independently from links reliability. Such a tree could be the minimum latency tree defined in [19]. We name this comparison tree SPT. SPT is any covering tree built without taking into account the reliability of the system components (links & processes). To compare the impact of MRT and SPT in our data replication solution, we define an evaluation method that consists of measuring the cost needed to completely hide the impact of the system unreliability. Indeed, to converge towards the same replication scheme as in a reliable environment, a simple idea is to resend each request until receiving an acknowledgement (ACK) for it. The number of retransmissions depends on the reliability of the link through which the request should be sent. In a reliable link, an ACK is received after the first request, in which case no retransmission is needed.

To show the benefit of using our MRT as the overlay of our communication model, we compare the retransmission needed when using both MRT and SPT to hide the environment unreliability. This comparison regards the communication cost, the number of messages retransmitted and its corresponding percentage as the portion of sent messages that represents the retransmissions.

Simulation configuration. To evaluate our solution, we conducted a set of experiments for various network configurations with 100 processes connected randomly with an average connectivity of 8, i.e., 8 direct neighbors. To simplify our results interpretation, we varied the links configuration L_i while assuming that processes are reliable i.e, $\forall p_i : P_i = 0$. At each experiment, we simulate the replication of one object and we assign to each process a fixed number of read and write requests that it submits periodically. The initial copy of the object to replicate is held by the node that initiates the replication, which is chosen randomly among the system nodes. We then measured the number of propagated messages (initial messages and retransmitted messages) at the convergence, i.e., when R stabilizes. In our simulation, we also assume that ACK messages are subject to loss.

5.1 AR Benefit

Figure 1 (a) shows the communication cost (in terms of all routed messages) of our solution when using both MRT and SPT at the convergence. The communication cost includes the number of requests submitted and the number of retransmitted messages (request or ACK) necessary to hide the unreliability. In Figure 1 (a), we show the average communication cost over executions, the worst and the best cases that our executions sample detected. In the corresponding executions we assigned, at each process, a periodic number of *Reads* in [0, 50] and a periodic number *Writes* in [0, 50]. When using MRT our solution induces a lower communication cost. Globally this cost increases as the link unreliability increases. This shows that as the reliability worsens, more message retransmission is needed. Note that the number of original requests (not including retransmissions) to route at the convergence is different for MRT and SPT since the generated replication scheme R is different. Indeed, the form of the tree influences significantly the resulting replication scheme as each tree has a

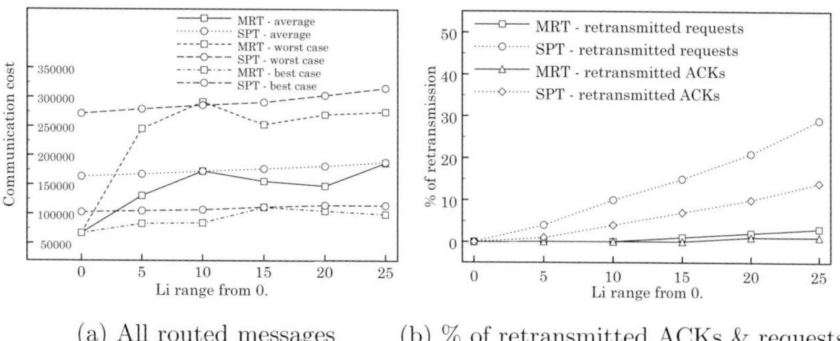

(a) All routed messages (b) % of retransmitted ACKs & requests

Fig. 1. Communication cost

different origin of requests reaching its replication scheme. For this reason, hereafter, we focus our evaluations on the induced retransmission to hide the system unreliability.

Figure 1 (b) shows the percentage of request messages and ACK messages retransmitted to ensure that every inter-nodes message (e.g., forwarded requests) is received when using the MRT and SPT trees. As shown in Figure 1 (b), this percentage is lower when using MRT than when using SPT. This difference increases as the links unreliability is in a larger range to reach the 15% when the links unreliability L_i is in [0 - 25%].

5.2 Varying Read/Write Pattern

In this section, we vary the range of fixed *Reads* and *Writes* assigned to nodes in order to evaluate their impacts on the retransmission cost. The indicated values of *Reads* and *Writes* in figures below represent the lower bound of a range of size 10. Figure 2 shows the percentage of requests retransmission needed to hide the system unreliability using MRT and SPT. As noticeable the variation of the Read/Write pattern has no impact on the percentage of retransmitted requests.

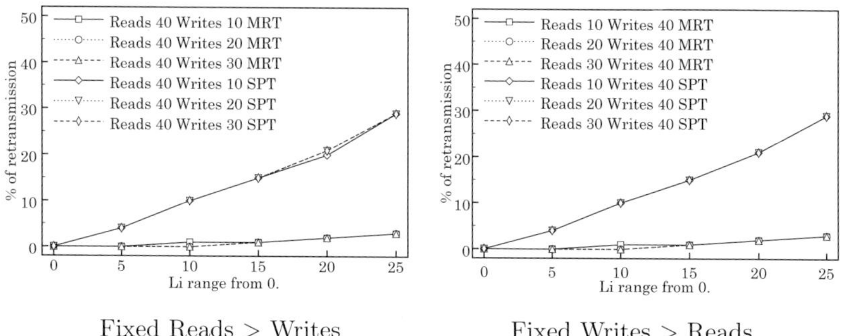

Fixed Reads > Writes Fixed Writes > Reads

Fig. 2. % of retransmitted requests to hide unreliability

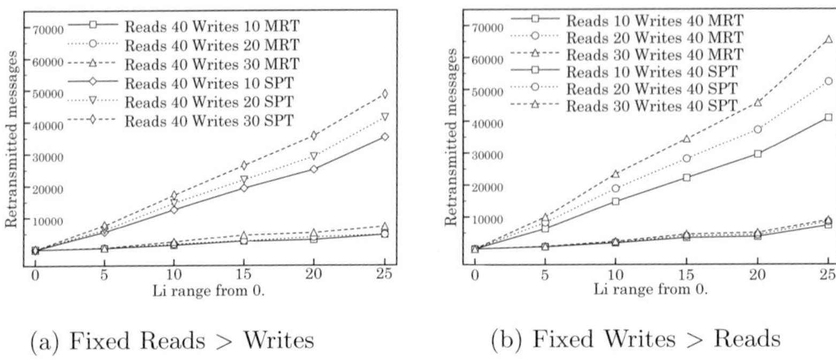

(a) Fixed Reads > Writes (b) Fixed Writes > Reads

Fig. 3. Number of retransmitted messages

This is however not the case for their corresponding number of retransmissions shown in Figure 3. The number of retransmissions needed to completely hide the links unreliability when using SPT is much higher than the one needed when using MRT. This difference increases as the links unreliability increases to reach 7 times more retransmissions when the links unreliability is in [0 - 25%]. Using both overlays structure, the number of retransmissions increases as the number of requests ($Reads + Writes$) increases. For the same fixed number of periodic $Reads$ at each process, when the number of periodic $Writes$ increases it induces more messages retransmissions than when we increase the number of $Reads$ while fixing the number of $Writes$. This is due to a different replication scheme at the convergence. In Figure 3 (a), we have more $Reads$ than $Writes$, which implies a large replication scheme through which each submitted $Write$ is broadcast. The larger the replication scheme, the higher is the cost induced by the broadcast of $Writes$. And as the global number of $Writes$ increases, this cost increases. In Figure 3 (b), we have more $Writes$ than $Reads$, which implies a small replication scheme. For the same number of global $Writes$, when the number of $Reads$ increases the number of propagated messages increases by the propagation of each $Read$ through a tree branch until reaching the replication scheme. This is less costly than a broadcast through a large replication scheme.

6 Related Work

Research efforts have considered distributed adaptive data replication systems from many different angles. Many previous works on adaptive data replication have considered user requests of some sort as the parameter to adapt to. In doing so, most of them formalize the problem as a cost function to optimize. The precise interpretation of adaptive data replication depends on the cost function. In [19], the replica management adapts to the read-write pattern. Based on the approach defined in [19], in this paper, our objective is to adapt to the read-write pattern while coping with an unreliable environment. While handling process and link failure-recovery the approach proposed [19] does not take into account the environment unreliability in its communication model. In [19], the link or process failure is handled by switching the execution in a failure mode where the replication scheme of each object is a singleton named *primary processor*. The execution then returns to the normal mode when the failed component (link or process) recovers. In our paper, however, the environment unreliability is adressed in a preventive way since our communication model selects apriori the most reliable paths to serve the requests routing in order to minimize the component failures risk. In addition, when a failure happens, our replication scheme is simply reconnected instead of being reduced to a singleton. The environment unreliability in this context was adressed also in a preventive way in [10] but in a different manner than ours. The approach defined in [10], distributes replicas in locations (or servers) whose failures are not correlated in order to mitigate the impact of correlated, shared-component failures. The adaptiveness of this approach relies on the proposed placement strategy as the number of replicas to place is assumed to be fixed by the storage system. Similarly to our paper, some works [17,18] integrate a communication model for their adaptive data-replication solutions. Other works have also adapted to the read-write pattern, e.g., in [7,16]. In addition to the read-write pattern, several objectives were defined to dictate the replication strategy. The approach defined in [7] also takes into consideration storage costs and node capacity to serve requests. In [11], a protocol dynamically replicates data so as to improve the client-server proximity without overloading any of the replica holder. In [3], the replica allocation aims to balance loads in terms of CPU and disk utilization in order to increase the system throughput. In [14], the replica placement strategy aims to minimize a cost function taking into account the average read latency, the average write latency and the amount of bandwidth used for consistency enforcement. Contrary to the distributed definition of the replication scheme of this paper, in [14] the definition of the placement configuration is done at a central server (the origin server) evaluating the cost of different possible configurations to select the best placement configuration yielding the least cost. Similarly, in [12] the placement configuration changes are decided at a central site based on statistics about the accesses data measured in a distributed manner. The starting point in [12] is also different than the one defined in this paper. That is, in [12], at system start-up, a full copy is available at each edge server. Throughout execution, a self-optimization algorithm is triggered periodically.

7 Conclusion

This paper proposed an adaptive replica placement solution for unreliable environments. The adaptiveness of this solution is at the replica placement level and the request routing level. In our evaluation, we showed that the unreliability impact on our replica placement algorithm could be hidden with a minor cost using our reliable communication model. However real-world deployment in WAN suffers from other contraints. One major constraint to be considered is the limited memory and CPU power at processes. Thus, assuming that each process has a global knowledge prevents this solution to scale in such environment.

References

1. Amir, Y., Tutu, C.: From total order to database replication. In: Proceedings of ICDCS, pp. 494–503. IEEE, Los Alamitos (2002)
2. Bernstein, P.A., Goodman, N., Wong, E., Reeve, C.L., Rothnie Jr., J.B.: Query processing in a system for distributed databases (sdd-1). ACM Trans. Database Syst. 6(4) (1981)
3. Elnikety, S., Dropsho, S.G., Zwaenepoel, W.: Tashkent+: memory-aware load balancing and update filtering in replicated databases. In: Euro. Sys., pp. 399–412 (2007)
4. Garbinato, B., Pedone, F., Schmidt, R.: An adaptive algorithm for efficient message diffusion in unreliable environments. In: Proceedings of IEEE DSN (2004)
5. Holliday, J., Agrawal, D., El Abbadi, A.: The performance of database replication with group multicast. In: Proceedings of FTCS, pp. 158–165. IEEE Computer Society Press, Los Alamitos (1999)
6. Jannotti, J., Gifford, D.K., Johnson, K.L., Kaashoek, M.F., O'Toole Jr., J.W.: Overcast: Reliable multicasting with an overlay network. In: Proceedings of OSDI (October 2000)
7. Kalpakis, K., Dasgupta, K., Wolfson, O.: Optimal placement of replicas in trees with read, write, and storage costs. IEEE Trans. Parallel Distrib. Syst. 12(6) (2001)
8. Kemme, B., Bartoli, A., Babaoglu, Ö.: Online reconfiguration in replicated databases based on group communication. In: DSN, pp. 117–130 (2001)
9. Kostic, D., Rodriguez, A., Albrecht, J., Bhirud, A., Vahdat, A.: Using random subsets to build scalable network services. In: Proceedings of USITS (March 2003)
10. MacCormick, J., Murphy, N., Ramasubramanian, V., Wieder, U., Yang, J., Zhou, L.: Kinesis: A new approach to replica placement in distributed storage systems. ACM Transactions on Storage (TOS) (to appear)
11. Rabinovich, M., Rabinovich, I., Rajaraman, R., Aggarwal, A.: A dynamic object replication and migration protocol for an internet hosting service. In: ICDCS (1999)
12. Serrano, D., no-Martínez, M., Jiménez-Peris, P.R., Kemme, B.: An autonomic approach for replication of internet-based services. In: SRDS, Washington, DC, USA, pp. 127–136. IEEE Computer Society, Los Alamitos (2008)
13. Serrano, D., Patiño-Martínez, M., Jiménez-Peris, R., Kemme, B.: Boosting database replication scalability through partial replication and 1-copy-snapshot-isolation. In: PRDC, pp. 290–297 (2007)
14. Sivasubramanian, S., Alonso, G., Pierre, G., van Steen, M.: GlobeDB: Autonomic data replication for web applications. In: Proc. of the 14th International World-Wide Web Conference, Chiba, Japan, pp. 33–42 (May 2005)
15. Stonebraker, M.: The design and implementation of distributed ingres. In: The INGRES Papers (1986)
16. Tsoumakos, D., Roussopoulos, N.: An adaptive probabilistic replication method for unstructured p2p networks. In: OTM Conferences, vol. (1) (2006)
17. van Renesse, R., Birman, K.P., Hayden, M., Vaysburd, A., Karr, D.A.: Building adaptive systems using ensemble. Softw., Pract. Exper. 28(9) (1998)
18. Vaysburd, A., Birman, K.P.: The maestro approach to building reliable interoperable distributed applications with multiple execution styles. TAPOS 4(2) (1998)
19. Wolfson, O., Jajodia, S., Huang, Y.: An adaptive data replication algorithm. ACM Trans. Database Syst. 22(2) (1997)
20. Wolfson, O., Milo, A.: The multicast policy and its relationship to replicated data placement. ACM Trans. Database Syst. 16(1) (1991)

FT-OSGi: Fault Tolerant Extensions to the OSGi Service Platform*

Carlos Torrão, Nuno Carvalho, and Luís Rodrigues

INESC-ID/IST
carlos.torrao@ist.utl.pt, nonius@gsd.inesc-id.pt, ler@ist.utl.pt

Abstract. The OSGi Service Platform defines a framework for the deployment of extensible and downloadable Java applications. Many of the application areas for OSGi have significant dependability requirements. This paper presents and evaluates FT-OSGi, a set of extensions to the OSGi Service Platform that allow to replicate OSGi services. FT-OSGi supports replication of OSGi services, including state-transfer among replicas, supports multiple replication strategies, and allows to apply a different replication strategy to each OSGi service.

1 Introduction

The OSGi Service Platform [1] (Open Services Gateway initiative) defines a component-based platform for applications written in the Java^{TM} programming language. The OSGi framework provides the primitives that allow applications to be constructed from small, reusable and collaborative components. It was developed with several applications in mind, including ambient intelligence, automotive electronics, and mobile computing. Furthermore, its advantages made the technology also appealing to build flexible Web applications [2].

Many of the application areas of OSGi have availability and reliability requirements. For instance, in ambient intelligence applications, reliability issues have been reported as one of the main impairments to user satisfaction [3]. Therefore, it is of utmost importance to design fault-tolerance support for OSGi.

This paper presents FT-OSGi, a set of fault tolerance extensions to the OSGi service platform. Our work has been inspired by previous work on fault-tolerant component systems such as Delta-4 [4], FT-CORBA [5,6,7] and WS-Replication [8], among others. Our solution, however, targets problems that are specific for the OSGi platform. More precisely, the proposed solution enriches the OSGi platform with fault tolerance by means of replication (active and passive) in an almost transparent way to the clients, keeping the same properties already provided by the OSGi platform.

A prototype of FT-OSGi was implemented. This prototype leverages on existing tools, such as R-OSGi [9] (a service that supports remote accesses to OSGi

* This work was partially supported by the FCT project Pastramy (PTDC/EIA/72405/2006).

R. Meersman, T. Dillon, P. Herrero (Eds.): OTM 2009, Part I, LNCS 5870, pp. 653–670, 2009.

services) and the Appia group communication toolkit [10] (for replica coordination). The resulting FT-OSGi framework can be downloaded from sourceforge[1]. The paper also presents an experimental evaluation of the framework, that measures the overhead induced by the replication mechanisms.

The remaining of the paper is structured as follows. The Section 2 describes the related work. The Section 3 presents the FT-OSGi extensions, describing its architecture, system components, how such components interact and how the proposed extensions are used by applications. The Section 4 presents an evaluation of FT-OSGi. Finally, the Section 5 concludes the paper and points to future research.

2 Related Work

This section makes a brief overview of OSGi and of the fault-tolerance techniques more relevant to our work. Then we overview previous work on fault-tolerant distributed component architectures from which the main ideas were inherited, including Delta-4, FT-CORBA and WS-Replication. Finally, we refer previous research that has addressed specifically the issue of augmenting OSGi with fault-tolerant features.

OSGi. The OSGi Framework forms the core of the OSGi Service Platform [1], which supports the deployment of extensible and downloadable applications, known as *bundles*. The OSGi devices can download and install OSGi bundles, and remove them when they are no longer required. The framework is responsible for the management of the bundles in a dynamic and scalable way. One of the main advantages of the OSGi framework is the support for the bundle "hot deployment", i.e., the support to install, update, uninstall, start or stop a bundle while the framework is running. At the time of writing of this paper, it is possible to find several implementations of the OSGi specification, such as Apache Felix [11], Eclipse Equinox [12] and Knopflerfish [13]. In many application areas, bundles provide services with availability and reliability requirements. For instance, in ambiance intelligence application, a bundle can provide services to control the heating system. Therefore, it is interesting to search for techniques that allow such services to be deployed and replicated in multiple hardware components, such that the service remain available even in the presence of faults.

Fault tolerance. A dependable computing system can be developed by using a combination of the following four complementary techniques: fault prevention, fault tolerance, fault removal and fault forecasting. This paper focus on fault-tolerance. Fault tolerance in a system requires some form of redundancy [14], and replication is one of the main techniques to achieve it. There are two main replication techniques [15]: passive and active replication. In passive replication, also known as primary-backup, one replica, called the *primary*, is responsible for processing and respond to all invocations from the clients. The remaining replicas,

[1] http://sourceforge.net/projects/ft-osgi

called the *backups*, do not process direct invocations from the client but, instead, interact exclusively with the primary. The purpose of the backups is to store the state changes that occur in the primary replica after each invocation. Furthermore, if the primary fails, one of the backup replicas will be selected (using some leader election algorithm previously agreed among all replicas) to play the role of new primary replica. In the active replication, also called the state-machine approach, all replicas play the same role thus there is no centralized control. In this case, all replicas are required to receive requests, process them and respond to the client. In order to satisfy correctness, requests need to be disseminated using a total-order-multicast primitive (also known as atomic multicast). This replication technique has the limitation that the operations processed by the replicas need to be deterministic (thus the name, state-machine).

Object replication. There is a large amount of published work on developing and replicating distributed objects. The Delta-4 [4] architecture was aimed at the development of fault-tolerant distributed systems, offering a set of support services implemented using a group-communication oriented approach. To the authors' knowledge, Delta-4 was one of the first architectures to leverage on group communication and membership technologies to implement several object replication strategies. Delta-4 supported three types of replicated components: *active replication*, *passive replication*, and *semi-active replication* (the later keeps several active replicas but uses a primary-backup approach to make decisions about non-deterministic events).

Arjuna [16] is an object-oriented framework, implemented in C++, that provides tools for the construction of fault-tolerant distributed applications. It supports atomic transactions controlling operations on persistent objects. Arjuna objects can be replicated to obtain high availability. Objects on Arjuna can be replicated either with passive and active replication. Passive replication in Arjuna is implemented on top of a regular Remote Procedure Call (RPC). Failure recovery is done with the help of a persistence storage.

The Common Object Request Broker Architecture (CORBA) [17] is a standard defined by the Object Management Group (OMG), which provides an architecture to support remote object invocations. The main component of the CORBA model is the Object Request Broker (ORB), which act as intermediary in the communication between a client object and a server object, shielding the client from differences in programming languages, platform and physical location. That communication of clients and servers is over the TCP/IP-based Internet Inter-ORB Protocol (IIOP). Several research projects have developed techniques to implement fault-tolerant services in CORBA [5,6,7] eventually leading the design of the FT-CORBA specification [18]. All implementations share the same design principles: they offer fault-tolerance by replicating CORBA components in a transparent manner for the clients. Different replication strategies are typically supported, including active replication and primary-backup. To facilitate inter-replica coordination, the system use some form of group-communication services [19]. To implement recovery mechanisms, CORBA components must be responsible to recover, when demanded, the three kinds of state present in every replicated CORBA object: application state, ORB state (maintained by the ORB) and infrastructure state (maintained by the Eternal [5]). To enable the capture and recover of the application state is necessary that CORBA objects implement Checkpointable interface that contains methods to retrieve (get_state()) and assign (set_state()) the state for that object.

The WS-Replication [8] applies the design principles used in the development of Delta-4 and FT-CORBA to offer replication in the Web Services architecture. It allows client to access replicated services whose consistency is ensured using a group communication toolkit that has been adapted to execute on top of a web-service compliant transport (SOAP).

OSGi replication. To the best of our knowledge, no previous work has addressed the problem of offering fault-tolerance support to OSGI applications in a general and complete manner, although several efforts have implemented one form or another of replication in OSGi. Thomsen [20] presents a solution to eliminate the single point of failure of OSGi-based residential gateways, using a passive replication based technique. However, the solution is specialized for the gateways. In a similar context, but with focus in the services provided through OSGi Framework, Heejune Ahn et al. [21] presents a proxy-based solution, which provides features to monitor, detect faults, recover and isolate a failed service from other service. Consequently, this solution adds four components to the OSGi Framework: proxy, policy manager, dispatcher and monitor. A proxy is constructed for each service instance, with the purpose of controlling all the calls to that service. The monitor is responsible for the state checking of each service. Finally, the dispatcher decides and routes the service call to the best implementation available with the help of the policy manager. In this work, Heejune Ahn et al. only provide fault tolerance to a stateless service, therefore, the service internal state and persistent data are not recovered.

3 FT-OSGi

This section presents FT-OSGi, a set of extensions to the OSGi platform to improve the reliability and availability of OSGi applications. This section shows the services provided by such extensions and how their are implemented, describing the several components of FT-OSGi and how these components interact.

3.1 Provided Services

The FT-OSGi provides fault tolerance to OSGi applications. This is done by replicating OSGi services in a set of servers. To access the services, the client application communicates with the set of servers in a transparent way. The services can be stateless or stateful. State management must be supported by the application programmer: in order to maintain the replicated state, the service must implement two methods, one for exporting its state (`Object getState()`) and another for updating its state (`void setState(Object state)`). The FT-OSGi extensions support three types of replication: active, eager-passive and

Table 1. Examples of configuration options for the proposed architecture

Configuration	Replication	Reply	State	Broadcast	Views
A	Active	First	Stateless	Total regular	Partitionable
B	Passive	First	Stateless	Reliable regular	Primary
C	Active	Majority	Stateful	Total uniform	Primary
D	Passive	First	Stateful	Reliable uniform	Primary

lazy-passive. The strategy used for replication of an OSGi service is chosen at configuration time, and different services with different replication strategies can coexist in the same FT-OSGi domain.

Replication is supported by group communication. Each replicated service may use a different group communication channel or, for efficiency, share a group with other replicated services. For instance, if two OSGi services are configured to use the same replication strategy, they can be installed in the same group of replicas. This solution has the advantage of reducing the number of control messages exchanged by the group communication system (for instance, two replicated services may use the same failure detector module).

When a service is replicated, multiple replies to the same request may be generated. There is a proxy installed in the client that collects the replies from servers and returns only one answer to the application, filtering duplicate replies and simulating the operation of a non-replicated service. The FT-OSGi proxy supports three distinct modes for filtering the replies from servers. In the `wait-first` mode, the first received reply is received and returned immediately to the client, all the following replies are discarded. In the `wait-all` mode, the proxy waits for all the replies from the servers, compares them and returns to the client one reply, if all the replies are equal. If there is an inconsistency in the replies, the proxy raises an exception. Finally, the `wait-majority` returns to the client as soon as a majority of similar replies is received. Distribution and replication is hidden from the clients, that always interact with a local instance of the OSGi framework. Thanks to this approach, the semantic of the OSGi events is maintained. All the events generated by the services and the OSGi framework itself are propagated to the clients.

Table 1 shows some examples of how to configure FT-OSGi applications. It is possible to configure the replication strategy, the filtering mode of server replies, and the operation of the group communication service.

3.2 System Architecture and Components

Figure 1 depicts the FT-OSGi architecture, representing the client and server components. Each node has an instance of the OSGi platform, the R-OSGi extension for distribution, and the FT-OSGi component. The main difference between a client and a server is the type of services installed. The servers maintain the services that will be used by clients. The clients contain proxies that represent locally the services that are installed in the servers. When a client needs to

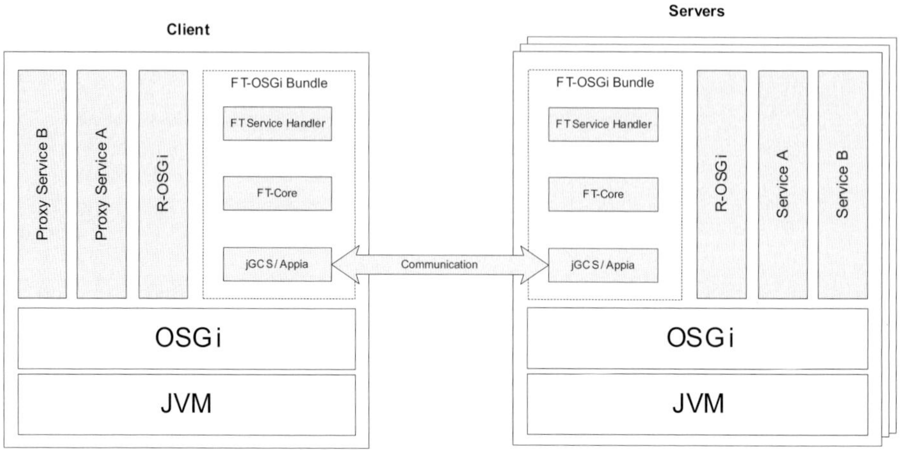

Fig. 1. Architecture of a server and a client

access a service that is installed remotely (in a replicated server), a local proxy is created to simulate the presence of that service.

The FT-OSGi is composed of several building blocks to support the communication between the nodes of the system and to support the consistency between replicas. The building blocks used to support communication and consistency are the R-OSGi and a Group Communication Service (GCS), that are described in the next paragraphs:

R-OSGi. R-OSGi [9] is a platform capable of distributing an OSGi application through several nodes in a network. R-OSGi is layered on top of the OSGi platform in an almost transparent way to applications, being possible to run any OSGi application in the R-OSGi platform with only minor changes on stateful services. The R-OSGi layer uses proxies to represent a service that is running in a remote node. To discover the services that are running in remote nodes, R-OSGi uses the *Service Location Protocol* (SLP) [22]. For each service that is advertised by a node in SLP, when another node needs that service, it creates locally a proxy to represent that service. When the application invokes a method in the proxy, that proxy will issue a remote method invocation in a transparent way to applications.

Group communication service. A Group Communication Service (GCS) provides two complementary services: (*i*) a membership service, that provides information about the nodes that are in the group and generates view changes whenever a member joins, leaves or is detected as failed, and (*ii*) a group communication channel between the nodes that belong to the group membership. The FT-OSGi uses a generic service (jGCS) [19] that can be configured to use several group communication toolkits, such as Appia [10] or Spread [23]. The prototype presented in this paper uses Appia, a protocol composition framework to support communication, implemented in the Java language. The main goal of Appia is to provide high flexibility when composing communication protocols in a stack, and to build protocols in a generic way for reusing them in different stacks. Appia contains a set of protocols that implement view synchrony, total order, primary views, and the possibility to create open and closed groups. An open group is a group of nodes that can send and receive messages from nodes that do not belong to the group. This particular feature is very important to FT-OSGi. It is also important to our system that the GCS used gives the possibility to chose the message ordering guarantees (regular FIFO for passive replication or total order for active replication), the reliability properties (regular or uniform broadcast) and the possibility to operate in a partitionable or non-partitionable group. Appia has also a service that maintains information about members of a group in a best effort basis. This service is called *gossip* service and allows the discovery of group members (addresses) from nodes that do not belong to the group.

On top of the previously described services, the following components were built to provide fault tolerance to OSGi services:

FT Service Handler. This component provides the information about the available services on remote nodes that can be accessed by the local node. In particular, it provides (for each service) FT-OSGi configuration options, such as, for instance, the replication strategy used and how replies are handled by the proxy.

FT-Core. This component is responsible for maintaining the consistency among all the replicas of the service. It also hides all the complexity of replication from the client applications. The FT-Core component is composed by four sub-components that are described next: the *Appia Channel Factory*, the *Client Appia Channel*, the *Server Appia Channel* and the *Replication Mechanisms*. The *Appia Channel Factory* component is responsible for the definition of the replication service for an OSGi service. Each OSGi service is associated with a group of replicas, which is internally identified by an address in the form `ftosgi://<GroupName>` (this address is not visible for the clients). The group of replicas support the communication between the replicas of the OSGi service and communication between the client and the group of replicas. The client is outside the group and uses the open group functionality supported by Appia. The communication between replicas uses view synchrony (with total order in the case of active replication). For each one of these communication types, an Appia channel is created. The channel to communicate among replicas is created when a service is registered with fault tolerance properties and is implemented in the *Server Appia Channel* component. The communication channel between clients and service replicas is created when some client needs to access a replicated service and is implemented in the *Client Appia Channel* component. Finally, the *Replication Mechanisms* component implements the replication protocols used in FT-OSGi (active and passive replication) and is responsible for managing the consistency of the replicated services. This component is also responsible for the recovery of failed replicas and for the state transfer to new replicas that dynamically join the group. For managing recovery and state transfer, this components uses the membership service provided by the GCS.

3.3 Replication Strategies

The FT-OSGi extensions support three different types of replication: *active replication*, *eager passive replication* and *lazy passive replication*. The three strategies are briefly described in the next paragraphs.

Active replication. This replication strategy follows the approach of standard active replication [15], where each and every service replica processes invocations from the clients. When some replica receives a new request, it atomic broadcasts the request to all replicas. All the replicas execute the same requests by the same global order. One limitation of this replication strategy is that it can only be applied to deterministic services.

Eager passive replication. In this case only one replica, the primary, deals with invocations from the clients [15]. The primary replica is the same for all services belonging to a replica group. The backup replicas receive state updates from the primary replica for each invocation of stateful services. The primary replica only replies to the client after broadcasting the state updates to the other replicas. Attached to the state update message, the backup also receives the response that will be sent by the primary replica to the client. This allows the backup to replace the primary and resend the reply to the client, if needed.

Lazy passive replication. This replication strategy follows the same principles of *eager passive replication*. However, the reply to the client is sent immediately, as soon as the request is processed. The state update propagation is done in background, after replying to the client. This strategy provides less fault tolerance guarantees, but is faster and many applications do not require strong guarantees.

3.4 Replica Consistency

The group of service replicas is dynamic, which means that it supports the addition and removal of servers at runtime. It also tolerates faults and later recovery of the failed replica. The following paragraphs describe the techniques used to manage dynamic replica membership.

Leader election. The replication protocols implemented in the *Replication Mechanisms* component need a mechanism for leader election for several reasons. In the case of passive replication, leader election is necessary to choose the primary replica, executing the requests and disseminating the updates to the other replicas. In the case of active replication, leader election is used to choose the replica that will transfer the state to replicas that are recovering or are joining the system. The leader election mechanism can be trivially implemented on top of jGCS/Appia because upon any membership change, all processes receive an ordered set of group members. By using this feature, the leader can be deterministically attributed by choosing the group member with the lower identification that also belonged to the previous view.

Joining New Servers. When one or more replicas join an already existing group, it is necessary to update the state of the incoming replicas. The state transfer starts when there is a view change and proceeds as follows. If there are new members joining the group, all replicas stop processing requests. The replica elected as primary (or leader) sends its state to all the replicas, indicating also the number of new replicas joining in the view. The state transfer also contains the services configurations for validity check purposes. When a joining replica receives the state message, it validates the service configurations, and updates its own state, it broadcasts an acknowledgment to all the members of the group. Finally, when all the group members receive a number of acknowledgments equal to the number of joining replicas, all resume their normal operation. During this process, three types of replica failures can occur: *i*) new joined replica failure; *ii*) primary (or leader) replica failure; *iii*) another (not new, neither primary) replica failure. To address the first type of failure, the remaining replicas will decrement, for each new joined replica that fails, the number of expected acknowledgments. The second type of failure only requires an action when the primary replica fails before sending successfully the state to all new joined replicas. In this case, the new primary replica sends the state to the replicas. This solution tries to avoid sending unnecessary state messages. Regarding the third type of failure, these failures do not affect the process of joining new servers.

Recovering From Faults. Through the fault detector mechanisms implemented on top of jGCS/Appia, it is possible for FT-OSGi to detect when a server replica fails. FT-OSGi treats a failure of a replica in the same way treats an intent leave of a replica from the group membership. When a replica fails or leaves, some approach is necessary to maintain the system running. If the failed or leaving replica was the leader replica (also known as primary replica for both passive replication strategies), it is necessary to run the leader election protocol to elect a new replica to play that role. Otherwise, the remain replicas just remove from the group membership the failed replica.

3.5 Life Cycle

This section describes how the FT-OSGi components are created on both the client and the group of servers. The FT-OSGi uses the SLP [22] protocol to announce the set of services available in some domain. The replication parameters are configured using Java properties. This feature allows to read the parameters, for instance, from a configuration file contained in a service. The replication parameters are the group name that contain the replicated service, the type of replication, the group communication configuration, among others.

When a new replica starts with a new service, it reads the configuration parameters for replication and creates an instance of *Server Appia Channel* with the specified group name. It creates also an instance of the *FT-Core* and *Replication Mechanisms* components with the specified replication strategy. Finally, the replica registers the service in SLP, specifying the service interface and that it can be accessed using the address `ftosgi://<GroupName>`. New replicas will

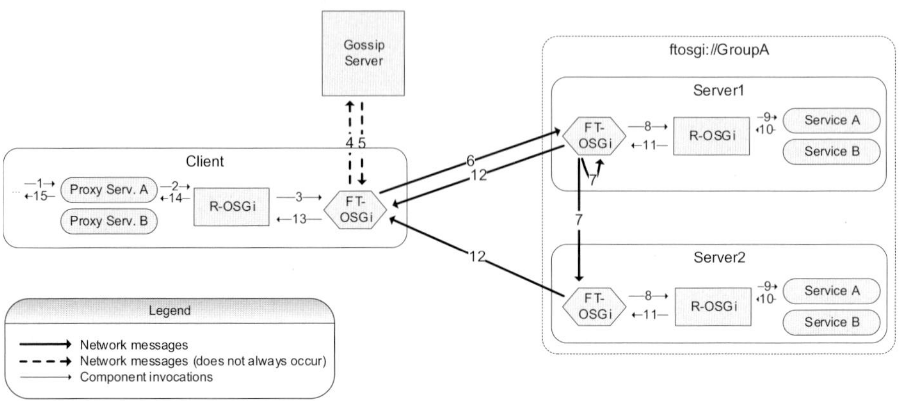

Fig. 2. Interaction between components when using active replication

also create an instance of *Server Appia Channel*, *FT-Core* and *Replication Mechanisms*, but they will join the already existing group.

The client starts by executing a query to SLP, asking for a reference that implements a service with a specified interface. If the service exists in the system, the SLP returns the address where the service can be found, in the form `ftosgi://<GroupName>`. In a transparent way to the client application, FT-OSGi creates a proxy that will represent locally the service and an instance of *Client Appia Channel* to send messages (requests) to the group of replicas of the service. After creating these components, a reference of the proxy is returned to the client application.

At this stage, replicas are deployed by scripts on behalf of the system administrator. As future work we plan to implement service factories that can create replicas on demand and provide support to components that perform the autonomic management of the membership. The next section describes how the several FT-OSGi components interact when a client invokes a replicated service.

3.6 Interaction between Components

The interaction between the system components depends on the replication strategy used by the service. For simplicity reasons, we will only illustrate the operation of active replication. The operation of passive replication is very similar, with the obvious differences of the replication protocol: only the primary executes the request, if the service is stateful, the primary reliable broadcasts a state update to all the replicas, and only the primary replies to the client.

The Figure 2 depicts the interaction between the several components of the FT-OSGi extensions. The client wants to invoke a method provided by Service A, which is replicated in Group A, and it already has an instance for the proxy that represents locally the Service A. The client starts by invoking that method on the local proxy (step 1). The service is actually deployed remotely, so the proxy invokes a Remote Method Invocation on R-OSGi (step 2). The original communication channel of R-OSGi was re-implemented to use an Appia channel, instead of a TCP connection. So, R-OSGi is actually using FT-OSGi to send the requests, through the *Client Appia Channel* component (steps 3 and 6). If the client does not have cached at least one address of the members of Group A, it queries the *gossip* service (steps 4 and 5). This request to the *gossip* service is also done periodically, in order to maintain the cache updated and is resent until it is successfully received by one of the servers. When one of the servers receives the client request, it atomically broadcasts the request to all the servers on Group A, using the *Server Appia Channel* component (step 7). This ensures that all servers execute the requests in the same global total order. For each request that is delivered to each replica by the atomic broadcast primitive, the replica delivers that request to the local instance of R-OSGi, that will call the method on the Service A and obtain a response (steps 8 to 11). Notice that these 4 steps are made on all the replicas. All the replicas reply to the client (step 12) that filters the duplicate replies and returns one reply to R-OSGi (step 13). Finally, R-OSGi replies to the proxy, that will return to the client application (steps 14 and 15).

3.7 Programing Example

This section illustrates how a client application and a service are implemented in the FT-OSGi architecture. We will start by showing how to implement and configure a service. The Listing 1 shows a typical *HelloWorld* example implemented as an OSGi service. The service implements an interface (`HelloService`) with two methods that are deterministic (lines 1 to 7). After implementing the service, it must be configured and registered in the OSGi platform. This is done in the class `Activator`, where it can be seen that the service is configured to use active replication, it is stateless, uses primary views, and it belongs to the group with the following address `ftosgi://GroupA` (lines 12 to 18). The registration of the service in the OSGi platform makes the service available to its clients and is done after the configuration process (line 19).

The same listing also shown an example of how a client application can obtain the instance of the replicated `HelloService` previously registered by the servers. First of all, in the class `Activator`, the application starts to obtain the local service `FTFramework` (lines 27 to 32). This `FTFramework` service is responsible to abstract the interaction with the SLP. Using that service, the application obtains the address `ftosgi://GroupA` (line 33), corresponding to the address where is located the `HelloService`. Notice that this address is actually in an opaque object of type `URI`, so it could be also an address of a non-replicated service (for instance, a R-OSGi service). Afterwards, with that address, the application can request the service instance (lines 34 and 35), which if it is successfully executed will create and register a service proxy of the `HelloService` in this local OSGi instance. Then, the proxy instance is returned to the client application, and that instance can be used like any other OSGi service. In this example an invocation of the method `speak()` is executed (line 40), which follows the invocation procedure of an actively replicated service, like it was described in the Section 3.6.

Listing 1. Example code

```
// server code
public class HelloServiceImpl implements HelloService {
 public String speak() {
  return "Hello World!";
 }
 public String yell() { return ("Hello World!".toUpperCase().concat("!!!")); }
}
public class Activator implements BundleActivator {
 private HelloService service;
 public void start(BundleContext context) throws Exception {
  service = new HelloServiceImpl();
  Dictionary<Object, Object> properties = new Hashtable<Object, Object>();
  properties.put(RemoteOSGiService.R_OSGi_REGISTRATION, Boolean.TRUE);
  properties.put(FTServiceTypes.FT_ROSGi_REGISTRATION, Boolean.TRUE);
  properties.put(FTServiceTypes.FT_ROSGi_FT_SERVICE_ID, "HelloService");
  properties.put(FTServiceTypes.FT_ROSGi_FT_TYPE, FTTypes.ACTIVE_STATELESS);
  properties.put(FTServiceTypes.FT_ROSGi_FT_GROUPNAME, "GroupA");
  properties.put(FTServiceTypes.FT_ROSGi_PRIMARY_VIEW, Boolean.TRUE);
  context.registerService(HelloService.class.getName(), service, properties);
 }
 public void stop(BundleContext context) throws Exception { service = null; }
}

// client code
public class Activator implements BundleActivator {
 public void start(BundleContext context) throws Exception {
  final ServiceReference ftRef = context.getServiceReference(FTFramework.class.getName());
  if (ftRef == null) {
    System.out.println("No FTFramework found!");
    return;
  }
  FTFramework ftFramework =(FTFramework)context.getService(ftRef);
  URI helloURI = ftFramework.getFTServiceURI(HelloService.class.getName());
  HelloService helloService =
   (HelloService)ftFramework.getFTService(HelloService.class.getName(), helloURI);
  if(helloService == null) {
   System.out.println("No HelloService found!");
   return;
  } else {
    System.out.println("Response: " + helloService.speak()); // Can start use service
  }
 }
 public void stop(BundleContext context) throws Exception {}
}
```

3.8 Some Relevant OSGi Implementation Details

This section focus in presenting the main issues that emerge by the replication of an OSGi service.

OSGi Service ID. Each OSGi service registered in an OSGi instance has a service id (`SID`) attributed by the OSGi framework itself. This SID is a `Long` object and identifies the service in an unique manner. R-OSGi uses this SID to identify remote services through different nodes since that SID is unique in each node. By extending that concept with an replicated service through several different nodes, the SID does not identify uniquely each service replica in all nodes, because the SID can be attributed differently in each node by the local OSGi instance for the replicated services. To solve this issue, FT-OSGi defines a replicated service id (`RSID`), which is defined by the service developer in the same way as the other service configurations, through service properties. The RSID was defined as a `String` object to let the developer choosing a more descriptive id, allowing unlimited and name space ids. The integration of RSID with R-OSGi is transparent, FT-OSGi always converts each RSID to the local OSGi SID in each replicated service interaction.

Filtering Replicated Events. The OSGi event mechanisms support the publish-subscribe paradigm. When replicating services, different replicas may publish multiple copies of the same event. These copies need to be filtered, to preserve the semantics of a non-replicated service. FT-OSGi addresses this issue using a similar approach as for filtering replies in active replication, i. e., the FT-OSGi component in the client is responsible to filter repeated events from the servers. The difficulty here is related with the possibility of non-deterministic events generation by different replicas. In a non-replicated system with R-OSGi, an unique id is associated with a event. In a replicated system is difficult to ensure the required coordination to have different replicas assign the same identifier to the replicated event. Therefore, the approach followed in FT-OSGi consists in explicitly comparing the contents of every event, ignoring the local unique id assigned independently by each replica. This approach avoids the costs associated with the synchronization of replicas for generating a common id for each event.

4 Evaluation

This section presents an evaluation of the FT-OSGi extensions, that will focus on the overhead introduced by replication. For these tests, an OSGi service with a set of methods was built, that (*i*) receive parameters and return objects with different sizes, generating requests with a growing message size, and (*ii*) have different processing times. The response time on FT-OSGi was measured with several replication strategies and compared it with R-OSGi, which is distributed but not replicated.

4.1 Environment

The machines used for the tests are connected by a 100Mbps Ethernet switch. The tests run in three FT-OSGi servers and one client machine. The servers have two Quad core processors Intel Xeon E5410 @ 2.33 Ghz and 4 Gbytes of RAM memory. One of the machines was also responsible for hosting the *gossip* service, which is a light process that does not affect the processing time of the FT-OSGi server. The client machine has one Intel Pentium 4 @ 2.80 Ghz (with Hyper-threading) processor and 2 Gbytes of RAM memory. All the server machines are running the Ubuntu Linux 2.6.27-11-server (64-bit) operating system and the client machine is running the Ubuntu Linux 2.6.27-14-server (32-bit) operating system. The tests were made using the Java Sun 1.6.0_10 virtual machine and the OSGi Eclipse Equinox 3.4.0 platform.

All the tests measure the time (in milliseconds) between the request and the reply of a method invocation on a OSGi service. In R-OSGi, the test was performed by invoking a service between a client and a server application. In FT-OSGi, different configurations were considered. The client issues a request to a group of 2 and 3 replicas. Group communication was configured using reliable broadcast in the case of passive replication and atomic broadcast (reliable broadcast with total order) in the case of active replication. Both group communication configurations used primary view membership.

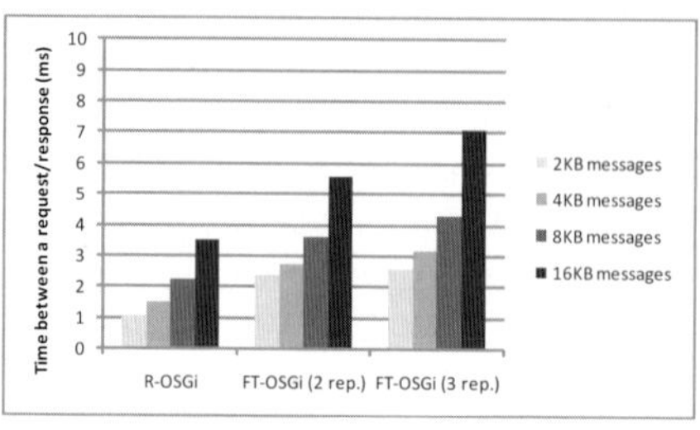

(a) Eager passive, no execution time.

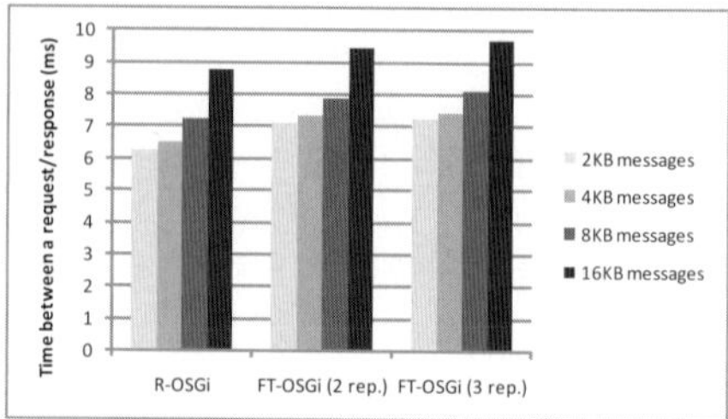

(b) Eager passive, execution time of $5ms$.

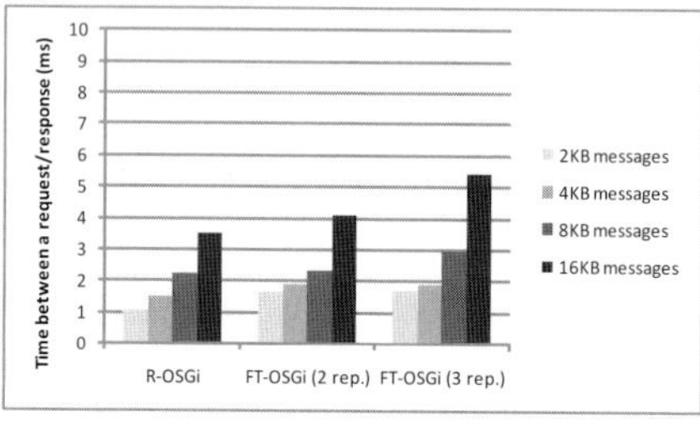

(c) Lazy passive, no execution time.

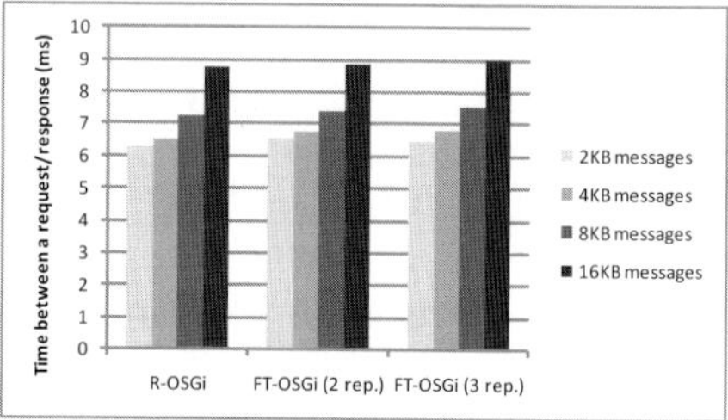

(d) Lazy passive, execution time of $5ms$.

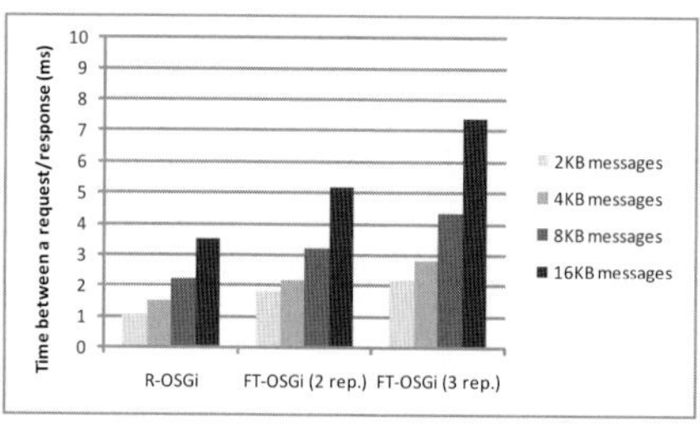

(e) Active, no execution time.

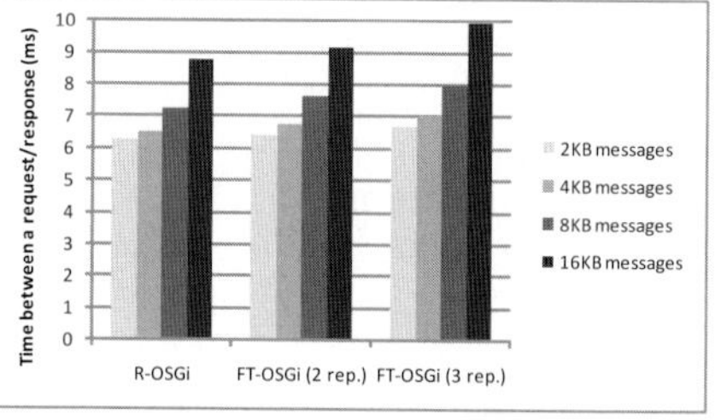

(f) Active, execution time of $5ms$.

Fig. 3. Replication overhead on different replication strategies

4.2 Replication Overhead

This section presents the replication overhead with different replication strategies, message sizes, execution times, and number of replicas. All tests for the active replication strategies of Figure 3 were executed using the *wait-first* reply filtering mode. The tests for both eager and lazy passive replication strategies were executed using a stateful service with 32 bytes of state. It was measured the response time with message sizes of 2 KBytes, 4 KBytes, 8 KBytes and 16 KBytes. Figure 3 shows the overhead of replication on active and passive replication. In the tests with no execution time, all the delays are due to remote method invocation and inter-replica coordination. The overhead of the replicated service is due to the extra communication steps introduced to coordinate the replicas. In the case of R-OSGi, where the service is located in another machine, but it is not replicated, there are two communication steps: request and reply. When using FT-OSGi, there are two extra communication steps for coordinating the replicas. In the case of the eager passive replication (Figure 3(a)), there is an additional overhead due to the dissemination of the new service state to the backup replicas. As expected, the lazy passive replication is the one with lower overhead (Figure 3(c)). It can also be observed that the message size has a similar impact on both the R-OSGi and FT-OSGi. On the other hand, as expected, adding extra replicas causes the overhead to increase. The tests presented in Figure 3 with an execution time of the invoked method of $5ms$, show that the overhead is smaller in all replication strategies, meaning that the execution time dominate the overhead of replication.

4.3 Response Filtering Modes on Active Replication

The Figure 4 shows the overhead in the case of active replication with the three reply filtering modes: *wait-first*, *wait-majority* and *wait-all*. The tests were configured to call a service method that takes $2ms$ to execute. The performance of a replicated system with two and three replicas was compared with R-OSGi (no replication). As it can be observed, the *wait-first* filtering mode does not introduce a large overhead when compared with R-OSGi. This can be explained by the fact that most of the requests are being received by the same node that orders the messages in the total order protocol. When the tests are configured to use the *wait-majority* and *wait-all* modes, the delay introduced by the atomic broadcast primitive is more noticeable. The results also show that the response time increases with the number of replicas.

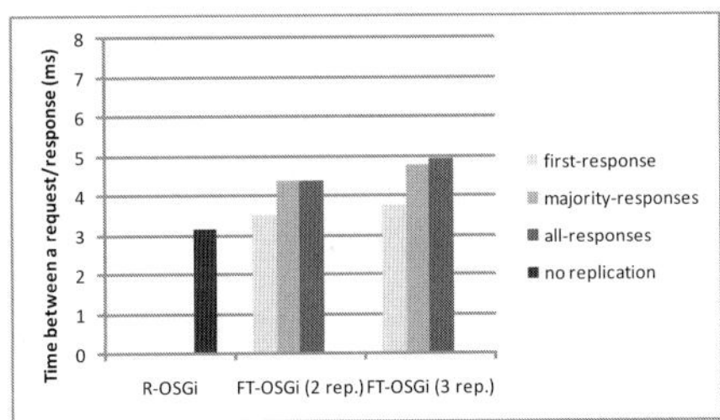

Fig. 4. Response time on active replication with the 3 filtering modes

4.4 Effect of State Updates on Passive Replication

Figure 5 presents the results when using passive replication with a stateful service. We measured the response time with a method execution time of $2ms$, and with different service state sizes of 32 bytes, 2 Kbytes, 4 Kbytes, 8 Kbytes and 16 Kbytes. The Figure 5(a) depicts the response times for the eager passive replication, where the state size has a direct impact in the replication overhead. On the other hand, in the lazy passive replication (Figure 5(b)), since the state transfer is made in background, in parallel with the response to the client, the state size has no direct impact on the response time.

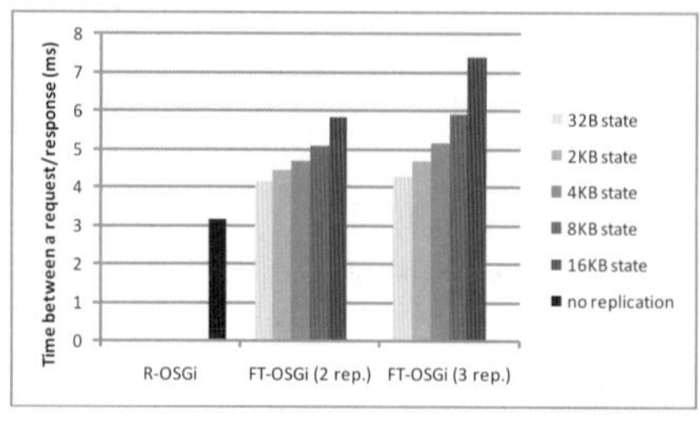

(a) Eager passive replication.

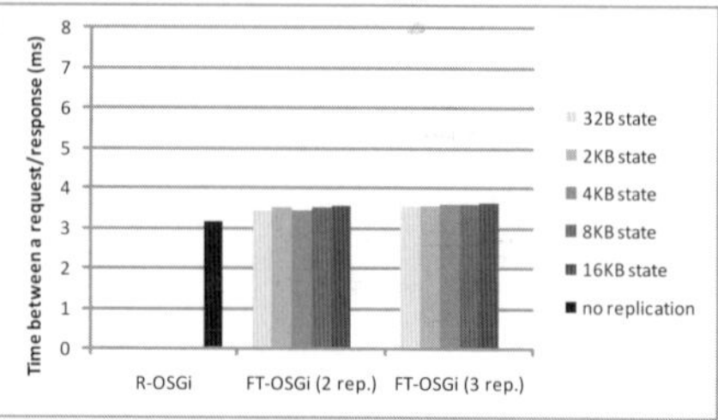

(b) Lazy passive replication.

Fig. 5. Passive replication with different state sizes

4.5 Failure Handling Overhead

Failure handling in FT-OSGi is based on the underlying view-synchronous group communication mechanisms. When a replica fails, the failure is detected and the replica expelled from the replica group. FT-OSGi ensures the availability of the service as long as a quorum of replicas remains active (depending on the reply filtering strategy, a single replica may be enough to ensure availability). Failure handling is performed in two steps. The first step consists in detecting the failure, which is based on timeouts. This can be triggered by exchange of data associated with the processing of requests or by a background heartbeat mechanism. When a failure is detected, the group communication protocol performs a view change, that requires the temporary interruption of the data flow to ensure view synchronous properties.

To assess the overhead of these mechanisms we artificially induced the crash of a replica. A non-leader replica is responsible for sending a special multicast control message that causes another target replica to crash. Since every replica receives the special message almost at the same time, we can measure the time interval between the crash and the moment when a new view, without the crashed replica, is received (and the communication is re-established). Additionally, we have also measure the impact of the crash on the client, by measuring the additional delay induced by the failure in the execution of the client request. We have repeated the same experience 10 time and made an average of the measured results.

The time to install a new group view as soon as a crash has been detected is, on average, $7ms$. The time to detect the failure depends on the Appia configuration. In the standard distribution, Appia is configured to operate over wide-area networks, and timeouts are set conservatively. Therefore, failure detection can take as much as $344ms$. By tuning the system configuration for a LAN setting, we were able to reduce this time to $73ms$. The reader should notice that the failure detection time has little impact on the client experience. In fact, while the replica failure is undetected, the remaining replicas may continue to process (and reply to) client requests. Therefore, the worst case perceived impact on the client is just the time to install a new view and, on average, much smaller than that, and in the same order of magnitude of other factors that may delay a remote invocation. As a result, in all the experiments, there were no observable differences from the point of view of remote clients between the runs where failures were induced and failure-free runs.

5 Conclusions and Future Work

This paper presents and evaluates FT-OSGi, a set of extensions to the OSGi platform to provide fault tolerance to OSGi applications. In FT-OSGi, each service can be configured to use active or passive replication (eager and lazy) and different services can coexist in the same distributed system, using different replication strategies. These extensions where implemented in Java and are available as open source software.

As future work we plan to implement the autonomic management of the group of replicas of each service. This will allow automatic recovery of failed replicas. We also plan to extend this work to support OSGi services that interact with external applications or persistent storage systems. Finally, the current version of FT-OSGi uses a naive approach to disseminate the state of the objects among the replicas. We intent to improve the current implementation by propagating only the JVM Heap changes. This approach will also allow that OSGi application can be integrated in FT-OSGi without needing any changes.

Acknowledgments. The authors wish to thank João Leitão by his comments to preliminary versions of this paper.

References

1. OSGi Alliance: Osgi service platform core specification (April 2007), `http://www.osgi.org/Download/Release4V41`
2. Spring Source: Spring Dynamic Modules for OSGi (2009), `http://www.springsource.org/osgi`
3. Kaila, L., Mikkonen, J., Vainio, A.M., Vanhala, J.: The ehome - a practical smart home implementation. In: Proceedings of the workshop Pervasive Computing @ Home, Sydney, Australia (May 2008)
4. Powell, D.: Distributed Fault Tolerance: Lessons from Delta-4. IEEE Micro", 36–47 (Feburary 1994)
5. Narasimhan, P., Moser, L., Melliar-Smith, P.: Eternal - a component-based framework for transparent fault-tolerant corba. Software Practice and Experience 32(8), 771–788 (2002)
6. Felber, P., Grabinato, B., Guerraoui, R.: The Design of a CORBA Group Communication Service. In: Proceedings of the 15th IEEE SRDS, Niagara-on-the-Lake, Canada, pp. 150–159 (October 1996)
7. Baldoni, R., Marchetti, C.: Three-tier replication for ft-corba infrastructures. Softw. Pract. Exper. 33(8), 767–797 (2003)
8. Salas, J., Perez-Sorrosal, F., no-Martínez, M.P., Jiménez-Peris, R.: Ws-replication: a framework for highly available web services. In: WWW 2006: Proc. of the 15th int. conference on World Wide Web, pp. 357–366. ACM, New York (2006)

9. Rellermeyer, J., Alonso, G., Roscoe, T.: R-osgi: Distributed applications through software modularization. Middleware, 1–20 (2007)
10. Miranda, H., Pinto, A., Rodrigues, L.: Appia, a flexible protocol kernel supporting multiple coordinated channels. In: Proceedings of the 21st International Conference on Distributed Computing Systems, pp. 707–710. IEEE, Los Alamitos (2001)
11. Apache Foundation: Apache felix, `http://felix.apache.org/`
12. Eclipse Foundation: Equinox, `http://www.eclipse.org/equinox/`
13. Knopflerfish Project: Knopflerfish, `http://www.knopflerfish.org/`
14. Nelson, V.P.: Fault-tolerant computing: Fundamental concepts. IEEE Computer 23(7), 19–25 (1990)
15. Guerraoui, R., Schiper, A.: Software-based replication for fault tolerance. Computer 30(4), 68–74 (1997)
16. Parrington, G.D., Shrivastava, S.K., Wheater, S.M., Little, M.C.: The design and implementation of arjuna. Technical report (1994)
17. Object Management Group: Corba: Core specification, 3.0.3 ed, OMG Technical Committee Document formal/04-03-01 (March 2004)
18. Narasimhan, P.: Transparent Fault Tolerance for CORBA. PhD thesis, Dept. of Electrical and Computer Eng., Univ. of California (1999)
19. Carvalho, N., Pereira, J., Rodrigues, L.: Towards a generic group communication service. In: Proc. of the 8th Int. Sym. on Distributed Objects and Applications (DOA), Montpellier, France (October 2006)
20. Thomsen, J.: Osgi-based gateway replication. In: Proceedings of the IADIS Applied Computing Conference 2006, pp. 123–129 (2006)
21. Ahn, H., Oh, H., Sung, C.: Towards reliable osgi framework and applications. In: Proceedings of the 2006 ACM symposium on Applied computing, pp. 1456–1461 (2006)
22. Guttman, E.: Service location protocol: automatic discovery of ip network services. Internet Computing, IEEE 3(4), 71–80 (1999)
23. Amir, Y., Danilov, C., Stanton, J.: A low latency, loss tolerant architecture and protocol for wide area group communication. In: Proceedings of the International Conference on Dependable Systems and Networks, DSN (June 2000)

A Component Assignment Framework for Improved Capacity and Assured Performance in Web Portals

Nilabja Roy, Yuan Xue, Aniruddha Gokhale, Larry Dowdy, and Douglas C. Schmidt

Electrical and Computer Science Department Vanderbilt University

Abstract. Web portals hosting large-scale internet applications have become popular due to the variety of services they provide to their users. These portals are developed using component technologies. Important design challenges for developers of web portals involve (1) determining the component placement that maximizes the number of users/requests (capacity) without increasing hardware resources and (2) maintaining the performance within certain bounds given by service level agreements (SLAs). The multitude of behavioral patterns presented by users makes it hard to identify the incoming workloads.

This paper makes three contributions to the design and evaluation of web portals that address these design challenges. First it introduces an algorithmic framework that combines bin-packing and modeling-based queuing theory to place components onto hardware nodes. This capability is realized by the Component Assignment Framework for multi-tiered internet applications (CAFe). Second, it develops a component-aware queuing model to predict web portal performance. Third, it provides extensive experimental evaluation using the Rice University Bidding System (RUBiS). The results indicate that CAFe can identify opportunities to increase web portal capacity by 25% for a constant amount of hardware resources and typical web application and user workloads.

1 Introduction

Emerging trends and challenges. Popular internet portals, such as eBay and Amazon, are growing at a fast pace. These portals provide many services to clients, including casual browsing, detailed reviews, messaging between users, buyer and seller reviews, and online buying/bidding. Customers visiting these web portals perform various types of behaviors, such as casual browsing, bidding, buying, and/or messaging. The broader the variety of services and user behavior a web portal supports, the harder it is to analyze and predict performance.

Performance of web portals can be quantified by average response time or throughput, which are functions of incoming load. In turn, the load generally corresponds to the arrival rate of user sessions or the concurrent number of clients handled by the portal. As the number of clients increases, performance can degrade with respect to customer's response time.

Service level agreements (SLAs) could be provided to bound the performance, such as an upper bound on response time. In such a case, a web portal needs to meet performance within the bounds given by the SLA. The capacity of such a web portal is

R. Meersman, T. Dillon, P. Herrero (Eds.): OTM 2009, Part I, LNCS 5870, pp. 671–689, 2009.

defined as the maximum number of concurrent user sessions or the maximum request arrival rate handled by applications with performance within the SLA bound.

The goal of portal designers is to deploy the given web portal to maximize its capacity, *i.e.*, maximize the number of users in the system while maintaining the average response time below the SLA limit. In turn, this goal helps maximize the revenue of the portal since more users can be accommodated. Portal capacity is generally proportional to the hardware, *i.e.*, more/better hardware means more/better capacity. Web portal designers are constrained by system procurement and operational costs, however, which yields the following questions: (1) for a fixed set of hardware, can web portals serve more clients and is the current hardware utilized in the optimal way?

To answer these questions, it is important to understand the way contemporary web portals are implemented. Component-oriented development (such as J2EE or .NET) is generally used to develop web portals by using the "divide and conquer" method. Each basic functionality (such as a business logic or a database query) is wrapped within a component, such as a Java class. Several components are then composed together to implement a single service, such as placing bids in an auction site or booking air tickets in a travel site.

A service is realized by deploying the assembly of components on the hardware resources. Maximizing the utilization of resources and meeting client SLA requirements can be achieved by intelligently placing the components on the resources. For example, colocating a CPU-intensive component with a disk-intensive component may yield better performance than two CPU-intensive components together.

Application placement in clustered servers has been studied extensively [1,2,3,4,5,6]. For example, [4,5,6] estimate resource usage of components and place components onto nodes by limiting their sum of resource usage within available node resource. They do not check response time of applications. On the other hand, [1] checks the response time while placing components. Their model is based on linear fitting and the effect of colocating components is estimated approximately. Other work [2,3] checks response time while provisioning resources using queuing theory.

In most related work, however, the functional granularity is modeled at the tier level, not at the component level. The downside of using the coarser granularity is that resources can remain unutilized in nodes where components could be configured, but tiers may not. If the models are component aware, available resources in the nodes can be better utilized by proper placement. Algorithms become more complicated, however, since the solution space is increased.

Solution approach → Intelligent component placing to maximize capacity. This paper presents the *Component Assignment Framework for multi-tiered internet applications* (CAFe), which is an algorithmic framework for increasing capacity of a web portal for a fixed set of hardware resources by leveraging the component-aware design of contemporary web portals. The goal of CAFe is to create a deployment plan that maximizes the capacity of the web portal so their performance remains within SLA bounds. It consists of a mechanism to predict the application performance coupled with an algorithm to assign the components onto the nodes.

CAFe complements related work by (1) introducing the concept of a performance bound through a SLA that acts as an additional constraint on the placement problem, which requires estimating and evaluating application performance against the SLA bound at every step, (2) creating service- and component-aware queuing models to estimate component resource requirements that predict overall service performance, (3) devising an algorithmic framework that combines a queuing model with a bin-packing algorithm to iteratively compute component placement while maximizing capacity of the application, and (4) showing how to balance resource usage across nodes in the network to deliver better performance.

Paper organization. The rest of the paper is organized as follows: Section 2 discusses an example to motivate intelligent component placement and formulate the problem CAFe is solving; Section 3 examines CAFe's algorithmic framework in detail; Section 4 empirically evaluates the performance of CAFe; Section 5 compares CAFe with related work; and Section 6 presents concluding remarks.

2 Motivating the Need for CAFe

This section presents an example to motivate intelligent component placement and formulate the problem CAFe is solving.

2.1 Motivating Example

The motivating example is modified from [7] to show how intelligently mapping the components of a web portal to available hardware nodes increases performance by several factors. The portal is an online bidding system (similar to eBay) that provides multiple services, such as browsing, bidding, and creating auctions. The portal is structured as a 3-Tier application with web-server, business logic, and database tiers.

Each service in the web portal consists of three components, one in each tier of the application. For example, the functionality of creating auctions is implemented by a service that is composed of three components: (1) a web service component that handles incoming requests, (2) a business tier component that implements application logic, and (3) a database component to make database calls. The web portal is deployed using a default deployment strategy with each tier placed on a single node. The portal performance is analyzed using an analytical model presented in [7].

The response times are given in Table 2, 2nd column. Most services have an unduly high response time, *e.g.*, Place Bid has a response time of 30 secs, which is unacceptable in an online bidding system. The CPU and disk utilizations of each node is given in Table 1a. Table 1a shows that the DB Server is disproportionately loaded. In particular, the DB Server disk is overloaded, whereas the Business Tier (BT) Server disk is

Table 1. The Utilization of resources

(a) Original Deployment

Server	CPU %	Disk %
Web Server	0.510	0.568
BT Node	0.727	0.364
DB Server	0.390	0.997

(b) After Reallocation

Server	CPU %	Disk %
Web Server	0.510	0.568
BT Node	0.822	0.554
DB Server	0.294	0.806

underloaded. One solution would be to move some components from the DB Server to the Business Tier Server so disk usage is more uniform across the nodes.

The database component for the "Login" service is chosen at random to move to the BT node. To analyze the new deployment, the analytic model is changed and the new utilization of the resources are given in Tables 1b, which shows that the resources are more evenly utilized, *e.g.*, the DB_Server disk is now only 80% utilized. The corresponding response times are given in the Table 2, $3^{rd}, column$. The percent decrease in the response times are given in the right-most column. These results show that the response times are reduced significantly.

Table 2. The Response Times of the Services Before and After Reallocation

Services	Response Time	New Response Time	% Decrease
Home	0.086	0.086	0.00
Search	12.046	0.423	96.49
View Bids	6.420	0.530	91.78
Login	17.230	0.586	96.60
Create Auction	27.470	0.840	96.95
Place Bid	30.800	0.760	97.53

The analysis clearly shows that the response time of all services can be improved significantly by placing the components properly. Proper placement can be achieved via two approaches. First, component-aware analytical models can be used to evaluate the impact of co-located components. Traditional models have looked at performance from the granularity of a tier [2] or have overlooked the effect of co-locating components [1]. Second, resource utilization can be balanced between the various nodes to ensure the load on any one resource does not reach 100%. This approach can be aided by analyzing performance at the component level (since components help utilize available resources more effectively) and using a placement routine that maps components to nodes so that the resources are utilized uniformly and none utilize 100% of any resource. Section 3 shows how CAFe combines both approaches to develop a component mapping solution that improves web portal capacity.

2.2 Problem Formulation and Requirements

As discussed in Section 1, the problem CAFe addresses involves maximizing the capacity (user requests or user sessions) of a web portal for given hardware resources, while ensuring that the application performance remains within SLA bounds. This problem can be stated formally as follows: The problem domain consists of the set of n components C, $\{C_1, C_2, ...C_n\}$, the set of m nodes P $\{P_1, P_2...P_m\}$, and the set of k services $\{S_1, S_2...S_k\}$. Each component has Service Demand D $\{D_1, D_2...D_n\}$. Each service has response time RT $\{RT_1, RT_2...RT_k\}$. The capacity of the application is denoted by either the arrival rate, λ for each service $\{\lambda_1.....\lambda_k\}$ or the concurrent number of customers M $\{M_1, M_2, ...M_k\}$. $SU_{i,r}$ gives the utilization of resource r by component i. The SLA gives an upper bound on the response times of each service $k\{RT_{sla,1}...RT_{sla,k}\}$.

CAFe must therefore provide a solution that places the n components in C to the m nodes P such that the capacity (either λ or M) is maximized while the response time is within the SLA limit $RT < RT_sla$. To achieve this solution, CAFe must meet the following requirements:

Place components onto nodes to balance resource consumption in polynomial time. Application components must be placed onto the available hardware nodes such that application capacity is maximized while the performance is within the upper bound set by a SLA. Since this is an NP-Hard problem [8] it is important to find out efficient heuristics that can find good approximate solutions. Section 3.1 describes how CAFe uses an efficient heuristic to place the components onto the available nodes and also ensuring that the resource utilization is balanced.

Estimate component requirement and application performance for various placement and workload. To place the components in each node, the resource requirement of each component is required. Moreover, for each placement strategy, the application performance must be compared with the SLA bound. Both vary with workload and particular placement. We therefore need a workload- and component-aware way of component resource requirement and application performance estimation. Section 3.2 describes how CAFe develops an analytical model to estimate component resource requirement and application performance.

Co-ordinate placement routine and performance modeling to maximize capacity. For each particular placement, the application performance need to be estimated to check if it is within the SLA limit. Conversely, performance estimation can only be done when a particular placement is given. Placement and performance estimation must therefore work closely to solve the overall problem. Section 3.3 describes how CAFe designs a algorithmic framework to co-ordinate the actions of a placement routine and an analytical model in a seamless fashion.

3 CAFe: A Component Assignment Framework for Multi-tiered Web Portals

This section discusses the design of CAFe and how it addresses the problem and requirements presented in Section 2.2. CAFe consists of two components: the placement algorithm and analytical model, as shown in Figure 1. The input to CAFe includes the set of application components and their inter-dependancy, as shown by the box on the left in Figure 1 and the set of available hardware nodes shown on the right. The output from CAFe is a deployment plan containing the mapping of application components to nodes. This mapping will attempt to maximize the application capacity. The rest of this section describes each element in CAFe.

3.1 Allocation Routine

As mentioned in Section 2.2, there is a need to develop efficient heuristics to place components onto nodes. Placing components to hardware nodes can be mapped as a bin-packing problem [9], which is NP-Hard in the general case (and which is what our scenarios present). Existing bin-packing heuristics (such as first-fit and best-fit) can be used to find a placement that is near optimal.

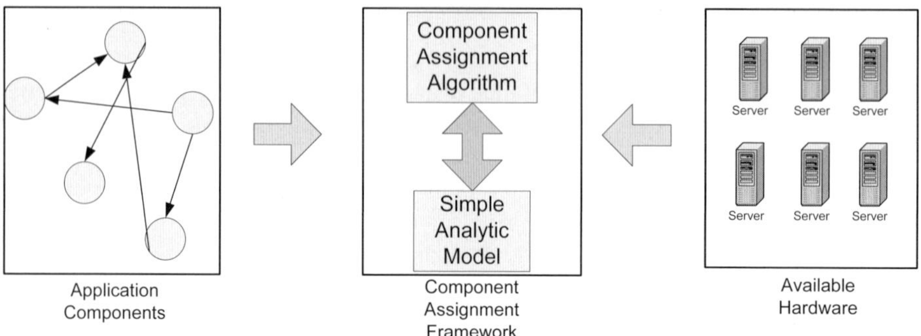

Fig. 1. The CAFe Component Assignment Framework Architecture

The motivating example in Section 2 describes the intuition behind the allocation routine, which is that performance increases by balancing the resource utilization between the various nodes and not allowing the load of any resource to reach 100%. We use worst-fit bin packing since it allocates items to bins by balancing the overall usage of each bin.

Algorithm 1. Allocate

C: Set of components
foreach *L:Set of Components that should remain local* **do**
  D← Sum the Service Demands of all components in L
  Replace all components in L with D in C
end
$DP = worst\_fit\_bin\_packing(C, P)$

Algorithm 1 gives the overall allocation routine, which is a wrapper around the worst-fit Algorithm 2. Algorithm 1 groups together components that are constrained to remain co-located in one machine. For example, they could be the database components that update the tables and need to be connected to the master instance of a database and hence msut be allocated to a single node. Algorithm 1 sums the Service Demands of the components that must be collocated and then replaces them by a hypothetical single component. A call to the $worst\_fit$ routine is made at the end of algorithm 1.

Algorithm 2. Worst-Fit Bin Packing

begin
  $Order\_Components(C)$ // Order the Components by resource requirement
  foreach $C_i \in C, 1 \leq i \leq |C|$ **do**
    // For all components
    $P_k \leftarrow$ The node with the maximum slack
    Place C_i on to P_k
  end
end

The worst-fit routine is given in Algorithm 2. The components are first ordered according to their resource requirements (Line 2). The algorithm then runs in an iterative fashion. At each step an item is allocated to a bin. All bins are inspected and the least utilized bin is selected.

3.2 Component- and Service-Aware Analytical Modeling

Section 2.2 also discusses the need for performance estimation to place components. This estimation process involves (1) predicting the resource requirements of each component for certain application loads, (2) predicting the response time of each service for a particular placement, and (3) computing the overall resource utilization of each node. As application components move among the various nodes, the performance of each service in the application will vary. Performance also depends upon the components that are collocated.

CAFe provides a queuing model that provides average case performance analysis of a system. It also models the interaction of collocated multiple components by modeling the queuing delay for resource contention. CAFe uses this information to transform a deployment plan produced by Algorithm 1 into a multiple class queuing model that maps each service as a class in the model. The components of a single class that are placed in the same node are considered as a single entity. Their Service Demands are summed together. After the model outputs the results, CAFe maps the performance parameters of each class onto the services.

Figure 2 shows the process of creating a model of the application. The input to such a process consists of the component placement map (mapping of application components to nodes) along with their Service Demands and the workload parameters, such as arrival rate of transactions and number of concurrent user sessions.

Depending upon the workload characteristics, a closed or an open model of the application is constructed. An open model assumes a continuous flow of incoming requests with a given average inter-arrival time between clients. A closed model assumes a fixed number of user sessions in steady state. The sessions are interactive and users make requests, then think for some time, and then make a subsequent request. If an application consists of independent requests arriving and being processed, it can be modeled as an open model. If there are inter-dependent sequences of requests coming from a single user, however, it must be modeled using a closed model.

These analytical model can be solved using standard procedures. In CAFe, the Mean Value Analysis (MVA) algorithm [7] is used to solve closed models, while an

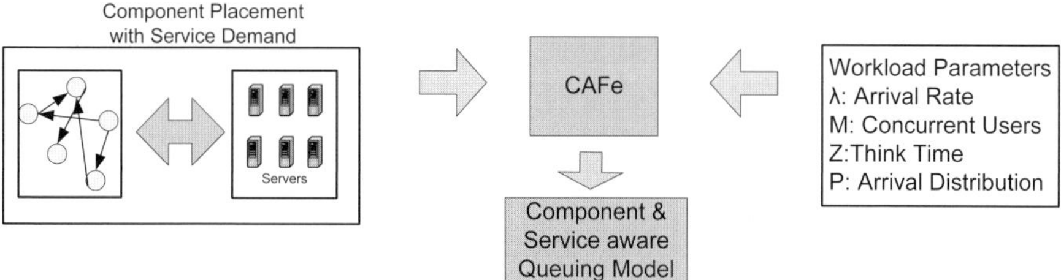

Fig. 2. Create Models of Application

algorithm based on the birth-death system is used to solve open models [7]. The solution to the analytical model provides the response times of the various services and also the utilization of the resources such as processor or disk usage in the various nodes.

3.3 Algorithmic Framework to Co-ordinate Placement and Performance Estimation

Section 2.2 also discusses the need for close co-ordination between placing the components and performance estimation. To meet this requirement CAFe provides a framework that standardizes the overall algorithm and defines a standard interface for communication between the placement and performance estimation. This framework also allows the configuration of other placement algorithms, such as integer programming and different analytical models like models based on worst case estimation. Different algorithms or analytical models can be configured in/out to produce results pertaining to the specific application domain or scenario.

CAFe uses an analytical model of the application and a placement routine to determine a mapping of the application components onto the available hardware nodes. CAFe attempts to maximize the capacity of the web portal, while ensuring that the response time of the requests is within the SLA bounds.

Algorithm 3. Component Assignment Framework

Input:
$C \leftarrow$set of N components to be deployed,
$D \leftarrow$set of Service Demands for all components, $D_{i,r} \leftarrow$Service Demand of component i on the device r
$P \leftarrow$set of K available nodes
RT_{sla} set of response time values for each service as specified by the SLA
Output:
Deployment plan $DP \leftarrow$ set of tuples mapping a component to a node,
M: Total Number of concurrent clients
$RT \leftarrow$set of response times for all components
RT_i: Total response time of service i
U_r: Total Utilization of each resource r in the nodes
$SU_{i,r}$: Utilization of resource r by component i
$SU \leftarrow$ set of resource utilization of all components
$Incr$: Incremental capacity at each step
$Init_Cap$:Initial Capacity
begin
  Intially, $DP = \{\}, M = Init_Cap, SU = U, RT_i = \sum_r D_{i,r}, incr = Incr$
  while $incr > 10$ **do**
    while $\nexists i : RT_i > RT_{sla,i}$ **do**
      // Check if any service RT is greater than SLA bound
      $DP = Allocate(SU,P)$ // Call Placement routine to get a placement
      $(RT,SU,U) = Model(M,D,DP)$// Call model to estimate performance for current placement
      $last\_M \leftarrow M$ // Save the previous capacity
      $M \leftarrow M + incr$ // Increment the capacity for the next iteration
    end
    // At least one service has Response Time greater than SLA bound for current capacity
    $M \leftarrow last\_M$ // Rollback to previous iteration's capacity
    $incr \leftarrow incr/2$ // Decrease incr by half
    $M \leftarrow M + incr$ // Now Increase capacity and repeat
  end
  // while($incr > 10$)
end

Algorithm 3 describes the component assignment framework. The algorithm alternates between invoking the placement algorithm and an evaluation using the analytic model. In each step, it increases the number of clients in the system by a fixed amount and rearranges the components across the nodes to minimize the difference in resource utilization. It then verifies that the response time is below the SLA limit using the analytical model and iterates on this strategy until the response time exceeds the SLA limit.

CAFe takes as input the details (such as the Service Demands of each component on each resource) of the N components to deploy. Service Demand is the amount of resource time taken by one transaction without including the queuing delay. For example, a single Login request takes 0.004 seconds of processor time in the database server. The Service Demand for Login on the database CPU then takes 0.004 seconds. As output, the framework provides a deployment plan(DP, estimated response (RT) time and total utilization (U) of each resource.

The initial capacity is an input ($Init\_Cap$, Line 2). The value of M is set equal to $Init\_Cap$. This capacity is an arrival rate for an open model or "number of users" for a closed model. The capacity (M) is increased in each step by an incremental step $incr$ (Line 9) which also can be parameterized ($Incr$). At each iteration, the response time of all services is compared with the SLA provided upper bound (inner while loop at Line 4).

Inside the inner loop, the framework makes a call to the Allocate module (Line 6), which maps the components to the nodes. This mapping is then presented to the Model (Line 7) along with the Service Demand of each component. The Model computes the estimated response time of each service and the utilization of each resource by each component. It also outputs the total utilization of each resource.

The inner loop of Algorithm 3 exits when response time of any service exceeds the SLA provided upper bound (*i.e.*, M reaches maximum capacity), at which point $incr$ is set to a lower value (one-half) and the algorithm continues from the previous value of M (Line 12). If the inner loop exits again, the value of $incr$ is lowered further (Line 13). The algorithm ends when the value of $incr$ is less than 10.

The output of the algorithm is that value of M, which yields the highest capacity possible and also a deployment plan (DP) that maps the application components onto the nodes. Though not provably optimal, the algorithm is a reasonable approximation. An optimal algorithm would require an integer programming routine [10] to obtain the mapping of the components to the nodes. Such an implementation would be NP-Hard, however, and thus not be feasible for large applications. CAFe therefore uses an intuitive heuristic based on the popular worst-fit bin packing algorithm [9].

4 Experimental Evaluation

4.1 Rice University Bidding System

This section describes our experimental evaluation of CAFe, which used the Java servlets version of the Rice University Bidding System (RUBiS) [11] to evaluate its effectiveness. RUBiS is a prototype of an auction site modeled after ebay that has the features of an online web portal studied in this paper. It provides three types of user sessions (visitor, buyer, and seller) and a client-browser emulator that emulates users behavior.

Table 3. Component Names for Each Service

Service Name	Home Page	Browse Page	Browse_Cat	Browse_Reg	Br_Cat_Reg
Business Tier	BT_H	BT_B	BT_BC	BT_BR	BT_BCR
DB Tier	_	_	DB_BC	DB_BR	DB_BCR

Service Name	Srch_It_Cat	Srch_It_Reg	View_Items	Vu_Usr_Info	Vu_Bid_Hst
Business Tier	BT_SC	BT_SR	BT_VI	BT_VU	BT_BH
DB Tier	DB_SC	DB_SR	DB_VI	DB_VU	DB_BH

A RUBiS session is a sequence of interactions for the same customer. For each customer session, the client emulator opens a persistent HTTP connection to the Web server and closes it at the end of the session. Each emulated client waits for a certain think time before initiating the next interaction. The next interaction is determined by a state transition matrix that specifies the probability to go from one interaction to another one. The load on the site is varied by altering the number of clients.

CAFe requires an analytical model of the application. Once the model is constructed and validated, it can be used in CAFe to find the appropriate component placement. The steps required to build the model are (1) compute Service Demand for each service provided for each customer type such as visitor or buyer and (2) build a customer behavior modeling graph of user interactions and calculate the percentage of requests for each service. For our experiments, a workload representing a set of visitor clients were chosen, so the workload consists of browsing by the users and is thus composed of read-only interactions. The components for each service in RUBiS is given in Table 3.

The RUBiS benchmark was installed and run on the ISISLab testbed (`www.isislab.vanderbilt.edu`) at Vanderbilt University using 3 nodes. One for the client emulators, one for the "Business Tier" and the other for "Database Tier". Each node has 2.8 GHz Intel Xeon processor, 1GB of ram, and 40GB HDD running Fedora Core 8.

4.2 Computing Service Demand

The Service Demand of each of the components must be captured to build an analytical model of the application. The RUBiS benchmark was run with increasing clients and its effect on various CPU, memory, and disk were noted. The memory and disk usages are shown in Figures 3a and 3b.

Disk usage is low ($\sim$0.2%) and memory usage was $\sim$40%. Moreover, these utilizations remained steady even as the number of clients are increased. Conversely, CPU usages increased as number of clients grew (the Actual line in the Figure 5a).

The results in Figures 3a and 3b show that CPU is the bottleneck device. The Service Demands were computed for the CPU and the disk. Since memory was not used fully, it is not a contentious resource and will not be used in the analytical model. Moreover, the CAFe placement routine ignores disk usage since it remains steady and is much less than CPU usage. The CAFe placement routine thus only uses one resource (CPU) to come up with the placement.

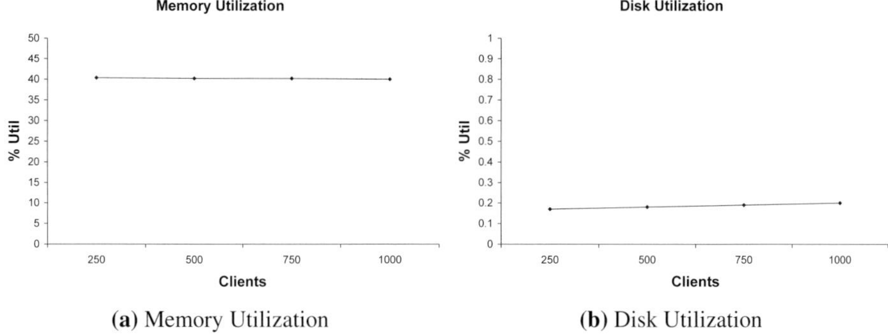

(a) Memory Utilization **(b)** Disk Utilization

Fig. 3. The Utilization of Memory and Disk for RUBiS Benchmark

Table 4. CPU Service Demand for Each Component

Service	Business Tier Component(secs)	DB Server Component(secs)	Description
home	0.002	0	Home Page
browse	0.002	0.0	Browse Main Page
browse_cat	0.0025	0.0005	Browse Categories
browse_reg	0.0025	0.0005	Browse Regions
br_cat_reg	0.003	0.0007	Browse Categories in Regions
Srch_it_cat	0.004	0.028	Search Items in Categories
Srch_it_reg	0.0021	0.027	Search Items in Regions
view_items	0.004	0.0009	View Items
vu_usr_info	0.003	0.001	View User Info
vu_bid_hst	0.004	0.004	View Bid History

RUBiS simplifies the calculation of Service Demand. It includes a client-browser emulator for a single client and makes requests on one service at a time. During the experiment, the processor, disk and memory usages were captured. After the experiment finished we used the Service Demand law [7] to calculate the Service Demand for that service. In some services (such as "Search Items in Categories") the Service Demand is load dependent. For such services the number of clients was increased and the Service Demands were measured appropriately.

The Service Demands of CPU for all the services measured in such a way are given in Table 4. Each service in RUBiS is composed of multiple components, with a component in the middle (Business) tier and one in the Database Tier. Each component has its own resource requirements or Service Demands.

4.3 Customer Behavior Modeling Graph

For the initial experiment, the workload was composed of visitor type of clients. A typical user is expected to browse across the set of services and visit different sections of the auction site. A transition probability is assumed for a typical user to move from one service to the other.

Table 5. Transition Probabilities Between Various Services

	home	browse	browse_cat	browse_reg	br_cat_reg	Srch_it_cat	Srch_it_reg	view_items	vu_usr_info	vu_bid_hst	view_items_reg	vu_usr_info_reg	vu_bid_hst_reg	Probabilities
home	0	0.01	0.0025	0.0025	0.0025	0.0025	0.0025	0.0025	0.0025	0.0025	0.0025	0.0025	0.0025	0.0026
browse	1	0	0.0075	0.0075	0.0075	0.0075	0.0075	0.0075	0.0075	0.0075	0.0075	0.0075	0.0075	0.0100
browse_cat	0	0.7	0	0	0	0	0	0	0	0	0	0	0	0.0070
browse_reg	0	0.29	0	0	0	0	0	0	0	0	0	0	0	0.0029
br_cat_reg	0	0	0	0.99	0	0	0	0	0	0	0	0	0	0.0029
Srch_it_cat	0	0	0.99	0	0	0.44	0	0.74	0	0	0	0	0	0.3343
Srch_it_reg	0	0	0	0	0.99	0	0.44	0	0	0	0.74	0	0	0.1371
view_items	0	0	0	0	0	0.55	0	0	0.8	0	0	0	0	0.2436
vu_usr_info	0	0	0	0	0	0	0	0.15	0	0.99	0	0	0	0.0747
vu_bid_hst	0	0	0	0	0	0	0	0.1	0.19	0	0	0	0	0.0386
view_items_reg	0	0	0	0	0	0	0.55	0	0	0	0	0.8	0	0.0999
vu_usr_info_reg	0	0	0	0	0	0	0	0	0	0	0.15	0	0.99	0.0306
vu_bid_hst_reg	0	0	0	0	0	0	0	0	0	0	0.1	0.19	0	0.0158

The various transition probabilities are given in Table 5.

Here element $p_{i,j}$ (at row i and column j) represents the probability of the ith service being invoked after the jth service is invoked. For example, a user in the web page "browse_cat"(browsing categories) has a 0.0025% chance of going to the "home" page and a 99% chance for moving on to "Search_it_cat"(searching for an item in a category).

The steady state probability (percentage of user sessions) for each service type is denoted by the vector π. The value of π_i denotes the percentage of user requests that invoke the the ith service. The vector π can be obtained by using a technique is similar to the one in [12]. Once computed, the amount of load on each service type can be calculated from the total number of user sessions. The rightmost column in Table 5 gives the steady state probabilities of each service.

4.4 Analytical Modeling of RUBiS Servlets

After the Service Demands and the steady state probability mix for each service is available, an analytical model of the application can be developed. The RUBiS benchmark assumes a client to carry out a session with multiple requests with think times in between. This type of a user behavior must be modeled with a closed model.

As soon as a client finishes, a new client takes its place. The average number of clients remains fixed. Figure 4 shows the analytical model of the RUBiS Servlets version. As mentioned in Section 4.2, the processor is the contentious resource. Each machine is represented by two queues, one for the CPU and the other for the disk.

Figure 4 also shows two queues for each of the two node in the deployment. The first node is the Business Tier, which also serves as the web server. The second node is the Database Server. The various client terminals are represented by delay servers. A delay server is a server that does not have a queue, so clients wanting to use the server can access it directly it without waiting. This design models user think times since as soon as a response to a previous request comes back, the user starts working on the next request.

Figures 5a and 5b compare the results predicted by the analytical model to the actual results collected from running the benchmark. The benchmark is run using progressively increasing number of clients for $250, 500, 750$ and $1,000$, respectively. The components are placed in the nodes using RUBiS's default strategy, which places all the Business Tier components in the Business Tier node and the entire database in the database server. The results in these figures show the model accurately predicts the response times of the services and the processor utilizations of the nodes. This model can therefore be used

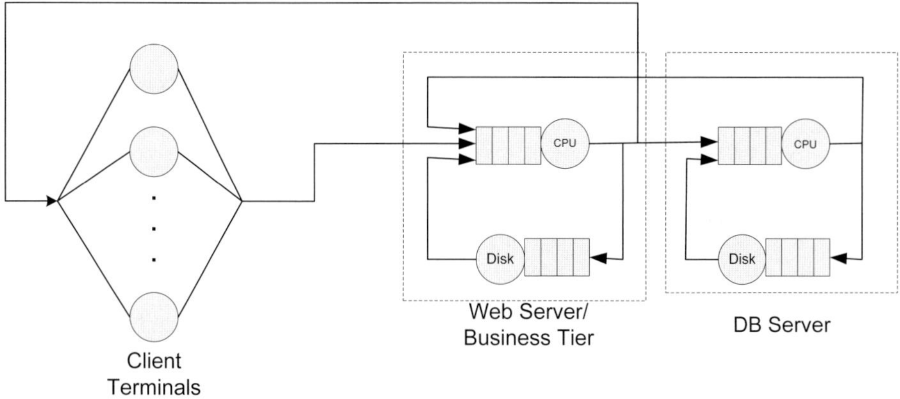

Fig. 4. Closed Queuing Model for Rubis Java Servlets Version

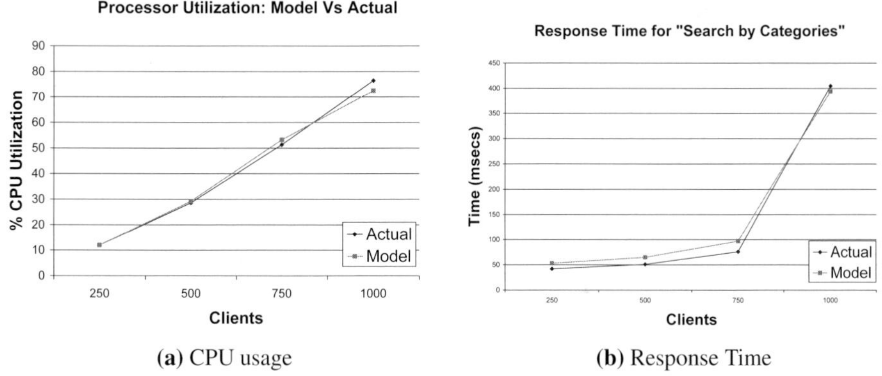

(a) CPU usage

(b) Response Time

Fig. 5. Validation of Analytical Model

by CAFe to find the placement of the components that optimizes the capacity of the deployment.

4.5 Application Component Placement

We now describe how CAFe iteratively places components onto hardware nodes. The SLA is assumed to have set an upper bound of 1 sec on the response time of all services. We use Algorithm 3 from Section 3, which considers CPU as the only resource since both memory and disk usage is minor compared to CPU usage, as described in Section 4.

The first iteration of Algorithm 3 uses the initial Service Demands for each application component. The Service Demands are given in Table 4. The set of all Service Demands and the available nodes (in this case 2) are used by the $Allocate$ Algorithm 1. This algorithm in turn invokes the $worst\_fit\_bin\_packing$ described in Algorithm 2, which places components on the two nodes.

As mentioned in Section 4, there are two nodes used for RUBiS benchmark: Business Tier Server (BT_SRV) and the Database Server (DB_SRV). The name of the nodes

Table 6. Component Placement and RUBiS Performance After Iteration 1

(a) Component Placement

BT_SRV		DB_SRV	
Component	CPU Util	Component	CPU Util
DB_SC	0.02783	DB_SR	0.02690
BT_VI	0.00405	BT_SC	0.00417
DB_BH	0.00400	BT_BH	0.00400
BT_UI	0.00300	BT_BCR	0.00325
BT_BC	0.00245	BT_BR	0.00253
BT_H	0.00200	BT_SR	0.00210
DB_UI	0.00100	BT_B	0.00200
DB_VI	0.00095	DB_BCR	0.00075
DB_BC	0.00055	DB_BR	0.00047

(b) Utilization and Response Time

Service	Business Tier Component%	DB Server Component%	Response Time
home	0.007	0.000	0.002
browse	0.029	0.000	0.002
browse_cat	0.025	0.005	0.004
browse_reg	0.010	0.002	0.004
br_cat_reg	0.014	0.003	0.005
Srch_it_cat	1.980	13.190	0.049
Srch_it_reg	0.380	5.260	0.041
view_items	1.940	0.490	0.006
vu_usr_info	0.480	0.120	0.005
vu_bid_hst	0.310	0.310	0.009

are given since the default deployment of RUBiS uses the BT_SRV to deploy all the business layer components and DB_SRV to deploy the database. In fact, such a tiered deployment is an industry standard [13].

Table 6a shows the placement of the components after the first iteration of Algorithm 3 in CAFe. The mapping of the components to the nodes, the total number of clients (100), and the Service Demands of the components are used to build the analytical model. It is then used to find the response time and processor utilization of the two servers, given in Table 6b. The response time of all the services is well below the SLA specified 1 sec. CAFe iterates and the processor utilization of each component found in the previous iteration is used in the *Allocate* routine.

In the second iteration, the *Allocate* Algorithm 1 produces the placement shown in Table 7.

In the third iteration, the number of clients, M is increase to 300. The placement computed by CAFe remains the same, however, and the response times of the two services "Search By Category" and "Search by Region" increase with each iteration as shown in Table 8.

Table 7. Iteration 2:Component Placement by Allocation Routine

BT_SRV		DB_SRV			
Component	CPU Util	Component	CPU Util	Component	CPU Util
DB_SC	13.19	DB_SR	5.26	BT_B	0.029
		BT_SC	1.97	BT_BC	0.025
		BT_VI	1.94	BT_BCR	0.014
		DB_VI	0.49	BT_BR	0.010
		BT_UI	0.48	BT_H	0.007
		BT_SR	0.39	DB_BC	0.005
		BT_BH	0.31	DB_BCR	0.003
		DB_BH	0.31	DB_BR	0.002
		DB_UI	0.12		

Table 8. Successive Iterations:Response Time of Each Service

Iteration	Clients	home	broBTe	broBTe_cat	broBTe_reg	br_cat_reg	Srch_it_cat	Srch_it_reg	view_items	vu_usr_info	vu_bid_hst
1	100	0.002	0.002	0.004	0.004	0.005	0.049	0.041	0.006	0.005	0.009
2	200	0.002	0.002	0.004	0.004	0.005	0.050	0.041	0.006	0.005	0.009
5	500	0.002	0.002	0.004	0.004	0.005	0.058	0.044	0.007	0.005	0.010
10	1000	0.002	0.002	0.005	0.005	0.006	0.088	0.049	0.007	0.006	0.011
15	1500	0.002	0.002	0.005	0.005	0.007	0.689	0.055	0.008	0.007	0.012
16	1600	0.003	0.003	0.006	0.006	0.007	1.119	0.057	0.008	0.007	0.012
17	1550	0.003	0.003	0.006	0.006	0.007	0.899	0.056	0.008	0.007	0.012
18	1575	0.003	0.003	0.006	0.006	0.007	1.011	0.057	0.008	0.007	0.012
19	1563	0.003	0.003	0.006	0.006	0.007	0.956	0.056	0.008	0.007	0.012

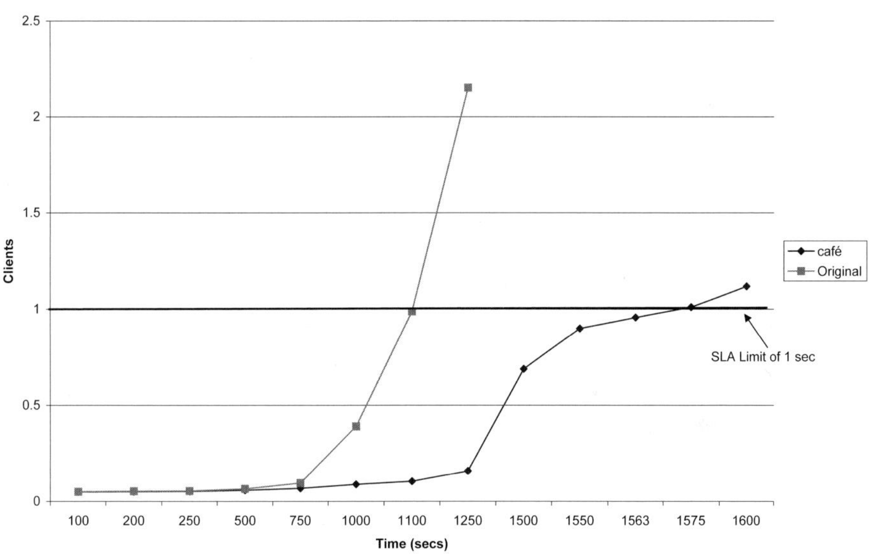

Fig. 6. Response Time with Increasing Clients

At the value of $M = 1600$, the response time of the service "Search by Category" crosses the SLA limit of 1 sec as shown in iteration 16 in Table 8. At that point *incr* variable in Algorithm 3 is reduced by half to 50 and M is reduced to the previous value of 1500. The algorithm continues from that point. Thus in iteration 17, value of M is 1550 In a similar way, for M equal to 1563, the response time of "Search by Category" is just below 1 sec (iteration 19). This response time is the maximum capacity of the application under a SLA response time of 1 sec. Figure 6 shows the comparison in the response time of the service "Search By Category," which is the bottleneck service.

4.6 Implementation of the CAFe Deployment Plan

We now describe how RUBiS uses the new deployment plan recommended by CAFe and empirically evaluate the performance improvement compared with the default tiered architecture used by RUBiS. This plan assigns all the Business Tier components in the BT_SRV and the entire database in the DB_SRV. The deployment suggested by CAFe is shown in Table 7, where component DB_SC is contained in one node and all the others are kept in the other node. The component DB_SC is the database component of the service "Search By Category," which is a read-only component that invokes a select query on the database.

One way to implement this assignment is to run a master instance of the database along with all the other components and run a slave instance of the database in the machine where DB_SC is run. The corresponding deployment is shown in Figure 7. In this figure there are two instances of the Database: the master instance is run in the machine BT_SRV and a slave instance is run in DB_SRV. All Business Tier components and the web server run in BT_SRV. These components make the database call on the master instance in BT_SRV. Only component DB_SC (which belongs to service "Search By Category") makes the database call to the slave instance (in DB_SRV). The component DB_SC is thus moved to DB_SRV, while all other components run in the BT_SRV.

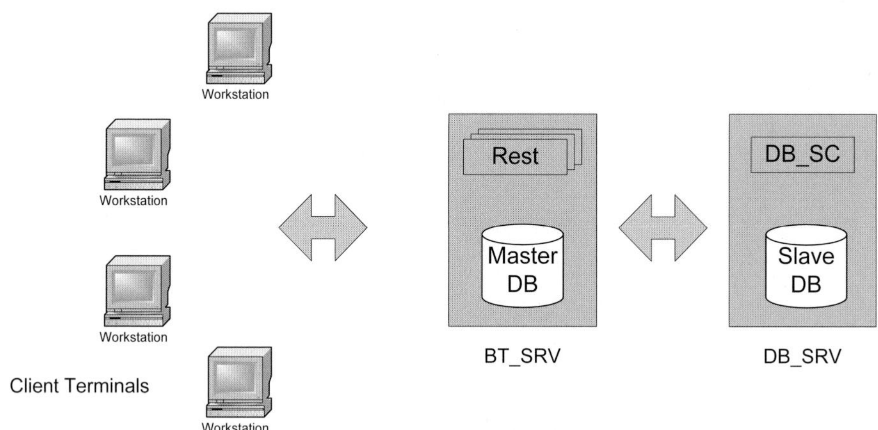

Fig. 7. Deployment of CAFe Suggested Assignment

Figure 8a shows the response times of the most loaded service "Search By Category" for the CAFe deployment. By comparison, the original response time with the 3-Tier deployment is also provided. The comparison shows that the CAFe deployment increases the capacity of the application. The solid line (Time = 1.0) parallel to the Client axis signifies the SLA limit of 1 sec. The response times for the 3-Tier deployment crosses the line just above 1,000 clients. In contrast, the CAFe deployment the response time graph crosses the line at just over 1250 clients, which provides an improvement of $\sim$25% in application capacity.

Figure 8b shows the processor utilization for the two cases. In the CAFe installation the DB_SRV is less loaded than in the 3-Tier deployment. The BT_SRV utilization also shows the CAFe installation uses more CPU time than in the 3-Tier installation. This result is expected since CAFe tends to balance out the component utilizations across the given machines.

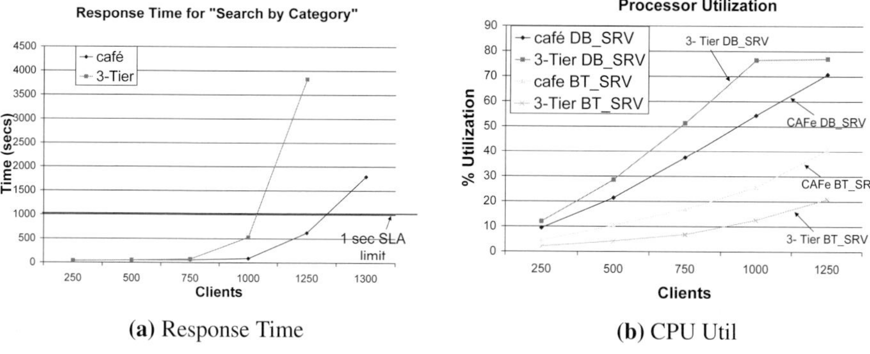

(a) Response Time (b) CPU Util

Fig. 8. Performance of CAFe Installation

CAFe's balancing is not perfect,however, since DB_SC (the database component of the "Search By Category" service) consumes more processor time than all other components. This result indicates that load balancing the DB_SRV on multiple components and moving the components to different machines may be advantageous.

5 Related Work

This section compares CAFe with related work in the area of system modeling and management.

Analytic models based on linear fitting. Stewart et. al. [1] proposed a profile-driven performance model for cluster based multi-component online services. They use this model to perform system management and implement component placement across nodes in the cluster. The main difference between CAFe and this work is that CAFe's modeling is based on queuing theory, whereas theirs is based on linear fitting. CAFe uses queuing theory since the impact of co-locating multiple components in the same node is better captured due to modeling of queuing delays. In [1]'s model, the impact of co-location is approximate.

Analytical modeling of multi-tiered applications have been pursued extensively in the research community. Closed queuing models of multi-tier internet application appear in [2,3] and [12]. Both model a single tier as a queue and do not have any concept of a component in their model. CAFe leverages the knowledge of the components in the system and generates a queuing model out of the component placement mapping. Thus CAFe uses components as the functional granularity which has the advantage of utilizing the available resources in the nodes in a much better way, as shown in Section 2. The work in [3] presents techniques to predict workload that could be used along with CAFe.

A framework for dynamic placement of clustered web applications is presented by Karve et. al. [4] and Kimbrel et. al. [5]. These approaches consider multiple resources that are load-dependent and load independent. They solve an optimization problem that attempts to change the component placement at run-time, while minimizing the number of changes. They characterize resource requirements of components using a simple model that calculates Service Demands of different requests. Urgaonkar et. al. [6] identify resource needs of application capsule or components by profiling them and uses it to characterise application Quality of Service (QoS) requirements. They also define an algorithm for mapping the application capsules on to the platforms or nodes available. CAFe differs from both these work in terms of its workload and performance modeling. CAFe defines a queuing model which can model the interaction of several components co-located in a node. It also models the behavior of clients using Customer Behavior Modeling Graphs to characterize incoming workload. The analytical models developed in CAFe can be used along with the algorithms presented in [4,5] or [6].

None of the work above (except [2]) has a concept of performance bound. CAFe introduces the performance bound through the concept of an SLA. The placement of the components is thus done to maximize capacity while ensuring that the performance remains within SLA bounds.

6 Concluding Remarks

This paper presented a *Component Assignment Framework for multi-tiered Internet Applications* (CAFe), which is a novel algorithmic framework for mapping components of a multi-tiered application onto hardware nodes in web portals. CAFe helps ensure that (1) the capacity of the application is potentially maximized and (2) response times remain within SLA prescribed bounds. CAFe complements research in the area of application placement by introducing the constraint of performance bounds. It also uses queuing-theoretic techniques to co-ordinate component placement and analytical modeling.

The paper also empirically evaluated CAFe against RUBiS, which is an industry-standard application benchmark. The experimental results showed how the CAFe Allocation algorithm can improve web portal performance by balancing the resource utilizations across various nodes. The performance improvement was 25%, which means that a representative web portal can handle more users without purchasing additional hardware. By using CAFe, therefore, earned revenue can potentially be increased by 25%.

References

1. Stewart, C., Shen, K.: Performance modeling and system management for multi-component online services. In: Proceedings of the 2nd conference on Symposium on Networked Systems Design & Implementation-Volume 2 table of contents, pp. 71–84. USENIX Association Berkeley, CA, USA (2005)
2. Urgaonkar, B., Pacifici, G., Shenoy, P., Spreitzer, M., Tantawi, A.: An analytical model for multi-tier internet services and its applications. SIGMETRICS Perform. Eval. Rev. 33(1), 291–302 (2005)
3. Urgaonkar, B., Shenoy, P., Chandra, A., Goyal, P.: Dynamic provisioning of multi-tier internet applications. In: Proceedings of Second International Conference on Autonomic Computing, ICAC 2005, pp. 217–228 (2005)

4. Karve, A., Kimbrel, T., Pacifici, G., Spreitzer, M., Steinder, M., Sviridenko, M., Tantawi, A.: Dynamic placement for clustered web applications. In: Proceedings of the 15th international conference on World Wide Web, pp. 595–604. ACM New York, NY (2006)
5. Kimbrel, T., Steinder, M., Sviridenko, M.I., Tantawi, A.: Dynamic application placement under service and memory constraints. In: Nikoletseas, S.E. (ed.) WEA 2005. LNCS, vol. 3503, pp. 391–402. Springer, Heidelberg (2005)
6. Urgaonkar, B., Shenoy, P., Roscoe, T.: Resource overbooking and application profiling in a shared Internet hosting platform (2009)
7. Menascé, D.A., Almedia, V.A.F., Dowdy, L.W.: Performance by design: Computer Capacity Planning by Example. Prentice Hall, Upper Saddle River (2004)
8. Urgaonkar, B., Rosenberg, A.L., Shenoy, P., Zomaya, A.: Application Placement on a Cluster of Servers. International Journal of Foundations of Computer Science 18(5), 1023–1041 (2007)
9. Coffman Jr., E., Garey, M.R., Johnson, D.S.: Approximation algorithms for bin packing: a survey (1996)
10. Schrijver, A.: Theory of linear and integer programming. Wiley, Chichester (1986)
11. Amza, C., Ch, A., Cox, A., Elnikety, S., Gil, R., Rajamani, K., Zwaenepoel, W.: Specification and Implementation of Dynamic Web Site Benchmarks. In: 5th IEEE Workshop on Workload Characterization, pp. 3–13 (2002)
12. Zhang, Q., Cherkasova, L., Mathews, G., Greene, W., Smirni, E.: R-capriccio: a capacity planning and anomaly detection tool for enterprise services with live workloads. In: Middleware 2007: Proceedings of the ACM/IFIP/USENIX 2007 International Conference on Middleware, pp. 244–265. Springer-Verlag New York, Inc., Heidelberg (2007)
13. Eckerson, W., et al.: Three Tier Client/Server Architecture: Achieving Scalability, Performance and Efficiency in Client Server Applications. Open Information Systems 10(1) (1995)

A Stability Criteria Membership Protocol for Ad Hoc Networks*

Juan Carlos García, Stefan Beyer, and Pablo Galdámez

Instituto Tecnológico de Informática, Universidad Politécnica de Valencia, 46022 Valencia, Spain
{juagaror,stefan,pgaldam}@iti.upv.es

Abstract. *Consensus* is one of the most common problems in distributed systems. An important example of this in the field of dependability is *group membership*. However, *consensus* presents certain impossibilities which are not solvable on asynchronous systems. Therefore, in the case of group membership, systems must rely on additional services to solve the constraints imposed on them by the impossibility of consensus. Such additional services exist in the form of *failure detectors* and *membership estimators*.

The contribution of this paper is the upper-level algorithm of a protocol stack that provides *group membership* for dynamic, mobile and partitionable systems, mainly aimed at mobile ad hoc networks. *Stability criteria* are established to select a subset of nodes with low failure probability to form stable groups of nodes. We provide a description of the algorithm and the results of performance experiments on the NS2 network simulator.

Keywords: ad hoc networks, distributed systems, consensus, group membership, stability criteria.

1 Introduction

Distributed agreement or *consensus* is a well-known problem in distributed systems, in which a set of nodes must agree on a value (or a set of values). Solving the consensus problem allows facing other closely related problems, such as group membership and provides the basis to implement dependability protocols, such as reliable group communication or replication protocols. Informally speaking, all processes of the system propose a value and must reach an unanimous and irrevocable decision about one of the proposed values.

However, the development of distributed protocols in asynchronous systems [1] is a difficult task due to many restrictions and impossibilities [2,3]. The authors of [4,5] show that the problem of consensus and membership cannot be solved in asynchronous distributed systems in which nodes can fail. In mobile ad hoc networks these problems are increased due to the inherently unreliable nature of ad hoc networks [6,7]. In this kind of

* This work has been partially supported by EU FEDER and Spanish MEC under grant TIN2006-14738-C02-01 and by EU FEDER and Spanish MICINN under grant TIN2009-14460-C03.

R. Meersman, T. Dillon, P. Herrero (Eds.): OTM 2009, Part I, LNCS 5870, pp. 690–707, 2009.

network, nodes can move, suffer delays in packet delivery and connect or disconnect at any moment. Furthermore, battery use and bandwidth limits, combined with the broadcast nature and the related tendency for network flooding of these wireless networks create further complications for protocols aiming at consensus or consensus approximations.

In order to avoid the theoretic impossibility of obtaining a deterministic solution, researchers have studied mechanisms that add a certain level of synchrony to the asynchronous model, or use probabilistic proposals of the algorithms. One of the most commonly used mechanism its to rely on an *unreliable failure detector* [8]. A similar solution, often employed in wireless networks, is the use of *collision detectors* [9].

Group membership is a particular case of the consensus problem. Group membership protocols install the same *system view* on all nodes that form the system. If the protocol detects new changes (a node wants to join or leave the network) it must start a *view change* process to install the new *system view* on the nodes of the group. This critical dependability service is typically used to build group communication middleware or replicated applications.

In this paper, we propose a group membership protocol for ad hoc networks that relies on a set of *stable* nodes provided by a *membership estimation* service. The members of the membership view are selected by *stability criteria* [10]. These parameters determine which nodes are the most reliable ones to form a membership group. As the system grows more stable, groups merge. We believe this model to be well suited for mobile environments for two main reasons: Firstly, the system copes well with the frequent occurrence of partitions inherent in MANETs, as it is likely that partitions occur along the less stable connections and individual stable groups remain intact. Secondly, in most dependable applications only stable nodes are of interest for hosting dependability services and, in general, for ensuring algorithms progress. For example, in the context of replication our membership service also serves as a replica placement algorithm, identifying stable nodes.

2 Related Work

A lot of bibliography on ad hoc networks exists. The characteristics of these networks and the problems they present are well described in [11,12]. Ad hoc networks technology may be used to deploy distributed applications. In [6,7] problems that appear in MANET's for distributed applications are discussed. Routing protocols are an essential service for this kind of networks and have a large impact on the performance of distributed algorithms, such as membership protocols. The most commonly used protocols are described in [13,14,15].

Consensus is one of the most common problems studied in distributed systems. There are many restrictions and impossibilities to be solved in asynchronous systems as shown in [5,4,3]. In order to help to solve consensus, additional mechanisms are often employed as *failure detectors*. Their properties are shown in [8,16]. A related solution for ad hoc networks, *collision detectors*, is proposed in [9] . Some *consensus* algorithms for ad hoc networks based on the above solutions are described in [17,18]. *Group membership* for partitionable wired networks is a consensus-based problem that has been widely studied in [19,20,21,22]. Finally, different solutions for the *MANET group membership* problem are studied in [23,24,25,26,27]. These works rely on neighbourhood or locality parameters to discover the members that can join to the group. Nevertheless, our proposal relies on other different criteria selected by the applications that choose the most stable nodes in the system. This allows to form "multi-hop" groups in a completely transparent way.

3 Model/Architecture

We use a protocol-layered architecture to provide *group membership* in ad hoc networks. This stack allows isolation of the different services that form the implementation of the final membership protocol. We observe three different services: the *routing service* is located at the bottom of the stack. Routing is an essential service in ad hoc networks, as it allows communication amongst the nodes of the network. It offers basic primitives to the upper level services to send/receive packets to/from other nodes. Although routing protocols do not usually offer full guarantees of delivery, they are capable to route packets in multi–hop scenarios. There are a wide range of available protocols, classified by their behavior and functionality. In our system we use the DSR *reactive* routing protocol. We use this pure-reactive routing protocol due we have made some optimizations to decrease its power consumption as we showed [28].

The *membership estimation service* is located at the middle of the stack. It monitorizes the system to provide an *estimated* composition. The service may provide some spurious faults or an incomplete view of the system, but it is an essential service since it allows to start communication with other discovered nodes of the system. Routing protocols do not usually offer *membership (estimation)* information. Furthermore, in our architecture this service also provides information about the stability conditions of nodes (criteria that nodes must fulfill to be considered part of a group). Basically, the algorithm works sending broadcast heartbeat messages with process estimation table every T_g time with a probability P_g. Processes that receive this messages update their estimation tables with the new information and decides to propagate the new changes with probability P_s. One process is considered failed if his heartbeat is not updated after T_f time. More information about the membership estimator and our underlying stack services can be found in [29,30].

Finally, the *membership service* is located at the top of the stack. This service installs the same *system view* (i.e. a subset of all nodes of the system) on each node that satisfies some *stability criteria*. The objective of these *stability criteria* is to select nodes that have low probability of failure or crash in the future ensuring that the *system view* does not include many unreliable nodes. We called *Membership Partition* to our *membership service*. This service can be used by an upper level protocol/application that is noticed about the strict view changes that appear in the system, managing and acting in consequence. For example, this service may be useful for an upper-level *group communication* protocol.

4 Stability Criteria

Our group membership protocol for ad hoc networks relies on a set of *stable* nodes provided by the low-level *membership estimation* service. This set of nodes is selected by a chosen *stability criterion*. This parameter determines which nodes are the most reliable ones to form a membership group. Ad hoc networks have inherent problems like the possibility of message loss and possible occurrence of node failures. We must choose a *stability criterion* that ensures that all nodes will be running during the new view installation process of the membership protocol in order to satisfy the *consensus* (all nodes have the same view) amongst them. Furthermore, the choice of the appropiate parameter will be influenced by the needs of the application that uses the membership protocol. Next we describe some parameters that we can use as stability criteria.

4.1 Stability Criteria Based on Time

One kind of parameter to ensure the *stability* is node activity time on the network. In our approach, we use a membership estimation service that provides a *view estimation* of the system. In this service, nodes interchange messages that contain information about them. If a node emits a lot of messages during the network lifetime there is high probability that the node does not suddenly disconnect or move out of reach and can form part of a *stable nodes* subset.

The time that a node is running (that is, the time that the membership estimator has detected its presence) can be used as *stability criterion*. Nevertheless, the choice of how long a node needs to be marked as *stable* is a difficult task. For example, if we choose a low value a lot of not fully reliable nodes may enter the system.In many ad hoc applications, nodes connect-disconnect frequently due to a variety of reasons. In certain cases, some nodes only require a brief connection to retrieve a system state and are switched off again to economize their power consumption. In other cases, nodes in movement may fastly transit a partition emitting few membership estimation packets. In this case the *stable set* of nodes may constanly suffer of changes and the strict membership protocol may install a lot of views that may contain untrusted nodes.

On the other hand, if we choose a large value for the time parameter, the inclusion of new nodes in the *strict system view* is delayed and these nodes may not participate inmediately in the system. This situation may be critical for certain problems. For example, a distributed application that replicates data amongst a few nodes that suffer from low battery level may need to reallocate the state of the system to other nodes. In this case, new nodes that enter in the system (and have a pre-supposed good level of battery power) may not be considered stable in time to participate in this proccess and the application may crash.

In what follows, we propose different choices to estimate the time value that may be used to select the set of stable nodes.

- $0 - Time$: this is the most basic parameter selection. In this case, we imediateley promote all nodes that the membership estimator service detects to the *stable subset* (at 0 seconds after its detection). In essence, the service acts as a classic membership protocol. New nodes that enter the system and are detected by the others are immediately included in the *stable view* of the system. This option provides most possible updated system view but we need be aware that we are likely to overload the membership protocol.
- $N - Time$: in this case a node is marked as *stable* if it has been at least N seconds in the system (that is, membership estimation service detected it N seconds ago).

The value of N can be tuned dynamically, adjusting it to possible variations of the network environment. For example if an excessive amount of view changes are observed N can be increased.
- $N(T_g/P_g) - Time$: finally, we can use a more complex *stability criterion* which involves two parameters of the *membership estimation* service. T_g refers to the time interval after which a node decides to propagate a message with a probability P_g. We can say that it is highly probable that every T_g/P_g time interval a node propagates a message. In other words we can use as a *stability criterion* nodes that approximately emit N or more packets since they are alive.

4.2 Stability Criteria Based on Distance

Nowadays, a lot of mobile devices incorporate a *Global Positioning System* or *GPS*, that allows calculating the position of the device on the Earth. Obviously, applications on such nodes can benefit from the location device, for example to obtain distances from the actual position to other positions. We may use this information to provide a *stable* set of nodes. The membership estimation can add the *position* data to the packets to be propagated along the system. This allows to know the last position of the node. We can define different uses of the *positioning* data to select the most *stable* nodes:

- **Based on a fixed or reference point:** certain application may require a group of nodes to be near a fixed point, that may be a physical point. If the nodes interchange position information they can calculate the distances to the reference-prefixed point.
- **Distance-Radio x:** if the membership estimator interchanges data about *position* it can select which nodes that are inside a x range of meters. This may express the nodes that are going to be in the radio range of the node, decreasing the probability of communication problems.
- **Moving from/to node:** finally, we can use the *position* data to calculate a node's most recent movement and direction. With this information, we can discard nodes that are moving away from the coverage area. In fact, there are optimizations for routing protocols that use this heuristic to decide to which node to route a packet.

In general, the precision of the above calculations will be strongly related to the data update frequency. One of the characteristics of Ad Hoc Networks is the possibility of node mobility. If we have a very dynamic system with low frequency of data refreshing the information can become obsolete quickly. As commented above, an additional device that provides the location is needed by the nodes.

4.3 Stability Criteria Based on Energy Level

One of the main constraints of ad hoc networks are the limited battery power of the participants. Usually, a heterogenous range of mobile devices form the system. These devices rely on a battery in order to provide the energy to nodes. Nodes may have a diverse range of battery types that differ in terms of capacity and durability.

We can define the next criteria:

- **Energy level of the devices:** the membership estimation service can propagate the power level of the different nodes, in order to determine the most *powerful* nodes.

This means that we can form a subset of *stable* nodes that includes only nodes that have at least a minimum level of energy.

- **Signal strength:** when a node receives a message it can measure the power of the signal strength. This measure may be important because the reduction of the normal (or latest) transmission's signal power may mean that the node is exiting from the radio coverage or the node may have low battery level.

4.4 Stability Criteria Based on Application Parameters

Finally, we can introduce a different criteria to select the most convenient nodes of the system. The application that is running on the system may select the nodes depending on its necessities. This may be employed to form a strict group for the particular application necessities.

5 Membership Protocol Description

A classic membership protocol tries to install the same view on a set of nodes of a group. Groups are formed by the nodes that explicitly join or leave the group. However, in ad hoc networks nodes join to form the network itself for a certain purpose. Due to dynamic nature of such a system (nodes can move, disconnect/reconnect or become unreachable) it is impractical to compose a *strict system view*. For this reason, we reduce the *strict system view* to the more reliable nodes in order to ensure a good performance of the upper-level applications and do not consume excessive bandwidth or power by the protocol. In this case, we change the definition of a group from the classic concept of nodes that want to join to a group (by means of *join/leave* calls) to the more reliable nodes, selected by stability criteria, forming implicit groups of stable nodes (without any explicit call to *join/leave* primitives).

The behavior of our proposed membership protocol (that we called *Membership Partition*) is based on the concept of a *partition* P. Every node that runs the protocol is part of a partition P. The view of a node is formed by the nodes that are in this partition; that is, $view_i = P_i$. The minimun size of a partition ($size(P_i)$) is 1 which represents an isolated group (node) that only contains the own node identifier $P_i = \{id\}$. Each node has an integer value called *heartbeat* that is incremented every time it tries to perform a new view installation process. Each view has a view identifier $< viewID >= \{id, heartbeat\}$ consisting of the node identifier that starts the new view installation and its own heartbeat. The heartbeat of the initiator node is increased at the begin of the view installation process.

The algorithm seeks to join groups of nodes to form a new partition P' with the new view identifier and containing the merge of the partitions. We define a partition as the set of nodes that have the same view identifier, that is, are forming a group.

5.1 Stable Nodes

The membership protocol relies on a membership estimation service that is responsible for maintaining an estimation (possibly different amongst all nodes) of the system composition, but also to provide information on which nodes are considered locally *stable* based on a certain set parameters. Each node has access to its own estimation and to the list of nodes it considers *stable*. We define the function $stable()$ which returns the set of stable nodes provided by the membership estimation. The Stable subset can be defined as $Stable_i = stable(Estimation_i)$. This set contains the nodes that satisfy the *stability criteria* on the local node.

5.2 Node Roles

When the membership protocol tries to install a new system view, nodes can adopt three different roles. The first role is the `initiator` node. This role is adopted by a node that detects a change in its set of *stable nodes*. This `initiator` is responsible for starting a new view installation process. The second role is `responder` node. This role is adopted by a node that receives a membership table request message from an `initiator`. Finally, the `nip` (node in partition) node role will be adopted by the rest of nodes in the view intallation process.

5.3 Inclusion Process

When a node detects the appearance of a new *stable* node it starts the partition merge or inclusion process. The detector node takes the `initiator` role and requests from the newly detected node its membership table (`MT_REQUEST` message). The newly detected node takes on the role of `responder` node. Once the `initiator` node has the membership table, it merges the two tables and generates the new proposal. This proposal is sent to the nodes in its partition (that take the `nip` role) and to `responder` node (`MT_PROPOSE` message). The responder propagates the proposal to its partition nodes (which also take the `nip` role).

Once the proposal has been received by each node in both partitions, these nodes respond with an *acknowledge* (or *no acknowledge*, it depends on his own decision about new proposal) message to their respective `initiator` or `responder` node. Both nodes wait to receive the messages from nodes in their own partition. If the `responder` node has collected enough *acknowledge* messages based on a *decision algorithm* it communicates this notice to the `initiator`. If the `initiator` node has been notified by the `responder` and it also has collected enough acknowledge messages from nodes in its partition, it then proceeds to propagate an `APPLY_TABLE_ACK` message to the nodes in its partition and to the `responder` node that re-sends the message to its own `nip` nodes. All nodes that receive this message will apply the proposal and establish it as the new view, ending the process. Finally, to avoid the possible coincidence that various nodes start the inclusion of a node in the group, we can add an additional probabilistic P_{change} parameter that reflects the probability to start the process every time that a new node is detected. This parameter aids to avoid some blocking conditions amongst nodes that simultaneously start the inclusion process.

5.4 Exclusion Process

The exclusion process is similar to the inclusion process with the exception that in this case the `responder` node role does not exist. The node that detects a node that is no longer *stable* sends a new *view* proposal to the nodes in its partition. These nodes accept (or not) the new proposal and reply to the `initiator` node . When the initiator node has collected enough *acknowledge* messages, it sends an `apply table` message to all nodes in the partition (`nip`), in order to apply the new proposal. As we showed above, in order to avoid the possible coincidence that various nodes start the exclusion of a node in the group we can use P_{change} parameter avoiding the blocking conditions amongst nodes.

5.5 Node States

The algorithm is defined by the phases or states that nodes transit during the view installation process. Nodes can be in one of these states:

- `NORMAL`: this state will be adopted by all nodes when they are not involved in an update view process.
- `PROPOSE_NIP`: this state will be adopted by the nodes that are not an `initiator` or `responder` node in the view installation process. Once these nodes receive a view proposal, the nodes enter this state storing the new proposal and recognizing the source node with a `PROPOSED_TABLE_ACK` message. Eventually, nodes return to `NORMAL` after applying the new proposal when an `APPLY_TABLE_ACK` message is received.
- `PROPOSE_I`: this state will be adopted by a node when it receives a response message of type `MT_REPLY`, which contains the membership table of the `responder`. The node sends the new proposal to all nodes in its partition and to the `responder` node. The node will stay in this state until nodes in its partition respond with an acknowledge message passing to a `DECIDE_I` state.
- `DECIDE_I`: this state will be adopted by the `initiator` node that has received a `PROPOSED_TABLE_ACK` message of all nodes in its partition. The node will be kept waiting for a `DECIDE_TABLE_ACK` message from the `responder`. After this event, the node changes back `NORMAL` state after sending an `APPLY_TABLE_ACK` message to all the nodes in its partition and to the `responder` node.
- `PROPOSE_R`: this state will be adopted by the `responser` node when it receives the new view proposal. Next, it propagates the new proposal to all the nodes in its partition and waits for their response.
- `DECIDE_R`: this state will be adopted by the `responder` node when it receives the responses of all nodes in its partition. In this state, the node waits for a `RESPONDER_APPLY_TABLE` message from the `initiator`. The node applies the proposal and sends an `APPLY_TABLE_ ACK` message to all the nodes in its partition. After that, it enters `NORMAL` state.

5.6 Timers

We established timers in order to determine if the process can continue its execution. If a timer expires and the protocol has not completed the phase, the process is aborted and nodes return to the `NORMAL` state. $\texttt{PROPOSE}_t$ is the maximum time that a node can be in a `PROPOSE` state, whereas $\texttt{DECIDE}_t$ is the maximum time that a node can be in a `DECIDE` state.

5.7 Consensus Amongst New Proposal

When the `responder` and `initiator` nodes send the proposal to the nodes in their partition, they may respond with two message types, *acknowledge* or *no acknowledge*. The `responder` and `initiator` execute a *consensus* protocol with received messages in order to continue or stop the process. We propose three kinds of *consensus* variants: **strict** in which all nodes must reply with an *acknowledge* message to continue the new view installation process. **Medium** at least X acknowledge messages must be received for the process to continue (usually $X = (nodesinpartition)/2$). Finally, **soft** requires that at least one node sends an *acknowledge* message, in order to continue.

5.8 Additional Functions

To ease the readability of the algorithm, we provide some functions that are used in different points of the algorithm's code:

- `setProposal()`: this function sets the received proposal identifier in the proposal variable states of the process.
- `isSameProposal()`: this function determines if the packet received is referred to the same proposal that process is attending.
- `decideProposal()`: this function evaluates the received proposal and determines an *acknoledge* or *not acknoledge* response. The decision amongst them may be configured by the application constrains.
- `decision()`: this function determines if the process has suficient *acknoledge* response to continue the view installation process.

5.9 Algorithm

The following is the pseudo-code of the algorithm:

```
init
   initiatorNode_=NULL; responderNode_=NULL;
   proposal_=EMPTY; responserDECIDEACK_=false;
   heartbeat=1; state_=NORMAL;
   viewID_={i, heartbeat}; view={i};
endInit

event newStableNode (node)
   heartbeat_++; responderNode_=node;
   responderDECIDEACK_=FALSE;
   send(responderNode_, MEMBERSHIP_TABLE_REQUEST);
endEvent

event unstableNode (node)
   heartbeat++;
   initiatorNode_=ownNode;
   state_=PROPOSE_I;
   deleteProcess_=true;
   responderDECIDEACK_=true;
   createNewProposalWithoutNode(node);
   sendProposalToNodesInPartition();
   if aloneInPartition_ then
      applyProposal();
      responderNode_=initiatorNode_=EMPTY;
      state_=NORMAL; responderDECIDEACK_=false;
   endif
endEvent


event receivedPacket (packet)
   if state_==NORMAL then
      if packet==MT_REQUEST then
         state_=PROPOSE_R;
         initiatorNode_=packet->initiatorID();
         proposalID_=(packet->initiatorID(),packet->HB());
```

```
      sendTo(initiatorNode_,MT_REPLY);
    endif

    if packet==MT_PROPOSE and isSameProposal(packet->IDView()) then
      state_=PROPOSE_NIP;
      setProposal(packet->proposal)
      initiatorNode_=packet->initiatorID();
      sendTo(packet->source(),decideProposal(ACK,NACK));
    endif
    if packet==MT_REPLY and responderNode_==packet->source() then
      initiatorNode_=packet->initiatorID();
      state_=PROPOSE_I;
      newProposal=mergePartitions();
      setProposal(newProposal);
      sendToPartitionNodes(newProposal);
      sendTo(responderNode_,MT_PROPOSE);
      if aloneInPartition_ then
        state=DECIDE_I;
    endif
  endif
  if state_==PROPOSE_NIP then
    if packet==APPLY_TABLE_ACK and isSameProposal(packet->IDView()) then
      applyProposal();
      responderNode_=initiatorNode_=empty; state_=NORMAL;
    endif
  endif
  if state_==PROPOSE_I then
    if packet==PROPOSED_TABLE_ACK and isSameProposal(packet->IDView()) then
      setMessageReceivedFromNode(packet->source());
      if decision()==true then
        state_=DECIDE_I;
    endif
    if packet==DECIDE_TABLE_ACK and responderNode_==packet->source() then
      responderDECIDEACK_=true;
  endif
  if state_==DECIDE_I then
    if responderDECIDEACK_==true or (packet==DECIDE_TABLE_ACK and responderNode_==packet->source()) then
      sendToPartitionNodes(APPLY_TABLE_ACK);
      applyProposal();
      if not deleteProcess_ then
        sendTo(responderNode_, RESPONDER_APPLY_TABLE);
      responderNode_=initiatorNode_=empty;
      state_=NORMAL; responderDECIDEACK_=false;
    endif
  endif
  if state_==PROPOSE_R then
    if packet==MT_PROPOSE and initiatorNode_==packet->source() then
      setProposal(packet->proposal);
      sendToPartitionNodes(packet->proposal);
      if aloneInPartition_ then
        state_=DECIDE_R;
        sendTo(initiatorNode_,DECIDE_TABLE_ACK);
      endif
    endif
    if packet==PROPOSED_TABLE_ACK and isSameProposal(packet->IDView()) then
      setMessageReceivedFromNode(packet->source());
      if decision()==true then
        state_=DECIDE_R;
        sendTo(initiatorNode_,DECIDE_TABLE_ACK);
      endif
    endif
  endif
  if state_==DECIDE_R then
    if packet==RESPONDER_APPLY_TABLE and packet->source()==initiatorNode_ then
      sendToPartitionNodes(APPLY_TABLE_ACK);
      applyProposal();
      responderNode_=initiatorNode_=empty; state_=NORMAL;
    endif
  endif
endEvent

event timerTimeout()
  initiatorNode_=responderNode_=empty;
  responderDECIDEACK_=false; state_=NORMAL;
endEvent
```

6 Experiments

We have performed a set of experiments to test our membership protocol which we call *Membership Partition*. First, we identify some scenarios and configurations that can help us to extract some conclusions on the energetic performance and bandwidth usage of the protocol.

General simulation configuration. We used the NS2 network simulator [31] to perform our tests. This tool implements the DSR routing protocol. The membership estimator and *Membership partition* protocols were implemented by us. We made a set of 20 simulations per configuration and scenario type, varying the number of nodes from 4 to 20. Simulation time was established to 300 seconds.

We identify two kinds of scenario sizes. *Single–hop* scenarios consist in a $100 \times 100\ \mathrm{m}^2$ flat grid. In this case, all nodes are inside radio signal strength of each other. These scenarios are used to simulate enclosed settings, such as meetings or museums in which network or visitor-guide applications are running. *Multi–hop* scenarios consist in a $400 \times 400\ \mathrm{m}^2$ flat grid. These scenarios are used to simulate "open-air" settings, in which network communication is needed.

Furthermore, for each of the above scenarios we use two different mobility patterns: *static* in which nodes do not move during the simulation time and *dynamic* in which nodes move following a Random WayPoint Model. In this case, nodes move randomly to a destination at a maximum speed of *3m/s* and stay there for a *pause time* of 25 seconds before changing speed and direction. We employethis model due we want to test the performance of our algorithm under bad scenarios (nodes may have random and "un-logical" movements that may cause different problems to establish the set of stable nodes). Finally, we set the values for the energy consumption model to be approximately the same as those of the real network interface *Lucent WaveLAN*.

Membership estimator configuration. The most important parameters of our *membership estimator* protocol are: P_g which indicates the probability that a node broadcasts its table after it increments its own heartbeat (we set this parameter at 0.7). P_s indicates the probability that a node that receives a message rebroadcast the updated table if there were changes (we set this parameter at 0.5). T_f indicates the time to consider a node as failed if its *heartbeat* is not updated after this time (we set this parameter at 4s). Finally, T_s indicates the time that a node waits to listen for packets during initialization (we set this parameter at 2s).

Membership Partition configuration. The *Membership partition* protocol needs to know which *stability criteria* must apply to determine which nodes can join the group. In the simulations we use the following criteria:

- `GOD CRITERION`: which emulates a *perfect* membership estimator. This means that we can provide an exact and precise information of the system composition at every moment without spurious mistakes. We can use it to model the behavior of a classic membership group protocol as it would be used in a wired environment. Unfortunately this membership estimator is impossible to implement in practice.
- `NO TIME CRITERION`: in this case we use a basic *Membership Partition* protocol to join every node that appears in the estimation without any restriction or condition. The strict membership protocol acts in this case as a classic membership group protocol over a wireless network. We use this configuration to compare our approach to that of classic membership services.
- `N TIME CRITERION`: in this case the *Membership Partition* protocol only selects nodes that have been in the membership estimation table at least N seconds.

- `N TG PG TIME CRITERION`: in this case the *Membership Partition* protocol only selects nodes that have been in the membership estimation table at least $N * (T_g/P_g)$ (T_g and P_g are membership estimator parameters) time seconds. That is, nodes have a high probability that they send about N packets.

In our experiments we employed (sorted from weak to strong stability conditions) the following configurations: *GOD CRITERION, NO TIME CRITERION, 5 TIME CRITERION, 5 TG PG TIME CRITERION, 10 TIME CRITERION* and *10 TG PG TIME CRITERION*. $Propose_i$ and $Decide_i$ timers are established to 1s. The timers are used to unblock the view installation process in case that one of the phases does not progress correctly, usually due to packet loss. In the simulations the *Membership Partition* protocol starts 5s after the *membership estimation* service, in order to initialize the system with some values in the tables. We employ a *soft* type decision for nodes that switch to an *unstable* state. This means that the group excludes a node when at least one node observes the transition to *unstable*.

7 Results

We have performed experiments to compare the power consumption of the *Membership Partition* protocol in different configurations. In Figures 1 and 3 we can see the results for each configuration of the static and dynamic single–hop scenarios.The results show that the use of *stability criteria* decreases the power consumption. If we use a strong *stability criterion* we restrict the conditions for a node to join the partition, decreasing the number of *inclusion processes*. This feature seems to be undesirable, but in a dynamic environment with strong constraints, such a trade-off may be beneficial, as the systems suffers less view changes and upper-level applications do not suffer delays in their progress due to the permanent process of inclusion/exclusion of nodes. Moreover, if we employ a strong *stability criterion* we ensure that groups are formed by the most reliable nodes of the system, allowing us to make a good choice for hosting reliable services.

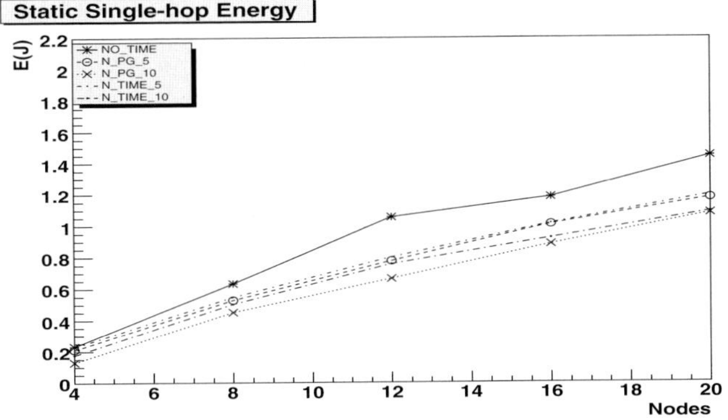

Fig. 1. Static Single–hop energy

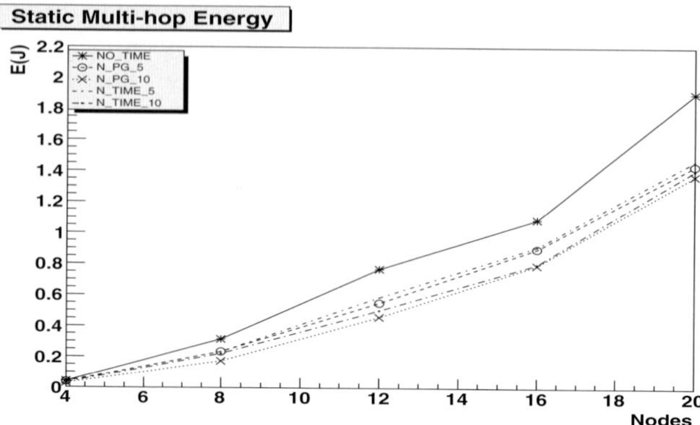

Fig. 2. Static Multi–hop energy

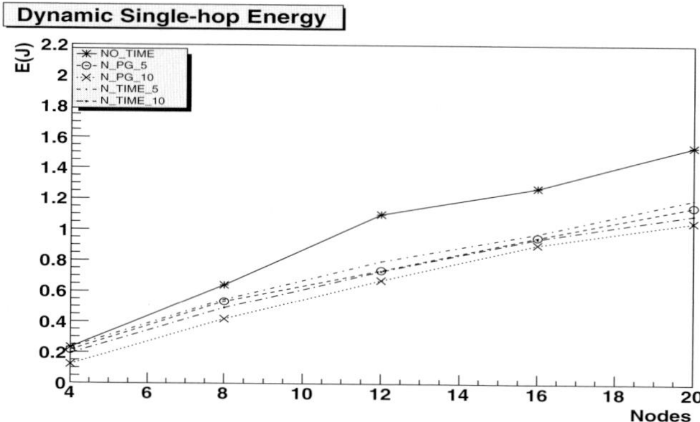

Fig. 3. Dynamic Single–hop energy

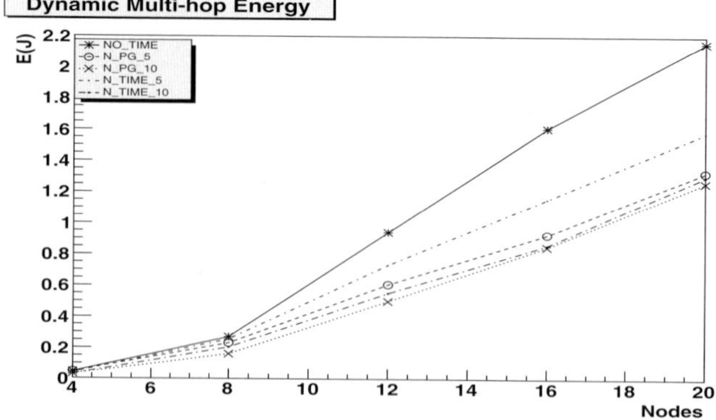

Fig. 4. Dynamic Multi–hop energy

We can see in Figures 2 and 4 the results for each configuration of the static and dynamic multi–hop scenarios. In this case, we observe that for a low number of nodes, we obtain lower power consumption than in single–hop scenarios, but a higher power consumption when the number of nodes increases. This is probably due to the possible existence of network partitions with a low number of nodes. If these occur, nodes that run *Membership Partition* form small partitions requiring less packets to interchange in every process. However, as the number of nodes grows reachability of nodes/partitions may increase, but the dynamicity causes the membership estimator to produce spurious failures due to the difficulty of maintaining its tables updated correctly. If we do not use any *stability criteria* (as *NO TIME*), we are essentially using a traditional membership service. We have included this experiment in the figures as a reference point to show the significant improvement in power consumption provided by our protocol.

To complement the results obtained in the study, we have performed simulations to compare the performance of the protocol in a single–hop scenario. We can model this

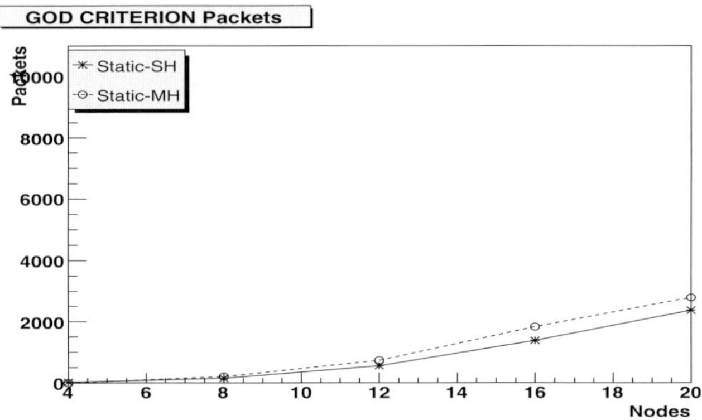

Fig. 5. GOD CRITERION Packets

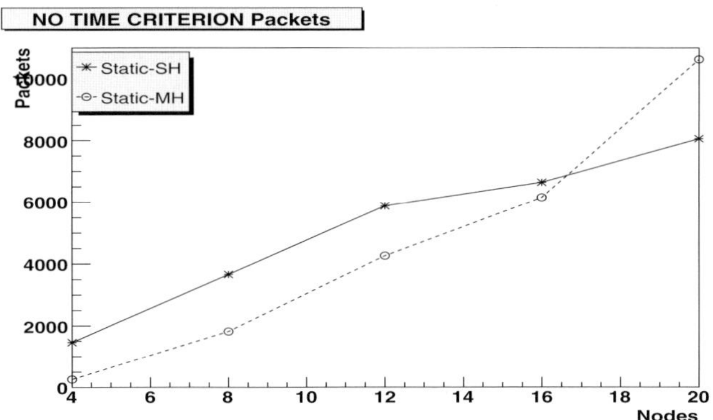

Fig. 6. NO TIME CRITERION Packets

scenario like a wired scenario in which the topology of the system is known in the case of the static experiments, whereas we can model the dynamic scenario as a conference room scenario where nodes enter and leave. We have performed experiments to compare the bandwidth usage (in terms of number of packets received/sent by a node during the simulation) of the *Membership Partition* protocol using the *NO TIME*, *N TG PG 10* and *N TIME 10* approaches comparing them with a *perfect system* (*GOD CRITERION*), that is a system that integrates a perfect *membership estimator* that retrieves the exact composition of the system without mistakes at any moment.

We can see in Figures 5, 6, 7 and 8 the results for the packets generated in each configuration. We observe that as expected *GOD CRITERION* experiments obtain the best results for all cases. This due to the system being fully connected and the algorithm rapidly forms a unique and big partition as the result of the merge of the small partitions and remains without changes during the simulation, decreasing the maintenance of the

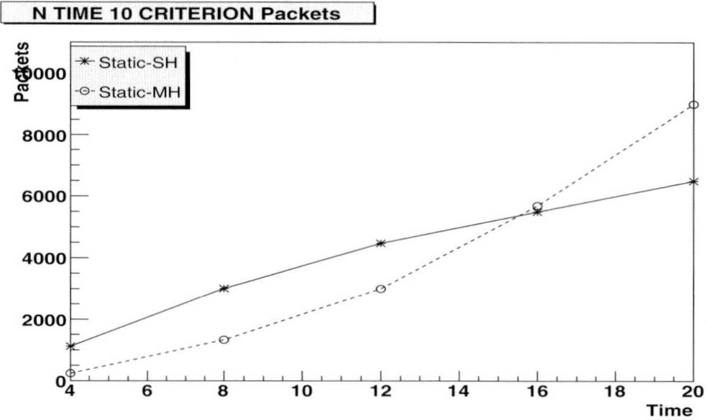

Fig. 7. N TIME 10 CRITERION Packets

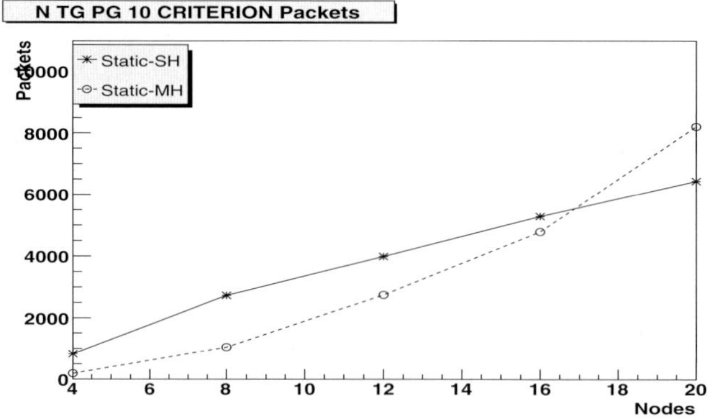

Fig. 8. N TG PG 10 CRITERION Packets

system (new view change processes) to the minimum. When we see the results for a more realistic environment (using data from a real membership estimator), we observe that the number of packets are higher. This due to the results provided by the *membership estimator* which cause a lot of spurious failures. Nevertheless, if we employ some restrictive *stability criteria* we can reduce the bandwidth usage of the protocol, as the system's groups are formed by *reliable* nodes.

Furthermore, we observe that in the *perfect system* in multi–hop scenarios more packets are generated than in single–hop scenarios, whereas in the other cases, multi–hop scenarios sent fewer packets than single–hop, when the number of nodes is low, but inverting this behavior when the number of nodes increases. When the number of nodes increase, the probability of isolation is reduced and nodes are more easily detected and updated by the *membership estimator*.

8 Conclusion and Future Work

In this work we presented a protocol stack which provides a group membership solution for ad hoc networks. In this protocol stack we find a strict membership protocol, a membership estimator protocol and a routing protocol. At the top of the layer strict membership generates a consistent view of the system that should be installed in a *stable* group of nodes. *Stable nodes* will be selected by the membership estimator protocol. The objective is to provide a subset of nodes of the system that satisfy certain *stability criterion*, in order to ensure the maximum reliability of the subset. Finally, at the bottom a routing protocol is used by the membership protocol to interchange the messages amongst nodes.

We believe that the selection of a reliable subset of nodes maximizes the probability that nodes do not fail during *view installation*. This reduces the *group membership/-consensus* problem in ad hoc networks. The subset of nodes is selected in function of a *stability criterion* that may be chosen for the particular necessities of the application. We presented results of experiments in which we observe that the use of *stability criteria* can significantly reduce the bandwidth and power consumption of nodes. In future work, we are planning to perform wider studies of particular application scenarios and aim to introduce further *stability criteria*.

References

1. Chockler, G., Keidar, I., Vitenberg, R.: Group communication specifications: a comprehensive study. ACM Computing Surveys 33(4), 427–469 (2001)
2. Schiper, A.: Practical impact of group communication theory. In: Schiper, A., Shvartsman, M.M.A.A., Weatherspoon, H., Zhao, B.Y. (eds.) Future Directions in Distributed Computing. LNCS, vol. 2584, pp. 1–10. Springer, Heidelberg (2003)
3. Fischer, M.J.: The consensus problem in unreliable distributed systems (a brief survey). Fundamentals of Computation Theory, 127–140 (1983)
4. Fischer, M.J., Lynch, N.A., Paterson, M.S.: Impossibility of distributed consensus with one faulty process. J. ACM 32(2), 374–382 (1985)
5. Chandra, T.D., Hadzilacos, V., Toueg, S., Charron-Bost, B.: On the impossibility of group membership. In: Proceedings of the 15th Annual ACM Symposium on Principles of Distributed Computing (PODC 1996), pp. 322–330. ACM, New York (1996)

6. Basile, C., Killijian, M., Powell, D.: A survey of dependability issues in mobile wireless networks. Technical report, LAAS CNRS Toulouse, France (2003)
7. Vollset, E., Ezhilchelvan, P.: A survey of reliable broadcast protocols for mobile ad-hoc networks. Technical Report CS-TR-792 (2003)
8. Chandra, T.D., Toueg, S.: Unreliable failure detectors for reliable distributed systems. J. ACM 43(2), 225–267 (1996)
9. Chockler, G., Demirbas, M., Gilbert, S., Newport, C., Nolte, T.: Consensus and collision detectors in wireless ad hoc networks. In: PODC 2005: Proceedings of the twenty-fourth annual ACM symposium on Principles of distributed computing, pp. 197–206. ACM Press, New York (2005)
10. Mostefaoui, A., Raynal, M., Travers, C., Patterson, S., Agrawal, D., Abbadi, A.E.: From static distributed systems to dynamic systems. In: SRDS 2005: Proceedings of the 24th IEEE Symposium on Reliable Distributed Systems, pp. 109–118. IEEE Computer Society, Los Alamitos (2005)
11. Baker, F.: An outsider's view of manet. Network Working Group Internet-Draft (2002)
12. Frodigh, M., Johansson, P., Larsson, P.: Wireless ad hoc networking, the art of networking without a network. Technical report, Ericsson Review No. 4 (2000)
13. Johnson, D.B., Maltz, D.A., Broch, J.: 5. In: DSR: The Dynamic Source Routing Protocol for Multihop Wireless Ad Hoc Networks, pp. 139–172. Addison-Wesley, Reading (2001)
14. Perkins, C.E., Royer, E.M.: Ad-hoc on-demand distance vector routing. In: WMCSA 1999: Proceedings of the Second IEEE Workshop on Mobile Computer Systems and Applications, vol. 90. IEEE Computer Society, Los Alamitos (1999)
15. Jacquet, P., Mühlethaler, P., Clausen, T., Laouiti, A., Qayyum, A., Viennot, L.: Optimized link state routing protocol for ad hoc networks. In: Proceedings of the 5th IEEE Multi Topic Conference (INMIC 2001), pp. 62–68 (2001)
16. Freiling, F.C., Guerraoui, R., Kouznetsov, P.: The failure detector abstraction. Technical report, Faculty of Mathematics and Computer Science, University of Manheim, Germany (2006)
17. Angluin, D., Fischer, M.J., Jiang, H.: Stabilizing consensus in mobile networks. In: Gibbons, P.B., Abdelzaher, T., Aspnes, J., Rao, R. (eds.) DCOSS 2006. LNCS, vol. 4026, pp. 37–50. Springer, Heidelberg (2006)
18. Wu, W., Yang, J., Raynal, M., Cao, J.: Design and performance evaluation of efficient consensus protocols for mobile ad hoc networks. IEEE Trans. Comput. 56(8), 1055–1070 (2007)
19. Dolev, D., Malki, D.: The transis approach to high availability cluster communication. Communications of the ACM 39, 64–70 (1996)
20. Moser, L.E., Melliar-smith, P.M., Agarwal, D.A., Budhia, R.K., Lingley-papadopoulos, C.A.: Totem: A fault-tolerant multicast group communication system. Communications of the ACM 39, 54–63 (1996)
21. Renesse, R.V., Birman, K.P., Maffeis, S.: Horus: A flexible group communication system. Communications of the ACM 39, 76–83 (1996)
22. Friedman, R., van Renesse, R.: Strong and weak virtual synchrony in horus. In: SRDS 1996: Proceedings of the 15th Symposium on Reliable Distributed Systems (SRDS 1996), vol. 140. IEEE Computer Society, Los Alamitos (1996)
23. Liu, J., Sacchetti, D., Sailhan, F., Issarny, V.: Group management for mobile ad hoc networks: design, implementation and experiment. In: MDM 2005: Proceedings of the 6th international conference on Mobile data management, pp. 192–199. ACM Press, New York (2005)
24. Mohapatra, P., Gui, C., Li, J.: Group communications in mobile ad hoc networks. Computer 37(2), 52–59 (2004)
25. Briesemeister, L., Hommel, G.: Localized group membership service for ad hoc networks. In: International Workshop on Ad Hoc Networking (IWAHN), pp. 94–100 (August 2002)

26. Roman, G.C., Huang, Q., Hazemi, A.: Consistent group membership in ad hoc networks. In: Proceedings of the 23rd international conference on Software engineering, pp. 381–388. IEEE Computer Society, Los Alamitos (2001)
27. Bollo, R., Le Narzul, J.P., Raynal, M., Tronel, F.: Probabilistic analysis of a group failure detection protocol. In: WORDS 1999: Proceedings of the Fourth International Workshop on Object-Oriented Real-Time Dependable Systems, Washington, DC, USA, vol. 156. IEEE Computer Society, Los Alamitos (1999)
28. García, J.C., Beyer, S., Galdámez, P.: Cross-layer cooperation between membership estimation and routing. In: SAC 2009: Proceedings of the ACM symposium on Applied Computing, pp. 8–15. ACM, New York (2009)
29. García, J.C., Banyuls, M.C., Galdámez, P.: Trading off consumption of routing and precision of membership. In: Proceedings of the 3rd International Conference Communications and Computer Networks, Marina del Rey, CA (USA), October 24–26, pp. 108–113. ACTA Press (2005)
30. García, J.C., Bañuls, M.-C., Beyer, S., Galdámez, P.: Effects of mobility on membership estimation and routing services in ad hoc networks. In: Stojmenovic, I., Thulasiram, R.K., Yang, L.T., Jia, W., Guo, M., de Mello, R.F. (eds.) ISPA 2007. LNCS, vol. 4742, pp. 774–785. Springer, Heidelberg (2007)
31. project, T.V.: The NS2 manual. Technical report, ISI (2004), http://www.isi.edu/nsnam/ns/ns-documentation.html

Proactive Byzantine Quorum Systems

Eduardo A.P. Alchieri[1], Alysson Neves Bessani[2],
Fernando Carlos Pereira[1], and Joni da Silva Fraga[1]

[1] DAS, Federal University of Santa Catarina - Florianópolis - Brasil
[2] University of Lisbon, Faculty of Sciences, LaSIGE - Lisbon - Portugal

Abstract. Byzantine Quorum Systems is a replication technique used to ensure availability and consistency of replicates data even in presence of arbitrary faults. This paper presents a Byzantine Quorum Systems protocol that provides atomic semantics despite the existence of Byzantine clients and servers. Moreover, this protocol is integrated with a protocol for proactive recovery of servers. In that way, the system tolerates any number of failures during its lifetime, since no more than f out of n servers fail during a small interval of time between recoveries. All solutions proposed in this paper can be used on asynchronous systems, which requires no time assumptions. The proposed quorum system read and write protocols have been implemented and their efficiency is demonstrated through some experiments carried out in the Emulab platform.

1 Introduction

Quorum systems [7] are fundamental tools used to ensure consistency and availability of data stored in replicated servers. Appart from its use in the construction of synchronization protocols (e.g., consensus), quorum-based protocols for register implementation are appealing due to their scalability and possibility of load balancing, since most operations does not need to be executed in all servers, but only in a subset of them (a quorum). The consistency of the stored data is ensured by the intersection between every quorum of the system. Quorum systems can be used to implement registers that provide read and write operations with several possible semantics (safe, regular or atomic) [8].

The concept of quorum systems was initially proposed for environments in which servers could be subject to crash faults [7]. Later, the model was extended to tolerate Byzantine faults [11]. However, the biggest challenge in quorum systems is how to design efficient protocols that tolerate malicious clients. The problem is that clients can execute some malicious actions to hurt system properties, e.g., sending an update only to some servers and not to a complete quorum [10]. This possibility of misbehaviour should not be discarded since quorum systems were developed to be used mainly in open systems such as the Internet, where there is a high probability of at least some clients being malicious.

The first protocols that tolerate Byzantine clients required at least $4f + 1$ servers to tolerate f faults (on servers) [11,12]. However, these protocols does not completely constraint faulty actions of malicious clients. There are some attacks

R. Meersman, T. Dillon, P. Herrero (Eds.): OTM 2009, Part I, LNCS 5870, pp. 708–725, 2009.

that still can be executed, e.g., a malicious client prepare several writes to be executed by another malicious client (colluder) after its exclusion of the system. More recently, these weakness were mitigated by the BFT-BC protocol [10], which requires only $3f + 1$ servers and allow clients to execute a write only after its previous write completes. The BFT-BC protocol relies on digital signatures (one of the biggest latency sources on Byzantine fault-tolerant protocols [4,6]) to constraint the actions of malicious clients.

In this paper we extend BFT-BC and propose a new Byzantine quorum system protocol in which the actions of malicious clients are constrained through threshold cryptography. The main benefit of our approach is **(1)** the use of a single public key for the whole storage service (instead of one public key per server), which simplifies the management of the system and make it more affordable in dynamic environments in which clients come and go, and **(2)** a considerable reduction on the size of message certificates, which makes our protocol much more efficient than BFT-BC when the number of faults tolerated f increases.

One important property of fault tolerant replicated systems is the bound f on the number of faulty servers that can be tolerated. This property can be a problem for long lived systems, since given a sufficient amount of time, an adversary can manage to compromise more than f servers and then impair the correctness of the system. To overcome this limitation, proactive recovery schemes [4,18] should be used. In a system in which this kind of technique is employed, each server is recovered periodically, in such a way that all servers are recovered in a bounded time interval. This mechanism allows an unlimited number of faults to be tolerated on the system lifetime, provided that no more than f faults occur during a recovery period. Another novel feature of our protocols is that the read/write algorithms were developed together with a proactive recovery scheme in order to make the register abstraction usefull in long lived systems. As far as we known, this is the first generic solution for register implementation that includes a proactive recovery scheme.

The contributions of this paper can be summarized as follows:

1. new quorum system read and write protocols that uses threshold cryptography to constraint the actions of malicious clients and ensure the consistency of the stored data;
2. a proactive recovery scheme for quorum systems that does not suffer from common weakness of previous efforts [18];
3. an experimental evaluation in which we compare our protocol with BFT-BC and shows the benefits (in terms of performance) of using threshold signatures instead of "common" public key criptography (e.g., RSA).

The paper is organized as follows. Section 2 presents our system model and the concept of Byzantine Quorum Systems, among other preliminary definitions used in this paper. Section 3 describes our proposal for Proactive Byzantine Quorum Systems. Section 4 presents an analytical analysis and some experiments realized with our protocols. Some related work are discussed in Section 5. Finally, Section 6 presents our final remarks and the next steps of this work.

2 Background

2.1 System Model

The system model consists of a set $C = \{c_1, c_2, ...\}$ of clients interacting with a set of n servers $S = \{s_1, ..., s_n\}$, which implement a quorum system with atomic semantic operations. Clients and servers have unique identifiers.

All the processes in the system (clients and servers) are prone to *Byzantine failures* [9]. A process that shows this type of failure may present any behavior: it may stop, omit transmission and delivery of messages, or arbitrarily deviate from its specification. A process that present this type of failure behavior is called faulty, otherwise it is called correct. However, faulty processes may be recovered, resuming a correct behavior again (proactive recovery). Furthermore, in this work we assume *fault independence* for processes, i.e., processes failures are uncorrelated. This assumption can be substantiated in practice using several types of diversity [14].

In terms of guarantees, the system remains correct while it presents a maximum of f faulty servers in a given time, being needed $n = 3f + 1$ servers in the system. Furthermore, an unlimited number of clients may be faulty.

We assume the asynchronous system model[1], where processes are connected by a network that may not send, delay, duplicate or corrupt messages. Moreover, time bounds for message transfers and local computations are unknown in the system. The only requirement for our protocols to terminate is that, if a process sends infinite times a message to another correct process, then this message will eventually be delivered in the receiver. To fulfill this requirement and to simplify the presentation of protocols, we consider that communications between processes are made through reliable and authenticated point-to-point channels.

Also, our protocols use threshold cryptography to constraint the actions of Byzantine clients and to ensure the integrity of stored data. Thus, we assume that in the startup of the system each server receives its partial key (a share of a service secret key [16]) that will be used in the preparation of partial signatures. A correct server never reveals its partial key. These keys are generated and distributed by a correct administrator that is only needed in the initialization of the system. The public key of the service, used to verify the signatures generated by the combination of signature shares (partial signatures) in this mechanism, is stored by the servers and is available to any process of the system.

We also assume the existence of a cryptographic hash function h resistant to collisions, so that any process is able to calculate the hash $h(v)$ from the value v. It is computationally infeasible to obtain two distinct values v e v' such that $h(v) = h(v')$.

Finally, to avoid replay attacks, some messages are tagged with *nonces*. We assume that clients do not choose repeated *nonces*, i.e., *nonces* already used.

---

[1] However, the proactive recovery procedure uses some mechanisms to guarantee that it starts and ends (Section 3.3).

2.2 Byzantine Quorum Systems

Byzantine Quorum Sytems [11], hereafter called simply as quorum systems, can be used to implement replicated data storage systems, ensuring consistency and availability of the stored data even in the presence of Byzantine faults in some replicas. Quorum systems algorithms are recognized for their good performance and scalability, once clients of these systems in fact access only a particular subset instead of all servers.

Servers in a quorum system are organized into subsets called quorums. Any two quorums have a nonempty intersection that contains a sufficient number of correct servers (ensuring consistency). Also, there is at least one quorum in the system that is formed only by correct servers (ensuring availability) [11]. Quorum systems are used to build register abstractions that provide read and write operations. The data stored on the register is replicated on the servers of the system. Each register stores a pair $\langle v, t\rangle$ with a value v of the stored data and an associated timestamp t.

The protocol presented in this paper is an extension of the BFT-BC [10], allowing proactive recovery of servers of the system. We choose this protocol due to its optimal resilience (it requires $n = 3f + 1$ to tolerate up to f faulty servers), strong semantics (implements an atomic register [8]) and tolerance to malicious clients.

In order to use a Byzantine quorum system with only $3f + 1$ servers it is required the stored data to be self-verifiable [11]. This is a direct consequence of the fact that the intersection between any two quorums on this system would have at least $f+1$ servers, and thus can contain only a single correct and updated server. Thus, clients correctly obtain the stored data, from this correct server, only if the register data is self-verifiable, i.e., it is possible to verify the integrity of the data stored on a single server without consulting other servers. In this sense, BFT-BC introduces a new way to make data self-verifiable: verification of a set of servers signatures. Other protocols, such as *f-dissemination quorum* system [11,12], are based on client signatures and therefore, they do not tolerate malicious clients[2].

Thus, to maintain its consistency semantics, BFT-BC uses a mechanism of completion proofs signed by servers at all phases of the protocol. To a client be able to enter in a new phase of the protocol, it is necessary that it presents a proof that it completed the previous phase. This proof, called certificate, comprises a set of signed replies collected from a quorum of $n - f = 2f + 1$ servers in the previous phase. For example, to a client write some value in the system it needs to have completed its previous write. By using this technique, BFT-BC employs 2/4 communication steps to execute read operations and 4/6 steps to execute write operations (Figure 1). The BFT-BC algorithm works as follows:

- **Read operations.** (Figure 1(a)) – the client requests a set of valid pairs $\langle v, t\rangle$ from a quorum of servers and selects that one with the highest timestamp $\langle v_h, t_h\rangle$. The operation ends if all returned pairs are equals, i.e., they

[2] The *f-masking quorum system* protocol [11,12] also tolerates malicious clients, but requires replication in $4f + 1$ servers, because it does not store self-verifying data.

have the same timestamp t_h and the same value v_h (this happens in executions without concurrency and failures). Otherwise, the client performs an additional phase of *write back* in the system and waits for confirmations until it can be assured that a quorum of servers has the most recent value v_h;

- **Write operations.** (Figure 1(b)) – the client obtains a set of timestamps from a quorum of servers (as in read operations) and then performs the preparation for its writing. In this phase it tries to obtain from a quorum of servers a set of proofs necessary to complete its current write operation. In case of success in the preparation, the client writes on a quorum of servers, waiting for confirmations. In an alternative scenario, the client may run an optimized protocol, performing in a single phase the both timestamp definition and write preparation (dotted lines in Figure 1(b)).

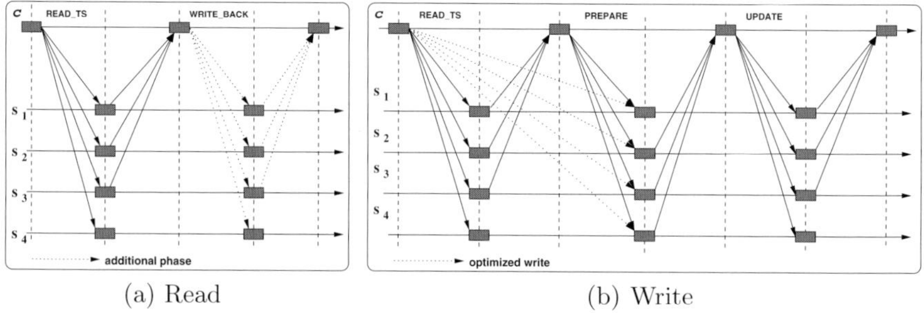

(a) Read (b) Write

Fig. 1. BFT-BC Read/Write Operations

2.3 Threshold Cryptography

The main mechanism used in this work is a threshold signature scheme (TSS) [16] by which it is possible to control actions of clients and to ensure the integrity of the data stored by servers. Its flexibility also facilitates the servers recovery procedure. In a scheme (n, k)-TSS, a trusted **dealer** initially generates n secret key shares (partial keys) $(SK_1, ..., SK_n)$, n verification keys $(VK_1, ..., VK_n)$, the group verification key VK and the public key PK used to validate signatures. Moreover, the dealer sends these keys to n different players, called **share holders**. Thus, each share holder i receives its partial key SK_i and its verification key VK_i. The public key and all verification keys are available for any part that composes the system.

After this initial setup, the system is able to generate signatures. To obtain a signature A to some data d, each share holder i generates its partial signature a_i (also called signature share) of d. Later, a **combiner** obtains at least k valid partial signatures $(a_1, ..., a_k)$ and builds the signature A through the combination of these k valid partial signatures. An important feature of this scheme is the impossibility of generating valid signatures with less than k valid partial signatures. This scheme is based on the following primitives:

- *Thresh_Sign*$(SK_i, VK_i, VK, data)$: function used by the share holder i to generate its partial signature a_i to *data* and the proofs v_i of validity of a_i, i.e.,$\langle a_i, v_i \rangle$;
- *Thresh_VerifyS*$(data, \langle a_i, v_i \rangle, PK, VK, VK_i)$: function used to verify if the partial signature a_i, obtained from the share holder i, is valid for *data*;
- *Thresh_CombineS*$(a_1, ..., a_k, PK)$: function used by the combiner to compose signature A from k valid partial signatures;
- *verify*$(data, A, PK)$: function used to verify if A is a valid signature of *data*.

In this work, we use the protocol proposed in [16], where it is proved that such a scheme is secure in the random oracles model [2], not being possible to forge signatures. This protocol represents a RSA [15] threshold signature scheme, i.e., the combination of partial signatures generates a RSA signature. In this model, the generation and verification of partial signatures is fully non-interactive, not being necessary to exchange messages to perform these operations.

3 Proactive Byzantine Quorum Systems

This section presents our approach to build a register that offers read and write operations, implemented through Byzantine quorum replication. Our protocol is based on BFT-BC [10] and therefore is able to tolerate malicious clients. The main feature introduced in our protocol is the possibility of proactive recovery of servers (Section 3.3), and thus it is called PBFT-BC (*proactive Byzantine fault-tolerance for Byzantine clients*). Moreover, PBFT-BC outperforms BFT-BC (Section 4) and also presents optimal resilience ($n = 3f + 1$).

A Byzantine quorum system should tolerate the presence of malicious clients, since they can perform the following actions to corrupt the system [10]: *(i)* write different values associated with the same timestamp; *(ii)* execute the protocol partially, updating the value stored by only some (few) servers; *(iii)* choose a very large timestamp and exhaust the timestamp space; and *(iv)* issue a large number of write requests and hand them off to a colluder who will run them after the bad client has been removed from the system.

Thus, protocols developed for this purpose should make impossible for malicious clients to perform these actions (or mask them). Our protocol uses a threshold signature scheme to control the actions performed by clients and to ensure the integrity of stored data. We consider the existence of a correct administrator who will perform the distribution function (dealer) of the scheme, where servers act as share holders and clients as combiners. The administrator generates the partial keys and distributes them to the servers. Also, all public information (public and verification keys) are sent to all servers. Thereafter, each client gets these public information from the servers by sending a request and waiting for $f+1$ identical replies. Notice that the administrator is only necessary in the initialization of the system, during the setup phase.

Our protocol uses quorums of $2f+1$ replicas and a $(n, 2f+1)$-TSS scheme, i.e., it is required a quorum of servers in order to generate a valid signature. Thus, all actions performed by clients must be authorized by a quorum of servers.

Clients need to prove that they are acting properly to move from one phase to another phase of the protocol. They do it by using certificates, which contain data indicating the validity of the actions that they are trying to execute and a signature that ensures the integrity of these data. This signature is generated by the $(n, 2f+1)$-TSS scheme ensuring that a quorum of servers approved its actions. Our protocols use two kinds of certificates:

Prepare Certificate: A client uses this certificate to prove that its writing was approved by at least a quorum of servers. On the other hand, servers use it to prove the integrity of the stored values. A prepare certificate pc has three fields: $pc.ts$ – timestamp of the proposed write; $pc.hash$ – hash of the value v proposed in the write; $pc.A$ – service signature, proving that at least a quorum of servers approve the write of v with timestamp $pc.ts$. A prepare certificate pc is valid only if its signature $pc.A$, for the tuple $\langle pc.ts, pc.hash\rangle$, is valid. This is determined by the operation $verify(\langle pc.ts, pc.hash\rangle, pc.A, PK)$.

Write Certificate: A client uses this certificate to prove that its last write has been completed. A write certificate wc has two fields: $wc.ts$ – timestamp of the completed write; $wc.A$ – service signature, proving that the client has done the write with timestamp $wc.ts$ in at least a quorum of servers. A write certificate wc is valid only if its signature $wc.A$, for $wc.ts$, is valid. This is determined by the operation $verify(wc.ts, wc.A, PK)$.

PBFT-BC can deal with multiple objects, since each of them has a distinct identifier. However, to simplify the presentation, we consider that servers store only a single register. Thus, each server i stores the following variables: **(1)** $data$ – value of the object; **(2)** P_{cert} – a prepare certificate valid for $data$; **(3)** P_{list} – a set of tuples $\langle c, ts, hash\rangle$ containing the client identifier c, the timestamp ts and the hash $hash$ of the value of a proposed write; **(4)** max_{ts} – timestamp of the latest write that i knows that was completed in at least a quorum of servers; **(5)** SK_i – partial key of i, that is used by the threshold signature scheme; **(6)** VK_i e VK – verification keys, that are used to generate proofs of validity of partial signatures; and **(7)** PK – service public key, that is used to validate certificates. Moreover, each client c uses the following variables: **(1)** W_{cert} – write certificate of the last write of c; **(2)** PK – service public key, that is used to validate certificates; and **(3)** VK e $VK_1, ..., VK_n$ – verification keys, that are used to validate partial signatures.

3.1 Read and Write Protocols

This section presents PBFT-BC's read and write protocols. We present the pseudocodes for each one of the operations and discuss their main features.

Clients should choose timestamps from different subsets in order to these protocols work properly. Thus, each client concatenates its unique identifier with a sequence number, i.e., $ts = \langle seq, id\rangle$. Timestamps are compared through their sequence number. If two timestamps have the same sequence number, then they are compared through their client identifier. Timestamps are incremented through the function $succ(ts, c) = \langle ts.seq + 1, c\rangle$.

Pseudocodes 1 and 2 present the write protocol executed by clients and servers, respectively. These pseudocodes represent the write version without optimizations, that demands three phases to complete a write operation.

Pseudocode 1. Protocol used by client c to write *value*.

- $w1.1$ Client c sends a message $\langle$READ_TS, $nonce\rangle$ to all servers.
- $w1.2$ c waits for a quorum $(2f + 1)$ of valid replies from different servers. A reply m_i from server i is valid if it is well-formed, i.e., $m_i = \langle$READ_TS_REPLY$, p, nonce\rangle$ where p is a valid prepare certificate (well-formed and the service signature is valid). Moreover, m_i should be correctly authenticated, i.e., its *nonce* matches the *nonce* used in step $w1.1$.
- $w1.3$ Among the prepare certificates received in step $w1.2$, c selects the certificate containing the highest timestamp, called P_{max}.
- $w2.1$ c sends a message $\langle$PREPARE$, P_{max}, ts, h(value), W_{cert}\rangle$ to all servers. Here $ts \leftarrow succ(P_{max}.ts, c)$, h is a hash function and W_{cert} is a write certificate of c's last write or *null* if this is c's first write.
- $w2.2$ c waits for a quorum $(2f + 1)$ of valid replies from different servers. A reply m_i from server i is valid if it is well-formed, i.e., $m_i = \langle$PREPARE_REPLY$, \langle ts_i, hash_i\rangle, \langle a_i, v_i\rangle\rangle$ where ts_i and $hash_i$ match the values sent in step $w2.1$ (ts and $h(value)$, respectively). Moreover, m_i is valid if $Thresh\_verifyS(\langle ts_i, hash_i\rangle, \langle a_i, v_i\rangle, PK, VK, VK_i)$ is *true*.
- $w2.3$ c combines the $2f + 1$ correct signature shares received in step $w2.2$, invoking $Thresh\_combineS(a_1, ..., a_{2f+1}, PK)$, and obtains the service signature A for the pair $\langle ts, h(value)\rangle$. Then, c forms a prepare certificate P_{new} for ts and $h(value)$ by using A.
- $w3.1$ c sends a message $\langle$WRITE$, value, P_{new}\rangle$ to all replicas.
- $w3.2$ c waits for a quorum $(2f+1)$ of valid replies from different servers. A reply m_i from server i is valid if it is well-formed, i.e., $m_i = \langle$WRITE_REPLY$, ts_i, \langle a_i, v_i\rangle\rangle$ where ts_i matches the value ts defined in step $w2.1$ and $Thresh\_verifyS(ts_i, \langle a_i, v_i\rangle, PK, VK, VK_i)$ is *true*.
- $w3.3$ c combines the $2f + 1$ correct signature shares received in step $w3.2$, invoking $Thresh\_combineS(a_1, ..., a_{2f+1}, PK)$, and obtains the service signature A for the timestamp ts. Then, c forms a write certificate W_{cert} for ts by using A. This certificate is used in c's next write.

In the first phase of the protocol the client defines the write timestamp and in the second phase it obtains a prepare certificate for this write operation. Afterwards, the write operation is definitely executed in the third phase. The progress of the protocol (i.e., for a client moves from one phase to another) is based on the use of certificates. The processing related with these certificates is one of the differences between PBFT-BC and BFT-BC (also, PBFT-BC provides protocols to recover servers periodically – Section 3.3). On PBFT-BC, a client obtains a certificate by waiting for a quorum of valid partial signatures (steps $w2.2$ e $w3.2$) and, then, combining them (steps $w2.3$ e $w3.3$), what results in a signature that is used to prove the validity of this certificate. Notice that to validate a certificate it is necessary to verify only one signature (service signature), differing from BFT-BC where a full quorum of signatures must be verified.

The most important phase of the write protocol is the second one. In this phase each server (Pseudocode 2) checks if: **(1)** the timestamp being proposed is correct; **(2)** the client is preparing just one write; **(3)** the value being prepared does not differ from a (possible) previous prepared value with the same timestamp; and **(4)** the client has completed its previous write. The item (1) is checked in the step $w2.1$. The items (2), (3) and (4) are checked in the steps $w2.2$

and $w2.3$, where each server uses its list of prepared writes (P_{list}). An important feature related with the use of this list is that a client is not able to prepare many write requests. Thus, a malicious client m is not able to prepare multiple write requests and hand them off to a colluder, that could execute them after m is removed from the system, what limits the damage caused by malicious clients.

Pseudocode 2. Write protocol executed at server i.

Upon receipt of $\langle \text{READ\_TS}, nonce \rangle$ **from** client c

- $w1.1$ i sends a reply $\langle \text{READ\_TS\_REPLY}, P_{cert}, nonce \rangle$ to c.

Upon receipt of $\langle \text{PREPARE}, P_c, ts, hash, W_c \rangle$ **from** client c

- $w2.1$ if the request is invalid or $ts \neq succ(P_c.ts, c)$, discard request without replying to c. A PREPARE request is invalid if either certificate P_c or W_c is invalid (not well-formed or the service signature is not valid).
- $w2.2$ if W_c is not $null$, set $max_{ts} \leftarrow max(max_{ts}, W_c.ts)$, and remove from P_{list} all entries e such that $e.ts \leq max_{ts}$.
- $w2.3$ if P_{list} contains an entry for c with a different ts or $hash$, discard the request without replying to c.
- $w2.4$ if $\langle c, ts, hash \rangle \notin P_{list}$ and $ts > max_{ts}$, add $\langle c, ts, hash \rangle$ to P_{list}.
- $w2.5$ i generates its signature share (partial signature): $\langle a_i, v_i \rangle \leftarrow Thresh\_sign(SK_i, VK_i, VK, \langle ts, hash \rangle)$.
- $w2.6$ i sends a reply $\langle \text{PREPARE\_REPLY}, \langle ts, hash \rangle, \langle a_i, v_i \rangle \rangle$ to c.

Upon receipt of $\langle \text{WRITE}, value, P_{new} \rangle$ **from** client c

- $w3.1$ if request is invalid or $P_{new}.hash \neq h(value)$, discard request without replying to c. A *write* request is invalid if the prepare certificate P_{new} is invalid (not well-formed or the service signature is not valid).
- $w3.2$ if $P_{new}.ts > P_{cert}.ts$, set $data \leftarrow value$ and $P_{cert} \leftarrow P_{new}$.
- $w3.3$ i generates its signature share (partial signature): $\langle a_i, v_i \rangle \leftarrow Thresh\_sign(SK_i, , VK_i, VK, P_{new}.ts)$.
- $w3.4$ i sends a reply $\langle \text{WRITE\_REPLY}, P_{new}.ts, \langle a_i, v_i \rangle \rangle$ to c.

Pseudocode 3. Read protocol executed at client c.

- $r1.1$ Client c sends a message $\langle \text{READ}, nonce \rangle$ to all servers.
- $r1.2$ c waits for a quorum ($2f + 1$) of valid replies from different servers. A reply m_i from server i is valid if it is well-formed, i.e., $m_i = \langle \text{READ\_REPLY}, value, p, nonce \rangle$ where p is a valid prepare certificate (well-formed and the service signature is valid) and $p.hash = h(value)$. Moreover, m_i should be correctly authenticated, i.e., its $nonce$ matches the $nonce$ used in step $r1.1$.
- $r1.3$ Among all replies received in step $r1.2$, c selects the reply with the prepare certificate containing the highest timestamp and returns the $value$ related with this reply. Also, if all timestamps obtained in step $r1.2$ are equals the read protocol ends.
- $r2.1$ Otherwise the client performs the write back phase for the highest timestamp. This is identical to phase 3 of writing (steps $w3.1$, $w3.2$ and $w3.3$), except that the client needs to send only to servers that are out of date, and it must wait only for enough responses to ensure that a quorum ($2f + 1$) of servers now have the updated information.

Pseudocodes 3 and 4 present the read protocol executed by clients and servers, respectively. Readings are completed in only one phase when there are no faults and concurrency on the system. However, an additional *writeback* phase [12] may be need, where clients write back the reading value in an enough number of servers, ensuring that the most recent information is stored in at least one quorum of servers.

Pseudocode 4. Read protocol executed at server i.

Upon receipt of $\langle \text{READ}, nonce \rangle$ **from** client c

r1.1 i sends a reply $\langle \text{READ\_REPLY}, data, P_{cert}, nonce \rangle$ to c.

The read protocol executed at servers is very simple. In these operations, servers send a reply containing the stored value and the certificate that proves the integrity of this data.

Correctness. The correctness conditions of PBFT-CUP, as the proofs that PBFT-CUP meets these conditions, are equals to those presented in [10], since these conditions are based on the validity of certificates and on the properties of quorum intersections. The atomicity of the register [8] is ensured by the write back phase of the read, i.e., writing back the read value ensures that all subsequent reads would read this value or a newer one. Wait-freedom is satisified due to the fact that all phases of the protocols require replies of only $n - f$ servers (a quorum), which are, by definition, always available on the system.

3.2 Optimizations

There are two optimizations that can be used to make the protocols more efficient in contention- and fault-free scenarios.

Avoiding the prepare phase of a write. As in BFT-BC (see Section 2.2), it is possible to agglutinate the functions of the phases 1 and 2 of the write protocol of PBFT-BC, reducing the number of communication steps from 6 to 4 in writes in which there are no faulty servers and no write contention. The idea is simple: if all timestamps read by a client c on the first phase of the write protocol are equal to ts, the obtained READ_TS_REPLY messages can be used as the prepare certificate for a timestamp $succ(ts, c)$. In order for this optimization to be used, the hash of the value to be written must be included both in READ_TS and READ_TS_REPLY messages.

Avoiding verification of partial signatures. Two of the most costly steps of the threshold signature scheme are: **(1)** verification of partial signatures (steps $w2.2$ and $w3.2$ of Pseudocode 1), executed by clients; and **(2)** generation of proofs of partial signatures validity (steps $w2.5$ and $w3.3$ of Pseudocode 2), executed by servers. In these steps, if there are no malicious servers in the system, the first quorum of replies received by clients will contain correct partial signatures that suffice to generate the service signature. So, we can change the algorithm to make the client first try to combine the first quorum of partial signatures received without verifying them. If some invalid partial signature is used in the combination, the service signature generated will be invalid for the data that the client is trying to obtain a signature. In this case, the client must wait for new partial signatures and make all possible combinations (using one quorum of partial signatures) until that it obtains a valid signature. Notice that, in the worst case, the client receives f invalid partial signatures in the first quorum of replies and must wait for more f replies in order to obtains a valid service signature. Moreover, as the system always has at least a quorum of correct servers, it is always possible to get a valid signature. In the fault-free case, this optimization drastically reduces the cryptographic processing time required in the write operations. Another advantage of this optimization is related with the proactive recovery of servers (Section 3.3), where their keys (verification keys and partial keys) are updated. Since clients do not verify the received partial signatures, they also do not need update the server verification key after the server recovery. Alternatively, to avoid many combinations in scenarios with failures, the client could execute the normal protocol when it does not obtain the correct service signature from the first quorum of replies.

3.3 Proactive Recovery

Most Byzantine fault-tolerant protocols consider that only a limited number of servers can fail during the system lifetime. However, many systems are designed to remain in operation for a long time (long lived systems), making this assumption problematic: given a sufficient amount of time, an adversary can manage to compromise more than f servers and corrupt the system.

To overcome this, we have developed a recovery protocol[3] that makes faulty replicas behave correctly again. Thus, PBFT-BC can tolerate any number of failures provided that only a limited number of faults f occurs concurrently within a small time interval, called **window of vulnerability**.

To recover a server, it is necessary execute the following actions: *(1)* reboot the system (both, hardware and configurations); *(2)* reload the code from a safe place (e.g., read-only memory); *(3)* recover the server state, that might have been corrupted; and *(4)* make obsolete any information that an attacker might have obtained (e.g., keys and vulnerabilities) from compromissed servers.

Additional Assumptions

To implement automatic recovery (without administrators), some additional assumptions are necessary. We use the same assumptions of [4]:

Key Pairs. Each process handles a pair of keys (public and private from an asymmetric cryptosystem). Private keys are known only to each owner, however

[3] As a Byzantine-faulty replica may appear to behave properly even when compromised, the recovery must be proactive, i.e., even correct servers must execute the recovery procedure.

all processes know all public keys (through certificates). These keys are only used to reestablish authenticated channels between processes, i.e., to share a secret.

Secure Cryptography. Each server has a secure cryptographic coprocessor that stores its private key. Thus, it can sign and decrypt messages without exposing the private key. This coprocessor also contains a counter that never goes backwards. The value of this counter is appended in each signed messages in order to avoid replay attacks. An example of co-processor that can be used to provide this service is the TPM (Trusted Platform Module) [19], already available in many commodity PCs.

Read-Only Memory. Each server stores the public keys of other servers, as well as the service public key PK, in some memory that survives failures without being corrupted. Moreover, a hash of the server code is also stored in this memory.

Watchdog Timer. Each server has a watchdog timer that periodically interrupts processing and hands control to a recovery monitor, which is stored in the read-only memory. An attacker is not able to change the rate of watchdog interruptions without having physical access to the machine.

Diversity in Time. The set of vulnerabilities of each replica is modified after each recovery. This ensures that an adversary will not use the same attack to compromise a server imediatelly after its recovery. This can be implemented through a combination of techniques [3] such as changing the operating system of the replica (to modify the set of vulnerabilities), using memory randomization (to modify the address layout of the server) and changing configuration parameters of the system.

Modified Protocol

The main change in the protocol is related with the windows of vulnerability. Each window has a number that is defined in increasing order. Servers append the number of the window of vulnerability in the replies, i.e., there is an additional parameter indicating in which window of vulnerability is the server. A client knows that servers have moved to the next window when it receives at least $f+1$ replies indicating this change in the system. Moreover, prepare certificates also have an additinal attribute to inform in which window it was created.

Sometimes, a client may restart one phase of a write operation concurrent with a recovery procedure. This can happen in phases 2 and 3, when a client is trying to obtain the service signature for a certificate. In fact, servers partial keys are updated by the recovery procedure, but a service signature is correctly generated only if all partial signatures are performed with partial keys of the same window of vulnerability. On the other hand, recovery does not affect read operations, unless, of course, while the server is rebooting and becomes unavailable for a short time.

The time to recover a server can be divided into three parts (Figure 2): T_r – time to restart the system; T_k – time to update keys (patial keys and session

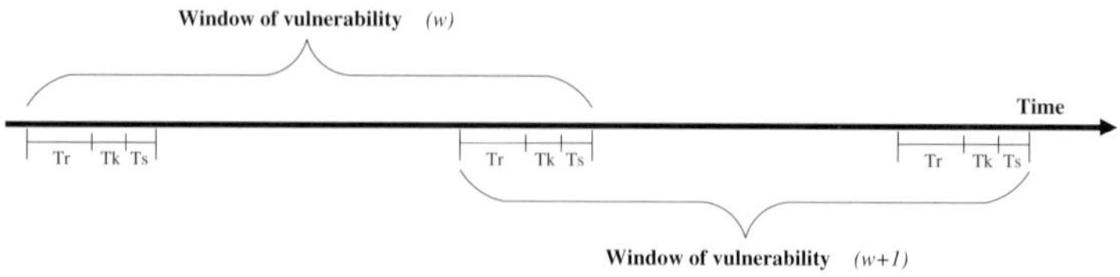

Fig. 2. Relationship between windows of vulnerability and recovery procedures

keys); and T_s – time to update the server state. A window of vulnerability starts at the begining of a recovery procedure and ends when the next recovery procedure ends (Figure 2). In this period, up to f out of $3f+1$ servers can fail.

It is impossible to develop protocols for proactive recovery in completely asynchronous systems [18], where we can not make any time assumption. Then, there is the problem of ensuring the periodic execution of the recovery procedure. Also, we must find ways to guarantee that this protocol ends.

To solve the first problem we use watchdogs timers that generate periodic interrupts. The second problem is much more complicated and should be solved in one of three ways: *(1)* assume (and justify this assumption!) that an adversary can not control the communication channels connecting two correct processes [22]; *(2)* use an hybrid distributed system, where the recovery procedure is performed in a synchronous and secure subsystem [17,5]; or *(3)* whether the recovery timeout expires, the server can contact an administrator that can take actions to allow the recovery to terminate [4]. Any of these methods can be incorporated in our system. The steps to recover a server are given bellow:

System Reboot. Each server restarts periodically when the watchdog timer goes off. Before rebooting, each server sends a message to inform other servers that it will perform the recovery, i.e., it is going to the next window of vulnerability. Any correct server that receives $f+1$ of these messages also restarts, even if its watchdog timer is not expired. This improves availability because the servers do not have to wait for their times to expire before changing to the next window of vulnerability. Moreover, the recovery monitor saves the state of the server ($data$ and P_{cert}) and its keys (partial key and verification key). Then, it reboots the system with correct code and restarts the server from the saved state. The integrity of the operating system and service code can be verified through the hashes of them stored in the read-only memory. If the copy of the code stored by the server is corrupt, the recovery monitor can fetch the correct code from other servers. Thus, it is guaranteed that the server code is correct and it did not lose its state. This state may be corrupted but the server must use it to process requests during the recovery phase in order to ensure the system properties when the recovering server is correct. Otherwise, the recovery procedure could cause more faults than the maximum tolerated by the system. But the recovering server could be faulty. So, the state should be updated, together with all confidential data that an attacker might have obtained (partial and session keys).

Keys Update. Session keys are updated as in [4]: the process i updates its session key used to authenticate a channel to other process j, by sending to j a secret that must be encrypted with the public key of j (thus, only j is able to access this secret) and signed with the private key of i (ensuring authenticity). So, a new session key is established to authenticate messages that i sends to j. These messages are signed by the secure coprocessor that appends the value of the counter (*count*) to avoid replay attacks, i.e., the *count* must be larger than the value of the last message that j has received from i. Moreover, the servers update its partial and verification keys through a proactive update protocol such as APSS [22]. In this procedure, the asymmetric keys of the servers are used to exchange confidential data.

State Update. Each server recovers its state by performing a normal read operation on a quorum of servers (considering himself). The value read is used to update the variables *data* and P_{cert}. Other variables are reset to the same value that they had in the boot of the system, i.e., P_{list} to a empty set and max_{ts} to null. These variables are used to avoid that malicious clients prepare many write operations. As these variables are "lost" in the recovery procedure, we use the following mechanism to control the actions of clients: as each prepare certificate contains the value w of the window of vulnerability, servers do not consider correct (in the third phase of the write) prepare certificates generated in previous windows of vulnerability. Thus, if a client receives replies from $f+1$ servers indicating that they have advanced to the next window of vulnerability, such client must restart the second phase of the write protocol and wait for responses from a quorum of servers that are in the new window. We could relax this requirement by accepting certificates generated in the previous window of vulnerability. In that case, a malicious client is able to prepare at most two (without optimization – 3 phases) or four (with optimization – 2 phases) write operations.

4 Evaluation

In this section, we analyze the PBFT-BC performance by comparing it with its precursor, the BFT-BC [10].

Analytical Evaluation. Both PBFT-BC and BFT-BC have read and write protocols with linear communication complexity ($O(n)$) and the same number of communication steps: 2 for reads (4 under write-contention) and 4 for writes (6 under write-contention). However, the messages of PBFT-BC are much smaller than the messages in BFT-BC due to the size of its certificates. A certificate of BFT-BC requires $2f+1$ signed messages from a quorum of servers, while in PBFT-BC, the certificate requires a single threshold signature. This represents an improvement by a a factor of $2f+1$ in terms of message size for PBFT-BC.

Even more important than lowering message sizes, are the bennefits of PBFT-BC in terms of the number cryptographic operations required by the protocols. Table 1 presents an analysis of the cryprographic costs to write a value in the system. As shown in the table, BFT-BC executes verifications in quadratic order ($O(f^2)$), while for PBFT-BC this cost is linear ($O(f)$). This happens because both protocols verify a quorum of certificates (first phase), where in BFT-BC

Table 1. Write protocol costs for a scenario without failures

Phase	BFT-BC		PBFT-BC	
	client	server	client	server
1ª	$4f^2 + 4f + 1$ verify	—	$2f + 1$ verify	—
2ª	$2f + 1$ verify	$4f + 2$ verify + 1 sign	1 comb of $2f + 1$ partial signatures	2 verify + 1 partial sign
3ª	$2f + 1$ verify	$2f + 1$ verify + 1 sign	1 comb of $2f + 1$ partial signatures	1 verify + 1 partial sign
Total Costs	$4f^2 + 8f + 3$ verify	$6f + 3$ verify + 2 sign	$2f + 1$ verify + 2 comb	3 verify + 2 partial sign
	$4f^2 + 14f + 6$ verify + 2 sign		$2f + 4$ verify + 2 partial sign + 2 comb	

each certificate is validated by a quorum of signatures and in PBFT-BC by only one signature. This cost is also reflected in read operations, where, discarding a possible write back phase, BFT-BC executes $4f^2 + 4f + 1$ verifications while PBFT-BC executes only $2f + 1$ verifications at client side.

Experimental Evaluation. To better ilustrate the bennefits of PBFT-BC when compared with BFT-BC we run some experiments to observe the latency of their read and write protocols. In particular, we are interested in observing the latency to execute read/write operations, since the effects caused by concurrency and failures is basically the same in both protocols.

We implemented both protocols using the Java programming language[4]. We evaluate the protocols without proactive recovery, since they can only impact the latency of operations executed during the recovery procedures.

In order to quantify this latency, some experiments were conducted in Emulab [20], where we allocate 11 *pc3000* machines (3.0 GHz 64-bit Pentium Xeon with 2GB of RAM and gigabit network cards) and a 100Mbs switched network. The network is emulated as a VLAN configured in a Cisco 4509 switch where we add a non-negligible latency of $10ms$ in the communications. The software installed on the machines was Red Hat Linux 6 with kernel 2.4.20 and Sun's 32-bit JRE version 1.6.0_02. All experiments were done with the Just-In-Time (JIT) compiler enabled, and run a warm-up phase to transform the bytecode into native code.

In the experiments, we use the standard 512-bits RSA signature in the BFT-BC and we configure the threshold signature scheme to generate RSA signatures of 512 bits in the PBFT-BC. The size of the objects, that were written and read from the system, was fixed in 1024 bytes. Then, we set the system with 4 ($f = 1$), 7 ($f = 2$) or 10 ($f = 3$) servers to analyze its scalability. We executed each operation 1000 times and obtained the mean time discarding the 5% values with greater variance.

Figure 3 presents the latency observed in the execution of each operation. We can see in this figure that PBFT-BC outperforms the performance of the BFT-BC. This happens mainly because the optimization in PBFT-BC write operations (verifications of partial signatures are avoided – Section 3.1) and

[4] For threshold cryptography, we adapt the library found at `http://threshsig.sf.net/`, which is an implementation of the protocol described in [16].

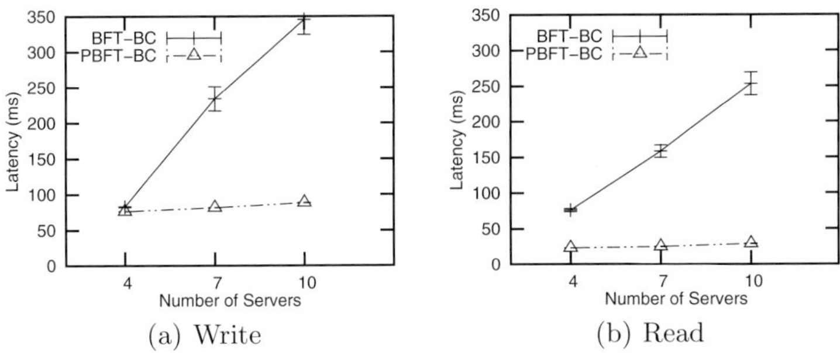

(a) Write (b) Read

Fig. 3. PBFT-BC X BFT-BC

due to the fact that PBFT-BC certificates have constant small size (only one signature), while BFT-BC certificates contains a quorum of signatures.

In fact, the total number of bytes exchanged by the client and each PBFT-BC server was approximately 2756 to write an object and 1466 to read an object, for all configurations of the system. On the other hand, using the BFT-BC protocol, these numbers increase to 5884/2229, 8120/2824 e 10292/3419 bytes for write/read operations in the system composed by 4, 7 or 10 servers, respectively. Other point to highlight is the low time need to produce a partial signature (aprox. 4.5ms) and to combine a quorum of these signatures (0.45ms for quorum of 3 servers). The time necessary to produce a standard RSA signature was approx. 1.5ms and each verification spent approx. 0.26ms, in both protocols.

These results clearly shows the benefits of using threshold signatures instead of signature sets to build certificate in quorum systems.

5 Related Work

There are many works that propose register implementation using quorum system protocols for different fault models (crash or malicious) and consistency semantics (safe, regular or atomic). The first protocols that tolerate malicious behavior of processes are presented in [11,12]. These works address two types of quorum systems: *(1) f-dissemination quorum system*, which does not allow the presence of malicious clients and requires $3f + 1$ servers; and *(2) f-masking quorum system*, which allows the presence of malicious clients but requires $4f + 1$ servers. Moreover, the *writeback* phase of the read protocol was introduced in [12]. This phase guarantees the atomic semantics of the operations performed on the system.

The quorum system more similar to PBFT-BC is the BFT-BC [10]. This system also tolerates malicious clients, requiring only $3f + 1$ servers. BFT-BC uses server signatures to guarantee the integrity of the stored data. The main difference of our work is the use of threshold cryptography to make data self-verifiable. This approach improves significantly the performance of the system, makes possible the update of the shares (partial keys) stored on the servers and facilitates the development of proactive servers recovery protocols since the clients keys do not need be changed.

Some protocols for proactive recovery of servers were proposed in [4,21,13]. Our protocol uses the same assumptions adopted in [4], that presents a protocol for active replication, but does not use threshold cryptography for signatures. Other works that use threshold cryptography are COCA [21], that implements a fault-tolerant online certification authority, and CODEX [13], that implements a distributed service for storage and dissemination of secrets. These works employ the APSS [22] proactive secret sharing protocol to update the shares of a service private key. Our protocol also employ the APSS protocol. However, the architectures of COCA and CODEX systems are significantly different from the adopted in our work. These systems use a server, called delegate, to presides over the processing of each client request. The use of delegates decrease the performance of the system since that are necessary more communication steps to perform an operation. Moreover, the additional assumptions for proactive recovery, assumed by COCA and CODEX, seem to be insufficient to guarantee the progress of the system [18]. The architecture Steward [1] also utilizes threshold cryptography, but it is used to guarantee the authenticity of a decision made by a set of processes.

6 Conclusions and Future Works

In this work we proposed a new protocol for Byzantine Quorum systems, the PBFT-BC, which tolerates malicious clients, presents optimal resilience and supplies a register with atomic semantics of operations. Moreover, we developed a protocol for proactive recovery of servers that implement the PBFT-BC. Also, we showed that PBFT-BC outperforms the performance of BFT-BC protocol, which implements a service with the same characteristics.

The next steps of this work will focus on extending this model to operate in a dynamic environment, where processes can join and leave the system at any time. In this direction, the use of threshold cryptography provides the necessary flexibility to make the PBFT-BC protocol adaptable to the changes that occur in the composition of the servers group.

Acknowledgments. Alchieri, Pereira and Fraga are supported by CNPq (Brazilian National Research Council). This work was partially supported by the FCT through the Multiannual and the CMU-Portugal Programmes.

References

1. Amir, Y., Danilov, C., Kirsch, J., Lane, J., Dolev, D., Nita-Rotaru, C., Olsen, J., Zage, D.: Scaling Byzantine fault-tolerant replication to wide area networks. In: Proc. of the International Conference on Dependable Systems and Networks, pp. 105–114 (2006)
2. Bellare, M., Rogaway, P.: Random oracles are practical: A paradigm for designing efficient protocols. In: Proc. of the 1st ACM Conference on Computer and Communications Security, November 1993, pp. 62–73 (1993)
3. Bessani, A., Daidone, A., Gashi, I., Obelheiro, R., Sousa, P., Stankovic, V.: Enhancing fault/intrusion tolerance through design and configuration diversity. In: Proc. of the 3rd Workshop on Recent Advances on Intrusion-Tolerant Systems (June 2009)

4. Castro, M., Liskov, B.: Practical Byzantine fault-tolerance and proactive recovery. ACM Transactions on Computer Systems 20(4), 398–461 (2002)
5. Correia, R., Sousa, P.: WEST: Wormhole-enhanced state transfer. In: Proc. of the DSN 2009 Workshop on Proactive Failure Avoidance, Recovery and Maintenance, PFARM (June 2009)
6. Dantas, W.S., Bessani, A.N., da Silva F.J., Correia, M.: Evaluating Byzantine quorum systems. In: Proc. of the 26th IEEE International Symposium on Reliable Distributed Systems (2007)
7. Gifford, D.: Weighted voting for replicated data. In: Proc. of the 7th ACM Symposium on Operating Systems Principles, December 1979, pp. 150–162 (1979)
8. Lamport, L.: On interprocess communication (part II). Distributed Computing 1(1), 203–213 (1986)
9. Lamport, L., Shostak, R., Pease, M.: The Byzantine generals problem. ACM Transactions on Programing Languages and Systems 4(3), 382–401 (1982)
10. Liskov, B., Rodrigues, R.: Tolerating Byzantine faulty clients in a quorum system. In: Proc. of the 26th IEEE International Conference on Distributed Computing Systems (June 2006)
11. Malkhi, D., Reiter, M.: Byzantine quorum systems. Distributed Computing 11(4), 203–213 (1998)
12. Malkhi, D., Reiter, M.: Secure and scalable replication in Phalanx. In: Proc. of 17th Symposium on Reliable Distributed Systems, pp. 51–60 (1998)
13. Marsh, M.A., Schneider, F.B.: CODEX: A robust and secure secret distribution system. IEEE Transactions on Dependable Secure Computing 1(1), 34–47 (2004)
14. Obelheiro, R.R., Bessani, A.N., Lung, L.C., Correia, M.: How practical are intrusion-tolerant distributed systems? DI-FCUL TR 06–15, Dep. of Informatics, University of Lisbon (September 2006)
15. Rivest, R.L., Shamir, A., Adleman, L.: A method for obtaining digital signatures and public-key cryptosystems. Communications of the ACM 21(2), 120–126 (1978)
16. Shoup, V.: Practical threshold signatures. In: Preneel, B. (ed.) EUROCRYPT 2000. LNCS, vol. 1807, pp. 207–222. Springer, Heidelberg (2000)
17. Sousa, P., Bessani, A.N., Correia, M., Neves, N.F., Verissimo, P.: Highly available intrusion-tolerant services with proactive-reactive recovery. IEEE Transactions on Parallel and Distributed Systems (to appear)
18. Sousa, P., Neves, N.F., Verissimo, P.: How resilient are distributed f fault/intrusion-tolerant systems? In: Proceedings of the International Conference on Dependable Systems and Networks - DSN 2005 (2005)
19. Trusted Computing Group. Trusted platform module web page (2009), `https://www.trustedcomputinggroup.org/groups/tpm/`
20. White, B., Lepreau, J., Stoller, L., Ricci, R., Guruprasad, S., Newbold, M., Hibler, M., Barb, C., Joglekar, A.: An integrated experimental environment for distributed systems and networks. In: Proc. of 5th Symposium on Operating Systems Design and Implementations (December 2002)
21. Zhou, L., Schneider, F., Van Rennesse, R.: COCA: A secure distributed online certification authority. ACM Transactions on Computer Systems 20(4), 329–368 (2002)
22. Zhou, L., Schneider, F.B., Van Renesse, R.: APSS: proactive secret sharing in asynchronous systems. ACM Transactions on Information and System Security 8(3), 259–286 (2005)

Model-Driven Development of Adaptive Applications with Self-Adaptive Mobile Processes

Holger Schmidt[1], Chi Tai Dang[2], Sascha Gessler[1], and Franz J. Hauck[1]

[1] Institute of Distributed Systems, Ulm University, Germany
{holger.schmidt,franz.hauck}@uni-ulm.de
[2] Multimedia Concepts and Applications, University of Augsburg, Germany
dang@informatik.uni-augsburg.de

Abstract. Writing adaptive applications is complex and thus error-prone. Our self-adaptive migratable Web services (*SAM-WS*s) already provide adaptation support in terms of location, available state, provided functionality and implementation in use. Yet, *SAM-WS*s still require developers implementing the adaptation logic themselves.

In this work, we present an approach to ease the implementation of adaptive applications with *SAM-WS*s. We introduce our concept of a self-adaptive mobile process (*SAMProc*), an abstraction for adaptive applications, and *SAMPEL*, an XML application to describe a *SAMProc*. We show a tool that automatically generates *SAM-WS*s adaptation code on the basis of the *SAMPEL* description. Then, we go even one step further by providing an Eclipse plug-in that allows automatic generation of the *SAMPEL* description on the basis of a graphic model. This enables generating a *SAM-WS* implementation with few clicks; developers have to write pure application logic only.

1 Introduction

Mobile and ubiquitous computing (UbiComp) [1] scenarios are characterised by a high heterogeneity of devices, such as personal digital assistents (PDAs), mobile phones and desktop machines. They face a very dynamic environment due to the fact that devices, users and even applications can potentially be mobile. For tapping the full potential of the environment, such scenarios require adaptive applications. For instance, such applications should be able to use as much resources as possible on powerful devices and only few resources on resource-limited devices. Additionally, they should be mobile in terms of migration to enable applications running on the best-fitting devices in the surroundings (e.g., to run on a specific user's device or on the most powerful device).

Our approach to tackle such scenarios is self-adaptive migratable Web services (*SAM-WS*s) [2]. We advocate that Web services will become a standard mechanism for communication in mobile and UbiComp scenarios due to the fact that Web services have already gained acceptance in standard environments to provide a heterogeneous communication model. This is supported by the fact

R. Meersman, T. Dillon, P. Herrero (Eds.): OTM 2009, Part I, LNCS 5870, pp. 726–743, 2009.

that there is already work on Web services providing reasonable communication between heterogeneous sensors [3]. Our *SAM-WS*s provide means to adapt themselves in terms of their location (i.e., weak service migration [4]), available state, provided functionality and implementation in use. At the same time, *SAM-WS*s maintain a unique service identity that allows addressing the Web service independent of its current location and adaptation (i.e., required to foster the collaboration between different *SAM-WS*–based applications). Although *SAM-WS*s provide a great flexibility for adaptive applications, developers have to manually implement the actual adaptation logic on their own.

In this work, we present a model-driven approach to ease the development of adaptive applications on the basis of *SAM-WS*s. Therefore, we build on our concept of a self-adaptive mobile process (*SAMProc*), which provides a novel abstraction for an adaptive application [5]. The basic idea is to describe the application as a *SAMProc* and to use this information to automatically generate the *SAM-WS* adaptation logic. As a novel description language, we present the self-adaptive mobile process execution language (*SAMPEL*), our new XML application to describe a *SAMProc*. Due to the fact that the business process execution language (BPEL) already provides means for orchestration of *standard* Web services, we implemented *SAMPEL* as a BPEL extension, which additionally supports describing *SAM-WS* behaviour regarding adaptation. We provide a tool, which automatically generates the adaptation logic of the corresponding *SAM-WS*; developers have to implement the pure application logic only. In comparison to related work [6,7], our approach is more lightweight because we generate node-tailored code which is not interpreted but executed at runtime. Additionally, we present an Eclipse plug-in that allows describing adaptive applications with a graphical notation. Modelling leads to an automatic generation of an appropriate *SAMPEL* description. In the overall process, our approach allows generating the adaptation logic of an adaptive application with only few clicks.

The rest of the paper is structured as follows. First, we introduce *SAM-WS* and an appropriate example application. In Section 3, we present our model-driven approach to develop adaptive applications with our *SAMProc* abstraction. After a discussion of related work in Section 4, we conclude and show future work.

2 Preliminaries

In the following, we introduce Web service basics and our *SAM-WS* extension. Then, we present a novel *mobile report* application, which acts as an exemplary adaptive application for the rest of the paper.

2.1 Web Services and Adaptive Web Service Migration

Web services are a common XML-based communication technology built upon standard Internet protocols [8]. They implement a service-oriented architecure (SOA) approach, in which functionality is provided by services only [9]. Web services are uniquely identified by uniform resource identifiers (URIs). The service interface and its protocol bindings are specified with the *Web services description language* (WSDL) [10]. WSDL binds the interface to a message protocol that is used to access the Web service. Therefore, Web services commonly use *SOAP* [11]. Due to the fact that Web services are built on top of XML technologies they are independent of platform and programming language. Thus, they perfectly suit heterogeneous environments.

We advocate that Web services will become a standard for communication in UbiComp. They have already gained acceptance in standard environments to provide heterogeneous communication and there is already work on Web services providing reasonable communication between heterogeneous sensors [3]. Yet, for tapping the full potential of UbiComp environments applications require adaptivity. In recent work [2], we proposed the concept of a *self-adaptive migratable Web service* (*SAM-WS*), which provides means for dynamic adaptation in terms of location (i.e., migration[1]), state, functionality and implementation.

For making Web services adaptive we introduce a *facet* concept. A facet represents a particular characteristic of the *SAM-WS* running on a particular *node* and comprising a particular *interface*, *implementation* and *state*. A *SAM-WS* adaptation can dynamically be applied at runtime by changing the location, interface, implementation or state of the *SAM-WS* (multiple concurrent changes are supported as well). A unique *SAM-WS* identity, which is used for continuously addressing the application, is maintained while adapting the *SAM-WS*.

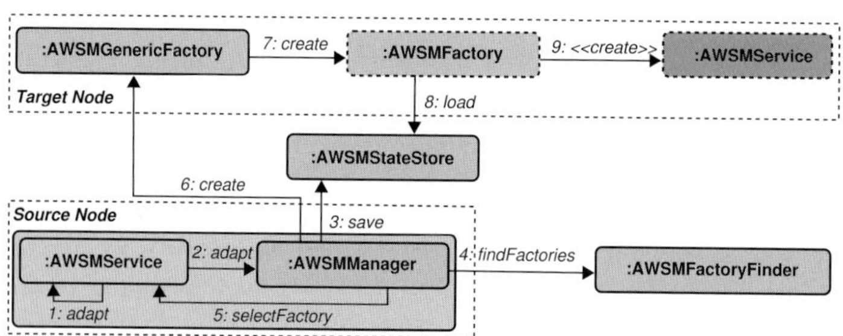

Fig. 1. Collaboration of adaptive Web service migration (*AWSM*) entities

Figure 1 shows our infrastructure services and their collaboration for the adaptation of a *SAM-WS*. For providing adaptation methods *SAM-WS*s have to implement our `AWSMService` interface. For easing application development the *SAM-WS* is able to use a generic service-internal `AWSMManager` entity[2], which manages the remaining adaptation steps (see Figure 1, step 2).

[1] *SAM-WS*s support weak migration [4]. Here, only application state is transferred but no execution-dependent state, such as values on the stack and CPU registers.

[2] Our *AWSM* prototype provides an `AWSMManager`. It is generic in the sense that it can be used within any *SAM-WS*.

The `AWSMManager` stores the active state[3] into an `AWSMStateStore` (step 3). The target location is determined by our `AWSMFactoryFinder` service[4]. The URI of an appropriate `AWSMGenericFactory` service is returned, which provides means to create *SAM-WS*s on a remote node. For allowing an application-specific selection of the best-fitting factory out of the list of appropriate factories returned from the factory finder we use a basic call-back mechanism (step 5). Then, the `AWSMGenericFactory` creates the desired target facet with the needed active state being loaded from the state store. If the code for the facet to create is unavailable at the target location, our infrastructure tries to load the code for a specific `AWSMFactory`, which is able to deploy the required *SAM-WS* facet (more details about our dynamic code loading feature can be found in [2]). In a last step, the old *SAM-WS* facet is removed.

Currently, application development support regarding *SAM-WS*s is quite limited. Our system provides generic adaptation code in terms of an abstract `AWSMServiceImpl` class for Java but the actual adaptation logic has to be implemented by the developer. Due to the fact that this is a non-trivial issue, there is still a need for further development support.

2.2 Example Application

Our example application supports crisis management by providing means for *spontaneous reporters* in the crisis surroundings to document the current situation (see Figure 2). Such a system is able to support rescue coordination centres with up-to-date information about the scene. Additionally, it can help afterwards with the investigation of causalities.

First, spontaneous reporters enter text, audio and video messages into a report application on their mobile device. The report is sent to *virtual first-aiders*, which undertake the task of reviewing the report. They prove the documentation and reject meaningless reports to disburden the *rescue coordination centre* where the accepted report is eventually presented.

The report application is implemented as a *SAM-WS*. With respect to this, unique reports are self-contained *SAM-WS* instances, which are adapted in terms of location to implement the mobile workflow. Additionally, our report application provides adaptivity in terms of programming language (due to heterogeneous environment with different hardware), functionality (each step in the mobile workflow needs different functionality) and state (e.g., anonymous reviewing: information about reporters should not be available at the virtual first-aiders but at the rescue coordination centre).

[3] Only implementation-independent state is externalised. It comprises only variables being interpretable by any possible implementation of a particular functionality. Moreover, we differentiate the overall *SAM-WS* state into active and passive state [2]. Thereby, active state is used within a particular *SAM-WS* facet while passive state is not. Passive state is stored in the `AWSMStateStore` and can be activated within another facet again.

[4] Due to the available universal description, discovery and integration (UDDI) mechanism for Web services [12], the `AWSMFactoryFinder` is implemented as a UDDI extension.

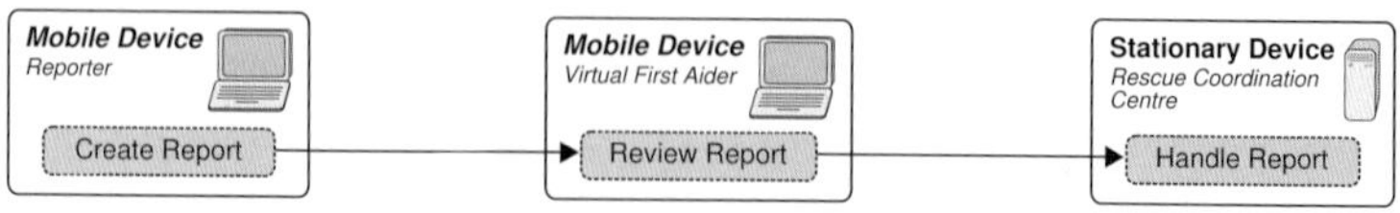

Fig. 2. Basic report application workflow

3 Building Adaptive Applications with SAMProc

The following sections present our model-driven approach to ease the implementation of adaptive applications with *SAM-WS*s.

3.1 Self-Adaptive Mobile Processes

In previous work [5], we introduced *SAMProc*:

> *A SAMProc can be seen as an ordered execution of services. It is able to adapt itself in terms of state, functionality and implementation to the current context and to migrate either for locally executing services or for accessing a particular context, while maintaining its unique identity.*

Thus, a *SAMProc* provides a high-level abstraction for adaptive applications. The idea is that developers should be able to model the application with its interactions and deployment aspects as a *SAMProc*. After a two-stage process (i.e., graphical model and its textual representation), a code generator automatically generates the application adaptation code, which is implemented as *SAM-WS*. Thus, only the pure application logic has to be provided by the developer.

In the following, we present *SAMPEL*, our textual *SAMProc* representation.

3.2 Self-Adaptive Mobile Process Execution Language

We introduce the *self-adaptive mobile process execution language* (*SAMPEL*) as our novel XML application, which is an extension of BPEL [13]. BPEL is an XML application being commonly used for describing business processes that are realised by Web services. Such a business process consists of the involved Web service interfaces, the interaction between them and a process state. Thus, BPEL is commonly characterised as a means for orchestration of Web services to build a business process. Such as Web services, BPEL uses WSDL to describe the involved Web service interfaces. A BPEL process itself is offered as a Web service. It is interpreted by BPEL engines, such as *ActiveBPEL* [14].

The way to describe processes with BPEL is suitable for describing *SAMProc*s as well. Yet, there are some issues why BPEL does not meet all requirements for *SAMProc*s. First, BPEL was particularly designed for business processes with focus on orchestration of Web services whereas *SAMProc*s rely on advanced concepts such as Web service facets and active process state. BPEL lacks support for these concepts. Additionally, BPEL processes are designed to be executed at a static location. Hence, BPEL does not provide the indispensable support for distribution aspects of *SAMProcs*. For instance, before migrating, a *SAMProc* has to select an appropriate location. Therefore, it needs context information about possible targets, such as available resources, being matched with its own context requirements. BPEL has to be extended for describing required context. Furthermore, current devices being used in UbiComp environments are highly resource-limited. Thus, it is in general not feasible to run a BPEL engine on these devices due to their high resource usage. Unlike BPEL, the *SAMPEL* process is not executed by a particular *SAMPEL* engine but used for node-tailored code generation. Additionally, *SAMPEL* provides support for the concepts of *SAMProcs*, in which the process is able to adapt itself according to the current execution platform. In Section 3.3, we present a code generator, which is able to automatically create code skeletons for all required implementations (i.e., Web service facets) of a *SAMPEL* process description. These code skeletons have already built-in support for process adaptation. Application developers have to implement the pure application logic only.

In the following subsections, we present *SAMPEL* in more detail. We attach particular importance to our extensions of BPEL.

Description Language. Like BPEL, a *SAMPEL* description is always paired with at least one WSDL description, which declares the *SAMProc* interfaces (i.e., interfaces of the Web service facets being implemented by the *SAMProc*).

Processes and Instances. A crucial difference between BPEL and *SAMPEL* is the conceptual view on a process. A BPEL process is an instance within a BPEL-engine and always has similar behaviour (e.g., starting with accepting a purchase order then communicating with involved Web services and eventually informing the purchaser about the shipping date). Unlike this, a *SAMPEL* process (i.e., *SAMProc*) is characterised by a highly dynamic behaviour since it can get adapted and migrated to another location at runtime, where it exists as an instance and handles user interactions (e.g., report application). Additionally, *SAMProcs* provide means to change the process state at runtime. Due to the inherent dynamics, a *SAMPEL* process is more functional oriented as opposed to a BPEL process (i.e., *SAMPEL* focuses on process functions instead of offering predefined process behaviour, such as in the purchase order example). This difference is reflected in the BPEL description by the placement of activities.

Figure 3 shows the process definition of a BPEL process with its activities. Activities specify the process behaviour, such as invoking a Web service and assigning a value to a variable. A BPEL process has activities in the main scope, whereas a *SAMPEL* process has not. Due to the functional oriented design, the activities of a *SAMPEL* process are basically determined by the activities within the method definitions (i.e., `eventHandler`, see below). Figure 4 shows the basic layout of a *SAMPEL* description. The `process` element contains all remaining parts of a *SAMProc*, which are explained in more detail in the following.

```
<process ...>
  ...
  ACTIVITIES
</process>
```

Fig. 3. Basic BPEL process description

```
<process ...>
  <partnerLinks>+
  ...
  </partnerLinks>
  <variables>?
  ...
  </variables>
  <correlationSets>
  ...
  </correlationSets>
  <eventHandlers>
  ...
  </eventHandlers>
</process>
```

Fig. 4. Basic *SAMPEL* process description

Scopes. A scope is a container for activities. As such, it is a structuring element that forms the control sequence of other elements. There are two kinds of scopes: the main scope and its sub-scopes. The main scope is implicitly defined by the `process` element and contains global variables, correlation sets and methods as shown in Figure 4. It must contain at least one method definition. Otherwise, the process has no activities. Sub-scopes (i.e., local scopes) can be defined by the `scope` element. As shown in Figure 5, sub-scopes must contain the activities to be executed and can contain elements that are used by the activities within the scope or its sub-scopes, such as local variables.

```
<scope>
  <partnerLinks>?
  ..
  </partnerLinks>
  <variables>?
  ..
  </variables>

  ACTIVITIES+
</scope>
```

Fig. 5. Basic *SAMPEL* scope description

PartnerLinks. Partner links are a *SAMPEL* concept inherited from BPEL. They allow declaring the communication endpoints of the *SAMPEL* process and its partner services. *SAMPEL* allows declaring partner links in the main scope as well as in sub-scopes. Figure 6 shows the concept of partner links illustrated for the report application. There, both processes—the reporter and the supervisor[5]—are represented by their *SAMPEL* description and WSDL description. Each process describes its communication endpoint by means of a partner link (i.e., `RLink` for the reporter process and `SLink` for the supervisor process). Each partner link relates to a partner link type of its WSDL description, which works as a bridge between partner links and a specific WSDL port type (i.e., contains the available operations). The same applies to the endpoints of other Web services (see Figure 6 for the partner link `RLink1` which links to the partner link type `SType` of the supervisor process). These definitions allow referring to communication partners within the process description.

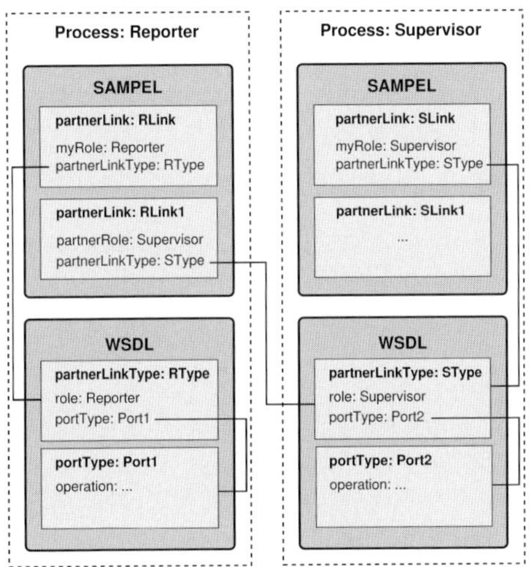

Fig. 6. Partner links

Figure 7 outlines the corresponding definition of partner links in a process description and a WSDL description. The upper part belongs to the *SAMPEL* description and the lower part to the corresponding WSDL description. A partner link must be defined inside a `partnerLinks` container element and is composed of a role, a name and the partner link type. The name of the partner link is referred within the process description to specify a communication endpoint for an activity. The role and partner link type refer to the WSDL description to select the WSDL port type for that communication endpoint. As shown in Figure 7, a partner link type has a name and must have at least one `role` element inside. A partner link from the process description points to a specific `role` element in the WSDL description by following the role and the name of the partner link type. Eventually, the `role` element refers to a WSDL port type.

[5] Here, we extended our report application with a particular supervisor entity that provides information about currently required information. This can highly increase the report quality.

```
<!-- SAMPEL -->
<process name="Report" xmlns:rpt="e1.wsdl">
  <partnerLinks>
    <partnerLink myRole="Reporter"
                 name="RLink1"
                 partnerLinkType="rpt:RType" />
  </partnerLinks>
</process>

<!-- WSDL -->
<definitions targetNamespace="e1.wsdl"
             xmlns:plnk="http://schemas.xmlsoap.org/ws/2003/05/partner-link/"...>
  ...
  <plnk:partnerLinkType name="ReportPLT">
    <plnk:role name="Reporter"
             portType="tns:ReportPortType" />
  </plnk:partnerLinkType>
</definitions>
```

Fig. 7. Partner link example

Variables. An important means for storing temporary values or maintaining the process state are variables. *SAMPEL* allows variable declarations in the main scope and sub-scopes with the `variable` element. The variables in the main scope are global variables and are treated as the implementation-independent process state (i.e., considered for migration; see Section 2.1). Variables within sub-scopes are local variables used by activities in the sub-scope and its nested sub-scopes. Figure 8 shows an example for two variable declarations with respect to the report application. A variable must be declared within the `variables` container and is composed of a variable name and a data type for the value. The data type can be declared either using an XML schema type (see Figure 8, line 2) or as a reference to a WSDL message type (line 3).

```
1 <variables>
2   <variable name="report" type="xsd:string" />
3   <variable name="recID" messageType="rpt:ID"/>
4   ...
5 </variables>
```

Fig. 8. Variable description example

Correlation Sets. SAMProcs are created at a particular location. Then, they are able to migrate to other locations. Thus, there can be multiple instances of the same *SAMProc* at the same location. For distinguishing between them, *SAMPEL* inherits the concept of correlation sets from BPEL to create a unique identifier (i.e., composed of two parts in the form `process-id.instance-id`). The process identifier identifies the *SAMPEL* description and the instance identifier identifies an actual process instance. The instance identifier is derived from one correlation set only. Therefore, it must be defined within the `process` element.

Figure 9 shows how to define a correlation set for identifying a report application instance within *SAMPEL* and WSDL. Correlation sets have to be defined

```
<!-- SAMPEL -->
<process ...>
  <correlationSets>
    <correlationSet name="ID"
     properties="rep:ReporterID rep:ReportNr" />
    ...
  </correlationSets>
  ...
</process>

<!-- WSDL -->
<definitions xmlns:vprop="http://docs.oasis-open.org/wsbpel/2.0/varprop" ...>
  ...
  <vprop:property name="ReporterID" type="xsd:int" />
  <vprop:propertyAlias propertyName="ReporterID" messageType="inMsg" part="ID" />
  ...
</definitions>
```

Fig. 9. Correlation set to identify report instance

in the process description inside a `correlationSets` element and consist of a name and properties. The name of a correlation set is referenced from within the process description, whereas the properties have to be defined in the corresponding WSDL description (properties must be mapped to message types defined in the WSDL description). This way, correlation to an instance can be determined from incoming messages. All properties have a name, are of particular XML schema type and can be assigned to a WSDL message type by means of a `propertyAlias` element. The property alias refers to a property and to a WSDL message type. If the WSDL message type has more than one part, the `part` attribute addresses the appropriate part.

Methods. The behaviour of *SAMPEL* processes is basically described by the activities in methods, which are specified with the `onEvent` element. In Figure 10, we specify a `setReport` method as part of our report application. An `onEvent` element requires several mandatory attributes: a reference to a partner link that declares the communication endpoint, the port type of the partner link, a name for the method and a message type that declares the input to the method. Beside these mandatory attributes, there is an optional `variable` attribute that implicitly creates a local variable in the scope of the method and fills that variable with the values submitted at method invocation. Each method must have a scope for activities and a reference to the correlation set to use (i.e., to address the right instance on the basis of the received message).

Distribution Aspects. Distribution aspects are specified with the `requires` element, which can be placed as part of an `onEvent` element (see Figure 10, line 11–13). It can also be placed before an activity and thereby effect only the following activity. Figure 10 shows a basic example where the scope is restricted to the reporter role. A `property` element needs a key and a value. We allow multiple properties inside the `requires` element, which are interpreted as follows. If two properties have different keys, then the values are linked in a logical

```
<eventHandlers>
  <onEvent partnerLink="ReporterPL"
          portType="ReporterPT"
          operation="setReport"
          messageType="ReportMsg"
          variable="Message" >
    <correlations>
      <correlation set="ReporterCS" />
    </correlations>

    <requires>
      <property key="role" value="Reporter" />
    </requires>

    <scope>
    ...
    </scope>
  </onEvent>
  ...
</eventHandlers>
```

Fig. 10. *SAMPEL*: method description for reporter

'and' manner. They are linked in a logical 'or' manner in case of the same keys. This forms a property set with key/value-pairs that restricts an activity. It is a flexible means to describe requirements regarding distribution. Unlike BPEL, this feature is unique to *SAMPEL*.

Basic Activities. Activities determine the behaviour of a process. Therefore, *SAMPEL* inherits most of the BPEL activities. Basic activities actually contribute to a process step and are essential elements, such as a variable copy operation and waiting for an answer.

Communication with other Web services is covered by the `invoke` activity (see Figure 11 for an invocation at the supervisor within our report application). For invoking a Web service, a partner link to the desired Web service has to be specified. Furthermore, the name of the operation must be provided and an optional port type can be specified. Parameters of the Web service invocation have to be specified with the `toPart` element. *Waiting for an invocation* (i.e., at server-side) can be established with a `receive` activity, which is similar to an `invoke` activity for the partner link, operation and port type.

Sending a reply to an invocation is an important activity for communication with other services. A `reply` element either corresponds to an `onEvent` or a `receive` element (there has to be a partner link to identify the corresponding element). Additionally, a variable containing the return value has to be provided.

Assigning a value to a variable is done by the `assign` activity copying a value to a destination variable. Within the `assign` element, multiple `copy` elements are allowed. Each `copy` element performs a copy operation to a declared variable.

An *explicitly waiting* activity is possible with the `wait` element. It allows specifying a blocking wait state either for a particular duration or until a given date and time. This can be used as part of polling sequences.

Extensible activities allow extending *SAMPEL* with custom activities. Figure 12 shows how to define an extensible activity by the `activity` element.

```
<invoke partnerLink="SupervisorPL" operation="getRequiredInfo" portType="SupervisorPT" >
  <toParts>
    <toPart part="ID" fromVariable="varId" />
  </toParts>
  <fromParts>
    <fromPart part="RequiredInfo" toVariable="varRequiredInfo" />
  </fromParts>
</invoke>
```

Fig. 11. *SAMPEL*: invocation at supervisor

```
<activity name="spellCheckReport" />
```

Fig. 12. *SAMPEL*: custom reporter activity

Here, we define an activity supporting reporters with spell-checker functionality. There is only one attribute allowed, which denotes the activity name. In a corresponding *SAM-WS* implementation, such an activity is mapped to abstract methods (supported by our code generator; see Section 3.3), which have to be implemented by application developers. Thus, developers are able to use advanced programming language features that cannot be specified with pure *SAMPEL*. In contrast to *SAMPEL*, this feature is not supported by BPEL.

Explicit middleware support for adaptation is realised by the `copy` and `adapt` elements. These elements are not supported by BPEL. The `copy` activity creates a copy of the instance and assigns it a new instance identifier. The `adapt` element contains a property set. According to the given adaptation properties the process is able to adapt in terms of location, state, functionality and implementation. Figure 13 shows an example that requests an adaptation of the report application. A corresponding *SAM-WS* implementation (see Section 3.3) is able to pass the property set to our *AWSM* platform, which automatically handles the required steps to implement the *SAMPEL* description.

```
<adapt>
  <property key="role" value="Reviewer" />
</adapt>
```

Fig. 13. *SAMPEL*: adaptation to reviewer

Structuring Activities. Structuring activities form the control sequence for basic activities and can be arbitrarily nested in order to build complex control sequences. The first sub-scope of a method represents the top-level structuring element for starting a control sequence. It contains basic activities and structuring activities. Any structuring activity can contain further sub-scopes. Basic activities can be executed in sequence by surrounding them with a `sequence` element. Execution in parallel can be performed with the `flow` element, which starts each containing activity at the same time and ends when the last activity has finished. *SAMPEL* also offers constructs for conditional execution as known from traditional programming languages. Figure 14 outlines the usage of an if/elseif/else-construct for our report application. The condition has to be an XPath expression that evaluates to a Boolean value. Other conditional execution constructs are loops described with the `while` and `repeatUntil` elements. Both evaluate an XPath expression to repeat the containing activities. The difference is that the `while` element stops as soon as the condition evaluates to a Boolean `false`, whereas the `repeatUntil` stops if the condition evaluates to a Boolean `true`. For a corresponding *SAM-WS* implementation, sequential and conditional activities can be mapped to corresponding programming language constructs, whereas parallel activities should use threads.

```
<if>
  <condition>string-length($report)&lt;=100</condition>
  <adapt><property key="Mem" value="1MB"/></adapt>
  <elseif>
    <condition>string-length($report)&lt;=1000</condition>
    <adapt><property key="Mem" value="5MB"/></adapt>
  </elseif>
  <else>
    <adapt><property key="Mem" value="10MB"/></adapt>
  </else>
</if>
```

Fig. 14. *SAMPEL*: conditional adaptation

3.3 Automatic Code Generation

For implementing the model-driven approach with our *SAMProc* concept, code skeletons for the *SAM-WS* implementations have to be automatically generated from the *SAMPEL* description (see Section 3.1). We provide a Java code generator for this task. It keeps pure application logic written by application developers separated from generated implementation skeletons by using abstract classes and inheritance. This allows developers extending and customising the implementations with the pure application logic.

Figure 15 shows the overall code generation process. The code generator uses the *SAMPEL* description and all referenced WSDL documents to generate skeletons of appropriate *SAM-WS* facets (see Section 2.1). In general, code generation for any programming language is possible but this paper focuses on Java.

First, the code generator determines the interface (i.e., a set of methods) and available state for each *SAM-WS* facet. Therefore, it analyses the property sets of the method definitions within the *SAMPEL* description (a method describes its property set with the `requires` element). The distinct sets out of all property sets of the method definitions determine the required facets. Thus, each facet provides a particular set of properties, which determine the methods building up the overall interface. For completing a specific facet, the respective active state is determined by identifying the global variables being used within the methods. Finally, an XML file is created that holds meta data about the facet and its implementation. This allows our *AWSM* platform to register implementations and take meta data into account for adaptation decisions.

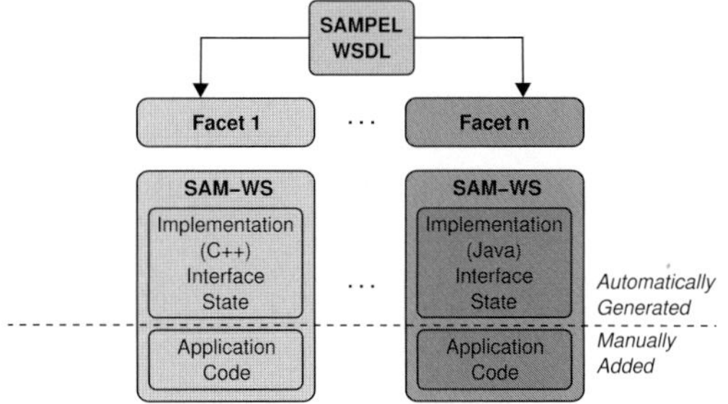

Fig. 15. *SAMPEL* code generator

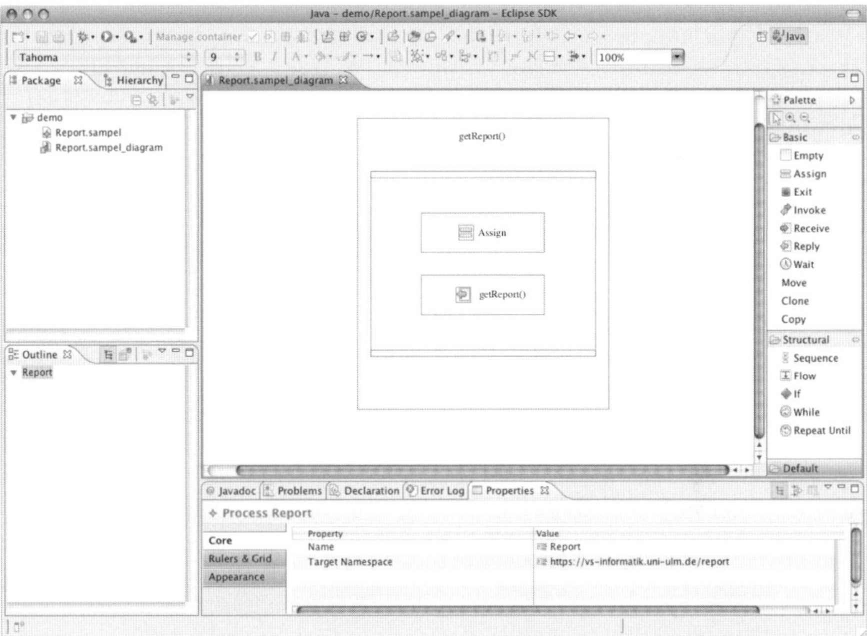

Fig. 16. *SAMPEL* diagram editor with a part of the report application model

For each facet, code generation starts with the main scope and recursively processes sub-elements, such as methods, sub-scopes and activities. Methods are implemented with conventional methods as provided by programming languages with the following exception. Due to the fact that *SAMPEL* methods allow activities after replying to a request, such as a `return` instruction, the instructions of each method are wrapped into an own Java thread, which continues with activities while the requested method can apply its `return` instruction.

Most basic activities are implemented using their direct programming language counterparts or with extensions, such as Apache Axis for Web service invocations. Extensible activities result in abstract methods that have to be implemented by application developers. For mapping the adaptation logic to a particular *SAM-WS* implementation, the property set of an adaptation request is passed through to our *AWSM* platform, which automatically manages the needed steps as described in Section 2.1. Structuring activities, such as conditional execution, are mapped to their direct programming language counterparts (e.g., `flow` elements are implemented as threads to achieve parallel execution).

Finally, the generated *SAM-WS* facet skeletons (i.e., abstract classes) contain basic adaptation support and programming-language–dependent realisations of the activities being specified with *SAMPEL*. This includes the adaptation logic as well. Thus, developers only have to add the pure application logic to implement their adaptive application on the basis of a *SAM-WS*.

To ease addressing of a specific *SAM-WS* with our *AWSM* platform, we implemented an addressing schema, in which the process name and the target location is sufficient. For this purpose, our code generator creates a *proxy Web service*, which should be deployed within the Web service container. The proxy receives all messages for a particular application (corresponds to a *SAMPEL* description) and routes them to the appropriate *SAM-WS* instance specified by the correlation set. Therefore, the generated *SAM-WS*s contain programme logic that automatically registers the particular instance at the proxy.

3.4 Modelling Self-Adaptive Mobile Processes

Overall, generating an adaptive application requires various XML documents. Paired with at least one WSDL description, an appropriate *SAMPEL* description provides the adaptation logic. For simplifying the generation of these XML documents, we developed an Eclipse plug-in to model adaptive applications with a graphical notation. Since Eclipse already provides a WSDL editor [15], our plug-in delivers a novel diagram editor that assists application developers in building *SAMPEL* descriptions by allowing them to drop and to combine activities onto a drawing canvas. During modelling, the editor provides further assistance by validating the structural and semantic correctness of the document. For initially creating a *SAMPEL* description we provide an Eclipse wizard. This wizard allows selecting different templates, which serve as scaffolds for common use cases.

The diagram editor is realised with the graphical modeling framework (GMF) [16], which provides a generative component and runtime infrastructure for graphical editors based on a structured data model (i.e., an Eclipse modeling framework (EMF) [17] *ecore model* derived from the *SAMPEL* XML schema). Since GMF separately manages data model, graphical representation and tooling definition, the editor is highly customisable and easy to extend.

Figure 16 shows the user interface of the diagram editor. The graphic notation is similar to the *business process modelling notation* (BPMN) [18] that allows modelling BPEL processes. All *SAMPEL* elements inherited from BPEL are represented with the direct BPMN counterpart. We only introduce new activities for explicit adaptation support (i.e., `move`, `copy`, `clone` and `adapt`). The diagram canvas represents the *SAMProc*. The palette on the right allows selecting differ-

Fig. 17. *SAMPEL* editor validation

ent tools to edit the process. In particular, there are tools for basic activities and structuring activities (see Section 3.2). Basic activities are essential elements of a process and are represented by a rectangular shape with a distinct icon and title. Since structuring activities form the control sequence for basic activities they are represented as titled rectangular containers for nested activities. Due to the fact that every change on the data model is performed using the GMF command framework, undo behaviour is seamlessly integrated.

The diagram itself does not display all required information to create a valid document. Otherwise, it would be too cluttered to be readable. Therefore, additional information can be captured and edited in the *Eclipse Properties View* by selecting the respective diagram element (see Figure 16).

The *Eclipse Problem View* displays errors and warnings being detected during validation (see Figure 17). Beside structural flaws, such as the fact that a scope has to contain at least one method definition, there are also various semantic correctness constraints. For instance, each process describes its communication endpoints by means of a partner link that has to relate to a predefined partner link type of its WSDL description. Those constraints are specified using the openArchitectureWare Check (oAW-Check) language [19], which is straightforward to use. Consequently, constraints can be extended with low efforts.

4 Related Work

There is related work in the area of mobile processes. For instance, *Ishikawa et al.* present a framework for mobile Web services, i.e., a synthesis of Web services and mobile agents [6]. Each mobile Web service has a process description on the basis of BPEL, which is used for interaction with other processes and for supporting migration decisions at runtime. This approach with its BPEL extension has similarities with *SAMProcs* and *SAMPEL*. Unlike our approach, it does not support adaptivity. Additionally, while the process description of mobile Web services is interpreted at runtime, we use *SAMPEL* for generating code. *Kunze et al.* follow a similar approach with DEMAC [7]. Here, the process description is a proprietary XML application being executed by the DEMAC process engine at runtime. Instead of using Web service and mobile agent concepts, plain process descriptions are transferred between process engines for achieving mobility. Unlike the DEMAC approach, we do not require a process execution engine on each device the platform is running. Additionally, we generate node-tailored code, which makes our approach more lightweight at runtime.

There is a lot of research regarding model-driven development of adaptive applications on the basis of distributed component infrastructures. Most of these systems, such as proposed by *MADAM* [20] and *Phung-Khac et al.* [21], allow modelling adaptation (i.e., dynamic component reconfiguration) decisions being executed at runtime. Unlike our approach, these frameworks are restricted to a custom component framework and do not support application migration.

Notations such as BPMN [18] define a business process diagram, which is typically a flowchart incorporating constructs suitable for process analysts and designers. Yet, the graphical notation for *SAMPEL* reflects the underlying structure of the language regarding its specific functional oriented design. This particularly suits application developers by emphasising the control-flow of an adaptive application.

5 Conclusion

In this work, we introduced *SAMPEL*, a novel XML application to describe adaptive applications on the basis of our *SAMProc* abstraction. Unlike related work, *SAMPEL* is not interpreted at runtime but used for generating the adaptation logic for our *SAM-WS*s. Application developers do not have to implement adaptation logic; they can focus on pure application logic. Furthermore, we allow modelling *SAMProc* adaptation with a graphic notation. Modelling with our Eclipse plug-in results in an automatic generation of an appropriate *SAMPEL* description. Thus, this work is a first step towards a model-driven development of tailored adaptive applications on the basis of *SAMProc*s.

If an application has many adaptation cases, our tool generates a lot of facets. For supporting the developer by generating as much code as possible we investigate adding Java code to custom *SAMPEL* activities. Such an approach is similar to *BPELJ* [22], which allows using Java code within BPEL descriptions.

References

1. Weiser, M.: The computer for the 21st century. Scientific American 265(3) (1991)
2. Schmidt, H., Kapitza, R., Hauck, F.J., Reiser, H.P.: Adaptive Web service migration. In: Meier, R., Terzis, S. (eds.) DAIS 2008. LNCS, vol. 5053, pp. 182–195. Springer, Heidelberg (2008)
3. Luo, L., Kansal, A., Nath, S., Zhao, F.: Sharing and exploring sensor streams over geocentric interfaces. In: GIS, pp. 1–10. ACM, New York (2008)
4. Fuggetta, A., Picco, G.P., Vigna, G.: Understanding code mobility. IEEE TSE 24(5), 342–361 (1998)
5. Schmidt, H., Hauck, F.J.: SAMProc: middleware for self-adaptive mobile processes in heterogeneous ubiquitous environments. In: MDS, pp. 1–6. ACM, New York (2007)
6. Ishikawa, F., Tahara, Y., Yoshioka, N., Honiden, S.: Formal model of mobile BPEL4WS process. IJBPIM 1(3), 192–209 (2006)
7. Kunze, C.P., Zaplata, S., Lamersdorf, W.: Mobile process description and execution. In: Eliassen, F., Montresor, A. (eds.) DAIS 2006. LNCS, vol. 4025, pp. 32–47. Springer, Heidelberg (2006)

8. W3C: Web services architecture (2004), http://www.w3.org/TR/ws-arch/
9. Barry, D.K.: Web Services and Service-Oriented Architectures. Elsevier, Amsterdam (2003)
10. W3C: Web services description language (WSDL) version 2.0 part 1: Core language (2007), http://www.w3.org/TR/wsdl20/
11. W3C: SOAP version 1.2 part 1: Messaging framework (2007), http://www.w3.org/TR/soap12-part1/
12. OASIS: Introduction to UDDI: Important features and functional concepts. Whitepaper, OASIS (2004)
13. OASIS: Web services business process execution language version 2.0 (2007)
14. Active Endpoints: ActiveBPEL open source engine project (2009), http://www.active-endpoints.com
15. Eclipse Foundation: Eclipse web tools platform (2009), http://www.eclipse.org/wtp
16. Eclipse Foundation: Graphical modeling framework (2009), http://www.eclipse.org/gmf
17. Eclipse Foundation: Eclipse modeling framework (2009), http://www.eclipse.org/emf
18. OMG: Business process modeling notation (BPMN), version 1.2. OMG Document formal/2009-01-03 (January 2009)
19. openArchitectureWare.org: openarchitectureware (2009), http://www.openarchitectureware.org
20. Geihs, K., Barone, P., Eliassena, F., Floch, J., Fricke, R., Gjorven, E., Hallsteinsen, S., Horn, G., Khan, M., Mamelli, A., Papadopoulos, G., Paspallis, N., Reichle, R., Stav, E.: A comprehensive solution for application-level adaptation. Software: Practice and Experience 39(4), 385–422 (2009)
21. Phung-Khac, A., Beugnard, A., Gilliot, J.-M., Segarra, M.-T.: Model-driven development of component-based adaptive distributed applications. In: SAC, pp. 2186–2191. ACM, New York (2008)
22. Blow, M., Goland, Y., Kloppmann, M., Leymann, F., Pfau, G., Roller, D., Rowley, M.: BPELJ: BPEL for Java. Whitepaper, BEA (2004)

An Architecture Independent Approach to Emulating Computation Intensive Workload for Early Integration Testing of Enterprise DRE Systems

James H. Hill

Department of Computer and Information Science
Indiana University/Purdue University at Indianapolis
Indianapolis, IN, USA
hillj@cs.iupui.edu

Abstract. Enterprise distributed real-time and embedded (DRE) systems are increasingly using high-performance computing architectures, such as dual-core architectures, multi-core architectures, and parallel computing architectures, to achieve optimal performance. Performing system integration tests on such architectures in realistic operating environments during early phases of the software lifecycle, *i.e.*, before complete system integration time, is becoming more critical. This helps distributed system developers and testers evaluate and locate potential performance bottlenecks before they become too costly to locate and rectify. Traditional approaches either (1) rely heavility on simulation techiques or (2) are too low-level and fall outside the domain knowledge distributed system developers and testers. Consequently, it is hard for distributed system developers and testers to produce realistic operating conditions for early integration testing of such systems.

This papers provides two contributions to facilitating early system integration testing of enterprise DRE systems. First, it provides a generalized technique for emulating computation intensive workload irrespective of the target architecture. Secondly, this paper illustrates how the emulation technique is used to evaluating different high-performance computing architectures in early phases of the software lifecycle. The technique presented in this paper is empirically and quantitatively evaluated in the context of a representative enterprise DRE system from the domain of shipboard computing environments.

1 Introduction

Emerging trends in enterprise distributed real-time and embedded systems. Enterprise distributed real-time and embedded (DRE) systems, such as mission avionics systems, shipboard computing environments, and traffic management systems, are increasingly using high-performance computing architectures [17,12], *e.g.*, multi-threaded, hyper-threaded, and multi-core processors. High-performance computing architectures help increase parallelization of computation intensive workload, which in turn can improve overall performance of enterprise DRE systems [17]. Furthermore, such computing architectures enable enterprise DRE systems to scale to the computation needs of next generation enterprise DRE systems, such as ultra-large-scale systems [13].

R. Meersman, T. Dillon, P. Herrero (Eds.): OTM 2009, Part I, LNCS 5870, pp. 744–759, 2009.

As enterprise DRE systems grow in both complexity and scale, it is becoming more critical to evaluate the system under development on different high-performance computing architectures early in the software lifecycle, *i.e*, before complete system integration time. This enables distributed system developers to determine which architecture is best for their needs. It also helps distributed system developers avoid the *serialized-phasing development problem* [19] where infrastructure- and application-level system entities are developed and validated in different phases of the software lifecycle, but fail to meet performance requirements when integrated and deployed together on the target architecture. Serialized-phasing development therefore makes is hard for distributed system developers and testers to identify potential performance bottlenecks before they become too costly to locate and rectify.

Existing techniques for evaluating and validating computation intensive systems on high-performance computing architectures early in the software lifecycle to overcome the serialized-phasing development problem rely heavily on simulation and/or analytical models [3, 4, 2, 5, 6]. Although this approach is feasible for predicting performance in early phases of the software lifecyle [18], such techniques cannot account for all the complexities of enterprise DRE systems (*e.g.*, the operating environment and underlying middleware. These complexities and others, such as arrival rates of events and thread/lock sychronization, are known to affect computation intensive workload and overall performance of such systems. Distributed system developers therefore need improved techniques for evaluating enterprise DRE systems on different high-performance computing architectures in early phases of the software lifecycle.

Solution approach → Abstracting computation intensive workload via emulation techniques. Emulation [20, 18] is an approach for constructing operational environments that are capable of generating realistic results. In the context of high-performance computing architectures, emulating computation intensive workload on the target architecture and operational environment enables distributed system developers and testers to construct realistic operational scenarios for evaluating different high-performance computing architectures. Moreover, it allows distributed system developers and testers to evaluate and validate system performance, such as latency, response time, and service time, when complexities of enterprise DRE systems are taken into account.

This paper describes a technique for emulating computation intensive workload to evaluate different high-performance computing architectures and evaluate and validate enterprise DRE system performance early in phases of the software lifecycle. The emulation technique presented in this paper abstracts away optimizations of high-performance computing architectures, such as caching and look-ahead processing, while letting the target architecture handle execution concerns, such as context-switching and parallel processing, which can vary between different high-performance architectures. The emulation technique is able to accurately emulate computation intensive workloads ranging from [1, 1000] msec irrespective of the underlying high-performance computing architecture. Experience gained from applying the emulation technique presented in this paper shows that it raises the level of abstraction for performance testing so that distributed system developers and testers focus more on locating potential performance bottlenecks instead of wrestling with low-level architectual concerns when constructing realistic operational environments.

Paper organization. The remainder of this paper is organized as follows: Section 2 introduces a case study for a representative enterprise DRE system; Section 3 presents the technique for emulating computation intensive workload independent of the underlying high-performance computing architecture; Section 4 presents empirical results for the emulation technique in the context of the case study; Section 5 compares this work with other related work; and Section 6 provides concluding remarks.

2 Case Study: The SLICE Scenario

The SLICE scenario is an representative enterprise DRE system from the domain of shipboard computing environments. It has be used in prior research and multiple case studies, such as evaluating system execution modeling tools [11, 8], conducting formal verification [9], and highlighting challenges of searching the deployment and configuration solution space of such systems [10]. Figure 1 shows an high-level model of the SLICE scenario.

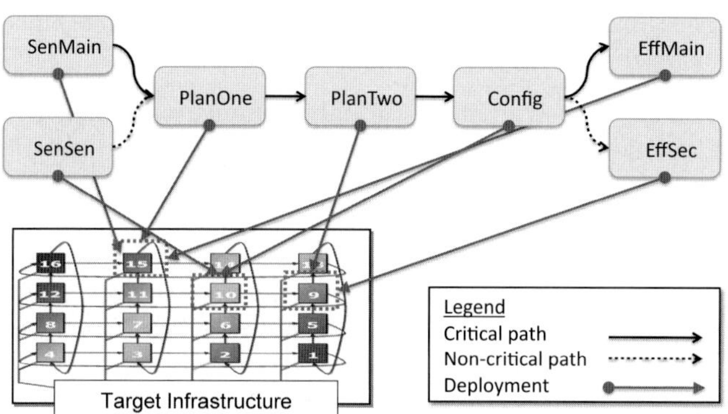

Fig. 1. High-level overview of the SLICE scenario

To briefly reiterate an overview of the SLICE scenario, Figure 1 illustrates that the SLICE scenario is composed of seven different component[1] instances named: (from left to right): *SenMain*, *SenSec*, *PlanOne*, *PlanTwo*, *Config*, *EffMain*, and *EffSec*. The directed lines between each component represents inter-communication between components, *e.g.*, sending/receiving an event. The SLICE scenario also has a *critical path* of execution, which is represented by the solid directed lines between components, that must be executed in a timely manner. Finally, each of the components in the SLICE scenario must be deployed (or placed) on one of three hosts in the target environment.

[1] In context of the SLICE scenario, a *component* is defined an abstraction that encapsulates common services that can be reused across different application domains.

The SLICE scenario is an application-level entity, however, the infrastructure-level entities it will leverage are currently under development. The SLICE scenario is therefore affected by the serialized-phasing development problem discussed in Section 1. To overcome the effects of serialized-phasing development, system developers are using system execution modeling tools to conduct system integration test for the SLICE scenario, such as evaluating end-to-end response time of its critical path, during early phases of the software lifecycle, *i.e.*, before complete system integration.

In particular, system developers are using the CUTS [11] system execution modeling tool, which is designed for component-based distributed systems, to conduct system integration tests during early phases of the software lifecycle, and to overcome the serialized-phasing development problem. CUTS uses domain-specific modeling languages [16] and emulation techniques to enable system integration testing during early phases of the software lifecycle on the target architecture using components that look and feel like their counterparts under development. Likewise, as the *real* components are developed, they can seamlessly replace the *faux* components to facilitate continuous system integration testing throughout the software lifecycle.

Similar to prior work [11], system developers intended to use CUTS to evaluate the end-to-end response time of the SLICE scenario's critical path. This time, however, system developers plan to extend their previous testing efforts and evaluate the end-to-end response time of the SLICE scenario's critical path on different high-performance computing architectures, such as multi-threaded, hyper-threaded, and multi-core architectures. Applying CUTS to evaluate the SLICE scenario on different high-performance computing architectures, however, revealed the following limitations:

- **Limitation 1. Inability to easily adapt to different high-performance computing architectures.** Different high-performance computing architectures have different hardware specifications, such as processor speed and number of processors. Different hardware specifications also affect the behavior of computation intensive workload. For example, execution time for two separate threads on a dual-core architecture will be less than execution time for the same threads on a single processor architecture—assuming there are no blocking affects, such as waiting for a lock, between the two separate threads of execution.

  System developers can use the high-performance computing architecture's hardware specification to model the behavior of computation intensive workload for emulation purposes. This approach, however, is too low-level and outside their knowledge domain. Likewise, such a model can be too costly to construct and validate, and can negatively impact overall emulation performance on the target architecture. System developers therefore need a technique that can easily adapt to different high-performance computing architecture and provide accurate emulation of CPU workload.

- **Limitation 2. Inability to adapt and scale to large number of computation resources.** System developers plan to leverage dynamic testbeds, such as Emulab [21], to conduct their integration tests. Emulab enables system developers to configure different topologies and operating systems to produce a realistic target environment for distributed system integration testing.

Although the SLICE scenario consists of three separate host, system developers intend to conduct scalability tests by increasing both the number of components and hosts in the SLICE scenario. System developers therefore need lightweight techniques that will enable them to rapidly include additional resources in their experiments, and still provide accurate emulation of computation intensive workload.

Due to these limitations it is hard for system developers of the SLICE scenario to evaluate the end-to-end response time of its critical path on different high-performance computing architecture. Moreover, this problem extends beyond the SLICE scenario and applies to other distributed systems that need to evaluate system performance on different (high-performance) computing architectures. The remainder of this paper, therefore, discusses a technique for overcoming the aforementioned limitations and improving CUTS to enable early system integration testing of computation intensive enterprise DRE systems on different high-performance computing architectures.

3 Architecture Independent Approach for Accurate Emulation of CPU Workload

This section discusses a technique for accurately emulating computation intensive workload independent of the underlying high-performance computing architecture. The approach abstracts CPU workload and focuses on overall execution time of the computation intensive workload under emulation.

3.1 Abstracting CPU Workload for Emulation

Conducting system integration test (1) during early stages of the software lifecycle, (2) on the target architecture, and (3) in the target environment enables distributed system developers to obtain realistic feedback for the system under development. Moreover, it enables distributed system developers to the locate potential performance bottlenecks so they can be rectified in a timely manner. The ability to locate such performance bottlenecks, however, depends heavily on the accuracy of such tests, *i.e.*, its behavior and workload, conducted during early phases of the software lifecycle.

In the context of high-performance computing architectures, it is possible to construct fine-grained models that will accurately emulate effects of their characteristics, such as enhanced performance due to CPU caching or look-ahead processing, or fetching instructions from memory. For example, Figure 2 illustrates an emulation/execution model for (a) fetching instructions from memory and (2) CPU caching effects.

As illustrated in Figure 2, in the case of (a) fetching instructions from memory, a portion of the computation intensive workload (*i.e.*, overall execution time) is attributed to fetching the instructions from memory. Likewise, in the case of (b) CPU caching effects, the portion of the overall execution time that would have been attributed to fetching CPU instructions from memory no longer exists.

Because high-performance computing architectures have many characteristics that can impact performance, it is not feasible to construct fine-grained emulation models that capture all effects—especially when conducting system integration tests during early phases of the software lifecycle. Instead, a more feasible approach is abstracting away such effects and focusing primarily on overall execution time, similar to profiling [7]. This is possible because as the same computation (re)occurs many times throughout the lifetime of a system, it will converge on an average execution time. The average execution time will therefore incorporate the effects from characteristics of its underlying high-performance computing architecture.

Figure 3 highlights abstracting the effects presented in Figure 2. As shown in Figure 3, instead of modeling each individual effect, the average execution time for the computation intensive workload is modeled. For example, (a) does not explicitly model the effects of fetching instructions from memory. Instead (a) models the average execution time of the computation intensive workload, which includes the execution time for occasionally fetching instructions from memory. The same holds to be true for (b), *i.e.*, instead of modeling performance gain from CPU caching effects, the average execution time for the computation intentsive workload is modeled, which includes such effects. The remainder of this section discusses how abstracting CPU effects is realized when emulating computation intensive workload for enterprise DRE systems.

3.2 Realizing Abstraction of Computation Intensive Workload in CUTS

In Section 3.1, a technique for abstracting the different CPU effects of high-performance computing architectures was discussed. The main thrust of the technique focuses on

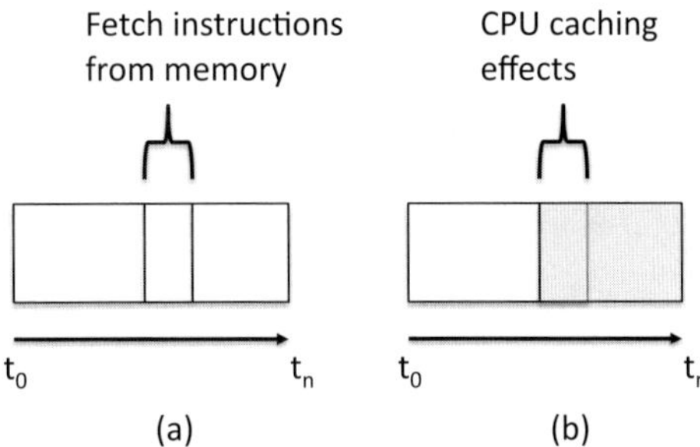

Fig. 2. Emulation/execution model to high-performance computing effects

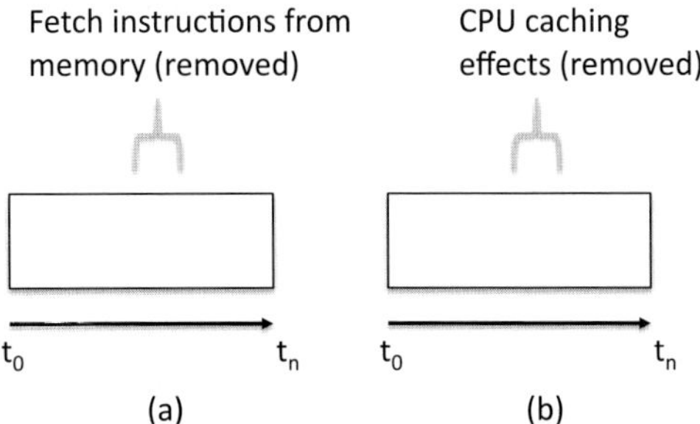

Fig. 3. Example of abstracting the emulation/execution model for computing effects

overall average execution time of a computation intensive workload (or emulated CPU workload), instead of modeling each individual CPU effect—especially when conducting system integration tests during early phases of the software lifecycle. This, in turn, simplifies emulating computation intensive workload for different high-performance computing architectures.

Since overall average execution time is the main focus of the abstraction technique, it is therefore feasibile to use an arbitrary computation to represent emulated CPU workload. The main challenge, however, is ensuring the arbitrary computation is capable of accurately representing different average execution times, *i.e*, scaling to different CPU (or computation) workloads. In CUTS (see Section 2), this challenge is resolved by using a calibration factor for an arbitrary CPU workload. The calibration factor is then used to scale the arbitrary CPU workload to different computational needs, *i.e.*, different average execution times. Algorithm 1 presents the algorithm for calibrating the CPU workload generator that realizes the abstraction technique in CUTS.

Algorithm 1. General algorithm for calibrating CPU workload generator that abstracts CPU effects.

```
procedure CALIBRATE(f, δ, max)
    f: target scaling factor for workload
    δ: acceptable error in calibration
    max: maximum number of attempts for calibration
    i ← 0
    bounds ← {0, MAX_INT}
    calib ← 0

    while i < max do
        while bounds[0] ≠ bounds[1] do
            calib ← (bounds[0] + bounds[1])/2
            t ← exec(computation, calib)

            if t > f + δ then
                bounds[1] ← calib
            else if t < f − δ then
                bounds[0] ← calib
            else
                break
            end if
        end while

        done ← VERIFY(calib)
        if done = true then
            return calib
        end if
    end while
end procedure
```

As illustrated in Algorithm 1, the goal of the calibration effort is to derive a calibration factor $calib$ that will achieve the scaling factor f (or time). As highlighted in Algorithm 1, the initial bounds of the calibration factor is defined as $[0, MAX\_INT]$. For each iteration at deriving the correct calibration factor for the CPU workload (or abstract computation), the median value is used as the potential calibration factor. If the calibration factor yields an execution time above the target scaling factor, then the current calibration factor becomes the upper bound. Likewise, if the calibration factor yields an execution time below the target scaling factor, then the current calibration factor becomes the lower bound.

This process continues until either (1) the calibration factor yields an execution time that is within acceptable error of the target scaling factor or (2) the lower and upper bounds are equal. In the case that the lower and upper bounds are equal and the calibration factor is not derived, the calibration effort is tried for up to max times. This can occur if there is background noise during the calibration exercise. If a calibration factor is derived, then it is verified that it will accurately generate execution times up to a user-defined execution time by scaling the calibration factor based on the number of iterations need to reach a target execution time. Finally, if the verification process fails, then the average error of the verification process for different execution times is used to adjust the current calibration factor for the next attempt.

By using the calibration exercise presented in Algorithm 1 it is possible to accurately emulate CPU workload independent of the underlying computational resources on different high-performance computing architectures, which has been realized in CUTS. Moreover, it enables emulation of computation intensive workload based on CPU time and (1) not "wall time" or (2) having to monitor how much time a given thread has currently executed on the CPU.

4 Evaluating Abstraction Technique for Emulating Computation Intensive Workload

The section presents the results for validating the computation intensive workload generator discussed in Section 3. This section also presents the results for applying the computation intensive workload generator to the SLICE scenario introduced in Section 2. It is necessary to validate the workload generator because it will ensure that system developers of the SLICE scenario are able to accurately emulate CPU workload for their early integration tests. Moreover, as components are collocated (*i.e.*, placed on the same host), the average execution time of its computation intensive workload is expected to be longer due to software/hardware contention [20, 18] than when the same components are deployed in isolation. It is therefore necessary to validate that the abstraction technique can produce such behavior/results.

The experiments described below (unless mentioned otherwise) were run in a representative target environment testbed at ISISlab (`www.isislab.vanderbilt.edu`). ISISlab is powered by Emulab software that configures network topologies and operating systems to produce a realistic target environment for distributed system integration testing. Each host in the experiment was an IBM Blade type L20, dual-CPU 2.8

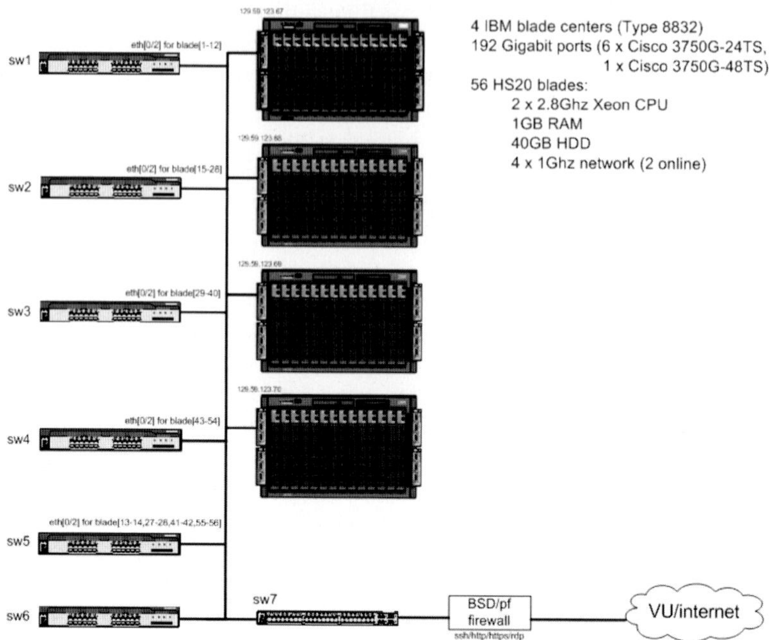

Fig. 4. ISISlab at Vanderbilt University

GHz processor with 1 GB RAM configured with the Fedora Core 6 operating system. Figure 4 shows a representative illustration of ISISlab at Vanderbilt University.

4.1 Validating the Calibration and Emulation Technique

Determining the upper bound of the emulation technique discussed in Section 3.2 will enable distributed system developers to understand how much CPU workload can accurately be guaranteed before the emulation becomes unstable. When calibrating the CPU workload generator using Algorithm 1 in Section 3.2 to determine this upper bound, all tests were conducted in an isolated environment on the target host. This prevents any interference from other process that many be executing in parallel on the target host.[2] Figure 5 highlights results of the calibration exercise when the scaling factor f is 20000 usec, acceptable error δ is 100 usec, and *max* is 10.

As illustrated in Figure 5, once the CPU workload generator is calibrated using Algorithm 1, the emulation technique is able to accurately generate computation intensive workloads up to 1000 msec (or 1 second) as illustrated in the upper graph. The lower graph illustrates the standard error for each of the execution times during verification process where the most error occurred between [70, 150] msec. After 1000 msec of computation, the emulation becomes unstable is not able to accurately emulate the

[2] It is hard to produce a completely isolated environment because a host may have background services running. It is assumed that such services, however, are negligible and do not interfere with the calibration exercise.

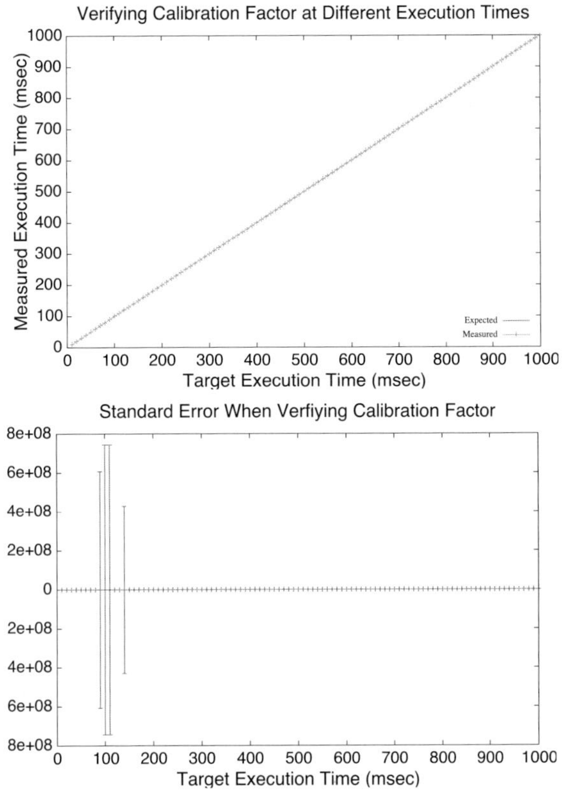

Fig. 5. Calibration results for the CUTS CPU workload generator

computation intensive workload (not shown in Figure 5). It is believed that the inaccuracy after 1000 msec is attributed to the fact that is hard to guarantee long running processes will occupy the CPU without real-time scheduling capabilities.

The calibration and verification results in presented in Figure 5 took only 1 try. This means that distributed system developers that who want to evaluate system performance have to ensure their CPU workload is less than 1000 msec, which is well above the upper bound for many short running computations of enterprise DRE systems, such as in the SLICE scenario. More importantly, they have a simple technique that will accurately emulate computation intensive workload without modeling low-level architecture concerns (*i.e.*, addresses Limitation 1 in Section 2).

Validating the calibration technique on multiple homogeneous hosts. The emulation technique presented in Section 3.2 and validated above enables system developers to accurately emulate computation intensive workload up to 1000 msec. This emulation technique, however, works only on the host on which the calibration was performed. Distributed system developers intend to use ISISlab to conduct early system integration tests of the SLICE scenario. Executing calibration test of each of the hosts in ISISlab, however, is hard and time-consuming because it requires isolated access to each host in a resource-sharing environment.

Distributed system developers of the SLICE scenario therefore hypothesize that it is possible to use the same calibration for different homogeneous hosts, or architectures. If this hypothesis is true, then it will reduce the complexity of managing and configuring integration testbeds. For example, if distributed system developers elect to use integration testbeds like ISISlab, then they have to only calibrate the CPU workload generator once on each class of hosts.

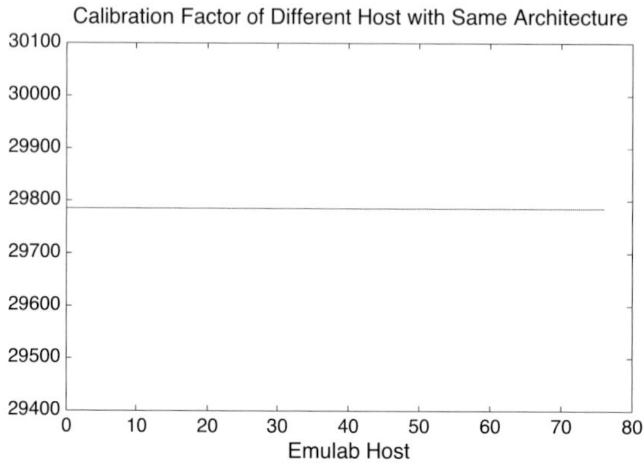

Fig. 6. Validating calibration technique on different hosts with same architecture

Figure 6 illustrates the results for calibrating the CPU workload generator on 77 different hosts in Emulab (`www.emulab.com`). Emulab was used for this experiment because it has more hosts for testing than ISISlab. Each host in Emulab used for this experiment was a Dell PowerEdge 2850s with a single 3 GHz processor, 2 GB of RAM, 2 x 10,000 RPM 146 GB SCSI disks configured with a Linux 2.6.19 Uniprocessor kernel. As illustrated in Figure 6, each host in the experiment yielded the same calibration factor. The distributed system developer's hypothesis was therefore true (*i.e.*, addresses Limitation 2 in Section 2). This also implied that different host with the same architecture and configuration can use the same calibration and consistantly generate the same accurate CPU workload for different average execution times ranging between [1, 1000] msec.

Validating the emulation technique against multiple threads of execution. Distributed system developers understand that as components of the SLICE scenario are collocated, their the execution time (or service time) for handling events will increase. This is due to having separate threads of execution for processing each event and hardware/software contention on the host. Distributed system developers of the SLICE scenario expect to experience similar behavior when using the CPU workload generator for early integration testing. They therefore hypothesize that the emulation technique realized in CUTS's CPU workload generator will produce execution times greater than the expected execution time depending on the number of threads executing on the host.

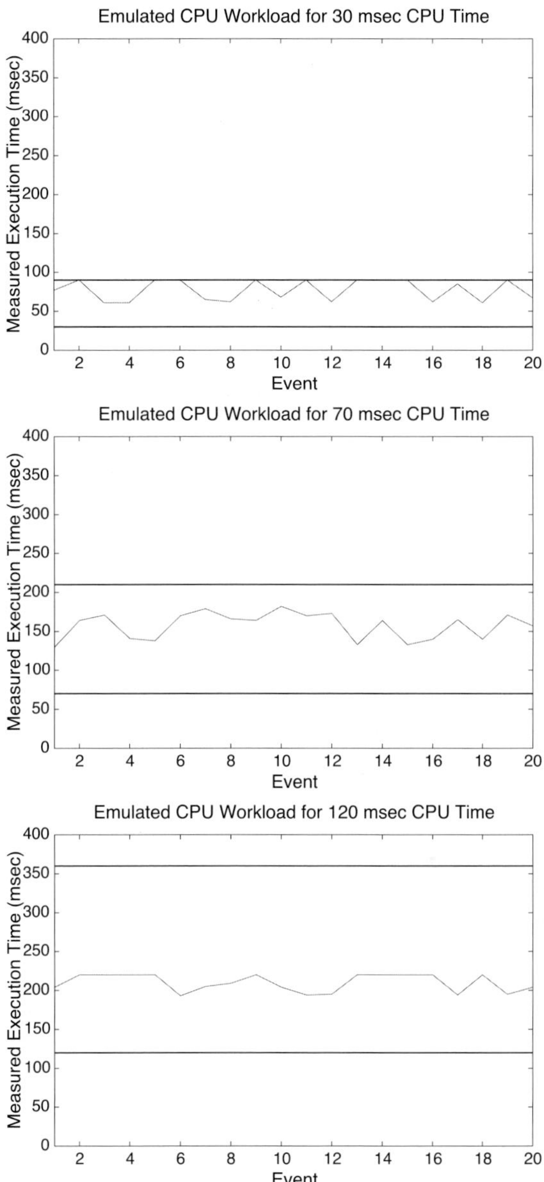

Fig. 7. Measured CPU workload for three threads of execution on same processor

Figure 7 presents the results for emulation 3 different threads of execution to single processor. As highlighted in Figure 7, each thread has an expected execution time: 30 msec (top), 70 msec (middle), and 120 msec (bottom). Figure 7 illustrates, however, that the measured execution time is greater than the specified execution time. This is because the CUTS's CPU workload generator is emulating CPU time instead of wall time. Moreover, if software performance engineering [18] techniques are taken into account, then the measured execution time for the emulation is bounded by Equation 1:

$$S_i < D_i < n \times S_i \quad (1)$$

where D_i is the measured service time (or service demand), n is the number of threads executing on the host, and S_i is the expected execution time (or service time) of the thread.

Because of the results presented in Figure 7, distributed system developer's hypothesis about the behavior of CUTS's CPU workload generator for multiple threads of execution was correct. More importantly, they have an emulation technique that will accurately emulate CPU workload, and produce realistic results when collocating components of the SLICE scenario.

4.2 Evaluating Performance of the SLICE Scenario

In Section 4.1, distributed system developers validated the emulation technique and capabilities of CUTS's CPU workload generator. Moreover, results showed that CUTS's CPU workload generator produces realistic behavior in environments with multiple threads of execution, such as collocating components of the SLICE scenario. Distributed system developers therefore plan to use the CUTS's CPU workload generator to evaluate the critical path of the SLICE scenario on different high-performance computing architectures.

In particular, distributed system developers plan to evaluate the improvement in performance of the critical path for the SLICE scenario when all components are collocated on the same host using different high-performance computing architectures. This will enable them to understand the side-effects of such architectures and can provide valuable insist in the future, such as determining how many hosts will be needed to ensure optimal performance of the SLICE scenario. They therefore used a single host in ISISlab to conduct several experiments.

Table 1 presents results of a single experiment that measured average end-to-end response time of the SLICE scenario's critical path. The experiment was executed on three different configurations/architectures of a single node is ISISlab. As highlighted in Table 1, the dual-core with hyper-threading had the best performance of the three, which distributed system developers of the SLICE scenario expected. More importantly, however, the results presented in Table 1 show that CUTS's CPU workload generator enabled distributed system developers of the SLICE scenario to construct realistic experiments and observe the effects of different high-performance computing architecture configurations during early phases of the software lifecycle.

Table 1. Emulation results of SLICE scenario on different architectures

Architecture	Avg. Exec. Time (msec)
Single	98.09
Dual-core	87.61
Dual-core (hyper-threaded)	65.8

5 Related Work

MinneSPEC [15] and Biesbrouck et. al [1] present benchmark suites for generating CPU workload. It is designed to generate computation workload for new high-performance computing architectures being simulated. The CPU workload generation technique discussed in this paper differs from MinneSPEC and Biesbrouck's work in that it abstracts away CPU characteristics that their benchmarks target in its computation workload. Moreover, the emulation technique discussed in this paper is designed to evaluate application-level performance and MinneSPEC and Biesbrouck's work is designed to evaluate architecture-level performance. It believed, however, that the computation workload in MinneSPEC and Biesbrouck's work can be used as the abstract computation workload for the emulation technique presented in this paper.

Jeong et. al [14] present a technique for emulating CPU-bound workload in the context designing database system benchmarks. Their CPU workload emulation technique uses a simple arithmetic computation that is continuously executed based on the number of times specified in the benchmark. The emulation technique in this paper extends their approach by representing similar arithmetic computations as abstract computations and accurately emulating them from [0,1000] msec. This enables distributed system developers to construct more controlled experiments when conducting system integration test during early phases of the software lifecycle.

6 Concluding Remarks

Evaluating enterprise DRE system performance on different high-performance computing architectures during early phases in the software lifecycle enables distributed system developers to make critical choices about system design and the target architecture before it is too late to change. Moreover, it enables them to identify performance bottlenecks before they become to costly to locate and rectify. This paper therefore presented a technique for emulating computation intensive workload independent of the underlying high-performance computing architecture during early phases of the software lifecyle. Based on the experience gained from applying the emulation technique for computation intensive enterprise DRE system, the following is a list of lessons learned:

- **Abstraction via emulation techniques simplifies construction of early integration test scenarios.** Instead of wrestling with low-level architecture concerns, distributed system developers focus on the overall execution times of the enterprise DRE system. This enables developers to create more realistic early system integration test with little effort, and concentrate more on evaluation the system under development on its target architecture.
- **Lack of details makes it hard to compare computation workload across heterogeneous architectures.** Different architectures have different characteristics, such as processor speed and type. Moving a computation from one architecture to another will yield different relative performance. Future work therefore includes enhancing the emulation techniques so computation intensive workloads have valid relative execution times when migrated between heterogeneous architectures.

- **Abstraction reduces the complexity of accurately emulating computation intensive workload.** This is because it does not rely on low-level, tedious, error-prone, and non-portable techniques, such as constantly sampling the clock and querying hardware (or software) for thread execution time.

CUTS and the CPU workload generator discussed in this paper are available for download in open-source format at the following location:
`www.dre.vanderbilt.edu/CUTS`.

Acknowledgements

Acknowledgements are given to Gautam Thaker from Lockheed Martin Advanced Technology Labs (ATL) in Cherry Hill, NJ. His knowledge of emulating CPU workload for enterprise DRE systems and critique on the technique and algorithm presented in this paper is greatly appreciated.

References

1. Biesbrouck, M.V., Eeckhout, L., Calder, B.: Representative Multiprogram Workloads for Multithreaded Processor Simulation. In: IEEE 10th International Symposium on Workload Characterization, September 2007, pp. 193–203 (2007)
2. Bohacek, S., Hespanha, J., Lee, J., Obraczka, K.: A hybrid systems modeling framework for fast and accurate simulation of data communication networks. In: Proceedings of ACM SIGMETRICS 2003 (June 2003)
3. Buck, J.T., Ha, S., Lee, E.A., Messerschmitt, D.G.: Ptolemy: A Framework for Simulating and Prototyping Heterogeneous Systems. In: International Journal of Computer Simulation, Special Issue on Simulation Software Development Component Development Strategies, April 4 (1994)
4. Carzaniga, A., Rosenblum, D.S., Wolf, A.L.: Design and Evaluation of a Wide-Area Event Notification Service. ACM Transactions on Computer Systems 19(3), 332–383 (2001)
5. de Lima, G.A., Burns, A.: An optimal fixed-priority assignment algorithm for supporting fault-tolerant hard real-time systems. IEEE Transactions on Computers 52(10), 1332–1346 (2003)
6. Haghighat, A., Nikravan, M.: A Hybrid Genetic Algorithm for Process Scheduling in Distributed Operating Systems Considering Load Balancing. In: Proceedings of Parallel and Distributed Computing and Networks (February 2005)
7. Hauswirth, M., Diwan, A., Sweeney, P.F., Mozer, M.C.: Automating Vertical Profiling. In: Proceedings of the 20th Annual ACM SIGPLAN Conference on Object-Oriented Programming, Systems, Languages, and Applications (OOPSLA 2005), pp. 281–296. ACM Press, New York (2005)
8. Hill, J.H., Gokhale, A.: Model-driven Engineering for Early QoS Validation of Component-based Software Systems. Journal of Software (JSW) 2(3), 9–18 (2007)
9. Hill, J.H., Gokhale, A.: Model-driven Specification of Component-based Distributed Real-time and Embedded Systems for Verification of Systemic QoS Properties. In: Proceeding of the Workshop on Parallel, Distributed, and Real-Time Systems (WPDRTS 2008), Miami, FL (April 2008)

10. Hill, J.H., Gokhale, A.: Towards Improving End-to-End Performance of Distributed Real-time and Embedded Systems using Baseline Profiles. In: Software Engineering Research, Management and Applications, SERA 2008 (2008); Special Issue of Springer Journal of Studies in Computational Intelligence 150(14), 43–57 (2008)
11. Hill, J.H., Slaby, J., Baker, S., Schmidt, D.C.: Applying System Execution Modeling Tools to Evaluate Enterprise Distributed Real-time and Embedded System QoS. In: Proceedings of the 12th International Conference on Embedded and Real-Time Computing Systems and Applications, Sydney, Australia (August 2006)
12. Hill, M.D.: Opportunities Beyond Single-core Microprocessors. In: Proceedings of the 14th ACM SIGPLAN symposium on Principles and practice of parallel programming, pp. 97–97. ACM Press, New York (2008)
13. Institute, S.E.: Ultra-Large-Scale Systems: Software Challenge of the Future. Technical report, Carnegie Mellon University, Pittsburgh, PA, USA (June 2006)
14. Jeong, H.J., Lee, S.H.: A Workload Generator for Database System Benchmarks. In: Proceedings of the 7th International Conference on Information Integration and Web-based Applications & Services, September 2005, pp. 813–822 (2005)
15. KleinOsowski, A., Lilja, D.J.: MinneSPEC: A New SPEC Benchmark Workload for Simulation-Based Computer Architecture Research. IEEE Computer Architecture Letters 1(1), 7 (2002)
16. Lédeczi, Á., Bakay, Á., Maróti, M., Völgyesi, P., Nordstrom, G., Sprinkle, J., Karsai, G.: Composing Domain-Specific Design Environments. Computer 34(11), 44–51 (2001)
17. Masters, M.W., Welch, L.R.: Challenges For Building Complex Real-time Computing Systems. Scientific International Journal for Parallel and Distributed Computing 4(2) (2001)
18. Menasce, D.A., Dowdy, L.W., Almeida, V.A.F.: Performance by Design: Computer Capacity Planning By Example. Prentice Hall PTR, Upper Saddle River (2004)
19. Rittel, H., Webber, M.: Dilemmas in a General Theory of Planning. Policy Sciences, 155–169 (1973)
20. Smith, C., Williams, L.: Performance Solutions: A Practical Guide to Creating Responsive, Scalable Software. Addison-Wesley Professional, Boston (2001)
21. White, B., Lepreau, J., Stoller, L., Ricci, R., Guruprasad, S., Newbold, M., Hibler, M., Barb, C., Joglekar, A.: An integrated experimental environment for distributed systems and networks. In: Proc. of the Fifth Symposium on Operating Systems Design and Implementation, pp. 255–270. USENIX Association, Boston (2002)

Managing Reputation in Contract-Based Distributed Systems*

Roberto Baldoni, Luca Doria, Giorgia Lodi, and Leonardo Querzoni

University of Rome "La Sapienza",
Via Ariosto 25, 00185, Rome, Italy
{baldoni,doria,lodi,querzoni}@dis.uniroma1.it

Abstract. In industry practice, bilateral agreements are established between providers and consumers of services in order to regulate their business relationships. In particular, Quality of Service (QoS) requirements are specified in those agreements in the form of legally binding contracts named *Service Level Agreements* (SLA). Meeting SLAs allows providers to be seen in the eyes of their clients, credible, reliable, and trustworthy. This contributes to augment their reputation that can be considered an important and competitive advantage for creating potentially new business opportunities.

In this paper we describe the design and evaluation of a framework that can be used for dynamically computing the reputation of a service provider as specified into SLAs. Specifically, our framework evaluates the reputation by taking into account two principal factors; namely, the run time behaviors of the providers within a specific service provision relationship, and the reputation of the providers in the rest of the system. A feedback-based approach is used to assess these two factors: through it consumers express a vote on the providers' behavior according to the actions the providers undertake.

We have carried out an evaluation that aimed at showing the feasibility of our approach. This paper discusses the principal results we have obtained by this evaluation.

Keywords: Service Level Agreements, reputation management, trust, voting mechanism, service providers, service consumers.

1 Introduction

Typically, bilateral agreements are established between providers and consumers of services. These agreements are used to regulate service provision relationships as they include all the rights and obligations the involved parties have to fulfill in order to successfully delivery and exploit the services.

It is a common industry practice for service providers to guarantee Quality of Service (QoS) requirements in the service provision relationship; these requirements are specified in the form of legally binding contracts termed Service Level

* This research is partially funded by the EU project CoMiFin (Communication Middleware for Financial Critical Infrastructures).

R. Meersman, T. Dillon, P. Herrero (Eds.): OTM 2009, Part I, LNCS 5870, pp. 760–772, 2009.

Agreements (SLAs). Meeting SLAs allows providers to be seen in the eyes of their clients, credible, reliable, and trustworthy, thus building a "reputation capital" which is a competitive advantage for them as it implies potentially new business opportunities. However, when providers fail in meeting their obligations, consumers can be led up to decrease their judgment on the reputation of the providers and, in the worse case, break business relationships because of their full untrustworthiness.

Augmenting and consolidating reputation become thus crucial requirements of many organizations participating in distributed IT systems. To motivate this statement we consider here an IT scenario that has recently gained particular attention: the protection of financial critical infrastructures [1]. Today, an increasing amount of financial services are provided also over public mediums such as Internet; this leads those services and the supporting IT infrastructures to be exposed to a variety of coordinated and massive Internet-based attacks and frauds that cannot be effectively faced by any single organization. Cross-domain interactions spanning different administrative organization boundaries are indeed in place in the financial context, and contribute to the born of the so-called *global financial ecosystem*. Protecting the financial ecosystem against the above Internet-based attacks requires global cooperation among ecosystem participating entities to monitor, share and react to emerging threats and attacks, and collectively offer the computational power to discover and contain those threats. For doing so, relationships among those entities are to be properly managed by establishing SLAs that regulate the sharing and monitoring capabilities of the cooperative environment. If those SLAs are effectively met, the financial entities can be seen reliable and trustworthy in their working environment, thus augmenting (or consolidating) their credibility (i.e., their reputation).

Therefore, we can state that reputation is an essential and strategic asset for many organizations and cannot be neglected. In particular, due to its importance, in this paper we argue that reputation has to be treated as other common QoS requirements such as service availability, and reliability and included into SLAs in the form of functions used for determining it.

The main contribution of this paper is then the design and evaluation of a framework that provides the necessary mechanisms for dynamically defining the reputation of service providers. In the reputation function evaluation, service consumers consider two main factors; namely, the run time behaviors the providers have with them in the service provision relationship, i.e., what we call *the local history* (see below), and the reputation of the providers in other business relationships. In particular, a feedback-based approach is used to assess these two factors: through it consumers express a vote on the providers' behavior according to the actions the providers undertake. These votes can be taken as input by other consumers, as specified into SLAs.

We have evaluated the feasibility of our approach in a scenario in which a service provider is involved in more than one SLA-based business relationship with many different service consumers. Depending on the impact of the two factors earlier mentioned, as specified by the SLAs, our evaluation shows that service consumers can dynamically and differently adapt their judgement on the provider reputation.

The rest of this paper is structured as follows. Section 2 presents a scenario of reference we have used in order to design our reputation management framework. Section 3 describes the framework that allows service providers and consumers to define reputation in SLA-based distributed systems. Section 4 discusses a case study we have identified in order to validate our framework, and presents a number of evaluations we have carried out for this purpose. Section 5 compares and contrasts our approach with relevant related work. Finally, Section 6 concludes the paper.

2 Reference Scenario

Figure 1 illustrates the scenario of reference we use in order to describe the principal functionalities of our reputation management framework.

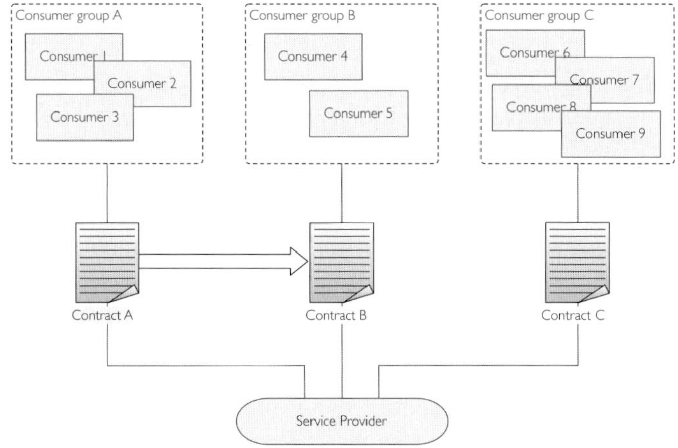

Fig. 1. Reference Scenario: an SLA-based distributed system

As shown in this Figure, we consider a distributed system in which a service provider is involved in a variety of business relationships for the provision of a specific service to different service consumers. The relationships are regulated through contracts, which we call SLAs from now on; each SLA defines different rights and obligations among the involved parties. Therefore, the distributed system can be conveniently thought of as consisting of a potentially large number of groups of consumers; these groups are formed based on the different SLA-based relationships established with the provider (the groups A, B, and C in Figure 1).

In particular, different SLAs may define different QoS requirements for which consumers are willing to pay. Included into these requirements, we propose to specify also the reputation of the provider in terms of an initial reputation value (see below) and the functions that are used to compute and update it according to the actions the provider undertake during the service provision.

In our scenario, the reputation defined into an SLA can be used as input for other SLAs, thus enabling links among SLAs and, consequently among groups of consumers. In Figure 1 this is represented by the arrow that links contract A with contract B for example. The binding among SLAs permits different consumers to take advantage of the knowledge on the provider reputation in other business relationships in order to enforce their trust in their counterpart. This knowledge can be crucial in some cases as it might be used for also determining how fast or slow can the reputation judgement towards service providers be changed. For example, if a service provider does not always behave correctly in the eyes of its clients, its reputation in other business relationships can count more and influence the final reputation evaluation.

Note that the scenario earlier described can be generalized to a distributed system with a large number of different service providers.

3 SLA-Based Reputation Framework

Owing to the reference scenario described in the previous section, our reputation management framework provides service consumers with a mean to compute the provider reputation. This is done by taking in input the SLA that regulates the business relationship among them, and that includes the rules for determining that reputation.

Thus, let p be a service provider involved in more than one SLA-based relationship with different consumers; we claim that in general its reputation is a function that can be defined as follows:

$$R_{p_j} = f(history_j, \{R_{p_i}, ..., R_{p_k}\} \setminus \{R_{p_j}\}) \tag{1}$$

$$\text{where} \quad i, j, k \in \{A, B, C, ...\}$$

To clarify the use of this function, we consider the specific scenario illustrated in Figure 1 and the reputation of the service provider from the point of view of the consumers in group B.

In this case, our framework takes in input the SLA_B, parses it, and determines the R_{p_B} that can be used by service consumers in group B in order to enforce their trust in the provider. To compute the reputation, two principal elements are taken into account: (i) the behavior of p in the SLA_B-based relationship: this is what we call *local history*; that is; the evaluation of the p's actions carried out in that specific relationship with consumers in group B, (ii) and the reputation that p has in group A, as SLA_B specifies that R_{p_A} is to be used (this is represented by the arrow in Figure 1). Note that, if more than one bind exists for SLA_B with other SLAs, the reputation function above is able to take into consideration all the reputation functions derived by the linked SLAs.

In the example of Figure 1 and using 3 above, we obtain that R_{p_B}=$f(history_B, \{R_{p_A}\})$ where $R_{p_A} = f(history_A)$. In case of group A, no links with other SLAs are established and thus the reputation of p in group A is based on the local history, only. Hence, our framework is capable of differentiating the reputation of a same provider in different consumers' groups based on the different SLAs in the distributed system.

The local history is in turn a function that is computed, as in case of 3 above, by our framework as follows:

$$history_j = g(init\_rep\_value, [facts_1(p), ..., facts_k(p)]) \qquad (2)$$

$$\text{where} \quad j \in \{A, B, C, ...\}$$

In other words, the local history is defined by considering an initial reputation value of the provider, which is included into the SLA, and an evaluation of the actions performed by p over the entire duration of the SLA (we call these evaluations *facts* in formula 2). These facts are the input of a feedback-based mechanism through which the different consumers belonging to a group can judge the provider's behavior based on the facts they have observed in that group.

The initial reputation is a value in the range of [0,1] and represents the initial credibility of the provider in the eyes of the different consumers. This credibility can be influenced by many factors: a previous business experience with the provider, the type of service the provider supply to the consumers, the level of QoS the provider guarantees to the consumers.

To make an example, we might think that some consumers have a previous long running business relationship with a specific provider so that their initial reputation judgement towards the provider can be high; that is, close to 1; in contrast, there can be other consumers that do not have any previous business experience with the provider so that they can be reluctant to consider the provider credible, and wish to evaluate its behavior during the duration of the contract in order to define its reputation.

The initial reputation value is then dynamically adjusted at run time by the local history function according to the judgements the consumers express on the provider p's actions. An example of a way to compute the reputation is described in the next section.

4 Adapting the Framework to an Example Scenario

In this section we show to the reader how the previously introduced framework can be adopted in a possible usage scenario, and which behaviour can be expected from it.

4.1 Scenario Description

Our possible usage scenario assumes a set of 310 clients distributed in 5 different usage groups. Clients of a group signed a contract with a service provider that can send continuous news updates on an argument of interests for the clients. All the five groups represent clients that signed similar contracts with different SLAs. Figure 2 reports this scenario. Note that, some SLAs are linked; for example,

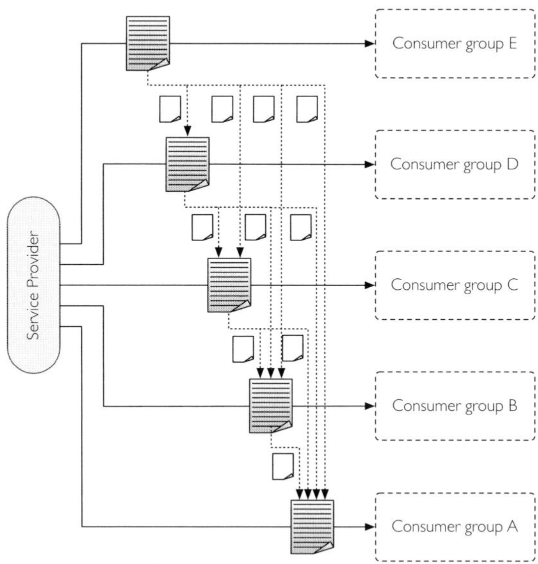

Fig. 2. An example scenario

users of group B are going to use reputation calculated at groups C, D and E and export their value to group A. Each group contains a different amount of clients: E has 10 members, D has 20 members, C has 40 members, B has 80 members, and, finally, A has 160 members.

The usage of a group reputation by another group, like group D using the calculated reputation by group E, can be seen as a sort of "service" offered by the latter to the former group. This service too can be regulated by a contract embedding various SLAs, among which, a reputation value can be present. This means that a group could repute differently reputation values provided by different groups. This "inter-group" contracts are shown in white in Figure 2.

Reputation as an SLA is defined within contracts by the two functions f and g introduced in section 3, that is the function that computes the reputation value (let us call it reputation function) and the history function. It is important to note that in each contract it is possible to introduce these functions in different ways in order to model different behaviours, but we just present one possible implementation of them. What is more, only for simplicity reasons we have used the same functions in all the contracts, but as said before we can imagine a more general situation where in each contract is introduced a different implementation of reputation and history functions.

Each client in the system maintains two different data structures for each subscribed service: the service providers reputation table and the groups reputation table. In the first table there is an entry for the service provider and a reputation number related to it; the second table contains an entry for each group whose reputation towards the service provider will be used by the client.

When a client in group x interacts with the service provider, it computes the following reputation function:

$$f_x = \frac{\sum_{i=1}^{n} R_i \cdot T_i + R_x}{n+1} \tag{3}$$

where R_i represents the reputation value provided by group i, $T_{i,x}$ is the reputation of group i with respect to clients in grup x (that is the value in the i-th entry of the groups reputation table), R_x is the current reputation of the service provider with respect to the client and, finally, n is the number of reputation values read from other groups. More specifically functions f for the different groups of our example scenario are defined as follows: $f_E = R_E$, $f_D = (R_E \cdot T_{E,D} + R_D)/2$, $f_C = (R_E \cdot T_{E,C} + R_D \cdot T_{D,C} + R_C)/3$, $f_B = (R_E \cdot T_{E,B} + R_D \cdot T_{D,B} + R_C \cdot T_{C,B} + R_B)/4$ and $f_B = (R_E \cdot T_{E,A} + R_D \cdot T_{D,A} + R_C \cdot T_{C,A} + R_B \cdot T_{B,A} + R_A)/5$.

The value returned by the function is the new reputation the client will assign to the service provider after each interaction with it. Note that, since interactions are assumed to be deterministic in this example scenario (e.g. we assume that all interactions with the service provider happens with the same order for all the clients in a group) and the method used to calculate f is deterministic, all clients in group x agree on the reputation value for the service provider.

After some interactions we suppose that a client can check through a set of facts the exact behaviour of the service provider. In the considered example the client could verify through a set of external information sources if the news sent since then by the service provider fulfills the requirements expressed in the contract. This can be modelled by defining the function g.

In our example scenario g is defined in such a way that the current reputation value for the service provider on a client is modified after each verification and increased if the verification confirms the expectations of the client ($R_x \geq 0.5$ and correct behaviour by the service provider or $R_x < 0.5$ and bad behaviour by the service provider), and reduced otherwise. The same type of update is executed on the reputation values for other groups too: if the reputation value read from a group confirms the verified behaviour of the service provider, the corresponding reputation value is increased, otherwise it is decreased. More specifically we used the following function:

$$g = (-1)^s \cdot K \cdot \frac{1}{\sigma\sqrt{2\pi}} e^{-\frac{(x-0.5)^2}{2\sigma^2}} \tag{4}$$

where K is a constant real multiplicative factor, σ is the standard deviation and s is parameter that has value 0 if the verification confirms the reputation value or value 1 otherwise. In our example scenario we assumed $K = 0.005$ and $\sigma = 0.15$. The rationale behind this formula is that reputation values in the middle of the range are not useful to take decisions; therefore, the formula is designed to quickly polarize the reputation value either towards a large or small value and then maintain the reputation in this state despite possible sources of fluctuations.

4.2 Preliminary Experimental Evaluation

Here we show some preliminary results obtained through an experimental study aimed at characterizing the behaviour of the two functions in the aforementioned scenario. The study was realized by a simple simulator that mimics the behaviour of our system by letting (i) the provider interact with all the clients, (ii) the clients calculate their reputation values and (iii) the clients adapt their values after validating the result of the previous interaction. A number of different interactions is realized in subsequent step, while the reputation values on the clients are observed. Note that our tests do not take into account all the problems involved with exchanging reputation values in a real distributed setting where clock are not perfectly synchronized, communications happening through network links can suffer from delays and failures, etc. While these implementation aspects are fundamental in order to make our system actually deployable, in this paper we focussed our attention on the potentialities offered in the proposed context by our framework, thus we limited our evaluation to more high level aspects.

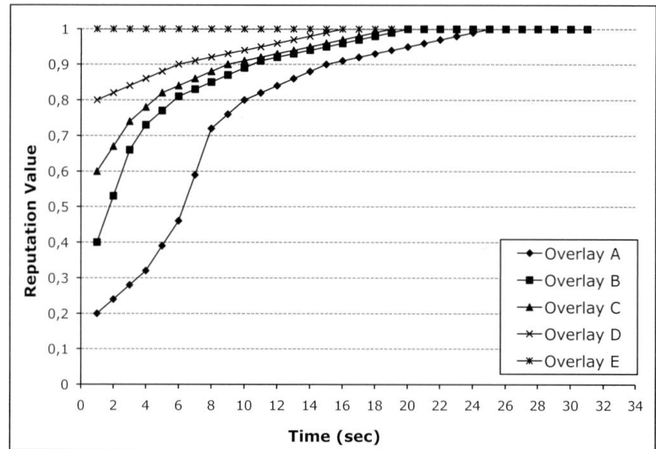

Fig. 3. Evolution of reputation when the provider behaves in a good manner

Figure 3 reports the behaviour of reputation values in a setting where the provider is supposed to always interact with clients completely fulfilling the requirements expressed in the contracts. Clients in different group start with different reputation levels ranging from 0.2 for group A to 1.0 for group E. The curves report the value of reputation calculated in different groups versus time, assuming that a new interaction with the provider happens every second. As the graph shows all reputation values converge toward value 1.0 driven by both the verification of the provider behaviour and the reputation values read from other groups.

Figure 4 shows the behaviour of reputation values in the opposite setting where the provider is supposed to always violate the SLAs defined in the contracts. As the graph shows, in this case all reputation values converge toward value 0.0 but do so with widely different behaviours. Clearly, clients in group E, starting from a reputation value 1.0, tend to maintain an higher reputation value. However, as time passes by, also the reputation of clients in E is driven toward 0 by the subsequent verification of the provider behaviour.

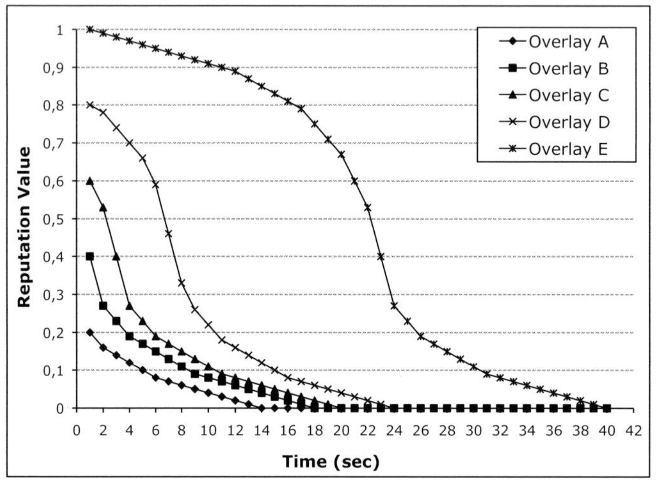

Fig. 4. Evolution of reputation when the provider behaves in a bad manner

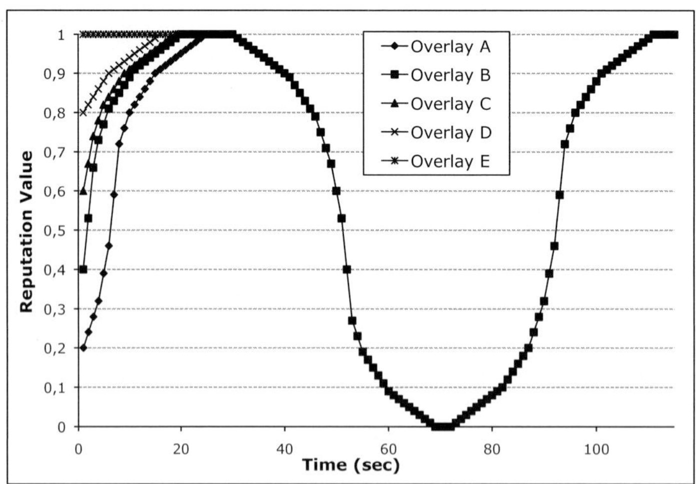

Fig. 5. Evolution of reputation when the provider behaves alternatively in a good and bad manner

More interestingly, Figure 5 shows the results obtained in a setting where the provider behaves alternatively in good and bad ways. The graph shows how the reputation values adapts to the provider behaviour in a consistent way: all groups agree on a same reputation value during the whole execution of the test, showing how the proposed adaptation of the framework to the scenario of interest is able to produce a shared picture of the provider reputation, regardless of the specific contracts signed with different clients.

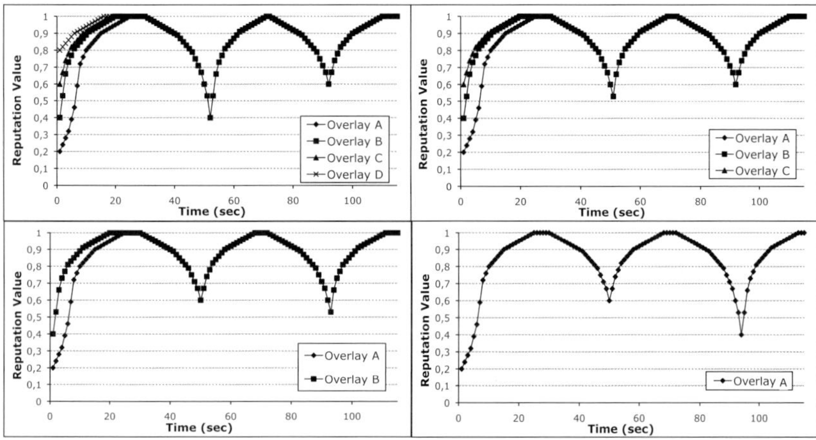

Fig. 6. Evolution of reputation towards group E (up/left), group D (up/right), group C (bottom/left) and group B (bottom/right)

We also observed, considering the same setting, how reputation among groups vary during the test. Figure 6 shows in four different graphs the average reputation of groups E, D, C and B respectively (reputation of group A is not considered in our example case). Clearly the reputation of groups show cuspids whenever that group changes its opinion with respect to the behaviour of the provider, i.e. when its reputation crosses the 0.5 "border" value. The time at which this crossing happens has an impact on the minimum value of the group reputation: the sooner is the better from this point of view as other groups will not be forced to decrease too much the reputation value.

5 Related Work

We have identified two principal macro research areas in which our work can be located; namely, reputation management systems and contract(SLA)-based distributed systems.

In this section we compare and contrast a number of works that fall into these two categories in order.

Reputation management is a body of research that has been recently widely investigated in the literature. This is proven by a large number of works that can be cited in the context of this paper [3], [8], [7], [5], [6], [4], [9]. For the sake of conciseness, we do not describe and contrast all of them in detail; rather we discuss here those works we believe are the closest to ours and from which we have derived a number of design principles and recommendations.

An important and successful example of reputation management is the online auction system eBay [2]. In eBay, buyers and sellers can rate each other after each transaction and the overall reputation of a participant is the sum of the ratings over a certain time period. This permits to build a user profile that might be used for future interactions with that user. Similarly, our approach exploits such an information that is however formally specified and used into SLAs with the service providers in order to compute their reputation value.

In [3] the authors describe a reputation system named EigenTrust that allows peers in a P2P enviroment to decrease the number of downloads of inauthentic files based on a unique global trust value assigned to each peer in the system. The value of trust is computed by taking into account the history of uploads. In essence, the EigenTrust system peers can choose from whom they can download files, thus effectively identifying and isolating malicious peers.

In [8] the authors propose an approach to P2P security where servents keep track and share with others information regarding the reputation of their peers. The reputation sharing mechanism is based on a distributed polling algorithm through which resource requestors can evaluate the reliablity of providers before initiating the download of files in a P2P system.

In [7] the authors describe a probability-based adaptive reputation mechanism by which newcomers will be trusted based on system's trust-probability that is dynamically changed according to the actions undertaken by the newcomers. In particular, the system presented in [7] relies on a group-based P2P environment in which users are organized into groups and one user belongs to only one group The groups are created based on the physical topology, the interest locality and so on. When a user joins the network a new reputation value is computed that is based on an initial probability; this probability is adjusted at run time according to the behaviors of all the new users.

All these works share a number of similarities with our approach. In all cases, a feedback-based mechanism is enabled in order to assess the reputation of participants of the system. In particular, in most cases a local reputation value is used as we use the local history function to compute the reputation of the provider locally to a specific group of consumers. In addition, all the works above are used in P2P environments for computing the reputation of peers for file sharing purposes. Typically, in these systems there is no contract used to regulate relationships among peers; the system is "flat" and the reputation is not controlled and differentiated on the basis of the type and requirements specified for the service provision. Only in [7] the authors design their reputation mechanism in a group-based P2P system. However, the groups are not formed based on contracts established among provider and consumers of services as we propose.

Owing to this latter observation, we have investigated a number of researchers that fall into the contract-based systems category in order to highlight the main differences from the reputation management standpoint. Once again, we report in this paper those that allowed us to derive important design principles for the development of our framework.

Specifically, in [10] the authors describe a distributed mechanism for managing the load in a federated system. The mechanism is based on a pairwise contracts negotiated offline between participants. Contracts set tightly bounded prices fro migrating each unit of load between two participants and specify the set of tasks that each is willing to execute on behalf of the other.

In [11] the authors describe an SLA-driven clustering mechanism for dynamic resource management. Specifically, they propose a middleware architecture that is capable of dynamically change the amount of clustered resources assigned to hosted applications on-demand, so as to meet application QoS requirements specified within SLAs.

In [12] the authors introduce a number of definition; namely, trust, reputation, trustworthiness in service-oriented environments. In addition they provide the readers with a definition of the trust and reputation relationships that can be created in such environments. Based on these definitions, they introduce a graphical notation for representing those relationships.

In [13] the authors present a contract framework in which accounting policies can be specified. The participants' trustworthiness is measured against these policies, thus ensuring that resources are allocated rationally. In their system the contract represents an action that is to be performed, and simultaneously provides a high-level description of the action. A subjective trust model is one of the foundation of their contract architecture: the belief, disbelief and uncertainty functions are used to represents the apparent trustworthiness of a participant; these functions are subjectively determined by each participant, based on their experiences.

From the assessment of the above works we can notice that most of them do not consider reputation as a requirement to be included in contracts. Only [13] seems to be the closest work to ours. However, to the best of our knowledge, the authors do not foresee a system in which contracts can be related one another by taking as input reputation policies specified in other contracts.

6 Concluding Remarks

In this paper we have described a framework that allows service consumers of a distributed system to assess the reputation of service providers in an SLA-based service provision relationship. In particular we have presented an example of usage of our framework in which groups of service consumers are able to dynamically adapt an initial reputation value specified into SLAs according to the observations on the behavior of the provider. We have carried out a preliminary evaluation of our framework that aims at showing how the reputation judgement of groups of consumers is dynamically adjusted in case the provider is reliable, malicious, or behaves sometimes in a reliable way and sometimes maliciously.

Future works include the design of a complete implementation of our framework that is able to deal with asynchronous settings and failures in the distributed system.

References

1. The CoMiFin project (2009), http://www.comifin.eu
2. Ebay, http://www.ebay.com
3. Kamvar, P.S., Schlosser, M.T., Garcia-Molina, H.: The EigenTrust Algorithm for Reputation Management in P2P Networks. In: ACM Proc. of WWW 2003 Budapest, Hungary, May 20-24 (2003)
4. Gupta, M., Judge, P., Ammar, M.: GossipTrust for Fast Reputation Aggregaion in Peer-to-Peer Networks. IEEE Transactions on Knowledgement and adata Engineering (February 2008)
5. Zhou, R., Hwang, K., Min, C.: A Reputation System for Peer-to-Peer Networks, In: Proc. of NOSSDAV 2003, Monterey, Califoria, USA, June 1-3 (2003)
6. Zhu, Y., Shen, H.: TrustCode: P2P Reputation-Based Trust Management Using Network Coding. In: Sadayappan, P., Parashar, M., Badrinath, R., Prasanna, V.K. (eds.) HiPC 2008. LNCS, vol. 5374, pp. 378–389. Springer, Heidelberg (2008)
7. Sun, L., Jiao, L., Wang, Y., Cheng, S., Wang, W.: An adaptive group-based reputation system in peer-to-peer networks. In: Deng, X., Ye, Y. (eds.) WINE 2005. LNCS, vol. 3828, pp. 651–659. Springer, Heidelberg (2005)
8. Cornelli, F., Damiani, E., De Capitani di Vimercati, S., Paraboschi, S., Samarati, P.: Choosing Reputable Servents in a P2P Network. In: ACM Proc. of WWW 2002, Honolulu, Hawaii, USA, May 7-11 (2002)
9. Procaccia, A.D., Bachrach, Y., Rosenschein, J.S.: Gossip-Based Aggregation of Trust in Decentralized Reputation Systems. In: Autonomous Agents and Multi-Agent Systems, January 22 (2009)
10. Balazinska, M., Balakrishnan, H., StoneBraker, M.: Contract-Based Load Management in Federated Distributed Systems. In: 1st Symposium on Networked Systems Design and Implementation, San Francisco, CA, USA (March 2004)
11. Lodi, G., Panzieri, F., Rossi, D., Turrini, E.: SLA-driven Clustering of QoS-aware Application Servers. IEEE Transactions on Software Engineering 33(3), 186–197 (2007)
12. Chang, E., Dillon, T.S., Hussain Khadeer, F.F.: Trust and Reputation Relationships in Service-Oriented Environments. In: Proc. of the Third International Conference on Information technology and Applications, ICITA 2005 (2005)
13. Shand, B., Bacon, J.: Policies in Accountable Contracts. In: Proc. of the Third International Workshop on Policies for Distributed Systems and Networks, POLICY 2002 (2002)

A Distributed Approach to Local Adaptation Decision Making for Sequential Applications in Pervasive Environments

Ermyas Abebe and Caspar Ryan

School of Computer Science & IT, RMIT University,
Melbourne, Victoria, Australia
{ermyas.abebe,caspar.ryan}@rmit.edu.au

Abstract. The use of adaptive object migration strategies, to enable the execution of computationally heavy applications in pervasive computing spaces requires improvements in the efficiency and scalability of existing local adaptation algorithms. The paper proposes a distributed approach to local adaptation which reduces the need to communicate collaboration metrics, and allows for the partial distribution of adaptation decision making. The algorithm's network and memory utilization is mathematically modeled and compared to an existing approach. It is shown that under small collaboration sizes, the existing algorithm could provide up to 30% less network overheads while under large collaboration sizes the proposed approach can provide over 900% less network consumption. It is also shown that the memory complexity of the algorithm is linear in contrast to the exponential complexity of the existing approach.

1 Introduction

With increasingly capable devices emerging in mobile computing markets and high speed mobile Internet permeating into current infrastructure, there is a growing need to extricate complex computing tasks from stationary execution domains and integrate them into the growing mobility of day to day life. Given the computationally heavy nature of many existing software applications, such a transition requires either the creation of multiple application versions for devices with different capabilities, or for mobile computing devices, such as smartphones and PDAs, to possess the resource capabilities of typical desktop computers. Despite the increasing capabilities being built into mobile devices, the majority of these pervasive computing machines are still incapable of handling the intensive resource demands of existing applications, and it is likely that the gap between mobile and fixed device capability will remain as both types of hardware improve. Therefore, an effective way to bridge the gap between computing resource availability is for applications to adapt dynamically to their environment.

Application Adaptation refers to the ability of an application to alter its behavior at runtime to better operate in a given execution scenario. For instance, an application might reduce its network communication in the event of an increase in bandwidth cost or a detected drop in bandwidth [1] or it might spread out its memory consumption across multiple nodes due to lack of resources [2-6].

R. Meersman, T. Dillon, and P. Herrero (Eds.): OTM 2009, Part I, LNCS 5870, pp. 773–789, 2009.

Adaptive object migration [2-6] is one such strategy in which a client distributes application objects to one or more remote nodes for execution. The approach allows constrained devices to execute applications with resource requirements exceeding their capabilities. Furthermore, the strategy allows the client to obtain improved performance and extended application functionality by utilizing externally available computing resources.

Adaptation entails incumbent overheads caused by the need to monitor and communicate available resources and application behavior. Additionally the computation of the runtime placement of objects or object clusters to suitable nodes incurs resource costs that could outweigh potential gains. These overheads limit the applicability of existing adaptation approaches in pervasive space. While the constrained nature of the devices in pervasive environments require that an adaptation process compute optimal decisions with minimal computation resources, the heterogeneity and indeterminate size of the collaboration, require that the solution scale adequately to diverse collaboration environments and application behavior.

This paper proposes a novel distributed adaptation decision making algorithm to improve the *efficiency* and *scalability* of adaptive object migration. The most relevant work is then compared in terms of its network and memory utilization using mathematical modeling and simulation. The results show that the proposed approach can provide more than 900% less network overhead under large collaboration sizes while maintaining linear memory complexity, compared with the exponential complexity of the existing approach. Finally, the paper outlines future work in terms of implementing and empirically evaluating the optimality of the adaptation decisions themselves.

The rest of this paper is organized as follows: Section 2 assesses existing work on adaptive object migration and identifies the most relevant work in the context of pervasive environments. Section 3 discusses the proposed distributed adaptation decision making algorithm. Section 4 mathematically models and comparatively evaluates the proposed approach to existing work. And finally Section 5 provides a summary and conclusion and outlines future work.

2 Literature Review

Object migration as a form of code mobility is not new. Early work [7-10] focused on manual migration of threads to remote nodes for objectives such as application behavior extension, load balancing and performance improvement. In contrast, the dynamic placement of objects in response to environmental changes or application requirements has been a focus of more recent research.

Adaptive object migration refers to the reactive and autonomous placement of objects on remote nodes based on a runtime awareness of the collaboration environment and application behavior. The approach allows applications to effectively utilize external computing resources to improve performance and extend application behavior while enabling devices to run applications with resource demands exceeding their capabilities.

Works on adaptive object migration differentiate between parallel and sequential applications. Adaptation for parallel applications [2, 5, 11] has typically focused on cluster and grid computing environments with objectives such as load balancing, performance improvement and improved data locality. Adaptive placement in parallel applications generally involves the co-location of related threads or *activities*, to minimize remote calls, and the distribution of unrelated ones to reduce resource contention. Work on sequential application adaptation has focused on more diverse computing environments including pervasive spaces. With objectives such as performance improvement and load balancing, decision making in sequential application adaptation involves matching objects to available computing nodes while minimizing inter-object network communication and improving overall utility. Sequential application adaptation [3, 4, 6, 12] presents more challenges as there is less explicit division in the units of distribution. As most work on adaptation in pervasive spaces involves sequential applications, this paper identifies relevant work in that domain alone. Nevertheless, although the algorithm is proposed in the context of sequential application adaptation, it is expected that the concepts of distributed decision making would also be applicable for parallel application adaptation.

Adaptive Offloading techniques [6, 12] and the approach presented by Rossi and Ryan [3] are works on adaptation for sequential applications specifically targeting mobile environments. *Adaptive offloading* techniques generally assume the existence of a single mobile device and a dedicated unconstrained surrogate node to serve as an offloading target. Adaptations are unidirectional in which computation is offloaded from the constrained device to the surrogate node. The approach presented by Rossi and Ryan [3] discusses adaptation occurring within *peer-to-peer* environments in which any node could potentially be constrained at some point during the collaboration. The approach presented by Rossi and Ryan [3] is more applicable to pervasive spaces in which spontaneous collaborations are formed from heterogeneous devices. Our adaptation algorithm will hence be presented as an extension to the Rossi and Ryan [3] algorithm. The paper hereon refers to the algorithm presented by Rossi and Ryan as the *existing* algorithm and the proposed approach as the *distributed* algorithm.

2.1 Adaptation Process

Generally, application adaptation approaches involve 3 basic components: 1. a metrics collection component, responsible for monitoring application behavior and available computing resources within the collaboration, 2. a decision making component responsible for determining object-to-node placements based on information from the metrics collection component and 3. an object migration component responsible for moving selected objects to their designated targets.

Based on the site at which decision making occurs, Ryan and Rossi [13], identify two types of adaptation processes, *Global (centralized)* and *Local (decentralized)* adaptation.

Global Adaptation: In *Global or Centralized Adaptation,* decision making is performed by a single dedicated and unconstrained machine. Other nodes within the collaboration periodically communicate their environment and application metrics [14] to this central node. The metrics pushed out by a node, includes the resource usage measurements of the device and the individual objects in its address space. When a node within the collaboration runs out of resources, the central node computes a new object distribution topology for the collaboration and offloads computation from the constrained device. The object topology is computed with the objective of providing optimum load balance and performance improvement while minimizing inter-object network communication.

While *Global Adaptation* allows for the near optimal placements of objects-to-nodes, the computation costs of computing an object topology for even a simple scenario with a few objects in small collaborations can be prohibitively expensive [13]. While Ryan and Rossi [13] discuss the possible use of Genetic Algorithms as a solution to reduce this cost, the approach would still be computationally expensive compared to a decentralized approach to decision making *(discussed in the next section).* Another disadvantage of *Global Adaptation* is the need for a reliable and unconstrained node for decision making. This presents a central point of failure and limits its applicability within ad-hoc collaborations of constrained devices.

Local Adaptation: In *Local or Decentralized Adaptation*, decision making is computed on individual nodes. Resource metrics of each node are periodically communicated to every other node within the collaboration. Unlike Global Adaptation, the metrics propagation includes only the collaboration's resource availability and not the software metrics of objects [14] (e.g. number of method invocations, method response times etc.). When a node runs out of resources, it computes an adaptation decision based on the information it maintains about the collaboration and the metrics of the objects in its memory space.

As the adaptation decisions are computed by considering only a subset of the overall object interactions, the decisions made are not as optimal as the centralized approach. However, such an approach removes the central point of failure, and offers a more scalable approach to adaptation decision making.

Figure 1 shows pseudo-code for the basic decision making computation of a local adaptation algorithm computed on individual machines, as presented by Rossi and Ryan [3]. The *evaluate* function computes a score determining the suitability of placing each mobile object, *o* on each remote target node, *n*. The object-to-node match with the highest score is selected for migration. The process is repeated until either all objects are migrated or an object-to-node match that can achieve a minimum threshold is not available. The score of an object placement (placement of object *o* to node *n*) considers two objectives: *resource offloading* from the constrained node, and *performance improvement* of the application. The *resource offloading* score evaluates the degree to which the source node's load can be mitigated and the load difference with the target node reduced whereas the *performance* score evaluates the degree of response time improvement achievable through the migration of an object to a target. Note that this serves as a basic example which can be readily extended to include more diverse goals such as battery life preservation, reliability etc.

In the context of pervasive spaces the local adaptation algorithm in Figure 1, presents a few shortcomings. As resource availability is inherently dynamic, suboptimal decisions could be made based on out-of-date information about the collaboration; therefore avoiding this problem requires nodes to communicate their environment metrics more frequently. The $O(N^2)$ message complexity[1], *where N is the number of nodes within the collaboration*, for one collaboration wide communication, increases the network overheads and limits the scalability of the approach. The storage and maintenance of this information also requires additional processor and memory resources.

---

[1] Message Complexity, in this context, is based on the number of messages sent and received by each node within the collaboration.

```
do
{
     maxScore = 0.5
     maxObject = null, maxNode = null
     for each mobile object o in local node do
       for each remote node n do
         score = evaluate(o, n)
         if (score > maxScore) then
            maxScore = score
            maxObject = o
            maxNode = n
         end if
       end for
     end for
     if (maxScore > 0.5) then
      move maxObject to maxNode
     end if
}
while (maxScore > 0.5)
```

Fig. 1. Local Adaptation Algorithm basic flow, Rossi and Ryan [3]

With a worst case complexity of $O(NM^2)$, *where M is the number of mobile objects*, decision making is computationally expensive for mobile devices in this environment, exacerbated by the decision being made on the already constrained adapting node. In response, nodes need to set lower constraint thresholds for triggering adaptation so that enough resources to compute adaptation decisions are reserved. This lowered threshold in turn results in increased adaptation throughout the collaboration causing additional network overheads and reduced application performance.

Given the diversity of the pervasive environment, adaptation algorithms need to consider additional metrics, such as location, battery usage, reliability etc. This requires adapting nodes to apply different score computation logic for different devices based on the metrics relevant to each target node, again introducing additional computation complexities to the adaptation logic.

In summary, while the approach is feasible in small, stable, homogenous collaborations, the above constraints limit the scalability under medium to large scale, heterogeneous and dynamic pervasive collaborations.

3 Distributed Approach

This section proposes a distributed decision making algorithm to improve the *efficiency* and *scalability* of the *existing* local adaptation algorithm discussed in section 2. The new *distributed* approach presents an alternative that avoids the need to periodically communicate environmental metrics throughout the collaboration, and partially distributes the decision making process to reduce the associated computation costs on the adapting node. Furthermore, the algorithm reduces the network and memory utilization costs that arise from increased collaboration size and heterogeneity.

The approach involves each node connecting to a multicast address through which all adaptation related communication will take place. The use of multicast groups for communication allows nodes to delegate the responsibility of maintaining an awareness of the collaborating nodes to external network devices (e.g. routers, switches). Once connected, each node monitors its own environment metrics and the metrics of the objects within its memory space, however unlike existing approaches this information is not communicated to other nodes. The anatomy of the *distributed* approach is described below and illustrated in Figure 2:

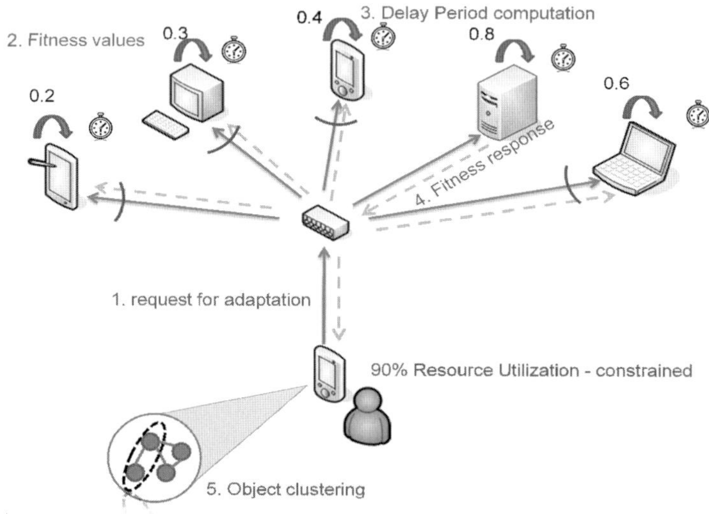

Fig. 2. Distributed adaptation decision making algorithm

1. Adaptation Request: When a node is resource constrained it multicasts an *adaptation request, R,* containing its own environment metrics and synoptic information about the objects in its memory space. The message complexity is *O(N),* since a single request is sent and *N-1* messages are received. The request sent by the node includes the following information:

- *Node URI:* This includes the IP address of the node and the port number of the middleware runtime.
- *Environmental Metrics (E):* This includes the memory, processor, network utilization and capacity of the source node (adapting node).
- *Synoptic information about objects:* The least resource utilization consumed by an object $ru^{O_{min}}$, and the total resource utilization consumed by all mobile objects within the node, $ru^{O_{total}}$ are recorded for each metric type (memory, network, and processor).
- *Minimum Threshold Value, k:* The minimum score that can result in migration. (similar to the score threshold identified by Rossi and Ryan [3])

2. Fitness Value Computation: Each target node receives the adaptation request and determines its own suitability for adaptation by computing a *fitness score*. This *fitness score* represents the amount of resources the source can offload to the target, thereby reducing the load disparity between the two nodes; as well as the response time improvement the target can provide. Additionally, candidate nodes which are already constrained could choose not to compute adaptation decisions hence reducing the amount of global processor cycles consumed for an adaptation. The fitness value computation is further discussed in section 3.1.

3. Delay Period Computation: Each node compares its *fitness score* against the minimum threshold, *k*, sent by the source. If a node can offer a better score, it computes a delay period which is inversely proportional to its score, providing fitter nodes with shorter delay periods. The node would then wait for the computed time period before reporting its fitness score to the source. The benefit of this approach is two-fold: firstly, it reduces the possibility of a multicast *sender-storm* in which the source might be inundated by fitness responses; and secondly, it allows for the suppression of further fitness reports once the fittest candidate has responded *(see 4. below)*, thereby reducing network costs. Though such a delay scheme would introduce latency in the decision making process, it occurs parallel to the application's execution and hence is unlikely to reduce application performance. Note that since the evaluation of the *distributed* adaptation algorithm presented in this paper involves mathematical modeling and simulation, the implementation specifics of the delay period computation are not explored. Nevertheless, although there is no *prima facie* reason to assume such an implementation would be problematic, the comparative network utilization simulation in section 4 considers both cases in which a delay period is, and is not, used.

4. Reporting the Fitness Score: Once the delay period of a candidate expires, it multicasts a response to the collaboration. The response is received by the source, and every other candidate within the collaboration. Each candidate compares the offered fitness score against its own and if the offered fitness value is greater, the candidate cancels its delay timer and need not reply to the source. However, if the target node can offer a better fitness value, it will continue to wait for its delay period to expire before communicating its fitness score in the same procedure, thus giving precedence to any other node which could be *fitter* than it. Enforcing the delay timer on every response, assures that the source receives responses from only the fittest candidate(s), hence minimizing resource costs and avoiding a *sender-storm* problem. The response that is multicast by a candidate includes the following information:

- *Node URI:* This includes the IP address of the candidate and the port number of its middleware runtime.
- *Environment Metrics(E):* The capacity and utilization of each resource metric type (memory, processor, network) of the target node
- *Fitness Score, (S):* The fitness score computed by the target node, which is a value between 0 and 1. (discussed in section 3.1)
- *Required Object Metrics Value, (ru^O):* The ru^O is a value between the $ru^{O_{\min}}$ and $ru^{O_{\text{total}}}$ value sent by the source (*step 1*), which describes the resource utilization of an imaginary object cluster for which the specified fitness score can be achieved *(discussed further in section 3.1)*. The source node will group objects so that the overall resource consumption of the group matches this specified value *(discussed below)*.

5. Clustering of Objects: The source node listens to fitness score multicasts for a pre-computed period of time. This *wait* period is based on the amount of time it would take a node achieving the minimum threshold score *k (discussed in 1 above)* to respond. If multiple fitness scores have been received by the time the source's *wait* period expires, the node selects the best offered score. The source node then groups objects within its memory space so as to meet the criteria, ru^O, required by the selected candidate. The identified object cluster would be a subset of the mobile objects within the memory space of the constrained source node. Once the grouping of objects is complete, the source migrates the cluster to the candidate node. The migration of object clusters instead of individual objects reduces object spread, and decreases inter-object network communication cost, thereby improving the optimality of adaptive decision making when compared to the *existing* approach. As clustering would be done to meet a single constraint or resource type *(discussed in section 3.1)*, a linear computation cost of $O(M)$ can be assumed in which objects are incrementally grouped, based on sorted metrics values. Another approach to object clustering could be to use runtime partitioning approaches [12]. However, since the main focus of this paper is to compare the network and memory consumption of the new algorithm to an existing approach, the processing complexities of object clustering techniques is beyond the scope of this paper.

3.1 Fitness Score Computation

The fitness score is at that heart of the *distributed* algorithm. It determines the degree of suitability of each candidate to the adaptation request, and guides the clustering performed on the source node. Like the algorithm proposed by Rossi and Ryan [3] the computation considers two objectives: *resource offloading* from the constrained node, and *performance improvement* of the application. The *resource offloading* score for each metric type, evaluates the degree to which the load of the source node can be mitigated and the load disparity with the target node reduced. The *performance improvement* score evaluates the degree of response time improvement that can be attained by the migration. The objectives considered can be easily extended to include more diverse goals such as battery life preservation, reliability etc. Though the details of fitness computation are a part of the algorithm's *effectiveness* and hence beyond the scope of this paper, as a proof of concept, the approaches and mathematical derivations of the computation of *resource offloading* scores for each metric type (memory, processor, network) have been presented as an online appendix to this paper [15]. While the *existing* approach computes a *resource offloading* score for individual objects, the *distributed* approach has the flexibility of attaining higher scores by grouping mobile objects together. The candidate computes an ideal ru^O value, which is the resource utilization of a hypothetical object cluster for which it can offer the most ideal score for each metric type. The individual scores achieved for each metric are then aggregated to compute a final *fitness score* as is shown in the online appendix in [15].

By allowing candidate nodes to compute their own fitness value, the approach removes the need to periodically communicate environmental metrics. This reduces network communication costs and limits other resource overheads associated with storing and maintaining collaboration information on every node. Consequently, the resource utilization of the *distributed* approach is linear to the number of nodes (discussed further in section 4), compared to the exponential cost of the *existing* approach, making it more scalable with regard to collaboration size.

By dividing and offloading the tasks of the adaptation process, the space and time complexity of decision making is reduced, making adaptation more feasible and efficient for the local constrained machine. This distribution further reduces resource contention between the executing application and the adaptation engine. Additionally, application objects are able to execute locally until higher constraint thresholds are reached, avoiding the need to maintain low thresholds so as to reserve resources for adaptation computation as is the case in the *existing* approach. Consequently this higher threshold limit reduces the number of adaptations and adaptation related overheads within the collaboration. The delegation of score computation to the remote nodes also allows individual nodes to easily factor in additional metrics into the decision computation process without further overheads on the adapting node. For instance a mobile device within the collaboration could factor in its battery life when computing a fitness score. This allows for increased diversity in the collaboration environment, as would be necessary for adaptation in pervasive environments.

As adaptive decisions are made using *live metrics* as opposed to *cached metrics*, the possibility of making suboptimal adaptation based on out-of-date metrics is minimized. The approach also lends itself to future work involving object clustering based on coupling information to obtain more efficient object topologies.

It is worth noting however, that the *distributed* approach could result in longer adaptation decision making times as a result of network latency and the delay scheme discussed in section 3. However, since adaptation decisions are performed parallel to the executing application, it is not expected to have a direct effect on the application's performance. Furthermore, it is expected that for large scale collaborations adapting computationally heavy applications, the execution of adaptation decisions on a local constrained device might take longer than the *distributed* approach which leverages externally available resources. This however relates to the performance aspect of our algorithm and is beyond the scope of this paper.

4 Evaluation

The utility of an adaptation algorithm is determined by two factors: The *efficiency* (resource utilization and performance) of the adaptation process and the *effectiveness* of the adaptation decision outcome or the optimality of the adapted object topology. Evaluating the latter requires that different application and collaboration behaviors be considered. This is because of the difference in granularity for which adaptation decisions are made and the difference in decision making location of the two algorithms. Hence an empirical study to determine the difference in decision optimality of the two algorithms would require the use of various application benchmark suites and the use of multiple collaboration environment settings. Though preliminary results of such a study are positive, further discussion of the *effectiveness* of the algorithms is left to future work.

In order to provide a comparative evaluation of both algorithms with respect to network and memory utilization, the algorithms are mathematically modeled under various environmental scenarios. Both the maximum degree with which one algorithm outperforms the other, as well as the specific environmental scenarios for which each algorithm provides comparatively lower resource utilization, are identified.

4.1 Environmental Settings

In order to compare the resource utilization of both algorithms under diverse environmental settings, a range of possible values for the variables identified in Equation 1-4 are identified and their feasibility constraints and relationships discussed.

In order to evaluate the resource utilization of the algorithms independently from their adaptation outcome, the adaptive decisions made by both algorithms are assumed to be the same. This means that both algorithms would compute the same adaptive object placements and hence incur similar object migration and inter-object communication costs under the same collaboration settings.

Table 1. Variables influencing network utilization and memory utilization in both algorithms

Variables	Constraints
Number of nodes *(N)*	$4 \leq N \leq 50$
Execution Time *(ET)*	$600s \leq ET \leq 86400s$
Frequency of propagation *(f)*	$0.0017/_s \leq f \leq 0.1/_s$
Number of adaptations *(Na)*	$Na \leq Np$
Number of propagations *(Np)*	$1 \leq Np \leq 8640, Np = ET * f$
Number of fitness values *(Nf)*	$Nf \leq N - 1$
Fitness report size *(F)*	$F = 2E$
Environment metrics size *(E)*	$200Bytes \leq E \leq 10000Bytes$
Adaptation request size *(R)*	$R = 2.5E$

Number of Nodes *(N)*: In order to observe the scalability of both algorithms with regards to collaboration size, a range of possible collaboration sizes ranging from a small scale collaboration of 4 nodes, to a large scale collaboration of 50 nodes, are considered.

Execution Time *(ET)*: The overall time spent executing and adapting a given application, ranging from a brief collaboration of 10 minutes up to 24 hours.

Frequency of Metrics Propagation *(f)*: This variable is applicable to the *existing* algorithm and refers to the frequency of propagating environmental metrics throughout the collaboration. A periodic metrics propagation is assumed for simplicity, though node triggered propagations based on degree of metrics change could be used [14]. The greater the fluctuation in resource availability within an environment the more frequently metrics propagations needs to occur. The higher value of 0.1/s (propagation every 10 seconds) would be more applicable to dynamic heterogeneous environments.

Number of Metrics Propagations *(Np= ET*f)*: The number of times a collaboration wide metrics communication occurs in the *existing* algorithm is the product of the *Execution Time (ET)* and the propagation *frequency (f)*. High number of propagations means either long executing applications, high frequency of metrics propagation or both. Hence a high number of propagations would be expected when executing long running applications in dynamic environments.

Number of Adaptations *(Na)*: The number of adaptations cannot exceed the total number of metrics propagations that occur within a given period. This is because the event in which a node exceeds its resource utilization constraint thresholds, hence requiring an adaptation, would be a less frequent occurrence than a metrics propagation which reports resource changes of much lesser degree. The reason behind this theoretical upper bound is that for an adaptation to occur in the existing algorithm, the adapting node would need to first have information about the collaboration. This implies that at least one collaboration wide propagation needs to occur prior to any adaptation. As every adaptation would cause notable resource utilization differences within the environment, it needs to be followed by a subsequent propagation. This means that for every adaptation there would be at least one more number of propagations. As the adaptation decisions and number of adaptations performed by both algorithms are assumed to be the same in this evaluation, it follows that the upper bound of *Na* for both algorithms would be *Np*.

Number of Fitness Reports Returned *(Nf)*: While the number of fitness values returned for each adaptation request could vary, for simplicity we assume a single *Nf* value in which a fixed number of nodes reply to every adaptation request. We consider a theoretical upper bound in which *Nf* is equal to the number of candidates, for a situation in which no functional delay scheme exists. The delay scheme, discussed in section 2, would significantly reduce the number of fitness responses by allowing only the fittest nodes to respond.

Fitness Report Message Size *(F)*: As discussed in section 3, the report communicated to candidates consists of more information than the simple environment metrics, E, propagated by the *existing* algorithm. Specifically, F is two times the message size of the *existing* algorithm.

Adaptation Request Message *(R)*: Similarly, it is determined that the adaptation request is 2.5 times the environmental metrics message size of the *existing* algorithm.

4.2 Network Utilization

The cost complexity of network utilization is assumed to be the total number of bytes sent and received by all nodes within the collaboration. Hence a unicast of an environment metrics message, E, from one node to the entire collaboration would cost $2(N-1)E$ whereas a multicast of the same information would cost $N \times E$.

Equation 1 models the total network utilization consumed by the existing algorithm of Rossi and Ryan [3] during a complete collaboration session. It is given as the sum of: the total bytes of all metrics propagations; the network utilization of each object due to *distributed* placement; each object migration during every adaptation; and any external network utilizations of the application *(e.g. http request etc.)*.

$$\mathrm{Nu(existing)} = 2\mathrm{Np}(\mathrm{E}(\mathrm{N}^2 - \mathrm{N})) + \sum_{i=1}^{O} \mathrm{Nu_{ext}}(\mathrm{O_i}) + \sum_{i=1}^{O}\sum_{j=1}^{m_i}(\mathrm{Nu_j} * \mathrm{NI_j}) + \sum_{i=1}^{Na}\sum_{j=1}^{O^i_{mo}}(\mathrm{SOS_{ij}} + \mathrm{ECS_j}) \tag{1}$$

Np = Number of propagations
E = Serialized size of the environment metrics.
N = Number of nodes within collaboration
O = Number of objects
m_i = *Number of methods of object i*
NI_j= *Number of invocations of method j.*
Nu_j= *Network utilization of method j [6]*

Na= Number of adaptations
SOSij=Serialized volume of object i during adaptation j.
ECS = Executable Codes size
O^i_{mo}=*number of migrated objects during adaptation i*

The network utilization of the *distributed* algorithm is the byte sum of: the adaptation requests sent; the adaptation reports responded for each adaptation; each object's network utilization due to *distributed* placement; each object migration on every adaptation; and any external network utilizations of the application *(e.g. http request etc.)(NU_{ext}).*

$$\mathrm{Nu(distributed)} = \mathrm{Na(nR)} + \sum_{i=1}^{Na} \mathrm{Nf_i}\,(\mathrm{N} \times \mathrm{F}) + \sum_{i=1}^{O} \mathrm{NU_{ext}(O_i)} + \sum_{i=1}^{O}\sum_{j=1}^{m_i} (\mathrm{Nu_{ij}} * \mathrm{NI}) + \sum_{i=1}^{Na}\sum_{j=1}^{O^i_{mo}} (\mathrm{SOS_{ij}} + \mathrm{ECS_j}) \qquad (2)$$

To visualize the network utilization of each algorithm under varying numbers of: metrics propagations (*Np*)*;* adaptations (*Na*); and collaboration sizes (*N*), a constant upper bound for the number of fitness values returned is first assumed (*Nf=N-1*). Setting *Nf* to its upper bound for each adaptation models the scenario in which no functional delay scheme exists in the new algorithm, thereby providing a worst case comparison, in terms of network utilization, of the *distributed* algorithm and the *existing* approach.

Figure 3 & Figure 4 show a 3D region of the values of *Na, Np* and *N* for which the network utilization of the *existing* algorithm would be better than the *distributed* approach and vice-versa. The region is determined based on a simple 3D *regionplot* performed using the computational tool Mathematica [16] for the ty $Nu(existing) < Nu(distributed)$ and vice-versa, under the constraints specified in section 4.1. Note that the costs incurred by external application network utilization, object placement and object migration cancel out as they are assumed to be the same for both algorithms.

Figure 3, shows that the *existing* algorithm provides comparatively better network utilization in small collaboration sizes and very high numbers of adaptations ($Na \geq 65\%Np$). In larger collaboration sizes, the *existing* algorithm provides better performance only at increasingly higher numbers of adaptations where $Na \cong Np$. Computing for the maximum disparity between the two algorithms within this region, shows that the *existing* algorithm could offer as much as 35% less network utilization when NA=NP.

Figure 4, shows the complementary region, under which the *distributed* algorithm provides better resource utilization. The algorithm provides better network utilization under a greater portion of the environment settings. While the algorithm provides better network utilization under relatively less numbers of adaptations, in small collaborations, it is able to outperform the *existing* approach even under high numbers of

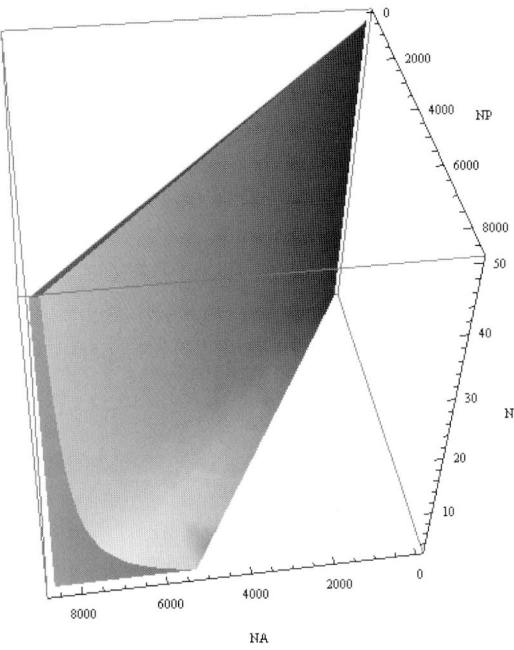

Fig. 3. The region of N, Na and Np for which the *existing* approach provides better network utilization. $\boldsymbol{Nu(existing) < Nu(distributed)}$

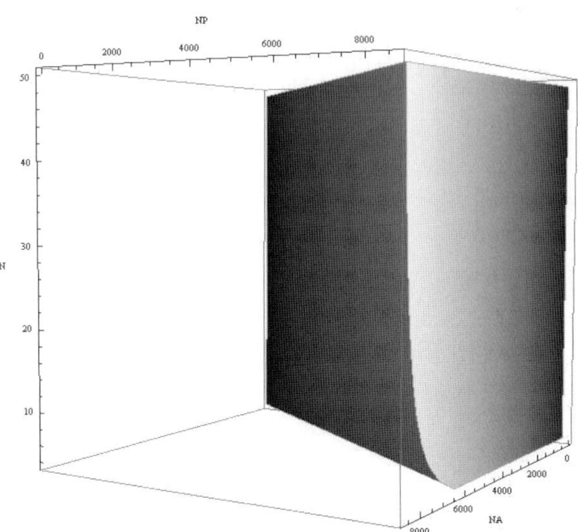

Fig. 4. The region of N, Na and Np for which the *distributed* algorithm provides better network utilization over the *existing* approach. $\boldsymbol{Nu(distributed) < Nu(existing).}$

adaptations in larger collaboration environments. Computing for the maximum degree of network utilization difference between the two algorithms under this region shows that the proposed *distributed* approach can provide over 900% less network utilization under relatively low number of adaptations (Na<10%Np).

In practice, a scenario in which the number of adaptations within the collaboration is as high as that favoring the *existing* algorithm is unlikely even in extremely dynamic computing environments. Such a high adaptation count suggests an object topology in a constant state of flux, thus implying that adaptation decisions are performed without benefits being gained, whilst adaptation overheads are still incurred. Furthermore an event in which a node exceeds its resource utilization threshold, thus requiring an adaptation, would be a far less frequent occurrence than metrics propagation occurring for lesser degrees of change in resources.

The above figures show that even without a delay scheme to reduce the number of fitness reports, the *distributed* algorithm provides better network utilization for a greater range of possible environmental scenarios and is more scalable with regard to the collaboration size and application execution duration.

The effect of a delay scheme on the results discussed above is shown in Table 2, wherein it is evident that more effective delay schemes reduce the maximum degree with which the *existing* algorithm outperforms the *distributed* solution until a delay period which is 55% effective (i.e. only 45% of the collaborating nodes reply on each adaptation) results in the *distributed* algorithm always outperforming the *existing* one. This can also be visualized as the shrinking of the region identified by Figure 3 and the expansion of the region shown in Figure 4.

Table 2. The effect of a delay scheme on the network utilization disparity between the algorithms

Delay Scheme % effectiveness	Maximum Degree with which *existing* can outperform *distributed*
0%	2.16 GB
5%	181.44 MB
25%	43.2 MB
50%	25.9 MB
55%	-19.44KB
75%	-54.90 KB
95%	-2.28 MB

4.3 Memory Utilization

Memory consumption is defined as the number of bytes stored by each algorithm for the duration of the collaboration. This excludes temporary memory resident data such as adaptation request information from other nodes, or storage of serialized environmental metrics before propagation. The assumption also disregards memory utilization of the middleware as it would not bear upon the comparison of the individual algorithms.

The global memory consumed in the collaboration by the *existing* algorithm of Rossi and Ryan [3] shown in Equation 3, is the sum of the memory consumed by: environment metrics of the entire collaboration stored on every node; the metrics information stored about each object; memory utilization of each object; and the memory utilization of the middleware framework.

Similarly, the total memory consumed by the *distributed* algorithm, shown in Equation 4, is the sum of: the environmental metrics of each node; the metrics information stored about each object; memory utilization of each object; and memory utilization of the middleware framework.

$$\mathrm{Mu(existing)} = \mathrm{N^2E} + \sum_{i=1}^{O} \mathrm{Mu}(O_i) + \sum_{i=1}^{O} \mathrm{Mu}(B_i) \tag{3}$$

$$\mathrm{Mu(distributed)} = \mathrm{NE} + \sum_{i=1}^{O} \mathrm{Mu}(O_i) + \sum_{i=1}^{O} \mathrm{Mu}(B_i) \tag{4}$$

Equation 3 models the memory utilization of the *existing* approach. Equation 4 models the memory utilization of the *distributed* algorithm.

O=Number of mobile objects
E = Environmental metrics size
N = number of nodes
Mu(x) = Average Memory utilization of x
B_i =Object metrics of object i

Equation 3 and 4 show that while the *existing* algorithm has a global memory complexity of $O(N^2)$ for storing collaboration information on every node, the *distributed* approach has a more favorable memory utilization complexity of *O(N)*. Figure 5 shows that the difference between the memory utilization of both algorithms increases exponentially with the increase in number of nodes and linearly with the increase in environmental metrics size. The *distributed* algorithm hence provides increasingly better memory utilization with an increase in the heterogeneity and size of the collaboration.

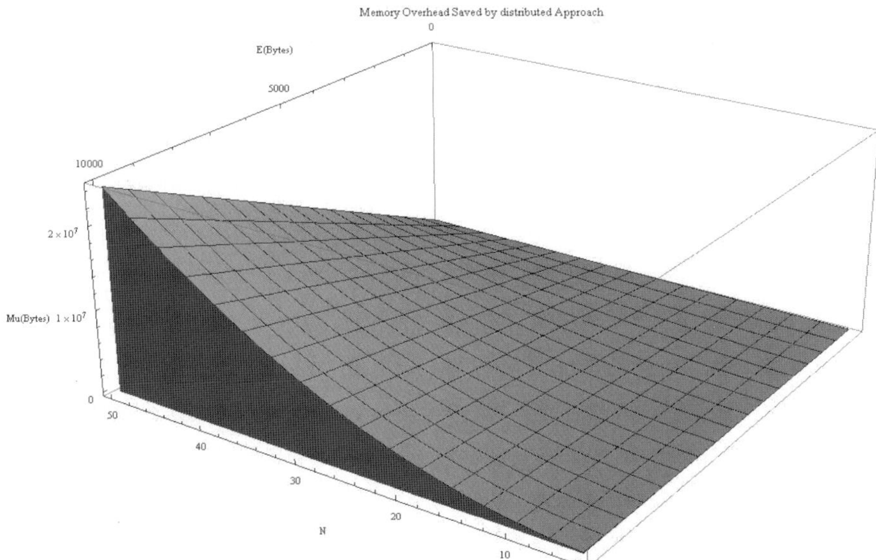

Fig. 5. Memory overhead saved by the *distributed* algorithm compared to the *existing* algorithm

5 Summary and Conclusion

The paper has proposed a *distributed* approach to local adaptation decision making to improve the *efficiency* and *scalability* of *existing* approaches. The network and memory utilization of the approach was mathematically modeled and compared to an *existing* algorithm. The results showed the *distributed* approach to be more scalable with regards to collaboration size and diversity of the computation environment, as well as offering over 900% better network utilization in large scale dynamic collaborations while still maintaining a linear memory utilization complexity.

Future work will focus on implementing and empirically evaluating the *effectiveness* of the algorithm's adaptation outcomes for the objective of application performance improvement. The use of a clustering technique, and an effective delay period will also be investigated and their performance impact analyzed.

References

1. Kim, M., Copeland, J.A.: Bandwidth sensitive caching for video streaming application. In: IEEE International Conference on Communications, 2003. ICC 2003 (2003)
2. Hütter, C., Moschny, T.: Runtime Locality Optimizations of Distributed Java Applications. IEEE Computer Society Press, Washington (2008)
3. Rossi, P., Ryan, C.: An Empirical Evaluation of Dynamic Local Adaptation for Distributed Mobile Applications. In: Proc. of 2005 International Symposium on Distributed Objects and Applications (DOA 2005), Larnaca, Cyprus (2005)
4. Ryan, C., Westhorpe, C.: Application Adaptation through Transparent and Portable Object Mobility in Java. In: Proc. of 2004 International Symposium on Distributed Objects and Applications (DOA 2004), Larnaca, Cyprus (2004)
5. Felea, V., Toursel, B.: Adaptive Distributed Execution of Java Applications. In: 12th Euromicro Conference on Parallel, Distributed and Network-Based Processing, PDP 2004 (2004)
6. Gu, X., et al.: Adaptive Offloading for Pervasive Computing. IEEE Pervasive Computing 3(3), 66–73 (2004)
7. Tilevich, E., Smaragdakis, Y.: J-orchestra: Automatic java application partitioning. In: Magnusson, B. (ed.) ECOOP 2002. LNCS, vol. 2374, pp. 178–204. Springer, Heidelberg (2002)
8. Philippsen, M., Zenger, M.: JavaParty Transparent remote objects in Java. In: Proc. ACM 1997 PPoPP Workshop on Java for Science and Engineering Computation (1997)
9. Fahringer, T.: JavaSymphony: A System for Development of Locality-Oriented Distributed and Parallel Java Applications. In: Cluster 2000. IEEE Computer Society Press, Los Alamitos (2000)
10. Garti, D., et al.: Object Mobility for Performance Improvements of Parallel Java Applications. Journal of Parallel and Distributed Computing 60(10), 1311–1324 (2000)
11. Sakamoto, K., Yoshida, M.: Design and Evaluation of Large Scale Loosely Coupled Cluster-based Distributed Systems. In: Li, K., Jesshope, C., Jin, H., Gaudiot, J.-L. (eds.) NPC 2007. LNCS, vol. 4672, pp. 572–577. Springer, Heidelberg (2007)
12. Ou, S., Yang, K., Liotta, A.: An adaptive multi-constraint partitioning algorithm for offloading in pervasive systems. In: Fourth Annual IEEE International Conference on Pervasive Computing and Communications, 2006. PerCom 2006 (2006)

13. Ryan, C., Rossi, P.: Software, performance and resource utilisation metrics for context-aware mobile applications. In: 11th IEEE International Symposium in Software Metrics (2005)
14. Gani, H., Ryan, C., Rossi, P.: Runtime Metrics Collection for Middleware Supported Adaptation of Mobile Applications. In: International Workshop on Adaptive and Reflective Middleware, ACM Middleware, 2006, Melbourne, Australia (2006)
15. Abebe, E., Ryan, C.: Online Appendix: Decision Computation Calculations for a Distributed Approach to Local Adaptation (2009), `http://goanna.cs.rmit.edu.au/~eabebe/DOA2009/Abebe_Ryan_Appendix.pdf`
16. Wolfram Research, Wolfram Mathematica 7 (2009)

Author Index

Printing: Mercedes-Druck, Berlin
Binding: Stein+Lehmann, Berlin